Principles of Development

Lewis Wolpert is Professor of Biology as Applied to Medicine at University College London, UK. He is the author of *The Triumph of the Embryo*, *A Passion for Science*, and *The Unnatural Nature of Science* and is a well-known broadcaster and journalist.

Rosa Beddington (deceased) was formerly Head of the Division of Mammalian Development at the National Institute for Medical Research, London, UK.

Thomas Jessell is Professor of Biochemistry and Molecular Biophysics, a member of the Center for Neurobiology and Behaviour, and a Howard Hughes Medical Institute investigator at the College of Physicians and Surgeons of Columbia University, New York, USA. He is an author of *Principles of Neural Science* and *Essentials of Neural Science and Behaviour*.

Peter Lawrence is in the Cell Biology Division at the Medical Research Council Laboratory of Molecular Biology, Cambridge, UK. He is the author of *The Making of a Fly*.

Elliot Meyerowitz is Professor in the Division of Biology at the California Institute of Technology, Pasadena, USA.

Jim Smith is Chairman of the Wellcome/CRC Institute of Cancer and Developmental Biology, Cambridge, UK.

Principles of Development

SECOND EDITION

Lewis Wolpert

Rosa Beddington

Thomas Jessell

Peter Lawrence

Elliot Meyerowitz

Jim Smith

OXFORD

UNIVERSITY PRESS

OXFORD

UNIVERSITY PRESS

Great Clarendon Street, Oxford OX2 6DP

Oxford University Press is a department of the University of Oxford
It furthers the University's objective of excellence in research, scholarship,
and education by publishing worldwide in

Oxford New York

Athens Auckland Bangkok Bogotá Buenos Aires Cape Town
Chennai Dar es Salaam Delhi Florence Hong Kong Istanbul Karachi
Kolkata Kuala Lumpur Madrid Melbourne Mexico City Mumbai Nairobi
Paris São Paulo Shanghai Singapore Taipei Tokyo Toronto Warsaw

with associated companies in Berlin Ibadan

Oxford is a registered trade mark of Oxford University Press
in the UK and in certain other countries

Published in the United States
by Oxford University Press Inc., New York

Library of Congress Cataloging in Publication Data
(Data applied for)

ISBN 0-19-879291-3 (Pbk)
ISBN 0-19-924939-3 (Hbk)

Typeset by RefineCatch Ltd, Broad Street, Bungay, Suffolk
Printed in Spain by Bookprint Ltd, Barcelona

Preface

Developmental biology is at the core of all biology. It deals with the process by which the genes in the fertilized egg control cell behavior in the embryo and so determine its pattern, its form, and much of its behavior. The progress in developmental biology in recent years with the application of advances in cell and molecular biology has been remarkable, and an enormous amount of information is now available.

In this second edition, we have included many recent advances, for example in the understanding of somite formation, and this new material is complemented by over 30 additional illustrations. Sections on the development of the heart, the vascular system, and teeth have been added, and we have given more attention, for example, to stem cells, signal transduction, and evolution.

Principles of Development is designed for undergraduates and the emphasis is on principles and key concepts. Central to our approach is that development can be best understood by understanding how genes control cell behavior. We have assumed that the students have some very basic familiarity with cell biology and genetics, but all key concepts, such as the control of gene activity, are explained in the text.

Conscious of the pressures on students, we have tried to make the principles as clear as possible and to provide numerous summaries, in both words and pictures. The illustrations in this book are a special feature and have been carefully designed and chosen to illuminate both experiments and mechanisms.

We have resisted the temptation to cover every aspect of development and have, instead, focused on those systems that best illuminate common principles. Indeed, a theme that runs throughout the book is that universal principles govern the process of development. At all stages, what we included has been guided by what we believe undergraduates should know about development.

We have thus concentrated our attention on vertebrates and *Drosophila*, but not to the exclusion of other systems, such as the nematode and the sea urchin, where they best illustrate a concept. An important feature of our book is the inclusion of the development of plants, which is usually neglected in textbooks. There have been striking advances in plant developmental biology in recent times, and some unique and important features have emerged. As knowing the basic features of the embryology of the main organisms used to study development is essential for an understanding of molecular mechanisms, we have introduced embryology at an early stage.

Whereas our emphasis has been on the laying down of the body plans and organ systems, such as limbs and the nervous system, we have also included later aspects of development, including growth and regeneration. The book concludes with a consideration of evolution and development.

In providing references, our prime concern has been to guide the students to helpful papers rather than to give credit to all the scientists who have made major contributions: to those whom we have neglected, we apologize.

For this new edition we welcome Jim Smith as a co-author and say good-bye to Jeremy Brockes, with thanks for all his help. Each chapter has also been reviewed by a number of experts (see page xix), to whom we give thanks. As with the first edition, I did all the writing—and I mean writing—in consultation with my co-authors; it was typed by Maureen Maloney and my revisions were deciphered, edited, and incorporated by our editor Eleanor Lawrence, whose expertise and influence pervades the book. Most of the new illustrations were brilliantly created or adapted by Matthew McClements, who created the illustrations for the first edition.

We are indebted to Jonathan Crowe, John Grandidge, and the illustrators at Oxford University Press, our new publishers, for their help and patience throughout the preparation of this new edition. My thanks to Vitek Tracz and Peter Newmark of the Current Science Group, without whom the first edition of this book would never have been started, let alone completed.

Finally, and sadly, we are acutely conscious of the loss of Rosa Beddington, who died last year, after a long illness. She was helpful to the end and we miss her terribly.

London
September 2001

L. W.

Contents

List of headings

Chapter 11: Development of the nervous system

Chapter 12: Germ cells and sex

Text acknowledgements

Many thanks to the following who kindly reviewed various parts of the book.

Chapter 1

Vernon French University of Edinburgh.
Sir John Gurdon Wellcome CRC Institute, Cambridge.
Andrew Murray University of California, San Francisco.
Daniel St Johnston Wellcome CRC Institute, Cambridge.

Chapter 2

Susan Darling University College London.
Brigid Hogan Vanderbilt University School of Medicine.
Nigel Holder King's College London.
Andrew Murray University of California, San Francisco.
Jonathan Slack University of Bath.
Daniel St. Johnston Wellcome CRC Institute, Cambridge.

Chapter 3

Les Dale University College London.
Richard Harland University of California, Berkeley.
Brigid Hogan Vanderbilt University School of Medicine.
Nigel Holder King's College London.
Hal Krinder University of Connecticut.
Janet Rossant Samuel Lunenfield Research Institute, Toronto.
Claudia Stern University College London.

Chapter 4

Richard Harland University of California, Berkeley.
Brigid Hogan Vanderbilt University School of Medicine.
Nigel Holder King's College London.
Hal Krinder University of Connecticut.
Janet Rossant Samuel Lunenfield Research Institute, Toronto.
Claudio Stern University College London.

Chapter 5

Kathryn Anderson University of California, Berkeley.
Peter Bryant University of California, Irvine.
Herbert Jackle Max-Planck Institute für Biophysikalische Chemie, Gottingen.
Daniel St Johnston Wellcome CRC Institute, Cambridge.

Chapter 6

Jeff Hardin University of Wisconsin, Madison.
James Priess Fred Hutchinson Cancer Center, Seattle.
Paul Sternberg California Institute of Technology, Pasadena.
Jeff Williams University College London.

Chapter 7

David Meinke Oklahoma State University.
Scott Poethig University of Pennsylvania.
Ian Sussex University of California, Berkeley.

Chapter 8

Marianne Bronner-Fraser University of California, Irvine.
Charles Ettensohn Carnegie Mellon University, Pittsburgh.
Raymond Keller University of California, Berkeley.
Malcolm Steinberg Princeton University.

Chapter 9

Margaret Buckingham Institut Pasteur, France.
David Latchman University College London Medical School.
Nicole Le Douarin Institut d'Embryologie Cellulaire et Moléculaire du CNRS et du Collège de France, Nogent-sur-Marne.
Roger Patient The Randall Institute, King's College London.
Fiona Watt Imperial Cancer Research Fund, London.

Chapter 10

Stephen Cohen European Molecular Biology Laboratory, Heidelburg.
Uma Thesleff Institute of Biotechnology, University of Helsinki.
Cheryl Tickle University of Dundee.
Andrew Tomlinson College of Physicians and Surgeons of Columbia University, New York.

Chapter 11

Michael Bate University of Cambridge.
Steven Easter University of Michigan.
Andrew Lumsden United Medical and Dental School, Guy's Hospital, London.
Jack Price SmithKline Beecham, Harlow, Essex.
Guy Tear United Medical and Dental School, Guy's Hospital, London.

Chapter 12

Denise Barlow The Netherlands Cancer Institute, Amsterdam.
Tom Cline University of California, Berkeley.
Jonathan Hodgkin MRC Laboratory of Molecular Biology, Cambridge.
Anne McLaren Wellcome CRC Institute, Cambridge.
Michael Whitaker Newcastle University.

Chapter 13

Hans Bode University of California, Irvine.
Susan Bryant University of California, Irvine.
Vernon French University of Edinburgh.
David Stocum Indiana University-Purdue University, Indianapolis.

Chapter 14

Michael Ashburner University of Cambridge.
Charles Brook University College London.
Peter Bryant University of California, Irvine.
Tom Kirkwood University of Manchester.
Stuart Reynolds University of Bath.
J. R. Tata National Institute for Medical Research, London.

Chapter 15

Michael Akam Wellcome CRC Institute, Cambridge.
Michael Coates University College London.
John Gerhart University of California, Berkeley.
Cliff Tabin Harvard Medical School, Boston.
Peter Holland University of Reading.

General

Brigid Hogan Vanderbilt University, School of Medicine,
David Kimelman University of Washington, Seattle.
Charles Rutherford Virginia Polytechnic Institute and State University, Blacksburg.
Nancy Wall Lawrence University.

Figure acknowledgements

Chapter 1

Fig. 1.15 illustration after Moore, K.L.: *Before we are Born: Basic Embryology and Birth Defects, 2nd edition*. Philadelphia: W.B. Saunders, 1983.

Chapter 2

Fig. 2.3 top photograph reproduced with permission from Alberts, B., Bray, D., Lewis, J., Raff, M., Roberts, K., Watson, J.D.: *Molecular Biology of the Cell, 3rd edition*. New York: Garland Publishing, 1994.

Fig. 2.5 photographs reproduced with permission from Kessel, R.G., Shih, C.Y.: *Scanning Electron Microscopy in Biology: A Student's Atlas of Biological Organization*. London, Springer-Verlag, 1974. © 1974 Springer-Verlag GmbH & Co. KG.

Fig. 2.6 illustration after Balinsky, B.I.: *An Introduction to Embryology*. Fourth edition. Philadelphia, W.B. Saunders, 1975.

Fig. 2.8 photograph reproduced with permission from Hausen, P., Riebesell, M.: *The Early Development of Xenopus laevis*. Berlin: Springer-Verlag, 1991.

Fig. 2.11 top photograph reproduced with permission from Kispert, A., Ortner, H., Cooke, J., Herrmann, B.G.: **The chick Brachyury gene: developmental expression pattern and response to axial induction by localized activin.** *Dev. Biol.* 1995, **168**:406–415.

Fig. 2.13 illustration adapted, with permission, from Balinsky, B.I.: *An Introduction to Embryology, 4th edition*. Philadelphia, W.B. Saunders, 1975. © 1975 Saunders College Publishing.

Fig. 2.17 illustration after Patten, B.M.: *Early Embryology of the Chick*. New York, Mc Graw-Hill, 1971.

Fig. 2.18 photographs reproduced with permission from Kispert, A., Ortner, H., Cooke, J., Herrmann, B.G.: **The chick Brachyury gene: developmental expression pattern and response to axial induction by localized activin.** *Dev. Biol.* 1995, **168**:406–415.

Fig. 2.19 illustration after Patton, B.M.: **The first heart beat and the beginning of embryonic circulation.** *American Scientist* 1951, **39**:225–243.

Fig. 2.20 top photograph reproduced with permission from Bloom, T.L.: **The effects of phorbol ester on mouse blastomeres: a role for protein kinase C in compaction?** *Development* 1989, **106**:159–171 Published by permission of The Company of Biologists Ltd.

Fig. 2.22 illustration after Hogan, B., Beddington, R., Costantini, F., Lacy, E.: *Manipulating the Mouse Embryo: A Laboratory Manual, 2nd edition*. New York: Cold Spring Harbor Laboratory Press, 1994.

Fig. 2.23 illustration adapted, with permission, from McMahon, A.P.: **Mouse development. Winged-helix in axial patterning.** *Curr. Biol.* 1994, 4:903–906.

Fig. 2.24 illustration after Hogan, B., Beddington, R., Costantini, F., Lacy, E.: *Manipulating the Mouse Embryo: A Laboratory Manual, 2nd edition*. New York: Cold Spring Harbor Laboratory Press, 1994.

Fig. 2.25 illustration after Kaufman, M.H.: *The Atlas of Mouse Development*. London: Academic Press, 1992.

Fig. 2.26 top photograph reproduced with permission from Kimmel, C.B., Ballard, W.W., Kimmel, S.R., Ullmann, B., Schilling, T.F.: **Stages of embryonic development of the zebrafish.** *Dev. Dynamics* 1995. **203**:253–310. © 1995 Wiley-Liss, Inc.

Fig. 2.27 photographs reproduced with permission from Kessel, R.G., Shih, C.Y.: *Scanning Electron Microscopy in Biology: A Student's Atlas of Biological Organization*. London, Springer-Verlag, 1974. © 1974 Springer-Verlag GmbH & Co. KG.

Fig. 2.29 top photograph reproduced with permission from Turner, F.R., Mahowald, A.P.: **Scanning electron microscopy of Drosophila embryogenesis. I. The structure of the egg envelopes and the formation of the cellular blastoderm.** *Dev. Biol.* 1976, 50:95–108. Middle photograph reproduced with permission from Turner, F.R., Mahowald, A.P.: **Scanning electron microscopy of Drosophila melanogaster embryogenesis. III. Formation of the head and caudal segments.** *Dev. Biol.* 1979, **68**:96–109.

Fig. 2.32 left photograph reproduced with permission from Turner, F.R., Mahowald, A.P.: **Scanning electron microscopy of Drosophila melanogaster embryogenesis. II. Gastrulation and segmentation.** *Dev. Biol.* 1977, **57**:403–416. Center photograph reproduced with permission from Alberts, B., Bray, D., Lewis, J., Raft, M., Roberts, K., Watson, J.D.: *Molecular Biology of the Cell, 3rd edition*. New York, Garland Publishing, 1994.

Fig. 2.38 illustration after Sulston, J.E., Schierenberg, E., White, J.G., Thomson, J.N.: **The embryonic cell lineage of the nematode Caenorhabditis elegans.** *Dev. Biol.* 1983, **100**:64–119.

Fig. 2.41 photographs reproduced with permission from Meinke, D.W.: **Seed development in Arabidopsis thaliana.** In *Arabidopsis*. Edited by Meyerowitz, E.M., & Somerville, C.: New York, Cold Spring Harbor Laboratory Press, 1994:pp253–295.

Chapter 3

Fig. 3.38 photograph reproduced with permission from Smith W.C., Harland, R.M.: **Expression cloning of noggin, a new dorsalizing factor localized to the Spemann organizer in Xenopus embryos.** *Cell* 1992, 70:829–840. © 1992 Cell Press.

Chapter 4

Box 4A illustration after Coletta, P.L., Shimeld, S.M., Sharpe, P.T.: The molecular anatomy of Hox gene expression. *J. Anat* 1994, **184**:15–22.

Fig. 4.9 illustration after Johnson, R.L., Laufer, E., Riddle, R.D., Tabin, C.: **Ectopic expression of Sonic hedgehog alters dorsal-ventral patterning of somites.** *Cell* 1994, 79:1165–1173.

Fig. 4.13 illustration after Burke, A.C., Nelson, C.E., Morgan, B.A., Tabin, C.: **Hox genes and the evolution of vertebrate axial morphology.** *Development* 1995, **121**: 333–346.

Fig. 4.20 illustration after Mangold, O.: Über die induktionsfahigkeit der verschiedenen bezirke der neurula von urodelen. *Naturwissenschaften* 1933, **21**:761–766.

Fig. 4.21 illustration after Kelly, O.G., Melton, D.A.: **Induction and patterning of the vertebrate nervous system.** *Tr. Genet.* 1995, **11**:273–278.

Fig. 4.22 illustration after Kintner, C.R., Dodd, J.: **Hensen's node induces neural tissue in *Xenopus* ectoderm. Implications for the action of the organizer in neural induction.** *Development* 1991, **113**:1495–1505.

Fig. 4.23 illustration after Doniach, T., Phillips, C.R., Gerhart, J.C.: **Planar induction of antero-posterior pattern in the developing central nervous system of *Xenopus laevis*.** *Science* 1992, **257**:542–545.

Fig. 4.24 illustration adapted, with permission, from Lumsden, A.: **Cell lineage restrictions in the chick embryo hindbrain.** *Phil. Trans. Roy. Soc. Lond. B* 1991, **331**:281–286.

Fig. 4.25 illustration adapted, with permission, from Lumsden, A.: **Cell lineage restrictions in the chick embryo hindbrain.** *Phil. Trans. Roy. Soc. Lond. B* 1991, **331**:281–286.

Fig. 4.27 illustration after Krumlauf, R.: **Hox genes and pattern formation in the branchial region of the vertebrate head.** *Tr. Genet.* 1993, **9**:106–112.

Fig. 4.28 photograph reproduced with permission from Lumsden, A., Krumlauf, R.: **Patterning the vertebrate neuraxis.** *Science* 1996, **274**:1109–1115 (image on front cover). © 1996 American Association for the Advancement of Science.

Chapter 5

Fig. 5.5 photograph reproduced with permission from Griffiths, A.J.H., Miller, J.H., Suzuki, D.T., Lewontin, R.C., Gelbart, W.M.: *An Introduction to Genetic Analysis, 6th edition.* New York: W.H. Freeman & Co., 1996.

Fig. 5.6 photograph reproduced with permission from Griffiths, A.J.H., Miller, J.H., Suzuki, D.T., Lewontin, R.C., Gelbart, W.M.: *An Introduction to Genetic Analysis, 6th edition.* New York: W.H. Freeman & Co., 1996.

Fig. 5.12 illustration after González-Reyes, A., Elliott, H., St. Johnston, D.: **Polarization of both major body axes in *Drosophila* by *gurken-torpedo* signalling.** *Nature* 1995, **375**:654–658.

Fig. 5.20 illustration after Lawrence, P.: *The Making of a Fly.* Oxford: Blackwell Scientific Publications, 1992.

Fig. 5.24 photograph reproduced with permission from Lawrence, P.: *The Making of a Fly.* Oxford: Blackwell Scientific Publications, 1992.

Fig. 5.25 illustration after Lawrence, P.: *The Making of a Fly.* Oxford: Blackwell Scientific Publications, 1992.

Box 5B illustration after Lawrence, P.: *The Making of a Fly.* Oxford: Blackwell Scientific Publications, 1992.

Fig. 5.32 illustration after Lawrence, P.: *The Making of a Fly.* Oxford: Blackwell Scientific Publications, 1992.

Fig. 5.38 illustration after Lawrence, P.: *The Making of a Fly.* Oxford: Blackwell Scientific Publications, 1992. Photograph reproduced with permission from Bender, W., Akam, A., Karch, F., Beachy, P.A., Peifer, M., Spierer, P., Lewis, E.B., Hogness, D.S.: **Molecular genetics of the bithorax complex in *Drosophila melanogaster*.** *Science* 1983, **221**:23–29 (image on front cover). © 1983 American Association for the Advancement of Science.

Chapter 6

Fig. 6.2 illustration after Sulston, J.E., Schierenberg, E., White, J.G., Thompson, J.N.: **The embryonic cell lineage of the nematode *Caenorhabditis elegans*.** *Dev. Biol.* 1983, **100**:69–119.

Fig. 6.3 photograph reproduced with permission from Strome, S., Wood, W.B.: **Generation of asymmetry and segregation of germ-line granules in early *C. elegans* embryos.** *Cell* 1983 **35**:15–25. © 1983 Cell Press.

Fig. 6.4 illustration after Sulston, J.E., Schierenberg, E., White, J.G., Thompson, J.N.: **The embryonic cell lineage of the nematode *C. elegans*.** *Dev. Biol.* 1983, **100**:69–119.

Fig. 6.5 photographs reproduced with permission from Wood W.B: **Evidence from reversal of handedness in C. elegans embryos for early cell interactions determining cell fates.** *Nature* 1991, **349**:536–538. © 1991 Macmillan Magazines Ltd.

Fig. 6.6 illustration after Mello, C.C., Draper, B.W., Priess, J.R.: **The maternal genes *apx-1* and *glp-1* and establishment of dorsal-ventral polarity in the early *C. elegans* embryo.** *Cell* 1994, **77**:95–106.

Fig. 6.9 illustration after Bürglin, T.R., Ruvkun, G.: **The *Caenorhabditis elegans* homeobox gene cluster.** *Curr. Opin. Gen. Devel* 1993, **3**:615–620.

Fig. 6.19 photograph reproduced with permission from Corbo, J.C., Levine, M., Zeller, R.W.: **Characterization of a notochord-specific enhancer from the Brachyury promoter region of the ascidian, *Ciona intestinalis*.** *Development* 1997, **124**:589–602. Published by permission of The Company of Biologists Ltd.

Fig. 6.21 illustration after Conklin, E.G.: **The organization and cell lineage of the ascidian egg.** *J. Acad. Nat. Sci. Philadelphia* 1905, **13**:1–119.

Fig. 6.22 illustration after Nakatani, Y., Yasuo, H. Satoh, N., Nishida, H.: **Basic fibroblast growth factor induces notochord formation and the expression of *As-T*, a *Brachyury* homolog, during ascidian embryogenesis.** *Development* 1996, **122**:2023–2031.

Fig. 6.23 top photograph reproduced with permission from Early, A.E., Gaskell, M.J., Traynor, D., Williams, J.G.: **Two distinct populations of prestalk cells within the tip of the migratory Dictyostelium slug with differing fates at culmination.** *Development* 1993, **118**:353–362. Published by permission of The Company of Biologists Ltd. Middle panel, photograph reproduced with permission from Jermyn K., Traynor, D., Williams, J.: **The initiation of basal disc formation in *Dictyostelium discoideum* is an early event in culmination.** *Development* 1996, **122**:753–760. Published by permission of The Company of Biologists Ltd.

Chapter 7

Fig. 7.1 illustration after Scheres, B., Wolkenfelt, H., Willemsen, V., Terlouw, M., Lawson, E., Dean, C., Weisbeek, P.: **Embryonic origin of the *Arabidopsis* primary root and root meristem initials.** *Development* 1994, **120**:2475–2487.

Fig. 7.2 illustration after Mayer, U., Torres-Ruiz, R.A., Berleth, T., Misera, S., Jurgens, G.: **Mutations affecting body organization in the *Arabidopsis* embryo.** *Nature* 1991, **353**:402–407.

Fig. 7.6 illustration after Alberts, B., Bray, D., Lewis, J., Raff, M., Roberts, K., Watson, J.D.: *Molecular Biology of the Cell, 2nd edition.* New York: Garland Publishing, 1989.

Fig. 7.9 illustration after Alberts, B., Bray, D., Lewis, J., Raff, M., Roberts, K., Watson, J.D.: *Molecular Biology of the Cell, 2nd edition.* New York: Garland Publishing, 1989.

Fig. 7.11 illustration after Steeves, T.A., Sussex, I.M.: *Patterning in Plant Development*. Cambridge: Cambridge University Press, 1989.

Fig. 7.13 illustration after McDaniel, C.N., Poethig, R.S.: **Cell lineage patterns in the shoot apical meristem of the germinating maize embryo**. *Planta* 1988, **175**:13–22.

Fig. 7.14 illustration after Irish, V.F.: **Cell lineage in plant development**. *Curr. Opin. Gen. Devel.* 1991, 1:169–173.

Fig. 7.15 illustration after Sachs, T.: *Pattern Formation in Plant Tissues*. Cambridge: Cambridge University Press, 1994; p.133.

Fig. 7.17 top panel, illustration after Poethig, R.S., Sussex, I.M.: **The cellular parameters of leaf development in tobacco: a clonal analysis**. *Planta* 1985, **165**:170–184. Bottom panel, illustration after Sachs, T.: *Pattern Formation in Plant Tissues*. Cambridge: Cambridge University Press, 1994.

Fig. 7.20 illustration after Scheres, B., Wolkenfelt, H., Willemsen, V., Terlouw, M., Lawson, E., Dean, C., Weisbeek, P.: **Embryonic origin of the *Arabidopsis* primary root and root meristem initials**. *Development* 1994, **120**:2475–2487.

Fig. 7.22 photograph reproduced with permission from Meyerowitz, E.M., Bowman, J.L., Brockman, L.L., Drews, G.N., Jack, T., Sieburth, L.E., Weigel, D.: **A genetic and molecular model for flower development in *Arabidopsis thaliana***. *Development Suppl.* 1991, pp157–167. Published by permission of The Company of Biologists Ltd.

Fig. 7.23 illustration after Coen, E.S., Meyerowitz, E.M.: **The war of the whorls: genetic interactions controlling flower development**. *Nature* 1991, **353**:31–37.

Fig. 7.24 photographs reproduced with permission from Meyerowitz, E.M., Bowman, J.L., Brockman, L.L., Drews, G.N., Jack, T., Sieburth, L.E., Weigel, D.: **A genetic and molecular model for flower development in *Arabidopsis thaliana***. *Development Suppl.* 1991, pp157–167. Published by permission of The Company of Biologists Ltd. (left panel); center panel from Bowman, J.L., Smyth, D.R., Meyerowitz, E.M.: **Genes directing flower development in *Arabidopsis***. *Plant Cell* 1989, 1:37–52. Published by permission of The American Society of Plant Physiologists.

Fig. 7.26 illustration after Dennis, E., Bowman, J.L.: **Manipulating floral identity**. *Curr. Biol* 1993, 3:90–93.

Fig. 7.27 illustration after Meyerowitz, E.M., Bowman, J.L., Brockman, L.L., Drews, G.N., Jack, T., Sieburth, L.E., Weigel, D.: **A genetic and molecular model for flower development in *Arabidopsis thaliana***. *Development Suppl.* 1991, pp157–167.

Fig. 7.28 illustration after Coen, E.S., Meyerowitz, E.M.: **The war of the whorls: genetic interactions controlling flower development**. *Nature* 1991, **353**:31–37.

Fig. 7.31 photograph reproduced with permission from Coen, E.S., Meyerowitz, E.M.: **The war of the whorls: genetic interactions controlling flower development**. *Nature* 1991, **353**:31–37. © 1991 Macmillan Magazines Ltd.

Fig. 7.32 illustration after Drews, G.N., Goldberg, R.B.: **Genetic control of flower development**. *Trends Genet.* 1989 **5**:256–261.

Chapter 8

Fig. 8.3 photographs reproduced with permission from Steinberg, M.S., Takeichi, M.: **Experimental specification of cell sorting, tissue spreading, and specific spatial patterning by quantitative differences in cadherin expression**. *Proc. Natl Acad. Sci./Dev. Biol* 1994, **91**:206–209. © 1994 National Academy of Sciences.

Fig. 8.6 illustration after Strome, S.: **Determination of cleavage planes**. *Cell* 1993, **72**:3–6.

Fig. 8.9 photograph reproduced with permission from Bloom T.L.: **The effects of phorbol ester on mouse blastomeres: a role for protein kinase C in compaction?** *Development* 1989, **106**:159–71.

Fig. 8.12 illustration after Coucouvanis, E., Martin, G.R.: **Signals for death and survival: a two-step mechanism for cavitation in the vertebrate embryo**. *Cell* 1995, **83**:279–287.

Fig. 8.16 illustration after Odell, G.M., Oster, G., Alberch, P., Burnside, B.: **The mechanical basis of morphogenesis. I. Epithelial folding and invagination**. *Dev. Biol.* 1981, **85**:446–462.

Fig. 8.18 photographs reproduced with permission from Leptin, M., Casal, J., Grunewald, B., Reuter, R.: **Mechanisms of early *Drosophila* mesoderm formation**. *Development Suppl.* 1992, pp23–31. Published by permission of The Company of Biologists Ltd.

Fig. 8.20 illustration after Balinsky, B.I.: *An Introduction to Embryology, 4th edition*. Philadelphia, W.B. Saunders, 1975.

Fig. 8.21 illustration after Hardin, J., Keller, R.: **The behavior and function of bottle cells during gastrulation of *Xenopus laevis***. *Development* 1988, 103:211–230.

Fig. 8.24 photograph reproduced with permission from Smith, J.C., Cunliffe, V., O'Reilly, M-A.J., Schulte-Merker, S., Umbhauer, M.: ***Xenopus Brachyury***. *Semin. Dev. Biol.* 1995, 6:405–410. © 1995 by permission of the publisher, Academic Press Ltd., London.

Fig. 8.31 illustration after Schoenwolf, G.C., Smith, J.L.: **Mechanisms of neurulation: traditional viewpoint and recent advances**. *Development* 1990 109:243–270.

Fig. 8.34 photograph reproduced with permission from Merrill, J.B., Santos, L.L.: **A scanning electron micrographical overview of cellular and extracellular patterns during blastulation and gastrulation in the sea urchin, *Lytechinus variegatus***. In *The Cellular and Molecular Biology of Invertebrate Development*. Edited by Sawyer, R.H. and Showman, R.M. University of South Carolina Press, 1985; pp3–33.

Fig. 8.39 illustration after Alberts, B., Bray, D., Lewis, J., Raft, M., Roberts, K., Watson, J.D.: *Molecular Biology of the Cell, 2nd edition*. New York: Garland Publishing, 1989.

Fig. 8.42 photographs reproduced with permission from Priess, J.R., Hirsh, D.I.: ***Caenorhabditis elegans morphogenesis*: the role of the cytoskeleton in elongation of the embryo**. *Dev. Biol* 1986, 117:156–173. © 1986 Academic Press.

Fig. 8.44 photographs reproduced with permission from Tsuge, T., Tsukaya, H., Uchimaya, H.: **Two independent and polarized processes of cell elongation regulate leaf blade expansion in *Arabidopsis thaliana* (L.) Heynh**. *Development* 1996, **122**:1589–1600. Published by permission of The Company of Biologists Ltd.

Chapter 9

Fig. 9.1 illustration after Friederich, E., Prignault, E., Arpin, M., Louvard, D.: **From the structure to the function of villin, an actin-binding protein of the brush border**. *BioEssays* 1990, **12**:403–408.

Fig. 9.3 illustration after Tijian, R.: **Molecular machines that control genes**. *Sci. Amer.* 1995, **272**:54–61.

Fig. 9.7 illustration after Alberts B., Bray, D., Lewis, J., Raff, M., Roberts, K., Watson, J.D.: *Molecular Biology of the Cell, 2nd edition*. New York: Garland Publishing, 1989.

Fig. 9.9 illustration after Alberts, B., Bray, D., Lewis, J., Raft, M., Roberts, K., Watson, J.D.: *Molecular Biology of the Cell, 2nd edition.* New York: Garland Publishing, 1989.

Fig. 9.ll illustration after Alberts, B., Bray, D., Lewis, J., Raft, M., Roberts, K., Watson, J.D.: *Molecular Biology of the Cell, 2nd edition.* New York: Garland Publishing, 1989.

Fig. 9.16 illustration after Metcalf, D.: **Control of granulocytes and macrophages: molecular, cellular, and clinical aspects.** *Science* 1991, 254:529–533.

Fig. 9.19 illustration after Crossley, M., Orkin, S.H.: **Regulation of the β-globin locus.** *Curr. Opin. Genet. Dev.* 1993, 3:232–237.

Fig. 9.22 illustration after Doupe, A.J., Landis, S.C., Patterson, P.H.: **Environmental influences in the development of neural crest derivatives: glucocorticoids, growth factors, and chromaffin cell plasticity.** *J. Neurosci.* 1985, 5:2119–2142.

Fig. 9.26 illustration after Janeway, C.A., Travers, P.: *Immunobiology: The Immune System in Health and Disease, 3rd edition.* London: Current Biology/Garland, 1997.

Fig. 9.27 illustration after Janeway, C.A., Travers, P.: *Immunobiology: The Immune System in Health and Disease, 3rd edition.* London: Current Biology/Garland Publishing, 1997.

Fig. 9.32 illustration after Okada, T.S.: *Transdifferentiation.* Oxford: Clarendon Press, 1992.

Fig. 9.33 illustration after Doupe, A.J., Landis, S.C., Patterson, P.H.: **Environmental influences in the development of neural crest derivatives: glucocorticoids, growth factors, and chromaffin cell plasticity.** *J. Neurosci.* 1985, 5:2119–2142.

Chapter 10

Box 10A top illustration after Meinhardt, H., Gierer, A.: **Applications of a theory of biological pattern formation based on lateral inhibition.** *J. Cell Sci.* 1974, 15:321–346.

Fig. 10.9 photograph reproduced with permission from Cohn, M.J., Izpisúa-Belmonte, J.C., Abud, H., Heath, J.K., Tickle, C.: **Fibroblast growth factors induce additional limb development from the flank of chick embryos.** *Cell* 1995, 80:739–746. © 1995 Cell Press.

Fig. 10.21 photograph reproduced with permission from Garcia-Martinez, V., Macias, D., Gañan, Y., Garcia-Lobo, J.M., Francia, M.V., Fernandez-Teran, M.A., Hurle, J.M.: **Internucleosomal DNA fragmentation and programmed cell death (apoptosis) in the interdigital tissue of embryonic chick leg bud.** *J. Cell Sci.* 1993, 106:201–208. Published by permission of The Company of Biologists Ltd.

Fig. 10.23 illustration after French, V., Daniels, G.: **Pattern formation: the beginning and the end of insect limbs.** *Curr. Biol.* 1994, 4:35–37.

Fig. 10.26 photograph reproduced with permission from Nellen, D., Burke, R., Struhl, G., Basler, K.: **Direct and long-range action of a dpp morphogen gradient.** *Cell* 1996, 85:357–368. © 1996, Cell Press.

Fig. 10.29 photograph reproduced with permission from Zecca, M., Basler, K., Struhl, G.: **Direct and long-range action of a wingless morphogen gradient.** *Cell* 1996, 87:833–844. © 1996 Cell Press.

Fig. 10.31 illustration after Bryant, P.J. **The polar coordinate model goes molecular.** *Science* 1993, 259:471–472

Chapter 11

Fig. 11.3 photograph reproduced with permission from Skeath, J.B., Doe, C.Q.: **The achaete-scute complex proneural genes contribute to neural precursor specification in the** *Drosophila* **CNS.** *Curr. Biol.* 1996, 6:1146–1152.

Fig. 11.5 illustration after Jan, Y.N., Jan, L.Y.: **Genes required for specifying cell fates in Drosophila embryonic sensory nervous system.** *Trends Neurosci.* 1990, 13:493–498.

Fig. 11.6 illustration after Campuzano, S., Modolell, J.: **Patterning of the** *Drosophila* **nervous system: the achaete-scute gene complex.** *Trends Genet.* 1992, 8:202–208.

Fig. 11.8 illustration after Guo, M., Jan, L.Y., Jan, Y.N.: **Control of daughter cell fates during asymmetric division: interaction of Numb and Notch.** *Neuron* 1996, 17:27–41.

Fig. 11.18 illustration after Rakic, P.: **Mode of cell migration to the superficial layers of fetal monkey neocortex.** *J. Comp. Neurol.* 1972, 145:61–83.

Fig. 11.22 illustration after Lawrence, P *The Making of a Fly.* Oxford Blackwell Scientific Publications, 1992.

Fig. 11.24 illustration after Alberts, B., Bray, D., Lewis, J., Raff, M., Roberts, K., Watson, J.D.: *Molecular Biology of the Cell, 2nd edition.* New York: Garland Publishing, 1989.

Fig.11.28 illustration after O'Connor, T.P., Duerr, J.S., Bentley, D.: **Pioneer growth cone steering decisions mediated by single filopodial contacts** *in situ. J. Neurosci.* 1990, 10:3935–3946.

Fig. 11.31 illustration after Tessier-Lavigne, M., Placzek, M.: **Target attraction: are developing axons guided by chemotropism?** *Trends Neurosci.* 1991, 14:303–310.

Fig. 11.32 photograph reproduced with permission from Serafini, T., Colamarino, S.A., Leonardo, E.D., Wang, H., Beddington, R., Skarnes, W.C., Tessier-Lavigne, M.: **Netrin-1 is required for commissural axon guidance in the developing vertebrate nervous system.** *Cell* 1996, 87:1001–1014. © 1996 Cell Press.

Fig. 11.39 illustration after Davies, AM.: **Neurotrophic factors: switching neurotrophin dependence.** *Curr. Biol* 1994, 4:273–276.

Fig. 11.40 illustration after Kandell, E.R., Schwartz, J.H., Jessell, T.M.: *Principles of Neural Science, 3rd edition.* New York: Elsevier Science Publishing Co., Inc., 1991.

Fig. 11.44 illustration after Goodman, C.S., Shatz, C.J.: **Developmental mechanisms that generate precise patterns of neuronal connectivity.** *Cell Suppl.* 1993, 72:77–98

Fig. 11.45 illustration after Goodman, C.S., Shatz, C.J.: **Developmental mechanisms that generate precise patterns of neuronal connectivity.** *Cell Suppl.* 1993, 72:77–98

Fig. 11.46 illustration after Kandell, E.R., Schwartz, J.H., Jessell, T.M.: *Essentials of Neural Science and Behavior.* Norwalk, Connecticut: Appleton & Lange, 1991.

Fig. 11.47 illustration after Goodman, C.S., Shatz, C.J.: **Developmental mechanisms that generate precise patterns of neuronal connectivity.** *Cell Suppl.* 1993, 72:77–98

Chapter 12

Fig. 12.2 illustration after Goodfellow, P.N., Lovell-Badge, R.: *SRY* and sex determination in mammals. *Ann. Rev. Genet.* 1993, 27:71–92.

Fig. 12.3 illustration after Higgins, S.J., Young, P., Cunha, G.R.: **Induction of functional cytodifferentiation in the epithelium of tissue recombinants II. Instructive induction of Wolffian duct epithelia by neonatal seminal vesicle mesenchyme.** *Development* 1989, **106**:235–250.

Fig. 12.7 illustration after Cline, T.W.: **The *Drosophila* sex determination signal: how do flies count to two?** *Trends Genet.* 1993 **9**:385–390.

Fig. 12.11 illustration after Clifford, R., Francis, R., Schedl, T.: **Somatic control of germ cell development.** *Semin. Dev. Biol* 1994, 5:21–30.

Fig. 12.16 illustration after Wylie, C.C., Heasman, J.: **Migration, proliferation, and potency of primordial germ cells.** *Semin. Dev. Biol.* 1993, 4:161–170.

Fig. 12.22 illustration after Alberts, B., Bray, D., Lewis, J., Raft, M., Roberts, K., Watson, J.D.: *Molecular Biology of the Cell, 2nd edition.* New York: Garland Publishing, 1989.

Chapter 13

Fig. 13.11 photographs reproduced with permission from Pecorino, L.T. Entwistle A., Brockes, J.P.: **Activation of a single retinoic acid receptor isoform mediates proximodistal respecification** *Curr. Biol.*1996, 6:563–569.

Fig. 13.13 illustration after French, V., Bryant, P.J., Bryant, S.V.: **Pattern regulation in epimorphic fields.** *Science* 1976, 193:969–981.

Fig. 13.14 illustration after French, V., Bryant, P.J., Bryant, S.V.: Pattern regulation in epimorphic fields. *Science* 1976, 193:969–981.

Fig. 13.15 photograph reproduced with permission from Müller, W.A.: **Diacylglycerol-induced multihead formation in Hydra.** *Development* 1989, **105**:309–316. Published by permission of The Company of Biologists Ltd.

Fig. 13.21 photograph reproduced with permission from Müller, W.A.: **Diacylglycerol-induced multihead formation in *Hydra.*** *Development* 1989, **105**:309–316. Published by permission of The Company of Biologists Ltd.

Chapter 14

Fig. 14.3 illustration after Edgar, B.A., Lehman, D.A., O'Farrell, P.H.: **Transcriptional regulation of *string (cdc25):* a link between develop-**mental programming and the cell cycle. *Development* 1994, **120**:3131–3143.

Fig. 14.5 photograph reproduced with permission from Harrison, R.G.: *Organization and Development of the Embryo.* New Haven: Yale University Press, 1969. © 1969 Yale University Press.

Fig. 14.6 illustration after Gray, H.: *Gray's Anatomy.* Edinburgh: Churchill-Livingstone, 1995.

Fig. 14.9 illustration after Walls, G.A.: **Here today, bone tomorrow.** *Curr. Biol.* 1993, 3:687–689.

Fig. 14.16 illustration after Tata, J.R.: **Gene expression during metamorphosis: an ideal model for post-embryonic development.** *BioEssays* 1993, **15**: 239–248.

Fig. 14.17 illustration after Tata, J.R.: **Gene expression during metamorphosis: an ideal model for post-embryonic development.** *BioEssays* 1993, **15**: 239–248.

Chapter 15

Fig. 15.5 illustration after Larsen, W.J.: *Human Embryology.* New York: Churchill Livingstone, 1993.

Fig. 15.6 illustration after Romer, A.S.: *The Vertebrate Body.* Philadelphia: W.B. Saunders, 1949.

Fig. 15.8 photographs reproduced with permission from Sordino, P., van der Hoeven, F., Duboule, D.: **Hox gene expression in teleost fins and the origin of vertebrate digits.** *Nature* 1995, **375**:678–681. © 1995 Macmillan Magazines Ltd.

Fig. 15.9 illustration after Coates, M.I.: **Limb evolution: fish fins or tetrapod limbs–a simple twist of fate?** *Curr. Biol.* 1995, 5:844–848.

Fig. 15.17 illustration after Akam, M.: **Hox genes and the evolution of diverse body plans.** *Phil, Trans. Roy. Soc. Lond. B* 1995, **349**:313–319.

Fig. 15.18 illustration after Ferguson, E.L.: **Conservation of dorsaventral patterning in arthropods and chordates.** *Curr. Opin. Genet. Dev.* 1996, **6**:424–431.

Fig. 15.19 illustration after Gregory, W.K.: *Evolution Emerging.* New York: Macmillan, 1957.

History and basic concepts

1

- ■ The origins of developmental biology
- ■ A conceptual tool kit

"Ideas about us have changed a lot; but now we have some reliable rules."

The development of multicellular organisms from a single cell—the fertilized egg—is a brilliant triumph of evolution. During embryonic development, the egg divides to give rise to many millions of cells, which form structures as complex and varied as eyes, arms, heart, and brain. This amazing achievement raises a multitude of questions. How do the cells arising from division of the fertilized egg become different from each other? How do they become organized into structures such as limbs and brains? What controls the behavior of individual cells so that such highly organized patterns emerge? How are the organizing principles of development embedded within the egg, and in particular within the genetic material—DNA? Much of the excitement in developmental biology today comes from our growing understanding of how genes direct these developmental processes, and genetic control is one of the main themes of this book. There are, however, thousands of genes involved in controlling development, and we will focus only on those that appear to have key roles and illustrate general principles.

One of the main tasks of early **embryogenesis**, the development of an embryo from the fertilized egg, is to lay down the overall body plan of the organism and we shall see that different organisms solve this fundamental problem in several ways. The focus of this book is mainly on animal development—that of vertebrates such as frogs, birds, fish, and mammals, and of a selection of invertebrates, such as the sea urchin, ascidians, and, above all, the fruit fly *Drosophila melanogaster* (Fig. 1.1) and the nematode worm *Caenorhabditis elegans*. It is in these last two organisms that our understanding of the genetic control of development is most advanced. We shall look briefly at some aspects of plant development, which differs in some respects from that of animals but involves similar principles.

The development of individual organs such as the vertebrate limb, the insect eye, and the nervous system illustrates multicellular organization and tissue differentiation at later stages in embryogenesis, and we consider

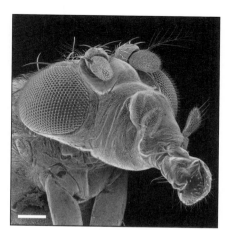

Fig. 1.1 Scanning electron micrograph of the head of an adult *Drosophila melanogaster*. Scale bar = 0.1 mm.

Photograph by D. Scharfe, from Science Photo Library.

Fig. 1.2 Photograph of a lizard, the South Eastern five-lined skink, after it has released its tail in defense. This species can deliberately shed its tail as a technique to avoid capture by predators; and then regenerate it. A piece of discarded tail can be seen below the skink.

Photograph from Oxford Scientific Films.

Fig. 1.3 Photograph of the adult South African claw-toed frog, *Xenopus laevis.* Scale bar = 1 cm.

Photograph courtesy of J. Smith.

some of these systems in detail. We also deal with the development of sexual characteristics. The study of developmental biology, however, goes well beyond the development of the embryo. We also need to understand how some animals can regenerate lost organs (Fig. 1.2), and how post-embryonic growth of the organism is controlled, a process which includes metamorphosis and aging. Taking a wider view, we finally consider how developmental mechanisms have evolved and how they constrain the very process of evolution itself.

One might ask whether it is necessary to cover so many different organisms and developmental systems in order to understand the basic features of development? The answer at present is yes. Developmental biologists do indeed believe that there are general principles of development that apply to all animals, but that life is too wonderfully diverse to find all the answers in a single organism. As it is, developmental biologists have tended to focus their efforts on a relatively small number of animals, chosen originally because they were convenient to study and amenable to experimental manipulation or genetic analysis. This is why some creatures, such as the frog *Xenopus laevis* (Fig. 1.3), the nematode *Caenorhabditis*, and the fruit fly *Drosophila*, have such a dominant place in developmental biology, and are encountered again and again in this book. Indeed, it is very encouraging that so few systems need to be studied in order to understand animal development. Similarly, the thale-cress, *Arabidopsis thaliana*, can be used as a model plant to consider the basic features of plant development.

One of the most exciting and satisfying aspects of developmental biology is that understanding a developmental process in one organism can help to illuminate similar processes elsewhere, for example in organisms much more like ourselves. Nothing illustrates this more dramatically than the influence our understanding of *Drosophila* development, and especially its genetic basis, has had throughout developmental biology. In particular, the identification of genes controlling early embryogenesis in *Drosophila* has led to the discovery of related genes being used in similar ways in the development of mammals and other vertebrates. Such discoveries encourage us to believe in the emergence of general developmental principles.

Frogs have long been a favorite organism for studying development because their eggs are large, and their embryos are robust, easy to grow in a simple culture medium, and relatively easy to experiment on. The South African claw-toed frog *Xenopus* (sometimes mistakenly called the 'South African clawed toad') is the model organism for many aspects of vertebrate development, and the main features of its development (Box 1A, p. 4) serve to illustrate some of the basic stages of development in all animals.

In the rest of this chapter we first look at the history of **embryology**—as the study of developmental biology has been called for most of its existence. The term developmental biology itself is of rather recent origin. We then introduce some key concepts that are used repeatedly in studying and understanding development.

The origins of developmental biology

Many questions in embryology were first posed hundreds, and in some cases thousands, of years ago. Appreciating the history of these ideas helps us to understand why we approach developmental problems in the way that we do today.

1.1 Aristotle first defined the problem of epigenesis and preformation

A scientific approach to explaining development started with Hippocrates in Greece in the 5th century BC. Using ideas that were current at the time, he tried to explain development in terms of the principles of heat, wetness, and solidification. About a century later the study of embryology advanced when the Greek philosopher Aristotle formulated a question that was to dominate much thinking about development until the end of the 19th century. Aristotle addressed the problem of how the different parts of the embryo were formed. He considered two possibilities: one was that everything in the embryo was preformed from the very beginning and simply got bigger during development; the other was that new structures arose progressively, a process he termed epigenesis (which means 'upon formation') and likened metaphorically to the 'knitting of a net'. Aristotle favored epigenesis and his conjecture was correct.

Aristotle's influence on European thought was enormous and his ideas remained dominant well into the 17th century. The contrary view to epigenesis, namely that the embryo was preformed from the beginning, was championed anew in the late 17th century. Many could not believe that physical or chemical forces could mold a living entity like the embryo. Against the contemporaneous background of belief in the divine creation of the world and all living things, they believed that all embryos had existed from the beginning of the world, and that the first embryo of a species must contain all future embryos.

Even the brilliant 17th century Italian embryologist Marcello Malpighi could not free himself from preformationist ideas. While he provided a remarkably accurate description of the development of the chick embryo, he remained convinced, against the evidence of his own observations, that the embryo was already present from the very beginning (Fig. 1.4). He argued that at very early stages the parts were so small that they could not be observed, even with his best microscope. Yet other preformationists believed that the sperm contained the embryo and some even claimed to be able to see a tiny human—a homunculus—in the head of each human sperm (Fig. 1.5).

The preformation/epigenesis issue was the subject of vigorous debate throughout the 18th century. But the problem could not be resolved until one of the great advances in biology had taken place—the recognition that living things, including embryos, were composed of cells.

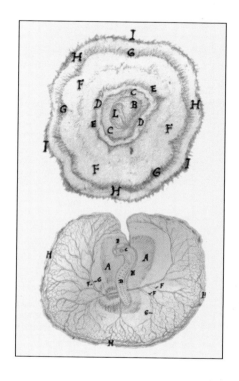

Fig. 1.4 Malpighi's description of the chick embryo. The figure shows Malpighi's drawings, made in 1673, depicting the early embryo (top), and at 2 days' incubation (bottom). His drawings accurately illustrate the shape and blood supply of the embryo. Reprinted by permission of the President and Council of the Royal Society.

Box 1A Basic stages of *Xenopus laevis* development

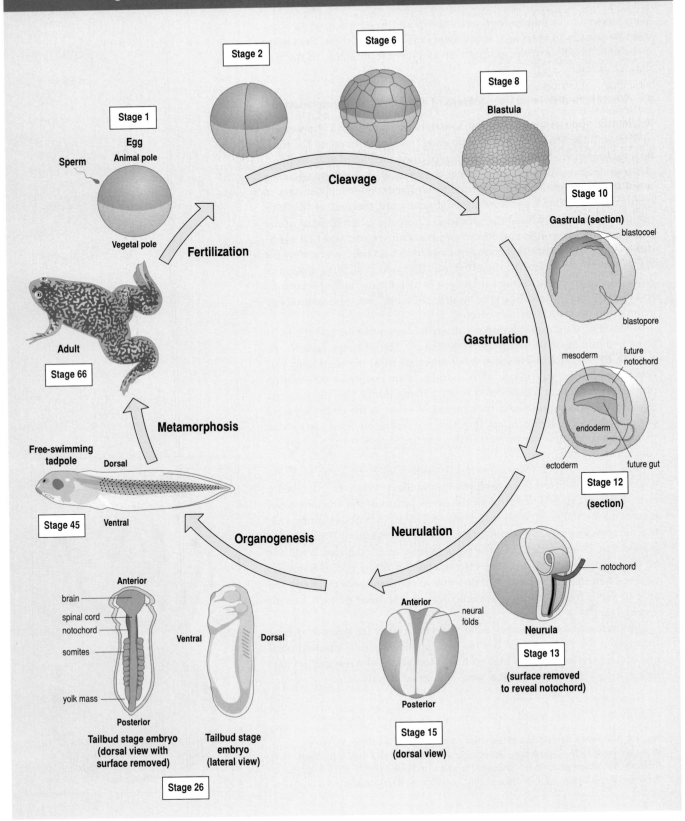

Although vertebrate development is very varied, there are a number of basic stages that can be illustrated by following the development of the frog *Xenopus laevis*, which is a favorite organism for experimental embryology. The unfertilized egg is a large cell. It has a pigmented upper surface (the **animal pole**) and a lower region (the **vegetal pole**) characterized by an accumulation of yolk granules. So even at the beginning, the egg is not uniform; in subsequent development, cells from the animal half become the anterior (head) end of the embryo.

After fertilization of the egg by a sperm, and the fusion of male and female nuclei, **cleavage** begins. Cleavages are mitotic divisions in which cells do not grow between each division, and so with successive cleavages the cells become smaller. After about 12 division cycles, the embryo, now known as a **blastula**, consists of many small cells surrounding a fluid-filled cavity (the **blastocoel**) above the larger yolky cells. Already, changes have occurred within the cells and they have interacted with each other so that some future tissue types—the **germ layers**—have become partly specified. The future **mesoderm** for example, which gives rise to muscle, cartilage, bone, and other internal organs like the heart, blood, and kidney, is present in the blastula as an equatorial band. Adjacent to it is the future **endoderm**, which gives rise to the gut, lungs, and liver. The animal region will give rise to **ectoderm**, which forms both the epidermis and the

nervous system. The future endoderm and mesoderm, which are destined to form internal organs, are still on the surface of the embryo. During the next stage—**gastrulation**—there is a dramatic rearrangement of cells; the endoderm and mesoderm move inside, and the basic body plan of the tadpole is established. Internally, the mesoderm gives rise to a rod-like structure (the **notochord**), which runs from the head to the tail, and lies centrally beneath the future nervous system. On either side of the notochord are segmented blocks of mesoderm called **somites**, which will give rise to the muscles and vertebral column, as well as the dermis of the skin.

Shortly after gastrulation, the ectoderm above the notochord folds to form a tube (the **neural tube**), which gives rise to the brain and spinal cord—a process known as **neurulation**. By this time, other organs, such as limbs, eyes, and gills, are specified at their future locations, but only develop a little later, during **organogenesis**. During organogenesis, specialized cells such as muscle, cartilage, and neurons differentiate. Within 48 hours the embryo has become a feeding tadpole with typical vertebrate features. Because the timing of each stage can vary, depending on environmental conditions, developmental stages in *Xenopus* and other embryos are often denoted by stage numbers rather than by hours of development

1.2 Cell theory changed the conception of embryonic development and heredity

The cell theory developed between 1820 and 1880 by, among others, the German botanist Matthias Schleiden and the physiologist Theodor Schwann, was one of the most illuminating advances in biology, and had an enormous impact. It was at last recognized that all living organisms consist of cells, which are the basic units of life, and which arise only by division from other cells. Multicellular organisms such as animals and plants could then be viewed as communities of cells. Development could not therefore be based on preformation but must be epigenetic, because during development many new cells are generated by division from the egg, and new types of cell are formed. A crucial step forward in understanding development was the recognition in the 1840s that the egg itself is but a single, albeit specialized, cell.

An important advance was the proposal by the 19th century German biologist August Weismann that the offspring does not inherit its characteristics from the body (the soma) of the parent but only from the **germ cells**—egg and sperm—and that the germ cells are not influenced by the body that bears them. Weismann thus drew a fundamental distinction between germ cells and **somatic cells** or body cells (Fig. 1.6). Characteristics acquired by the body during an animal's life cannot be transmitted to the germ cells. As far as heredity is concerned, the body is merely a carrier of germ cells. As the English novelist and essayist Samuel Butler put it: "A hen is only an egg's way of making another egg".

Work on sea urchin eggs showed that after fertilization the egg contains two nuclei, which eventually fuse; one of these nuclei belongs to the egg while the other comes from the sperm. Fertilization therefore results in an

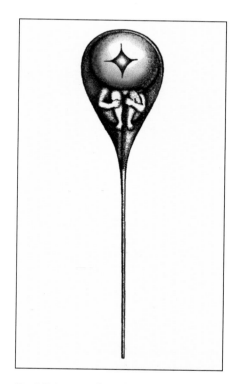

Fig. 1.5 Some preformationists believed that a homunculus was curled up in the head of each sperm. An imaginative drawing, after Nicholas Hartsoeker (1694).

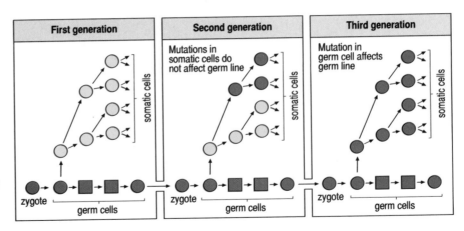

Fig. 1.6 The distinction between germ cells and somatic cells. In each generation germ cells give rise to both somatic cells and germ cells, but inheritance is through the germ cells only. Changes that occur due to mutation in somatic cells can be passed on to their daughter cells but do not affect the germ line.

egg carrying a nucleus with contributions from both parents, and it was concluded that the cell nucleus must contain the physical basis of heredity. The climax of this line of research was the eventual demonstration, toward the end of the 19th century, that the chromosomes within the nucleus of the **zygote** (the fertilized egg) are derived in equal numbers from the two parental nuclei, and the recognition that this provided a physical basis for the transmission of genetic characters according to laws developed by the Austrian botanist and monk Gregor Mendel. The constancy of chromosome number from generation to generation in somatic cells was found to be maintained by reduction division (**meiosis**) in germ cells. The **diploid** precursors to the germ cells contain two copies of each chromosome, one maternal and one paternal. This number is halved by meiosis during formation of the gametes, so that each **haploid germ cell** contains only one copy of each chromosome. The diploid number is restored at fertilization.

1.3 Mosaic and regulative development

Once it was recognized that the cells of the embryo arose by cell division from the zygote, the question of how cells became different from one another emerged. With the increasing emphasis on the role of the nucleus, in the 1880s Weismann put forward a model of development in which the nucleus of the zygote contained a number of special factors or **determinants** (Fig. 1.7). He proposed that as the fertilized egg underwent division (cleavage), these determinants would be distributed unequally to the daughter cells and so would control the cells' future development. The fate of each cell was therefore predetermined in the egg by the factors it would receive during cleavage. This type of model was termed 'mosaic', as the egg

Fig. 1.7 Weismann's theory of nuclear determination. Weismann assumed that there were factors in the nucleus that were distributed asymmetrically to daughter cells during cleavage and directed their future development.

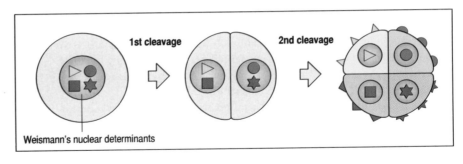

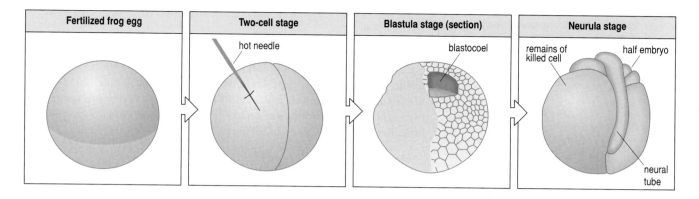

could be considered to be a mosaic of discrete localized determinants. Central to Weismann's theory was the assumption that early cell divisions must be **asymmetric divisions**, so that the daughter cells would be quite different from each other as a result of unequal distribution of nuclear components.

In the late 1880s, initial support for Weismann's ideas came from experiments carried out independently by the German embryologist Wilhelm Roux, who experimented with frog embryos. Having allowed the first cleavage of a fertilized frog egg, Roux destroyed one of the two cells with a hot needle and found that the remaining cell developed into a well-formed half-larva (Fig. 1.8). He concluded that the "development of the frog is based on a mosaic mechanism, the cells having their character and fate determined at each cleavage".

But when Roux's fellow countryman, Hans Driesch, repeated the experiment on sea urchin eggs, he obtained quite a different result (Fig. 1.9). He wrote later:

"But things turned out as they were bound to do and not as I expected; there was, typically, a whole gastrula on my dish the next morning, differing only by its small

Fig. 1.8 Roux's experiment to investigate Weismann's theory of mosaic development. After the first cleavage of a frog embryo, one of the two cells is killed by pricking it with a hot needle; the other remains undamaged. At the blastula stage the undamaged cell can be seen to have divided as normal into many cells that fill half of the embryo. The development of the blastocoel is also restricted to the undamaged half. In the damaged half of the embryo, no cells appear to have formed. At the neurula stage, the undamaged cell has developed into something resembling half a normal embryo.

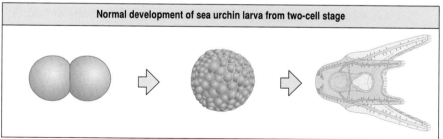

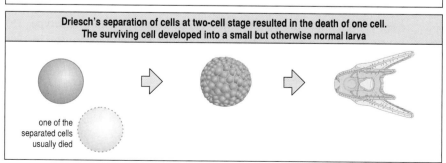

Fig. 1.9 The outcome of Driesch's experiment on sea urchin embryos, which first demonstrated the phenomenon of regulation. After separation of cells at the two-cell stage, the remaining cell develops into a small, but whole, normal larva. This contradicts Roux's earlier finding that if one of the cells of a two-cell frog embryo is damaged, the remaining cell develops into a half-embryo only (see Fig. 1.8).

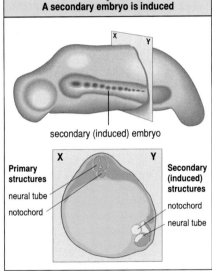

Dorsal lip of blastopore grafted from an unpigmented species of newt to the blastocoel roof of a pigmented species

Triton cristatus gastrula

dorsal lip of blastopore

blastocoel

Triton taeniatus gastrula

A secondary embryo is induced

secondary (induced) embryo

Primary structures

neural tube

notochord

Secondary (induced) structures

notochord

neural tube

Fig. 1.10 The dramatic demonstration by Spemann and Mangold of induction of a new main body axis by the organizer region in the early amphibian gastrula. A piece of tissue (yellow) from the dorsal lip of the blastopore of a newt (*Triton cristatus*) gastrula is grafted to the opposite side of a gastrula from another, pigmented, newt species (*Triton taeniatus*, pink). The grafted tissue induces a new body axis containing neural tube and somites. The unpigmented graft tissue forms a notochord at its new site (see section in lower panel) but the neural tube and the other structures of the new axis have been induced from the pigmented host tissue.

size from a normal one; and this small but whole gastrula developed into a whole and typical larva."

Driesch had completely separated the cells at the two-cell stage and obtained a normal but small larva. That was just the opposite of Roux's result, and was the first clear demonstration of the developmental process known as **regulation**. This is the ability of the embryo to develop normally even when some portions are removed or rearranged. An explanation of Roux's experiment is given in Section 3.2.

1.4 The discovery of induction

Although the concept of regulation implied that cells must interact with each other, the central importance of cell–cell interactions in embryonic development was not really established until the discovery of the phenomenon of **induction**, in which one tissue directs the development of another, neighboring, tissue.

The importance of induction and other cell–cell interactions in development was proved dramatically in 1924 when Spemann and his assistant Hilde Mangold carried out their famous organizer transplant experiment in amphibian embryos. They showed that a partial second embryo could be induced by grafting one small region of a newt embryo onto a new site on another embryo (Fig. 1.10). The grafted tissue was taken from the dorsal lip of the **blastopore**—the slit-like invagination that forms where gastrulation begins on the dorsal surface of the amphibian embryo (see Box 1A). This small region they called the **organizer**, since it seemed to be ultimately responsible for controlling the organization of a complete embryonic body. For their discovery, Spemann received the Nobel Prize for Physiology or Medicine in 1935, one of only two ever given for embryological research. Sadly, Hilde Mangold had died earlier, in an accident, and so could not be honored.

1.5 The coming together of genetics and development

During much of the early part of this century there was little connection between embryology and genetics. When Mendel's laws were rediscovered in 1900 there was a great surge of interest in mechanisms of inheritance, particularly in relation to evolution, but less so in relation to development. Genetics was seen as the study of the transmission of hereditary elements from generation to generation, whereas embryology was the study of how an individual organism develops and, in particular, how cells in the early embryo became different from each other. Genetics seemed, in this respect, to be irrelevant to development.

An important concept that eventually helped to link genetics and embryology was the distinction between **genotype** and **phenotype**. This was first put forward by the Danish botanist Wilhelm Johannsen in 1909. The genetic endowment of an organism—the genetic information it acquires from its parents—is the genotype. Its visible appearance, internal structure, and biochemistry at any stage of development is the phenotype. While the genotype certainly controls development, environmental factors interacting with the genotype influence the phenotype. Despite having identical genotypes, identical twins can develop considerable differences in their phenotypes as they grow up (Fig. 1.11), and these tend to become more evident with age. The problem of development could now be posed in terms of the relationship between genotype and phenotype; how the genetic endowment becomes 'translated' or 'expressed' during development to give rise to a functioning organism.

The coming together of genetics and embryology was a slow and tortuous process. Little progress was made until the nature and function of the genes were much better understood. The discovery in the 1940s that genes encode proteins was a major turning point. As it was already clear that the properties of a cell are determined by the proteins it contains, the fundamental role of genes in development could at last be appreciated. By controlling which proteins were made in a cell, genes could control the changes in cell properties and behavior that occurred during development.

Fig. 1.11 The difference between genotype and phenotype. These 'identical twins' have the same genotype because one fertilized egg split into two during development. Their slight difference in appearance is due to nongenetic factors, such as environmental influences.

Photograph courtesy of José and Jaime Pascual.

Summary

The study of embryonic development started with the Greeks more than 2000 years ago. Aristotle put forward the idea that embryos were not contained completely preformed in miniature within the egg, but that form and structure emerged gradually, as the embryo developed. This idea was challenged in the 17th and 18th centuries by those who believed in preformation, the idea that all embryos that had been, or ever would be, had existed from the beginning of the world. The emergence of the cell theory in the 19th century finally settled the issue in favor of epigenesis, and it was realized that the sperm and egg are single, albeit highly specialized, cells. Some of the earliest experiments showed that very early sea urchin embryos are able to regulate, that is to develop normally even if cells are removed or killed. This established the important principle that development must depend at least in part on communication between the cells of the embryo. Direct evidence for the importance of cell–cell interactions came from the organizer graft experiment carried out by Spemann and Mangold in 1924, showing that the cells of the amphibian organizer region could induce a new partial embryo from host tissue when transplanted into another embryo. The role of the genes in controlling development has only been fully appreciated in the past 30 years and the study of the genetic basis of development has been made much easier in recent times by the techniques of molecular biology.

A conceptual tool kit

Development into a multicellular organism is the most complicated fate a single living cell can undergo; in this lies both the fascination and the challenge of developmental biology. Yet only a few basic principles are needed to start to make sense of developmental processes. The rest of this chapter is devoted to introducing these key concepts. These principles are encountered repeatedly throughout the book, as we look at different organisms and developmental systems, and should be regarded as a conceptual tool kit, essential for embarking on a study of development.

Genes control development by controlling where and when proteins are synthesized, and the number involved is of the order of many thousands— even a reasonable approximation is not known. Their activity sets up complex networks of interactions between different proteins and between

Fig. 1.12 Light micrograph of *Xenopus* eggs after four cell divisions. Scale bar = 1 mm.

Photograph courtesy of J. Slack.

proteins and genes within cells, and hence to interactions between cells. It is these interactions that determine how the embryo develops, and so no developmental process can be attributed to the function of a single gene or single molecule. The amount of information on molecular processes in development is now enormous. We will make no attempt to describe this information in detail, but will concentrate on processes that provide insight into the mechanisms of development and illustrate general principles.

1.6 Development involves cell division, the emergence of pattern, change in form, cell differentiation, and growth

Development is essentially the emergence of organized structures from an initially very simple group of cells. It is convenient to distinguish five main developmental processes, even though in reality they overlap with and influence one another considerably.

Fertilization is followed by a period of rapid cell division where the egg divides into a number of smaller cells (Fig. 1.12). These divisions are known as cleavage divisions and, unlike the cell divisions that take place during cell proliferation and growth of a tissue, there is no increase in cell mass between each division. The cell cycle during cleavage consists simply of phases of DNA replication, mitosis, and cell division with no intervening stage of cell growth. The cleavage stage of embryogenesis thus rapidly divides the embryo into a number of cells, each containing a copy of the genome.

Pattern formation is the process by which a spatial and temporal pattern of cell activities is organized within the embryo so that a well-ordered structure develops. In the developing arm, for example, pattern formation is the process that enables cells to 'know' whether to make the upper arm or fingers, and where the muscles should form. There is no single universal strategy or mechanism of patterning; rather, it is achieved by a variety of cellular and molecular mechanisms in different organisms and at different stages of development.

Pattern formation initially involves laying down the overall **body plan**— defining the main **axes** of the embryo so that the anterior and posterior ends, and the dorsal and ventral sides of the body are specified. One can distinguish at least one main body axis in all multicellular organisms. In animals this refers to the axis that runs from 'head' to 'tail' (antero-posterior). In plants, the main axis runs from the growing tip to the roots. Many animals also have a distinguishable front and back, which defines another axis (dorso-ventral). A striking feature of these axes is that they are almost always at right angles to one another. The two axes may therefore be thought of as making up a coordinate system (Fig. 1.13).

The next stage in pattern formation in animal embryos is allocation of cells to the different germ layers—the ectoderm, mesoderm, and endoderm (Box 1B). During further pattern formation, cells of these germ layers acquire different identities so that organized spatial patterns of cell differentiation emerge, such as the arrangement of skin, muscle, and cartilage in developing limbs, and the arrangement of neurons in the nervous system. In the earliest stages of pattern formation, differences between cells are not easily detected and probably consist of subtle chemical differences caused by a change in activity of a very few genes.

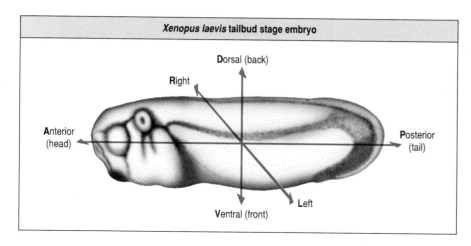

Fig. 1.13 The main axes of a developing embryo. The antero-posterior axis and the dorso-ventral axis are at right angles to one another, as in a coordinate system.

The third important developmental process is change in form, or **morphogenesis**. Embryos undergo remarkable changes in three-dimensional form—you need only look at your hands and feet. At certain stages in development, there are characteristic and dramatic changes in form, of which **gastrulation** is the most striking. Almost all animal embryos undergo gastrulation, during which the gut is formed and the main body plan emerges. During gastrulation, cells on the outside of the embryo move inwards and, in animals such as the sea urchin, gastrulation even transforms a hollow spherical blastula into a gastrula with a hole through the middle—the gut (Fig. 1.14). Morphogenesis in animal embryos can also involve extensive cell migration. Most of the cells of the human face, for

Box 1B Germ layers

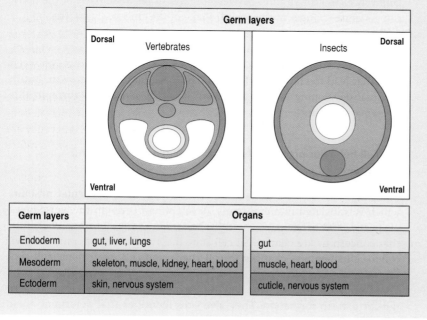

The concept of germ layers is useful to distinguish between regions of the early embryo that give rise to quite distinct types of tissues. It applies to both vertebrates and invertebrates. All the animals considered in this book, except for the coelenterate *Hydra*, are triploblasts, with three germ layers: the endoderm, which gives rise to the gut and its derivatives, such as the liver and lungs in vertebrates; the mesoderm, which gives rise to the skeleto-muscular system, connective tissues, and other internal organs such as the kidney and heart; and the ectoderm, which gives rise to the epidermis and nervous system. These are specified early in development. The boundaries between the different layers can be fuzzy and there are notable exceptions. The neural crest in vertebrates, for example, is ectodermal in origin but gives rise both to neural tissue and to some skeletal elements, which would normally be considered mesodermal in origin.

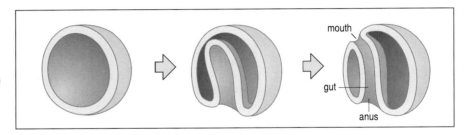

Fig. 1.14 Gastrulation in the sea urchin. Gastrulation transforms the spherical blastula into a structure with a hole through the middle, the gut. The left-hand side of the embryo has been removed.

example, are derived from cells that migrated from the **neural crest**, which originates on the back of the embryo.

The fourth developmental process we must consider here is **cell differentiation**, in which cells become structurally and functionally different from each other, ending up as distinct cell types, such as blood, muscle, or skin cells. Differentiation is a gradual process, cells often going through several divisions between the time they start differentiating and the time they are fully differentiated (when some cell types stop dividing altogether). In humans, the fertilized egg gives rise to at least 250 clearly distinguishable types of cell.

Pattern formation and cell differentiation are very closely interrelated, as we can see by considering the difference between human arms and legs. Both contain exactly the same types of cell—muscle, cartilage, bone, skin, and so on—yet the pattern in which they are arranged is clearly different. It is essentially pattern formation that makes us different from elephants and chimpanzees.

The fifth process is **growth**—the increase in size. In general there is little growth during early embryonic development and the basic pattern and form of the embryo is laid down on a small scale, always less than a few millimeters in extent. Subsequent growth can be brought about in a variety of ways: cell multiplication, increase in cell size, and deposition of extracellular materials such as bone and shell. Growth can also be morphogenetic. Differences in growth rates between organs, or between parts of the body, can generate changes in the overall shape of the embryo (Fig. 1.15).

These five developmental processes are neither independent of each other nor strictly sequential. In very general terms, however, one can think of pattern formation in early development specifying differences between cells that lead to changes in form, cell differentiation, and growth. But in any real developing system there will be many twists and turns in this sequence of events.

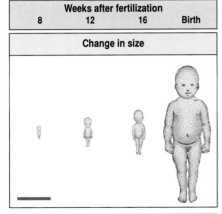

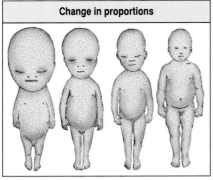

Fig. 1.15 The human embryo changes shape as it grows. From the time the body plan is well established at 8 weeks until birth the embryo increases in length some tenfold (upper panel), while the relative proportion of the head to the rest of the body decreases (lower panel). As a result, the shape of the embryo changes. Scale bar = 10 cm. After Moore, K.L.: 1983.

1.7 Cell behavior provides the link between gene action and developmental processes

Gene expression within cells leads to the synthesis of particular proteins which is translated into embryonic development through the consequent properties and behavior of the cells. The main categories of cell behavior that concern us are changes in cell state (i.e. the pattern of gene activity), cell-to-cell signaling, changes in cell shape and cell movement, cell proliferation, and cell death.

Changing patterns of gene activity during early development are essential for pattern formation. They give cells identities that determine their future behavior and lead eventually to their final differentiation. And, as we saw in the example of induction by the Spemann organizer, the capacity of cells to influence each other's fate by producing and responding to signals is crucial for development. By their movement or change in shape, cells generate the physical forces that bring about morphogenesis (Fig. 1.16). The curvature of a sheet of cells into a tube, as happens in *Xenopus* and other vertebrates during formation of the neural tube, is the result of contractile forces generated by cells changing shape at certain positions within the cell layer. Cells also carry adhesion molecules on their surface, which hold them together as tissues and can also guide the migration of cells such as the neural crest cells of vertebrates, which leave the neural tube to form many structures elsewhere in the body.

Later in development, growth involves cell proliferation, which can also influence final form, as parts of the body grow at different rates. **Programmed cell death**, or **apoptosis**, is also an intrinsic part of the developmental process; cell death in developing hands and feet helps to form the fingers and toes from a continuous sheet of tissues. We can therefore describe and explain developmental processes in terms of how individual cells and groups of cells behave. Because the final structures generated by development are themselves composed of cells, explanations and descriptions at the cellular level can provide an account of how these adult structures are formed.

Since development can be understood at the cellular level, we can pose the question of how genes control development in a more precise form. We can now ask how genes are controlling cell behavior. The many possible ways in which a cell can behave therefore provide the link between gene activity and the morphology of the adult animal—the final outcome of development. Cell biology provides the means by which the genotype becomes translated into the phenotype.

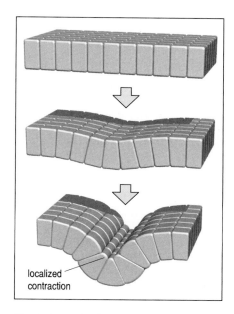

Fig. 1.16 Localized contraction of particular cells can cause a whole sheet of cells to fold. Contraction of a line of cells at their apices due to the contraction of cytoskeletal elements causes a furrow to form in a sheet of epidermis.

1.8 Genes control cell behavior by controlling which proteins are made by a cell

What a cell can do is determined very largely by the proteins present within it. The hemoglobin in red blood cells enables them to transport oxygen; skeletal muscle cells are able to contract because they contain an arrangement of the contractile apparatus of myosin, actin, tropomyosin, and other muscle-specific proteins. All these are very special, or 'luxury', proteins that are not involved in what are considered to be the housekeeping activities of the cell that keep it alive and functioning. Housekeeping activities are common to all cells and include the production of energy and all the intermediate biochemical pathways involved in the breakdown and synthesis of molecules necessary for the life of the cell. Although there are some variations, both qualitative and quantitative, in housekeeping proteins in different cells, they are not important players in development. In development we are concerned primarily with those luxury or tissue-specific proteins that make cells different from one another.

Genes control development mainly by determining which proteins are made in which cells and when. In this sense they are passive participants in the process, compared with the proteins for which they code, which are the agents that directly determine cell behavior. Whether a particular protein is synthesized in a cell requires first that its gene is switched on—that the

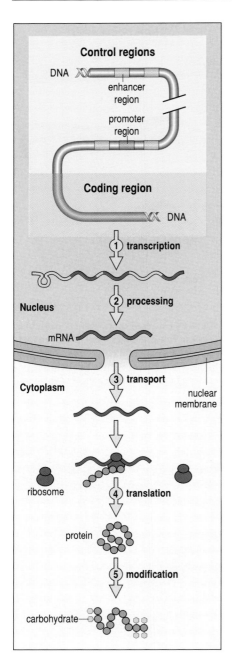

Fig. 1.17 Gene expression and protein synthesis. A protein-coding gene comprises a stretch of DNA that contains a coding region, which contains the instructions for making the protein, and adjacent control regions—promoter and enhancer regions—at which the gene is switched on or off. The promoter region is the site at which RNA polymerase binds and starts transcribing. The enhancer regions may be thousands of base pairs distant from the promoter. Transcription of the gene into RNA (1) may be either stimulated or inhibited by transcription factors that bind to promoter and enhancer regions. The RNA formed by transcription is spliced to remove introns (yellow) and processed within the nucleus (2) to produce mRNA that is exported to the cytoplasm (3) for translation into protein at the ribosomes (4). Control of gene expression and protein synthesis occurs mainly at the level of transcription but can also occur at later stages. For example, mRNA may be degraded before it can be translated. If it is not translated immediately it may be stored in inactive form in the cytoplasm for translation at some later stage. Some proteins require post-translational modification (5) to become biologically active.

gene is being **transcribed** into RNA. The RNA is eventually **translated** into protein. It should also be recognized that, as a result of RNA processing, the RNA transcript may be spliced in different ways to give rise to different **messenger RNAs (mRNAs)**; thus the same gene may produce different proteins with different properties. Translation does not automatically follow transcription, because protein production can also be controlled at later stages in gene expression. Fig. 1.17 shows the main stages in gene expression at which the production of a protein can be controlled. For example, mRNA may be degraded before it can be exported from the nucleus. Even if it reaches the cytoplasm, its translation may be inhibited there. In the eggs of many animals, preformed mRNA is prevented from being translated until after fertilization.

Even if a gene has been transcribed and the mRNA translated, the protein may still not be able to function. Many newly synthesized proteins require further **post-translational modification** before they acquire biological activity. Examples are the digestive enzymes trypsin and chymotrypsin, which only have enzymatic activity after the protein chain is cut and fragments of the protein removed. But, as a rough guide, if a gene is being transcribed, it is reasonable to expect the protein for which it codes to be present in the cell. Untranslated sequences in RNA can also have an important regulatory role. Also, some genes, such as those for ribosomal RNA, do not code for proteins.

Both housekeeping and luxury genes encode proteins that are involved directly in cell function, that is, enzymes, receptor proteins, growth factors, and the structural components of cells. Of particular importance in development are genes that encode **transcription factors** or gene regulatory proteins, proteins that are involved in activating or repressing transcription. Transcription factors act by binding to the control regions of genes (see Fig. 1.17) or by interacting with other DNA-binding proteins.

An intriguing question is how many genes out of the total genome are involved in the development of an embryo? This is not an easy estimate to make. In a few particularly well-studied cases we have rough estimates of the minimum number of genes involved in a particular aspect of development. In the early development of *Drosophila*, about 60 genes are known to be directly involved in pattern formation up to the time of segmentation—that is, when the embryo becomes divided into segments. In the nematode *Caenorhabditis*, some 50 genes are known to be needed to specify a small reproductive structure known as the vulva. All these are quite small numbers compared to the thousands of genes that are active at the same time; some of these are essential to development in that they are necessary for maintaining life, but provide no or little information for influencing development. Recent studies have shown that many genes change their activity during development, and since the total number of genes in the nematode and *Drosophila* is around 19,000 and 13,000 respectively, the total number of genes involved in development is many thousands, probably not less than 10,000 in vertebrates. For flies and vertebrates, a guess at the total number of developmental genes could range from 10,000 to 20,000, respectively. This number can be compared with the 40,000 genes thought to be present in the human genome. One should always remember that by alternative splicing a single gene can code for more than one protein.

1.9 Differential gene activity controls development

All the somatic cells in an embryo are derived from the fertilized egg by successive rounds of mitotic cell division. Thus, with rare exceptions, they all contain identical genetic information, the same as that in the zygote. The differences between cells must therefore be generated by differences in gene activity which lead to the synthesis of different proteins. Turning the correct genes on or off in the correct cells at the correct time becomes the central issue in development. Since the activity of a gene is determined by the transcription factors that bind to its control regions, these control regions and the factors that activate or inhibit them are a fundamental aspect of the developmental process.

As all the key steps in development reflect changes in gene activity, one might be tempted to think of development simply in terms of mechanisms for controlling gene expression. But this would be highly misleading. For gene expression is only the first step in a cascade of cellular processes that lead via protein synthesis to changes in cell behavior and so direct the course of embryonic development. To think only in terms of genes is to ignore crucial aspects of cell biology, such as change in cell shape, that may be initiated at several steps removed from gene activity. In fact, there are very few cases where the complete sequence of events from gene expression to altered cell behavior has been worked out. The route leading from gene activity to a structure such as the five-fingered hand may be tortuous.

1.10 Development is progressive and the fate of cells becomes determined at different times

As embryonic development proceeds, the organizational complexity of the embryo becomes vastly increased over that of the fertilized egg. Many different cell types are formed, spatial patterns emerge, and there are major changes in shape. All this occurs more or less gradually, depending on the particular organism. But, in general, the embryo is first divided up into a few broad regions, such as the future germ layers (mesoderm, ectoderm, and endoderm). Subsequently, the cells within these regions have their fates more and more finely determined. Mesoderm, for example, becomes differentiated into muscle cells, cartilage cells, bone cells, the fibroblasts of connective tissue, and the cells of the dermis of the skin. **Determination**

Normal fate	Region B not determined	Region B determined	Region B specified

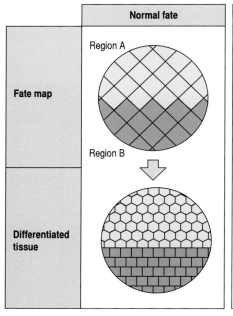 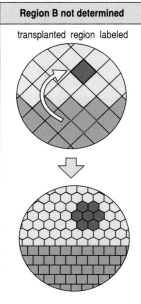 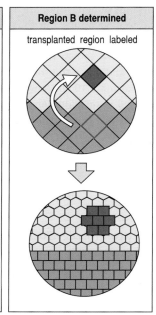 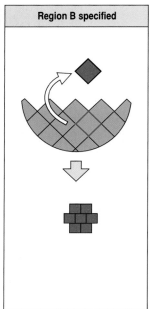

Fig. 1.18 The distinction between cell fate, determination, and specification. In this idealized system, regions A and B differentiate into two different sorts of cell, depicted as hexagons and squares. The fate map (first panel) shows how they would normally develop. If cells from region B are grafted into region A and now develop as A-type cells, the fate of region B has not yet been determined (second panel). By contrast, if region B cells are already determined when they are grafted to region A, they will develop as B cells (third panel). Even if B cells are not determined, they may be specified, in that they will form B cells when cultured in isolation from the rest of the embryo (fourth panel).

implies a stable change in the internal state of a cell, and an alteration in the pattern of gene activity is assumed to be the initial step.

It is important to understand clearly the distinction between the normal fate of a cell at any particular stage, and its state of determination. The **fate** of a group of cells merely describes what they will normally develop into. By marking cells of the early embryo one can find out, for example, which ectodermal cells will normally give rise to the nervous system, and of those, which to the retina in particular. However, that in no way implies that those cells can only develop into a retina, or are already determined or committed to doing so.

A group of cells is called **specified** if, when isolated and cultured in the neutral environment of a simple culture medium away from the embryo, they develop more or less according to their normal fate (Fig. 1.18). For example, cells at the animal pole of the amphibian blastula are specified to form ectoderm, and will form epidermis when isolated. Cells that are specified in this technical sense need not yet be determined, for influences from other cells can change their normal fate; if tissue from the animal pole is put in contact with cells from the vegetal pole, the animal pole tissue will form mesoderm instead of epidermis. At a later stage of development, however, the cells in the animal region have become determined as ectoderm and their fate cannot then be altered. Tests for specification rely on culturing the tissue in a 'neutral' environment lacking any inducing signals, and this is often difficult to achieve.

The state of determination of cells at any particular stage can be demonstrated by transplantation experiments. At the blastula stage of the amphibian embryo, one can graft the ectodermal cells that give rise to the eye into the side of the body and show that the cells develop according to their new position; that is, into mesodermal cells like those of the notochord and somites (Fig. 1.19). At this early stage, their potential for development is much greater than their normal fate. However, if the same operation is done at a later stage, then the future eye region will form

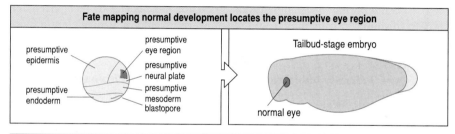

Fate mapping normal development locates the presumptive eye region

presumptive epidermis

presumptive eye region
presumptive neural plate
presumptive mesoderm
blastopore

presumptive endoderm

Tailbud-stage embryo

normal eye

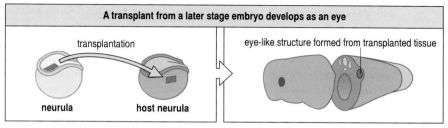

The presumptive eye region of a gastrula, transplanted into the trunk of a neurula embryo forms structures typical of that region

transplantation

gastrula   host neurula

tissues from host   somitic tissue

tissues from transplant

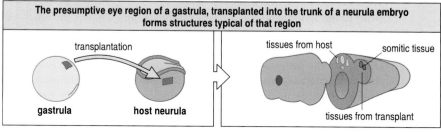

A transplant from a later stage embryo develops as an eye

transplantation

neurula   host neurula

eye-like structure formed from transplanted tissue

Fig. 1.19 Determination of the eye region with time in amphibian development. If the region of the gastrula that will normally give rise to an eye is grafted into the trunk region of a neurula (middle panel), the graft forms structures typical of its new location, such as notochord and somites. If, however, the eye region from a neurula is grafted into the same site (bottom panel), it develops as an eye-like structure, since at this later stage it has become determined.

structures typical of an eye. At the earlier stage the cells were not yet determined as eye cells, whereas later they had become so.

It is a general feature of development that cells in the early embryo are less narrowly determined than those at later stages; with time, cells become more and more restricted in their developmental potential. We assume that determination involves a change in the genes that are expressed by the cell and that this change fixes or restricts the cell's fate, thus reducing its developmental options.

We have already seen how, even at the two-cell stage, the cells of the sea urchin embryo do not seem to be determined. Each thus has the potential to generate a whole new larva (see Section 1.3). Embryos like these, where the potential of cells is much greater than that indicated by their normal fate, are termed **regulative**. Vertebrate embryos are among those capable of considerable regulation. In contrast, those embryos where, from a very early stage, the cells can develop only according to their early fate are termed mosaic. As we saw in Section 1.3, this terminology has a long history. It describes eggs and embryos that develop as if their pattern of future development is laid down very early, even in the egg, as a 'mosaic' of different molecules. The different parts of the embryo then develop quite independently of each other. In such embryos, cell interactions may be quite limited. The demarcation between these two strategies of development is not always sharp, and in part reflects the time when determination occurs; it occurs much earlier in mosaic systems.

The difference between regulative and mosaic embryos reflects the relative importance of cell–cell interactions in each system. The occurrence of regulation absolutely requires interactions between cells, for how else

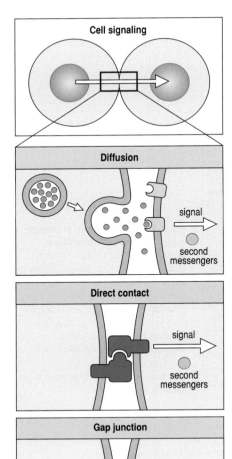

Cell signaling

Diffusion

signal

second messengers

Direct contact

signal

second messengers

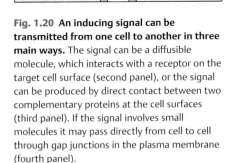

Gap junction

Fig. 1.20 An inducing signal can be transmitted from one cell to another in three main ways. The signal can be a diffusible molecule, which interacts with a receptor on the target cell surface (second panel), or the signal can be produced by direct contact between two complementary proteins at the cell surfaces (third panel). If the signal involves small molecules it may pass directly from cell to cell through gap junctions in the plasma membrane (fourth panel).

could normal development take place and deficiencies be recognized and restored? A truly mosaic embryo, however, would in principle not require such interactions. No purely mosaic embryos are known to exist.

1.11 Inductive interactions can make cells different from each other

Making cells different from one another is central to development. There are numerous examples in development where a signal from one group of cells influences the development of an adjacent group of cells. This is known as induction, and the classic example is the action of the Spemann organizer in amphibians (see Section 1.4). Inducing signals may be propagated over several or even many cells, or be highly localized. The inducing signal from the amphibian organizer affects many cells, whereas other inducing signals may pass from one cell to its immediate neighbor.

There are three main ways in which inducing signals may be passed between cells (Fig. 1.20). First, the signal is transmitted through the extracellular space, usually by means of a secreted diffusible molecule. Second, cells may interact directly with each other by means of molecules located on their surface. In both these cases, the signal is generally received by receptor proteins in the cell membrane and is subsequently relayed through intracellular signaling systems to produce the eventual cellular response. Third, the signal may pass from cell to cell directly through gap junctions. These are specialized protein pores in the apposed plasma membranes, which provide direct channels of communication between the cytoplasm of adjacent cells through which small molecules can pass.

In the case of signaling by a diffusible molecule or by direct contact, the signal is received at the cell membrane. If it is to alter gene expression in the nucleus, the signal has to be transmitted from the membrane to the cell's interior. This process is known generally as **signal transduction**, and is carried out by relays of intracellular signaling molecules that are activated when the extracellular signaling molecule binds to its receptor. These intracellular signaling proteins and small-molecule 'second messengers' interact with one another to transmit the signal onward in the cell. Regulation of protein activity by phosphorylation is an important component of most signaling pathways. Different signals received by a cell can be integrated by the interaction of the different signaling pathways.

A further important feature of induction is whether or not the responding cell is **competent** to respond to the inducing signal. This **competence** may depend on, for example, the presence of the appropriate receptor and transducing mechanism, or on the presence of the particular transcription factors needed for gene activation. A cell's competence for a particular response can change with time; the Spemann organizer can induce changes in the cells it affects only during a restricted time window.

In embryos, it seems that small is generally beautiful where signaling and pattern formation are concerned. Whenever a pattern is being specified, the size of the group of cells involved is barely, if ever, greater than 0.5 mm in any direction; that is, some 50 cell diameters. Many patterns are specified on a much smaller scale and involve just tens or a few hundred cells. This means that the inducing signals involved in pattern formation reach over distances of the order of only ten times a cell diameter. The final organism may be very big, but this is almost entirely due to growth of the basic pattern.

1.12 **The response to inductive signals depends on the state of the cell**

Inductive signals can alter how the induced cells develop. They can thus be regarded as providing the cells with instructions as how to behave. It is important to realize that the response to inductive signals is entirely dependent on the current state of the cell. It not only has to be competent to respond but the number of possible responses is usually very limited. An inductive signal can only select one response from a small number of possible cellular responses. All inductions and signals are not really instructive but are essentially selective. A truly instructive signal would be one that provided the cell with entirely new information and capabilities, by providing it with, for example, new DNA or proteins, which is not thought to occur in development.

As an analogy, consider a juke-box containing a hundred records. When one selects a record to be played, the machine has not been given any new information; rather, one of its repertoire of records has simply been selected. It would be quite different if a new record were added. That would be to provide new information, and would be equivalent to introducing a completely new gene or protein into a cell, which rarely occurs during development. Like the juke-box, a cell's behavior can be changed only by external signals within the constraints provided by its current state.

Because signals are essentially selective and depend on the state of the cell, different signals can activate a particular gene at different stages of development. Genes can be turned on and off repeatedly during development. Much of the complexity of development is the result of changes in cell behavior consequent on a signal from another cell. Having received a signal, a cell can develop autonomously, that is without new signals from another cell, for some time.

The emphasis on inductive signals acting by selection has a further important implication for biological economy. The same signal can be used to elicit different responses in different cells. A particular signaling molecule, for example, can act on several types of cell, evoking a characteristic and different response from each, depending on their developmental history. As we see in future chapters, evolution has been quite lazy with respect to this aspect of development; once a suitable set of signaling molecules has been assembled, its members are used again and again.

1.13 **Patterning can involve the interpretation of positional information**

One general mode of pattern formation can be illustrated by considering the patterning of a simple non-biological model—the French flag (Fig. 1.21). The French flag has a simple pattern: one-third blue, one-third white, and one-third red, along just one axis. Moreover, the flag comes in many sizes but always the same pattern, and thus can be thought of as mimicking the capacity of an embryo to regulate. Given a line of cells, any one of which can be blue, white, or red, and given also that the line of cells can be of variable length, what sort of mechanism is required for the line to develop the pattern of a French flag?

One solution is for the cells to acquire **positional information**. That is, the cells acquire an identity, or **positional value**, that is related to their position along the line with respect to the boundaries at either end. After they have acquired their positional values, the cells interpret this

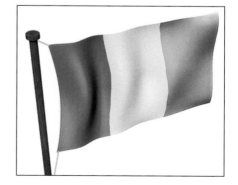

Fig. 1.21 The French flag.

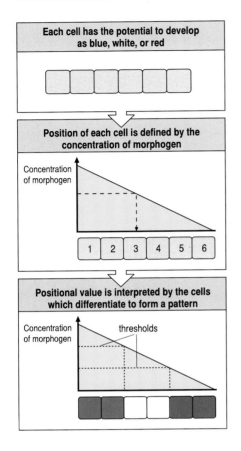

Fig. 1.22 The French flag model of pattern formation. Each cell in a line of cells has the potential to develop as blue, white, or red. The line of cells is exposed to a concentration gradient of some substance and each cell acquires a positional value defined by the concentration at that point. Each cell then interprets the positional value it has acquired and differentiates into blue, white, or red, according to a predetermined genetic program, thus forming the French flag pattern. Substances that can direct the development of cells in this way are known as morphogens. The basic requirements of such a system are that the concentration of substance at either end of the gradient must remain different from each other but constant, thus fixing boundaries to the system. Each cell must also contain the necessary information to interpret the positional values. Interpretation of the positional value is based upon different threshold responses to different concentrations of morphogen.

information by differentiating according to their genetic program. Those in the left-hand third of the line will become blue, those in the middle third white, and so on.

Pattern formation using positional information implies at least two distinct stages: first the positional value has to be specified with respect to some boundary, and then it has to be interpreted. The separation of these two processes has an important implication: it means that there need be no set relation between the positional values and how they are interpreted. In different circumstances, the same set of positional values could be used to generate the Italian flag or another pattern. How positional values will be interpreted will depend on the particular genetic instructions active in the group of cells and will be influenced by their developmental history.

Cells could have their position specified by a variety of mechanisms. The simplest is based on a gradient of some substance. If the concentration of some chemical decreases from one end of a line of cells to the other, then the concentration of that chemical in any cell along the line effectively specifies the position of the cell with respect to the boundary (Fig. 1.22). A chemical whose concentration varies, and which is involved in pattern formation, is called a **morphogen**. In the case of the French flag, we assume a source of morphogen at one end and a sink at the other, and that the concentrations of morphogen at both ends are kept constant but are different from each other. Then, as the morphogen diffuses down the line, its concentration at any point effectively provides positional information. If the cells can respond to **threshold concentrations** of the morphogen—for example, above a particular concentration the cells develop as blue, while below this concentration they become white, and at yet another lower concentration they become red—the line of cells will develop as a French flag (see Fig. 1.22). Thresholds can represent the amount of morphogen that must bind to receptors to activate an intracellular signaling system, or concentrations of transcription factors required to activate particular genes. The use of threshold concentrations of transcription factors to specify position is most beautifully illustrated in the early development of *Drosophila*, as we shall see in Chapter 5.

The French flag model, simple though it is, illustrates two important features of development in the real world. The first is that, even if the length of the line varies, the system regulates and the pattern will still form correctly, given that the boundaries of the system are properly defined by keeping the different concentrations of morphogen constant at either end. The second is that the system could also regenerate the complete original pattern if it were cut in half, provided that the boundary concentrations were re-established. It is therefore truly regulative. We have discussed it here as a one-dimensional patterning problem, but the model can easily be extended to provide two-dimensional patterning (Fig. 1.23).

1.14 Lateral inhibition can generate spacing patterns

Many structures, such as the feathers on a bird's skin, are found to be more or less regularly spaced with respect to one another. Such spacing could occur by the mechanism of **lateral inhibition** (Fig. 1.24). Given a group of cells that all have the potential to differentiate in a particular way, for example as feathers, it is possible to regularly space the cells that will form feathers by a mechanism in which cells that begin to form feathers inhibit the adjacent cells from doing so. This is reminiscent of the spacing of trees in a forest being caused by competition for sunlight and nutrients. In embryos, lateral inhibition is often the result of the differentiating cell secreting an inhibitory molecule that acts locally on the nearest cell neighbors to prevent them developing in a similar way.

1.15 Localization of cytoplasmic determinants and asymmetric cell division can make cells different from each other

Positional specification is just one way in which cells can be given a particular identity. A separate mechanism is based on **cytoplasmic localization** and asymmetric cell division (Fig. 1.25). Asymmetric divisions are so called because they result in daughter cells having properties different from each other, independently of any environmental influence. The properties of such cells therefore depend on their **lineage** or line of descent, and not on environmental cues. Although some asymmetric cell divisions are unequal divisions in that they produce cells of different sizes, this is not usually the most important feature in animals; it is the unequal distribution of cytoplasmic factors that makes the division asymmetric. An alternative way of making the French flag pattern from the egg would be to have chemical differences (representing blue, white, and red) distributed in the egg in the form of determinants that foreshadowed the French flag. When the egg underwent cleavage, these cytoplasmic determinants would become distributed amongst the cells in a particular way and a French flag would develop. This would require no interactions between the cells, which would have their fates determined from the beginning.

Although such extreme examples of mosaic development are not known in nature, there are many cases where eggs or cells divide so that some cytoplasmic determinant becomes unequally distributed between the two daughter cells and they develop differently. This happens at the first cleavage of the nematode egg, and defines the antero-posterior axis of the embryo. The germ cells of *Drosophila* are also specified by cytoplasmic determinants, in this case contained in the cytoplasm located at the posterior end of the egg. However, daughter cells most often become different because of signals from other cells or from their extracellular environment, rather than because of the unequal distribution of cytoplasmic determinants.

One way in which cells can be generated or renewed is by a **stem cell**, which is a cell capable of repeated division and which produces daughter cells that can either remain stem cells or differentiate into a variety of cell

Fig. 1.23 Positional information could be used to generate an enormous variety of patterns. A good example, as shown here, is where the people seated in a stadium each have a position defined by their row and seat numbers. Each position has an instruction about which colored card to hold up, and this makes the pattern. If the instructions were changed, a different pattern would be formed.

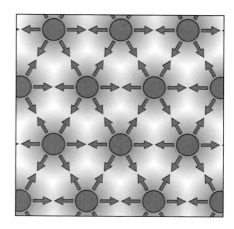

Fig. 1.24 Lateral inhibition can give a spacing pattern. If developing structures produce an inhibitor which prevents the formation of any similar structures in the area adjacent to them, the structures may become evenly spaced.

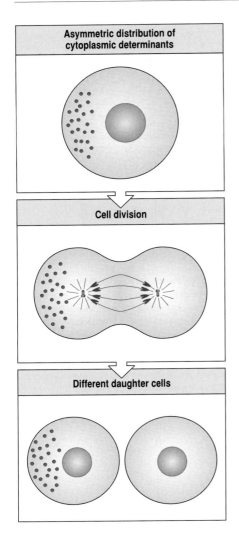

Asymmetric distribution of cytoplasmic determinants

Cell division

Different daughter cells

Fig. 1.25 Cell division with asymmetric distribution of cytoplasmic determinants. If a particular molecule is distributed unevenly in the parent cell, cell division will result in its being shared unequally between the cytoplasm of the two daughter cells. The more localized the cytoplasmic determinant is in the parental cell, the more likely it will be that one daughter cell will receive all of it and the other none, thus producing a distinct difference between them.

types. This is achieved in one of two main ways. cells. In one type of strategy the stem cell divides asymmetrically, so that one daughter remains a stem cell while the other daughter cell undergoes cell differentiation (Fig. 1.26). This difference in behavior may be due to cytoplasmic determinants or to external signals. In the other strategy, the stem cell gives rise to daughter cells that on average give rise to one stem cell and one committed to differentiation. Thus asymmetry of fate in this second system is achieved on a population basis rather than at the level of the division of a single cell.

1.16 The embryo contains a generative rather than a descriptive program

All the information for embryonic development is contained within the fertilized egg. So how is this information interpreted to give rise to an embryo? One possibility is that the structure of the organism is somehow encoded as a descriptive program in the genome. Does the DNA contain a full description of the organism to which it will give rise: is it a blueprint for the organism? The answer is no. The genome contains instead a program of instructions for making the organism—a generative program—in which the cytoplasmic constituents of eggs and cells are essential players along with the genes.

A descriptive program such as a blueprint or a plan describes an object in some detail, whereas a generative program describes how to make an object. For the same object the programs are very different. Consider origami, the art of paper folding. By folding a piece of paper in various directions it is quite easy to make a paper hat or a bird from a single sheet. To describe in any detail the final form of the paper with the complex relationships between its parts is really very difficult, and not of much help in explaining how to achieve it. Much more useful and easier to formulate are instructions on how to fold the paper. The reason for this is that simple instructions about folding have complex spatial consequences. In development, gene action similarly sets in motion a sequence of events that can bring about profound changes in the embryo. One can thus think of the genetic information in the fertilized egg as equivalent to the folding instructions in origami; both contain a generative program for making a particular structure.

A further distinction can be made between the embryo's genetic program and the developmental program of a particular cell or group of cells. The genetic program refers to the totality of information provided by the genes,

Stem cells

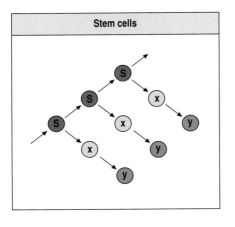

Fig. 1.26 Stem cells. Stem cells (S) are cells that both renew themselves and give rise to differentiated cell types. Thus, a daughter cell arising from stem cell division may either develop into another stem cell or give rise to a quite different type of cell (X), which can then differentiate further (Y).

whereas a developmental program may refer only to that part of the genetic program that is controlling a particular group of cells. As the embryo develops, different parts acquire their own developmental programs as a result of cell–cell interactions and the activities of selected sets of genes. Each cell in the embryo thus has its own developmental program, which may change as development proceeds.

1.17 The reliability of development is achieved by a variety of means

The development of embryos is remarkably consistent and reliable; one need only look at the similar lengths of your legs, each of which develops quite independently for some 15 years (Chapter 10). Embryonic development needs to be reliable in the sense that the adult organism can function properly. For example, both wings of a bird must be very similar in size and shape if it is to be able to fly satisfactorily. How such reliability is achieved is an important problem in development.

Development needs to be reliable in the face of fluctuations that occur either within the embryo or in the external environment. Internal fluctuations include small changes in the concentrations of molecules, and also changes due, for example, to mutations in genes not directly linked to the development of the organ in question. External factors that could perturb development include temperature and environmental chemicals.

One of the central problems of reliability relates to pattern formation. How are cells specified to behave in a particular way in a particular position? Two ways in which reliability is assured are apparent redundancy of mechanism and negative feedback. **Redundancy** is when there are two or more ways of carrying out a particular process; if one fails for any reason another will still function. It is like having two batteries in your car instead of just one. True redundancy—such as having two identical genes in a haploid genome with identical functions—is rare except for cases such as ribosomal RNA, where there can be hundreds of similar genes. Apparent redundancy, on the other hand, where a given process can be specified by several different mechanisms, is probably one of the ways that embryos can achieve such precise and reliable results. It is like being able to draw a straight line with either a ruler or a piece of taut string. This type of situation is not true redundancy, but may give the impression of being so if one mechanism is removed and the outcome is still apparently normal.

Negative feedback also has a role in ensuring consistency; here, the 'end-product' of a process inhibits an earlier stage and thus keeps the level of product constant. The classic example is in metabolism, where the end-product of a biochemical pathway inhibits one of the enzymes that act early in the pathway. Yet another reliability mechanism relates to the complexity of the networks of gene activity that operate in development. There is evidence that these networks are robust and relatively insensitive to small changes in, for example, the rates of the individual processes involved.

1.18 The complexity of embryonic development is due to the complexity of cells themselves

Cells are, in a way, more complex than the embryo itself. The reason is that the network of interactions between proteins and DNA within any individual cell contains many more components and is very much more complex than the interactions between the cells of the developing embryo. Some of the basic cell activities that are involved in bringing about the development of the embryo, such as cell division, response to cell signals, and cell motility are discussed above. Each of these activities involves inter-actions between numerous proteins that vary both with time and with location in the cell. Cell division for example, is a highly complex cell bio-logical program that takes place over time, with a set order of stages, and also requires precise spatial structural organization at mitosis. Signal transduction is another good example. When a cell receives a signal at its cell membrane from another cell, it results in a cascade of protein–protein interactions that may lead, for example, to an ion channel being opened in the membrane, to genes being turned on or off, or to a change in cell behavior such as migration.

Any given cell is expressing hundreds, if not thousands, of different genes at any given time. Much of this gene expression may reflect an intrinsic program of activity, independent of external signals. It is this complexity that determines how cells respond to the signals they receive; how a cell responds to a particular signal depends on its internal state. This state can reflect the cell's developmental history—cells have good memories—and so different cells can respond to the same signal in very different ways. We shall see many examples of the same signals being used over and over again by different cells at different stages of embryonic development, with different biological outcomes.

We have at present only a fragmentary picture of how all the genes and proteins in a cell, let alone a developing embryo, interact with one another. But new technologies are now making it possible to detect the simul-taneous activity of hundreds of genes in a given tissue. How to interpret this information, and make biological sense of the patterns of gene activity revealed, is a massive task for the future.

Summary

Development results from the coordinated behavior of cells. The major processes involved in development are cell division, pattern formation, morphogenesis or change in form, cell differentiation, cell migration, cell death, and growth. Genes control cell behavior by controlling where and when proteins are synthesized, and thus cell biology provides the link between gene action and developmental processes. During development, cells undergo changes in the genes they express, in their shape, in the signals they produce and respond to, in their rate of proliferation, and in their migratory behavior. All these aspects of cell behavior are controlled largely by the presence of specific proteins; gene activity controls which proteins are made. Since the somatic cells in the embryo generally contain the same genetic information, the changes that occur in development are controlled by the differential activity of selected sets of genes in different groups of cells. The genes' control regions are fundamental to this process. Development is progressive and the fate of cells becomes determined at different times. The potential for development of cells in the early embryo is usually much greater than their normal fate, but this potential becomes more restricted as development proceeds. Inductive interactions, involving signals from one tissue or cell to another, are one of the main ways of changing cell fate and directing development. Asymmetric cell divisions, in which cytoplasmic components are unequally distributed to daughter cells, can also make cells different. One widespread means of pattern generation is through positional information; cells first acquire a positional value with respect to boundaries and then interpret their positional values by behaving in different ways. Developmental signals are more selective than instructive, choosing one or other of the developmental pathways open to the cell at that time. The embryo contains a generative, not a descriptive, program—it is more like the instructions for making a structure by paper-folding than a blueprint. A variety of mechanisms, including apparent redundancy and negative feedback, are involved in making development remarkably reliable. The complexity of development lies within the cells.

SUMMARY TO CHAPTER 1

All the information for embryonic development is contained within the fertilized egg—the diploid zygote. The genome of the zygote contains a program of instructions for making the organism. In the working out of this developmental program, the cytoplasmic constituents of the egg and of the cells it gives rise to are essential players along with the genes. Strictly regulated gene activity, by controlling which proteins are synthesized, and where and when, directs a sequence of cellular events that brings about the profound changes that occur in the embryo during development. The major processes involved in development are cell division, pattern formation, morphogenesis, cell differentiation, cell migration, cell death, and growth. All of these processes can be influenced by communication between the cells of the embryo. Development is progressive, with the fate of cells becoming more precisely specified as development proceeds. Cells in the early embryo usually have a much greater potential for development than is evident from their normal fate, and this enables embryos to develop normally even if cells are removed, added, or transplanted to different positions. Cells' developmental potential becomes much more restricted as development proceeds. Many genes are involved in controlling the complex interactions that occur during development, and reliability is achieved in a variety of ways. Thousands of genes control the development of animals and plants, and a full understanding of the processes involved is far from being achieved. While basic principles and certain well-studied developmental systems are quite well understood, there are still many gaps.

REFERENCES

The origins of developmental biology

Cole, F.J.: *Early Theories of Sexual Generation.* Oxford: Clarendon Press, 1930.

Hamburger, V.: *The Heritage of Experimental Embryology: Hans Spemann and the Organizer.* New York: Oxford University Press, 1988.

Needham, J.: *A History of Embryology.* Cambridge: Cambridge University Press, 1959.

Sander, K.: **"Mosaic work" and "assimilating effects" in embryogenesis: Wilhelm Roux's conclusions after disabling frog blastomeres.** *Roux's Arch. Dev. Biol.* 1991, **200**: 237–239.

Sander, K.: **Shaking a concept: Hans Driesch and the varied fates of sea urchin blastomeres.** *Roux's Arch. Dev. Biol.* 1992, **201**: 265–267.

Wilson, E.B.: *The Cell in Development and Heredity.* New York: Macmillan, 1896.

Wolpert, L.: **Evolution of the cell theory.** *Phil. Trans. Roy. Soc. Lond. B* 1995, **349**: 227–233.

A conceptual tool kit

Graveley, B.R.: **Alternative splicing: increasing diversity in the proteomic world.** *Trend Genet.* 2001, **17**: 100–107.

Jordan, J.D., Landau, E.M., Lyengar, R: **Signaling networks: the origins of cellular multitasking.** *Cell* 2000, **103**: 193–200.

Roberts, K., *et al.*: *Essential Cell Biology: An Introduction to the Molecular Biology of the Cell.* New York: Garland Publishing, 1998.

Wolpert, L.: **Do we understand development?** *Science* 1994, **266**: 571–572.

Wolpert, L.: **One hundred years of positional information.** *Trends Genet.* 1996, **12**: 359–364.

Model systems

2

- Model organisms: vertebrates
- Model organisms: invertebrates
- Model systems: plants
- Identifying developmental genes

"In order to understand how we build things, you must first just watch us do it."

Although the development of a wide variety of species has been studied at one time or another, a relatively small number of organisms provide most of our knowledge about developmental mechanisms. We can thus regard them as models for understanding the processes involved. Sea urchins and frogs were the main animals used for the first experimental investigations at the beginning of the century (see Chapter 1) because their developing embryos are both easy to obtain and, in the case of the frog, sufficiently large and robust for relatively easy experimental manipulation, even at quite late stages. Among vertebrates, the frog *Xenopus*, the mouse, the chick, and more recently the zebrafish, are the main model systems now studied. Among invertebrates, the fruit fly *Drosophila* and the nematode worm *Caenorhabditis elegans* are presently the focus of most attention because a great deal is known about their developmental genetics and they can also be deliberately genetically modified.

The reasons for these choices are partly historical—once a certain amount of research has been done on one animal it is more efficient to continue to study it rather than start at the beginning again with another species—and partly a question of ease of study and biological interest. Each species has its advantages and disadvantages as a developmental model. The chick embryo, for example, has long been studied as an example of vertebrate development because fertile eggs are easily available, the embryo withstands experimental microsurgical manipulation very well, and it can be cultured outside the egg. A disadvantage, however, is that little is known about the chick's developmental genetics. However, we know a great deal about the genetics of the mouse, although the mouse is more difficult to study in some ways as development takes place entirely within the mother. Many developmental mutations have been identified in the mouse, and it is also amenable to genetic modification by transgenic techniques (see Box 3A, p. 68). It is also the best experimental model we have for studying mammalian development, including that of humans. The zebrafish (*Brachydanio rerio*) is a very recent addition to the select list of vertebrate model systems; it is easy to breed in large numbers, the embryos are transparent and so cell divisions and tissue movements can be followed visually, and it has great potential for genetic investigations.

Although the genetics of the fruit fly *Drosophila melanogaster* have been studied since the beginning of the 20th century, research on its development has come into its own only in relatively recent times, with the advent of molecular biological techniques that can exploit the immense fund of knowledge of fruit fly genetics. The nematode worm's prominence as a model developmental system is even more recent; it relies on the simplicity of the organism—the embryo has fewer than 1000 cells (not counting the germ cells)—the ability to follow development cell by cell in the transparent embryo, and its invariant cell lineages. The ancestry of every cell in a nematode can be traced back to the zygote through a series of invariant cell divisions. In addition, the nematode is amenable to genetic analysis and genetic modification.

Among plants, the small crucifer *Arabidopsis thaliana* has an increasingly important role as a developmental model for flowering plants, especially in relation to the genetic basis of plant development.

Although this book focuses mainly on the organisms covered in this chapter, and mainly on animal models, there are many other animals and plants whose development is of great interest and has been studied in some detail; some of these, for example molluscs and tunicates, are considered in later chapters. By comparing developmental mechanisms in a variety of organisms it becomes possible to identify those mechanisms that are conserved; that is, mechanisms that are used in different groups of organisms. Comparative studies suggest that, despite immense differences in the details, it is likely that the most basic mechanisms of development are similar in all animals and are derived from the earliest animal ancestors. Thus, elucidation of a developmental process in one animal is often of great help in understanding development in another.

Before embarking on a consideration of the mechanisms of embryonic development, one must first be familiar with the stages that embryos pass through. It is essential to understand clearly how their structure, or morphology, changes during development to give rise to the larval or adult form. This chapter takes you through the main developmental stages of a number of model systems, and introduces some essential terminology. Although this aspect of development is largely descriptive, its importance cannot be underestimated.

The developmental cycles of the main model organisms are dealt with only in outline here; the details and mechanisms of developmental processes will be covered in later chapters. Here, you will find background information and terminology that will be needed when developmental mechanisms are considered later, so regard this chapter also as one to refer back to. Since a major aim of developmental biology is to identify those genes that control development, we finally look briefly at some basic strategies for obtaining and screening developmental mutants.

Model organisms: vertebrates

All vertebrate embryos pass through a similar set of developmental stages. After fertilization, the zygote undergoes cleavage, during which the embryo divides into a number of smaller cells without any increase in overall mass. This is followed by gastrulation, in which cell movements result in the germ layers (see Box 1B, page 11) moving into the correct places for further development. At the end of gastrulation, the **ectoderm** covers the embryo, and the **mesoderm** and **endoderm** have moved inside. One of the earliest mesodermal structures that can be recognized is the rod-shaped notochord, which forms along the antero-posterior axis of the body. This later becomes incorporated into the vertebral column. The vertebral column and the muscles of the trunk and limbs develop from blocks of mesodermal tissue—somites—that form in an antero-posterior sequence on either side of the notochord. All vertebrates form these structures, although the details of their earlier development differ between the vertebrate classes. The brain and spinal cord are derived from ectoderm directly above the notochord that forms the neural tube. The vertebrate 'body plan' is illustrated in Fig. 2.1.

Fig. 2.2 shows the differences in form and shape of a range of early vertebrate embryos. Most of the apparent differences in the development of different vertebrates up to and including gastrulation are related to the nutrition of the embryo. The overall form of the embryo is affected by the amount of yolk in the egg, and the structures the embryo has to develop to utilize this nutrient source. In the case of mammals, whose eggs have no yolk, the extra-embryonic structures of the placenta have to develop. But after gastrulation, all vertebrate embryos pass through a **phylotypic stage**—at which they all more or less resemble each other (see Fig. 2.2) and show the specific features of chordate embryos such as the notochord, somites, and neural tube.

Before describing the development of some model vertebrates, a general point must be made in relation to the staging of development. Amphibians, for example, can develop quite normally over a range of temperatures, but the rate of development changes considerably at different temperatures. Therefore one needs a way of charting development by stages rather than by the time after fertilization. For amphibians, this is provided by tables describing normal development, in which different developmental stages are identified by their main features and given a number. A stage 10 *Xenopus* embryo, for example, refers to an embryo at a very early stage of gastrulation. Similar tables of normal development describe stages in the chick embryo; they are required here since one does not always know the time of fertilization or when the fertilized egg has been placed in the incubator. For mouse embryos, which develop in a much more constant environment, staging is again by reference to the structure of the embryo; somite number is often used as an indication of developmental stage. For earlier stages of mouse development, before the somites have formed, time is often expressed as days *post coitum*, that is, days after mating.

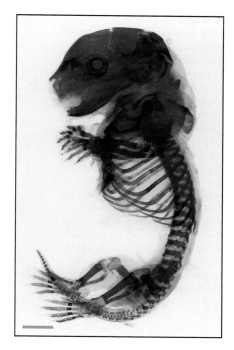

Fig. 2.1 The skeleton of a mouse embryo illustrates the vertebrate body plan. The skeletal elements in this embryo have been stained with a dye. The vertebral column, which develops from blocks of somites, is divided up into cervical (neck), thoracic (chest), lumbar (lower back), and sacral (hip and lower) regions. The paired limbs can also be seen. Scale bar = 1 mm.

Photograph courtesy of M. Maden.

Fig. 2.2 Vertebrate embryos go through a similar phylotypic stage, but the embryos show considerable differences in form before gastrulation. The top row shows representative vertebrate embryos, in cross-section, at the stage corresponding roughly to the *Xenopus* blastula (left panel) just before gastrulation commences. The main determinant of tissue arrangement is the amount of yolk (yellow) in the egg. The mouse embryo (third from left) at this stage has implanted into the uterine wall and thus has already developed some extra-embryonic tissues required for implantation. The mouse embryo proper is a small cup-shaped blastoderm at the center of these structures, seen here in cross-section as a U-shaped epithelial layer. The bottom row shows all of these embryos after gastrulation, at around the phylotypic stage, when they all more-or-less resemble each other and show the characteristic vertebrate features.

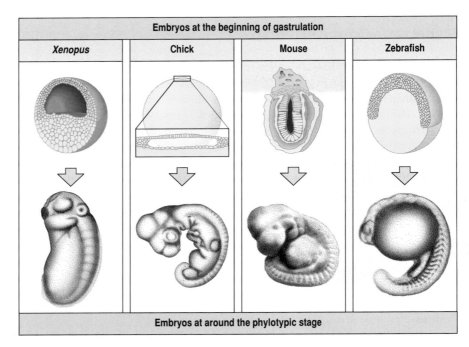

2.1 Amphibians: *Xenopus laevis*

The amphibian most commonly used for developmental work nowadays is the African claw-toed frog *Xenopus laevis*, which is able to develop normally in tap water. The life cycle of *Xenopus* is shown in Fig. 2.3 (opposite). Much classical embryology was, however, done on embryos of newts and salamanders. These belong to an order of amphibians different from that of *Xenopus*, and their development is not the same in some details. A great advantage of *Xenopus* is that its fertilized eggs are easy to obtain; females and males need only be injected with the human hormone chorionic gonadotropin and put together overnight. Eggs can also be fertilized in a dish by adding sperm to eggs released after hormonal stimulation of the female. The embryos of *Xenopus* are extremely hardy and, like those of many other amphibians, are highly resistant to infection after microsurgery. The description of amphibian development in this chapter will focus almost exclusively on *Xenopus*. The large eggs of amphibians (1–2 mm in diameter) are invaluable for experimental manipulation. It is also easy to culture fragments of early *Xenopus* embryos in a simple, chemically defined solution.

The mature *Xenopus* egg has a dark, pigmented **animal region** and a pale, yolky, and heavier **vegetal region** (Fig. 2.4). Before fertilization, the egg is enclosed in a protective **vitelline membrane**, which is embedded in a gelatinous coat. Meiosis is not yet complete and, while the first meiotic division has resulted in a small cell—a **polar body**—forming at the animal pole, the second meiotic division is completed only after fertilization, when the second polar body also forms at the animal pole (Box 2A).

At fertilization, one sperm enters the egg in the animal region. The egg completes meiosis and the egg and sperm nuclei then fuse to form the diploid zygote nucleus. The vitelline membrane lifts off the egg surface and

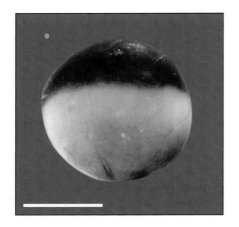

Fig. 2.4 The unfertilized egg of *Xenopus*. The surface of the animal half (top) is pigmented and the paler, vegetal half of the egg is heavy with yolk. Scale bar = 1 mm.

Photograph courtesy of J. Smith.

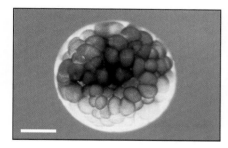

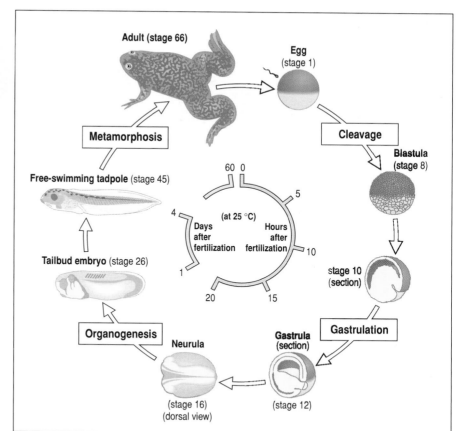

Fig. 2.3 Life cycle of the African claw-toed frog *Xenopus laevis*. The numbered stages refer to standardized stages of *Xenopus* development. More stages can be seen in the larger figure in Box 1A, p. 4. The photographs show: an embryo at the blastula stage (top, scale bar = 0.5 mm); a tadpole at stage 45 (middle, scale bar = 1 mm); and an adult frog (bottom, scale bar = 1 cm).

Photographs courtesy of J. Slack (top, from Alberts, B., et al.: 1994) and J. Smith (middle and bottom).

Box 2A Polar body formation

Polar bodies are small cells formed by meiosis during the development of an oocyte into an egg. In this highly schematic illustration, the segregation of only one pair of chromosomes is shown for simplicity. There are two cell divisions associated with meiosis, and one daughter from each division is almost always very small compared with the other, which becomes the egg—hence the term polar bodies for these smaller cells.

The timing of meiosis in relation to the development of the oocyte varies in different animals, and in some species meiosis is completed and the second polar body formed only after fertilization. In general, polar body formation is of little importance for later development, but in some animals the site of formation is a useful marker for the embryonic axes.

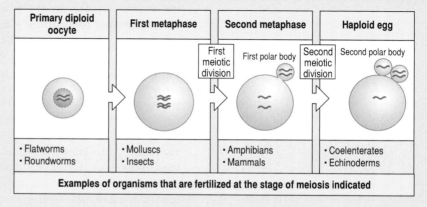

Primary diploid oocyte	First metaphase	Second metaphase	Haploid egg
• Flatworms • Roundworms	• Molluscs • Insects	• Amphibians • Mammals	• Coelenterates • Echinoderms

Examples of organisms that are fertilized at the stage of meiosis indicated

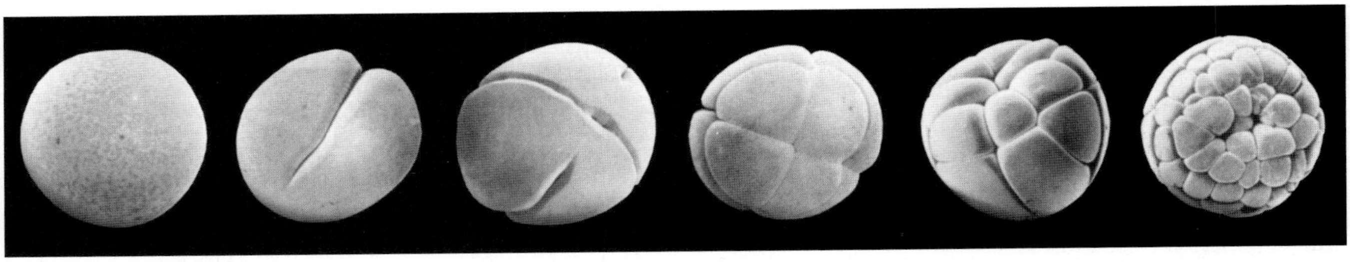

Fig. 2.5 Cleavage of the *Xenopus* embryo. The *Xenopus* embryo undergoes successive cleavages at intervals of about 20 minutes.

Photographs courtesy of R. Kessel, from Kessel, R.G., et al.: 1974.

within about 15 minutes the egg has rotated within it under the influence of gravity, so that the heavier, yolky, vegetal region is now downward. About an hour after fertilization, there is a rotation of the egg cortex—a gel-like superficial layer beneath the plasma membrane—which moves relative to the cytoplasm beneath it. As we shall see in Chapter 3, cortical rotation determines the future dorsal side of the *Xenopus* embryo, which develops opposite the site of sperm entry.

The first cleavage occurs along the animal–vegetal axis within 90 minutes of fertilization, and divides the embryo into equal halves (Fig. 2.5). Further cleavages follow rapidly at intervals of about 20 minutes. The second cleavage is also along the animal–vegetal axis but at right angles to the first. The third cleavage is equatorial, at right angles to the first two, and divides the embryo into four animal cells and four larger vegetal cells. The cells deriving from cleavage divisions are often called **blastomeres**. Continued cleavage results in formation of smaller and smaller blastomeres, since there is no growth between cell divisions. Cells at the vegetal pole are larger than those at the animal pole. Inside this spherical mass of cells, a fluid-filled cavity—the blastocoel—develops in the animal region and the embryo is now called a blastula.

At the end of blastula formation the *Xenopus* embryo has gone through about 12 cell divisions and is made up of several thousand cells. At the blastula stage, the mesodermal and endodermal germ layers that give rise to internal structures are located in the equatorial and vegetal regions and are essentially on the outside of the embryo, while the ectoderm, which will eventually cover the whole of the embryo, is still confined to the animal region (Fig. 2.6, first panel). The belt of tissue around the equator which, as we shall see, plays a crucial part in future development, is known as the **marginal zone**. At this stage, the blastula is in the form of a hollow sphere with radial symmetry. The cell movements of gastrulation convert this into a three-layered structure with clearly recognizable antero-posterior and dorso-ventral axes and bilateral symmetry.

Gastrulation involves extensive cell movements and rearrangement of the tissues of the blastula so that they become located in their proper positions in relation to the overall body plan of the animal. Because it involves changes in form in three dimensions, gastrulation can be quite difficult to visualize. Gastrulation is initiated by a small slit-like infolding—the **blastopore**—that forms on the surface of the blastula on the dorsal side (see Fig. 2.6, second panel). Once gastrulation has started, the embryo is known as a **gastrula**. The layers of future endoderm and mesoderm in the marginal zone move inside the gastrula through the dorsal lip of the blastopore and converge and extend along the antero-posterior axis beneath the ectoderm, while the ectoderm spreads downward to cover the whole

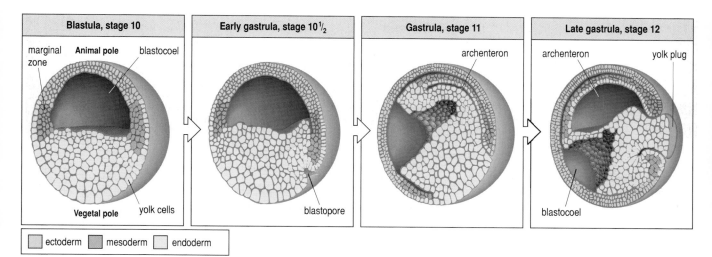

| Blastula, stage 10 | Early gastrula, stage 10½ | Gastrula, stage 11 | Late gastrula, stage 12 |

marginal zone · **Animal pole** · blastocoel

Vegetal pole · yolk cells

blastopore

archenteron

archenteron · yolk plug

blastocoel

☐ ectoderm ☐ mesoderm ☐ endoderm

embryo. The layer of dorsal endoderm is closely applied to the mesoderm; the space between it and the yolky vegetal cells is known as the **archenteron** (see Fig. 2.6, third panel) and is the precursor of the gut cavity. The inward movement of endoderm and mesoderm eventually spreads to form a complete circle around the blastopore.

By the end of gastrulation, the blastopore has closed, the dorsal mesoderm lies beneath the dorsal ectoderm, and the lateral mesoderm begins to spread in a ventral direction on either side; the inner surface of the archenteron becomes completely covered by a layer of endoderm, forming the gut. At the same time, the ectoderm has spread to cover the whole embryo by a process known as **epiboly**. There is still a large amount of yolk present, which provides nutrients until the larva—the tadpole—starts feeding.

During gastrulation, the mesoderm in the dorsal region develops into two main structures, the notochord and the somites. The notochord is a stiff, rod-like structure that forms along the dorsal midline and eventually becomes incorporated into the vertebrae. The somites form by segmentation of the mesoderm lying immediately either side of the notochord. Somites are formed in pairs, and segmentation proceeds in an antero-posterior direction.

Gastrulation is succeeded by neurulation—the formation of the neural tube, the early embryonic precursor of the central nervous system. While the notochord and somites are developing, the neural plate ectoderm above them begins to develop into the neural tube and the embryo is then called a **neurula**. The early sign of neural development is the formation of the **neural folds**, which form on the edges of the **neural plate**. These rise up, fold toward the midline and fuse together to form the **neural tube**, which sinks beneath the epidermis (Fig. 2.7). The anterior neural tube gives rise to the brain; further back, the neural tube overlying the notochord will develop into the spinal cord.

A section taken across the middle of the body shows the internal structure of the *Xenopus* embryo just after the end of neurulation (Fig. 2.8). Now that the germ layers are in place they begin to develop into specific tissues. The main structures that can be recognized at this stage are the neural tube, the notochord, the somites, the **lateral plate mesoderm**, and the endoderm lining the gut. By this stage, different parts of the somites can be distinguished: the most dorsal region has formed the **dermatome**, which

Fig. 2.6 Gastrulation in amphibians. The blastula (first panel) contains several thousand cells and there is a fluid-filled cavity, the blastocoel, beneath the cells at the animal pole. Gastrulation begins (second panel) at the blastopore, which forms on the dorsal side of the embryo. Future mesoderm and endoderm of the marginal zone move inside at this site through the dorsal lip of the blastopore, the mesoderm ending up sandwiched between the endoderm and ectoderm in the animal region (third panel). The tissue movements create a new internal cavity—the archenteron—which will become the gut. Endoderm in the ventral region also moves inside through the ventral lip of the blastopore (fourth panel) and will eventually completely line the archenteron. At the end of gastrulation the blastocoel has considerably reduced in size. After Balinsky, B.I.: 1975.

Sagittal section	Dorsal view of embryo	Transverse section

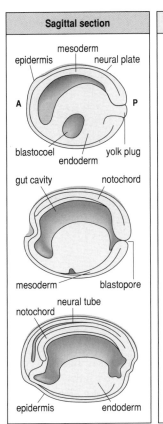

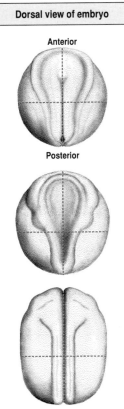

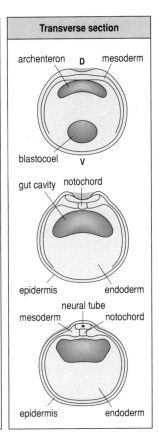

Fig. 2.7 Neurulation in amphibians.
Top row: the notochord begins to form in the midline. At the same time the neural plate develops neural folds. Middle and bottom rows: the neural folds come together in the midline to form the neural tube, from which the brain and spinal cord will develop. During neurulation, the embryo elongates along the antero-posterior axis. The left panel shows sections through the embryo in the planes indicated by the red dotted lines in the center panel. The center panel shows dorsal surface views of the amphibian embryo. The right panel shows sections through the embryo in the planes indicated by the blue dotted lines in the center panel.

Fig. 2.8 A cross-section through a stage 22 *Xenopus* embryo just after gastrulation and neurulation are completed. The germ layers are now all in place for future development and organogenesis. The most dorsal parts of the somites have already begun to differentiate into the dermatome, which will give rise to the dermis. Scale bar = 0.2 mm.

Photograph from Hausen, P., Riebesell, M.: 1991.

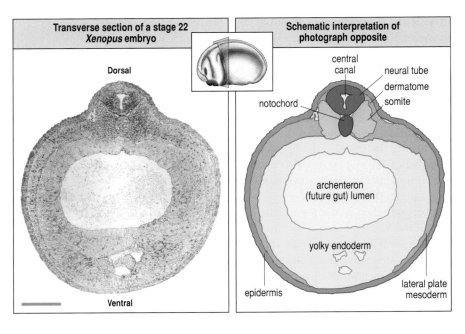

Transverse section of a stage 22 *Xenopus* embryo	Schematic interpretation of photograph opposite

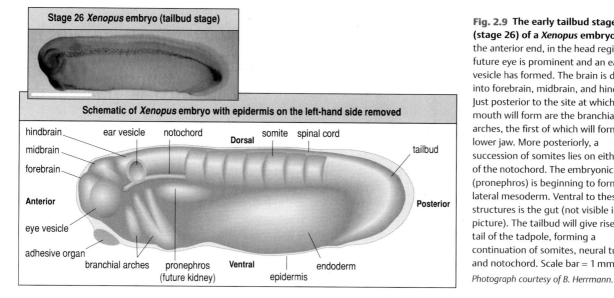

Fig. 2.9 The early tailbud stage (stage 26) of a *Xenopus* embryo. At the anterior end, in the head region, the future eye is prominent and an ear vesicle has formed. The brain is divided into forebrain, midbrain, and hindbrain. Just posterior to the site at which the mouth will form are the branchial arches, the first of which will form the lower jaw. More posteriorly, a succession of somites lies on either side of the notochord. The embryonic kidney (pronephros) is beginning to form from lateral mesoderm. Ventral to these structures is the gut (not visible in this picture). The tailbud will give rise to the tail of the tadpole, forming a continuation of somites, neural tube, and notochord. Scale bar = 1 mm.

Photograph courtesy of B. Herrmann.

will give rise to the dermis. The rest of the somite gives rise to the vertebrae and to the muscles of the trunk. The unsegmented lateral plate mesoderm, lying lateral and ventral to the somites, gives rise to tissues of the heart and kidney, as well as to the gonads and gut muscle, while the most ventral mesoderm gives rise to the blood-forming tissues. The endoderm lining the gut will bud off organs such as the liver and lungs.

The embryo now begins to look something like a tadpole and we can recognize the main vertebrate features (Fig. 2.9). At the anterior end the brain is already divided up into a number of regions, and the eye and ear have begun to develop. There are also three branchial arches, of which the most anterior one will form the lower jaw. More posteriorly, the somites and notochord are well developed. The post-anal tail of the tadpole is formed last. It develops from the tailbud which, at the dorsal lip of the blastopore, gives rise to the continuation of notochord, somites, and neural tube.

Many other internal structures in vertebrates are formed from **neural crest cells**. They come from tissue at the tip of the neural folds, and after neural tube fusion detach and migrate as single cells between the mesodermal tissues. Neural crest gives rise to a remarkable variety of tissues, including the sensory and autonomic nervous systems, the bones of the skull, and pigment cells. They provide the exception to the general rule that ectodermal cells form either nervous system or epidermis, since they also give rise to cartilage.

After organogenesis is completed, the mature tadpole hatches out of its jelly covering and begins to swim and feed. Later, the tadpole larva will undergo metamorphosis to give rise to the adult frog; the tail regresses and the limbs form.

2.2 Birds: the chicken

Avian embryos are very similar to those of mammals in the morphological complexity of the embryo and the general course of embryonic development, but are easier to obtain and observe. Many observations and

Structure of the fertilized hen's egg when laid

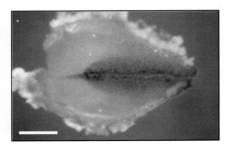

egg shell
egg white
blastoderm (embryo)
yolk
air sac
shell membranes   vitelline membrane
yolk balancer

Fig. 2.10 The development of the hen egg at the time of laying. Cleavage begins after fertilization while the egg is still in the oviduct. The albumen (egg white) and shell are added during the egg's passage down the oviduct. At the time of laying the embryo is a disc-shaped cellular blastoderm lying on top of a massive yolk, which is surrounded by the egg white and shell.

manipulations can be carried out simply by opening the egg, but the embryo can also be cultured outside the egg. This is particularly convenient for some experimental microsurgical manipulations and investigation of the effects of chemical compounds. The later development of a chick embryo is similar to that of a mouse embryo, and so provides a valuable complement to studies of mouse embryology.

The egg is fertilized and begins to undergo cleavage while still in the hen's oviduct. The cytoplasm and nucleus of the fertilized egg is confined to a small patch, several millimeters in diameter, lying on a large mass of yolk. Cleavage in the oviduct results in the formation of a disc of cells called a **blastodisc** or **blastoderm**. During a 20 hour passage down the oviduct, the

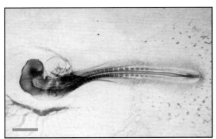

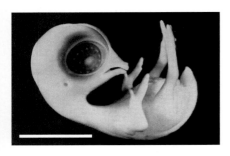

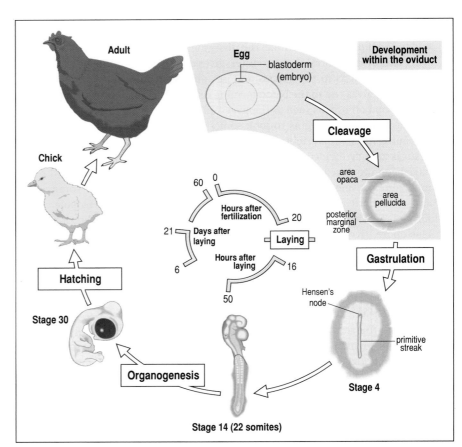

Fig. 2.11 Life cycle of the chicken. The egg is fertilized in the hen and by the time it is laid, cleavage is complete and a cellular blastoderm lies on the yolk. After gastrulation, the primitive streak forms. Regression of Hensen's node is associated with somite formation. The photographs show: the primitive streak surrounded by the area pellucida (top, scale bar = 1 mm); a stage 14 embryo (50–53 hours after laying) with 22 somites (the head region is well defined and the transparent organ adjacent to it is the ventricular loop of the heart) (middle, scale bar = 1 mm); a stage 35 embryo, about 8½–9 days after laying, with a well-developed eye and beak (bottom, scale bar = 10 mm).

Top photograph courtesy of B. Herrmann, from Kispert, A., et al.: 1995.

egg becomes surrounded by albumen (egg white), the shell membranes, and the shell (Fig. 2.10 opposite). At the time of laying, the blastoderm, which is analogous to the amphibian blastula, is composed of some 60,000 cells. The chick developmental cycle is shown in Fig. 2.11 (opposite).

After the egg is laid, cleavage continues with the formation of the furrows. The early cleavage furrows extend downward from the surface of the cytoplasm but do not completely separate the cells, whose ventral faces initially remain open to the yolk. Cleavage results in a circular blastoderm several cells thick. Its central region, which overlies a cavity, is translucent and is known as the **area pellucida**, in contrast to the outer region, which is in the darker **area opaca** (Fig. 2.12). Between the area pellucida and the yolk is a cavity—the subgerminal space—and a layer of cells called the **hypoblast** now develops over the yolk. The hypoblast cells come from two sources: the **posterior marginal zone**, which lies at the junction between the area opaca and area pellucida at the posterior of the embryo, and the

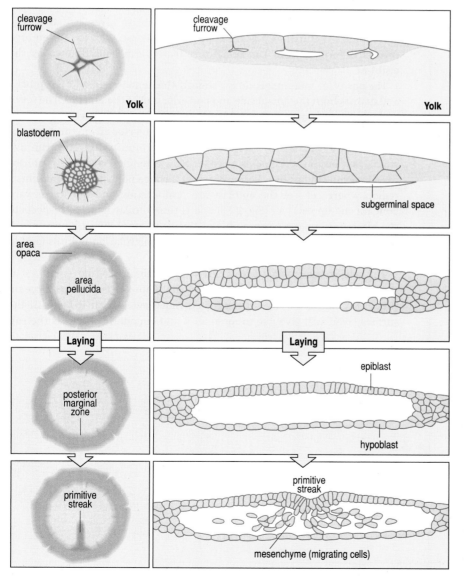

Fig. 2.12 Cleavage and epiblast formation in the chick embryo. By the time the egg is laid, cleavage has divided the small area of egg cytoplasm free from yolk into a disc-shaped cellular blastoderm. The first cleavage furrows extend downward from the surface of the egg cytoplasm and initially do not separate the blastoderm completely from the yolk. In the cellular blastoderm the central area overlying the subgerminal space is called the area pellucida and the marginal region the area opaca. The hypoblast forms as a layer of cells overlying the yolk and will give rise to extra-embryonic structures, while the upper layers of the blastoderm—the epiblast—give rise to the embryo proper.

Fig. 2.13 Ingression of mesoderm and endoderm during gastrulation in the chick embryo. Gastrulation begins with the formation of the primitive steak, a region of proliferating and migrating cells, which elongates from the posterior marginal zone. Future mesodermal and endodermal cells migrate through the primitive streak into the interior of the blastoderm. During gastrulation, the primitive streak extends about half-way across the area pellucida (see Fig. 2.12). At its anterior end an aggregation of cells known as Hensen's node forms. As the streak extends, cells of the epiblast move toward the primitive streak (arrows), move through it, and then outward again underneath the surface to give rise to the mesoderm and endoderm internally, the latter displacing the hypoblast. Adapted from Balinsky, B.I., et al.: 1975.

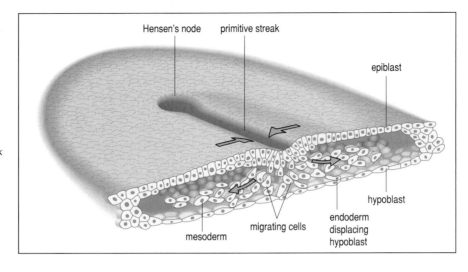

Fig. 2.14 Regression of Hensen's node. After extending about half-way across the blastoderm, the primitive streak begins to regress, with Hensen's node moving in a posterior direction as the head fold and neural plate begin to form. As the node moves backward, the notochord develops in the area anterior to it and somites begin to form on either side of the notochord.

overlying cells of the blastoderm. The hypoblast gives rise to extra-embryonic structures such as the stalk of the yolk sac, whereas the embryo proper is formed from the remaining blastoderm cells, known as the **epiblast**.

The posterior marginal zone is a slightly thickened region of the epiblast and defines both the dorsal side and posterior end of the embryo. The onset of gastrulation is marked by the development of the **primitive streak**. This is the forerunner of the antero-posterior axis; it develops from the posterior marginal zone and is fully extended by 16 hours after laying. The primitive streak first becomes visible as a denser strip extending from the posterior marginal zone to just over half-way across the area pellucida. It represents a region where cells of the epiblast are proliferating and moving inward beneath the upper layer (Fig. 2.13) and is thus similar in some respects to the blastopore region of amphibians. Unlike in amphibians however, cell proliferation and growth in size occur during gastrulation in birds (and in mammals).

The cells in the epiblast of the posterior marginal zone move forward as the streak extends across the area pellucida. Those cells that converge on the streak, move through it, and then move away from it beneath the surface layer, will give rise to mesoderm and endoderm, whereas the sur-

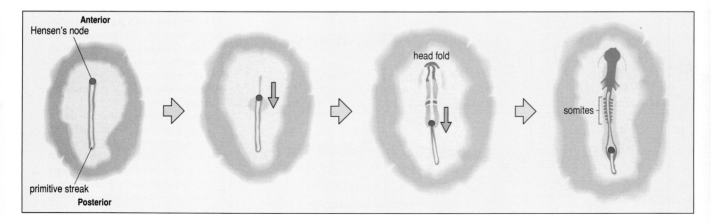

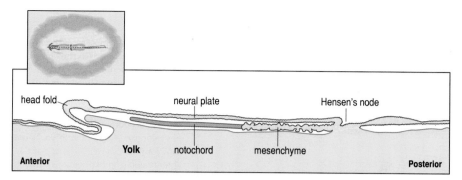

head fold   neural plate   Hensen's node

Yolk   notochord   mesenchyme

Anterior   **Posterior**

Fig. 2.15 Head fold and notochord formation during node regression in the chick embryo. The diagram shows a sagittal section through the chick embryo (inset, dorsal view) at the stage of head fold formation as Hensen's node starts to regress. As the node regresses, the notochord starts to form anterior to it; the undifferentiated mesenchyme on either side of the notochord will form somites.

face layer of the epiblast gives rise to the ectoderm. The prospective endoderm displaces the hypoblast and the mesoderm forms a layer between ectoderm and endoderm.

During gastrulation, the area pellucida changes from circular to pear shaped, and at the anterior end of the primitive streak a condensation of cells known as **Hensen's node** is formed. Once most of the mesoderm and endoderm has moved inward, the primitive streak begins to regress, Hensen's node moving toward the posterior end of the embryo (Fig. 2.14). The head region of the embryo is demarcated anterior to the node by the head fold, an infolding of the blastoderm composed of both ectoderm and endoderm. Cells from Hensen's node give rise to the notochord and contribute to the somites during node regression (Fig. 2.15). As the node regresses in a posterior direction, notochord and somites form immediately anterior to it (Fig. 2.16); by 25 hours after laying, about seven pairs of somites have formed. Somite formation progresses in the posterior direction at a rate of about one pair of somites per hour, with somites on each side of the notochord being formed together. The somites are formed in the presomitic mesoderm, which lies between Hensen's node and the last-formed somite.

As the notochord is formed, the neural tube begins to develop as a pair of folds on either side of the midline of the neural plate ectoderm above it. Unlike in *Xenopus*, where the neural tube forms at the same time along the whole length of the midline, folding and closure of the chick embryo neural tube starts at the anterior end and proceeds in a posterior direction (Fig. 2.17). The folds fuse in the dorsal midline and cells detach from the neural crest on either side of the site of fusion. At the same time the head fold develops, separating the head from the surface of the epiblast. Accompanying neurulation and development of the head fold, the embryo also folds on the ventral side to form the gut. This brings the two heart rudiments together to form one organ lying ventral to the gut. The further development of the mesoderm is rather similar to that of *Xenopus*, the somites giving rise to the vertebrae, the axial and limb muscles, and the dermis. By 2 days after laying, the embryo has reached the 20-somite stage (Fig. 2.18).

By 3 days after laying, 40 somites have formed, the head is well developed, the heart is formed, and the limbs are beginning to develop. Blood vessels and blood islands, where hematopoiesis is occurring, have developed in the extra-embryonic tissues; the vessels connect up with those of the embryo to provide a circulation with a beating heart.

At this stage, the embryo turns on its side and the head is strongly flexed.

Fig. 2.16 Scanning electron micrograph of chick early somites and neural tube. There are blocks of somites adjacent to the neural tube and the notochord lies beneath it. The lateral plate mesoderm flanks the somites. Scale bar = 0.1 mm.

Photograph courtesy of J. Wilting.

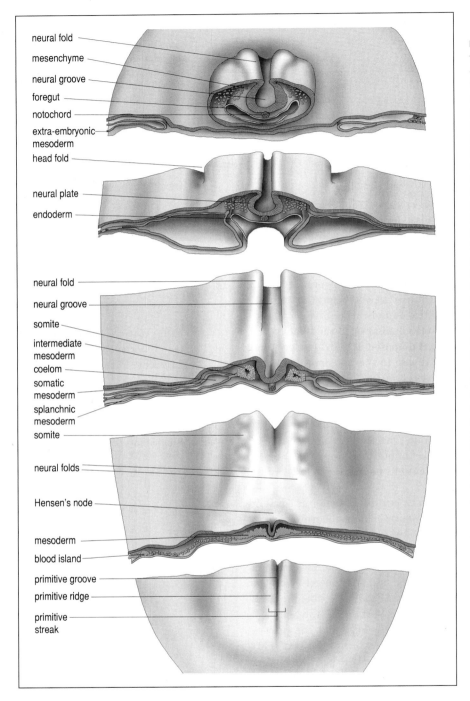

labels on figure:
neural fold
mesenchyme
neural groove
foregut
notochord
extra-embryonic mesoderm
head fold
neural plate
endoderm
neural fold
neural groove
somite
intermediate mesoderm
coelom
somatic mesoderm
splanchnic mesoderm
somite
neural folds
Hensen's node
mesoderm
blood island
primitive groove
primitive ridge
primitive streak

Fig. 2.17 Development of the neural tube and mesoderm in the chick embryo. Once the notochord has formed, neurulation begins, following notochord formation in an anterior to posterior direction. The figure shows a series of sections along the antero-posterior axis of a chick embryo. Neural tube formation is well advanced at the anterior end (top two sections), where the head fold has already separated the future head from the rest of the blastoderm and the ventral body fold has brought endoderm from both sides of the body together to form the gut. During neurulation, the neural plate changes shape: neural folds rise up on either side and form a tube when they meet in the midline. The mesenchymal mesoderm in this region will give rise to head structures. Further back (middle sections), in the future trunk region of the embryo, notochord and somites have formed and neurulation is starting. At the posterior end, behind Hensen's node (bottom section), notochord formation, somite formation, and neurulation have not yet begun. The mesoderm internalized through the primitive streak starts to form structures appropriate to its position along the antero-posterior and dorso-ventral axes. For example, in the future trunk region, the intermediate mesoderm will form the mesodermal parts of the kidney, and the splanchnic mesoderm will give rise to the heart. The body fold will continue down the length of the embryo, forming the gut and also bringing paired organ rudiments that initially form on each side of the midline (e.g. those of the heart and dorsal aorta) together to form the final organs lying ventral to the gut. Blood islands, from which the first blood cells are produced, form from the ventral-most part of the lateral mesoderm. After Patten, B.M.: 1971.

The embryo gets its nourishment through extra-embryonic membranes (Fig. 2.19), which also provide protection. The fluid-filled **amniotic sac** provides mechanical protection; a **chorion** surrounds the whole embryo and lies just beneath the shell; an **allantois** both receives excretory products and provides the site of oxygen and carbon dioxide exchange; and a **yolk sac** surrounds the yolk.

In the remaining time before hatching, eyes develop from the optic

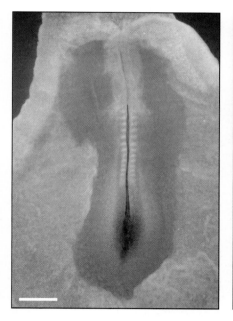 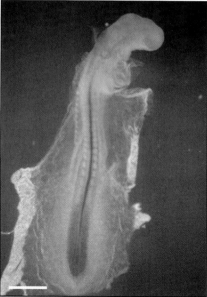 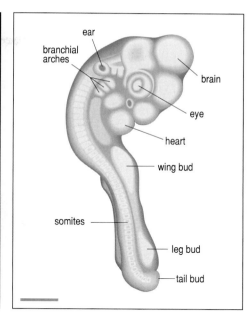

vesicles, and the inner ear develops from the otic vesicles. The embryo grows in size, the internal organs develop, the wings, legs, and beak are formed, and down feathers grow on the wings and body. The chick hatches 21 days after the egg is laid.

2.3 Mammals: the mouse

The mouse has a life cycle of 9 weeks, from fertilization to mature adult (Fig. 2.20), which is relatively short for a mammal, and is one of the reasons that the mouse has become a model organism for vertebrate development. Another is that it is amenable to both classical genetic analysis and to the generation of mutants by genetic modification. But like all mammals, the

Fig. 2.18 Development of the chick embryo. Left panel: at the 13-somite stage. At the anterior end (top), the head fold has formed. The dark region at the posterior end is Hensen's node. The somites can be seen on either side of the notochord as blocks of white tissue. Between the node and the last-formed somite is mesoderm that will segment into somites. Center panel: at the 20-somite stage. Right panel: at the 40-somite stage. Development of the head region and the heart are quite well advanced, and the wing and leg buds are present as small protrusions. Scale bars = 1 mm.
Photographs courtesy of B. Hermann, from Kispert, A., et al.: 1995.

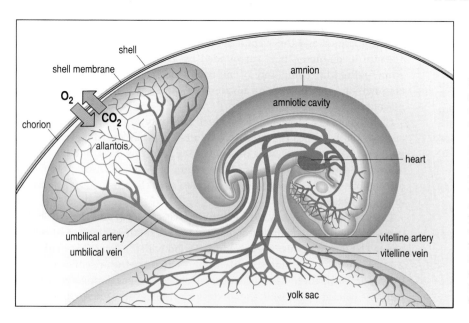

Fig. 2.19 The extra-embryonic structures and circulation of the chick embryo. A chick embryo at the same stage as that shown in the third panel in Fig. 2.18 is depicted *in situ*. The embryo has turned on its side, and its heart is beating. The yolk is surrounded by the yolk sac membrane. The vitelline vein takes nutrients from the yolk sac to the embryo and the blood is returned to the yolk sac via the vitelline artery. The umbilical artery takes waste products to the allantois and the umbilical vein brings oxygen to the embryo. The amnion and fluid-filled amniotic cavity provide a protective chamber for the embryo. After Patten, B.M.: 1951.

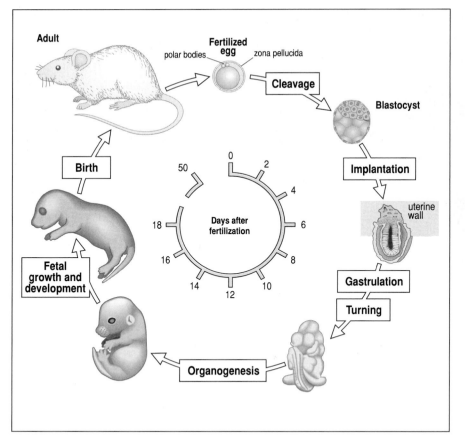

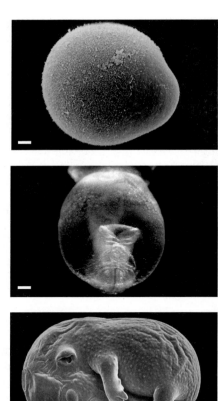

Fig. 2.20 The life cycle of the mouse. The egg is fertilized in the oviduct, where cleavage also takes place before implantation of the blastocyst in the uterine wall at 5 days after fertilization. Gastrulation and organogenesis then take place over a period of around 7 days and the remaining 6 days before birth are largely a time of overall growth. After gastrulation the mouse embryo undergoes a complicated movement known as 'turning' in which it becomes surrounded by its extra-embryonic membranes (not shown here). The photographs show (from top): a fertilized mouse egg just before the first cleavage (scale bar = 10 μm); anterior view of a mouse embryo at 8 days after fertilization (scale bar = 0.1 mm); and a mouse embryo at 14 days after fertilization (scale bar = 1 mm).

Photographs courtesy of: T. Bloom (top, from Bloom, T.L.: 1989); N. Brown (middle); and J. Wilting (bottom).

mouse embryo develops inside the mother, and so is not easily accessible for experimental manipulation or continuous observation, although it can be cultured outside the mother for short periods. The mouse is the mammalian model system that is most often used to help us understand human development.

Fertilization of the egg takes place internally in the oviduct; meiosis is then completed and the second polar body forms. The egg is small, about 100 mm in diameter. It is surrounded by a protective external coat, the **zona pellucida**, composed of mucopolysaccharides and glycoproteins. Mammalian embryos rely on nutrients obtained from the mother via the placenta.

Cleavage takes place in the oviduct and only after 4½ days does the embryo implant into the uterine wall after being released from the zona pellucida. Gastrulation takes place over the next few days, and by 10 days after fertilization all the organs have begun to develop. During the next 9 days before birth, organogenesis continues, as in the chick, and the embryo grows in size.

Early cleavages are very slow compared with *Xenopus* and chick, the first occurring about 24 hours after fertilization and subsequent cleavages at about 12-hour intervals. They produce a solid ball of cells, the **morula** (Fig. 2.21). At the eight-cell stage the blastomeres increase the area of cell surface in contact with each other in a process called compaction. After compaction, the cells are polarized; their exterior surfaces carry microvilli whereas

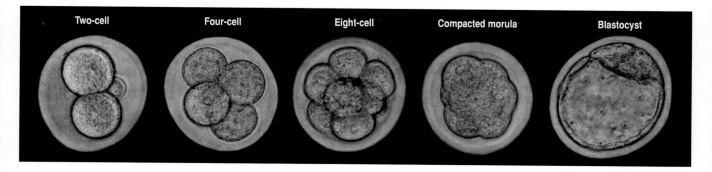

| Two-cell | Four-cell | Eight-cell | Compacted morula | Blastocyst |

their inner surfaces are smooth. Further cleavages are somewhat variable and are both radial and tangential, so that by the equivalent of the 32-cell stage the morula contains about 10 internal cells and more than 20 outer cells.

A special feature of mammalian development is that the early cleavages give rise to two groups of cells—the **trophectoderm** and the **inner cell mass**. The internal cells of the morula give rise to the inner cell mass and the outer cells to the trophectoderm. The trophectoderm will give rise to extra-embryonic structures such as the placenta, which provides a pathway for nutrition of the embryo via the mother, while the embryo proper develops from a small set of cells in the inner cell mass. At this stage (3½ days gestation) the embryo is known as a **blastocyst** (see Fig. 2.21). Fluid is pumped by the trophectoderm into the interior of the blastocyst, which causes the trophectoderm to expand and form a fluid-filled vesicle containing the inner cell mass at one end.

From 3½ to 4½ days gestation the inner cell mass becomes divided into two regions. The surface layer in contact with the fluid-filled cavity of the blastocyst becomes the **primitive endoderm**, and will contribute to extra-embryonic membranes, while the remainder of the inner cell mass—the **primitive ectoderm** or epiblast—will develop into the embryo proper as well as some extra-embryonic membranes. At this stage the embryo releases itself from the zona pellucida still surrounding it, and implants into the uterine wall.

The course of early post-implantation development of the mouse embryo from around 4½ days to 8½ days appears more complicated than that of the chick, partly because of the need to produce a larger variety of extra-embryonic membranes, and partly because the epiblast from which the embryo will develop is distinctly cup-shaped in the early stages. This is a peculiarity of mouse and other rodent embryos. The epiblasts of human and rabbit embryos, for example, are flat. In essence, however, the development of the mouse embryo proper is very similar to that of the chick.

The first 2 days of post-implantation development are shown in Fig. 2.22. At implantation, the cells of the mural trophectoderm (not the region in contact with the inner cell mass) replicate their DNA without cell division (endo-reduplication), giving rise to trophoblast giant cells which invade the uterus during implantation. The rest of the trophectoderm grows to form the ectoplacental cone and the **extra-embryonic ectoderm**, which both contribute to the formation of the placenta. Some cells from the primitive endoderm migrate to cover the whole inner surface of the mural trophectoderm. They become the parietal endoderm. The remaining primitive

Fig. 2.21 Cleavage in the mouse embryo. The photographs show the cleavage of a fertilized mouse egg from the two-cell stage through to the formation of the blastocyst. After the eight-cell stage, compaction occurs, forming a solid ball of cells called the morula, in which individual cell outlines can no longer be discerned. The internal cells of the morula give rise to the inner cell mass, which can be seen as the compact clump at the top of the blastocyst. It is from this that the embryo proper forms. The outer layer of the hollow blastocyst—the trophectoderm—gives rise to extra-embryonic structures.

Photographs courtesy of T. Fleming.

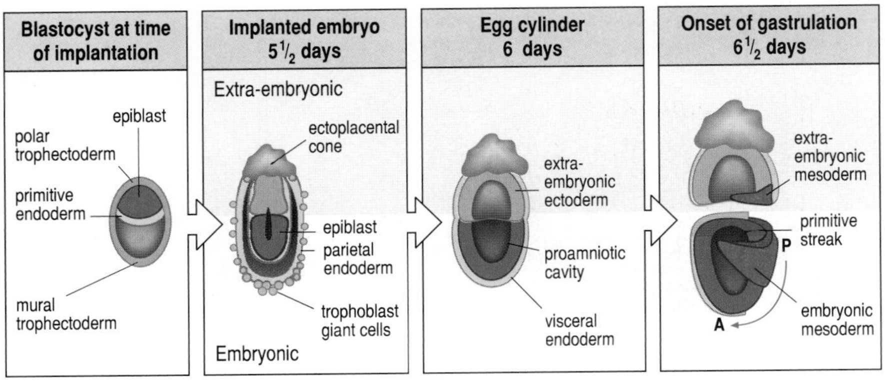

Fig. 2.22 Early post-implantation development of the mouse embryo. First panel: before implantation, the fertilized egg has undergone cleavage to form a hollow blastocyst, in which a small group of cells, the inner cell mass, will give rise to the embryo, while the rest of the blastocyst forms the trophectoderm, which will develop into extra-embryonic structures. At the time of implantation the inner cell mass divides into two regions: the primitive ectoderm or epiblast, which will develop into the embryo proper, and the primitive endoderm, which will contribute to extra-embryonic structures. Second panel: the polar trophectoderm in contact with the epiblast forms extra-embryonic tissues, the ectoplacental cone, and extra-embryonic ectoderm which contributes to the placenta. The mural trophectoderm gives rise to the trophoblast giant cells. The epiblast elongates and develops an internal cavity (proamniotic cavity) which gives it a cup-shaped form. Third panel: the cylindrical structure containing both the epiblast and the extra-embryonic tissue derived from the polar trophectoderm is known as the egg cylinder. The parietal endoderm and trophoblast giant cells are not shown in this or any subsequent figures. Fourth panel: the primitive streak appears at the posterior of the epiblast (P) and extends anteriorly (arrow) to the bottom of the cylinder. A is anterior. The egg cylinder has been sectioned across in this view so that the primitive streak on the interior surface of the 'cup' can be indicated. Epiblast cells passing through the streak will become mesoderm and endoderm. The endodermal layer is not shown on the right-hand side so that the mesoderm spreading over the outer surface of the epiblast can be seen. Note that, given the topology of the mouse embryo at this stage, the germ layers will appear inverted (ectoderm on the inner surface of the cup, endoderm on the outer) compared to the frog gastrula. After Hogan, B., *et al.*: 1994.

endoderm cells form the **visceral endoderm** which covers the elongating egg cylinder containing the epiblast.

By 6 days after fertilization, an internal cavity has formed inside the epiblast, which becomes cup-shaped—U-shaped when seen in cross-section (see Fig. 2.22, third panel). The embryo proper develops from this curved layer of epithelium, which at this stage contains about 1000 cells. The first easily visible sign of its future axis is at about 6½ days, when gastrulation begins with the formation of the primitive streak. The streak starts as a localized thickening at a point on the circumference of the cup; this is the future posterior end of the embryo. The inside of the cup is the future dorsal side of the embryo. Proliferating epiblast cells migrate through the primitive streak, and spread out laterally and anteriorly between the ectoderm and the visceral endoderm to form a mesodermal layer (Fig. 2.23). Some epiblast-derived cells enter the visceral endoderm layer and gradually displace it to form the definitive endoderm.

The development of the primitive streak in the mouse is similar to that in the chick; first it elongates towards the future anterior end of the embryo, with a condensation of cells at its anterior end corresponding to Hensen's node (Fig. 2.24). Cells migrating anteriorly through the node will form the notochord, and both the notochord and somites form anterior to it. Some migrating cells pass through the mesoderm to form an embryonic endodermal layer (the future gut), eventually displacing the visceral endoderm cells completely.

At around 8½ days, neural folds have started to form at the anterior end

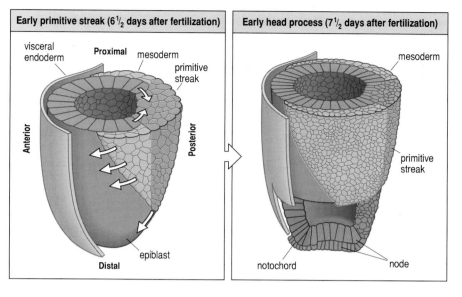

Fig. 2.23 Gastrulation in the mouse embryo. At the beginning of gastrulation, epiblast cells move through the primitive streak to give rise to the mesoderm and definitive endoderm (definitive endoderm is not shown in this diagram).

on the dorsal side of the embryo. In these final stages of gastrulation the embryo also undergoes an episode of complex folding, in which the embryonic endoderm—initially on the ventral surface of the embryo—becomes internalized to form the gut, while the heart and liver move into their final positions relative to the gut, and the head becomes distinct. The embryo then turns so that it becomes surrounded by its extra-embryonic membranes (Fig. 2.25). By 9 days, gastrulation is complete: the embryo has a distinct head and the forelimb buds are starting to develop. Organogenesis proceeds very much as in the chick embryo, at least in the initial stages.

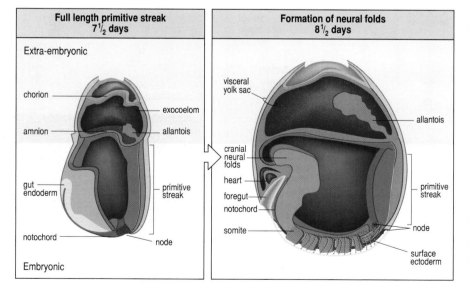

Fig. 2.24 Early development of the mouse embryo to the beginning of neurulation. Left panel: the primitive streak continues to extend. Further development of extra-embryonic structures involves the production of extra-embryonic mesoderm at the posterior end of the primitive streak. This eventually contributes to the amnion, the visceral yolk sac, and the allantois and chorion, which are important components of the placenta. Right panel: during the final stages of gastrulation, organogenesis commences in the anterior part of the embryo with the formation of the heart, the cranial neural folds, and the appearance of somites. After Hogan, B., et al.: 1994.

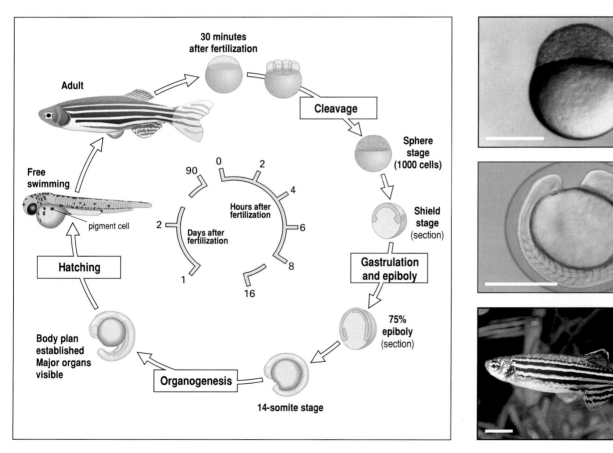

8½ days after fertilization	3–4 hours later	9 days after fertilization	9½ days after fertilization

(9 days after fertilization) allantois

(9½ days after fertilization) placenta / parietal yolk sac / visceral yolk sac / amnion / gut

Fig. 2.25 Turning in the mouse embryo. Between 8½ and 9½ days, the mouse embryo becomes entirely enclosed in the protective amnion and amniotic fluid. The visceral yolk sac, a major source of nutrition, surrounds the amnion and the allantois connects the embryo to the placenta. After Kaufman, M.H.: 1992.

2.4 Fishes: the zebrafish

The zebrafish is receiving increasing attention as a model for vertebrate development. Its two great advantages are its short life cycle of approximately 12 weeks (Fig. 2.26), which makes genetic analysis so much easier; and the transparency of the embryo, so that the fate of individual cells during development can be observed (see Fig. 2.26). The zebrafish egg is

Adult

30 minutes after fertilization

Cleavage

Free swimming

pigment cell

Hatching

90   0   2

Hours after fertilization

Days after fertilization

2   1   16   8   6   4

Body plan established Major organs visible

Organogenesis

14-somite stage

Sphere stage (1000 cells)

Shield stage (section)

Gastrulation and epiboly

75% epiboly (section)

Fig. 2.26 Life cycle of the zebrafish. The zebrafish embryo develops as a cup-shaped blastoderm sitting on top of a large yolk cell. It develops rapidly and by 2 days after fertilization the tiny fish, still attached to the remains of its yolk, hatches out of the egg. The top photograph shows a zebrafish embryo at the sphere stage of development, with the embryo sitting on top of the large yolk cell (scale bar = 0.5 mm). The middle photograph shows an embryo at the 14-somite stage, showing developing organ systems. Its transparency is useful for observing cell behavior (scale bar = 0.5 mm). The bottom photograph shows an adult zebrafish (scale bar = 1 cm).

Photographs courtesy of C. Kimmel (top, from Kimmel, C.B., et al.: 1995), N. Holder (middle), and M. Westerfield (bottom).

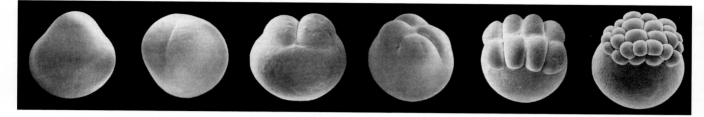

about 0.7 mm in diameter, with the cytoplasm and nucleus at the animal pole sitting upon a large mass of yolk. After fertilization, the zygote undergoes cleavage but, as in the chick, cleavage does not extend into the yolk and results in a mound of blastomeres perched above the yolk. The first five cleavages are all vertical, and the first horizontal cleavage gives rise to the 64-cell stage about 2 hours after fertilization (Fig. 2.27).

Further cleavage results in a blastoderm with a single outer layer of flattened cells, known as the outer enveloping layer, and a deep layer of more rounded cells, overlying the yolk (Fig. 2.28). The yolk syncytial layer is an extra-embryonic layer that arises from marginal blastomeres which, during early blastula stage, collapse onto the yolk cell, forming a ring around the blastodisc edge. The blastoderm expands in a vegetal direction by the spreading process known as epiboly, which we have met already in the *Xenopus* gastrula, to cover the yolk cell. By about 5½ hours after fertilization it has spread half-way to the vegetal pole. Gastrulation then begins, with the prospective endodermal and mesodermal cells of the deep layer turning inward at the margin of the blastoderm, a process known as involution. These cells migrate toward the future dorsal side, the tissue converging toward the midline of the embryo and extending at the same time as the embryo elongates in an antero-posterior direction. The future mesoderm and endoderm come to lie beneath the ectoderm. Gastrulation in the zebrafish has many features in common with gastrulation in *Xenopus*, but one difference is that involution occurs all around the periphery of the blastoderm at about the same time. By 9 hours the notochord becomes distinct, and gastrulation is complete by 10 hours. Neurulation and somite formation then follow.

Over the next 12 hours the embryo elongates, and the rudiments of the primary organ systems become recognizable. Somites appear anteriorly at about 10½ hours, and new ones are formed at intervals of initially 2, then 3 hours; by 18 hours, 18 somites are present. The nervous system develops

Fig. 2.27 Cleavage of the zebrafish embryo is initially confined to the animal (top) half of the embryo.

Photographs courtesy of R. Kessel, from Kessel, R.G., et al.: 1974.

Fig. 2.28 Epiboly and gastrulation in the zebrafish. At the end of the first stage of cleavage the zebrafish embryo is composed of a cluster of blastomeres sitting on top of the yolk. With further cleavage and spreading out of the layers of cells (epiboly), the upper half of the yolk becomes covered by a cup-shaped blastoderm. Gastrulation occurs by involution of cells in a ring around the edge of the blastoderm. The involuting cells converge on the dorsal midline to form the body of the embryo encircling the yolk.

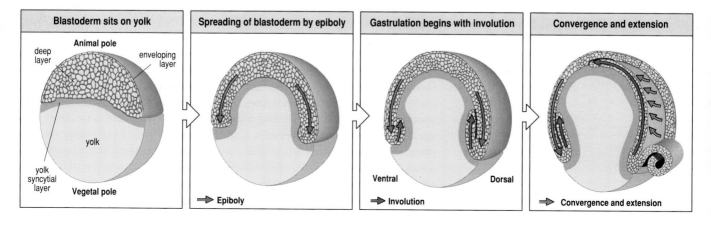

rapidly. Optic vesicles, which give rise to the eyes, can be distinguished at 12 hours as bulges from the brain and, by 18 hours, the body starts to twitch. At 48 hours the embryo hatches, and the young fish begins to swim and feed.

Summary

While the early development of different vertebrates can differ considerably, they all first undergo cleavage to form a blastula-like structure. The mouse and other mammals are special, since so much of the early mammalian embryo gives rise to extra-embryonic structures whereas the embryo proper derives from just a few cells of the inner cell mass. In all vertebrate embryos, gastrulation is followed, or is accompanied in its late stages, by neurulation—the formation of the neural tube. During gastrulation there is extensive cell movement so that the three germ layers—ectoderm, mesoderm, and endoderm—take up their appropriate positions. The mesoderm immediately on either side of the notochord forms somites, while the ectoderm lying above the notochord forms the neural tube, which develops into the brain and spinal cord. In the chick and mouse, complex extra-embryonic structures involved in nutrition, gas exchange, secretion, and mechanical protection are formed.

Model organisms: invertebrates

Although the development of the many different kinds of invertebrates is very diverse, certain features are not only common to the development of most invertebrates but are also seen in vertebrate development. These include cleavage, formation of a blastula or blastoderm, and gastrulation. Compared with vertebrate embryos, the number of cells in some invertebrate embryos is small and, as in the nematode for example, they have a stereotyped pattern of cleavage in which the fate of each cell can be identified.

2.5 The fruit fly *Drosophila melanogaster*

The wealth of genetic studies on *Drosophila* development, together with the feasibility of combining genetic and microsurgical manipulation, have made this small fly one of the best understood developmental systems. The entire genome of *Drosophila melanogaster* has been sequenced and is thought to contain around 13,600 genes. The life cycle of *Drosophila* is shown in Fig. 2.29.

The *Drosophila* egg is sausage shaped and the future anterior end is easily recognizable by the micropyle, a nipple-shaped structure in the tough external coat surrounding the egg. Sperm enter the anterior end of the egg through the micropyle. After fertilization and fusion of the sperm and egg nuclei, the zygote nucleus undergoes a series of rapid mitotic divisions, one about every 9 minutes but, unlike most animal embryos, there is no cleavage of the cytoplasm. The result is a **syncytium** in which many nuclei are present in a common cytoplasm (Fig. 2.30); the embryo essentially remains a single cell during its early development. After nine divisions the nuclei move to the periphery to form the **syncytial blastoderm**. This is equivalent to the blastula or blastoderm stage of other animals. Shortly afterwards,

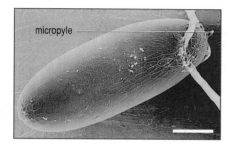

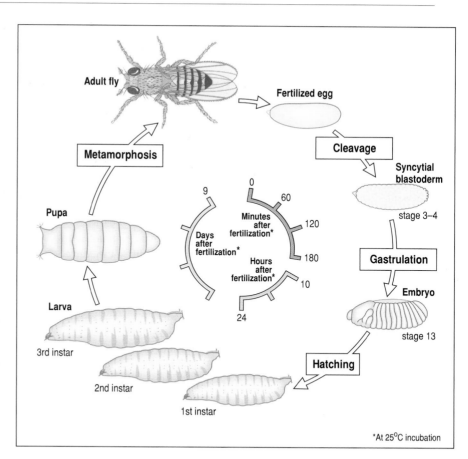

membranes grow in from the surface to enclose a nucleus and form cells, and the blastoderm becomes truly cellular after about 13 mitoses. Not all the nuclei give rise to the cells of the cellular blastoderm; some 15 or so end up at the posterior end of the embryo and develop into **pole cells**, which later give rise to germ cells, that is, sperm or eggs. Because of the formation of a syncytium, even large molecules such as proteins can diffuse between nuclei during the first 3 hours of development and, as we shall see in Chapter 5, this is of great importance to early *Drosophila* development.

All the future tissues are derived from the single epithelial layer of the cellular blastoderm. For example, prospective mesoderm is located in the most ventral region, while the future midgut derives from two regions of prospective endoderm, one at the anterior and the other at the posterior end of the embryo. Endodermal and mesodermal tissues move to their future positions inside the embryo during gastrulation, leaving ectoderm as the outer layer (Fig. 2.31). Gastrulation starts at about 3 hours after fertilization when the future mesoderm in the ventral region invaginates to form a furrow along the ventral midline. The mesodermal cells are initially internalized by formation of a mesodermal tube in a process rather similar to neural tube formation in vertebrates. The mesoderm cells then separate from the surface layer of the tube and migrate under the ectoderm to internal locations, where they later give rise to muscle and other connective tissues.

Fig. 2.29 Life cycle of *Drosophila melanogaster.* After cleavage and gastrulation the embryo becomes segmented and hatches out as a feeding larva. The larva grows and goes through two molts (instars), eventually forming a pupa that will metamorphose into the adult fly. The photographs show scanning electron micrographs of: a *Drosophila* egg before fertilization (top). The sperm enters through the micropyle. The dorsal filaments are extra-embryonic structures; a *Drosophila*, 2nd instar larva (middle); and a *Drosophila* pupa (bottom). Scale bars = 0.1 mm.

Photographs courtesy of F. R. Turner (top, from Turner, F.R., et al.: 1976; middle, from Turner, F.R., et al.: 1979).

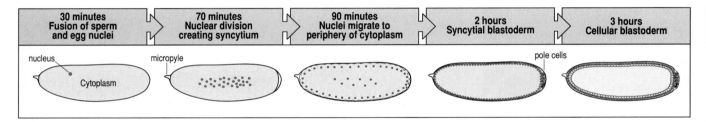

| 30 minutes Fusion of sperm and egg nuclei | 70 minutes Nuclear division creating syncytium | 90 minutes Nuclei migrate to periphery of cytoplasm | 2 hours Syncytial blastoderm | 3 hours Cellular blastoderm |

Fig. 2.30 Cleavage of the *Drosophila* embryo. After fusion of the sperm and egg nuclei, there is rapid nuclear division but no cell walls form, the result being a syncytium of many nuclei in a common cytoplasm. After the ninth division the nuclei move to the periphery to form the syncytial blastoderm. After about 3 hours, cell walls develop, giving rise to the cellular blastoderm. About 15 pole cells, which will give rise to germ cells, form a separate group at the posterior end of the embryo. Times given are for incubation at 25°C.

In insects, as in all arthropods, the main nerve cord lies ventrally, rather than dorsally as in vertebrates. Shortly after the mesoderm has invaginated, ectodermal cells of the ventral region that will give rise to the nervous system leave the surface individually and form a layer of **neuroblasts** between the mesoderm and the outer ectoderm. At the same time, two tube-like invaginations develop at the sites of the future anterior and posterior midgut. These grow inward and eventually fuse to form the endoderm of the midgut, while ectoderm is dragged inward behind them at each end to form the foregut and the hindgut. The outer ectoderm layer develops into the epidermis. There are no cell divisions during gastrulation, but once it is completed, cells start to divide again. The cells of the epidermis only divide twice before they secrete a cuticle.

Also during gastrulation, the ventral blastoderm or **germ band**, which comprises the main trunk region, undergoes germ band extension, which drives the posterior trunk regions round the posterior end and onto what was the dorsal side (Fig. 2.32). The germ band later retracts as embryonic development is completed. At the time of germ band extension the first external signs of **segmentation** can be seen. A series of evenly spaced grooves form more or less at the same time and these demarcate **parasegments**, which later give rise to the **segments** of the larva and adult. Parasegments and segments are out of register, so that a segment is formed by the posterior region of one parasegment and the anterior region of the next. There are 14 parasegments: three contribute to mouth parts of the head, three to the thoracic region and eight to the abdomen.

The larva (Fig. 2.33) hatches about 24 hours after fertilization, but the different regions of the larval body are well defined several hours before that. The head is a complex structure, largely hidden from view before the larva hatches. Structures associated with the most anterior region of the head are called the acron. At the posterior end, the most terminal structures are called the telson. Between these extremities, three thoracic

mesoderm	nervous system
amnioserosa	yolk
gut	epidermis ● germ line

Fig. 2.31 Gastrulation in *Drosophila*. Gastrulation begins when the future mesoderm invaginates in the ventral region, first forming a furrow and then an internalized tube. The cells then leave the tube and migrate internally under the ectoderm. The nervous system comes from cells which leave the surface of the ventral blastoderm and form a layer between the ventral ectoderm and mesoderm. The gut forms from two invaginations at the anterior and posterior end that fuse in the middle. The midgut region is endoderm, while the foregut and hindgut are of ectodermal origin.

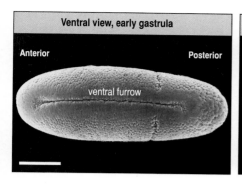

Ventral view, early gastrula

Anterior · Posterior

ventral furrow

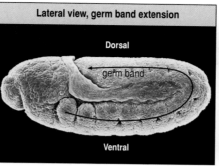

Lateral view, germ band extension

Dorsal

germ band

Ventral

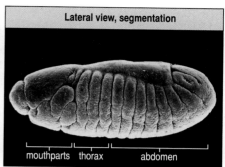

Lateral view, segmentation

mouthparts · thorax · abdomen

Fig. 2.32 Gastrulation, germ band extension, and segmentation in the *Drosophila* embryo. Gastrulation involves the future mesoderm moving inside through the ventral furrow. During gastrulation, the ventral blastoderm (the germ band) extends, driving the posterior trunk region onto the dorsal side and segmentation now takes place. Later the germ band shortens. Scale bar = 0.1 mm.

Photographs courtesy of F. Turner (left from Turner, F.R., et al.: 1977; middle from Alberts, B., et al.: 1994).

segments and eight abdominal segments can be distinguished by specializations in the cuticle secreted by the epidermis. On the ventral side of each segment are denticle belts and other cuticular structures characteristic of each segment. As the larva feeds and grows, it molts, shedding its cuticle. This occurs twice, each stage being called an **instar**.

The *Drosophila* larva has neither wings nor legs; these and other organs emerge when the larva undergoes hormone-induced **metamorphosis** after the third instar. These structures are, however, already present in the larva as **imaginal discs**, small sheets of prospective epidermal cells derived from the cellular blastoderm and usually containing about 40 cells each. These discs grow throughout larval life and form folded sacs of epithelia to accommodate their increase in size. There are imaginal discs for each of the six legs, two wings, and the two halteres (balancing organs), and for the genital apparatus, eyes, antennae, and other adult head structures (Fig. 2.34). In the segments of the abdomen there are groups of about ten

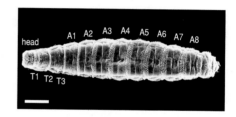

head · A1 A2 A3 A4 A5 A6 A7 A8 · T1 T2 T3

Fig. 2.33 Ventral view of a *Drosophila* larva. T1 to T3 are the thoracic segments and A1 to A8, the abdominal segments. The characteristic pattern of denticles can be seen in the anterior region of each abdominal segment. Scale bar = 0.1 mm.

Photograph courtesy of F.R. Turner.

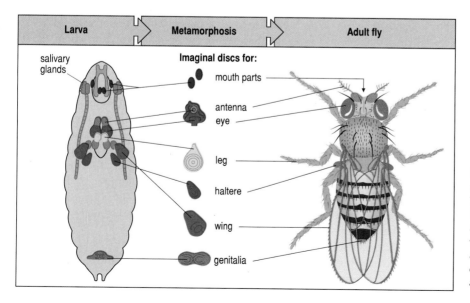

Larva · Metamorphosis · Adult fly

salivary glands

Imaginal discs for:

mouth parts

antenna

eye

leg

haltere

wing

genitalia

Fig. 2.34 Imaginal discs give rise to adult structures at metamorphosis. The imaginal discs in the *Drosophila* larva are small sheets of epithelial cells; at metamorphosis they give rise to a variety of adult structures. The abdominal cuticle comes from groups of histoblasts located in each larval abdominal segment.

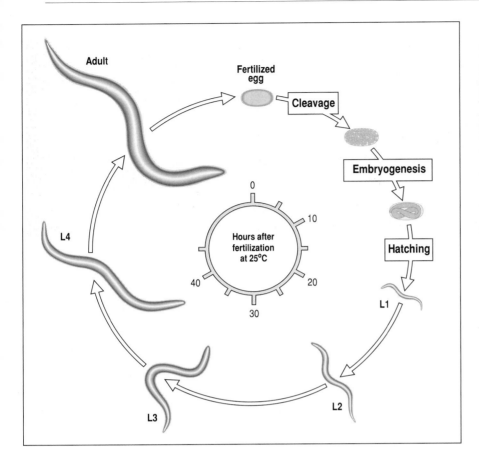

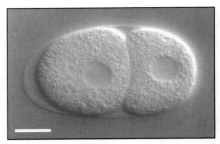

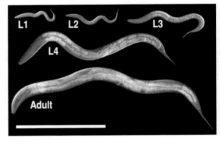

Fig. 2.35 Life cycle of the nematode
Caenorhabditis elegans. After cleavage and
embryogenesis there are four larval stages (L1–
L4) before the sexually mature adult develops.
Adults of *C. elegans* are usually hermaphrodite,
although males can develop. The photographs
show: the two-cell stage (top, scale
bar = 10 μm); an embryo after gastrulation with
the future larva curled up (middle, scale
bar = 10 μm); and the four larval stages and
adult (bottom, scale bar = 0.5 mm).

Photographs courtesy of J. Ahringer.

histoblast cells. These cells do not divide but will contribute to the epidermis of the adult at metamorphosis.

2.6 The nematode *Caenorhabditis elegans*

The free-living soil nematode *Caenorhabditis elegans*, whose life cycle is
shown in Fig. 2.35, is now used as a major model organism in developmental biology. Its genome has been sequenced and contains about 19,000
genes. Its advantages are its suitability for genetic analysis, its small number of cells (558 in the first larval stage) and their invariant lineage, and
the transparency of the embryo that allows the formation of each cell to
be observed. Its genome has also been completely sequenced. *Caenorhabditis elegans* has a simple anatomy and the adults are about 1 mm long and
just 70 mm in diameter. Nematodes can be grown on agar plates in large
numbers and early larval stages can be stored frozen and later resusci-
tated. This nematode reproduces primarily by self-fertilization of adult
hermaphrodites, although males can develop under special conditions.
Embryonic development is rapid, the larva hatching after 15 hours at
20°C, though maturation through larval stages to adulthood takes about
50 hours.

The nematode egg is small, only 50 μm in diameter. Polar bodies are
formed after fertilization. Before the male and female nuclei fuse, there is
what appears to be an abortive cleavage, but after fusion of the nuclei true

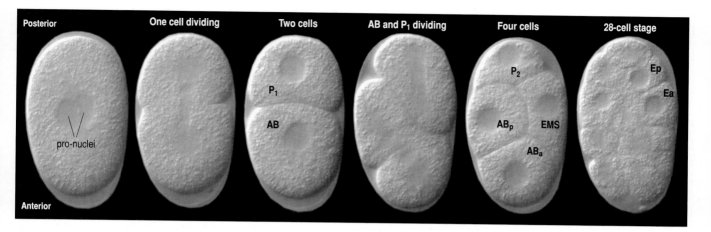

cleavage begins (Fig. 2.36). The first cleavage is asymmetric and generates an anterior AB cell and a smaller posterior P_1 cell. At the second cleavage, AB divides to give AB_a anteriorly and AB_p posteriorly, while P_1 divides to give P_2 and EMS. At this stage the main axes can already be identified, since P_2 is posterior and AB_p dorsal. Further cleavage of the AB cells gives rise mainly to hypodermis (the outer layers of the worm), neurons, and muscle. We concentrate here on the fates of the P_2 and EMS cells (Fig. 2.37). EMS divides into E and MS. E gives rise to the gut, whereas MS gives rise to muscle, glands, and neurons. P_2 divides to give P_3 and C. C forms muscle, hypodermis, and neurons, and P_3 divides into P_4 and D. D gives rise to muscle and P_4 gives rise to the germ cells. All these cells undergo a further well-defined pattern of cell divisions. Gastrulation starts at the 28-cell stage, when the descendants of the E cell that will form the gut move inside. Not

Fig. 2.36 Cleavage of the *C. elegans* embryo. After fertilization, the pronuclei of the sperm and egg fuse. The egg then divides into a large anterior AB cell and a smaller, posterior P_1 cell. At the next cell division, AB divides into AB_a and AB_p, while P_1 divides into P_2 and EMS. The EMS cell divides into the E cells which give rise to the intestine, and MS cells (not labeled here). Each of the cells will continue to divide within these groups.

Photographs courtesy of J. Ahringer.

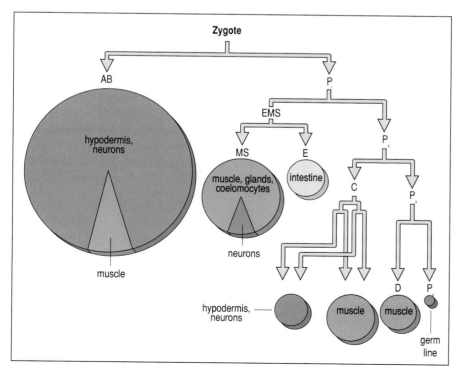

Fig. 2.37 Cell lineage and cell fate in the early *C. elegans* embryo. The fertilized egg divides into an anterior AB cell and a posterior P_1 cell. The AB cell gives rise to hypodermis (the outer layers of the embryo), neurons, and some muscle. The P_1 cell divides to give EMS and P_2. The EMS cell then divides to give cells MS and E which develop respectively into muscle, glands, and coelomocytes, and into the gut. Further divisions of the P lineage are rather like stem cell divisions, with one daughter of each division (C and D) giving rise to a variety of tissues while the other (P_2 and P_3) continues to act as a stem cell. Eventually, P_4 gives rise to the germ cells.

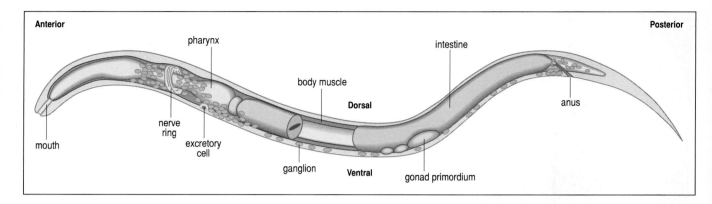

Fig. 2.38 *Caenorhabditis elegans* **larva at the L1 stage (20 hours after fertilization).** The vulva will form from the gonad primordium.

all cells that are formed during embryonic development survive; programmed death of specific cells is an integral feature of nematode development.

The newly hatched larva (Fig. 2.38), while similar in overall organization to the mature adult, is sexually immature and lacks a gonad and its associated structures, such as the vulva, which are required for reproduction. Post-embryonic development takes place during a series of four molts. While the newly hatched larva contains 558 nuclei, the adult hermaphrodite contains 959 somatic cell nuclei in addition to a variable number of germ cells. The emphasis is on nuclei rather than cells because some cells are syncytial and have several nuclei. The additional cells in the adult are derived largely from precursor blast cells (P cells) that are distributed along the body axis. Each of these blast cells founds an invariant lineage involving between one and eight cell divisions. The vulva, for example, is derived from blast cells P_5, P_6, and P_7. One may think of post-embryonic development in the nematode as the adding of adult structures to the basic larval plan.

Summary

Like vertebrates, invertebrate embryos undergo cleavage, form a blastula-like structure, and then gastrulate so that the endoderm and mesoderm migrate into the embryo to take up their correct location, with the ectoderm on the outside. There can be considerable differences in these processes. In insects, the early embryo is a synctium with several thousand nuclei, forming a superficial layer on the outside of the embryo. This layer later forms a cellular blastoderm of several thousand cells. The nematode is an example of an embryo with an invariant cell lineage, and the total number of cells is small.

Model systems: plants

The study of developmental biology has been dominated by animals, but plant development is now receiving increasing attention. There are important differences between plant and animal development, the most obvious being the absence of cell migration and tissue movements, so that cell division and cell expansion play a large part in morphogenesis. There is nothing in plants that corresponds to gastrulation. A particular feature of

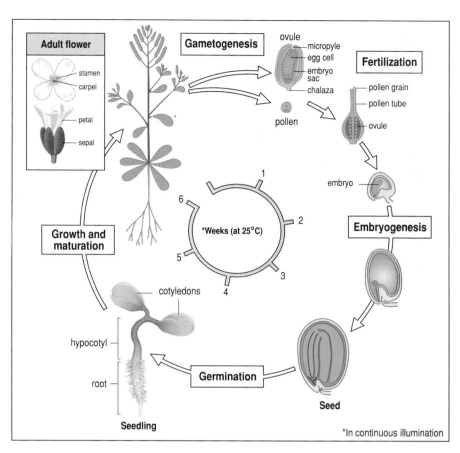

Fig. 2.39 Life cycle of *Arabidopsis*. In flowering plants, egg cells are contained separately in ovules inside the carpels. Fertilization of an egg cell by a male nucleus from a pollen grain takes place inside the ovary. The egg then develops into an embryo contained within the ovule coat, forming a seed. *Arabidopsis* is a dicotyledon, and the mature embryo has two wing-like cotyledons (storage organs) at the apical (shoot) end of the main axis—the hypocotyl—which contains a shoot meristem at one end and a root meristem at the other. Following germination, the seedling develops into a plant with roots, a stem, leaves, and flowers. The photograph shows five mature *Arabidopsis* plants.

plants is the development of all adult structures from the **meristems**—groups of undifferentiated cells set aside in shoot and root tips.

2.7 *Arabidopsis thaliana*

The equivalent of *Drosophila* in the study of plant development is the small annual crucifer *Arabidopsis thaliana*, often called thale cress, which is well suited to genetic studies. Its genome has been sequenced and contains about 25,000 genes. Its life cycle is shown in Fig. 2.39. This annual flowering plant develops as a small ground-hugging rosette of leaves, from which a branched flowering stem is produced with an inflorescence at the end of each branch.

Each flower (Fig. 2.40) consists of four sepals surrounding four white petals; inside the petals are six stamens, which contain the male pollen, and a central ovary of two carpels which contain ovules. Each ovule contains an egg cell. Following fertilization, the embryo develops inside the ovule, taking about 2 weeks to form a mature seed. Flower buds are usually visible on a young plant 3–4 weeks after seed germination. The complete life cycle is thus about 6–8 weeks. In the ovule, the fertilized egg cell is surrounded by specialized nutritive tissue, the **endosperm**, which provides the food source for embryonic development.

The early embryo is a mass of small, undifferentiated cells that becomes differentiated into three main tissues: the outer epidermis; prospective vascular tissue, which runs through the center of the main axis and

Fig. 2.40 An individual flower of *Arabidopsis*. Scale bar = 1 mm.

Fig. 2.41 *Arabidopsis* embryonic development. Light micrographs (Nomarski optics) of cleared wild-type seeds of *Arabidopsis thaliana*. The cotyledons can already be seen at the heart stage. The embryo proper is attached to the seed coat through a filamentous suspensor. Scale bar = 20 μm.

Photographs courtesy of D. Meinke, from Meinke, D.W.: 1994.

cotyledons; and the ground tissue that surrounds it. The **cotyledons** (seed leaves) are storage organs, developed by the embryo, that will provide nutrition for the germinating seedling. Monocotyledons, such as maize, have a single cotyledon; dicotyledons, such as *Arabidopsis*, have two. **Apical meristems**, composed of undifferentiated cells capable of continued division, develop at each end of the main axis and give rise to the root and shoot of the seedling.

The ovule containing the embryo matures into a seed, which remains dormant until suitable external conditions trigger germination. The early stages of germination and seedling growth rely on food supplies stored in the cotyledons. The shoot and root elongate and emerge from the seed. Once the shoot emerges above ground it starts to photosynthesize and forms the first true leaves at the shoot apex. About 4 days after germination the seedling is a self-supporting plant. All the adult structures such as leaves, stem, flowers, and roots are derived from the apical meristems.

The egg cell is fertilized by pollen from the male sex organs, the stamens. A pollen grain deposited on the carpel surface grows a tube that penetrates the carpel and delivers two haploid pollen nuclei to an ovule. One nucleus fertilizes the egg cell while the other fuses with the two so-called polar nuclei, which go on to develop into the triploid endosperm. Early cell division produces an embryo composed of two parts; the embryo proper, and the suspensor, which attaches the embryo to maternal tissue and is a source of nutrients (Fig. 2.41). Initial patterns of cleavage up to the 16-cell stage are highly reproducible. At the octant (eight-cell) stage the embryo proper and suspensor are clearly distinct. Even at this early stage it is possible to make a fate map for the major regions of the seedling. The upper tier of cells gives rise to the cotyledons, which will provide nutrition, the next tier is the origin of the hypocotyl, and the region of the suspensor where it joins the embryo gives rise to the root. At the 16-cell stage the epidermal layer, or dermatogen, is established. Slightly later, at the globular stage, the future vascular and ground tissue can be identified. With further cell divisions the heart stage is reached, in which the two cotyledons appear as wing-like structures. Further expansion of the cotyledons and hypocotyl takes place to give the embryo its form in the mature seed. The future shoot apical meristem is a small cluster of cells between the

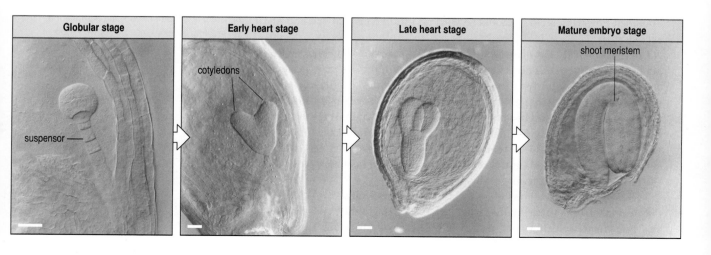

cotyledons and remains quiescent until germination. The embryo is now enclosed in a seed coat and awaits germination.

Meristem activity is entirely post-embryonic. Cell division in the shoot meristem results in both apical growth and the initiation of leaves. Leaf initiation is first indicated by a localized swelling—a leaf primordium—on the apical meristem; this gradually forms a protrusion that gives rise to the leaf. Flowering involves the transformation of a vegetative shoot meristem into reproductive mode. In *Arabidopsis*, the apical shoot meristem becomes an **inflorescence meristem**, throwing off individual **floral meristems**, each of which develops into a single flower. Instead of leaves, a floral meristem gives rise in sequence to sepals, petals, stamens, and carpels.

Summary

The annual thale cress, *Arabidopsis thaliana*, has a complete life cycle of about 6–8 weeks. Post-embryonic growth is entirely from meristems composed of undifferentiated cells capable of continued division, which develop at each end of the main axis and give rise to the root and shoot of the seedling, and all structures of the adult plant. Early patterns of cleavage in the embryo are highly reproducible and it is possible to make a fate map of the main regions of the seedling at a very early stage. In the adult plant, the shoot apical meristem gives rise to stem and leaves before becoming transformed into an inflorescence meristem. This produces floral meristems, each of which develops into a single flower. A floral meristem gives rise in sequence to sepals, petals, stamens, and carpels.

Identifying developmental genes

A major goal of developmental biology is to understand how genes control embryonic development, and to do this one must first identify those genes critically involved in controlling development. This task can be approached in a variety of ways, depending on the organism involved, but the general starting point is the identification of mutations that alter development in some specific and informative way. In the following sections general strategies are outlined for obtaining, and screening for, such mutations; other methods will be described in later chapters. Techniques for identifying genes that control development and detecting and manipulating their expression in the organism will also be described elsewhere, along with techniques for genetic manipulation.

Only some of our model organisms are suitable for genetic analysis. In spite of their importance, amphibian embryos cannot be used for genetic studies, as their breeding period is far too long and their genetics are virtually unknown. The situation with birds is only marginally better, but using techniques of direct DNA analysis, developmental genes are beginning to be identified in these organisms by their sequence similarity to well-characterized developmental genes from organisms such as *Drosophila* and mice. In general, when an important developmental gene has been identified in one animal, it has proved very rewarding to consider whether a **homologous gene** (that is, one with a certain minimum degree of nucleotide sequence similarity that indicates descent from a common ancestral gene) is present and is acting in some developmental capacity in other animals. As we shall see in Chapter 4, this approach has, for example, identified a hitherto unsuspected class of genes in vertebrates that affect segmental patterning along part of the antero-posterior axis. These were identified by their homology with genes that control antero-posterior patterning in early *Drosophila* development.

2.8 Developmental genes can be identified by rare spontaneous mutation

All the organisms dealt with in this book are sexually reproducing diploids so that their **somatic cells** contain two copies of each gene (with the exception of those on sex chromosomes). *Xenopus* is different, however, as it is tetraploid, so has double the number of chromosomes compared to diploids. One copy (**allele**) of each gene is contributed by the male parent and the other by the female. For many genes there are several different 'normal' alleles present in the population, which leads to the variation in normal phenotype one sees within any sexually reproducing species. Occasionally, however, a mutation will occur spontaneously in a gene and there will be marked change, usually deleterious, in the phenotype of the organism.

Many of the genes that affect development have been identified by spontaneous mutations that disrupt their function and produce an abnormal phenotype. Such mutations are rare, however, and to produce more it has been necessary to carry out large-scale mutagenesis experiments using chemical mutagens or X-irradiation and to screen for developmental mutations, as described in Section 2.9. Mutations are classified broadly according to whether they are dominant or recessive (Fig. 2.42). **Dominant** and **semi-dominant** mutations are those that produce a distinctive phenotype when the mutation is present in just one of the alleles of a pair; that is, it exerts an effect in the **heterozygous** state. By contrast, **recessive** mutations such as *white* in *Drosophila* alter the phenotype only when both alleles of a pair carry the mutation; that is, when they are **homozygous**.

In general, dominant mutations are more easily recognized, particularly

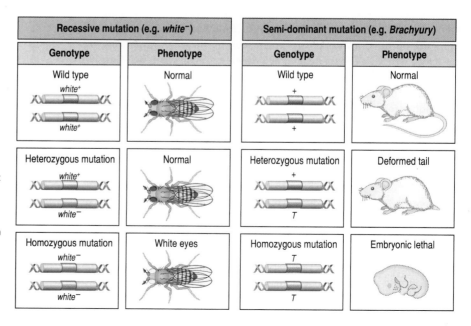

Fig. 2.42 Types of mutations. Left: a mutation is recessive when it only has an effect in the homozygous state, that is, when both copies of the gene carry the mutation. Right: by contrast, a dominant or semi-dominant mutation produces an effect on the phenotype in the heterozygous state; that is, when just one copy of the mutant gene is present. A plus sign denotes wild type, and a minus sign recessive, *T* is the mutant form of the gene *Brachyury*.

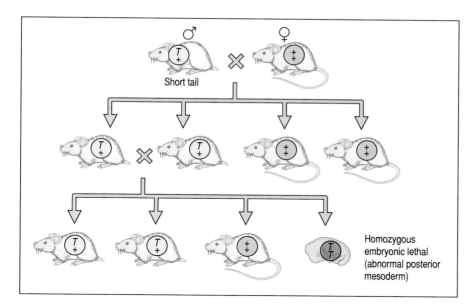

Fig. 2.43 Genetics of the semi-dominant mutation *Brachyury* (*T*) in the mouse. A male heterozygote carrying the *T* mutation merely has a short tail. When mated with a normal (wild type, +) female some of the offspring will also be heterozygotes and have short tails. Mating two heterozygotes together will result in some of the offspring being homozygous (*T/T*) for the mutation, resulting in a severe and lethal developmental abnormality in which the posterior mesoderm does not develop.

if they affect gross anatomy or coloration, provided that they do not cause the early death of the embryo in the heterozygous state. However, truly dominant mutations are rare. A mutation in the mouse gene *Brachyury*, which is involved in mesoderm formation, is a classic example of a semi-dominant mutation and was originally identified because mice hetero-zygous for this mutation (symbolized by *T*) have short tails. When the mutation is homozygous it has a much greater effect, and embryos die at an early stage, indicating that the gene is required for normal embryonic development (Fig. 2.43). In the *Brachyury* mutation, the development of pos-terior mesoderm is affected; the anatomical defect seen in heterozygotes is relatively minor because the normal copy of the gene also present is able to compensate to some extent. Once breeding studies had confirmed that the *Brachyury* trait was due to a single gene, the gene could be mapped to a location on a particular chromosome by classical genetic mapping tech-niques. The gene has now been cloned; that is, isolated in a pure form that can be sequenced or used for other molecular studies. Identifying recessive mutations is more laborious, as the heterozygote has a phenotype identical to a normal wild-type animal, and a carefully worked-out breeding program is required to obtain homozygotes. Identifying potentially lethal recessive developmental mutations requires careful observation and analysis in mammals, as the homozygotes may die unnoticed inside the mother.

Very rigorous criteria must be applied to identify those mutations that are affecting a genuine developmental process and not just affecting some vital but routine housekeeping function without which the animal cannot survive. One simple criterion for a developmental mutation is embryonic lethality, but this also catches mutations in genes involved in housekeeping functions. Mutations that produce abnormal patterns of embryonic devel-opment are much more promising candidates for true developmental mutations.

2.9 Developmental genes can be identified by induced mutation and screening

Valuable though spontaneous mutations have been in the study of development, suitable mutations are rare. Many more developmental genes have been identified by inducing random mutations in a large number of organisms by chemical treatments or irradiation with X-rays, and then screening for mutants of developmental interest. The aim, where possible, is to treat a large enough population so that, in total, a mutation is induced in every gene in the genome. This sort of approach can best be used in organisms that breed rapidly and can be obtained and treated conveniently in very large numbers.

Zebrafish offer a potentially very valuable vertebrate system for large-scale mutagenesis because large numbers can be handled, and the transparency and large size of the embryos makes it easier to identify developmental abnormalities. However, unlike the case of *Drosophila* described later, there are as yet no genetic means of eliminating unaffected individuals automatically. This means that all the progeny of a cross have to be examined visually.

A screening program using zebrafish involves breeding for three generations (Fig. 2.44). Male fish treated with a chemical mutagen are crossed

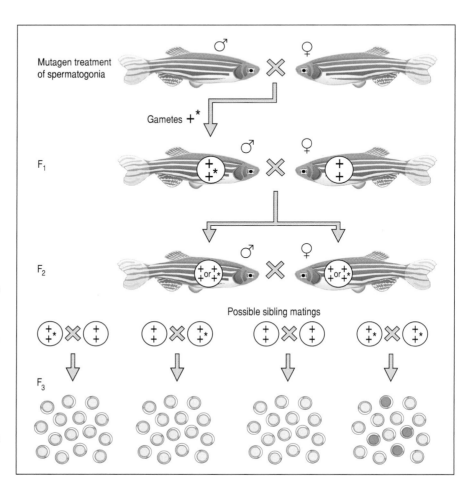

Fig. 2.44 Genetic screening to produce homozygous mutant zebrafish embryos. Male fish are treated with a mutagen and mated with wild-type females. In the F$_1$ offspring each individual is heterozygous for a different mutated gene. Further mating with wild-type females results in 50% of each F$_2$ family carrying the same mutation. The siblings within each family are mated and the embryos examined for developmental abnormalities. Embryos homozygous for the induced mutation will be found in the offspring of 25% of the matings. A plus sign with an asterisk indicates the induced mutation.

with wild-type females; their F_1 male offspring are crossed again with wild-type females, and the female and male siblings from each of these crosses are themselves crossed. The offspring from each of these pairs are examined separately for homozygous mutant phenotypes. If the F_1 fish carry a mutation, then in 25% of the F_2 matings two heterozygotes will mate and 25% of their offspring will be homozygous for the mutation. Zebrafish can also be made to develop as haploids by fertilizing the egg with sperm heavily irradiated with ultraviolet light. This allows one to detect early-acting recessive mutations without having to breed the fish to obtain homozygous embryos.

Many of the developmental mutations that have led to our present understanding of early *Drosophila* development came from a brilliantly successful screening program that searched the *Drosophila* genome systematically for mutations affecting the patterning of the early embryo. Its success was recognized with the award of a Nobel Prize for Physiology or Medicine to Edward Lewis, Christiane Nüsslein-Volhard, and Eric Wieschaus in 1995, only the second given for developmental biology.

In this screening program thousands of flies were mutagenized with a chemical mutagen, and were bred and screened according to the strategy described in Box 2B (p. 62). Given the number of progeny involved, it was important to devise a strategy that would reduce the number of flies that had to be examined to find a mutation. So the search was restricted to mutations on just one chromosome at a time. As described in Box 2B, the program also incorporated a means of identifying flies homozygous for the male-derived chromosomes carrying the induced mutations and, most importantly, also included a way of automatically deleting from the population flies that could not be carrying a mutated chromosome.

A phenotypic character of particular value in screening for patterning mutants in *Drosophila* is the stereotyped denticle pattern of the larval segments, irregularities in which enable alterations in pattern to be quickly recognized visually. In this way the key genes involved in patterning the early *Drosophila* embryo were first identified. They were subsequently mapped and many have now been cloned. Screening was also successful because the mutations that were inevitably produced in housekeeping genes were mostly rescued by the actions of maternal housekeeping genes; that is, those genes acting in the mother that provide many housekeeping functions to the egg. Indeed, it is necessary to do special screens for maternal-effect genes; that is, genes acting in the mother that are involved in patterning the egg during oogenesis. A similar approach has been used to induce developmental mutants in mice, but screening is much more difficult and is limited by both the enormous numbers of mice needed and the difficulty of screening for embryonic phenotype while the embryo is still inside the mother.

The nematode is also amenable to large-scale screening, and that is why its developmental genetics are advanced despite the fact that it has only been studied for a relatively short time. *Caenorhabditis* is the subject of an intensive program of genetic analysis and the complete DNA sequence of its genome is now available. Around 19,000 protein-coding genes have been identified, which is about half the number expected to be found in the human genome. Fewer than 200 genes have been identified that control cell fate and patterning.

In *Arabidopsis*, mutagenesis is most frequently performed not on the

Box 2B Mutagenesis and genetic screening strategy for identifying developmental mutants in *Drosophila*

The mutagen ethyl methane sulfonate (EMS) was applied to large numbers of male flies homozygous for a recessive mutation on the selected chromosome. The chromosome marked in this way is designated 'a' in the figure. A recessive mutation is chosen that gives adult flies an easily distinguishable but viable phenotype when homozygous (such as the mutation *white⁻*, see Fig. 2.42)

The treated males, who now produce sperm with a variety of induced mutations on the 'a' chromosomes (*a**), were crossed with untreated females that carried different mutations (*DTS* and *b*) on their two 'a' chromosomes, but were otherwise wild type. These mutations track the untreated female-derived chromosomes and automatically eliminate all embryos carrying two female-derived chromosomes in subsequent generations. *DTS* is a dominant temperature-sensitive mutation that causes death of the fly when the incubation temperature is raised to 29˚C. *b* is a non-developmental lethal recessive so that any flies homozygous for this female-derived chromosome will die as normal-looking embryos and be automatically eliminated. The female flies also carried a balancer chromosome (not shown) that prevented recombination at meiosis; this is to prevent recombination between male-derived and female-derived chromosomes in females. There is no recombination at meiosis in male *Drosophila*.

To identify the new recessive mutations (*a**) caused by the chemical treatment, a large number of the heterozygous males arising from this first cross were again outcrossed to *DTS/b* females. Of the offspring of each individual cross, only *a*/b* flies survive when placed at 29˚C; all other combinations die. The surviving siblings were then intercrossed and the offspring of each cross screened for patterning mutants. There are three possible outcomes: flies homozygous for the induced mutation *a** (which will also be homozygous for the original mutation marking chromosome a in males); heterozygous *a** flies; and homozygous b flies (which die as embryos).

If *a** is indeed a patterning mutation which is lethal in the larva, then the culture tube in which the cross is made will

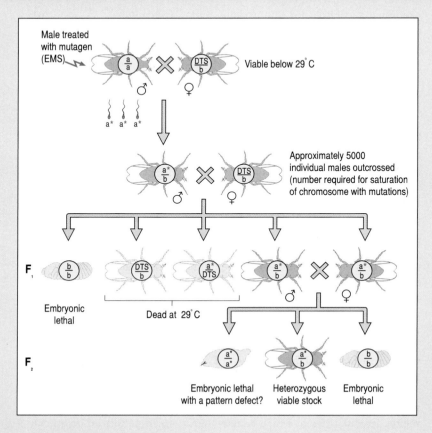

contain no adult flies with the original male phenotype, say white-eyed. Therefore tubes containing white-eyed flies can be discarded immediately, as the induced mutation *a** they carry must have allowed them to develop to adulthood even when homozygous and is therefore likely to be of little interest. If no white-eyed flies are present in a tube, then the homozygous *a*/a** embryos may have died or been arrested in their development as a result of abnormal development, and the mutation is of potential interest. The embryos of larvae from this cross can then be examined for pattern defects, as in the example illustrated here. The phenotypically wild-type adults in this tube are heterozygous for *a** and are used as breeding stock to study the mutant further. This whole program has to be repeated for each of the four chromosome pairs of *Drosophila*.

gametes but on the seed. If the mutant cell ends up in the apical meristem it has a chance of contributing to the sex cells when the flower is formed, and can thus give rise to gametes containing the mutation. Since *Arabidopsis* is self-fertilizing, both homozygotes and heterozygotes for the mutation can be obtained in the next generation.

Summary

Genes controlling development can be identified by mutations that affect embryonic development. It is easier to recognize dominant as distinct from recessive mutations, as the latter have a distinguishable phenotype only in homozygotes. In some animals, such as *Drosophila*, recessive mutations in many genes controlling early development have been identified by chemically inducing mutations in large populations of flies, breeding them, and screening the embryos. Zebrafish offer the possibility of similar large-scale screening for developmental genes in a vertebrate.

SUMMARY TO CHAPTER 2

Our knowledge and understanding of development depends on a rather small number of intensively studied model organisms, most of which are animals. These model systems represent quite a wide range of organisms, each with their own advantages and disadvantages for studying development. In spite of the differences between various animals, there are broad similarities in their early development. In all the animals studied here, cleavage of the fertilized egg leads to a multicellular blastula or blastoderm stage; this is followed by gastrulation, during which cells of the three germ layers—endoderm, mesoderm, and ectoderm—become located in the correct position for future development of the animal body. In most embryos, future endoderm and mesoderm are located initially on the outside of the embryo and move inside during gastrulation.

In vertebrates, gastrulation is followed by a distinct episode of neurulation, in which the future central nervous system is formed. During gastrulation and neurulation the overall shape of the embryo is altered from a sphere or disc of cells into that of a recognizable embryo, with antero-posterior and dorso-ventral axes. One of the main differences between embryos relates to their rate of development and the number of cells present at different stages.

In plants, there is virtually no cell movement and nothing that corresponds to gastrulation. The shape of the early plant embryo is molded by patterns of cell division. Unlike animals, in which the late embryo can be thought of as a miniature of the larva or adult, all adult plant structures are derived at a later stage from the shoot and root meristems.

A variety of techniques can be used for identifying developmental genes. One powerful and successful strategy has been to induce large numbers of mutations in a population by chemical treatment, and then to screen the offspring for developmental mutations. Such programs have contributed many of the known developmental mutations in *Drosophila*, *Caenorhabditis*, and zebrafish.

GENERAL REFERENCES

Bard, J.B.L.: *Embryos. Color Atlas of Development*. London: Wolfe, 1994.

Carlson, B.M.: *Patten's Foundations of Embryology*. New York: McGraw-Hill, Inc., 1996.

SECTION REFERENCES

2.1 Amphibians: *Xenopus laevis*

Hausen, P., Riebesell, H.: *The Early Development of Xenopus laevis*. Berlin: Springer-Verlag, 1991.

Nieuwkoop, P.D., Faber, J.: *Normal Tables of Xenopus laevis*. Amsterdam: North Holland, 1967.

2.2 Birds: the chicken

Bellairs, R., Osmond, M.: *An Atlas of Chick Development*. London: Academic Press, 1998.

Hamburger, V., Hamilton, H.L.: **A series of normal stages in the development of a chick.** *J. Morph*. 1951, **88**: 49–92.

Lillie, F.R.: **Development of the Chick: An Introduction to Embryology.** New York: Holt, 1952.

Patten, B.M.: *The Early Embryology of the Chick*. New York: McGraw-Hill, 1971.

2.3 Mammals: the mouse

Hogan, H., Beddington, R., Costantini, F., Lacy, E.: *Manipulating the Mouse Embryo. A Laboratory Manual* (2nd edn). New York: Cold Spring Harbor Laboratory Press, 1994.

Kaufman, M.H.: *The Atlas of Mouse Development*. London: Academic Press, 1992.

2.4 Fishes: the zebrafish

Kimmel, C.B., Ballard, W.W., Kimmel, S.R., Ullmann, B., Schilling, T.F.: **Stages of embryonic development of the zebrafish**. *Dev. Dynam.* 1995, **203**: 253–310.

Westerfield, M. (ed.): *The Zebrafish Book; A Guide for the Laboratory Use of Zebrafish (Brachydanio rerio)*. Eugene, Oregon: University of Oregon Press, 1989.

2.5 The fruit fly *Drosophila melanogaster*

Ashburner, M.: *Drosophila. A Laboratory Handbook*. New York: Cold Spring Harbor Laboratory Press, 1989.

2.6 The nematode *Caenorhabditis elegans*

Ruvkun, G., Hobart, O.: **The taxonomy of developmental control in** *Caenorhabditis elegans*. *Science* 1998, **282**: 2033–2040.

Sulston, J.E., Scherienberg, E., White, J.G., Thompson, J.N.: **The embryonic cell lineage of the nematode** *Caenorhabditis elegans*. *Dev. Biol.* 1983, **100**: 64–119.

Sulston, J.: **Cell lineage**. In *The Nematode Caenorhabditis elegans*. Edited by Wood, W.B. New York: Cold Spring Harbor Laboratory Press, 1988: 123–156.

Wood, W.B.: **Embryology**. In *The Nematode Caenorhabditis elegans*. Edited by Wood, W.B. New York: Cold Spring Harbor Laboratory Press, 1988: 215–242.

2.7 *Arabidopsis thaliana*

Mansfield. S.G., Briarty, L.G.: **Early embryogenesis in** *Arabidopsis thaliana*. **II. The developing embryo**. *Can. J. Bot.* 1991, **69**: 461–476.

Meyerowitz, E.M.: *Arabidopsis—a useful weed*. *Cell* 1989, **56**: 263–269.

2.8 Developmental genes can be identified by rare spontaneous mutation
&
2.9 Identification of developmental genes by induced mutation and screening

Driever, W., Solnica-Krezel, L., Schier, A.F., Neuhauss, S.C.F., Malicki, J., Stemple, D.L., Stainier, D.Y.R., Zwartkruis, F., Abdelilah, S., Rangini, Z., Belak, J., Boggs, C.: **A genetic screen for mutations affecting embryogenesis in zebrafish**. *Development* 1996, **123**: 37–46.

Gelbart, W.M., Griffiths, A.J.F., Lewontin, R.C., Miller, J.H., Suzuki, D.T.: *An Introduction to Genetic Analysis* (5th edn). New York: W.H. Freeman and Co., 1995.

Haffter, P., Granato, M., Brand, M., Mullins, M.C., Hammerschmidt, M., Jiang, Y-J., Heisenberg, C-P., Kelsh, R.N., Fabian, C., Nüsslein-Volhard, C.: **The identification of genes with unique and essential functions in the development of the zebrafish,** *Danio rerio*. *Development* 1996, **123**: 1–36.

Mullins, M.C., Hammerschmidt, M., Haffter, P., Nüsslein-Volhard, C.: **Large scale mutagenesis in the zebrafish: in search of genes controlling development in a vertebrate**. *Curr. Biol.* 1994, **4**: 189–202.

Patterning the vertebrate body plan I: axes and germ layers

3

- Setting up the body axes
- The origin and specification of the germ layers

" When young, we needed to decide who would be at the front and who at the back. Thus, we were assigned to these main groups. All this required many conversations and even some messages from the outside. "

All vertebrates, despite their many outward differences, have a similar basic body plan. The defining vertebrate structures comprise the segmented vertebral column surrounding the spinal cord, with the brain at the anterior end enclosed in a bony or cartilaginous skull. These prominent structures mark the **antero-posterior axis**, the main body axis of vertebrates. The head is at the anterior end of this axis, followed by the trunk with its paired appendages—limbs in terrestrial vertebrates and fins in fish—and terminating in many vertebrates in a post-anal tail. In addition, the vertebrate body has a distinct **dorso-ventral** (back–belly) polarity, with the mouth defining the ventral side. The antero-posterior and dorso-ventral axes define the left and right sides of the animal, and internal organs such as the heart and liver are asymmetrically arranged.

The overall similarity of the body plan in all vertebrates suggests that the developmental processes that establish it are similar in different animals. All vertebrate embryos do indeed pass through a common stage known as the **phylotypic stage**, in which the head is distinct and a neural tube runs along the dorsal midline, under which runs the notochord, flanked on either side by the mesodermal somites. At this stage, most vertebrate embryos are very similar, though differing in size and some details. Features special to different groups, such as beaks, wings, and fins, appear later (Fig. 3.1).

There are, however, considerable differences between vertebrate embryos at the pre-gastrula stages of development, particularly in relation to how and when the axes are set up, and how early patterning is established. These differences are related mainly to the very different modes of reproduction amongst vertebrates: yolk provides all the nutrients for amphibian, bird, and fish development. Mammalian eggs, by contrast, are small and non-yolky, and the embryo is nourished initially by fluids in the oviduct and uterus and then through the placenta; this requires the development of specialized extra-embryonic structures at a very early stage.

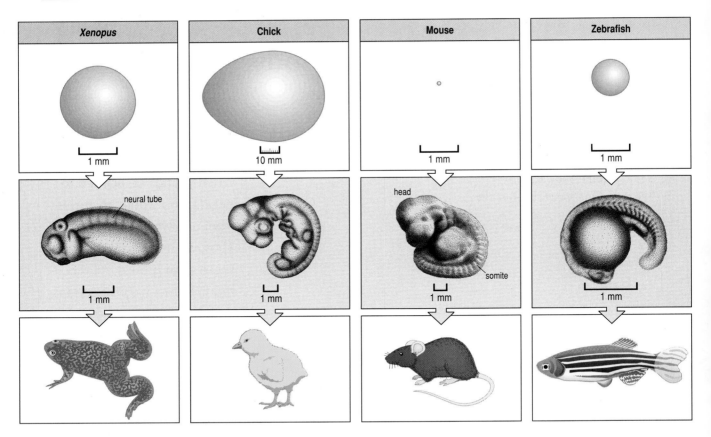

Fig. 3.1 All vertebrate embryos pass through a common phylotypic stage. The eggs of the frog (*Xenopus*), chick, mouse, and zebrafish are very different in size (top row), and their early development (not shown) is rather different, yet they all pass through an embryonic stage at which they look rather similar (middle row). This is the phylotypic stage, when the body axis has developed, and neural tube, somites, notochord, and head structures are present. After this stage their development diverges again. Paired appendages, for example, develop into fins in fish, and wings and legs in the chick (bottom row).

These reproductive differences affect the mechanisms used to specify the axes, the time at which they are established, and what cells of the earliest 'embryo' forms the extra-embryonic structures rather than the embryo proper.

In this and the following chapter, we concentrate on the setting-up of the main body axes, the induction and patterning of the mesoderm, and the induction and early patterning of structures along the antero-posterior axis, including the nervous system. This takes the vertebrate embryo up to the stage at which it has become recognizably vertebrate, with somites, notochord, and neural tube. More detailed consideration of the development of particular organs and structures, such as the limbs, the head, and the nervous system itself, is covered in later chapters.

We deal here mainly with three vertebrates whose early development has been particularly well studied: amphibians are represented by the frog *Xenopus*, birds by the chicken, and mammals by the mouse. We also refer to the zebrafish, which is becoming an increasingly important model organism for developmental studies as genetic analyses confirm inferences drawn from other vertebrates and provide new insights. The development of *Xenopus* inevitably provides the main focus, as this is the organism in which the setting-up of the body plan is best understood.

To consider early vertebrate development in an integrated way, we divide it into three main stages, and at each stage the similarities and differences between organisms are discussed. First, we consider the setting-up of the main body axes—the antero-posterior and the dorso-ventral axes—and to what extent these axes are already present in the egg or are specified by external signals. A central issue in the early development of any animal is the role of maternal factors in the egg. To what extent is the very early embryo already patterned as a result of maternal factors laid down in the egg during its development in the mother's ovary? Here we must draw a distinction between **maternal genes** and **zygotic genes**. Maternal genes are expressed in the mother during the development of the egg, and affect subsequent embryonic development through maternal factors—proteins and mRNAs—laid down in the egg during **oogenesis**. Maternal genes can control not only which proteins and mRNAs are put into the egg but also how they are distributed within the egg. Zygotic genes, on the other hand, are expressed in the developing embryo itself from the embryo's own genome. The second stage covers the specification of the three germ layers: the endoderm, which gives rise to the gut, and its derivatives such as liver and lungs; the mesoderm, which forms the notochord, skeletal structures, muscle, connective tissue, kidney, heart, and blood, as well as some other tissues; and the ectoderm, which gives rise to the epidermis, the brain and spinal cord, and neural crest. The third stage to be considered is the patterning of the germ layers, particularly the mesoderm, and the early patterning of the nervous system.

Although these three stages follow each other roughly in developmental time, there are no sharp boundaries between them, and the later stages, in particular, overlap considerably. The first two stages, along with the early patterning of the mesoderm along the dorso-ventral axis, will be the subject of this chapter, while Chapter 4 deals with patterning along the antero-posterior axis and the final emergence of the characteristic vertebrate body plan.

Setting up the body axes

Different vertebrates use quite different strategies to set up the primary embryonic axes. We begin with amphibians, in which the establishment of the axes is by far the best understood, and then compare this strategy with those of birds and mammals. As well as antero-posterior and dorso-ventral polarity, vertebrates also have bilateral symmetry, with many structures occurring in pairs, with one on either side of the midline, and we shall see how this bilateral symmetry is set very early on in *Xenopus*. We then briefly consider the intriguing question of how the left–right asymmetry or 'handedness' of a number of internal organs might be determined.

3.1 The animal–vegetal axis of *Xenopus* is maternally determined

The *Xenopus* egg possesses a distinct polarity even before it is fertilized, and this polarity influences the pattern of cleavage. One end of the egg, the **animal pole**, which sits uppermost, has a heavily pigmented surface, whereas most of the yolk is located toward the opposite end, the **vegetal pole**, which is unpigmented (see Fig. 2.4). The pigment itself has no role in development but is a useful marker for the developmental differences between the animal and vegetal halves of the egg. The animal half contains the nucleus, which is located close to the animal pole. The planes of early cleavages are related to the **animal–vegetal axis**. The first plane of cleavage is parallel with the axis and often defines a plane corresponding to the

Box 3A Intercellular signals

Proteins that are known to act as signals between cells during development belong to seven main families. Some of these families, such as the fibroblast growth factors (FGF), were originally identified because they were essential for the survival and proliferation of mammalian cells in tissue culture, hence the term 'growth factor'. The members of the seven families are either secreted or are part of the cell's surface, and they provide intercellular signals in both vertebrates and invertebrates at many stages of development.

The signal molecules listed in the table all act by binding to receptors on the surface of a target cell. This produces a signal that is passed, or transduced, across the cell membrane by the receptor to the intracellular biochemical signaling pathways. Each type of signal has a corresponding receptor or set of receptors, and only cells with the appropriate receptors on their surface can

respond. The intracellular pathways can be complex and involve numerous different proteins. The ultimate response of the cell to the growth factors active during development is a change in gene expression, with specific genes being switched on or off.

Some signal proteins, such as those of the transforming growth factor-β family (TGF-β), act as dimers; two molecules form a complex that binds to and activates the receptor, which is itself a dimer. In some cases, the active form of the protein signal is a heterodimer, made up of two different members of the same family. Secreted signals can diffuse short distances through tissue and set up concentration gradients. Some signal molecules, however, remain bound to the cell surface and thus can only interact with receptors on a cell that is in direct contact. One such is Delta protein, which is a membrane-bound protein that interacts with the Notch receptor protein on an adjacent cell.

Family	Receptors	Examples of roles in development
Fibroblast growth factor (FGF) Ten mammalian FGFs; FGF-1 to FGF-10 and eFGF	Receptor tyrosine kinases	Induction of spinal cord; signal from apical ridge in vertebrate limb
Epidermal growth factors (EGF)	EGF receptor	Insect eye
Transforming growth factor-β (TGF-β) Large family, which includes activin, Vg-1, bone morphogenetic proteins (BMPs), nodal (mouse), decapentaplegic (*Drosophila*)	Receptors associated with a cytoplasmic serine-threonine protein kinase. Receptors act as dimers	Mesoderm induction in *Xenopus*; patterning of dorso-ventral axis and imaginal discs in *Drosophila*
Hedgehog Hedgehog in insects, Sonic hedgehog and Indian hedgehog in vertebrates	Patched	Positional signal in vertebrate limb and neural tube, and insect wing and leg discs
Wingless (Wnt) Wingless in insect, various Wnt proteins in vertebrates	Frizzled	Dorso-ventral axis specification in *Xenopus*; insect segment and imaginal disc specification
Delta and **Serrate**	Notch	Inhibitory signal in nervous system
Ephrins	Ephrin receptors	Vertebrate nervous system

antero-posterior axis of the embryo, the second divides the embryo into ventral and dorsal regions, whereas the third cleavage is at right angles to the axis and divides the embryo into animal and vegetal halves (see Fig. 2.5).

Differences in the localization of maternally provided mRNAs and proteins are present along the animal–vegetal axis of the egg before cleavage. Messenger RNAs for housekeeping proteins such as histones are abundant throughout the unfertilized egg, but there are also smaller amounts of localized mRNAs that are presumed to encode proteins with developmental roles.

The proteins encoded by several of these mRNAs belong to developmentally important families of signaling molecules, and are strong candidates for signals involved in specifying early polarity and inducing the mesoderm. One such maternal mRNA encodes the signaling protein Vg-1, which is a member of the transforming growth factor-β (TGF-β) family

(Box 3A). Vg-1 mRNA is localized at the vegetal half of the egg (Fig. 3.2), where its presence can be detected by *in situ* hybridization and autoradiography (Box 3B, p. 70). Vg-1 mRNA is synthesized during early oogenesis and becomes localized in the vegetal cortex of fully grown oocytes. It then moves into the vegetal cytoplasm prior to fertilization. Its localization is determined by the 3′ untranslated region of the mRNA. An mRNA encoding another signaling protein, Xwnt-11, is also localized in the vegetal hemisphere. The Wnt family of vertebrate proteins are related to, and named after, the wingless protein of *Drosophila* and the Int protein in mice, which is involved in retroviral integration. They are important signal proteins in pattern formation in the fly and other organisms. Another maternal mRNA localized to the vegetal hemisphere of the mature *Xenopus* oocyte encodes the transcription factor VegT. VegT mRNA is only translated after fertilization.

It should be emphasized that the axes of the tadpole are not directly comparable to those of the fertilized egg. The animal–vegetal axis of the egg is certainly related to the antero-posterior axis of the tadpole, as the head forms from the animal region. However, just where the head will form is not determined until after the second main body axis—the dorsoventral axis—is fixed, after fertilization. The precise position of the future antero-posterior axis thus depends on the specification of the dorso-ventral axis.

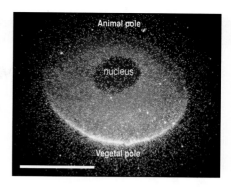

Fig. 3.2 Distribution of mRNA for the growth factor Vg-1 in the amphibian egg. *In situ* hybridization with a radioactive probe for maternal Vg-1 mRNA shows its localization (yellow) in the vegetal region. Scale bar = 1 mm. *Photograph courtesy of D. Melton.*

3.2 The dorso-ventral axis of an amphibian embryo is determined by the site of sperm entry

The spherical unfertilized egg of *Xenopus* is radially symmetric about the animal–vegetal axis, and this symmetry is broken only when the egg is fertilized. Sperm entry sets in motion a series of events that define the dorso-ventral axis of the gastrula (see Section 2.1 for a general outline of *Xenopus* development), with the dorsal side forming more or less opposite the sperm entry point.

Within 40 minutes of fertilization, changes in the egg become distinguishable opposite the site of sperm entry. The plasma membrane and the cortex—a gel-like layer of actin filaments and associated material about 5 μm thick beneath the membrane—rotates about 30° relative to the rest of the cytoplasm, which remains stationary. This **cortical rotation** is driven by microtubules and is toward the site of sperm entry, the vegetal cortex opposite the entry site moving toward the animal pole (Fig. 3.3).

Fig. 3.3 The future dorsal side of the amphibian embryo develops opposite the site of sperm entry. After fertilization (first panel), the cortical layer just under the cell membrane rotates toward the site of sperm entry, moving over the cytoplasm within. This movement results in the dorsal side being defined by the formation of the Nieuwkoop or signaling center (second panel). Later, the Spemann organizer and blastopore will develop in a region just above it (third panel). The fourth panel shows a tailbud stage embryo after gastrulation and neurulation. V = ventral; D = dorsal; A = anterior; P = posterior.

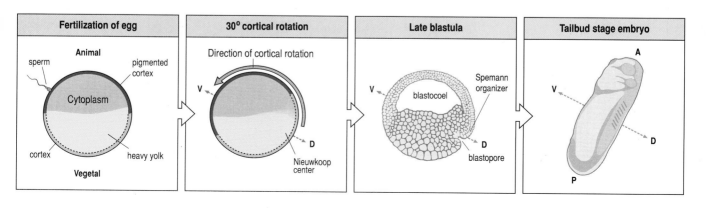

Box 3B *In situ* detection of gene expression

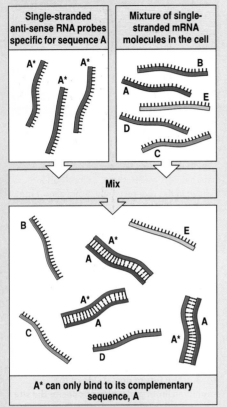

Single-stranded anti-sense RNA probes specific for sequence A

A* A* A*

Mixture of single-stranded mRNA molecules in the cell

B A E D C

Mix

B A* A E A* A C A* A A A* D

A* can only bind to its complementary sequence, A

In order to understand how gene expression is guiding development, it is essential to know exactly where and when particular genes are active. Genes are switched on and off during development and patterns of gene expression are continually changing. Several powerful techniques are available that show where a gene is being expressed within a tissue or within a whole early embryo.

One set of techniques uses *in situ* **hybridization** to detect the mRNA that is being transcribed from a gene. If an anti-sense RNA probe shares complementary sequence with a length of mRNA being transcribed in the cell, it will hybridize (pair tightly) to the mRNA. A nucleic acid probe of the appropriate sequence can therefore be used to locate its complementary mRNA in a tissue slice or a whole embryo. The probe may be labeled in various ways—with a radioactive isotope, a fluorescent dye, or an enzyme for histochemical localization—to enable it to be detected. Radioactively labeled probes are detected by autoradiography, as illustrated in the bottom panels, whereas probes labeled with colored dyes are observed directly. Enzyme-labeled probes are mixed with a substrate that produces a localized colored product. The probes can be applied to both tissue sections and whole-mount preparations.

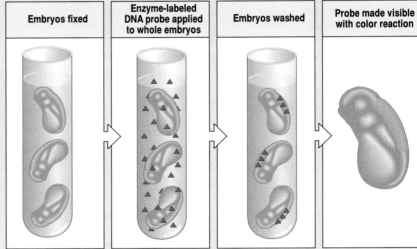

| Embryos fixed | Enzyme-labeled DNA probe applied to whole embryos | Embryos washed | Probe made visible with color reaction |

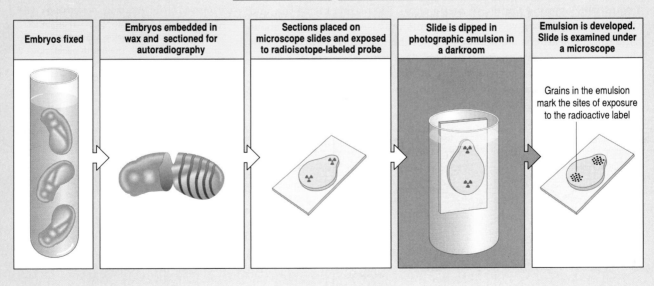

| Embryos fixed | Embryos embedded in wax and sectioned for autoradiography | Sections placed on microscope slides and exposed to radioisotope-labeled probe | Slide is dipped in photographic emulsion in a darkroom | Emulsion is developed. Slide is examined under a microscope |

Grains in the emulsion mark the sites of exposure to the radioactive label

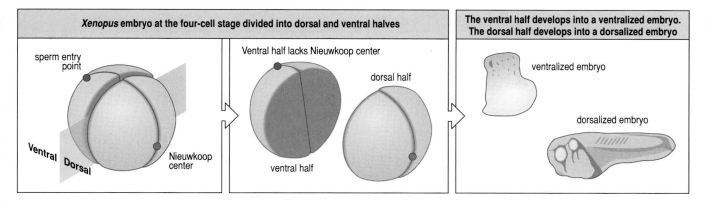

The crucial developmental consequence of cortical rotation is to cause the dorsal side of the embryo to acquire some sort of identity, and this leads to the formation of a **signaling center** in the vegetal region on the side opposite sperm entry. A signaling center is a localized region of the embryo that exerts a special influence on surrounding cells and can thus determine how they will develop. This particular signaling center is known as the **Nieuwkoop center**, after the Dutch embryologist Pieter Nieuwkoop, who discovered it. The Nieuwkoop center sets the initial dorso-ventral polarity in the blastula.

The Nieuwkoop center exerts its influence on dorso-ventrality from a very early stage. The first cleavage usually passes through the point of sperm entry and so divides the egg, and also the Nieuwkoop center, into left and right halves, defining the plane of bilateral symmetry of the animal body. The second cleavage divides the egg into ventral and dorsal halves. The importance of the center is shown by experiments that divide the embryo into two at the four-cell stage in such a way that one half contains the Nieuwkoop center and the other does not. The half containing the center will develop most structures, although the resulting embryo will lack some ventral regions. The half without the Nieuwkoop center will develop much more abnormally, with little relevance to the cell's normal fate, producing a distorted, radially symmetric **ventralized** embryo that lacks all dorsal and anterior structures (Fig. 3.4).

The influence of the center can also be seen dramatically in experiments in which cells from the region of the Nieuwkoop center of a 32-cell *Xenopus* embryo are grafted into the ventral side of another embryo. This gives rise to a twinned embryo with two dorsal sides (Fig. 3.5), whereas grafting ventral cells to the dorsal side has no effect. Signals from the Nieuwkoop center are required for the future development of all dorsal and anterior structures. Cells from the graft do not contribute to the mesodermal and neural tissues of the new axes, but do contribute to endodermal tissues, emphasizing their organizing abilities.

We can now understand the result of Roux's classic experiment (see Fig. 1.8) in which he destroyed one cell of a frog embryo at the two-cell stage and got a half-embryo rather than a half-sized whole embryo. The crucial feature of his experiment, it turns out, was that the killed cell remained attached, but the embryo did not 'know' it was dead. The plane of first cleavage had already passed through the Nieuwkoop center. The remaining living cell therefore developed as just one half-embryo and had just one half

Fig. 3.4 The Nieuwkoop center is essential for normal development. If a *Xenopus* embryo is divided into a dorsal and a ventral half at the four-cell stage, the dorsal half containing the Nieuwkoop center develops as a dorsalized embryo lacking a gut, while the ventral half, which has no Nieuwkoop center, is ventralized and lacks both dorsal and anterior structures.

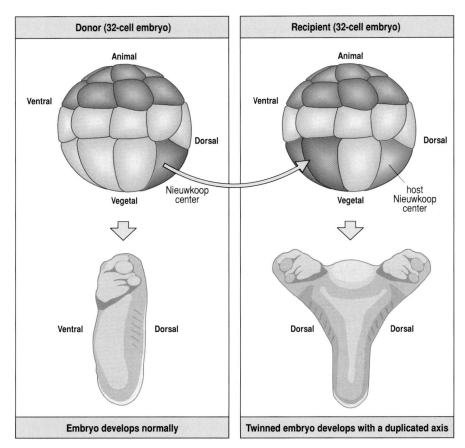

Donor (32-cell embryo)   **Recipient (32-cell embryo)**

Animal

Ventral   Ventral

Dorsal   Dorsal

Nieuwkoop center   host Nieuwkoop center

Vegetal   Vegetal

Ventral   Dorsal   Dorsal   Dorsal

Embryo develops normally   **Twinned embryo develops with a duplicated axis**

Fig. 3.5 The Nieuwkoop center can specify a new dorsal side. Grafting vegetal cells containing the Nieuwkoop center from the dorsal side to the ventral side of a 32-cell *Xenopus* blastula results in the formation of a second axis and the development of a twinned embryo. The grafted cells signal but do not contribute to the second axis.

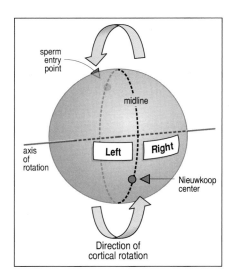

Fig. 3.6 Bilateral symmetry in amphibians results from cortical rotation. Cortical rotation toward the point of sperm entry defines the midline of the embryo, since the Nieuwkoop center lies on this midline and will give rise to the left and right sides.

of the center. If the killed cell had been separated from the living blastomere, the embryo could then have regulated and developed into a small-sized whole embryo. We now look at the relation between cortical rotation and the specification of the Nieuwkoop center.

3.3 The Nieuwkoop center is specified by cortical rotation

How sperm entry triggers cortical rotation is as yet unknown, but it probably causes a signal to be generated and passed to the egg's cytoskeleton. During cortical rotation the cortex slides with respect to the rest of the cytoplasm; this involves the interaction of the cortex with parallel arrays of microtubules located in the cytoplasm of the vegetal region. The rotation probably affects the whole embryo, but the key result is that it brings about a localized change in the vegetal region on the side of the egg opposite the point of sperm entry. This change specifies the Nieuwkoop center in the vegetal region on that side, just below the equator, possibly because the vegetal cortex can now interact with cytoplasm nearer to the animal pole.

As well as specifying the dorsal side, cortical rotation can also establish the plane of bilateral symmetry of the future embryo. The rotation toward the point of sperm entry is greatest in the plane defined by the point of sperm entry. This is the future midline and contains the Nieuwkoop center (Fig. 3.6). The first cleavage usually passes through the midline, dividing the egg into two initially symmetrical halves.

Blocking cortical rotation, and thus the formation of the Nieuwkoop center, leads to highly abnormal development. Cortical rotation can be prevented by irradiating the vegetal side of the egg with ultraviolet (UV) light, which disrupts the microtubule array responsible for the movement. Embryos developing from such treated eggs are ventralized; they are deficient in structures normally formed on the dorsal side, and develop excessive amounts of blood-forming mesoderm, a tissue normally most abundant at the embryo's ventral midline. With increasing doses of radiation, both dorsal and anterior structures are lost and the ventralized embryo appears as little more than a small, distorted cylinder. It has developed in the same way as an isolated ventral half of a four-cell embryo that lacks a Nieuwkoop center (see Fig. 3.4).

UV-irradiated eggs can be rescued by establishing a new Nieuwkoop center. This can be done by re-orienting the eggs (and thus mimicking cortical rotation) after irradiation. They can also be rescued at a slightly later stage by a direct graft of dorsal cells (as in Fig. 3.5) containing the Nieuwkoop center from another embryo at the 32-cell stage. This shows that it is not the cortical rotation itself but the specification of a Nieuwkoop center that is crucial. By establishing a Nieuwkoop center, both these treatments specify a dorsal side and enable the embryo to continue its development normally.

Whereas UV irradiation ventralizes the embryo, treatment with lithium chloride **dorsalizes** it, promoting the formation of dorsal and anterior structures at the expense of ventral and posterior structures. UV irradiation and lithium chloride probably produce their effects by interfering in some way either with the proteins involved in establishing the dorso-ventral axis or with their distribution. Some of these proteins are considered in the next section.

3.4 Maternal proteins with dorsalizing and ventralizing effects have been identified

Cortical movement results in an increase in the concentration of some proteins on the dorsal side. Some localized maternal proteins can both act like an additional Nieuwkoop center in a normal embryo, and rescue UV-irradiated eggs. One of these is β-catenin, whose mRNA is present in the egg. β-Catenin protein accumulates on the dorsal side of the egg following cortical rotation, and eventually moves from the cytoplasm into nearby nuclei. By the 16-cell stage, β-catenin protein can be detected in nuclei on the dorsal side (Fig. 3.7). The transcription factor siamois, which is involved in specifying the dorsal region, is expressed in the region where β-catenin levels are high; this is thought to be the result of the intranuclear β-catenin acting as a transcriptional regulator to help switch on the *siamois* gene. The dorsalizing effects of β-catenin can be seen when it is injected into ventral vegetal cells; this results in formation of a second main body axis (Fig. 3.8). The preferential accumulation of β-catenin on the dorsal side is thought to be due to prevention of its degradation. β-Catenin degradation involves a large protein complex that contains the protein kinase, glycogen synthase kinase-3 (GSK-3). This promotes the phosphorylation of β-catenin and targets it for degradation by cellular proteases. On the prospective ventral side of the embryo, β-catenin is degraded, but on the dorsal side, the activity of this complex appears to be inhibited, β-catenin is not degraded, and it

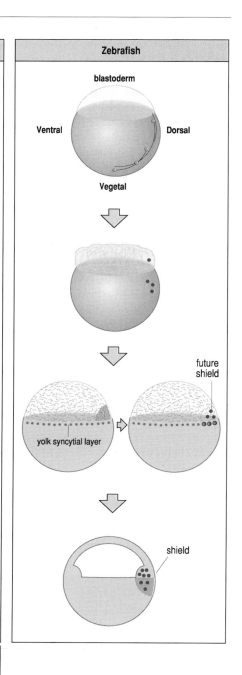

Fig. 3.7 The establishment of the dorsal organizer and axis formation in *Xenopus* and zebrafish embryos. In *Xenopus*, cortical rotation moves dorsal determinants towards the future dorsal side of the embryo, creating a large region where, starting at the 32-cell stage, β-catenin moves from the cytoplasm into nuclei. The region in which β-catenin is nuclear defines the blastula organizer and expresses the transcription factor siamois. At the gastrula stage, siamois and other factors further define the vegetal head organizer and marginal trunk organizer. In zebrafish, dorsal determinants are transported towards the future dorsal side of the embryo and enter the blastoderm. At the mid-blastula transition, β-catenin is translocated into dorsal yolk syncytial layer nuclei. In both species, the dorsal organizer forms where β-catenin is translocated into the nuclei. After: Kodjabachian, L., Lemaire, P.: 1998.

accumulates to high levels. β-Catenin is a component of the Wnt signal transduction pathway, as shown in Fig. 3.9, but it is unlikely that Wnt signals themselves are required for the initial specification of the axis. It seems that components of the pathway are moved along microtubules to the dorsal side of the fertilized egg following cortical rotation; when they are activated, the action of the GSK-3 complex is blocked, thus leading to the accumulation of β-catenin on the dorsal side only. Lithium acts by blocking GSK-3 activity.

One of the main roles of the Nieuwkoop center is to specify another key

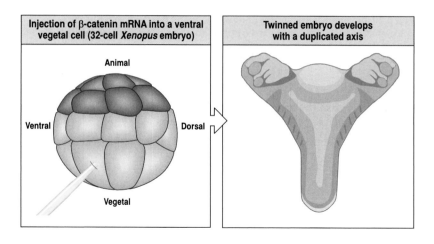

Injection of β-catenin mRNA into a ventral vegetal cell (32-cell *Xenopus* embryo)

Animal

Ventral — Dorsal

Vegetal

Twinned embryo develops with a duplicated axis

Fig. 3.8 Induction of a new dorsal side by injection of β-catenin mRNA. Injection of mRNA encoding β-catenin into ventral vegetal cells can specify a new Nieuwkoop center at the site of injection, leading to a twinned embryo. Some other vegetally localized maternal mRNAs, such as *Vg-1*, also have similar effects.

dorsal signaling center—the **Spemann organizer**—which arises just above the Nieuwkoop center at the late blastula–early gastrula stage (see Fig. 3.3). As we shall see, signals originating in the Spemann organizer are involved in further patterning along both the antero-posterior and dorso-ventral axes of the embryo, and in inducing the central nervous system.

In the zebrafish embryo the animal–vegetal axis is clear from an early stage, but how the dorso-ventral axis is specified is not known. The embryonic shield, which forms at the dorsal margin, is the zebrafish organizer (see Fig. 3.7). The vegetal yolk cell specifies the position of the shield. A

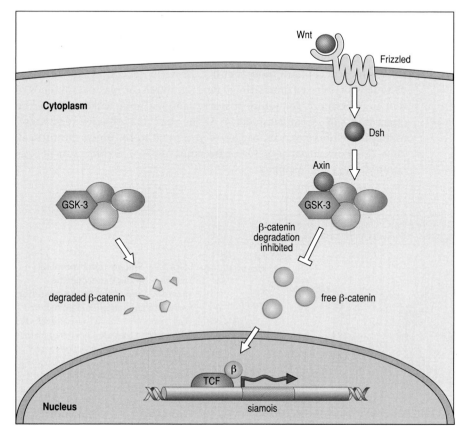

Fig. 3.9 β-Catenin is a component of the Wnt pathway. When a Wnt protein binds to its receptor, Frizzled, a signal is transduced across the cell membrane to the intracellular signaling protein Dishevelled (Dsh). Activated Dsh interacts with the protein axin to prevent a protein complex containing glycogen synthase kinase (GSK-3) preparing β-catenin (β) for degradation. Stabilized free β-catenin enters the nucleus where it forms a complex with the transcription factor TCF and activates expression of target genes such as *siamois*.

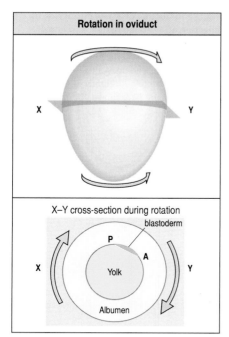

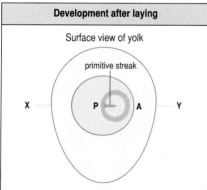

Fig. 3.10 Gravity defines the antero-posterior axis of the chick. Rotation of the egg in the oviduct of the mother results in the blastoderm being tilted in the direction of rotation, though it tends to remain uppermost. The posterior marginal zone (P) develops at that side of the blastoderm which is uppermost and initiates the primitive streak. A = anterior.

signal localized at the vegetal pole of the yolk cell is essential for organizer induction and for the development of the nervous system and trunk. β-Catenin accumulates in the dorsal yolk syncytial layer, suggesting a similar process to that in *Xenopus*, but there is no cortical rotation, only cytoplasmic streaming (see Section 2.4 for a general outline of zebrafish development).

3.5 The antero-posterior axis of the chick blastoderm is set by gravity

Because there is so much yolk in a hen's egg, cleavage is restricted to a thin layer of cytoplasm and the chick embryo starts off as a disc of cells—the **blastoderm**—sitting on the top of the yolk (see Section 2.2 for a general outline of chick development). Like the amphibian blastula, the chick blastoderm is initially radially symmetric. This symmetry is broken when the posterior end of the embryo is specified. The future posterior end becomes evident soon after the egg is laid, when a denser area of cells appears at one side of the blastoderm; this is the **posterior marginal zone** (see Fig. 2.13). From this region, the **primitive streak** will develop. The streak defines the position of the antero-posterior axis within the blastoderm. It also defines the axis of bilateral symmetry and thus the dorso-ventral axis.

The position of the posterior marginal zone, and thus of the posterior end of the antero-posterior axis, is specified by gravity. During its passage through the hen's uterus, which takes about 20 hours, the fertilized egg moves pointed end first and rotates slowly around its long axis, each revolution taking about 6 minutes. Cleavage has already started at this stage, so the blastoderm contains thousands of cells when the egg is laid. The egg is obliquely tilted in the gravitational field, although the blastoderm tends to remain uppermost. The future posterior marginal zone will develop at the uppermost side of the blastoderm. As the shell and albumen (egg white) rotate, the embryo and its weight of yolk tend to return to the vertical. Thus, the future blastoderm region becomes tipped in the direction of rotation of the egg (Fig. 3.10).

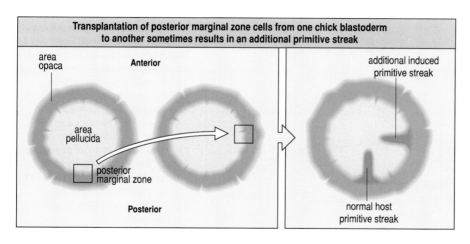

Fig. 3.11 The posterior marginal zone of the chick specifies the posterior end of the antero-posterior axis. Grafting posterior marginal zone cells to another site in the marginal zone can result in the formation of an extra primitive streak, which defines a new antero-posterior axis. This does not always occur. Usually the more advanced of the streaks is the only one to develop because it inhibits the development of the other.

The posterior marginal zone can be thought of as an organizing center analogous in some way to the Nieuwkoop center in *Xenopus*, since it too can induce a new axis: if a fragment of the posterior marginal zone is grafted to another part of the marginal zone, it may induce a new primitive streak (Fig. 3.11). In general, however, only one axis develops in the grafted embryos—either the host's normal axis or one induced by the graft. This suggests that the more advanced of the two organizing centers inhibits streak formation elsewhere. A chick gene related to *Vg-1* of *Xenopus* is expressed at the site of primitive streak formation. If cells expressing this Vg-1 protein are grafted to another part of the marginal zone, they can induce a complete new primitive streak, thus simulating grafts from the posterior marginal zone.

3.6 The axes of the mouse embryo are recognizable early in development

The very earliest stages of development of a mammalian egg differ considerably from those of either *Xenopus* or the chick, as the mammalian egg contains no yolk (see Section 2.3 for a general outline of mouse development). Thus, the extra-embryonic structures of the placenta, which nourishes the embryo, must develop. There is no clear sign of polarity in the mouse egg, nor evidence for localized maternal factors affecting later development. However, the site of the second polar body and the point of sperm entry may define axes in the fertilized egg that are related to future axes in the blastocyst.

The early cleavages do not follow a well-ordered pattern. Some cleavages are parallel to the egg's surface, so that a solid ball of cells (the **morula**) is formed with outer and inner cell populations (Fig. 3.12). By the 32-cell stage, the mouse embryo has developed into a **blastocyst**, a hollow sphere of epithelium containing a small mass of some 10–15 cells attached at one end. These cells are the **inner cell mass**, from which the embryo proper and some extraembryonic structures will develop, whereas the outer

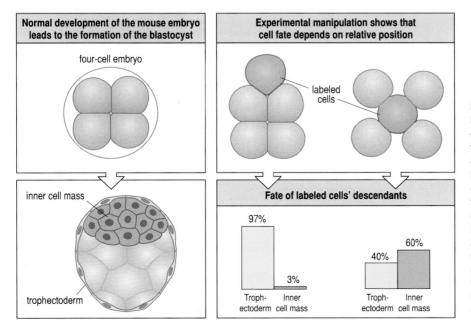

Fig. 3.12 The specification of the inner cell mass of a mouse embryo depends on the relative position of the cells with respect to the inside and outside of the embryo. If labeled blastomeres from a four-cell mouse embryo are separated and combined with unlabeled blastomeres from another embryo, it can be seen that blastomeres on the outside of the aggregate more often give rise to trophectoderm, 97% of the labeled cells ending up in that layer. The reverse is true for the origin of the inner cell mass, blastomeres on the inside giving rise most often to this.

epithelium will form the **trophectoderm**, which gives rise to extra-embryonic structures associated with implantation and formation of the placenta.

The specification of cells as either inner cell mass or trophectoderm depends on their relative positions in the cleaving embryo. Determination of their fate occurs only after the 32-cell stage, and during earlier stages all the cells seem to be equivalent in their ability to give rise to either tissue. The most direct evidence for the effect of position comes from taking individual cells (blastomeres) from disaggregated four-cell or eight-cell embryos, labeling them, and then combining the labeled blastomeres in different positions with respect to unlabeled blastomeres from another embryo. If the labeled cells are placed on the outside of a group of unlabeled cells they usually give rise to trophectoderm; if they are placed inside, so that they are surrounded by unlabeled cells, they more often give rise to inner cell mass (see Fig. 3.12). If a whole embryo is surrounded by other blastomeres, it too can become part of the inner cell mass of a giant embryo. Aggregates composed entirely of either 'outside' or 'inside' cells of early embryos can also develop into normal blastocysts, showing that there is no specification of these cells, other than by their position, at this stage.

Although the mouse embryo is highly regulative, there is evidence for establishment of axes at a very early stage, although these axes can be re-established if development is disrupted. The position of the polar body defines an 'animal–vegetal' axis in the zygote, with the polar body at the animal pole. The mouse embryo is essentially a spheroidal ball of cells up to the blastocyst stage when, following the specification of the inner cell mass and trophectoderm, a blastocoel cavity develops asymmetrically within the blastocyst, leaving the inner cell mass attached to one part of the trophectoderm (Fig. 3.13). The blastocyst now has one distinct axis running from the site where the inner cell mass is attached—the **embryonic pole**—to the

Fig. 3.13 The axes of the early mouse embryo. As shown in the right-hand panel, at the blastocyst stage, about 4 days, the inner cell mass is confined to the embryonic, or dorsal, region and this defines an embryonic–abembryonic axis (which relates geometrically, although not in terms of cell fate, to the dorsal–ventral axis of the future epiblast). The inner cell mass (ICM) is oval, and thus has an axis of bilateral symmetry, which is related to the site of the polar body (see photo; the red line indicates the axis of bilateral symmetry). The asterisks indicate the location of the polar body. The green cells at either end of the axis have been injected with a long-lasting marker RNA for green fluorescent protein to trace their fate.

Photos from Weber, R.J. et al.: 1999.

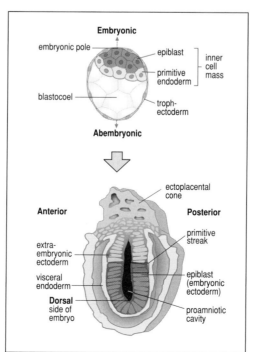

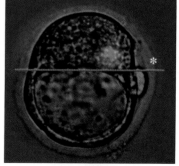

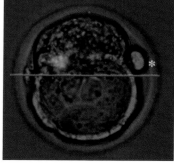

opposite end (the **embryonic-abembryonic axis**, with the blastocoel occupying most of the abembryonic half). This axis corresponds to the dorso-ventral axis (see Section 2.3), but only in a geometrical sense, and is not related to specification of cell fate. It appears that the embryonic–abembryonic axis is related to the site of sperm entry. There is evidence that the site of sperm entry predicts the plane of the first cleavage and can define the embryonic and abembryonic halves of the future blastocyst, with the sperm entry site coming to lie on the boundary between the embryonic and abembryonic halves.

The inner cell mass is oval, which provides the blastocyst with a second axis, an axis of bilateral symmetry, which relates spatially to the antero-posterior axis of the future embryo. The axis of bilateral symmetry is related to the site of the second polar body (Fig. 3.13). Nevertheless, the regulative powers of the embryo argue against any determinants in the egg specifying the axes. Rearranging or removing cells at the eight-cell stage has little effect on development; neither does removing cytoplasm from regions of the egg.

By about 4½ days after fertilization the inner cell mass has become differentiated into two tissues: primary or primitive endoderm on its blastocoelic surface (which will form extra-embryonic structures); and the **epiblast** within, from which the embryo and some extra-embryonic structures will develop. The blastocyst implants in the uterine wall, and the trophectoderm at the embryonic pole proliferates to form the ectoplacental cone, producing extra-embryonic ectoderm that pushes the inner cell mass across the blastocoel. A cavity—the proamniotic cavity—is then formed within the epiblast as its cells proliferate. The mouse epiblast also known as the **embryonic ectoderm**, is now an epithelial sheet and is formed into a cup shape, U-shaped in section (Fig. 3.13). During this process there is considerable cell mixing, and it is not possible at this early stage to identify individual cells that will become dorsal or ventral. It should be noted that the cup-shaped epiblast is a peculiarity of rodents. The epiblast in other mammals, humans and rabbits for example, is flat.

The first sign of the antero-posterior axis of the embryo can be seen at 5½ days after fertilization, about 24 hours before gastrulation begins. At 6½ days, epiblast cells begin to form a primitive streak, which is equivalent to gastrulation in the frog. Primitive streak formation occurs at the posterior pole of the embryo. The streak elongates in an anterior direction and has a node at its anterior end, which is analogous to the organizer of other vertebrates. But while transplantation of this node can induce secondary axes, these invariably lack the most anterior regions, including the brain. These anterior structures are in part induced by the extra-embryonic **anterior visceral endoderm**, into which ventral visceral endoderm cells have moved (Fig. 3.14). Before streak formation this region expresses homeodomain transcription factors such as Hex, Otx-2, Lim-2, and goosecoid. Removal of the region early in gastrulation leads to loss of markers of prospective forebrain. Another extra-embryonic structure, the extra-embryonic ectoderm, is involved in setting up the posterior to anterior axis in the epiblast.

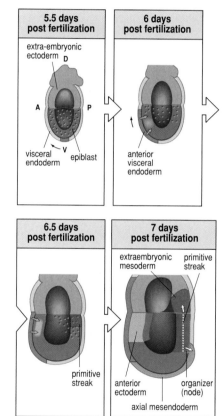

Fig. 3.14 Cell movement in the formation of the antero-posterior axis of the mouse embryo. At around 5.5 days, before primitive streak formation, ventral visceral endoderm (green) expresses the gene *Hex* and begins to move anteriorly. At 6.5 days, the primitive strea begins to form at the posterior end of the epiblast and then extends anteriorly with a well defined node. The anterior visceral endoderm induces anterior ectoderm.

3.7 Specification of left–right handedness of internal organs requires special mechanisms

Vertebrates are bilaterally symmetric about the midline of the body for many structures, such as eyes, ears, and limbs. But, while the vertebrate body is outwardly symmetric, most internal organs are in fact asymmetric with respect to the left and right sides. In mice, for example, the heart is on the left side, the right lung has more lobes than the left, the stomach and spleen lie to the left, and the liver has a single left lobe. This handedness of organs is remarkably consistent, but there are rare individuals, about one in 10,000 in humans, who have the condition known as **situs inversus**, a complete mirror-image reversal of handedness. Such people are generally asymptomatic even though all their organs are reversed.

Specification of left and right is fundamentally different from specifying the other axes of the embryo, as left and right have meaning only after the antero-posterior and dorso-ventral axes have been established. If one of these axes is reversed, then so too will be the left–right axis (it is for this reason that handedness is reversed when you look in a mirror: your dorso-ventral axis is reversed, and hence left becomes right and *vice versa*). While the molecular mechanism by which organ handedness is specified remains a mystery, one suggestion is that asymmetry at the molecular level is converted to an asymmetry at the cellular and multicellular level. If that were so, the asymmetric molecules or molecular structure would need to be oriented with respect to both the antero-posterior and dorso-ventral axes.

3.8 Organ handedness in vertebrates is under genetic control

In the early chick embryo, several genes are expressed asymmetrically with respect to Hensen's node (see Section 2.2), which is at the anterior end of the primitive streak. Grafting experiments show that asymmetry is initially specified in the tissue adjacent to the node, and is then acquired by the node itself. The gene *Sonic hedgehog*, whose protein product is implicated in a variety of developmental processes (see Section 4.2), is expressed on the left side only of Hensen's node. The TGF-β family member activin and its receptor are expressed on the right side and may repress *Sonic hedgehog* expression on this side. On the left side, Sonic hedgehog protein induces the expression of *nodal* (Fig. 3.15). The expression of the gene *lefty* in the

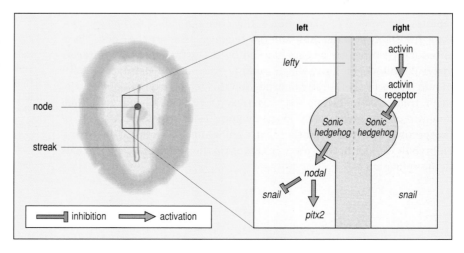

Fig. 3.15 Pathways determining left–right asymmetry in the chick embryo. Activation of an activin receptor on the right side leads to the inhibition of *Sonic hedgehog* expression in the adjacent part of the node. *Sonic hedgehog* is still being expressed in the left half of the node and this activates the gene *nodal*, which in turn leads to activation of the gene *pitx2* and inhibition of *snail*. *pitx2* and *snail* encode transcription factors. Expression of *lefty* in the left half of the primitive streak provides a midline barrier to these signals. This is a simplified version of a much more complex set of interactions.

left half of the primitive streak produces a barrier that prevents signals crossing the midline. If this pattern of *nodal* expression is made symmetric by placing a pellet of cells secreting Sonic hedgehog protein on the right-hand side, then organ asymmetry is randomized. There is a complex pattern of gene expression that involves both gene induction and suppression, and the asymmetric pattern of gene expression is somewhat different in the mouse, chick, and *Xenopus*. However, both nodal protein and the transcription factor Pitx2 are expressed on the left side in all three species. Ectopic expression of Pitx2 on the right can lead to randomization of symmetry.

In *Xenopus*, localized injection of processed Vg-1 protein into the right side of an early embryo randomizes organ asymmetry, suggesting that asymmetrical distribution of Vg-1 in the Nieuwkoop center could underlie left–right asymmetry. The generation of asymmetry has been suggested to involve orientation of an asymmetric molecule, which specifies a small differential between left and right sides, but there is no direct evidence for this.

In mice, the *iv* gene is involved in specifying organ handedness, by means as yet unknown. In animals homozygous for a mutant *iv* allele, the handedness of the organs is reversed in half of the animals (Fig. 3.16), implying that the specification of handedness has become random in these mutants. The mutation therefore affects not the generation of asymmetry itself but the mechanism by which it is normally consistently biased to one side. These mutant mice quite often show **heterotaxis**, the condition where organs of normal and inverted asymmetry are present in the same animal. This suggests that the generation of asymmetry in different organs may occur independently. The *iv* gene codes for a dynein, a motor protein that moves along microtubules and is also involved in the movement of cilia. The gene has now been renamed *left–right dynein (lrd)*. Another mutation, *inv*, causes the complete reversal of handedness in mice; the *inv* gene codes for a protein containing ankyrin repeats, which suggests an involvement with the cytoskeleton.

In humans, situs inversus sometimes occurs with Kartagener's syndrome, which is a recessive defect. As in the *iv* mice, handedness becomes random; that is 50% of those with the syndrome have altered asymmetry. In individuals with this syndrome, the cilia that line the surface of respiratory organs, such as the lungs, are nonfunctional and do not beat, and so these

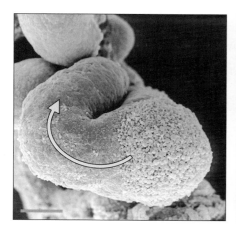 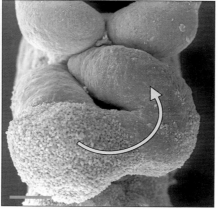

Fig. 3.16 Left–right asymmetry of the mouse heart is under genetic control. Each photograph shows a mouse heart viewed anteriorly after the loops have formed. The normal asymmetry of the heart results in it looping to the left, as indicated by the arrow (left panel). 50% of mice that are homozygous for the mutation in the *iv* gene have hearts that loop to the right (right panel). Scale bar = 0.1 mm.

Photographs courtesy of N. Brown.

individuals suffer from respiratory problems. The immotile cilia lack the motor protein dynein, which is essential for their movement. Dynein has other roles in cells, where it is associated with microtubules, and so microtubules and other cytoskeletal structures may have a role in generating asymmetry, as they themselves are asymmetric structures. There is evidence that in the normal embryo, ciliated cells in the node at the end of the primitive streak produce a leftward flow of extra-embryonic fluid, but that in *lrd* mutant mice the cilia in the node are immotile. Thus cilia might provide an early mechanism for specifying left–right asymmetry in humans. However, this mechanism probably does not operate in chick or *Xenopus*.

Summary

Setting-up the body axes in vertebrates involves maternal factors, external influences, and cell–cell interactions. In the amphibian embryo, maternal factors determine the animal–vegetal axis, which approximately corresponds to the antero-posterior axis, whereas the dorso-ventral axis is specified by the site of sperm entry and the resulting cortical rotation, which leads to the establishment of the Nieuwkoop center. In chick embryos, the dorso-ventral axis is specified at cleavage in relation to the yolk while the setting of the antero-posterior axis involves gravity, which determines the side of the blastoderm at which the posterior marginal zone, and thus the primitive streak, will form. Specification of the axes in the mouse embryo does not involve any maternal component. They are established later in tissues descended from the inner cell mass, which gives rise to primitive endoderm as well as the epiblast. The dorso-ventral axis is related to the position of the inner cell mass on the trophectoderm, while the antero-posterior axis is related to the site of polar body formation. The generation of the consistent left–right organ asymmetry found in vertebrates is under genetic control and involves cell–cell interactions.

Summary: vertebrate axis determination		
	Dorso-ventral axis	**Antero-posterior axis**
Xenopus	sperm entry point and cortical rotation. Dorsal side and Nieuwkoop center form on side opposite to sperm entry. Caterin on dorsal side	specified by Nieuwkoop center
Chick	axis of bilateral symmetry	gravity
Mouse	interaction between inner cell mass and trophectoderm	intercellular interactions?
Zebrafish	yolk cell specifies dorsal side	shield specifies the animal region

The origin and specification of the germ layers

We have seen in the preceding sections how the main axes are laid down in various vertebrate embryos. We now focus on the earliest patterning of the embryo with respect to these axes: the specification of the three germ layers—endoderm, mesoderm, and ectoderm—and their further diversification.

All the tissues of the body are derived from these three germ layers. The mesoderm becomes subdivided into cells that give rise to notochord, to muscle, to heart and kidney, and to blood-forming tissues, amongst others. The ectoderm becomes subdivided into cells that give rise to the epidermis and that develop into the nervous system. The endoderm gives rise to the gut and organs such as the lungs. We first look at the **fate maps** (see Section 1.10) of early embryos of different vertebrates, which tell us which tissues the different regions of the embryo give rise to. We then consider how the germ layers are specified and subdivided, with the main focus on *Xenopus*, in which these processes are best understood and where some of the genes and proteins involved have been identified.

3.9 A fate map of the amphibian blastula is constructed by following the fate of labeled cells

Examination of the *Xenopus* blastula at the 32-cell stage gives no indication of how the different regions will develop, but individual cells can be identified with respect to the animal–vegetal and dorso-ventral axes. By following the fate of individual cells, or groups of cells, we can make a map on the blastula surface showing the regions that will give rise to, for example, somites, brain, spinal cord, or gut. The fate map shows where the tissues of each germ layer normally come from, but it indicates neither the full potential of each region for development nor to what extent its fate is already specified or determined in the blastula. While we know what each of the early cells will give rise to, the embryo does not. Early vertebrate embryos have considerable capacity for regulation when pieces are removed or transplanted to different parts of the same embryo (see Section 1.10). This implies considerable developmental plasticity at this early stage and also that the actual fate of cells is heavily dependent on the signals they receive from neighboring cells.

One way of making a fate map is to stain various parts of the surface of the early embryo with a lipophilic dye such as diI, and observe where the labeled region ends up. Individual cells can also be labeled by injection of stable high molecular weight molecules such as rhodamine-labeled dextran, which cannot pass through cell membranes and so are restricted to the injected cell and its progeny; as rhodamine fluoresces red in UV light, the rhodamine dextran can be easily detected under a UV microscope. Fig. 3.17 shows a *Xenopus* embryo labeled for fate mapping with the green-fluorescing dye fluorescein-dextran-amine. In *Xenopus* and the zebrafish there is no cell growth, and so the label does not become diluted as development proceeds.

The fate map of the late *Xenopus* blastula (Fig. 3.18) shows that the yolky vegetal region, which occupies the lower third of the spherical blastula, gives rise to most of the endoderm. The yolk, which is present in all cells,

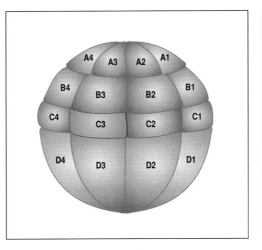

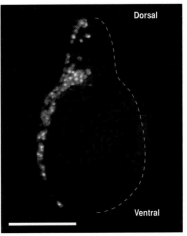

Fig. 3.17 Fate mapping of the early *Xenopus* embryo. Left panel: a single cell in the embryo, C3, is labeled by injecting fluorescein-dextran-amine, which fluoresces green under UV light. Right panel: a cross-section of the embryo, made at the tailbud stage, shows that the labeled cell has given rise to mesoderm cells on one side of the embryo. Scale bar = 0.5 mm.

Photograph courtesy of L. Dale.

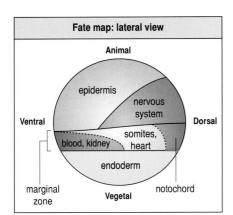

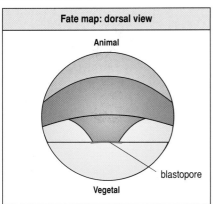

Fig. 3.18 Fate map of a late *Xenopus* blastula. The ectoderm gives rise to the epidermis and nervous system. Along the dorso-ventral axis the mesoderm gives rise to notochord, somites, heart, kidneys, and blood. Note that blood can also form in more dorsal regions. In *Xenopus*, although not in all amphibians, there is also endoderm (not shown here) overlying the mesoderm in the marginal zone.

provides all the nutrition for the developing embryo, and is gradually used up as development proceeds. At the other pole, the animal hemisphere becomes ectoderm, which becomes further diversified into epidermis and the future nervous tissue. The mesoderm forms a belt-like region, known as the **marginal zone**, around the equator of the blastula. In *Xenopus*, but not in all amphibians, a thin outer layer of presumptive endoderm overlies the presumptive mesoderm in the marginal zone.

The fate map of the blastula makes clear the function of gastrulation. At the blastula stage, the endoderm which gives rise to the gut is on the outside and so must move inside. Similarly, mesodermal tissues that will form internal tissues and organs such as muscle, bone, heart, kidneys, and blood, are on the outside of the embryo. During gastrulation, the marginal zone moves into the interior through the dorsal blastopore, which lies above the Nieuwkoop center. The fate map of the mesoderm (see Fig. 3.18) shows that it becomes subdivided along the dorso-ventral axis of the blastula. The most dorsal mesoderm gives rise to the notochord, followed, as we move ventrally, by somites (which give rise to muscle tissue), lateral plate (which contains heart and kidney mesoderm), and blood islands (tissue where hematopoiesis first occurs in the embryo). There are also important differences between the future dorsal and ventral sides of the animal hemisphere: the epidermis comes mainly from the ventral side of the animal hemisphere, whereas the nervous system comes from the dorsal side. The epidermis spreads to cover the whole of the embryo after neural tube formation.

The terms dorsal and ventral in relation to the fate map can be somewhat confusing, because the fate map does not correspond exactly to a neat set of axes at right angles to each other. As a result of cell movements during gastrulation, cells from the dorsal side of the blastula give rise to some ventral parts of the anterior end of the embryo, such as the head, as well as to dorsal structures, and will also form some other ventral structures, such as the heart. The ventral region gives rise to ventral structures in the anterior part of the embryo but will also form some dorsal structures posteriorly. This is why the 'dorsalized' embryos described in Section 3.4 have overdeveloped anterior structures and lack posterior regions.

The fate map of the *Xenopus* blastula has reasonably well-defined boundaries. However, there is some local cell movement and mixing of cells as the embryo develops and deeper cells move to the surface. The fate map that one can obtain in *Xenopus* in no way implies that the fate of the cells in the early embryo is fixed, but rather reflects the stereotyped pattern of tissue movements that carry cells to their positions in the later embryo.

In contrast to the fate map, which gives no indication of the actual differences between cells at the time they are labeled, a **specification map** gives some indication of such differences (see Section 1.10). A specification map of the blastula is constructed by culturing small pieces of the blastula in a simple culture medium and observing what tissues they form. The specification map of the *Xenopus* blastula corresponds quite well to some features of the fate map, but there are important differences, particularly in the ectodermal and mesodermal regions (Fig. 3.19). No neural tissue develops from explants of cells from the animal half of the blastula, and no muscle develops from any but the most dorsal-most mesodermal fragments. This shows that the ectoderm has not yet become specified as prospective neural cells and prospective epidermal cells, and that prospective muscle has not yet been specified within the mesoderm. Nevertheless, the specification map shows that there are already important regional differences in cell states at the blastula stage.

3.10 The fate maps of vertebrates are variations on a basic plan

Fate maps of the early embryos of chick, mouse, and zebrafish have been prepared using techniques essentially similar to those used for *Xenopus*: cells in the early embryo are labeled and their fate followed.

A fate map of the chick embryo cannot be made at the early blastoderm stage that corresponds roughly to the *Xenopus* blastula. This is partly because so much of the chick embryo comes from the posterior marginal zone, which is still a very small region of the total blastoderm at this stage. Unlike *Xenopus*, there is considerable cell proliferation and growth in the chick embryo during primitive streak formation and gastrulation. There are also extensive cell movements both before and during the emergence of the primitive streak (see Fig. 2.13) and gastrulation. Once the primitive streak has formed, the picture becomes clearer, and presumptive endoderm, mesoderm, and ectoderm can be mapped (Fig. 3.20).

At the stage shown in Fig. 3.20, the blastoderm has become a three-layered structure. Cells have ingressed through the primitive streak into the interior to form mesodermal and endodermal layers. Most of the cells that now form the outer surface of the blastoderm are prospective ectoderm and will form neural tube and epidermis, but there are still regions of the outer blastoderm that will move through the streak and give rise to mesoderm. Hensen's node, an aggregation of cells at the anterior end of the streak, is prospective mesoderm; as the node regresses it leaves cells behind that will form the notochord and also contribute to the somites. In the mesodermal layers of the blastoderm, the mesoderm lying along the antero-posterior midline will give rise to somites and is surrounded by cells that will form the lateral plate mesoderm and structures such as the heart and kidney. In the lowest layer of the embryo, closest to the yolk, the presumptive endoderm is surrounded by cells that will form extra-embryonic structures.

In the case of the mouse, most cells from the inner cell mass of embryos less than 3½ days old can give rise to many different embryonic tissues, as

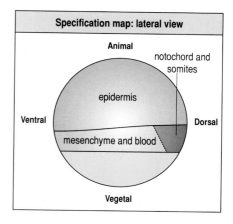

Fig. 3.19 Specification map of a *Xenopus* late blastula. The specification map is constructed from the results of experiments showing how isolated fragments of blastula develop in a simple culture medium.

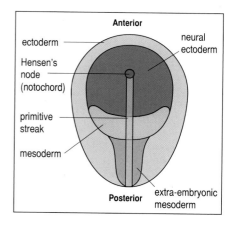

Fig. 3.20 Fate map of a chick embryo when the primitive streak has fully formed. The diagram shows a view of the dorsal surface of the embryo. Almost all the endoderm has already moved through the streak to form a lower layer, so is not shown.

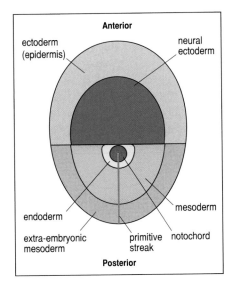

Fig. 3.21 Fate map of a mouse at the late gastrula stage. The embryo is depicted as if the 'cup' has been flattened and is viewed from the dorsal side. At this stage the primitive streak is at its full length.

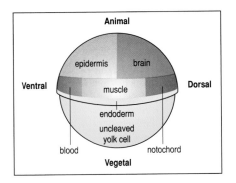

Fig. 3.22 Fate map of zebrafish at the early gastrula stage. The three germ layers come from the blastoderm, which sits on the lower hemisphere composed of an uncleaved yolk cell. The endoderm comes from the margin of the blastoderm and some has already moved inside.

well as to some extra-embryonic structures, such as the visceral and parietal endoderm, and so at this stage a fate map like that of *Xenopus* cannot be constructed. At about 4–4½ days, the inner cell mass gives rise to an outer layer of cells, the primitive endoderm (see Fig. 3.13). The cells lying between the primitive endoderm and the polar trophectoderm comprise the embryonic (primitive) ectoderm or epiblast. Although the primitive endoderm gives rise to extra-embryonic structures only, the embryonic ectoderm gives rise to all of the embryo proper and all the extra-embryonic mesodermal structures.

At 6–7 days gestation the mouse epiblast becomes transformed into the three germ layers by the formation of a primitive streak and gastrulation. Gastrulation in the mouse is essentially very similar to gastrulation in the chick, but at this stage the mouse epiblast is folded into a cup, which makes the process more difficult to follow. A detailed fate map of this stage has been established by tracing the descendants of single cells that have been labeled by injection with a dye. There is, however, extensive cell mixing and cell proliferation in the epiblast. Descendants of a single cell can spread widely and give rise to cells of different germ layers, so that only about 50% of the labeled clones have progeny in only one germ layer.

The fate map obtained is basically similar to that of the chick at the primitive streak stage, making allowances for the fact that the mouse epiblast is cup shaped, in contrast to the sheet-like chick epiblast (Fig. 3.21). The node forms at the anterior end of the primitive streak in the mouse embryo, and gives rise to the notochord and part of the somites. The middle part of the streak gives rise mainly to lateral plate mesoderm, while the posterior part of the streak provides the extra-embryonic mesoderm of the amnion, visceral yolk sac, and allantois.

In the zebrafish embryo there is extensive cell mixing during the transition from blastula to gastrula, and so it is not possible to construct a reproducible fate map at cleavage stages. In this the zebrafish resembles the mouse. The zebrafish late blastula comprises a cup-shaped blastoderm of deep cells and a thin overlying layer, sitting on top of a large yolk cell. The overlying layer is largely protective and is eventually lost. At the beginning of gastrulation, the fate of deep-layer cells, from which all the cells of the embryo itself will come, is correlated with their position in respect of the animal pole. Cells at the margin of the blastoderm give rise to the endoderm, cells slightly further toward the animal pole to mesoderm, while ectodermal cells come from the blastoderm nearest the animal pole (Fig. 3.22).

A fate map for each of the germ layers has also been constructed: in the ectoderm, for example, forebrain structures come from a region near the animal pole, while hindbrain structures come from nearer the margin. In the mesoderm, the future notochord is located on the dorsal side whereas presumptive blood-forming tissue is located ventrally. In general terms, the fate map of the zebrafish is rather similar to that of an amphibian, if one imagines the vegetal region of the amphibian blastula being replaced by one large yolk cell.

The fate maps of the different vertebrates are thus similar when one looks at the relationship between the germ layers and the site of ingression of cells at gastrulation (Fig. 3.23). The differences are due mainly to the yolkiness of the different eggs, which determines the pattern of cleavage and influences the shape of the early embryo. The similarity in relationship

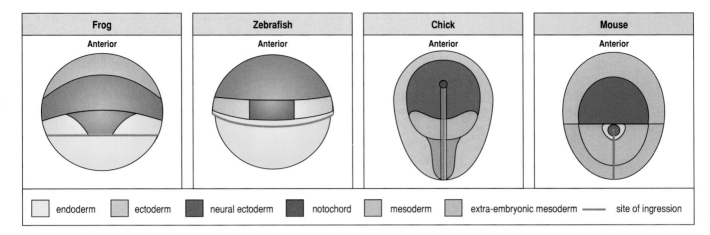

Frog	Zebrafish	Chick	Mouse
Anterior	Anterior	Anterior	Anterior

■ endoderm   ■ ectoderm   ■ neural ectoderm   ■ notochord   ■ mesoderm   ■ extra-embryonic mesoderm   —— site of ingression

between the germ layers implies that similar mechanisms must be involved in their specification. Fate maps are not specification maps, and they do not reflect the full potential for development of the cells of these early embryos. At the late blastula and early gastrula stages, when these maps are made, vertebrate embryos are still capable of considerable regulation.

3.11 Cells of early vertebrate embryos do not yet have their fates determined

All early vertebrate embryos have considerable powers of regulation (see Section 1.3) when parts of the embryo are removed or rearranged. Experiments in which cells from an early *Xenopus* blastula are transplanted to a different position on another blastula, and then develop in accordance with their new position, show that at this stage the fates of cells can be altered. Such experiments show that at the blastula, and even later, stages, many cells are not yet determined or specified (see Section 1.10); their potential for development is greater than their position on the fate map suggests.

Fragments of a fertilized *Xenopus* egg that are only one-fourth of the normal volume develop into more-or-less normally proportioned but small embryos. There must therefore be a patterning mechanism involving cell interactions that can cope with such differences in size. There are, however, limits to the capacity for regulation. Isolated animal and vegetal halves of an eight-cell *Xenopus* embryo do not develop normally; dorsal halves of eight-cell embryos regulate to produce a reasonably normal embryo, but early ventral halves do not. Instead they make an abnormal embryo lacking anterior and dorsal structures (see Fig. 3.4) and with much less muscle than the fate map would suggest. As we saw in Section 3.2, these results are linked to the presence or absence of a Nieuwkoop center in the fragments and thus the presence or absence of early cell–cell signals.

The early embryo's powers of regulation reflect the state of determination of individual cells. The state of determination of cells, or of small regions of an embryo, can be studied by transplanting them to a different region of a host embryo and seeing how they develop (see Section 1.10). If they are already determined, they will develop according to their original position. If they are not yet determined, they will develop in line with their new position. This can be shown experimentally by introducing a single labeled cell from a *Xenopus* blastula into the blastocoel of a later-stage host

Fig. 3.23 The fate maps of vertebrate embryos at comparable developmental stages. In spite of all the differences in early development, the fate maps of vertebrate embryos at stages equivalent to a late blastula or early gastrula show similarities. All maps are shown in a dorsal view. The future notochord mesoderm occupies a central dorsal position. The neural ectoderm lies adjacent to the notochord, with the rest of the ectoderm anterior to it. The mouse fate map depicts the late gastrula stage. The future ectoderm of the zebrafish is on its ventral side.

embryo and following its fate. The transferred cell divides, and during gastrulation its progeny become distributed to different parts of the embryo.

In general, cells in transplants made from early blastulas are not yet determined; their progeny will differentiate according to the signals they receive at their new location. So cells from the vegetal pole, which would normally form endoderm, can contribute to a wide variety of other tissues such as muscle or nervous system, when grafted at an early stage. Similarly, early animal pole cells, whose normal fate is epidermis or nervous tissue, can form endoderm or mesoderm. With time, cells gradually become determined, so that similar cells taken from later-stage blastulas and early gastrulas develop according to their fate at the time of transplantation.

Early mouse embryos can regulate to achieve the correct size. Giant embryos formed by aggregation of several embryos in early cleavage stages can achieve normal size within about 6 days by reducing cell proliferation. The mouse embryo retains considerable capacity to regulate until late in gastrulation. Even at the primitive streak stage, up to 80% of the cells of the epiblast can be destroyed with the drug mitomycin C and the embryo can still recover and develop with relatively minor abnormalities. The early chick embryo also has remarkable powers of regulation and fragments of the blastoderm can give rise to whole embryos.

Further evidence for regulation in mammals comes from twinning. Twins can result from a separation of the cells at the two-cell stage, but twinning can also occur in humans at a stage as late as 7 days of gestation, when the primitive streak has already started to form. Since early vertebrate embryos show considerable capacity for regulation and many of the cells are not determined, this implies that cell–cell communication must determine cell fate.

To study this question, one can create **chimeric** mice—that is, mice that are mosaics of cells with two different genetic constitutions—by fusing two embryos. Chimeras made from a normal embryo and one that is genetically similar, but has a mutated version of a single gene, can be used to find out whether the effects of the gene are **cell-autonomous** or **non-autonomous**. If only the mutant cells exhibit the mutant phenotype and are not 'rescued' by the normal cells, the gene is acting cell-autonomously. This means that the product of the gene is acting solely within the cell in which it is made, and is not influencing other cells. The cells of a black mouse, for example, remain black when put into a white mouse, and they do not change the white cells to black; they are thus autonomous with respect to pigmentation (Fig. 3.24). In contrast, a gene is acting non-autonomously when either the mutant cells in the chimera appear to act normally or the normal cells start to show the mutant phenotype. Non-autonomous action is typically due to a gene product that is secreted by one cell and acts on the other.

The cells in the inner cell mass of the mouse embryo are not yet determined. We have already seen this in relation to early mouse development, where cells of the inner cell mass and the trophectoderm are specified purely by their relative position on the inside or the outside of the embryo (see Fig. 3.12). The cells of the inner cell mass itself are pluripotent up to 4½ days after fertilization—they can give rise to many cell types. If 4½-day cells are injected into the inner cell mass of a 3½-day blastocyst, they can contribute to all tissues of the embryo, including the germ cells. This provides another way of producing chimeric mice. Lines of embryonic stem cells (ES cells) derived from cells of the inner cell mass will behave like inner cell

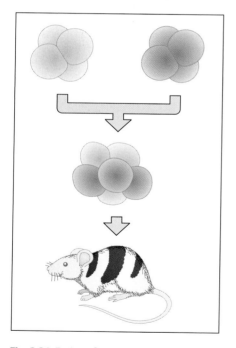

Fig. 3.24 Fusion of mouse embryos gives rise to a chimera. If an eight-cell stage embryo of an unpigmented strain of mouse is fused with a similar embryo of a pigmented strain, the resulting embryo will give rise to a chimeric animal, with a mixture of 'pigmented' and 'unpigmented' cells. The distribution of the different cells in the skin gives this chimera a stripy coat.

Box 3C Transgenic mice

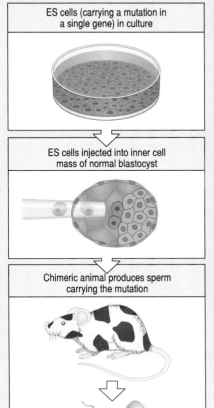

ES cells (carrying a mutation in a single gene) in culture

ES cells injected into inner cell mass of normal blastocyst

Chimeric animal produces sperm carrying the mutation

When studying the role of a particular gene in development, it is an enormous advantage to be able to study the effects of a mutation in that gene. One way of obtaining an animal with the desired mutation is simply to wait for it to turn up in the population, but in vertebrates the wait may be very long indeed. Developmental mutations, in particular, are rarely identified. In mice, however, it is possible to produce animals with a particular mutant genetic constitution using transgenic techniques. One technique involves the injection of DNA into the nucleus of the fertilized egg.

Another means of producing transgenic mice with a desired mutation is by introducing **embryonic stem cells** (**ES cells**) carrying the mutation into the blastocyst. ES cells are cultured cells derived from the inner cell mass; they can be maintained in culture indefinitely and grown in large numbers. Inner cell mass cells introduced into the inner cell mass of another embryo will populate all of the mouse's tissues and will contribute to the germ cells. For, example, if ES cells from a black-pigmented mouse are introduced into the inner cell mass of an embryo of a white mouse, the mouse that develops from this embryo will be a chimera of 'black' and 'white' cells. In the skin, this mosaicism is visible as patches of black and white hairs (see figure).

ES cells can be genetically manipulated in culture to produce mutant cells in which a particular gene or genes have been inactivated or new genes introduced. This technique is particularly powerful for creating loss-of-function mutations to ascertain the role of particular genes in development (see Box 4B, p. 122). Mutations that lead to the complete absence of the function of the gene are known as **gene knock-outs**. Some mutations do not lead to loss of function, but to a change in function.

Because the initial transgenic animals are a mixture of mutant and normal cells they may show few, if any, effects of the mutation. If they carry the mutant gene—the **transgene**—in their germ cells, however, interbreeding can produce a permanent line of non-chimeric transgenic animals in which the mutation is present in either the heterozygous or the homozygous state.

mass cells when injected into a host embryo. Mice with a novel genetic constitution—**transgenic** mice—can also be generated by the introduction of embryonic stem cells carrying particular mutations. If these contribute to the germ cells, a line of mice that carry the introduced mutation in all their cells can be bred (Box 3C and Box 4B, p. 122).

We next examine mechanisms for specifying the germ layers, focusing particularly on the induction of mesoderm in *Xenopus*, which is the best understood system in which this problem has been studied.

3.12 In *Xenopus* the mesoderm is induced by signals from the vegetal region and the endoderm is specified by maternal factors

The ability of the early *Xenopus* embryo to regulate implies that the cells are communicating with each other through extracellular signals. At the time the *Xenopus* egg is laid, there are, however, already differences along the animal–vegetal axis (see Section 3.1). Thus, early patterning involves both cell–cell signaling and localized maternal factors. When explants from different regions of the early blastula are cultured in a simple medium containing the necessary salts for ion balance, tissue from the region nearest the animal pole will form a ball of epidermal cells, while explants from the vegetal region are endodermal in their development. These results are in line with the normal fates of these regions. It is thus generally accepted that all the ectoderm and endoderm are specified by maternal factors in the egg. There is no evidence that any signals from other regions of the embryo are necessary for their initial specification. The mesoderm, however, is different, and its formation is dependent on signals from the endoderm.

The formation of mesoderm in amphibians is dependent on inducing signals produced by the vegetal region of the blastula. These convert a band of adjacent animal cells from an ectodermal fate to a mesodermal one (Fig. 3.25). The standard experiment for studying mesoderm induction is to take a small piece of tissue from the animal half of a blastula (the **animal cap**), which would normally produce only ectoderm, and place it in contact with tissue from the vegetal region. This combination is cultured for 3 days and examined for the presence of mesoderm. Mesoderm can be distinguished by its histology as, after 3 days culture, it may contain muscle, notochord, blood, and loose mesenchyme (connective tissue). It can also be identified by the typical proteins that cells of mesodermal origin may produce, such as muscle-specific actin.

Using these criteria, one finds that the animal portion of the combined explant not only forms epidermis, but also a substantial amount of

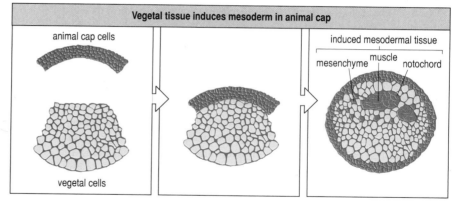

Fig. 3.25 Induction of mesoderm by the vegetal region in the *Xenopus* blastula. Top panels: explants of animal cap cells or vegetal cells on their own from a late blastula form only ectoderm or endoderm, respectively. Explants from the equatorial region, where animal and vegetal regions are adjacent, form mesodermal tissues (mesenchyme, blood cells like erythrocytes, notochord, and muscle), showing that mesoderm induction has taken place. The reason for differences between the mesodermal tissues formed by ventral and dorsal explants at this stage are explained later. Bottom panels: when pieces of animal and vegetal regions from an early blastula are combined and cultured for a few days, mesoderm is induced from the animal cap tissue. This mesoderm contains notochord, muscle, blood, and loose mesenchyme.

mesoderm (see Fig. 3.25). One can confirm that it is the animal cap cells, and not the vegetal cells, that are forming mesoderm, by pre-labeling the animal region of the blastula with a cell-lineage marker and showing that the labeled cells form the mesoderm. Clearly, the vegetal region is producing a signal or signals that can induce mesoderm. In the zebrafish, the mesoderm is induced by the yolk syncytial layer and transplanted yolk syncytial layer can induce both endoderm and mesoderm.

In contrast to the mesoderm, the endoderm in *Xenopus* is maternally specified in the vegetal region. The transcription factor VegT, which is translated from maternal mRNA localized in the vegetal region of the egg and inherited by the cells that develop from this region, has a key role in endoderm specification. Injection of VegT mRNA into animal cap cells induces expression of endoderm-specific markers; blocking the translation of VegT in the vegetal region, by injecting oligonucleotides which are antisense to the mRNA, results in a loss of endoderm. The zebrafish homolog of *VegT* is *spadetail*, which is not expressed maternally. Fish carrying mutant *spadetail* have a complex change in phenotype; there is loss of endoderm and mesoderm from the trunk, but the tail is relatively normal.

3.13 The mesoderm is induced by a diffusible signal during a limited period of competence

The explant system in which pieces of tissue are placed in contact with one another (described in Section 3.12) is well suited for experimental investigation of mesoderm induction and the animal cap cells' response. If the explanted animal and vegetal fragments are separated by a filter with pores too small to allow cell contacts to develop, induction still takes place. This suggests that the mesoderm-inducing signal is in the form of secreted molecules that diffuse across the extracellular space, and does not pass directly from cell to cell via cell junctions (see Fig. 1.20).

The distance over which the signal acts to induce muscle is small, about 80 μm, or four cell diameters in the blastula. This can be shown by blocking both cell movement and cell division in the animal cap explant with the drug cytochalasin; the boundary between the inducing vegetal tissue and the induced mesodermal tissue can then be clearly distinguished. Of course, the distance of 80 μm reflects only the response of the induced cells, and the signal may well be present further away, but at a concentration below that necessary for induction of muscle to occur, but high enough, perhaps, for other cell types to be specified.

The animal cap is **competent** to respond to the inducing signal only for a limited time. Using explanted tissues from embryos of different ages, it has been shown that mesoderm induction is almost complete by the time gastrulation starts. Only a short period of contact is required between the inducing vegetal region and the responding animal cap cells: 2 hours is sufficient to give some induction of muscle, and 5 hours contact leads to complete induction of mesoderm tissues. The animal cap loses its competence to respond about 11 hours after fertilization.

Differentiation of a mesodermal tissue, muscle, appears to depend on a **community effect** in the responding cells. A few animal cap cells placed on vegetal tissue will not be induced to express muscle-specific genes. Even when a small number of individual cells are placed between two groups of vegetal cells, induction does not occur. By contrast, larger aggregates of

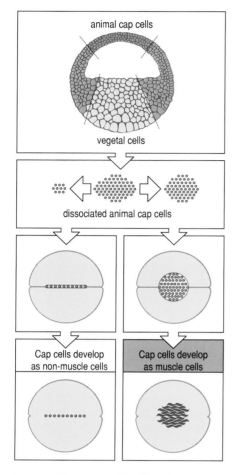

Fig. 3.26 The community effect. One or a small number of animal cap cells in contact with vegetal tissue are not induced to become mesodermal cells and do not begin to express mesodermal markers such as muscle-specific proteins. A sufficiently large number of animal cap cells, about 200, must be present for induction of muscle differentiation to occur.

Fig. 3.27 Timing of muscle gene expression is not linked to the time of mesoderm induction. Animal cap cells isolated from an early *Xenopus* blastula are competent to respond to mesoderm-inducing signals only for a period of about 7 hours, between 4 and 11 hours after fertilization. For expression to occur, exposure to inducer must be for at least 2 hours within this period. Irrespective of when the induction occurs within this competence period, muscle gene expression occurs at the same time—6 hours after fertilization.

animal cap cells respond by strongly expressing muscle-specific genes (Fig. 3.26). An explanation for this is that the induced cells produce a factor that has to reach a sufficiently high concentration for muscle differentiation to occur. This concentration is reached only when there are sufficient cells within a confined volume.

What then determines the extent of mesoderm induction during normal development? One possibility is that the inducing signal from the vegetal region forms a gradient with a threshold below which mesoderm induction does not occur. We look next at the mechanism controlling the temporal sequence of events following induction.

3.14 An intrinsic timing mechanism controls the time of expression of mesoderm-specific genes

Developmental events need to be coordinated both in space and time. Relatively little attention has been given to the timing of developmental processes, which is nevertheless of great importance. For example, following mesoderm induction a cascade of events leads eventually to gastrulation, and we need to understand how these are timed so that they occur in the correct sequence.

As a result of mesoderm induction, muscle-specific genes begin to be expressed in the mesoderm at the mid-gastrula stage. One might expect the timing of this gene expression to be closely coupled to the time at which the mesoderm is induced—but it is not. There is a period of about 7 hours at the blastula stage during which animal cells are competent to respond to a mesoderm-inducing signal. Mesoderm-specific gene expression always starts about 5 hours after the end of this time. For some induction to occur, the animal cap cells require exposure to inducing signal only for a period of about 2 hours. Irrespective of when during the 7-hour period of competence the cells are exposed to the 2-hour induction, the time at which muscle-specific gene expression starts remains the same (Fig. 3.27). Muscle gene expression can occur as early as 5 hours after induction, if induction occurs late in the period of competence, or as late as 9 hours after induction, if induction occurs early in the competent period. These results suggest that there is an independent timing mechanism by which the cells monitor the time elapsed since fertilization and then, provided they have been induced, express muscle-specific genes.

The time at which animal cells lose the ability to respond to the mesoderm-inducing signal also seems to be fixed by an intrinsic timing mechanism. This timing is unaffected by blocking cleavage or by the time

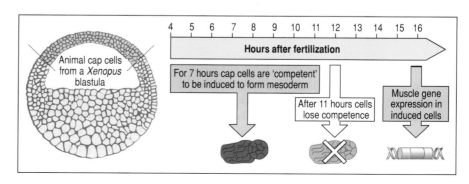

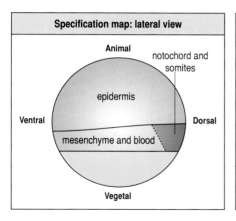

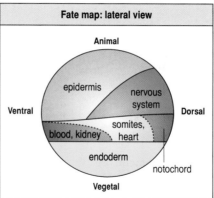

Fig. 3.28 The difference between the fate map and specification map of a *Xenopus* blastula. The fate of a region when isolated and placed in culture is shown on the specification map (left panel), while the normal fate of the blastula regions is shown in the fate map (right panel). There is a clear difference in specification of dorsal and ventral mesoderm. While the notochord's fate and specification maps correspond at this stage, that of the rest of the mesoderm is much more labile, and the specification of most of the somites and other mesodermal tissues has yet to occur. That involves signals from the region of the Spemann organizer, which acts just before and during gastrulation, as well as signals from the ventral region.

of the onset of zygotic gene expression at the mid-blastula transition (see Section 3.17). It also persists even if animal cap tissue is dissociated into single cells several hours before the normal time of transition and cultured so that the cells cannot communicate with each other. A timing mechanism in which the concentration of some protein increases or decreases to a threshold level, is ruled out as, surprisingly, no new protein synthesis is required. Possibly the timing mechanism is based on the breakdown of some protein or the synthesis of some other class of molecule.

An extended period of competence gives the embryo a certain latitude as to when mesoderm induction actually takes place. This means that the timing of the inductive signal does not have to be rigorously linked to the time when the animal region is competent.

3.15 Several signals induce and pattern the mesoderm in the *Xenopus* blastula

We can see from the blastula fate map (Fig. 3.28) that the mesoderm is divided into a number of regions along the dorso-ventral axis, with the notochord originating in the most dorsal region, and the blood-forming tissue most ventrally, although significant amounts of blood also form from dorsal regions. But from the specification map we see that, at the same blastula stage, only a small region on the dorsal side is specified as muscle, whereas the fate map shows that a great deal of muscle will come from more lateral and ventral regions. Thus, explants from the dorsal marginal zone of a blastula, taken after mesoderm induction has begun but before it is completed, behave much in line with their normal fate; they develop into notochord and muscle, and the explants even mimic gastrulation movements by converging and extending. By contrast, ventral and lateral marginal zone explants develop into mesenchyme and blood-forming tissue only (see Fig. 3.25). They do not give rise to any muscle, although their normal fate in the embryo is to form considerable amounts.

These results, together with other evidence discussed later, allow us to construct a model of mesoderm induction that involves at least four different signals. Induction by the vegetal region involves at least two sets of signals: one is a general mesoderm inducer, broadly specifying a ventral-type mesoderm, which can be considered the ground or default state; the second signal, from the Nieuwkoop center, acts simultaneously or a little later, and specifies the dorsal-most mesoderm that will contain the

Fig. 3.29 Four signals involved in mesoderm induction. Two signals originate in the vegetal region, the first from the ventral region (1) and the second on the dorsal side from the region of the Nieuwkoop center (2). The first signal specifies ventral mesoderm, and the dorsal signal specifies the Spemann organizer (O) and dorsal mesoderm. The third signal (3) emanates from the organizer and dorsalizes the adjacent mesoderm by inhibiting the ventralizing action of the fourth signal (4), which comes from the ventral region.

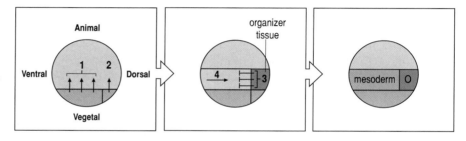

Spemann organizer and form notochord. There are then two further sets of signals, which pattern the ventral mesoderm along the dorso-ventral axis, subdividing it into prospective muscle, kidney, and blood. The third set of signals comes from the organizer region and modifies the ventralizing action of the fourth set of signals, which comes from the ventral region (Fig. 3.29). This model in no way implies that only four distinct signaling molecules are required, or indeed that all four signals are qualitatively different. It is quite possible that each 'signal' represents the actions of more than one molecule, or that different signals represent the same molecule acting at different concentrations.

3.16 Mesoderm-inducing and patterning signals are produced by the vegetal region, the organizer, and the ventral mesoderm

Direct evidence for at least two signals from the vegetal region is provided by comparing the inducing effects of dorsal and ventral vegetal regions (Fig. 3.30). Dorsal vegetal tissue containing the Nieuwkoop center induces notochord and muscle from animal cap cells, whereas ventral vegetal tissue induces mainly blood-forming tissue and little muscle. A minimum of two signals from the vegetal region can thus specify the broad differences between dorsal and ventral mesoderm. These signals are, however, insufficient to explain all of the patterning. In normal development, the ventral

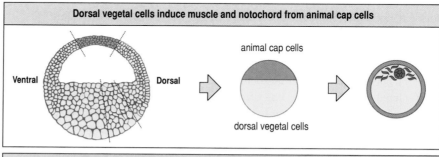

Fig. 3.30 Differences in mesoderm induction by dorsal and ventral vegetal regions. The dorsal vegetal region of the *Xenopus* blastula, which contains the Nieuwkoop center, induces notochord and muscle from animal cap tissues, while ventral vegetal cells induce blood and associated tissues. This is good evidence for different inducing signals coming from the dorsal and ventral vegetal regions.

mesoderm makes a major contribution to the somites, and hence to muscle, yet isolated explants of presumptive ventral mesoderm from an early blastula do not make muscle. Another signal or set of signals is required to pattern the ventral mesoderm further.

The third signal, which has a dorsalizing effect, originates in the Spemann organizer itself. Evidence for this signal comes from combining a fragment of dorsal marginal zone from a late blastula with a fragment of the ventral presumptive mesoderm. The ventral fragment will form substantial amounts of muscle, whereas ventral mesoderm isolated from an early blastula after induction by the vegetal region only will form mainly blood-forming tissue and mesenchyme. The fourth signal, which emanates from the ventral region of the embryo, ventralizes the mesoderm and interacts with the dorsalizing signal, which limits its influence.

A dramatic demonstration of the action of the third signal is to graft the Spemann organizer into the ventral marginal zone of an early gastrula (Fig. 3.31). The graft induces a complete new dorsal axis, and the result is a twinned embryo. This is the famous experiment, carried out by Hans Spemann and Hilde Mangold in the 1920s (see Fig. 1.10), which first identified this key signaling region. The Spemann organizer and the dorsal mesoderm are specified by the Nieuwkoop center in the presumptive endoderm, which lies just vegetal to it. We can now see that the dorsalizing effect of the Nieuwkoop center, observed in the experiments described in Section 3.2, is due to its induction of the Spemann organizer. We next consider some of the molecules that seem to be involved in mesoderm induction and patterning. Some of these are no longer derived from maternal sources, but are produced by zygotic gene expression—the expression of the embryo's own genes.

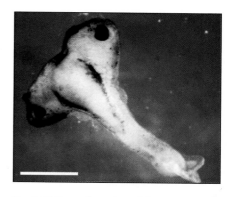

Fig. 3.31 Transplantation of the Spemann organizer can induce a new axis in *Xenopus*. The third set of signals required for mesoderm induction and patterning come from the Spemann organizer region. Their effect can be seen by transplanting the Spemann organizer into the ventral region of another gastrula. The resulting embryo has two distinct heads, one of which was induced by the Spemann organizer. The organizer therefore produces signals that not only pattern the mesoderm dorso-ventrally, but induce neural tissue and anterior structures. Scale bar = 1 mm.

Photograph courtesy of J. Smith.

3.17 Zygotic gene expression begins at the mid-blastula transition in *Xenopus*

The *Xenopus* egg contains quite large amounts of maternal mRNA, which is laid down during oogenesis. In addition, there are large amounts of stored proteins; there is, for example, sufficient histone protein for the assembly of more than 10,000 nuclei. On fertilization, the rate of protein synthesis increases 1½-fold, and during cleavage a large number of new proteins begin to be synthesized, as shown by two-dimensional electrophoresis of extracts of whole embryos. All these proteins are synthesized by translation of preformed maternal mRNA. There is, however, very little new mRNA synthesis until 12 cleavages have taken place and the embryo contains 4096 cells. This point is the so-called **mid-blastula transition** (although it actually happens in the late blastula, just before gastrulation starts). Transcription of the embryo's own genes begins at the mid-blastula transition, and paternal genes are transcribed for the first time in the life of the embryo.

The start of transcription coincides, more or less, with several other changes in the blastula. The first cleavages take place at regular 35-minute intervals, but at the 12th cleavage they become asynchronous as cells take different amounts of time to complete the next cell cycle (Fig. 3.32). At the same time, cells become more motile and can be seen to form small outpushings. It is the coincidence of all these events, which may not be causally related, that leads to the stage being referred to as the mid-blastula transition.

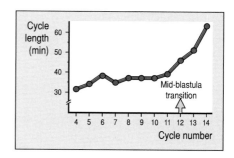

Fig. 3.32 Timing of the cell cycle during cleavage in *Xenopus*. While the cell cycles of early cleavages in *Xenopus* are short and synchronous, later cleavages are longer and asynchronous. The mid-blastula transition in *Xenopus* occurs at the 12th cleavage.

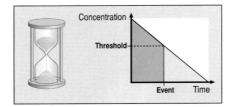

Fig. 3.33 Timing mechanism that could operate in development. A mechanism based on an analogy to an egg-timer could measure time to mid-blastula transition. The decrease in concentration of some molecule, such as a repressor, could occur with time, and the transition could occur when the repressor reaches a critically low threshold concentration. This would be equivalent to all the sand running into the bottom of the egg-timer.

How is the mid-blastula transition triggered? Suppression of cleavage, but not DNA synthesis, by cytochalasin B does not alter the timing of transcriptional activation, and so this is not linked directly to cell division. Neither are cell–cell interactions involved, as dissociated blastomeres undergo the transition at the same time as intact embryos. The key factor in triggering the mid-blastula transition seems to be the ratio of DNA to cytoplasm—the quantity of DNA present per unit mass of cytoplasm.

Direct evidence for this comes from increasing the amount of DNA artificially by allowing more than one sperm to enter the egg or by injecting extra DNA into the egg. In both cases, transcriptional activation occurs prematurely, suggesting that there may be some fixed amount of a general repressor of transcription present initially in the egg cytoplasm. As the egg cleaves, the amount of cytoplasm does not increase, but the amount of DNA does. The amount of repressor in relation to DNA gets smaller and smaller until there is insufficient to bind to all the available sites on the DNA and the repression is lifted. Timing of the mid-blastula transition thus seems to fit with a timing model of the hour-glass egg-timer type (Fig. 3.33). In such a model something has to accumulate, in this case DNA, until a threshold is reached. The threshold is determined by the initial concentration of the cytoplasmic factor, which does not increase.

3.18 Candidate mesoderm inducers have been identified in *Xenopus*

As we saw in Section 3.16, the vegetal region produces the mesoderm-inducing signals (signals 1 and 2 in Fig. 3.29). From the explant experiments outlined there, these mesoderm-inducing signals seem most likely to be secreted proteins. Two main approaches are used to test for mesoderm-inducing and patterning factors. One is to apply the candidate factor directly to isolated animal caps in culture. The other is to inject the mRNA encoding the suspected inducer into the animal pole cells of the early blastula.

By itself, the ability to induce mesoderm in culture does not prove that a particular protein is a natural inducer in the embryo. Rigorous criteria must be met before such a conclusion can be reached. These criteria include the presence of the protein in the right concentration, place, and time in the embryo; the demonstration that the appropriate cells can respond to the factor; and the demonstration that blocking the response prevents induction taking place. On all these criteria, however, the evidence for a key role for a member or members of the TGF-β family in mesoderm induction is rather good.

The most obvious candidates for the vegetal mesoderm-inducing signals are the maternal factors localized in the vegetal region of the egg. A prime candidate is Vg-1, a maternally expressed member of the TGF-β family, whose mRNA is localized in the vegetal region (see Section 3.1). Like all TGF-β family proteins, newly synthesized Vg-1 has to be proteolytically processed before it becomes active. Although the precursor protein is abundant in the vegetal region, the injection of neither its mRNA nor the precursor protein itself into the animal cap has any significant effect. This suggests that Vg-1 activity is regulated at the post-translational level and that the animal cap cells are unable to process the Vg-1 precursor efficiently.

Properly processed Vg-1 protein does indeed have a pronounced

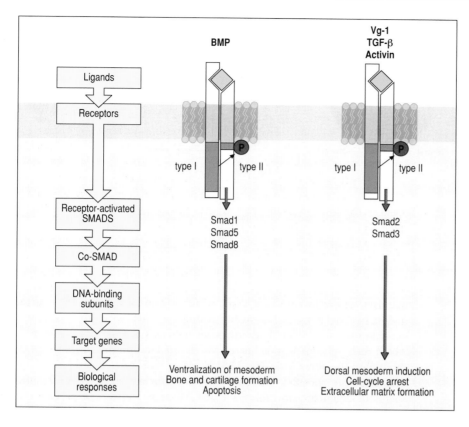

Fig. 3.34 Signaling by members of the TGF-β family of growth factors. Members of this family, such as Vg-1, BMP and activin, act at cell-surface receptors that are heterodimers of two different subunits. There are several different forms of type I and type II subunits and these combine to form distinctive receptors for different TGF-family members. Binding of ligand causes the initiation of intracellular signaling pathways involving the SMAD proteins. These pathways use slightly different intermediary molecules for each receptor and thus can lead to the activation of different sets of target genes. The biological response depends on the combination of activated target genes and the particular cellular environment.

mesoderm-inducing effect on animal cap cells. When a suitably engineered *Vg-1* RNA construct is injected into animal cap explants, the expression of mature active Vg-1 induces dorsal mesoderm. Vg-1 expression also rescues embryos ventralized by UV irradiation (see Section 3.3). Treatment of isolated animal caps with purified active Vg-1 protein induces embryo-like structures with clear axial organization and heads. Vg-1 is therefore a possible candidate for a mesoderm inducer. At high concentrations Vg-1 induces dorsal mesoderm, whereas at lower concentrations it induces ventral-type mesoderm; inhibition of its action leads to defects in dorsal mesoderm.

Another member of the TGF-β family—activin—also has mesoderm-inducing activity. Activin was isolated from the culture fluid of a *Xenopus* cell line because of its powerful inducing activity. The response of animal caps to purified activin is also concentration dependent: at higher concentrations notochord develops, together with muscle, whereas at lower concentrations only muscle is induced. Although activin-like activity can be detected in extracts from oocytes and early embryos, there is no evidence so far for maternal activin mRNA in the egg. Activin itself may therefore not be the primary inducing signal *in vivo*. Different members of the TGF-β family may bind and act through the same receptors, so the activin applied to the animal cap cells could be acting through the same pathway as, say, Vg-1 (Fig. 3.34).

Indeed, there is evidence that activin is not the natural mesoderm inducer. This function is probably carried out by other members of the TGF-β family which are encoded by zygotic genes of the *nodal* family. The

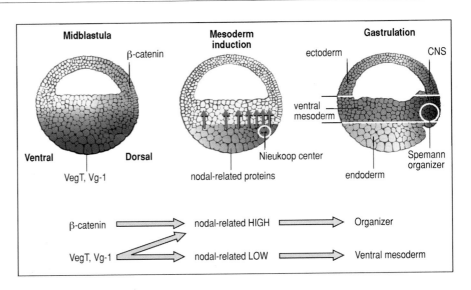

Fig. 3.35 A gradient in nodal-related proteins may provide the first two signals in mesoderm induction. VegT and Vg-1 are maternal proteins in the vegetal region that activate the transcription of nodal-related genes. The presence of β-catenin on the dorsal side results in a dorsal-to-ventral gradient in the nodal proteins. These induce mesoderm and, at high doses, specify the Nieuwkoop center, which in turn induces the Spemann organizer on the dorsal side.

maternal transcription factor VegT, which is also localized in the vegetal region (see Section 3.1), probably acting together with Vg-1, activates expression of *nodal*-related zygotic genes such as *Xnr*, and the gene *derriere*, which also encodes a member of the TGF-β family. The presence of β-catenin on the dorsal side (Fig. 3.35) increases *nodal*-related gene transcription and this specifies the Nieuwkoop center and, consequently, the Spemann organizer, on the dorsal side. These processes can provide the first two signals. If VegT is severely depleted, *nodal*-related genes and *derriere* are downregulated and almost no mesoderm develops, showing that VegT is crucial for mesoderm induction. Injection of mRNA for the *nodal*-related genes or *derriere* can, however, rescue mesoderm formation, suggesting that these are most likely to be the actual mesoderm inducers. The *nodal*-related genes rescue head, trunk, and tail mesoderm, but *derriere* rescues only trunk and tail. Activin is present at 50% of its normal level in VegT-depleted embryos but no mesoderm is formed; however, in a similar fashion to derriere protein, activin can partially rescue VegT-depleted embryos. Activin's role in mesoderm induction remains unclear.

A general approach to identifying signaling molecules is to block the response of a cell to a particular signal by preventing the receptors for that signal from being activated. If mesoderm induction is also prevented by this treatment, it is evidence that the signal proteins that can bind to the receptor in question are causally involved. A receptor for several TGF-β family growth factors is the activin type II receptor, which is expressed and uniformly distributed throughout the early *Xenopus* blastula. Receptors for proteins of the TGF-β family only function as dimers and their function can be blocked by the presence of a mutant subunit, which associates with a normal subunit to produce an inactive receptor (Fig. 3.36). If mRNA for a mutant receptor subunit is injected into the early *Xenopus* embryo, mesoderm formation is prevented. The presence of a mutant receptor subunit has the same effect as a **dominant-negative mutation** in the gene coding for the receptor; that is, it inhibits receptor function. This direct biochemical intervention is particularly useful in *Xenopus*, where there is at present no means of producing suitable genetic mutations.

These experiments show that proteins of the TGF-β family are involved in

Fig. 3.36 A mutant activin receptor blocks mesoderm induction. Receptors for factors of the TGF-β family function as dimers. Ligand binding causes dimerization of type I and type II receptors, which activates a serine–threonine kinase in the cytoplasmic region of the receptor. Receptor function can be blocked by introducing mRNA encoding a mutant receptor subunit that lacks most of the cytoplasmic domain, and so cannot function. It can bind ligand and form heterodimers with normal receptor subunits but cannot signal. It thus acts as a dominant-negative mutation of receptor function. When mRNA encoding the mutant receptor subunit is injected into cells of the two-cell *Xenopus* embryo, subsequent mesoderm formation is blocked. No mesoderm or axial structures are formed except for the cement gland, the most anterior structure of the embryo.

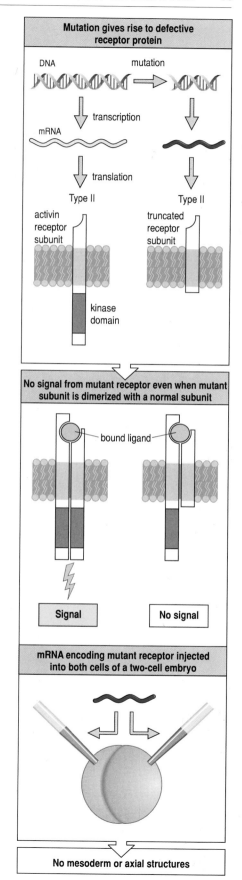

mesoderm induction, and could provide a key element of the first two signals, but they give no indication as to which proteins they are. Another protein involved in mesoderm induction is the *Xenopus* equivalent of fibroblast growth factor (FGF). This is present in the blastula, mainly in the animal hemisphere, and may be necessary to potentiate the response of animal cap cells to TGF-β-like molecules.

3.19 Mesoderm-patterning factors are produced within the mesoderm

A number of proteins seem to be involved in patterning the mesoderm along the dorso-ventral axis once it has been induced (Fig. 3.37). The gene *noggin*, which was identified during a screen for factors that could rescue UV-irradiated embryos, is expressed in the Spemann organizer (Fig. 3.38), the source of the third signal (see Section 3.16). Noggin is a secreted protein unrelated to any of the known growth factor families. *noggin* expression does not induce mesoderm in animal pole explants but can dorsalize explants of ventral marginal zone tissue, thus making it a good candidate for one of the third class of signals that patterns the mesoderm along the dorso-ventral axis. The proteins chordin and frizbee, also secreted by the organizer, are other proposed components of the third signal. Surprisingly, the action of all these signals is not on the cells themselves but on the fourth set of signals.

The fourth set of signals, emanating from the ventral region of the mesoderm, promotes ventralization of the mesoderm. Candidates for these signals are bone morphogenetic protein 4 (BMP-4), a member of the TGF-β family, and Xwnt-8. BMP-4 is expressed uniformly throughout the late *Xenopus* blastula, and Xwnt-8 is expressed in the future mesoderm. As gastrulation proceeds, BMP-4 is no longer expressed in dorsal regions. When the action of BMP-4 is blocked throughout by introducing a dominant-negative receptor, the embryo is dorsalized, with ventral cells now differentiating as both muscle and notochord. The secreted protein Xwnt-8, which is expressed in the future mesoderm, can also ventralize the embryo. How do the dorsalizing factors, represented by signal 3, and the ventralizing factors, represented by signal 4, interact? The action of the dorsalizing agents is not on the presumptive mesodermal cells themselves, but on the ventralizing factors. Noggin and chordin interact with BMP-4 and prevent it from binding to its receptor. In this way, a functional gradient of BMP-4 activity is set up across the dorso-ventral axis with its high point ventrally and little or no activity in the presumptive dorsal mesoderm. Frizbee, a secreted Wnt-binding protein, generates a similar ventral to dorsal gradient of Wnt activity by binding to Wnt proteins and preventing them acting on

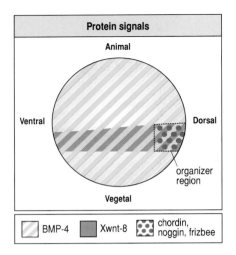

Protein signals

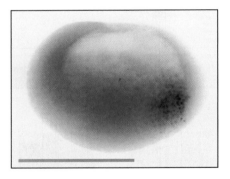

Fig. 3.37 Distribution of protein signals in the *Xenopus* blastula. The signals from the organizer block the action of BMP-4 and Xwnt-8.

Fig. 3.38 Expression of *noggin* in the *Xenopus* blastula. *noggin* expression is shown as the dark-staining area in the region of the Spemann organizer. Scale bar = 1 mm.

Photograph courtesy of R. Harland, from Smith, W.C., et al.: 1992.

the presumptive dorsal mesoderm. Fig. 3.39 summarizes the main proteins so far identified as mesoderm inducers and mesoderm-patterning factors in *Xenopus*.

As more proteins involved in mesoderm patterning are discovered, additional refinements in the signals are emerging. Secreted metalloproteases such as Xolloid, which is expressed in ventral regions, cleave chordin *in vitro* and block chordin activity *in vivo*. Xolloid is thought to act as a clearing agent for chordin, reducing the long-range diffusion of this protein and helping to maintain a gradient of chordin dorsalizing activity. Other proteins are thought to help maintain the gradient in BMP activity. The protein Twisted gastrulation (Tsg), which is localized to the ventral regions of the embryo during gastrulation, where BMP-4 signaling is at its peak, can dislodge BMP from fragments of chordin produced by Xolloid, thus preventing BMP's inactivation. As might be expected from its ventral expression, injection of *Tsg* mRNA ventralizes dorsal mesoderm, whereas inhibition of Tsg activity disrupts tail development. The secreted protein cerberus, which is produced in anterior endoderm, inhibits the actions of BMPs, nodal proteins and Wnt proteins; it inhibits mesoderm formation and is involved in the induction of anterior structures.

The absence of genetics in *Xenopus* is compensated for to some extent by genetic analysis in zebrafish. Mutagenesis screens have identified several genes in zebrafish similar to those involved in dorso-ventral patterning in *Xenopus* and, in general, confirm the conclusions obtained. *Swirl* and *snailhouse* code for BMP-2 and BMP-7. They are expressed, as in *Xenopus*, on the ventral side of the embryo and mutants that do not produce these proteins are dorsalized. The zebrafish gene *chordino* codes for chordin, which antagonizes the actions of BMP. *Chordino* mutants show an expansion of ventral and lateral mesoderm. *nodal* is expressed in a layer of cells at the boundary between endoderm and mesoderm (the **endo-mesoderm**), just above the yolk cell. Double mutations in the *nodal*-related genes *squint* and *cyclops* block nodal function. The mutant embryos lack both head and trunk mesoderm, but there is some mesoderm in the tail region. A similar result is obtained with mutations in *one-eye pinhead*, which is required for signaling by nodal.

Compared to *Xenopus*, we are largely ignorant of the mechanisms by which mesoderm is specified in the mouse and chick. In the chick, mesoderm specification occurs during primitive streak formation. The chick homolog of Vg-1 can induce a whole new axis when cells secreting it are grafted to the margin of an early chick blastoderm, before primitive streak formation, suggesting a role for Vg-1 in mesoderm induction similar to that in *Xenopus*. Chick epiblast isolated before streak formation will form some mesoderm containing blood vessels, blood cells, and some muscle, but no dorsal mesodermal structures such as notochord. Treatment of the isolated epiblast with activin, however, results in the additional appearance of notochord and more muscle, suggesting that TGF-β family members act as mesoderm-inducing and/or patterning signals in chick embryos as well as in *Xenopus*. Other experiments suggest that the full development of the axis, including anterior structures, requires the action of Wnt proteins as well as TGF-β family members.

In the mouse, the gene *nodal* is expressed in the primitive streak at the time of mesoderm formation. In mutants lacking nodal protein function, mesoderm does not form during gastrulation, suggesting a role for nodal in

Signals in early *Xenopus* development		
Factor	**Protein family**	**Effects**
Mesoderm induction		
Xnr-1	TGF-β family	
Xnr-2	TGF-β family	
Xnr-3	TGF-β family	
Derriere	TGF-β family	
Vg-1	TGF-β family	
Activin	TGF-β family	
Mesoderm patterning		
Bone morphogenetic protein (e.g. BMP-4)	TGF-β family	ventral mesoderm patterning
Xwnt-8	Wnt family	ventralizes mesoderm
Fibroblast growth factor (FGF)	FGF	ventral mesoderm induction
Noggin		dorsalizes—binds BMP-4
Chordin		dorsalizes—binds BMP-4
Frizbee		dorsalizes—binds BMP-4

Fig. 3.39 Signals in early *Xenopus* development.

mesoderm formation, as in *Xenopus*. Mice lacking either activin or the type II activin receptor still develop mesoderm, which suggests that neither activin nor the type II receptor are necessary for mesoderm development in mammals.

3.20 Mesoderm induction activates genes that pattern the mesoderm

The signals described in Sections 3.18 and 3.19 pattern the mesoderm by turning on groups of genes that control mesoderm differentiation. Going from dorsal to ventral, the most dorsal mesoderm will form the notochord, followed by somites, heart, and kidneys. A gene characteristic of mesoderm, and which is expressed early in prospective mesoderm, is *Brachyury*, which encodes a transcription factor. This gene was first identified in mice, where it is required for formation of most of the mesoderm, especially the posterior mesoderm (see Section 2.8). In all vertebrates, *Brachyury* is initially expressed throughout the presumptive mesoderm (Fig. 3.40), later becoming confined to the notochord, the dorsal-most derivative of the mesoderm, and to posterior mesoderm (the tailbud). The *Xenopus* homolog of *Brachyury* is switched on in presumptive ectoderm experimentally treated with mesoderm-inducing factors such as activin. The maintenance of *Brachyury* expression depends on the expression of a gene for a member of the FGF family, which is a direct target of *Brachyury*.

Injection of *Brachyury* mRNA into the *Xenopus* embryo and its resulting overexpression in presumptive ectoderm causes this to form ventral mesoderm; at high doses Brachyury causes formation of muscle. These results strongly suggest a key role for Brachyury in mesoderm patterning. This notion is further strengthened by the finding that the *no-tail* mutant in zebrafish, which results in an absence of posterior mesoderm, is due to a mutation in the zebrafish homolog of *Brachyury*.

Blocking the function of Brachyury leads to an inhibition of gastrulation movements because this causes downregulation of Xwnt-11, whose pattern of expression is similar. The regulation of gastrulation movements by

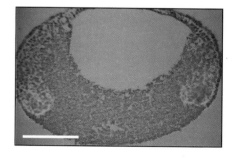

Fig. 3.40 Expression of *Brachyury* in the *Xenopus* blastula. A cross-section through the embryo along the animal–vegetal axis shows that *Brachyury* (red) is expressed in the future mesoderm. Scale bar = 0.5 mm.

Photograph courtesy of M. Sargent and L. Essex.

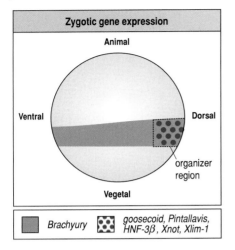

Zygotic gene expression

Animal

Ventral

Dorsal

organizer region

Vegetal

| ▨ *Brachyury* | ▨ *goosecoid, Pintallavis, HNF-3β, Xnot, Xlim-1* |

Fig. 3.41 Zygotic gene expression in a late *Xenopus* blastula. The expression domains of a number of zygotic genes that code for transcription factors correspond quite well to demarcations on the specification map. The gene *Brachyury* is expressed in a ring around the embryo corresponding quite closely to the future mesoderm. Several transcription factors are expressed in the region of the dorsal mesoderm that corresponds to the Spemann organizer.

Xwnt-11 is not by the wingless signaling pathway illustrated in Fig. 3.9, but by one that has been shown in *Drosophila* to affect the cytoskeleton and cell polarity.

On the dorsal side of the *Xenopus* embryo, the Nieuwkoop center induces the Spemann organizer in the dorsal mesoderm. The organizer is not only involved in patterning the dorso-ventral axis of the mesoderm but, as we shall see in Chapter 4, plays a role in patterning the antero-posterior axis of both the mesoderm and the nervous system. One of the first zygotic genes to be expressed in the organizer region is *goosecoid*, which was identified by screening a cDNA library made from RNA isolated from the *Xenopus* dorsal mesoderm region. *goosecoid* is a homeobox-containing gene (see Box 4A, p. 117), encoding a transcription factor with a homeodomain somewhat similar to that of both the gooseberry and bicoid proteins of *Drosophila*—hence the name. It is a zygotic gene that is expressed in the mesoderm after the mid-blastula transition.

In line with its presence in the organizer region, microinjection of *goosecoid* mRNA into the ventral region of the blastula mimics to some extent transplantation of the Spemann organizer (see Fig. 3.31), resulting in the formation of a secondary axis. Genes for other transcription factors are also expressed in the organizer region (Fig. 3.41). These include *Pintallavis* and *HNF-3β*, both of which code for proteins with so-called forkhead domains, and *Xnot* and *Xlim-1*, which code for homeodomain proteins. *Xnot* appears to have a role in the specification of notochord, which develops from the organizer mesoderm; overexpression of *Xnot* results in a notochord that is larger than normal. Its expression, like that of *Brachyury*, is induced by mesoderm inducers such as activin. A zebrafish homolog of *Xnot*, the gene *floating head*, has been identified as essential for formation of the zebrafish notochord. This gene is expressed in the presumptive notochord region and its mutation results in complete absence of the notochord and some increase in muscle.

3.21 Gradients in signaling proteins and threshold responses could pattern the mesoderm

Although possible secreted signaling proteins have been identified, it is still not clear how they turn on genes like *goosecoid* and *Brachyury* in the right place. One model for patterning the mesoderm, and other tissues, proposes that positional information is provided by a dorso-ventral gradient of a morphogen. Several of the proteins identified as possible mesoderm-patterning agents in *Xenopus*, such as the nodal-related proteins, are indeed expressed in a graded fashion. Just how gradients are set up with the necessary precision is not clear. Simple diffusion of a morphogen may play a part but more complex cellular processes are likely to be involved.

Experiments with activin provide an example of how a diffusible protein could pattern a tissue by turning on particular genes at specific threshold concentrations. Although activin itself may not be responsible for meso-derm patterning in this way *in vivo*, animal cap cells from a *Xenopus* blastula respond to increasing doses of activin by activation of different genes at different threshold concentrations. Increasing activin concentration by as little as 1½-fold results in a dramatic alteration in the pattern of marker proteins expressed and in the tissues that differentiate. For example, this small increase causes a change from homogeneous formation of muscle to formation of notochord. Increasing concentrations of activin can specify several different cell states that correspond to the different regions along the dorso-ventral axis. At the lowest concentrations of activin, only epidermis develops. Then, as the concentration increases, *Brachyury* is expressed, together with muscle-specific genes such as that for actin. With a further increase in activin, *goosecoid* is expressed, and this corresponds to the dorsal-most region of the mesoderm—the organizer (Fig. 3.42). Similar results can be obtained by injecting increasing quantities of activin mRNA. One can thus see how graded signals could activate transcription factors in particular regions and thus pattern tissues.

Further support for the idea that gradients and threshold concentrations of morphogen could pattern the mesoderm was provided by experiments in which vegetal tissues were injected with increasing amounts of activin mRNA. These were then placed in contact with an animal cap. The results showed that activin diffused into the animal cap. *Brachyury* was turned on at some distance from the source, while *goosecoid* was expressed nearest to the source, which fits with the idea of a concentration gradient, in this case of activin, turning on genes at a specific threshold concentration.

How do the cells distinguish between different concentrations of activin? Occupation of just 2% of the activin receptors is required to activate expression of *Brachyury*, while 6% occupation is required for *goosecoid* to be expressed. The link between the signal concentration and gene expression may not be so simple, however. There appear to be additional layers of intracellular regulation; cells expressing *goosecoid* at high activin concentrations also repress *Brachyury* expression, for example; this involves the action of the goosecoid protein itself together with other proteins.

Excellent evidence for a secreted morphogen acting at a distance and turning on genes at specific threshold concentrations comes from zebrafish. In these embryos, the *nodal*-related gene *squint* is involved in patterning the mesoderm. Injection of *squint* mRNA into a single cell of an early zebrafish embryo resulted in high-threshold genes being activated in adjacent cells, whereas low-threshold genes were activated in more distant cells. Experiments also ruled out a relay mechanism in this case.

We are now in a position to consider the final emergence of the typical vertebrate body plan. Further patterning of the germ layers occurs during gastrulation, along both the antero-posterior and dorso-ventral axes, and this is discussed in the next chapter.

Summary

Once the antero-posterior and dorso-ventral axes are established, one can begin to construct a fate map for the germ layers. There are strong similarities in the fate maps of the amphibian, zebrafish, chick, and mouse at later stages. Even though there is good evidence for the maternal specification of some regions such as the future endoderm in amphibians, the embryo can still undergo considerable regulation at the blastula stage. This implies that interactions between cells, rather than intrinsic factors, have a central role even in early amphibian development. This strategy is particularly pronounced in the mouse and chick, where there is no evi-

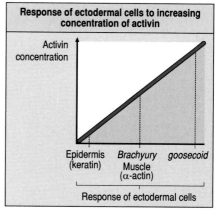

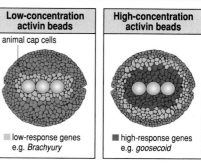

Fig. 3.42 Graded responses of early *Xenopus* tissue to increasing concentrations of activin. When animal cap cells are treated with increasing concentrations of activin, particular genes are activated at specific concentrations, as shown in the top panel. At intermediate concentrations of activin, *Brachyury* is induced, whereas *goosecoid*, which is typical of the organizer region, is only induced at high concentrations. If beads releasing a low concentration of activin are placed in the center of a mass of animal cap cells (lower left panel), expression of low-response genes such as *Brachyury* is induced immediately around the beads. With a high concentration of activin in the beads (lower right panel), *goosecoid* and other high-response genes are now expressed around the beads and the low-response genes farther away.

dence for any maternal specification, and it is position that determines cell fate.

In *Xenopus*, the mesoderm and some endoderm are induced from animal cap tissue by the vegetal region, which contains the Nieuwkoop center. Early patterning of the mesoderm can be accounted for by a four-signal model. The first signal is a general mesoderm inducer, specifying a ventral-type mesoderm. The second specifies dorsal mesoderm, including the organizer, while the third signal comes from the ventral side and ventralizes the mesoderm. The fourth signal originates from the organizer and establishes further pattern within the mesoderm by interacting with the third signal.

Protein growth factors such as members of the TGF-β family are excellent candidates for the natural mesoderm-inducing factors as well as for patterning the mesoderm. Other signaling proteins such as noggin and chordin inhibit the action of BMP-4, and so are involved in specifying the dorsal and ventral mesoderm. *Brachyury* and *goosecoid* are early mesodermally expressed genes encoding transcription factors and their pattern of expression may be specified by gradients in signaling proteins, the genes being turned on at particular threshold concentrations.

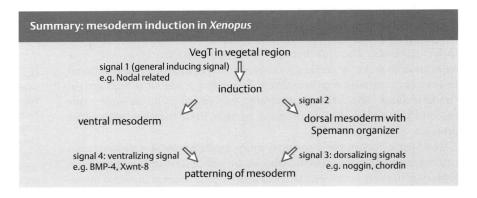

Summary: mesoderm induction in *Xenopus*

SUMMARY TO CHAPTER 3

All vertebrates have the same basic body plan. During early development, the antero-posterior and dorso-ventral axes of this body plan are set up. The mechanism is different in frog, chick, zebrafish, and mouse but can involve localized maternal determinants, external signals, and cell–cell interactions. This early patterning also establishes bilateral asymmetry. It is possible to construct a fate map in the early embryo for the three germ layers—mesoderm, endoderm, and ectoderm. The fate maps of the different vertebrates have strong similarities. At this early stage the embryos are still capable of considerable regulation and this emphasizes the essential role of cell–cell interactions in development. In *Xenopus*, at least four separate signals are involved in mesoderm induction and early patterning. Good candidates for these signals have been identified and include members of the TGF-β family. These signals activate mesoderm-specific genes such as *Brachyury* at particular concentrations and so their gradients could pattern the mesoderm. The summary table opposite lists all genes considered in this chapter in relation to *Xenopus*.

Summary: genes involved in patterning of axes and germ layers

Gene	Maternal/ Zygotic	Type of protein	Where expressed	Effects
activin	Z	TGF-β family	?	mesoderm induction
BMP-4	Z	transcription factor	late blastula	ventralizes mesoderm
Brachyury	Z	transcription factor	early mesoderm	mesoderm development
β-catenin	M	gene regulatory protein	egg	dorsalizing signal
cerberus	Z	secreted	vegetal egg	mesoderm inhibition
chordin	Z	secreted signal molecule	organizer	dorsalizes mesoderm
derriere	Z	TGF-β family	vegetal egg	mesoderm induction
fibroblast growth factor	Z	secreted signal molecule	blastula	ventral mesoderm induction
goosecoid	Z	transcription factor	organizer	organizer function
GSK-3	M	protein kinase	egg	suppresses dorsalizing signals
HNF-3β	Z	transcription factor	organizer	organizer development
noggin	M/Z	secreted	organizer	dorsalizes mesoderm
Pintallavis	Z	transcription factor	organizer	?
siamois	Z	transcription factor	dorsal blastula	dorsalizing signal
VegT	M	transcription factor	vegetal egg	induces endoderm and mesoderm signals
Vg-1	M	TGF-β family	vegetal egg	mesoderm induction
Xlim-1	Z	transcription factor	organizer	?
Xnot	Z	transcription factor	organizer	notochord specification
Xnr-1	Z	secreted	vegetal egg	mesoderm induction
Xnr-2	Z	secreted	vegetal egg	mesoderm induction
Xnr-4	Z	secreted	vegetal egg	mesoderm induction
Xwnt-11	M	Wnt family	vegetal egg	mesoderm induction
Xwnt-8	Z	Wnt family	propective mesoderm	ventralizes mesoderm

REFERENCES

3.1 The animal–vegetal axis of *Xenopus* is maternally determined

Foristall, C., Pondel, M., Chen, L., King, M.L.: **Patterns of localization and cytoskeletal association of two vegetally localized RNAs, Vg-1 and Xcat-2.** *Development* 1995, **121**: 201–208.

Kofron, M., Demel, T., Xanthos, J., Lohr, J., Sun, B., Sive, H., Osada, S-I., Wright, C., Wylie, C., Heasman, J.: **Mesoderm induction in *Xenopus* is a zygotic event regulated by maternal VegT via TGFb growth factors.** *Development* 1999, **126**: 5759–5770.

Weeks, D.L., Melton, D.A.: **A maternal mRNA localized to the vegetal hemisphere in *Xenopus* eggs codes for a growth factor related to TGF-β.** *Cell* 1987, **51**: 861–867.

3.2 The dorso-ventral axis of amphibian embryos is determined by the site of sperm entry

Gerhart, J., Danilchik, M., Doniach, T., Roberts, S., Browning, B., Stewart, R.: **Cortical rotation of the *Xenopus* egg: consequences for the antero-posterior pattern of embryonic dorsal development.** *Development* (Suppl.) 1989, 37–51.

3.4 Maternal proteins with dorsalizing and ventralizing effects have been identified

Dale, L.: **Vertebrate development: mutiple phases to endoderm formation.** *Curr. Biol.* 1999, **9**: R812–R815.

He, X., Saint-Jennet, J-P., Woodgett, J.R., Varmus, H.E., Dawid, I.B.: **Glycogen synthase kinase-3 and dorsoventral patterning in *Xenopus* embryos.** *Nature* 1995, **374**: 617–622.

Kessler, D.S., Melton, D.A.: **Induction of dorsal mesoderm by soluble, mature Vg1 protein.** *Development* 1995, **121**: 2155–2164.

Kodjabachian L., Lemaire P.: **Embryonic induction: is the Nieuwkoop centre a useful concept?** *Curr. Biol.* 1998, **8**: R918–R921.

Kodjabachian, L., Dawid, I.B., Toyama, R.: **Gastrulation in zebrafish: what mutants teach us.** *Dev. Biol.* 1999, **126**: 5309–5317.

Miller, J.R., Moon, R.T.: **Analysis of the signaling activities of localization mutants of beta-catenin during axis specification in *Xenopus*.** *J. Cell Biol.* 1997, **139**: 229–243.

Smith, J.: **T-box genes: what they do and how they do it.** *Trends Genet.* 1999, 15: 154–158.

Smith, W.C., Harland, R.M.: **Injected Xwnt-8 RNA acts early in *Xenopus* embryos to promote formation of a vegetal dorsalizing center.** *Cell* 1991, **67**: 753–765.

Sokol, S.Y.: **Wnt signaling and dorso-ventral axis specification in vertebrates.** *Curr. Opin. Genet. Dev.* 1999, **9**: 405–410.

Sokol, S., Christian, J.L., Moon, R.T., Melton, D.A.: **Injected Wnt RNA induces a complete body axis in *Xenopus* embryos.** *Cell* 1991, **67**: 741–752.

3.5 The antero-posterior axis of the chick blastoderm is set by gravity

Khaner, O., Eyal-Giladi, H.: **The chick's marginal zone and primitive streak formation. I. Coordinative effect of induction and inhibition.** *Dev. Biol.* 1989, **134**: 206–214.

Kochav, S., Eyal-Giladi, H.: **Bilateral symmetry in chick embryo determination by gravity.** *Science* 1971, **171**: 1027–1029.

Seleiro, E.A.P., Connolly, D.J., Cooke, J.: **Early developmental expression and experimental axis determination by the chicken Vg-1 gene.** *Curr. Biol.* 1996, **11**: 1476–1486.

3.6 The axes of the mouse embryo are recognizable early in development

Beddington, R.S.P., Robertson, E.J.: **Axis development and early asymmetry in mammals.** *Cell* 1999, **96**: 195–209.

Ciemerych, M.A., Mesnard, D., Zernicka-Goetz, M.: **Animal and vegetal poles of the mouse predict the polarity of the embryonic axis yet are nonessential for development.** *Development* 2000, **127**: 3467–3474.

Hillman, N., Sherman, M.I., Graham, C.: **The effect of spatial arrangement on cell determination during mouse development.** *J. Embryol. Exp. Morph.* 1972, **28**: 263–278.

Lewis, N.E., Rossant, J.: **Mechanism of size regulation in mouse embryo aggregates.** *J. Embryol. Exp. Morph.* 1982, **72**: 169–181.

Lu, C.C., Brennan, J., Robertson, E.J.: **From fertilization to gastrulation: axis formation in the mouse embryo.** *Curr. Opin. Genet. Dev.* 2001, **11**: 384–392.

Piotrowska, K., Zernicka-Goetz, M.: **Role for sperm in spatial patterning of the early mouse embryo.** *Nature* 2001, **409**: 517–521.

Weber, R.J., Pedersen, R.A., Wianny, F., Evans, M.J., Zernica-Goetz, M. **Polarity of the mouse embryo is anticipated before implantation.** *Development* 1999, **126**: 5591–5596.

3.7 Specification of left–right handedness of internal organs requires special mechanisms

Brown, N.A., Wolpert, L.: **The development of handedness in left/right asymmetry.** *Development* 1990, **109**: 1–9.

3.8 Organ handedness in vertebrates is under genetic control

Capdevila, J., Vogan, K.J., Tabin, C.J., Izpisúa Belmonte, J.C.: **Mechanisms of left-right determination in vertebrates.** *Cell* 2000, **101**: 9–21.

Cooke, J.: **Vertebrate left and right: finally a cascade, but first a flow.** *BioEssays* 1999, **21**: 537–541.

Jost, H.J.: **Diverse initiation in a conserved left/right pathway.** *Curr. Opin. Genet. Dev.* 1999, **9**: 422–426.

3.9 A fate map of the amphibian blastula is constructed by following the fate of labeled cells

Dale, L., Slack, J.M.W.: **Fate map for the 32 cell stage of *Xenopus laevis*.** *Development* 1987, **99**: 527–551.

Lane, M.C., Smith, W.C.: **The origins of primitive blood in *Xenopus*: implications for axial patterning.** *Development* 1999, **126**: 423–434.

3.10 The fate maps of vertebrates are variations on a basic plan

Beddington, R.S.P., Morgenstern, J., Land, H., Hogan, A.: **An *in situ* transgenic enzyme marker for the midgestation mouse embryo and the visualization of inner cell mass clones during early organogenesis.** *Development* 1989, **106**: 37–46.

Gardner, R.L., Rossant, J.: **Investigation of the fate of 4–5 day post-coitum mouse inner cell mass cells by blastocyst injection.** *J. Embryol. Exp. Morph.* 1979, **52**: 141–152.

Helde, K.A., Wilson, E.T., Cretehos, C.J., Grunwald, D.J.: **Contribution of early cells to the fate map of the zebrafish gastrula.** *Science* 1994, **265**: 517–520.

Kimmel, C.B., Warga, R.M., Schilling, T.F.: **Origin and organization of the zebrafish fate map.** *Development* 1990, **108**: 581–594.

Lawson, K.A., Meneses, J.J., Pedersen, R.A.: **Clonal analysis of epiblast fate during germ layer formation in the mouse embryo.** *Development* 1991, **113**: 891–911.

Stern, C.D.: **The marginal zone and its contribution to the hypoblast and primitive streak of the chick embryo.** *Development* 1990, **109**: 667–682.

Stern, C.D., Canning, D.R.: **Origin of cells giving rise to mesoderm and endoderm in chick embryo.** *Nature* 1990, **343**: 273–275.

3.11 Cells of early vertebrate embryos do not yet have their fates determined

Snape, A., Wylie, C.C., Smith, J.C., Heasman, J.: **Changes in states of commitment of single animal pole blastomeres of *Xenopus laevis*.** *Dev. Biol.* 1987, **119**: 503–510.

Wylie, C.C., Snape, A., Heasman, J., Smith, J.C.: **Vegetal pole cells and commitment to form endoderm in *Xenopus laevis*.** *Dev. Biol.* 1987, **119**: 496–502.

3.12 In *Xenopus* the mesoderm is induced by signals from the vegetal region and the endoderm is specified by maternal factors

Kimelman, D., Griffin, J.P.G.: **Vertebrate mesoderm induction and patterning.** *Curr. Opin. Genet. Dev.* 2000, **10**: 350–356.

3.13 The mesoderm is induced by a diffusible signal during a limited period of competence

Gurdon, J.B., Lemaire, P., Kato, K.: **Community effects and related phenomena in development.** *Cell* 1993, **75**: 831–834.

3.15 Several signals induce and pattern the mesoderm in the *Xenopus* blastula

Slack, J.M.W.: **Inducing factors in *Xenopus* early embryos.** *Curr. Biol.* 1994, **4**: 116–126.

3.17 Zygotic gene expression begins at the mid-blastula transition in *Xenopus*

Davidson, E.: *Gene Activity In Early Development.* New York: Academic Press, 1986.

Yasuda, G.K., Schübiger, G.: **Temporal regulation in the early embryo: is MBT too good to be true?** *Trends Genet.* 1992, **8**: 124–127.

3.18 Candidate mesoderm inducers have been identified in *Xenopus*

Amaya, E., Musci, T.J., Kirschner, M.W.: **Expression of a dominant negative mutant of the FGF receptor disrupts mesoderm formation in *Xenopus* embryos.** *Cell* 1991, **66**: 257–270.

De Robertis, E.M., Larrain, J., Oelgeschläger, M., Wessely, O.: **The** establishment of Spemann's organizer and patterning of the vertebrate embryo. *Nat. Rev. Genet.* 2000, **1**: 171–181.

Kemmati-Brivanlou, A., Melton, D.A.: **A truncated activin receptor inhibits mesoderm induction and formation of axial structures in *Xenopus* embryos.** *Nature* 1992, **359**: 609–614.

Kofron, M., Demel, T., Xanthos, J., Lohr, J., Sun, B., Sive, H., Osada, S-I., Wright, C., Wylie, C., Hesman, J.: **Mesoderm induction in *Xenopus* is a zygotic event regulated by maternal VegT via TGFbeta growth factors.** *Development* 1999, **126**: 5759–5770.

Massagué, J.: **How cells read TGF-β signals.** *Nat. Rev. Mol. Cell Biol.* 2000, **1**: 169–178.

Schier, A.F., Shen, M.M.: **Nodal signalling in vertebrate development.** *Nature* 2000, **403**: 385–389.

3.19 Mesoderm patterning factors are produced within the mesoderm

Cooke, J., Takado, S., McMahon, A.: **Experimental control of axial pattern in the chick blastoderm by local expression of *Wnt* and *activin*; the role of HNK-1 positive cells.** *Dev. Biol.* 1994, **164**: 513–527.

Dale, L.: **Pattern formation: A new twist to BMP signalling.** *Curr. Biol.* 2000, **10**: R671–R673.

Leyns, L., Bouwmeester, T., Kim, S-H., Piccolo, S., De Robertis, E.M.: **Frzb-1 is a secreted antagonist of Wnt signaling expressed in the Spemann organizer.** *Cell* 1997, **88**: 747–756.

Moon, R.T., Brown, J.D., Yang-Snyder, J.A., Miller, J.R.: **Structurally related receptors and antagonists compete for secreted Wnt ligands.** *Cell* 1997, **88**: 725–728.

Oelgeschläger, M., Larrain, J., Geissert, D., De Robertis, E.M.: **The evolutionarily conserved BMP-binding protein Twisted gastrulation promotes BMP signalling.** *Nature* 2000, **405**: 757–763.

Piccolo, S., Sasai, Y., Lu, B., De Robertis, E.M.: **Dorsoventral patterning in *Xenopus*: inhibition of ventral signals by direct binding of chordin to BMP-4.** *Cell* 1996, **86**: 589–598.

Schier, A.F.: **Axis formation and patterning in zebrafish.** *Curr. Opin. Genet. Dev.* 2001, **11**: 393–404.

Smith, J.: **Angles on activin's absence.** *Nature* 1995, **374**: 311–312.

Zimmerman, L.B., De Jesús-Escobar, J.M., Harland, R.M.: **The Spemann organizer signal noggin binds and inactivates bone morphogenetic protein 4.** *Cell* 1996, **86**: 599–606.

3.20 Mesoderm induction activates genes that pattern the mesoderm

Ang, S-L., Rossant, J.: **HNF-3β is essential for node and notochord formation in mouse development.** *Cell* 1994, **78**: 561–574.

Isaacs, H.V., Pownall, M.E., Slack, J.M.W.: **eFGF regulates *Xbra* expression during *Xenopus* gastrulation.** *EMBO J.* 1994, **13**: 4469–4481.

Schulte-Merker, S., Smith, J.C.: **Mesoderm formation in response to *Brachyury* requires FGF signalling.** *Curr. Biol.* 1995, **5**: 62–67.

Tada, M, Smith, J.C.: **Xwnt11 is a target of Brachyury: regulation of gastrulation movements via Dishevelled, but not through the canonical Wnt pathway.** *Development* 2000, **127**: 2227–2228.

Taira, M., Jamrich, M., Good, P.J., Dawid, L.B.: **The LIM domain-containing homeobox gene Xlim-1 is expressed specifically in the organizer region of *Xenopus* gastrula embryos.** *Genes Dev.* 1992, **6**: 356–366.

3.21 Gradients in signaling proteins and threshold responses could pattern the mesoderm

Chen, Y., Schier, AF.: **The zebrafish Nodal signal Squint functions as a morphogen.** *Nature* 2001, **411**: 607–609.

Green, J.B.A., New, H.V., Smith, J.C.: **Responses of embryonic *Xenopus* cells to activin and FGF are separated by multiple dose thresholds and correspond to distinct axes of the mesoderm.** *Cell* 1992, **71**: 731–739.

Gurdon, J.B., Harger, P., Mitchell, A., Lemaire, P.: **Activin signalling and response to a morphogen gradient.** *Nature* 1994, **371**: 487–492.

Gurdon, J.B., Standley, H., Dyson, S., Butler, K., Langon, T., Ryan, K., Stennard, F., Shimizu, K., Zorn, A.: **Single cells can sense their position in a morphogen gradient** *Development* 1999, **126**: 5309–5317.

Jones, C.M., Armes, N., Smith, J.C.: **Signaling by TGF-β family members: short-range effects of Xnr-2 and BMP-4 contrast with the long-range effects of activin.** *Curr. Biol.* 1996, **6**: 1468–1475.

Papin, C., Smith, J.C.: **Gradual refinement of activin-induced thresholds requires protein synthesis.** *Dev. Biol.* 2000, **217**: 166–172.

Reilly, K.M., Melton, D.A.: **Short-range signaling by candidate morphogens of the TGF-β family and evidence for a relay mechanism of induction.** *Cell* 1996, **86**: 743–754.

Patterning the vertebrate body plan II: the mesoderm and early nervous system

4

- Somite formation and patterning
- The role of the organizer region and neural induction

" Early on, we behaved as a single unit, as there were no barriers between us. Once our positions were known, we became separate and joined different groups. We only spoke to those in our own group. "

In the previous chapter we saw how the body axes are set up and how the three germ layers are initially specified in various vertebrate embryos. Although amphibian, fish, chick, and mouse embryos share some features at these stages, there are many significant differences. As we approach the phylotypic stage—the embryonic stage common to all vertebrates (see Fig. 2.2)—the similarity between vertebrate embryos becomes greater, and so we can consider the patterning of the vertebrate body plan in a general way. By the phylotypic stage, the embryo has undergone gastrulation, and the main axial structures characteristic of vertebrate embryos—somites, notochord, and neural tube—are well developed and already show signs of regional organization along both the antero-posterior and dorso-ventral axes. In this chapter we shall look at how this patterning is achieved. Each region, such as an individual somite, now develops largely independently.

During gastrulation, the germ layers—mesoderm, endoderm, and ectoderm—move to the positions in which they will develop into the structures of the larval or adult body. The antero-posterior body axis of the vertebrate embryo emerges clearly, with the head at one end and the future tail at the other (Fig. 4.1). In this chapter, we focus mainly on the patterning of the somitic mesoderm that forms the skeleton and muscles of the trunk,

Fig. 4.1 Rearrangement of the presumptive germ layers during gastrulation and neurulation in *Xenopus*. The mesoderm (pink and red), which is in an equatorial band at the blastula stage, moves inside to give rise to the notochord, somites, and lateral mesoderm (not shown). The endoderm (yellow) moves inside to line the gut. The neural tube (dark blue) forms and the ectoderm (light blue) covers the whole embryo. The antero-posterior axis emerges, with the head at the anterior end.

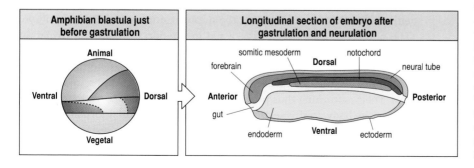

Amphibian blastula just before gastrulation	Longitudinal section of embryo after gastrulation and neurulation

Amphibian blastula just before gastrulation: Animal, Ventral, Dorsal, Vegetal

Longitudinal section of embryo after gastrulation and neurulation: somitic mesoderm, notochord, forebrain, Dorsal, neural tube, Anterior, Posterior, gut, endoderm, Ventral, ectoderm

and of the ectoderm that will develop into the future nervous system. The phenomenon of gastrulation and the action of the organizer region are crucial to establishing the vertebrate body plan (see Section 1.6), and will be discussed in this chapter in relation to their role in the patterning processes. A detailed discussion of the behavior of cells and tissues during gastrulation will, however, be deferred to Chapter 8.

After gastrulation, the part of the mesoderm that comes to lie along the dorsal side of the embryo, under the ectoderm, gives rise to the notochord and somites, and to a small amount of head mesoderm anterior to the notochord. During gastrulation, cells of the dorsal-most mesoderm (the organizer region) are internalized, and eventually form a rigid rod-like notochord along the dorsal midline, flanked on each side by blocks of somites, which are derived from cells lying on either side of the organizer region in the marginal zone mesoderm of the blastula (see Fig. 3.18). In vertebrates, the notochord is a transient structure, and its cells eventually become incorporated into the vertebral column. During neurulation, the neural tube is formed from the ectoderm overlying the notochord, and develops into the brain and spinal cord. The somites, now positioned on either side of the neural tube, give rise to the vertebrae and ribs, to the muscles of the trunk and limbs, and also contribute to the dermis of the skin. Neural crest cells migrate away from the neural tube and develop into a variety of tissues that include skeletal elements of the head, the sensory and autonomic nervous systems, and pigment cells.

Both the mesodermally derived structures along the antero-posterior axis of the vertebrate trunk and the ectodermally derived nervous system have a distinct antero-posterior organization. The vertebrae, for example, have characteristic shapes in each of the four anatomical regions: cervical, thoracic, lumbar, and sacral. In this chapter, we first examine the development of the somites and how they are patterned. We then deal with how their positional identity along the antero-posterior axis is specified. In later sections, we consider the function of the vertebrate organizer and the movements of gastrulation in establishing the antero-posterior organization of the embryo and its coordination with the dorso-ventral organization that we discussed in the previous chapter. Finally, the induction and early patterning of the nervous system will be discussed.

Somite formation and patterning

In Chapter 3, we discussed the early specification of the mesoderm and its patterning along the dorso-ventral axis. The fate maps of the various vertebrates (see Fig. 3.23) show that the notochord develops from the most dorsal region of the mesoderm, and somites from a more ventral region on either side. During gastrulation, the mesoderm moves inside the embryo. The notochord then develops as a rod in the dorsal midline, neurulation begins, and the prospective somitic mesoderm segments into blocks, which will eventually flank the neural tube. The somites give rise to the body and limb muscles, the cartilage that forms the vertebrae and ribs, and the dermis. Their patterning thus provides much of the antero-posterior organization of the body. In this section, we will look at the initial formation of the somites after gastrulation and how they are patterned.

4.1 Somites are formed in a well-defined order along the antero-posterior axis

In the chick embryo, somite formation occurs in the mesodermal region anterior to and just lateral to the regressing Hensen's node (see Fig. 2.15). Between the node and the most recently formed somite, there is an unsegmented region—the **pre-somitic mesoderm**—which will segment into about 12 somites, although the number differs in different vertebrates. Changes in cell shape and intercellular contacts in the pre-somitic mesoderm result in the formation of distinct blocks of cells—the somites. Somites are formed in pairs, one on either side of the notochord, with each pair of somites forming simultaneously. Somite formation begins at the anterior 'head' end and proceeds in a posterior direction.

The sequence of somite formation in the unsegmented region is unaffected by transverse cuts in the plate of pre-somitic mesoderm, suggesting that somite formation is an autonomous process and that, at this time, no signal specifying antero-posterior position or timing is involved. Even if a small piece of the unsegmented mesoderm is rotated through 180°, each somite still forms at the normal time, but with the sequence of formation running in the opposite direction to normal in the inverted tissue (Fig. 4.2). So, before somite formation begins, a molecular pattern that specifies the time of formation of each somite has already been laid down in the pre-somitic mesoderm, and the prospective identity of each somite is due to the temporal order in which they leave the pre-somitic mesoderm.

The cells that give rise to the somites originate in the epiblast on either side of the anterior primitive streak and move into it at gastrulation to form a population of somitogenic stem cells in and around Hensen's node and later in the tailbud. These stem cells divide, and those that remain in the stem-cell region continue to be self-renewing stem cells, but those that leave the node region as it regresses form the pre-somitic mesoderm. As new cells are being added to the pre-somitic mesoderm at the posterior end of the chick embryo, somites are forming at the anterior end (Fig. 4.3), left panel.

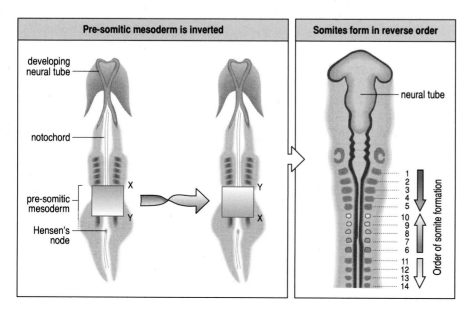

Fig. 4.2 The temporal order of somite formation is specified early in embryonic development. Somite formation in the chick proceeds in an antero-posterior direction. Somites form sequentially in the pre-somitic region between the last-formed somite and Hensen's node, which moves posteriorly. If the antero-posterior axis of the pre-somitic mesoderm is inverted through 180°, as shown by the arrow, the temporal order of somite formation is not altered—somite 6 still develops before somite 10.

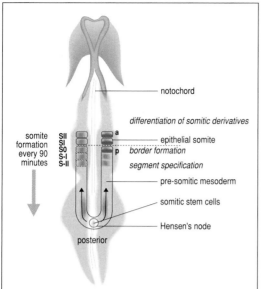

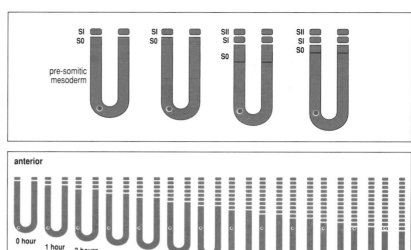

Fig. 4.3 Somite formation in the chick. As shown in the left-hand panel, the somites are generated successively from pre-somitic mesoderm, which is derived from somitic stem cells in the primitive streak. As pre-somitic cells are released into the posterior pre-somitic mesoderm, a new pair of somites buds from the anterior end every 90 minutes. SI, the most recently formed somite; SII, the last but one somite formed; S0, somite in the process of formation, whose boundaries are not yet set; S–I, S–II, blocks of pre-somitic cells that will form somites. At formation, each somite acquires antero-posterior polarity, after which it can respond to the signals that pattern it along the antero-posterior and dorso-ventral axes. The top right-hand panel shows stages in one of the cycles of *c-hairy1* expression (blue) that sweep from posterior to anterior of the pre-somitic mesoderm every 90 minutes. During each cycle, a given pre-somitic cell (red dot) experiences distinct phases when *c-hairy1* is expressed and when it is not expressed. The lower right-hand panel shows the progress of a pre-somitic cell (red dot) from the time it enters the pre-somitic mesoderm until it is incorporated into a somite. Somitic cells in the anterior somites will have experienced fewer cycles of *c-hairy1* expression than those in posterior somites, and this could define a 'clock' that is both linked to somite segmentation and 'tells' the somite its position along the antero-posterior axis.

A favored model is that the rate of somite formation is largely determined by an internal 'clock' in the pre-somitic mesoderm. This 'clock' is represented by periodic cycles of gene expression, such as that of the gene *c-hairy 1* in the chick embryo, whose expression sweeps from the posterior to the anterior end of the pre-somitic mesoderm with a period of 90 minutes, the time it takes for a pair of somites to form. In a newly formed somite, *c-hairy 1* expression becomes restricted to the posterior end of the somite, where it persists, while a new wave of *c-hairy 1* expression starts at the tail end of the pre-somitic mesoderm (Fig. 4.3).

The connection between thses oscillations and somite formation is not yet clear, but one of the proteins whose expression cycles is Lunatic fringe, which potentiates activity of the Notch–Delta signaling pathway. This pathway (Fig. 4.4) is widely involved in determining cell fate and delimiting boundaries, and is involved in setting the somite boundaries. Mice mutant for the Delta–Notch pathway often do not form somites, and if they do, the somites vary in size and are different on each side of the body.

One idea is that the timing and position of somite formation is determined by the interaction of the segmentation 'clock' with the growth factor FGF-8. This is expressed in the posterior part of the pre-somitic mesoderm and keeps it in a labile state in which it does not form somites. As the node regresses, any given pre-somitic cell will eventually move out

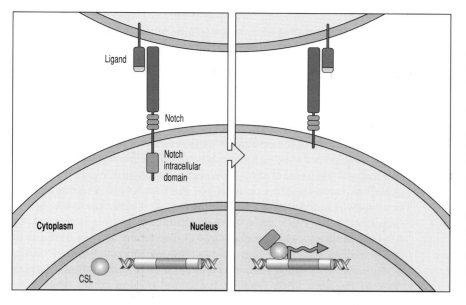

Fig. 4.4 The core Notch signaling pathway. Binding of a receptor protein of the Notch family to its membrane-bound ligand such as Delta or Serrate activates intracellular signaling from the receptor. This is thought to involve the enzymatic cleavage of the cytoplasmic tail of the receptor (Notch intracellular domain) and its translocation to the nucleus to bind and activate a transcription factor of the CSL family. The final outcome of this pathway is the activation of specific genes. Notch signaling is very versatile. In different organisms and in different developmental circumstances, activation of Notch switches different genes on or off.

of the posterior area (see Fig. 4.3), away from the influence of FGF-8. It has been proposed that when this transition occurs at a particular phase of a cell's 'clock' cycle, this starts the cellular changes leading to somite formation.

Somites differentiate into particular axial structures depending on their position along the antero-posterior axis. The anterior-most somites contribute to the skull, those posterior to them will form cervical vertebrae, and more posterior ones will develop as thoracic vertebrae with ribs. Specification by position has occurred before somite formation begins during gastrulation; if unsegmented somitic mesoderm from, for example, the presumptive thoracic region is grafted to replace the presumptive mesoderm of the neck region, it will still form thoracic vertebrae with ribs (Fig. 4.5). How then is the pre-somitic mesoderm patterned so that somites

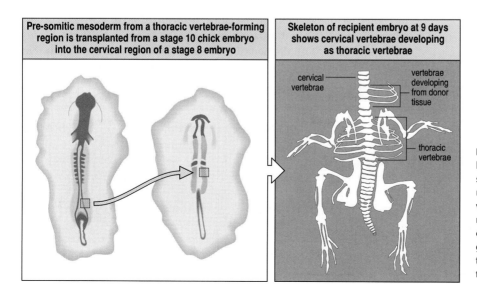

Fig. 4.5 The pre-somitic mesoderm has a positional identity before somite formation. Pre-somitic mesoderm that will give rise to thoracic vertebrae is grafted to an anterior region of a younger embryo that will develop into cervical vertebrae. The grafted mesoderm develops according to its original position and forms ribs in the cervical region.

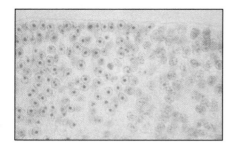

Fig. 4.6 Photograph of quail-chick chimeric tissue. The quail cells are on the left and the chick cells on the right.

Photograph courtesy of Nicole Le Douarin.

acquire their identity and form particular vertebrae? We consider this below, but first we deal with the patterning of the individual somite with respect to the different tissues it gives rise to.

4.2 The fate of somite cells is determined by signals from the adjacent tissues

The somites of the vertebrate embryo give rise to major axial structures: the cartilage cells of the embryonic axial skeleton—the vertebrae and ribs; all the skeletal muscles, including those of the limbs; and the dermis. The fate maps of particular somites have been made by grafting somites from a quail into a corresponding position in a chick embryo at a similar stage of development and following the fate of the quail cells. These can be distinguished from chick cells by their distinctive nuclei, which can be detected in histological sections (Fig. 4.6). The lateral and medial parts of chick somites are of different origins, and are brought together during gastrulation; the medial portion comes from cells in the primitive streak close to Hensen's node, whereas the lateral portion comes from more posterior cells.

Cells located in the dorsal and lateral regions of a newly formed somite make up the **dermomyotome**, which expresses the *Pax3* gene, a homeobox-containing gene of the *paired* family (see Box 4A, p. 117). The dermomyotome is made up of the **myotome**, which gives rise to muscle cells, and the **dermatome**, an epithelial sheet over the myotome which gives rise to the dermis. Cells from the medial region of the somite form mainly axial and back muscles, and express the muscle-specific transcription factor MyoD and related proteins, whereas lateral cells migrate to give rise to abdominal and limb muscles. The ventral part of the medial somite contains **sclerotome** cells that express the *Pax1* gene and migrate ventrally to surround the notochord and develop into vertebrae and ribs (Fig. 4.7).

Which cells will form cartilage, muscle, or dermis is not yet determined at the time of somite formation. Specification of these fates requires signals from tissues adjacent to the somite. This is clearly shown by experiments in which the dorso-ventral orientation of newly formed somites is inverted; they still develop normally. In the chick, determination of myotome occurs within hours of somite formation, whereas the future sclerotome is only determined later. Both the neural tube and notochord produce signals that

Fig. 4.7 The fate map of a somite in the chick embryo. The ventral medial quadrant (blue) gives rise to the sclerotome cells, which migrate to form the cartilage of the vertebrae. The rest of the somite—the dermomyotome—forms the dermatome and myotome, which give rise to the dermis and all the trunk muscles, respectively. The dermomyotome also gives rise to muscle cells that migrate into the limb bud.

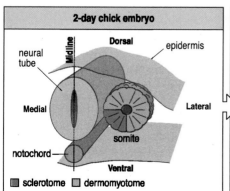
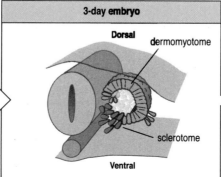
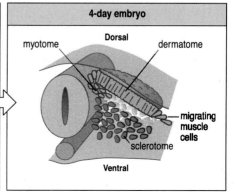

pattern the somite and are required for its future development. If the noto-chord and neural tube are removed, the cells in the somites undergo apoptosis; neither vertebrae nor axial muscles develop, although limb musculature still does.

The role of the notochord in specifying somitic cells has been shown by experiments in the chick, in which an extra notochord is implanted to one side of the neural tube, adjacent to the somite. This has a dramatic effect on somite differentiation, provided the operation is carried out on unseg-mented pre-somitic mesoderm: when the somite develops, there is an almost complete conversion to cartilage precursors (Fig. 4.8), suggesting that the notochord is an inducer of cartilage. The neural tube also has a cartilage-inducing effect on somites, which is mediated by the most ventral region of the tube, the floor plate (see Section 11.6). There is also evidence for a signal from the lateral plate mesoderm, which is involved in specify-ing the lateral part of the dermomyotome, and for a signal from the overlying ectoderm (Fig. 4.9).

Some of the signals that pattern the somite have been identified. In the chick, both the notochord and the floor plate express the gene *Sonic hedge-hog*, which encodes a secreted protein that seems to be a key molecule for positional signaling in a number of developmental situations (see Section 10.5). (We met *Sonic hedgehog* in Chapter 3 as a gene involved in the asym-metry of structures about the midline. There, this gene was being expressed at a quite different stage of development and in different tissues.) One model proposes that the signal generated by *Sonic hedgehog* specifies the ventral region of the somite and is required for sclerotome development. Signals from the dorsal neural tube and from the overlying non-neural ectoderm would specify the dorsal region. The TGF-β family member bone morphogenetic protein 4 (BMP-4) and secreted signaling proteins of the Wnt family (both of which we also encountered in Chapter 3), are good candidates for the lateral and dorsal signals, respectively. BMP-4 specifies dorsal cartilage. The Sonic hedgehog and BMP-4 signals are mutually antagonistic. These are signals that are used over and over again during development.

Regulation of the Pax homeobox genes in the somite by signals from the notochord and neural tube seems to be important in determining cell fate. *Pax3* is initially expressed in all cells that will form somites. Its expression is then modulated by the BMP-4 and Wnt family proteins so that it becomes confined to muscle precursors. It is then further downregulated in cells that differentiate as the muscles of the back, but remains switched on in the migrating presumptive muscle cells that populate the limbs. Mice that lack a functional *Pax3* gene—*Splotch* mutants—lack limb muscles. In the chick, *Pax1* has been implicated in the formation of the scapula, a key element in the shoulder girdle, part of which is contributed by somites. Unlike the *Pax1*-expressing cells of the vertebrae, which are of sclerotomal origin, the blade of the scapula is formed from dermomyo-tome cells of chick somites 17–24, whereas the head of the scapula is derived from lateral plate mesoderm. All the scapula-forming cells express *Pax1*.

Having seen how somites are formed and how their different regions are patterned after gastrulation, we now discuss the patterning of the pre-somitic mesoderm along the antero-posterior axis that gives each somite its individual character.

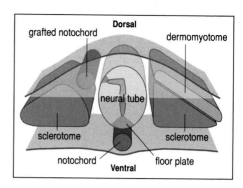

Fig. 4.8 A signal from the notochord induces sclerotome formation. A graft of an additional notochord to the dorsal region of a somite in a 10-somite embryo suppresses the formation of the dermomyotome from the dorsal portion of the somite, and induces the formation of sclerotome, which develops into cartilage. The graft also affects the shape of the neural tube.

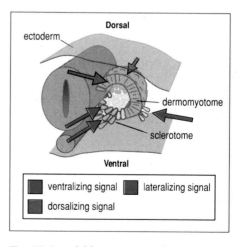

Fig. 4.9 A model for patterning of somite differentiation. The sclerotome is thought to be specified by a diffusible signal, probably the Sonic hedgehog protein, from the notochord and the floor plate of the neural tube (blue arrows). Signals from the dorsal neural tube and ectoderm (pink arrows) would specify the dermomyotome, together with lateral signals (green arrows) from the lateral plate mesoderm. After Johnson, R.L.: 1994.

4.3 Positional identity of somites along the antero-posterior axis is specified by Hox gene expression

The antero-posterior patterning of the mesoderm is most clearly seen in the differences in the vertebrae, each vertebra having well-defined anatomical characteristics depending on its location along the axis. The most anterior vertebrae are specialized for attachment and articulation of the skull, while the cervical vertebrae of the neck are followed by the rib-bearing thoracic vertebrae and then those of the lumbar region which do not bear ribs, and finally, those of the sacral and caudal regions. This antero-posterior patterning is a process distinct from the patterning of individual somites in respect of the cells that will produce muscle, cartilage, and dermis, and it occurs earlier, while the pre-somitic mesoderm is still unsegmented. Patterning of the skeleton along the body axis is based on the mesodermal cells acquiring a positional value that reflects their position along the axis and so determines their subsequent development. Mesodermal cells that will form thoracic vertebrae, for example, have different positional values from those that will form cervical vertebrae.

Patterning along the antero-posterior axis in all vertebrates involves the expression of a set of genes that specify positional identity along the axis. These are the **Hox genes**, members of the large family of **homeobox genes** that are involved in many aspects of development (Box 4A). The concept of positional identity, or positional value, has important implications for developmental strategy; it implies that a cell or a group of cells in the embryo acquires a unique state related to its position at a given time, and that this determines its later development (see Section 1.13).

Homeobox genes that specify positional identity along the antero-posterior axis were originally identified in the fruit fly *Drosophila* and it turned out that related genes are also involved in patterning the vertebrate axis. As we shall see in the final part of this chapter, patterning along the antero-posterior axis by Hox genes and other homeobox genes is not confined to mesodermal structures; the hindbrain, for example, is also divided into distinct regions.

All the homeobox genes whose functions are known encode transcription factors. The subset known as the Hox genes are the vertebrate counterparts of a cluster of homeobox genes in *Drosophila* that is involved in specifying the identities of the different segments of the insect body. These are discussed fully in Chapter 5. Most vertebrates have four separate clusters of Hox genes that are thought to have arisen by duplications of the genes within a cluster, and of the clusters themselves (see Box 4A). The zebrafish is unusual in having six. A particular feature of Hox gene expression in both insects and vertebrates is that the genes in each cluster are expressed in a temporal and spatial order that reflects their order on the chromosome.

A simple idealized model illustrates the key features by which a Hox gene cluster records positional identity. Consider four genes, 1, 2, 3, and 4, arranged along a chromosome in that order (Fig. 4.10). The genes are expressed in a corresponding order along the antero-posterior axis of a tissue. Thus, gene 1 is expressed throughout the tissue with its anterior boundary at the anterior end. Gene 2 has its anterior boundary in a more posterior position and expression continues posteriorly. The same principles apply to the two other genes. This pattern of expression defines four

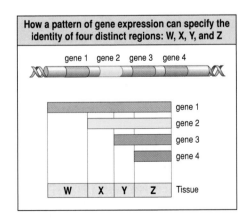

Fig. 4.10 Gene activity can provide positional values. The model shows how the pattern of gene expression along a tissue can specify the distinct regions W, X, Y, and Z. For example, only gene 1 is expressed in region W but all four genes are expressed in region Z.

Box 4A Homeobox genes

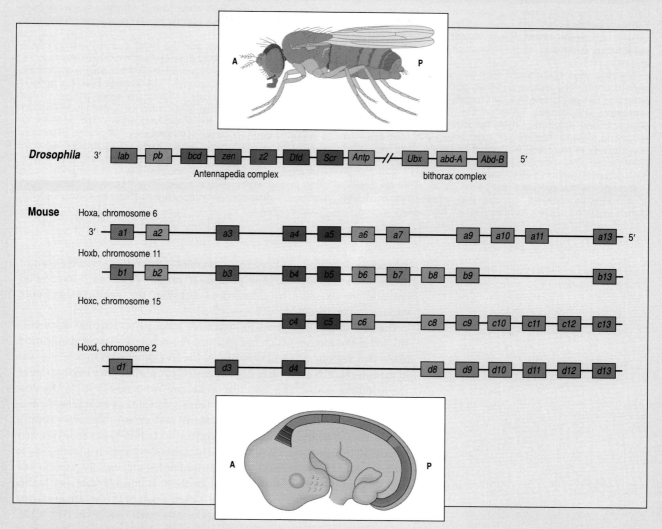

The homeobox gene family encodes a large group of transcription factors which all contain a similar DNA-binding region of around 60 amino acids called the homeodomain. The **homeodomain** contains a helix-turn-helix DNA-binding motif which is characteristic of many DNA-binding proteins. This domain is encoded by a DNA sequence of 180 base pairs termed the **homeobox**. Many homeobox genes are involved in development, and the homeobox was originally identified in genes that control patterning in *Drosophila* development.

The name 'homeobox' comes from the fact that mutations in some of these genes result in what is known as a **homeotic** transformation, in which one structure replaces another. For example, in one homeotic mutation in *Drosophila*, a segment in the fly's body that does not normally bear wings is transformed into an adjacent segment that does bear wings, resulting in a fly with four wings.

Clusters of homeotic genes involved in specifying segment identity were first discovered in the fruit fly *Drosophila*. Similar complexes of homeotic genes have been identified in many animals. In vertebrates, the related clusters are known as the Hox complexes, and the homeoboxes of the genes are related to the Antennapedia homeobox of *Drosophila*. In the mouse, there are four unlinked Hox complexes, designated Hoxa, Hoxb, Hoxc, and Hoxd, (originally called Hox1, Hox2, Hox3, and Hox4) located on chromosomes 6, 11, 15, and 2, respectively (see figure).

The vertebrate clusters have arisen by duplication of an ancestral cluster, possibly related to the single Hox cluster in the lancelet (amphioxus), a simple chordate. Thus, corresponding genes within the four clusters resemble each other closely. The original cluster is thought to have formed by gene duplication and divergence and all Hox genes thus resemble each other to some extent; the homology is most marked within the

contd

Box 4A contd

homeobox and less marked in sequences outside it. Genes that have arisen by duplication and divergence within a species are known as **paralogs**, and the corresponding genes in the different clusters (e.g. *Hoxa4, Hoxb4, Hoxc4, Hoxd4*) are usually known as a **paralogous subgroup**. In the mouse there are 13 paralogous groups.

The Hox gene clusters and their role in development are of ancient origin. The mouse and frog genes are similar to each other and to those of the fruit fly *Drosophila*, both in their coding sequences and in their order on the chromosome. In both *Drosophila* and vertebrates, these homeotic genes are involved in specifying regional identity along the antero-posterior axis. The Hox clusters in mice and in *Drosophila* (where they are called HOM genes) almost certainly arose by gene duplication in some common ancestor of vertebrates and insects.

Most genes that contain a homeobox do not, however, belong to a homeotic complex, nor are they involved in homeotic transformations. Other subfamilies of homeobox genes in

vertebrates include the **Pax genes**, which contain a homeobox typical of the *Drosophila* gene *paired*. All these genes encode transcription factors with various functions in development and cell differentiation.

The homeobox genes are the most striking example of a widespread conservation of developmental genes in animals. It is widely believed that there are common mechanisms underlying the development of all animals. This implies that if a gene is identified as having a central role in the development of one animal, it is worth looking to see where it is present in another animal and whether it has a similar function. This strategy of comparing genes by sequence homology has proved extremely successful in identifying genes involved in development in vertebrates. Numerous genes first identified in *Drosophila*, in which the genetic basis of development is far better understood than in any other animal, have proved to have counterparts involved in development in vertebrates. Illustration after Coletta, P., *et al.*: 1994.

distinct regions, coded for by the expression of different combinations of genes. If the amount of gene product is varied within each expression domain, for example by interactions between the genes, many more regions can be specified.

The role of the Hox genes in vertebrate axial patterning has been best studied in the mouse, which has four Hox clusters as a result of duplication of the Hox regions. As in all vertebrates, the Hox genes start to be expressed in mesoderm cells at an early stage of gastrulation when they begin to leave the primitive streak. The 'anterior' genes are expressed first. As the posterior pattern develops later, clearly defined patterns of Hox gene expression are most easily seen in the mesoderm and neural tube, after somite formation and neurulation, respectively (Fig. 4.11). Hox genes are expressed as gastrulation proceeds. Typically, the pattern of expression of each gene is characterized by a relatively sharp anterior border and, usually, a much less well defined posterior border. Although there is considerable overlap in expression, almost every region in the anterior part of the antero-posterior axis is characterized by a particular set of expressed Hox genes (Fig. 4.12). For example, the most anterior somites are characterized by expression of

Fig. 4.11 Hox gene expression in the mouse embryo after neurulation. The three panels show lateral views of 9½ days post-coitum embryos immunostained with antibodies specific for the protein products of the *Hoxb1, Hoxb4,* and *Hoxb9* genes. The arrowheads indicate the anterior boundary of expression of each gene within the neural tube. The position of the three genes within the Hoxb gene complex is indicated (inset). Scale bar = 0.5 mm.

Photographs courtesy of A. Gould.

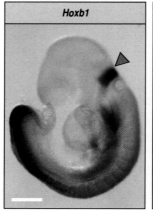
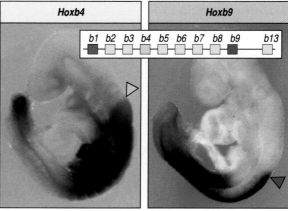

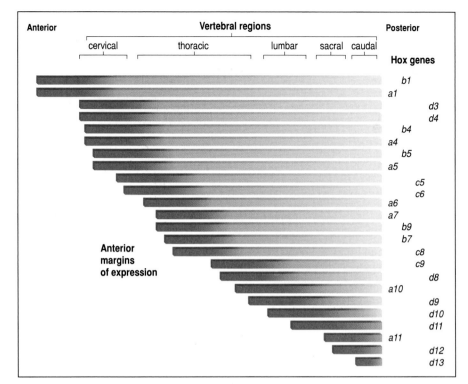

Fig. 4.12 Hox gene expression along the antero-posterior axis of the mouse mesoderm. The anterior border of each gene is shown by the dark red blocks. Expression usually extends backward some distance but the posterior margin of expression may be poorly defined. The pattern of Hox gene expression could specify the identity of the tissues at different positions. For example, the pattern of expression is quite different in anterior and posterior regions of the body axis.

genes *Hoxa1* and *Hoxb1*, and no other Hox genes are expressed in this region. By contrast, all the Hox genes are expressed in the most posterior regions. The Hox genes thus provide a code for regional identity. The most anterior expression of Hox genes is in the hindbrain; the more anterior regions of the vertebrate body—the anterior head, forebrain, and midbrain—are characterized by expression of homeobox genes such as *Emc* and *Otx*, and not by Hox genes.

If we focus on just one set of Hox genes, those of the Hoxa complex, we find that the most anterior border of expression in the mesoderm is that of *Hoxa1* in the posterior head mesoderm, while *Hoxa11*, the most posterior gene in the Hoxa cluster, has its anterior border of expression in the sacral region (see Fig. 4.12). This exceptional correspondence, or **co-linearity**, between the order of the genes on the chromosome and their order of spatial and temporal expression along the antero-posterior axis, is typical of all the Hox complexes. The genes of each Hox complex are expressed in an orderly sequence, with the gene lying most 3′ in the cluster being expressed the earliest and in the most anterior position. The correct expression of the Hox genes is dependent on their position in the cluster, and anterior genes must be expressed before more posterior genes.

Evidence that the Hox genes are involved in controlling regional identity comes from comparing their patterns of expression in mouse and chick with the well-defined anatomical regions—cervical, thoracic, and so on (Fig. 4.13). Hox gene expression corresponds nicely with the different regions. For example, even though the number of cervical vertebrae in birds (14) is twice that of mammals, the anterior boundaries of *Hoxc5* and *Hoxc6* gene expression in both chick and mouse lie on either side of the cervical/thoracic boundary. A correspondence between Hox gene

Fig. 4.13 Patterns of Hox gene expression in the mesoderm of chick and mouse embryos, and their relation to regionalization. The posterior margins of expression of Hox genes in the mesoderm vary along the axes. The vertebrae are derived from somites, 40 of which are shown. The vertebrae have characteristic shapes in each of the five regions: cervical (C), thoracic (T), lumbar (L), sacral (S), and caudal (Ca). Which somites form which vertebrae differs in chick and mouse. For example, thoracic vertebrae start at somite 20 in the chick, but at somite 12 in the mouse. The transition from one region to another corresponds with the pattern of Hox gene expression, so *Hoxc5* and *Hoxc6* are expressed on either side of the cervical and thoracic vertebral transition in both chick and mouse. Similarly, *Hoxd9* and *Hoxd10* are expressed at the transition between lumbar and sacral regions. After Burke, A.C.: 1995.

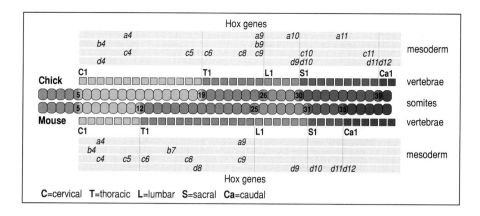

expression and region is also similarly conserved among vertebrates at other anatomical boundaries.

It must be emphasized that the summary picture of Hox gene expression given in Fig. 4.12 does not represent a 'snapshot' of expression at a particular time but rather the total overall pattern of expression. Some genes are switched on early and are then downregulated, while others are expressed considerably later; the most posterior Hox genes such as *Hoxd12* and *Hoxd13*, for example, are expressed in the post-anal tail, which develops later. Moreover, this summary picture reflects the general expression of the genes in embryonic regions; not all Hox genes expressed in a region are expressed in all the cells of that region. Nevertheless, the overall pattern suggests that the combination of Hox genes provides positional identity. In the cervical region, for example, each somite, and thus each vertebra, could be specified by a unique pattern of Hox gene expression.

As we saw in Fig. 4.5, grafting experiments show that the character of the somites is already determined in the pre-somitic mesoderm and that somitic tissue transplanted to other levels along the axis retains its original identity. It has also been shown to retain its original pattern of Hox gene expression. By contrast, transplanted lateral plate mesoderm takes on the Hox expression pattern of its new site. Hox genes providing positional specification in somites and lateral plate seem to be separate systems, though similar mechanisms may be involved.

We still do not know what switches on the Hox genes in the mesoderm during gastrulation, and how co-linear expression is ensured. In all vertebrates, the Hox genes begin to be expressed (the anterior-most genes being expressed first) at an early stage of gastrulation, when the mesodermal cells begin their gastrulation movements. If an 'early' Hoxd gene is relocated to the 5′ end of the Hoxd complex, its expression pattern then resembles that of the neighboring *Hoxd13*. This shows that the structure of the Hox complex is crucial in determining the pattern of Hox gene expression.

One way the antero-posterior pattern of Hox gene expression might be established in the somitic mesoderm is through linking gene activation to the time spent in the region adjacent to the node. Such a mechanism could provide the striking correspondence in the order of Hox genes on the chromosome with their spatial and temporal expression. One could propose a mechanism that starts opening up a Hox gene complex at its 3′ end in the stem-cell region, but which stops as the cells move into the more anterior mesoderm from which the somites segment. If the spread of activation of the Hox complex along the chromosome in the 5′ direction ceases when cells move out of the posterior region, then more of the complex can be expressed in posterior cells, which leave last, than in anterior cells, which leave first. Such a mechanism, which resembles that proposed for the patterning of the proximo-distal axis of the vertebrate limb (see Section 10.2), is entirely speculative at present. There is evidence that the segmentation clock coordinates a burst of Hox gene expression in cells that have left the posterior pre-somitic mesoderm and are approaching the position at which they will form a somite. This may be when the cells acquire their positional value. As we saw earlier, an FGF signal in the posterior pre-somitic mesoderm is likely to set the position of segment formation. Thus, the time of a somite's formation and its identity would be tightly co-ordinated by the interactions between FGF, the segmentation clock and the Hox genes.

If the Hox genes do provide positional values that determine a region's subsequent development, then if their pattern of expression is altered, one would expect morphological changes. This is indeed the case, as we see next.

4.4 Deletion or overexpression of Hox genes causes changes in axial patterning

In order to see how Hox genes control patterning, either their expression can be prevented by mutation, or they can be expressed ectopically, in abnormal positions. Hox gene expression can be eliminated from the developing mouse embryo by gene knock-out techniques (Box 4B, p. 122). Experiments along these lines have shown that the absence of a given Hox gene affects patterning in a way that accords with the idea that Hox gene activity provides the cells with positional identity. For example, mice in which the gene *Hoxa3* has been deleted show structural defects in the region of the head and thorax, where this gene is normally strongly expressed, and tissues derived from both ectoderm and mesoderm are affected. But the Hox genes seem to specify positional identity in rather complex ways. There is undoubtedly some apparent **redundancy** between the effects of some of the genes, and when one gene is removed, another may serve in its place. This can make it difficult to interpret the results of Hox gene inactivation. There is also interaction between the individual genes, and this can further complicate results. For example, with the mutated *Hoxa3* gene described above, more posterior axial structures, where the inactivated gene is also normally expressed, show no evident defects.

This observation illustrates a general principle of Hox gene expression, which is that more posteriorly expressed Hox genes tend to inhibit the action of the Hox genes normally expressed anterior to them; this phenomenon is known as **posterior dominance** or **posterior prevalence**. This means that a change in Hox gene expression usually affects the most anterior regions in which the gene is expressed, leaving posterior structures relatively unaffected. The effects of a Hox gene knock-out can also be tissue specific, so that certain tissues in which a Hox gene is normally expressed appear normal, while other tissues at the same position along the antero-posterior axis are affected. The apparent absence of an effect may be due to redundancy, with **paralogous genes** from another complex being able to compensate. For example, *Hoxb1* is expressed in the same

Box 4B Gene targeting: insertional mutagenesis and gene knockout

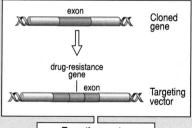

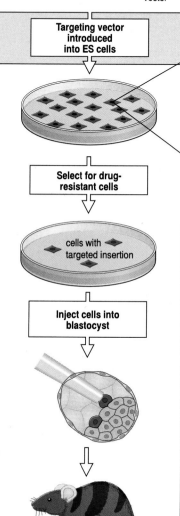

chimera
(mutant gene in germ line)

In order to study the function of a gene controlling development, it is highly desirable to be able to introduce an altered gene into the animal to see what effect this has. Mice into which an additional or altered gene has been introduced are known as **transgenic** mice. Two main techniques for generating transgenic mice are currently in use. One is to inject DNA containing the required gene directly into the nuclei of fertilized eggs; the other is to alter or add a gene to the genome of **embryonic stem cells** (ES cells) in culture, and then to inject the genetically altered cells into the blastocyst, where they become part of the inner cell mass.

ES cells can be genetically altered by techniques that can be used to create a mutation in a particular gene. A vector DNA molecule that is introduced into an ES cell by

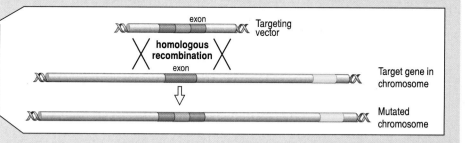

transfection will usually insert randomly in the genome. However, it is possible to tailor the vector DNA in such a way that only those DNA molecules that insert at a specific predetermined site by **homologous recombination**, and thus insert into and mutate a particular gene, will be selected. The DNA to be introduced must contain enough sequence homology with the target gene that it will insert within the target gene in at least a few cells in the culture, even though most insertions will be random. These mutated ES cells can then be introduced into the blastocyst, producing a transgenic mouse carrying a mutation in a known gene (see figure). The use of homologous recombination to inactivate a gene is known as **gene knock-out** when the animal is homozygous for the inactivated gene. Many mutations produced by this technique do not result in a knock-out, but the mutation alters gene function.

The enormous advantage of using ES cells over microinjection methods for generating transgenic mice is that it is possible to design a selection procedure to isolate just those rare cells in which the DNA has incorporated at the desired site. These can then be used to generate the chimeric embryo. The selection procedure is based on including certain genes for drug-resistance and drug-sensitivity in the DNA construct such that when the DNA inserts at the correct site, only the cells with the correctly targeted DNA can be selected.

The mutated ES cells are then introduced into the cavity of an early blastocyst, which is then returned to the uterus. They become incorporated into the inner cell mass and thus into the embryo, where they can give rise to germ cells and gametes. Once the mutant gene has entered the germ line, strains of mice heterozygous for the altered gene can be intercrossed to produce either viable homozygotes or homozygous lethals, depending on the gene involved, and the effect of completely inactivating and so knocking-out the gene can be examined.

A technique for targeting a gene knock-out to a specific tissue and/or at a particular time in development is provided by the *Cre-lox* system. The target gene is first 'loxed' by inserting a *loxP* sequence of 34 base pairs on either side of the gene. These transgenic mice are then crossed with another line of transgenic mice carrying the gene for the recombinase Cre. *loxP* sequences are recognized by Cre, which will excise all the DNA between the two *loxP* sites. In the offspring, if Cre is expressed in all cells, then all cells will excise the 'loxed' target, causing an ubiquitous knock-out of the target gene. However, if the gene for Cre is under the control of a tissue-specific promoter, so that, for example, it is only expressed in heart tissue, the target gene will only be excised in heart tissue (see figure opposite). If the Cre gene is linked to an inducible control region, it is also possible to induce excision of the target gene at will by exposing the mice to the inducing stimulus.

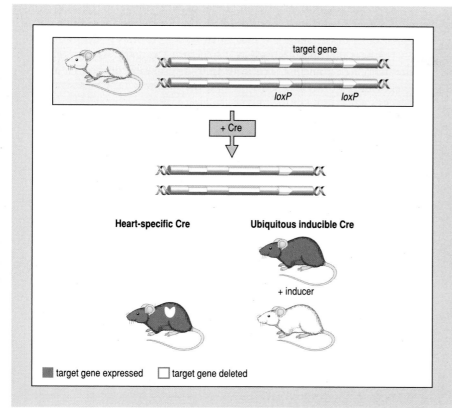

Heart-specific Cre   Ubiquitous inducible Cre

+ inducer

■ target gene expressed   □ target gene deleted

A significant number of knock-outs of a single gene result in mice developing without any obvious abnormality or with fewer and less severe abnormalities than might be expected from the normal pattern of gene activity. A striking example is that of *myoD*, a key gene in muscle differentiation. In *myoD* knock-outs, the mice are anatomically normal, although they do have a reduced survival rate. This could mean that other genes can substitute for some of the functions of *myoD*.

However, it is unlikely that any gene is without any value at all to an animal. It is much more likely that there is an altered phenotype in these apparently normal animals, which is too subtle to be detected under the artificial conditions of life in a laboratory. Redundancy is thus probably apparent rather than real. A further complication is the possibility that, under such circumstances, related genes with similar functions may increase their activity to compensate for the mutated gene.

region as *Hoxa1* (see Fig. 4.12), and so may be largely able to fulfill the function of an absent *Hoxa1* gene.

Loss of Hox gene function often results in **homeotic transformation—** the conversion of one body part into another, for example. This is the case in a knock-out mutation of *Hoxc8*. In normal embryos, *Hoxc8* is expressed in the thoracic and more posterior regions of the embryo from late gastrulation onward. Mice homozygous for mutant *Hoxc8* die within a few days of birth, and have abnormalities in patterning between the seventh thoracic vertebra and the first lumbar vertebra. The most obvious homeotic transformations are the attachment of an eighth pair of ribs to the sternum and the development of a 14th pair of ribs on the first lumbar vertebra (Fig. 4.14). Thus, the absence of *Hoxc8* modifies the development of some of the cells that would normally express it. Its absence gives them a more anterior positional value, and they develop accordingly. In mice in which *Hoxd11* is mutated, anterior sacral vertebrae are transformed into lumbar vertebrae. Another example of the homeotic transformation of a structure into one normally anterior to it can be seen in knock-out mutations of *Hoxb4*. In normal mice, *Hoxb4* is expressed in the mesoderm that will give rise to the axis (the second cervical vertebra), but not in that giving rise to the atlas (the first cervical vertebra). In *Hoxb4* knock-out mice, the axis is transformed into another atlas.

By contrast, abnormal expression of Hox genes in anterior regions that normally do not express them can result in transformations of anterior structures into structures that are normally more posterior. For example,

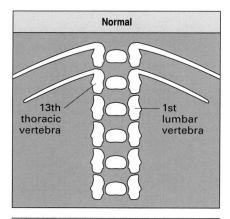

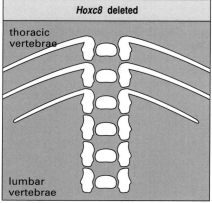

Fig. 4.14 Homeotic transformation of vertebrae due to deletion of *Hoxc8* in the mouse. In loss-of-function homozygous mutants of *Hoxc8*, the first lumbar vertebra is transformed into a rib-bearing thoracic vertebra. The mutation has resulted in the transformation of the lumbar vertebra into a more anterior structure.

when *Hoxa7*, whose normal anterior border of expression is in the thoracic region, is expressed throughout the whole antero-posterior axis, the basal occipital bone of the skull is transformed into a pro-atlas structure, normally the next most posterior skeletal structure. Overexpression of *Hoxa2* in the first chick branchial arch leads to transformation of the first arch cartilages such as Meckel's cartilage and the quadrate into second arch cartilages such as those of the tongue skeleton.

There are synergistic interactions between Hox genes of the same paralogous group. Thus, knock-outs of *Hoxa3* do not affect the first cervical vertebra—the atlas—or the basal occipital bone of the skull to which it connects, even though *Hoxa3* is expressed in the mesoderm that gives rise to these bones. However, knock-outs of *Hoxd3* (which is also expressed in this region) cause a homeotic transformation of the atlas into the adjacent basal occipital bone. A double knock-out of *Hoxa3* and *Hoxd3* results in complete deletion of the atlas. The complete absence of this bone in the absence of Hox gene expression suggests that one target of Hox gene action is the cell proliferation required to build such a structure from the somite cells. Although we do not know how the pattern of Hox gene expression is specified, it has been found that the pattern of vertebrae can be anteriorized by a secreted protein—growth/differentiation factor 11 (GDF-11), a member of the TGF-β family. Deletion of the gene for GDF-11 in mice results in an increase in thoracic vertebrae from 13 to 17 or 18, a small increase in the number of lumbar vertebrae and a very reduced tail. This shows that a secreted molecule can alter the pattern, and, as we see in the next section, a small diffusible molecule—retinoic acid—can also do so.

4.5 Retinoic acid can alter positional values

Retinoic acid, a derivative of vitamin A, is a small hydrophobic molecule with an important role in local signaling in vertebrate development. Like the steroid and thyroid hormones, it diffuses across the plasma membrane unaided and binds to intracellular receptors; the complex of receptor and retinoic acid then functions as a transcription factor. A variety of experiments has shown that retinoic acid can alter cells' positional values in limb development (see Section 10.5), and it can also have effects on the antero-posterior axis.

The developmental abnormalities induced by retinoic acid are apparently the result of its interference with the normal establishment of the Hox gene expression pattern. For example, treatment of early mouse embryos with retinoic acid results in homeotic transformation of vertebrae, both anterior and posterior transformations being induced, depending on the time of treatment. It is likely that this effect of retinoic acid is mediated in part by its actions on Hox gene expression, and studies of cells in culture show that the Hox genes can be induced by retinoic acid in a concentration-dependent manner. In *Xenopus*, the gene *Xlhbox6* (related to mouse *Hoxb9*) is normally expressed in posterior regions, but in embryos treated with retinoic acid, the expression of *Xlhbox6* extends anteriorly, and dorsal mesoderm and anterior structures are defective. As there is evidence for a gradient in retinoic acid along the antero-posterior axis in the mouse embryo, it could be important in activating Hox genes in normal antero-posterior patterning.

Summary

Somites are blocks of mesodermal tissue that are formed after gastrulation. They form sequentially in pairs on either side of the notochord, starting at the anterior end of the embryo. The somites give rise to the vertebrae, to the muscles of the trunk and limbs, and to the dermis of the skin. The pre-somitic mesoderm is patterned along its antero-posterior dimension before somite formation, and the first manifestation of this pattern is the expression of the Hox genes in the mesoderm. The somites are also patterned by signals from the notochord, neural tube, and ectoderm, which induce particular regions of each somite to give rise to muscle, cartilage, or dermis.

The regional character of the mesoderm that gives rise to somites is specified even before the somites form. The positional identity of the somites is specified by the combinatorial expression of genes of the Hox complexes along the antero-posterior axis, from the hindbrain to the posterior end, with the order of expression of these genes along the axis corresponding to their order along the chromosome. Mutation or overexpression of a Hox gene results, in general, in localized defects in the anterior parts of the regions in which the gene is expressed, and can cause homeotic transformations. We can think of Hox genes as providing positional information that specifies the identity of a region and its later development.

Although we have concentrated on the expression of Hox genes in the mesoderm, they are also expressed in a patterned way in the neural tube after its induction, and we shall return to this aspect of antero-posterior regionalization later. But next, we look at the role of the crucially important organizer region both in neural induction and in organizing the antero-posterior axis in vertebrate embryos.

Neural induction and the role of the organizer

The Spemann organizer of amphibians, the shield in zebrafish, Hensen's node in the chick, and the equivalent node region in the mouse all have a similar global organizing function in vertebrate development. They can, with the exception of the mouse node, all induce a complete body axis if transplanted to another embryo at an appropriate stage, and so are able to organize and coordinate both dorso-ventral and antero-posterior aspects of the body plan, as well as to induce neural tissue from ectoderm. In the mouse, the anterior visceral endoderm is required for the induction of anterior structures (see Section 3.6).

During gastrulation, the ectoderm lying along the dorsal midline of the embryo becomes specified as neural plate; during the subsequent stage of neurulation, it forms the neural tube, and then differentiates into the brain and spinal cord (see Section 2.1). The brain and spinal cord must develop in the correct relationship with other body structures, particularly the meso-dermally derived structures that give rise to the skeleto-muscular system. Thus, patterning of the nervous system must be linked to that of the mesoderm. In this part of the chapter, we consider the patterning of the neural tube up to shortly after its closure. Consideration of the hindbrain will take us to a later stage when that region becomes segmented and the neural crest cells have migrated.

The function of the organizer has been best studied in amphibians, and

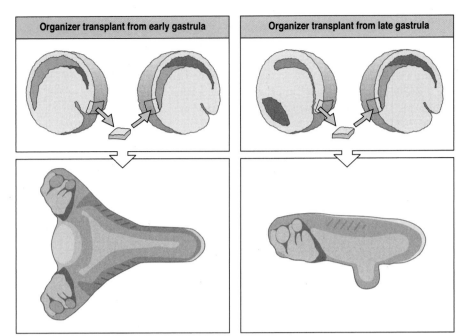

Fig. 4.15 The inductive properties of the organizer change during gastrulation. A graft of the organizer region, from the dorsal lip of the blastopore of an early frog gastrula to the ventral side of another early gastrula, results in the development of an additional anterior axis at the site of the graft (left panels). A graft from the dorsal lip region of a late gastrula to an early gastrula only induces formation of tail structures (right panels).

we have already emphasized its role in the dorso-ventral patterning of the mesoderm of *Xenopus* (see Sections 3.16 and 3.19). We now discuss its profound effects on the antero-posterior axis.

4.6 The vertebrate head is specified by signals different from those for the trunk

In amphibians, the action of the organizer is dramatically demonstrated in what is classically known as **primary embryonic induction**. If the Spemann organizer from an early gastrula (corresponding to the dorsal lip of the amphibian blastopore) is grafted to the ventral side of the marginal zone of another gastrula, the result is a twinned embryo (Fig. 4.15). This second embryo can have a well-defined head and trunk region and even a tail, but will be joined to the main embryo along the axis (see Fig. 1.10). A variety of other treatments, such as grafting dorsal vegetal blastomeres containing the Nieuwkoop center to the ventral side, produce a similar result (see Fig. 3.5), but what all these treatments have in common is that, directly or indirectly, they result in the formation of a new Spemann organizer. An important issue is whether or not there are separate organizers for the trunk and tail. A graft of an organizer from an early gastrula induces a complete body axis and central nervous system, while older organizers only induce posterior structures (Fig. 4.15). Is this a change in the quantity of inducing signals or are different signals involved?

The cells of the organizer region in *Xenopus* give rise during gastrulation to anterior endoderm, prospective head mesoderm (the pre-chordal mesoderm), and the notochord. Understanding how the organizer mesoderm organizes the overall pattern of the antero-posterior axis is not so straightforward, however. In amphibians, the inductive properties of the Spemann organizer region change throughout gastrulation. In transplantation experiments of the type described above, a dorsal lip taken from an early

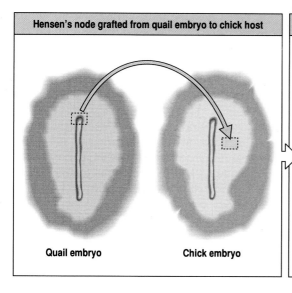
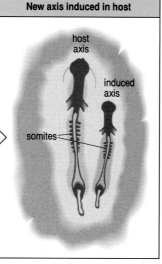

Hensen's node grafted from quail embryo to chick host

Quail embryo   Chick embryo

New axis induced in host

host axis
induced axis
somites

Fig. 4.16 Hensen's node can induce a new axis in avian embryos. When Hensen's node from a quail embryo is grafted to a position lateral to the primitive streak of a chick embryo at the same stage of development, a new axis forms at the site of transplantation. Histological examination shows that although some of the somites of this new axis are formed from the graft itself (quail tissue can easily be distinguished from chick tissue, see Fig. 4.6), others have been induced from host tissue that does not normally form somites.

gastrula induces a complete additional embryo, a dorsal lip from a mid-gastrula induces a trunk and tail but no head, while a dorsal lip from a late gastrula induces only a tail (see Fig. 4.15). As gastrulation proceeds, the antero-posterior axis becomes specified, and so at later stages cells in the blastopore induce only posterior structures.

The avian equivalent to the Spemann organizer is Hensen's node, the region at the anterior end of the primitive streak in the chick blastoderm. It contributes to notochord, pre-chordal mesoderm, somites and gut endoderm, and can induce an additional axis, complete with somites, if grafted beneath a chick epiblast at the head-process stage of development (Fig. 4.16). This stage is reached when elongation of the primitive streak is complete, the notochord (head process) has started to form anteriorly, but Hensen's node has not yet started to regress (see Fig. 2.14). Induction occurs if the graft is placed quite close to the streak; it can then induce normally non-axial mesoderm to form somites and other axial structures. When grafted to the area opaca, it can induce a whole neural axis. The node at the anterior region of the primitive streak in the mouse can induce a similar duplication of the mouse axis on transplantation, with the exception of the forebrain, which requires the anterior visceral endoderm.

The organizer is a complex structure whose constitution changes during gastrulation, and we have just seen that at early stages it induces heads, whereas later it induces tails. In the early gastrula of *Xenopus*, the vegetal half of the organizer region will induce genes that are expressed more anteriorly. The inducing regions in the early *Xenopus* embryo, as determined from the results of transplantation of small pieces of tissue from the dorsal marginal zone of the early gastrula, are shown in Fig. 4.17. In the chick, the node itself is specified by signaling, involving both TGF-β and Wnt family members, from a region in the middle of the primitive streak. Cells move into the node at an early stage and, as the node regresses, the cells divide and some leave the node region to give rise to the notochord and somites.

A number of genes are specifically expressed in the organizer of both *Xenopus* and the mouse (Fig. 4.18) and, as we saw in Section 3.19, some of

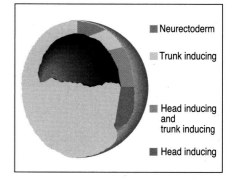

■ Neurectoderm
■ Trunk inducing
■ Head inducing and trunk inducing
■ Head inducing

Fig. 4.17 Inducing regions in the *Xenopus* organizer. Regions of head-inducing and trunk-inducing activity in the early *Xenopus* gastrula, as determined from transplantation experiments. Adapted from Schneider, V.A., Mercola, M.: 1999.

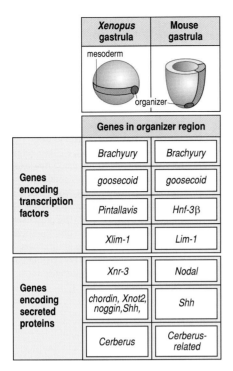

Xenopus gastrula	Mouse gastrula
mesoderm	
organizer	
Genes in organizer region	

	Xenopus gastrula	Mouse gastrula
Genes encoding transcription factors	Brachyury	Brachyury
	goosecoid	goosecoid
	Pintallavis	Hnf-3β
	Xlim-1	Lim-1
Genes encoding secreted proteins	Xnr-3	Nodal
	chordin, Xnot2, noggin, Shh,	Shh
	Cerberus	Cerberus-related

Fig. 4.18 Genes expressed in the Spemann organizer region of the *Xenopus* gastrula, and in Hensen's node in the mouse gastrula. There is a similar pattern of gene activity in the two animals, with homologous genes being expressed. *Shh = Sonic hedgehog.*

these genes have a role in patterning the mesoderm along the future antero-posterior axis. The secreted proteins chordin and noggin antagonize BMP and pattern the mesoderm. The gene *goosecoid* can cause a complete secondary axis to develop when its mRNA is injected into the ventral side of an early frog embryo. However, mice lacking goosecoid have a normal antero-posterior axis, so this gene cannot be solely involved in patterning along this axis.

As we saw in Section 3.6, the extra-embryonic anterior visceral endoderm in the mouse embryo has a key role in specifying anterior structures. In *Xenopus*, the cerberus protein, which is related to a protein expressed in mouse anterior visceral endoderm, is produced in the anterior endoderm and may be involved in the induction of the head region. If misexpressed, cerberus can induce ectopic heads, albeit with one eye only, with no trunk structures. It acts by binding to and blocking the action of BMPs, Wnts and nodal proteins; all three of these activities appear to be necessary for ectopic head induction. Down-regulation of Wnt signaling in anterior regions seems to be essential for forebrain and heart development. Other inhibitors of Wnts, such as dickopf-1 and frizbee, only induce ectopic heads if BMPs are inhibited at the same time. This shows that a complex set of signals is involved in specifying head and truck.

4.7 The neural plate is induced in the ectoderm

The induction of neural tissue from ectoderm was first indicated by the organizer transplant experiment in frogs, which is described in Fig. 4.15; in the secondary embryo that forms at the site of transplantation, a nervous system develops from the host ectoderm that would normally have formed ventral epidermis. This suggests that neural tissue can be induced from as yet unspecified ectoderm by signals emanating from the mesoderm of the organizer region. The requirement for induction is confirmed by experiments that exchange prospective neural plate ectoderm for prospective epidermis before gastrulation; the transplanted prospective epidermis develops into neural tissue, and the transplanted prospective neural tissue into epidermis (Fig. 4.19). This shows that the formation of the nervous system is dependent on an inductive signal.

An enormous amount of effort in the 1930s and 1940s was devoted to trying to identify the signals involved in neural induction in amphibians. Researchers were encouraged by the finding that a dead organizer region could still induce neural tissue. It seemed to be merely a matter of hard work to isolate the chemicals responsible. Alas, the search was fruitless, for it appeared that an enormous variety of substances were capable of varying degrees of neural induction. As it turned out, this was because newt ectoderm, the main experimental material used, seems to have a high propensity to develop into neural tissue on its own. This is not the case with *Xenopus* ectoderm, although prolonged culture of disaggregated ectodermal cells can result in their differentiation as neural cells. Whatever the nature of the signal, it seems that it is a molecule that can diffuse through a Nucleopore filter (which prevents cell contact but allows the passage of quite large molecules, such as proteins), and that contact lasting about 2 hours is required for induction to occur. The molecules responsible for neural induction have still not been definitively identified, although there are

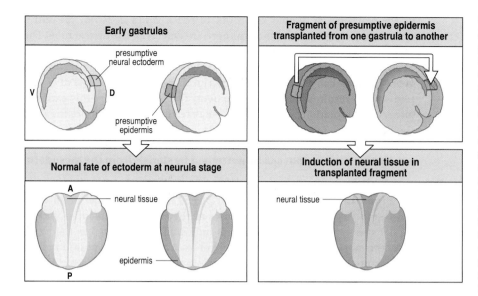

Fig. 4.19 The nervous system of *Xenopus* is induced during gastrulation. The left panels show the normal development fate of ectoderm at two different positions in the early gastrula. The right panels show the transplantation of a piece of ventral ectoderm, whose normal fate is to form epidermis, from the ventral side of an early gastrula to the dorsal side of another, where it replaces a piece of dorsal ectoderm whose normal fate is to form neural tissue. In its new location, the transplanted prospective epidermis develops not as epidermis but as neural tissue, and forms part of a normal nervous system. This shows that the ventral tissue has not yet been determined at the time of transplantation, and that neural tissue is induced during gastrulation.

some very strong candidates. However, contrary to previous ideas, inducing molecules do not act directly on the cells that will form neural tissue, but act instead by releasing the inhibition of neural tissue formation mediated by proteins such as the BMPs, in a manner reminiscent of the patterning of the early mesoderm.

A number of secreted proteins, including chordin, noggin, and cerberus, are BMP antagonists involved in neural induction. BMP-4 is pivotal in neural induction, as it inhibits cells from forming neural tissue; if BMP signaling is inhibited, neural tissue develops. One such inhibitor is noggin. As discussed in Section 3.19, noggin is secreted by the *Xenopus* organizer region and is probably one of the signals that dorsalizes the presumptive mesoderm. Noggin may also be a neural induction factor: if it is added at high concentration to isolated *Xenopus* blastula animal caps, neural markers are induced. Noggin also has the expression pattern and activity expected of a neural inducer. Another secreted protein, chordin, is expressed in the organizer-derived mesoderm underlying the future neural plate; it too has neuralizing activity and acts by binding to BMP-4 and preventing its action. The antagonistic actions of BMP-4 and chordin in neural induction could be similar to those seen in the dorso-ventral patterning of the mesoderm itself (see Section 3.19). The situation is somewhat different in the chick. There, BMP-4 does not inhibit neural development and, perhaps as a consequence, chordin cannot induce neural tissue in the chick, even though it is expressed in Hensen's node. But FGF is required for neural induction and Wnt signals can block this.

Neural induction is thus a complex multistep process and may begin before the organizer acts. Although the organizer contributes to neural induction, it may not be necessary for neural tissue to form, as a neural plate will still develop in frogs, birds, and mice if the organizer, or node, is excised during gastrulation. One might explain this effect by reformation of the node, but this has not been established. More probably, the potential to form neural tissue is already present in the ectoderm and is specified by earlier signals. Although *noggin* and *chordin* code for neural-inducing signals in the zebrafish, a double knock-out of these genes does not prevent a neural plate developing, although there are defects in the neural tube. The early developmental signal that confers a neural fate on the ectoderm of *Xenopus* most probably involves the Wnt/β-catenin signaling pathway, as Wnt signaling can block *BMP-4* transcription and synthesis of BMP-4 protein, and so result in neural development. Thus, signaling by this pathway after cortical rotation (see Section 3.4) may result in neural development on the dorsal side.

4.8 The nervous system can be patterned by signals from the mesoderm

Pieces of mesoderm taken from different positions along the antero-posterior axis of a newt neurula and placed in the blastocoel of an early newt embryo induce neural structures at the site of transplantation. Positional specificity in this induction is indicated by the fact that the structures that are formed correspond more or less to the original position of the transplanted mesoderm. Pieces of anterior mesoderm induce a head with a brain, whereas posterior pieces induce a trunk with a spinal cord (Fig. 4.20). Another indication of positional specificity in induction comes from the observation that pieces of the neural plate themselves induce similar regional neural structures in adjacent ectoderm when transplanted beneath the ectoderm of a gastrula. Indications that gene expression in the mesoderm may be influencing gene expression in the ectoderm come from the observation of coincident expression of several Hox genes in the notochord and in the pre-somitic mesoderm and ectoderm at the same position along the antero-posterior axis: *Xlhbox1* in *Xenopus* and *Hoxb1* in the mouse are examples of genes that have coincident expression of this type.

Hox gene expression cannot be detected in the anterior-most neural tissue of the mouse, but genes such as *Otx* and *Emc* are expressed anterior to the hindbrain. These genes encode homeodomain transcription factors and specify pattern in the anterior brain in a manner similar to the Hox genes more posteriorly. The *Drosophila orthodenticle* gene and the mouse *Otx* genes are homologous and provide a good example of the conservation of gene function during evolution. *orthodenticle* is expressed in the posterior region of the future *Drosophila* brain and mutations in the gene result in a greatly reduced brain. In mice, *Otx1* and *Otx2* are expressed in overlapping domains

Fig. 4.20 Induction of the nervous system by the mesoderm is region specific. Mesoderm from different positions along the dorsal antero-posterior axis of early newt neurulas induces structures specific to its region of origin when transplanted to ventral regions of early gastrulas. Anterior mesoderm induces a head with a brain (top panels), whereas posterior mesoderm induces a posterior trunk with a spinal cord ending in a tail (bottom panels). After Mangold, O.: 1933.

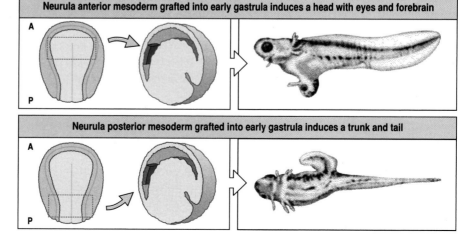

Neurula anterior mesoderm grafted into early gastrula induces a head with eyes and forebrain

A

P

Neurula posterior mesoderm grafted into early gastrula induces a trunk and tail

A

P

in the developing forebrain and hindbrain, and mutation in *Otx1* leads to brain abnormalities and epilepsy. Mice with a defective *Otx* gene can be partly rescued by replacing it with *orthodenticle*, even though the sequence similarity of the two proteins is confined to the homeodomain region. Human *Otx* can even rescue *orthodenticle* mutants in *Drosophila*.

In the mouse, the node, usually considered to be similar in function and properties to the organizer in chick and frog, can induce only posterior structures. In mouse embryos lacking HNF-3β, a transcription factor essential for node development, there is no node but a full range of neural tissues. In mice, the extra-embryonic anterior visceral endoderm is essential for induction of anterior structures, including the forebrain (see Section 3.6).

One model for neural ectoderm patterning suggests that qualitatively different mesodermal inducers are present at different positions along the antero-posterior axis. However, many experiments fit quite well with a simpler two-signal model of neural patterning (Fig. 4.21), in which differences are due to quantitative rather than qualitative differences in the inducing signal. In this model, the first, activating, signal is produced by the whole mesoderm and induces the ectoderm to become anterior neural tissue. The second signal transforms part of this tissue so that it acquires a more posterior identity. This latter, transforming, signal would be graded in the mesoderm with the highest concentration at the posterior end. Chordin and noggin are good candidates for the first activating signal and may act by inhibiting BMPs and Wnts (see Sections 4.6 and 4.7). FGF and Wnt3A are candidates for posteriorizing transforming signals.

In the chick, there is some evidence for the activation/transformation model outlined above. Young nodes induce a full central nervous system whereas older ones only induce more posterior ones. As the node emits strong posteriorizing signals, the prospective anterior neural tissue must be protected from these signals. This is achieved in the chick by the prospective forebrain territory moving anterior to the node under the cells of the spreading hypoblast.

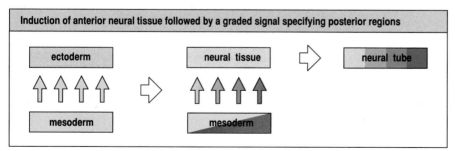

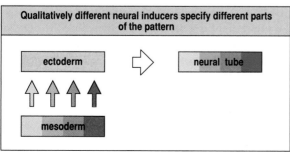

Fig. 4.21 Models of neural patterning by induction. Top panel: in the two-signal model, one signal from the mesoderm first induces anterior tissue throughout the corresponding ectoderm. A second, graded signal from the mesoderm then specifies more posterior regions. Bottom panel: in an alternative model, qualitatively different inducers are localized in the mesoderm. After Kelly, O.G., *et al.*: 1995.

Fig. 4.22 Hensen's node from a chick embryo can induce gene expression characteristic of neural tissue in *Xenopus* ectoderm. Tissues from different parts of the primitive streak stage of a chick epiblast are placed between two fragments of animal cap tissue (prospective ectoderm) from a *Xenopus* blastula. The induction in the *Xenopus* ectoderm of genes that characterize the nervous system is detected by looking for expression of mRNAs for neural cell adhesion molecule (N-CAM) and neurogenic factor-3 (NF-3), which are expressed specifically in neural tissue in stage-30 *Xenopus* embryos. Only transplants from Hensen's node induce the expression of these neural markers in the *Xenopus* ectoderm. (EF-1α is a common transcription factor expressed in all cells.) After Kintner, C.R., *et al.*: 1991.

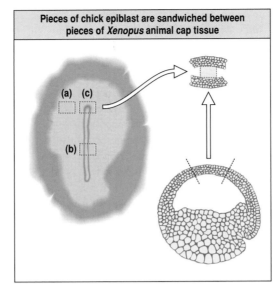

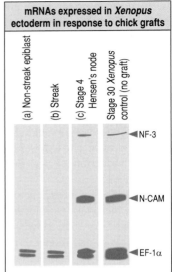

Mesoderm also induces neural tissue in chick and mouse embryos. Neural tissue can be induced in the chick epiblast—in both the area pellucida and the area opaca—by grafts from the primitive streak. Inductive activity is initially located in the anterior primitive streak and later, during regression of Hensen's node, becomes confined to the region just anterior to the node. By the four-somite stage inductive activity has disappeared; the competence of the ectoderm to respond disappears at about the same time.

Hensen's node from the chick embryo can induce neural gene expression in amphibian (*Xenopus*) ectoderm (Fig. 4.22), which is of great interest as it suggests that there has been an evolutionary conservation of inducing signals. Moreover, early nodes induce gene expression characteristic of anterior amphibian neural structures, whereas older nodes induce gene expression typical of posterior structures. These results are in line with the theory that the node is able to specify different antero-posterior positional values, and they also confirm the essential similarity of Hensen's node to the Spemann organizer. But the molecular mechanisms will be different, as chordin and BMP are not involved in the chick.

4.9 Signals that pattern the neural plate may travel within the neural plate itself

Development of neural tissue was originally thought to occur only if the mesoderm came to lie immediately beneath the ectoderm and in contact with it. It now appears that there is a second route by which the nervous system can be patterned. The classical one is the traditional vertical, or transverse, route from the mesoderm to the overlying ectoderm, whereas the other is planar, the signal being generated within the neural plate itself and travelling within the ectodermal sheet.

Evidence that signals within the plate may be important comes from explants of early *Xenopus* gastrulas that consist of dorsal mesoderm, including the Spemann organizer, and ectoderm that will normally give rise to

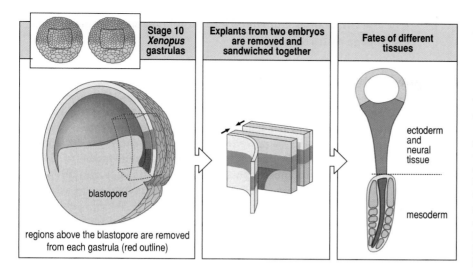

| Stage 10 *Xenopus* gastrulas | Explants from two embryos are removed and sandwiched together | Fates of different tissues |

regions above the blastopore are removed from each gastrula (red outline)

blastopore

ectoderm and neural tissue

mesoderm

Fig. 4.23 **Induction and patterning of the nervous system involves planar signals originating within the ectoderm.** Explants containing both mesoderm (pink and red) and prospective neural ectoderm (blue) are taken from the marginal zone of early gastrulas at the time when the mesoderm begins to invaginate. Two such explants are sandwiched together (to prevent curling of the strips), and cultured. In these explants, mesoderm lies posterior and in the same plane as the ectoderm, with only a small point of contact, rather than underneath the ectoderm and in contact with it along its whole length as it is in the embryo. Nevertheless, the explants differentiate into both neural tissue and mesoderm. The ectoderm and mesoderm converge and extend in the explants, and genes such as *engrailed-2* and *Krox-20*, which are normally expressed in region-specific patterns within neural tissue, are expressed. After Doniach, T., *et al.*: 1992.

neural tissue after induction. Two of these explants are cultured as a sandwich in such a way as to keep the strip flat and prevent it rounding up. Both ectoderm and mesoderm in such explants converge and extend as they do during normal gastrulation (see Section 8.10), and the ectoderm differentiates into neural tissue, as shown by the expression of the neural-specific cell adhesion molecule N-CAM (Fig. 4.23).

The neural ectoderm of such explants shows clear spatial patterning, as indicated by the correct order of expression of genes that are normally localized to specific regions along the antero-posterior axis of the neural plate, namely *engrailed-2, Krox-20, Xlhbox1*, and *Xlhbox6*. Moreover, two of the genes (*engrailed-2* and *Xlhbox1*) are also expressed in the mesoderm of the explants, as they are in the mesoderm of intact embryos. Thus, the expression of these two genes occurs independently in mesoderm and ectoderm without the close apposition of these two tissues that normally occurs in embryos *in vivo*. The most likely explanation of these results is that signals confined to the plane of either the ectoderm or mesoderm are both involved in patterning the ectoderm, perhaps in the form of a gradient with its high point at the site of the organizer. Within the developing nervous system, the early patterning of the hindbrain region has been best studied and will be considered below.

Good evidence for planar signals also comes from zebrafish mutants. In the absence of the signal delivered by nodal protein (see Section 3.18), no anterior or trunk mesoderm develops yet the neural plate is normally patterned. This must be due to signals within the plate itself, emanating both from the shield and from an anterior signaling region in the ectoderm, which patterns the forebrain. At the midbrain–hindbrain boundary there is also a signaling center that patterns the regions on either side; if grafted to the forebrain it induces a midbrain/hindbrain pattern locally. The midbrain–hindbrain boundary is defined by the posterior limit of expression of the gene *Otx2*, which is expressed in the anterior neural tube. The signaling molecules FGF-8 and Wnt-1 are subsequently expressed on either side of the border and pattern these regions.

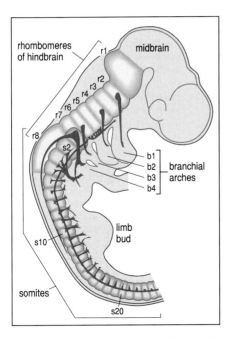

Fig. 4.24 The nervous system in a 3-day chick embryo. The hindbrain is divided into eight rhombomeres (r1 to r8). The positions of the cranial nerves III to XII are shown in green. b1 to b4 are the four branchial arches. b1 gives rise to the jaws. s = somites. Adapted from Lumsden, A.: 1991.

4.10 The hindbrain is segmented into rhombomeres by boundaries of cell-lineage restriction

Patterning of the posterior region of the head and the hindbrain involves segmentation of the neural tube along the antero-posterior axis. This does not occur elsewhere along the spinal cord, where the pattern of dorsal root ganglia and ventral motor nerves at regular intervals—one pair per somite—is imposed by the somites. In the chick embryo, three segmented systems can be seen in the posterior head region by 3 days of development: the mesoderm on either side of the notochord is subdivided into somites, the hindbrain (the rhombencephalon) is divided into eight **rhombomeres**, and the lateral mesoderm has formed a series of branchial arches which are populated by neural crest cells (Fig. 4.24).

Development of the head in the hindbrain region involves several interacting components. The neural tube gives rise both to the segmentally arranged cranial nerves that innervate the face and neck, and to neural crest cells, which in turn give rise both to peripheral nerves and to skeletal elements. In addition, the otic vesicle gives rise to the ear. The main skeletal elements of the head in this region develop from the first three branchial arches, into which neural crest cells migrate (see Section 2.2). For example, the first arch gives rise to the jaws, while the second arch develops into the bony parts of the ear. This region of the head is a particularly valuable model for studying patterning along the antero-posterior axis because of the presence of the numerous different structures ordered along it.

Immediately after the neural tube of the chick embryo closes in this region, the future hindbrain becomes constricted at evenly spaced positions to define eight rhombomeres (see Fig. 4.24). The cellular basis for these constrictions is not understood but may involve differential cell division or changes in cell shape. Whatever the underlying cause, it seems that the boundaries between rhombomeres are barriers of **cell-lineage restriction**; that is, once the boundaries form, cells and their descendants are confined within a rhombomere and do not cross from one side of a boundary to the other. Marking of individual cells shows that before the constrictions become visible, the descendants of a given labeled cell can populate two adjacent rhombomeres. After the constrictions appear, however, descendants of cells then within a rhombomere never cross the boundaries and are thus confined to a single rhombomere (Fig. 4.25). It seems that the cells of a rhombomere share some adhesive property that prevents them mixing with those of adjacent rhombomeres, and this involves ephrins, membrane-bound proteins that interact with cell-surface ephrin receptors (Eph) on adjacent cells and which can generate bidirectional signals (Fig. 4.26). Ephrin receptors and ephrins are thought to be separately expressed in alternating rhombomeres, thus preventing cell mixing at the boundaries. This implies that cells in each rhombomere may be under the control of the same genes, and that the rhombomere is a developmental unit. A rhombomere is thus behaving like a compartment, which is a common feature of insect development, as we see in Chapter 5, but seems to be rare in vertebrates.

The idea that each rhombomere is a developmental unit is supported by the observation that when an odd-numbered and an even-numbered rhombomere from different positions along the antero-posterior axis are placed

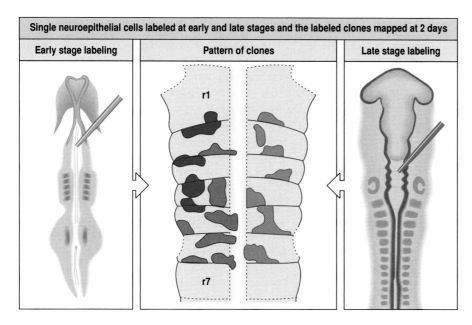

Single neuroepithelial cells labeled at early and late stages and the labeled clones mapped at 2 days

Early stage labeling | Pattern of clones | Late stage labeling

Fig. 4.25 Lineage restriction in rhombomeres of the embryonic chick hindbrain. Single cells are injected with a label (rhodamine dextran) at an early stage (left panel) or a later stage (right panel) of neurulation, and their descendants are mapped 2 days later. Cells injected before rhombomere boundaries form give rise to some clones that span two rhombomeres (dark red) as well as those that do not cross boundaries (red). Clones marked after rhombomere formation never cross the boundary of the rhombomere that they originate in (blue). Adapted from Lumsden, A.: 1991.

next to each other after a boundary between them has been surgically removed, a new boundary forms. No boundaries form when different odd-numbered rhombomeres are placed next to each other, however, suggesting that their cells have similar surface properties.

The division of the hindbrain into rhombomeres has functional significance in that each has a unique identity and this determines how each will develop. As we see below, their development is under the control of Hox gene expression. We first consider the behavior of the neural crest cells, which originate from the neural tube and initially migrate over the rhombomeres. They populate the branchial arches, subsequently giving rise to structures such as the lower jaw.

4.11 Neural crest cells have positional values

The cranial neural crest that migrates out from the rhombomeres of the dorsal region of the hindbrain also has a segmental arrangement. This has been revealed by labeling chick neural crest cells *in vivo* and following their migration pathways. The patterns of crest emergence and migration correlate closely with the rhombomere from which the crest cells come. Thus, branchial arches 1, 2, and 3 become populated by crest cells from rhombomeres 2, 4, and 6, respectively. In the chick, crest cells from rhombomeres 3 and 5 are apparently largely eliminated from further development by programmed cell death, or apoptosis.

The crest cells have already acquired a positional value before they begin to migrate. When crest cells of rhombomere 4 are replaced by cells from rhombomere 2 taken from another embryo, these cells enter the second branchial arch but develop into structures characteristic of the first arch, to which they would normally have migrated. This can result in the development of an additional lower jaw in the chick embryo.

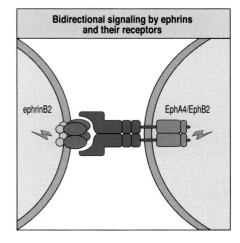

Fig. 4.26 Signaling by ephrins and their receptors. Ephrins binding to their receptors (Eph) can generate a bidirectional signal, in which the ephrin itself also generates a signal in the cell that carries it.

4.12 Hox genes provide positional identity in the hindbrain region

Hox gene expression provides a possible molecular basis for the positional identity of both the rhombomeres and the neural crest. Hox genes are expressed in the mouse embryo hindbrain in a well defined pattern, which closely correlates with the segmental pattern (Fig. 4.27). For example, *Hoxb3* has its most anterior region of expression at the border of rhombomeres 4 and 5, while *Hoxb2* has its anterior border at the border of rhombomeres 2 and 3 (Fig. 4.28). In general, the paralogous genes of the different Hox complexes have similar patterns of expression. It is clear that the three paralogous groups involved have different anterior margins of expression, paralog 1 (i.e. *Hoxa1*, *Hoxb1*, etc.) being most anterior, followed by paralogs 2 and 3. The pattern of Hox gene expression in the ectoderm and branchial arches at a particular position along the antero-posterior axis is similar to that in the neural tube and neural crest, and it may be that the crest cells induce their positional values in the overlying ectoderm during their migration.

Transplantation of rhombomeres from an anterior to a more posterior position alters the pattern of Hox gene expression so it becomes the same as that normally expressed at the new location. The signals responsible for this reprogramming originate from the neural tube itself and not the surrounding tissues. Studies of the control of Hox gene expression at the molecular level have provided some indication as to how their pattern of expression is controlled. For example, although the *Hoxb2* gene is expressed in the three contiguous rhombomeres 3, 4, and 5, its expression in rhombomeres 3 and 5 is controlled quite independently from its expression in rhombomere 4. The regulatory regions of the *Hoxb2* gene carry two separate

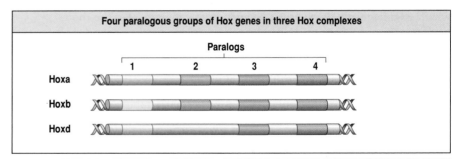

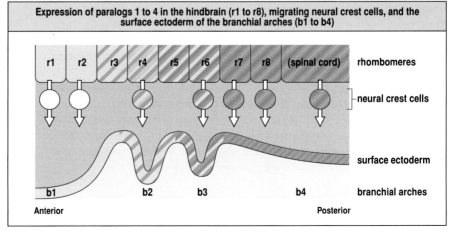

Fig. 4.27 Expression of Hox genes in the branchial region of the head. The expression of genes of three paralogous Hox complexes in the hindbrain (rhombomeres r1 to r8), neural crest, branchial arches (b1 to b4), and surface ectoderm is shown. *Hoxa1* and *Hoxd1* are not expressed at this stage. The arrows indicate the migration of neural crest cells into the branchial arches. Note the absence of neural crest migration from r3 and r5. After Krumlauf, R.: 1993.

enhancer elements which regulate its expression in these three rhombomeres. Expression in rhombomeres 3 and 5 is controlled through one of these enhancers, while expression in rhombomere 4 is controlled through the other (see Fig. 4.28). In rhombomeres 3 and 5, *Hoxb2* is activated in part by the zinc finger transcription factor encoded by the gene *Krox-20*, which is expressed in these rhombomeres but not in rhombomere 4. There are binding sites for the Krox-20 protein in the enhancer element that activates expression of *Hoxb2* in rhombomeres 3 and 5. How the spatially organized expression of transcription factors such as Krox-20 is achieved is not yet known. In the case of *Hoxb4*, whose anterior boundary of expression is at rhombomere 6, there is evidence that it is induced and localized by signaling both within the neural tube and from adjacent somites. There is also a role for retinoic acid in patterning the hindbrain, as an absence of retinoic acid leads to loss of hindbrain rhombomeres.

An experiment showing that Hox genes determine cell behavior in the rhombomeres comes from the misexpression of *Hoxb1* in an anterior rhombomere. Motor axons from a pair of rhombomeres project to a single branchial arch; rhombomere r2 projects to the first branchial arch while r4 projects to the second arch. *Hoxb1* is expressed in r4 but not in r2. If *Hoxb1* is misexpressed in an r2 rhombomere, this then sends axons to the second arch.

Gene knock-outs in mice have also shown that the Hox genes are involved in patterning of the hindbrain region, though the results are not always easy to interpret; knock-out of a particular Hox gene can affect different populations of neural crest cells in the same animal, such as those that will form neurons and those that will form skeletal structures. Knock-out of the *Hoxa2* gene, for example, results in skeletal defects in that region of the head corresponding to the normal domain of expression of the gene which extends from rhombomere 3 backward. Segmentation itself is not affected, but the skeletal elements in the second branchial arch, all of which come from neural crest cells derived from rhombomere 4, are abnormal. The usual elements, such as the stapes of the inner ear, are absent, but instead some of the skeletal elements normally formed by the first arch develop, such as Meckel's cartilage, which is a precursor element in the lower jaw. Thus, suppression of *Hoxa2* causes a partial homeotic transformation of one segment into another.

These observations, together with those described earlier in this chapter, show that during gastrulation the cells of vertebrates acquire positional values along the antero-posterior axis, and that this positional identity is encoded by the genes of the Hox complexes. Many of the anatomical differences between vertebrates are probably simply due to differences in the subsequent targets of Hox gene actions, which result in the emergence of different but homologous skeletal structures—the mammalian jaw or the bird's beak, for example. These principles will be elaborated further in Chapter 5, which deals with the development of the fruit fly *Drosophila*, the animal in which Hox-like genes were first identified, and the concept of their role in regional specification first formulated.

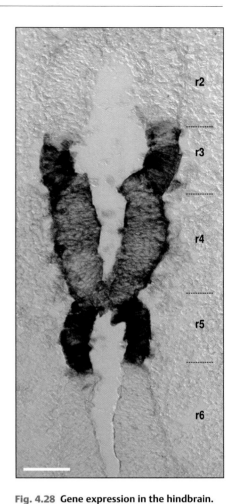

r2

r3

r4

r5

r6

Fig. 4.28 Gene expression in the hindbrain. The photograph shows a coronal section through the hindbrain of a 9½ days post-coitum mouse embryo, which is transgenic for two reporter constructs. The first construct contains the *lacZ* gene under the control of an enhancer from *Hoxb2*, which directs expression in rhombomeres 3 and 5 (revealed as blue staining). The second construct contains an alkaline phosphatase gene under the control of an enhancer from *Hoxb1*, which directs expression in rhombomere 4 (revealed as brown staining). A similar enhancer directing expression in rhombomere 4 exists for *Hoxb2*. Anterior is uppermost, and the positions of five of the rhombomeres are indicated (r2 to r6). Scale bar = 0.1 mm.

Photograph courtesy of J. Sharpe, from Lumsden, A., Krumlauf, R.: 1996.

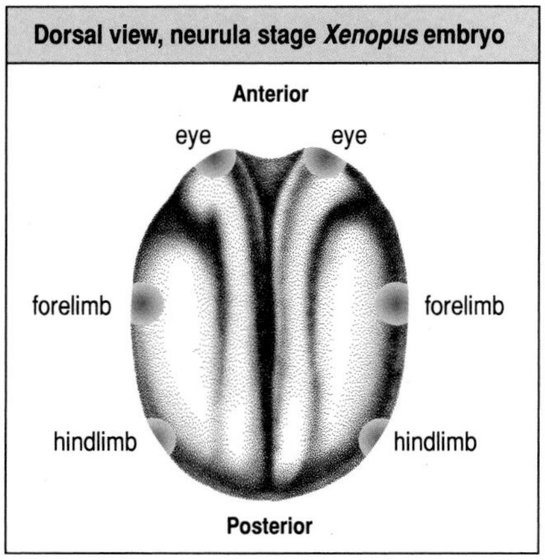

Fig. 4.29 The *Xenopus* embryo has become regionalized by the neurula stage. Various organs such as limbs, heart, and eyes will develop from specific regions (red) of the neurula after gastrulation is complete. Some of these regions, like the limb buds, are already determined at this stage and will not form any other structure. The boundaries of the regions are not sharply defined, however, and within each region or 'field' considerable regulation is still possible.

4.13 The embryo is patterned by the neurula stage into organ-forming regions that can still regulate

At the neurula stage, the body plan has been established and the regions of the embryo that will form limbs, eyes, heart, and other organs have become determined (Fig. 4.29).

This contrasts sharply with the blastula stage, at which time such determination has not yet occurred. The basic vertebrate phylotypic body plan is thus established during gastrulation. But although the positions of various organs are fixed, there is no overt sign yet of differentiation. Numerous grafting experiments have shown that the potential to form a given organ is now confined to specific regions. Each of these regions has, however, considerable capacity for regulation, so that if part of the region is removed a normal structure can still form. For example, the region of the neurula that will form a forelimb will, when transplanted to a different region, still develop into a limb. If part of a future limb region is removed, the remaining part can still regulate to develop a normal limb.

Summary

Patterning along both the antero-posterior and dorso-ventral axes is closely related to the action of the Spemann organizer and its morphogenesis during gastrulation. When grafted to the ventral side of an early gastrula, the Spemann organizer induces both a new dorso-ventral axis and a new antero-posterior axis, with the development of a second twinned embryo. In chick development, Hensen's node serves a function similar to that of the Spemann organizer, and it too can specify a new antero-posterior axis. In the mouse, the anterior visceral endoderm is necessary for forebrain induction.

The vertebrate nervous system, which forms from the neural plate, is induced both by early signals and by cells that give rise to the mesoderm and come to lie beneath prospective neural plate ectoderm during gastrulation. For those molecules that can induce neural tissue in *Xenopus*, such as the noggin protein, induction of neural differentiation is due to inhibition of BMP activity. Patterning of the neural plate, which includes the movement of signals within the plate itself, can partly be accounted for by a two-signal model: the ectoderm is first specified as anterior neural tissue and then a second set of signals, possibly graded, specifies more posterior structures.

The hindbrain is segmented into rhombomeres, with the cells of each rhombomere respecting their boundaries. Neural crest cells from the hindbrain populate specific mesodermal regions such as the branchial arches in a position-dependent fashion. A Hox gene code provides positional values for the rhombomeres and neural crest cells of the hindbrain region, while other genes specify more anterior regions. By the neurula stage, after gastrulation, the body plan has been established.

SUMMARY TO CHAPTER 4

The germ layers specified during blastula formation become patterned along the antero-posterior and dorso-ventral axes during gastrulation. The Spemann organizer in amphibians and its counterpart, Hensen's node, in chick and mouse embryos, are involved in the initial patterning that underlies the regionalization of the antero-posterior axis. Positional identity of cells along the antero-posterior axis is encoded by the combinatorial expression of genes of the four Hox complexes, which provide a code for regional identity. There is both spatial and temporal co-linearity between the order of Hox genes on the chromosomes and the order in which they are expressed along the antero-posterior axis of the embryo from the hindbrain backwards. Inactivation or overexpression of Hox genes can lead both to localized abnormalities and to homeotic transformations of one 'segment' of the axis into another, indicating that these genes are crucial in specifying regional identity. At the end of gastrulation, the basic body plan has been laid down and the nervous system induced. Specific regions of each somite give rise to cartilage, muscle, and dermis, and these regions are specified by signals from the notochord, neural tube, and epidermis. Induction and patterning of the nervous system involves both signals in the early embryo and from the underlying mesoderm, and planar signals arising within the neural plate itself. In the hindbrain, Hox gene expression provides positional values for both neural tissue and neural crest cells.

Summary: patterning of the vertebrate axial body plan

gastrulation and Spemann organizer activity

⇩

the Hox gene complexes are expressed along the antero-posterior axis

⇩

Hox gene expression establishes positional identity for mesoderm, endoderm, and ectoderm

mesoderm develops into notochord, somites, and lateral plate mesoderm

early signals and mesoderm induce neural plate from ectoderm

⇩

somites receive signals from notochord, neural tube, and ectoderm

mesoderm and planar ectodermal signals give regional identity to neural tube

somite develops into sclerotome and dermomyotome

rhombomeres and neural crest in the hindbrain are characterized by regional patterns of Hox gene expression

GENERAL REFERENCE

Duboule, D. (ed.): *Guidebook to the Homeobox Genes.* Oxford: Oxford University Press, 1994.

SECTION REFERENCES

4.1 Somites are formed in a well-defined order along the antero-posterior axis

Artavanis-Tsakonas, S., Rand, M.D., Lake, R.J.: **Notch signaling: cell fate control and signal integration in development.** *Science* 1999, **284**: 770–776.

Dale, K.J., Pourquié, O.: **A clockwork somite.** *BioEssays* 2000, **22**: 72–83.

Dubrulle, J., McGrew, M.J., Pourquié, O.: **FGF signaling controls somite boundary position and regulates segmentation clock control of spatiotemporal Hox gene activation.** *Cell* 2001, **106**: 219–232.

Jiang, Y-J., Aerne, B.L., Smithers, L., Hadon, C., Ish-Horowicz, D., Lewis, J.: **Notch signalling and the synchronization of the segmentation clock.** *Nature* 2000, **408**: 475–478.

Kieny, M., Mauger, A., Sengel P.: **Early regionalization of somitic mesoderm as studied by the development of the axial skeleton of the chick embryo.** *Dev. Biol.* 1972, **28**: 142–161.

Vasiliauskas, D., Stern, C.D.: **Patterning the embryonic axis:FGF signaling and how vertebrate embryos measure time.** *Cell* 2001, **106**: 133–136

4.2 The fate of somite cells is determined by signals from the adjacent tissues

Brand-Saberi, B., Christ, B.: **Evolution and development of distinct cell lineages derived from somites.** *Curr. Topics Dev. Biol.* 2000, **48**: 1–42.

Cossu, G., Tajbakhsh, S., Buckingham, M.: **How is myogenesis initiated in the embryo?** *Trends Genet.* 1996, **12**: 218–223.

Docktor, J., Ordahl, C.R.: **Dorsoventral axis determination in the somite: a re-examination.** *Development* 2000, **127**: 2201–2206.

Fan, C.M., Porter, J.A., Chiang, C., Chang, D.T., Beachy, P.A., Tessier-Lavigne, M.: **Long range sclerotome induction by Sonic hedgehog: direct role of the amino-terminal cleavage product and modulation by the cyclic AMP signaling pathway.** *Cell* 1995, **81**: 457–465.

Gossler, A., de Angelis, M.H.: **Somitogenesis.** *Curr. Topics Dev. Biol.* 1998, **38**: 225–287.

Huang, R., Zhi, Q., Patel, K., Wilting, J., Christ, B.: **Dual origin and segmental organisation of the avian scapula.** *Development* 2000, **127**: 3789–3794.

Mumm, J.S., Kopan, R.: **Notch signaling from the outside in.** *Dev. Biol.* 2000, **228**: 151–165.

Olivera-Martinez, I., Coltey, M., Dhouailly, D., Pourquié, O.: **Mediolateral somitic origin of ribs and dermis determined by quail-chick chimeras.** *Development* 2000, **127**: 4611–4617.

Pourquié, O., Fan, C-M., Coltey, M., Hirsinger, E., Watanabe, Y., Bréant, C., Francis-West, P., Brickell, P., Tessier-Lavigne, M., Le Douarin, N.M.: **Lateral and axial signals involved in avian somite patterning: a role for BMP-4.** *Cell* 1996, **84**: 461–471.

Selleck, M., Stern, C.D.: **Fate mapping and cell lineage analysis of Hensen's node in the chick embryo.** *Development* 1991, **112**: 615–626.

4.3 Positional identity of somites along the antero-posterior axis is specified by Hox gene expression

Burke, A.C., Nelson, C.E., Morgan, B.A., Tabin, C.: **Hox genes and the evolution of vertebrate axial morphology.** *Development* 1995, **121**: 333–346.

Duboule, D.: **Temporal colinearity and the phylotypic progression: a basis for the stability of a vertebrate Bauplan and the evolution of morphologies through heterochrony.** *Development* (Suppl.) 1994, 35–142.

Duboule, D.: **Vertebrate Hox gene regulation: clustering and/ or colinearity?** *Curr. Opin. Genet. Dev.* 1998, **8**: 514–518.

Godsave, S., Dekker, E.J., Holling, T., Pannese, M., Boncinelli, E., Durston, A.: **Expression patterns of Hoxb in the *Xenopus* embryo suggest roles in antero-posterior specification of the hindbrain and in dorso-ventral patterning of the mesoderm.** *Dev. Biol.* 1994, **166**: 465–476.

Hunt, P., Krumlauf, R.: **Hox codes and positional specification in vertebrate embryonic axes.** *Annu. Rev. Cell Biol.* 1992, **8**: 227–256.

Kondo, T., Duboule, D.: **Breaking colinearity in the mouse HoxD complex.** *Cell* 1999, **97**: 407–417.

Krumlauf, R.: **Hox genes in vertebrate development.** *Cell* 1994, **78**: 191–201.

McGinnis, W., Krumlauf, R.: **Homeobox genes and axial patterning.** *Cell* 1992, **68**: 283–302.

Nowicki, J.L., Burke, A.C.: *Hox* genes and morphological identity: axial versus lateral patterning in the vertebrate mesoderm. *Development* 2000, **127**: 4265–4275.

Vasiliauskas, D., Stern, C.D.: **Patterning the embryonic axis: FGF signaling and how vertebrate embryos measure time.** *Cell* 2001, **106**: 133–136.

4.4 Deletion or overexpression of Hox genes causes changes in axial patterning

Condie, B.G., Capecchi, M.R.: **Mice with targeted disruptions in the paralogous genes *Hoxa3* and *Hoxd3* reveal synergistic interactions.** *Nature* 1994, **370**: 304–307.

Duboule, D.: **Vertebrate Hox genes and proliferation: an alternative pathway to homeosis?** *Curr. Opin. Genet. Dev.* 1995, **5**: 525–528.

Favier, B., Le Meur, M., Chambon, P., Dollé, P.: **Axial skeleton homeosis and forelimb malformations in *Hoxd11* mutant mice.** *Proc. Natl Acad. Sci. USA* 1995, **92**: 310–314.

Grammatopoulos, G.A., Bell, E., Toole, L., Lumsden, A., Tucker, A.S.: **Homeotic transformation of branchial arch identity after hoxa2 overexpression.** *Development* 2000, **127**: 5355–5365.

Jegalian, B.G., De Robertis, E.M.: **Homeotic transformations in the mouse induced by overexpression of a human Hox 3.3 transgene.** *Cell* 1992, **71**: 901–910.

Le Mouellic, H., Lallemand, Y., Brulet, P.: **Homeosis in the mouse induced by a null mutation in the *Hox 3.1* gene.** *Cell* 1992, **69**: 251–264.

Rossant, J., Spence, A.: **Chimeras and mosaics in mouse mutant analysis.** *Trends Genet.* 1998, **14**: 358–363.

4.5 Retinoic acid can alter positional values

Conlon, R.A.: **Retinoic acid and pattern formation in vertebrates.** *Trends Genet.* 1995, **11**: 314–319.

Kessel, M., Gruss, P.: **Homeotic transformations of moving vertebrae and concomitant alteration of the codes induced by retinoic acid.** *Cell* 1991, **67**: 89–104.

Ruiz-i-Altaba, A., Jessell, T.: **Retinoic acid modifies mesodermal patterning in early *Xenopus* embryos.** *Genes Dev.* 1991, **5**: 175–187.

Sive, H.L., Cheng, P.F.: **Retinoic acid perturbs the expression of *Xhox.lab* genes and alters mesodermal determination in *Xenopus laevis*.** *Genes Dev.* 1991, **5**: 1321–1332.

4.6 The vertebrate head is specified by signals different from those for the trunk

Beddington, R.S.P., Robertson, E.H.: **Axis development and early asymmetry in mammals.** *Cell* 1999, **96**: 195–209.

Brickman, J.M., Jones, C.M., Clements, M., Smith, J.C., Beddington, R.S.P.: **Hex is a transcriptional repressor that contributes to anterior identity and suppresses Spemann organiser function.** *Development* 2000, **127**: 2303–2315.

Gad, J.M., Tam, P.P.L.: **The mouse becomes a dachshund.** *Curr. Biol.* 1999, **9**: R783–R786.

Griffin, K., Patient, R., Holder, N.: **Analysis of FGF function in normal and no tail zebrafish embryos reveals separate mechanisms for formation of the trunk and tail.** *Development* 1995, **121**: 2983–2994.

Jones, C.M., Broadbent, J., Thomas, P.Q., Smith, J.C., Beddington, R.S.P.: **An anterior signalling centre in *Xenopus* revealed by the homeobox gene XHex.** *Curr. Biol.* 1999, **9**: 946–954.

Joubin, K., Stern, C.D.: **Molecular interactions continuously define the organizer during the cell movements of gastrulation.** *Cell* 1999, **98**: 559–571.

Niehrs, C.: **Head in the Wnt: the molecular nature of Spemann's head organizer.** *Trends Genet.* 1999, **15**: 314–315.

Piccolo, S., Agius, E., Leyns, L., Bhattacharya, S., Grunz, H., Bouwmeester, T., De Robertis, E.M.: **The head inducer Cerberus is a multifunctional antagonist of Nodal, BMP and Wnt signals.** *Nature* 1999, **397**: 707–710.

Schneider, V.A., Mercola, M.: **Spatially distinct head and heart inducers within the *Xenopus* organizer region.** *Curr. Biol.* 1999, **9**: 800–809.

Yamaguchi, T.P.: **Heads or tails: Wnts and anterior–posterior patterning.** *Curr. Biol.* 2001, **11**: R713–R724.

4.7 The neural plate is induced in the ectoderm

Bachiller, D., Klingensmith, J., Kemp, C., Belo, J.A., Anderson, R.M., May, S.R., McMahon, J.A., McMahon, A.P., Harland, R.M., Rossant, J., De Robertis, E.M.: **The organizer factors Chordin and Noggin are required for mouse forebrain development.** *Nature* 2000, **403**: 658–661.

Harland, R.: **Neural induction.** *Curr. Opin. Genet. Dev.* 2000, **10**: 357–362.

Kemmati-Brivanlou, A., Melton, D.: **Vertebrate embryonic cells will become nerve cells unless told otherwise.** *Cell* 1997, **88**: 13–17.

Sasai, Y., Lu, B., Steinbesser, H., De Robertis, E.M.: **Regulation of neural induction by the Chd and BMP-4 antagonistic patterning signals in *Xenopus*.** *Nature* 1995, **376**: 333–336.

Streit, A., Stern, C.D.: **Neural induction: a bird's eye view.** *Trends Genet.* 1999, **15**: 20–24.

Streit, A., Berliner, A.J., Papanayotou, C., Sirulnik, A., Stern, C.D.: **Initiation of neural induction by FGF signalling before gastrulation.** *Nature* 2000, **406**: 74–78.

Wilson, P., Kemmati-Brivanlou, A.: **Induction of epidermis and inhibition of neural fate by BMP-4.** *Nature* 1995, **376**: 331–333.

Wilson, S.L., Rydström, A., Trimborn, T., Willert, K., Nusse, R., Jessell, T. M., Edlund, T.: **The status of Wnt signaling regulates neural and epidermal fates in the chick embryo.** *Nature* 2001, **411**: 325–329.

4.8 The nervous system can be patterned by signals from the mesoderm

Ang, S.L., Rossant, J.: **HNF-3b is essential for node and notochord formation in mouse development.** *Cell* 1994, **78**: 561–574.

Brocolli, V., Boncinelli, E., Wurst, W.: **The caudal limit of *Otx2* expression positions the isthmaic organizer.** *Nature* 1999, **401**: 164–168.

Doniach, T.: **Basic FGF as an inducer of antero-posterior neural pattern.** *Cell* 1995, **85**: 1067–1070.

Foley, A.C., Skromne, I., Stern, C.D.: **Reconciling different models of forebrain induction and patterning: a dual role for the hypoblast.** *Development* 2000, **127**: 3839–3854.

Kelly, O.G., Melton, D.A.: **Induction and patterning of the vertebrate nervous system.** *Trends Genet.* 1995, **11**: 273–278.

Kintner, C.R., Dodd, J.: **Hensen's node induces neural tissue in *Xenopus* ectoderm. Implications for the action of the organizer in neural induction.** *Development* 1991, **113**: 1495–1505.

Sasai, Y., De Robertis, E.M.: **Ectodermal patterning in vertebrate embryos.** *Dev. Biol.* 1997, **182**: 5–20.

Sharman, A.C., Brand, M.: **Evolution and homology of the nervous system: cross-phylum rescues of *otd/Otx* genes.** *Trends Genet.* 1998, **14**: 211–214.

Stern, C.D.: **Initial patterning of the central nervous system: how many organizers?** *Nature Rev. Neurosci.* 2001, **2**: 92–98.

Storey, K., Crossley, J.M., De Robertis, E.M., Norris, W.E., Stern, C.D.: **Neural induction and regionalization in the chick embryo.** *Development* 1992, **114**: 729–741.

4.9 Signals that pattern the neural plate may travel within the neural plate itself

Doniach, T., Phillips, C.R., Gerhart, J.C.: **Planar induction of antero-posterior pattern in the developing central nervous system of *Xenopus laevis*.** *Science* 1992, **257**: 542–545.

Ruiz-i-Altaba, A., Melton, D.: **Interaction between peptide growth factors and homeobox genes in the establishment of antero-posterior polarity in frog embryos.** *Nature* 1989, **341**: 33–38.

Sive, H.L., Hattori, K., Weintraub, H.: **Progressive determination during formation of antero-posterior axis in *Xenopus laevis*.** *Cell* 1989, **58**: 171–180.

4.10 The hindbrain is segmented into rhombomeres by boundaries of cell-lineage restriction

Klein, R.: **Neural development: bidirectional signals establish boundaries.** *Curr. Biol.* 1999, **9**: R691–R694.

Lumsden, A.: **Cell lineage restrictions in the chick embryo hindbrain.** *Phil. Trans. Roy. Soc. Lond. B* 1991, **331**: 281–286.

Xu, Q., Mellitzer, G., Wilkinson, D.G.: **Roles of Eph receptors and ephrins in segmental patterning.** *Phil. Trans. Roy. Soc. B* 2000, **355**: 993–1002.

4.11 Neural crest cells have positional values

Keynes, R., Lumsden, A.: **Segmentation and the origin of regional diversity in the vertebrate central nervous system.** *Neuron* 1990, **4**: 1–9.

4.12 Hox genes provide positional identity in the hindbrain region

Bell, E., Wingate, R.J., Lumsden, A.: **Homeotic transformation of rhombomere identity after localized Hoxb1 misexpression.** *Science* 1999, **284**: 2168–2171.

Grapin-Botton, A., Bonnin, M-A., McNaughton, L.A., Krumlauf, R., Le Douarin, N.M: **Plasticity of transposed rhombomeres: Hox gene induction is correlated with phenotypic modifications.** *Development* 1995, **121**: 2707–2721.

Hunt, P., Krumlauf, R.: **Hox codes and positional specification in vertebrate embryonic axes.** *Annu. Rev. Cell Biol.* 1992, **8**: 227–256.

Krumlauf, R.: **Hox genes and pattern formation in the branchial region of the vertebrate head.** *Trends Genet.* 1993, **9**: 106–112.

Nonchev, S., Maconochie, M., Vesque, C., Aparicio, S., Ariza-McNaughton, L., Manzanares, M., Maruthainar, K., Kuroiwa, A., Brenner, S., Charnay, P., Krumlauf, R.: **The conserved role of *Krox-20* in directing *Hox* gene expression during vertebrate hindbrain segmentation.** *Proc. Natl Acad. Sci. USA* 1996, **93**: 9339–9345.

Rijli, F.M., Mark, M., Lakkaraju, S., Dierich, A., Dolle, P., Chambon, P.: **A homeotic transformation is generated in the rostral branchial region of the head by disruption of *Hoxa2*, which acts as a selector gene.** *Cell* 1993, **75**: 1333–1349.

4.13 The embryo is patterned by the neurula stage into organ-forming regions that can still regulate

De Robertis, E.M., Morita, E.A., Cho, K.W.Y.: **Gradient fields and homeobox genes.** *Development* 1991, **112**: 669–678.

Development of the *Drosophila* body plan

5

- Maternal genes set up the body axes
- Polarization of the body axes during oogenesis
- Zygotic genes pattern the early embryo
- Segmentation: activation of the pair-rule genes
- Segment polarity genes and compartments
- Segmentation: selector and homeotic genes

> "From the beginning we knew where we were even though there were no boundaries between us. Later, we divided into separate groups and were given a proper address."

We are much more like flies in our development than you might think. Astonishing discoveries in developmental biology over the past ten years have revealed that many of the genes that control the development of the fruit fly *Drosophila* are similar to those controlling development in vertebrates, and indeed in many other animals. It seems that once evolution had found a satisfactory way of patterning animal bodies, it tended to use the same mechanisms and molecules over and over again with, of course, some important modifications.

Drosophila is the best understood of all developmental systems, especially at the genetic level, and although it is an invertebrate it has had an enormous impact on our understanding of the genetic basis of vertebrate development. We have already seen this with the *Hox* genes (see Box 4A, p. 117), which were first discovered in *Drosophila*. The pre-eminent place of *Drosophila* in modern developmental biology was recognized by the award of the 1995 Nobel Prize for Physiology or Medicine for work that led to a fundamental understanding of how genes control development in the fly embryo. This was only the second time that the Nobel Prize had been awarded for work in developmental biology. Many questions that still remain unanswered in vertebrates have been solved at the molecular level for *Drosophila*; these include the mechanisms of axis determination in the egg, and the identification and mechanism of action of key signaling circuits and transcriptional regulators in pattern formation. While insect and vertebrate development may seem to be very different, much has been learnt that can be applied to vertebrate development; indeed, many of the key genes in vertebrate development were originally identified as developmental genes in *Drosophila*. The protein-coding genome of *Drosophila* has been sequenced. It contains about 13,600 genes, surprisingly only twice the number in yeast and fewer than the 19,000 genes of the nematode *Caenorhabditis*.

Drosophila, like many other insects, hatches from the egg as a larva,

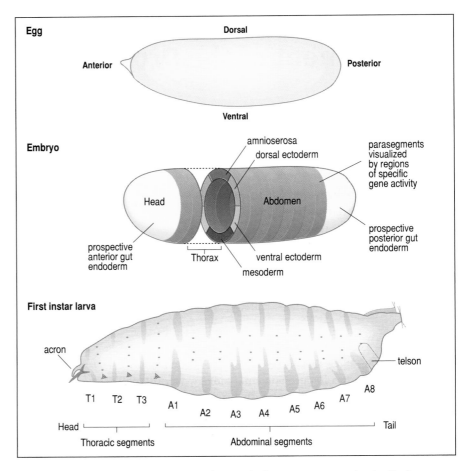

Fig. 5.1 Patterning of the *Drosophila* embryo. The body plan is patterned along two distinct axes. The antero-posterior and dorso-ventral axes are at right angles to each other and are laid down in the egg. In the early embryo, the dorso-ventral axis is divided into four regions: mesoderm (red), ventral ectoderm (yellow), dorsal ectoderm (orange), and amnioserosa (an extra-embryonic membrane, green). The ventral ectoderm gives rise to both ventral epidermis and neural tissue, the dorsal ectoderm to epidermis. The antero-posterior axis becomes divided into different regions that later give rise to the head, thorax, and abdomen. After the initial division into broad body regions, segmentation begins. The future segments can be visualized as transverse stripes by staining for specific gene activity; these stripes demarcate 14 parasegments, 10 of which are marked. The embryo develops into a segmented larva. By the time the larva hatches, the 14 parasegments have been converted into thoracic (T1–T3) and abdominal (A1–A8) segments, which are offset from the parasegments by one half segment. Different segments are distinguished by the patterns of bristles and denticles on the cuticle. Specialized structures, the acron and telson, develop at the head and tail ends, respectively.

which grows and subsequently undergoes metamorphosis into the adult (see Fig. 2.29). In this chapter, we look at how the basic body plan of the *Drosophila* larva is established. We see how the antero-posterior and dorso-ventral axes are determined, how the embryo becomes divided into a series of segments each with its unique identity, and how the mesoderm and ectoderm become specified. The first half of the chapter concentrates on the development of the embryo up to the stage at which it becomes segmented. In the second half of the chapter, we consider how the segments are patterned and acquire their unique identities. We leave until Chapter 10 the development of the imaginal discs—groups of cells set aside in the embryo that eventually give rise to adult structures such as wings and legs at metamorphosis. The imaginal discs provide continuity between the pattern of the larval body and that of the adult, even though the processes of metamorphosis intervene.

Like all animals with bilateral symmetry, the *Drosophila* larva is patterned along two distinct and largely independent axes: the antero-posterior and dorso-ventral axes, which are at right angles to each other. Along the antero-posterior axis the larva appears regularly segmented, and is divided into several broad anatomical regions. At the anterior end is the head, behind which are three thoracic segments followed by eight abdominal segments (Fig. 5.1). Each segment has its own unique character, as revealed by both its external cuticular structure and its internal organization. At each end of the larva are specialized structures—the acron at the head end and the telson at the tail end.

Early in embryogenesis, the dorso-ventral axis becomes divided up into four regions, which give rise to the dorso-ventral organization of the larval body. Organization along the antero-posterior and dorso-ventral axes of the early embryo develops more or less simultaneously, but is specified by independent mechanisms and by different sets of genes in each axis.

Early *Drosophila* development is peculiar to insects, as early patterning occurs within a multinucleate **syncytial blastoderm** formed by repeated rounds of nuclear division without any corresponding cytoplasmic division (see Fig. 2.30). Only after the beginning of segmentation does the embryo become truly multicellular. The absence of cells in the early *Drosophila* embryo represents an important difference from other organisms. At the syncytial stage, the whole embryo can be considered as a multinucleate single cell, and many proteins, including those that are not normally secreted from cells, such as transcription factors, can diffuse throughout the blastoderm and enter the nuclei. Concentration gradients, which can provide positional information for the nuclei, can thus be set up (see Section 1.13).

Early development is essentially two-dimensional; patterning occurs mainly in the blastoderm, the superficial layer of the embryo that consists first of nuclei and later of cells. But the larva is a three-dimensional object, with internal structures. This third dimension develops later, at gastrulation, when parts of the surface layer move into the interior to form the gut, the mesodermal structures that will give rise to muscle, and the ectodermally derived nervous system.

We start this chapter by looking at the appearance of the first level of antero-posterior and dorso-ventral organization in the syncytial embryo, before returning to the formation of the *Drosophila* egg to show how the positional information that organizes the early embryo is originally laid down in the developing oocyte by the mother.

Maternal genes set up the body axes

The earliest stage of *Drosophila* development is guided by preformed mRNAs and proteins that are synthesized and laid down in the egg by the mother fly. Several of these become localized at the ends of the egg while it is being formed in the ovary. The genes responsible for this maternal contribution are known as **maternal genes** as they must be expressed by the mother and not by the embryo; they are expressed in the tissues of the ovary during oogenesis. By contrast, **zygotic genes** are those required during the

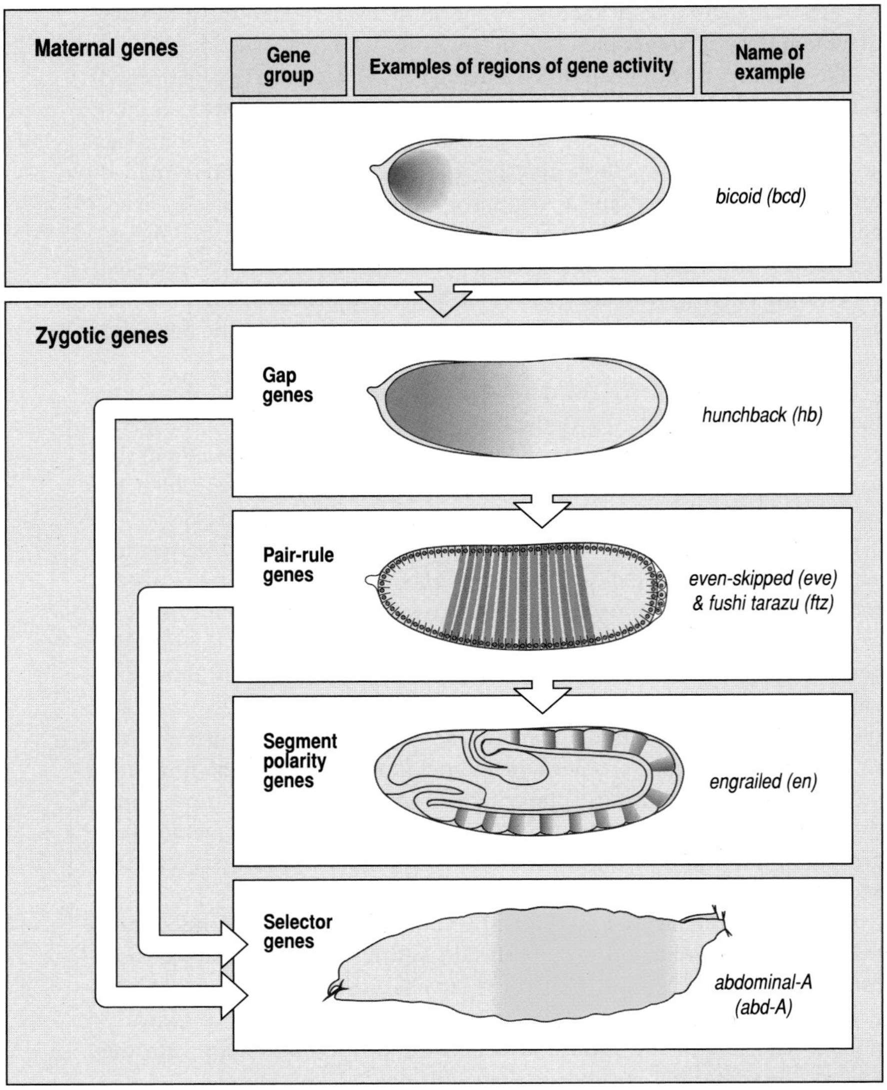

Fig. 5.2 The sequential expression of different sets of genes establishes the body plan along the antero-posterior axis. After fertilization, maternal gene products laid down in the egg, such as *bicoid* mRNA, are translated. They provide positional information which activates the zygotic genes. The four main classes of zygotic genes acting along the antero-posterior axis are the gap genes, the pair-rule genes, the segment polarity genes, and the selector, or homeotic, genes. The gap genes define regional differences that result in the expression of a periodic pattern of gene activity by the pair-rule genes, which define the parasegments and foreshadow segmentation. The segment polarity genes elaborate the pattern in the segments, and segment identity is determined by the selector genes. The functions of each of these classes of genes are discussed in this chapter.

development of the embryo; they are expressed in the nuclei of the embryo itself.

About 50 maternal genes in all are involved in setting up the two axes and a basic framework of positional information, which is then interpreted by the embryo's own genetic program. All later patterning, which involves expression of the zygotic genes, is built on this framework (Fig. 5.2). Maternal gene products establish the axes and set up regional differences along each axis in the form of spatial distributions of RNA and proteins. These proteins then activate zygotic genes in the nuclei at particular positions along both axes for the next round of patterning. The sequential activities of the maternal and zygotic genes pattern the embryo in a series of steps. Broad regional differences are established first, and these are then refined to produce a larger number of smaller developmental domains, each characterized by a unique profile of zygotic gene activity. Developmental genes act in a strict temporal sequence. They form a hierarchy of gene activity in which the action of one set of genes is essential for another set of genes to be activated, and thus for the next stage of development to occur. We first look at how maternal gene products specify the antero-posterior axis.

5.1 Three classes of maternal genes specify the antero-posterior axis

Maternal gene expression creates differences in the egg along the antero-posterior axis even before the egg is fertilized. These differences distinguish the future head and posterior ends of the adult. Maternal genes are identified by mutations that, when present in the mother, do not damage her but have effects on the development of her progeny. The roles of the maternal genes can be deduced from the effects of these **maternal-effect mutations** on the larva. The mutations fall into three classes: those that affect anterior regions; those that affect posterior regions; and those that affect both the terminal regions (Fig. 5.3). Mutations of genes in the anterior class, such as *bicoid*, lead to a reduction or loss of head and thoracic structures, and in some cases, their replacement with posterior structures. Posterior group mutations, such as *nanos*, cause the loss of abdominal regions, leading to a smaller than normal embryo; and those of the terminal class, such as *torso*, affect the specialized structures—the acron and telson—at the head and tail ends of the embryo. Each class of genes acts more or less independently of the others. The apparently idiosyncratic naming of genes in *Drosophila* usually reflects the attempts by the discoverer to describe the mutant phenotype. In this chapter we meet quite a number of gene names; all these are listed, together with their functions where known, in the table

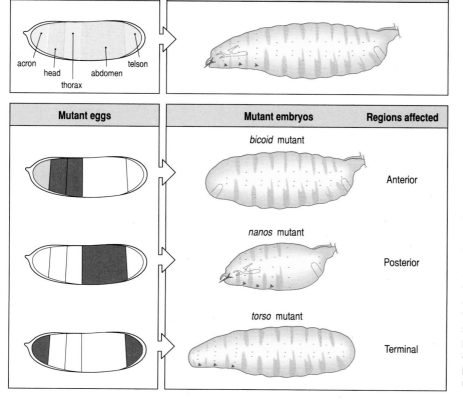

Fig. 5.3 The effects of mutations in the maternal gene system. Mutations in maternal genes lead to deletions and abnormalities in anterior, posterior, or terminal structures. The wild-type fate map shows which regions of the egg give rise to particular regions and structures in the larva. Regions that are affected in mutant eggs and which lead to lost or altered structures in the larva are shaded in red. In *bicoid* mutants there is a partial loss of anterior structures and the appearance of a posterior structure—the telson—at the anterior end. *nanos* mutants lack a large part of the posterior region. *torso* mutants lack both acron and telson.

at the end of this chapter. Of the 50 or so maternal genes, the products of four in particular—*bicoid, nanos, hunchback,* and *caudal*—become distributed along the antero-posterior axis and are crucial in establishing it.

Some of the maternal genes exert their effect on the follicle cells in the mother's ovary, which are derived from the mesoderm of the mother and form a bag which contains the oocyte and its nurse cells. We shall return to the crucial role of the follicle cells in patterning the oocyte later in this chapter.

5.2 The *bicoid* gene provides an antero-posterior morphogen gradient

In the unfertilized egg, *bicoid* mRNA is localized at the anterior end. After fertilization it is translated, and the bicoid protein diffuses from the anterior end and forms a concentration gradient along the antero-posterior axis. This provides the positional information required for further patterning along this axis. Historically, the bicoid protein gradient provided the first concrete evidence for the existence of the morphogen gradients in animals that had been postulated to control pattern formation (see Section 1.13).

The role of the *bicoid* gene was first elucidated by a combination of genetic and physical experiments on the *Drosophila* embryo. Female flies lacking *bicoid* gene expression produce embryos that have disrupted anterior segments and thus have no proper head and thorax (see Fig. 5.3). They also

Fig. 5.4 The *bicoid* gene is necessary for the development of anterior structures. Embryos whose mothers lack the *bicoid* gene lack anterior regions (second row). Transfer of anterior cytoplasm from wild-type embryos to *bicoid* mutant embryos causes some anterior structure to develop at the site of injection (third row). If wild-type anterior cytoplasm is transplanted to the middle of a *bicoid* mutant egg or early embryo, head structures develop at the site of injection, flanked on both sides by thoracic-type segments (fourth row). These results can be interpreted in terms of the anterior cytoplasm setting up a gradient of bicoid protein with the high point at the site of injection (see graphs, bottom left panel).

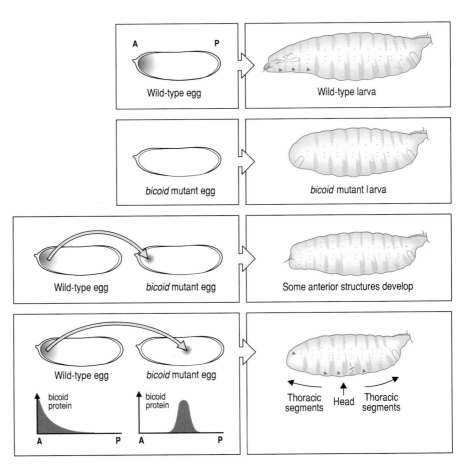

have a telson instead of an acron at the head end. In a separate line of investigation into the role of localized cytoplasmic factors in anterior development, normal eggs were pricked at the anterior ends and some cytoplasm allowed to leak out. The embryos that developed bore a striking resemblance to *bicoid* mutant embryos. This suggested that normal eggs have some factor(s) in the cytoplasm at their anterior end which is absent in *bicoid* mutant eggs. And indeed, *bicoid* mutant embryos can be partially rescued, in the sense that they will develop more normally, if anterior cytoplasm of wild-type embryos is injected into their anterior regions (Fig. 5.4 opposite). Moreover, if normal anterior cytoplasm is injected into the middle of a fertilized *bicoid* mutant egg, head structures develop at the site of injection and the adjacent segments become thoracic segments, setting up a mirror-image body pattern at the site of injection. The simplest interpretation of these experiments is that the *bicoid* gene is necessary for the establishment of the anterior structures because it establishes a gradient in some substance whose source and highest level are at the anterior end: this substance is the bicoid protein.

Using *in situ* hybridization (see Box 3B, p. 70), *bicoid* mRNA has been shown to be present in the anterior region of the unfertilized egg, where it is attached to the cytoskeleton. This mRNA is not translated until after fertilization. Staining with an antibody against the bicoid protein shows that the protein is absent from the unfertilized egg, but that after fertilization the mRNA is translated into protein which forms a gradient with the high point at the anterior end of the egg, at the site of its synthesis (Fig. 5.5). As the bicoid protein diffuses through the embryo, it also breaks down—it has a half-life of about 30 minutes—and this breakdown is important in establishing the antero-posterior concentration gradient.

The bicoid protein is a transcription factor and acts as a morphogen, as described in more detail in Section 5.11. It switches on certain zygotic genes at different threshold concentrations, so initiating a new pattern of gene expression along the axis. Thus, *bicoid* is a key maternal gene in early *Drosophila* development. The other maternal genes of the anterior group are mainly involved in the localization of *bicoid* mRNA to the anterior end of the egg during oogenesis and in the control of its translation.

5.3 The posterior pattern is controlled by the gradients of nanos and caudal proteins

For proper patterning along an axis both ends need to be specified, and bicoid protein defines only the anterior end of the antero-posterior axis. The posterior end is specified by the actions of at least nine maternal genes—the posterior group genes. Just as mutations in the *bicoid* gene result in larvae in which head and thoracic regions do not develop normally, mutations in the posterior group genes result in larvae in which abdominal development is abnormal. These mutant embryos are shorter than normal because there is no abdomen (see Fig. 5.3). One of the actions of the maternal posterior group genes, for example *oskar*, is to localize *nanos* mRNA at the extreme posterior pole of the unfertilized egg, as well as to specify the egg posterior germ plasm that gives rise to the germ cells. Like *bicoid* mRNA, *nanos* mRNA is translated after fertilization to give a concentration gradient of nanos protein, in this case with the highest level at the posterior end of the embryo.

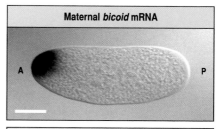

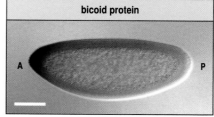

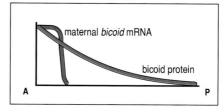

Fig. 5.5 The distribution of the maternal mRNA for *bicoid* in the egg and the gradient of bicoid protein after fertilization. Top panel: the mRNA is visualized by *in situ* hybridization. Middle panel: the bicoid protein is stained with a labeled antibody. Bottom panel: translation of *bicoid* mRNA and diffusion of bicoid protein from its site of synthesis produces an antero-posterior gradient of bicoid protein in the embryo. Scale bars = 0.1 mm.

Photographs courtesy of R. Lehmann, from Suzuki, D.T., et al.: 1996.

Fig. 5.6 Establishment of a maternal gradient in hunchback protein. Left panel: in the unfertilized egg, maternal *hunchback* mRNA (turquoise) is present at a relatively low level throughout the egg, whereas *nanos* mRNA (yellow) is located posteriorly. The photograph is an *in situ* hybridization showing the location of nanos mRNA (black). Right panel: after fertilization, *nanos* mRNA is translated and nanos protein blocks translation of *hunchback* mRNA in the posterior regions, giving rise to a shallow antero-posterior gradient in maternal hunchback protein. The photograph shows the graded distribution of nanos protein, detected with a labeled antibody.

Photographs courtesy of R. Lehmann, from Suzuki, D.T., et al.: 1996.

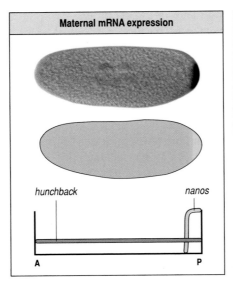

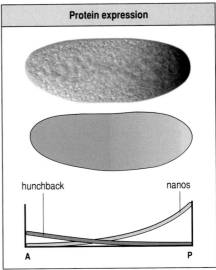

Unlike the bicoid protein, however, nanos protein does not act directly as a morphogen to specify the abdominal pattern. It has a quite different role. Its function is to suppress, in a graded way, the translation of the maternal mRNA of another gene, *hunchback*, so that a clear gradient of zygotically expressed hunchback protein can be subsequently established and act as a morphogen for the next stage of patterning. Maternal *hunchback* mRNA is uniformly distributed in the embryo, but the embryo's own *hunchback* genes are also activated at the anterior end of the embryo by the high concentrations of bicoid protein. To establish a clear antero-posterior gradient of hunchback protein, maternal *hunchback* translation has to be prevented, as otherwise there would be too high a concentration of hunchback protein in the posterior region. The nanos protein prevents maternal hunchback translation (Fig. 5.6), by binding to a complex of *hunchback* mRNA and the protein encoded by the posterior group gene *pumilio*. If maternal hunchback is completely removed from embryos, then nanos becomes completely dispensable, showing that it is needed only to remove the maternal hunchback protein.

The fourth maternal product crucial to establishing the posterior end of the axis is *caudal* mRNA. Like maternal *hunchback* mRNA, it is uniformly distributed throughout the egg at first. A posterior to anterior gradient of the caudal protein is established by inhibition of caudal protein synthesis by the bicoid protein. Because the concentration of bicoid protein is low at the posterior end of the embryo, caudal protein concentration is highest there. Mutations in the *caudal* gene result in abnormal development of abdominal segments.

Soon after fertilization, therefore, several gradients of maternal proteins have been established along the antero-posterior axis. Two gradients—bicoid and hunchback proteins—run in an anterior to posterior direction, while caudal protein is graded posterior to anterior. We next look at the quite different mechanism that specifies the two termini of the embryo.

5.4 The anterior and posterior extremities of the embryo are specified by cell-surface receptor activation

A third group of maternal genes specifies the structures at the extreme ends of the antero-posterior axis—the acron and the head region at the anterior end, and the telson and the most posterior abdominal segments at the posterior end. A key gene in this group is *torso*; mutations in *torso* can result in embryos developing neither acron nor telson (see Fig. 5.3). This indicates that the two terminal regions, despite their topographical separation, are not specified independently but use the same pathway.

The terminal regions are specified by an interesting mechanism that also involves a maternal gene product that has been localized to particular regions in the egg. Terminal specification is due to activation, at the two poles only, of a receptor protein encoded by the maternal gene *torso*, which then transmits a signal to the adjacent cytoplasm. The torso protein is uniformly distributed throughout the egg plasma membrane, but is only activated at the ends of the fertilized egg because the protein ligand, probably the trunk protein, that stimulates it is only present there. The ligand is laid down at these two sites in the vitelline envelope outside the egg plasma membrane during oogenesis.

Before fertilization, the ligand is immobilized within the vitelline envelope and cannot come into contact with the receptor. Only when development begins after fertilization is the ligand released into the perivitelline space where it can bind to torso protein. The ligand is only present in small quantities and so most becomes bound to torso at the poles, with little left to diffuse further away. In this way, a localized area of receptor activation is set up at each pole (Fig. 5.7). Stimulation of torso produces a signal that is transduced across the plasma membrane to the interior of the developing embryo. This signal directs the activation of zygotic genes at both poles, thus defining the two extremities of the embryo. The torso protein is one of a large group of transmembrane receptors known as receptor tyrosine kinases; they possess an intrinsic tyrosine protein kinase in the cytoplasmic portion of the receptor. This is activated when the extracellular part of the receptor binds its ligand, and transmits the signal onwards by phosphorylating cytoplasmic proteins.

The ingenious mechanism for setting up a localized area of receptor activation is not confined to determination of the terminal regions of the embryo, but is also used in setting up the dorso-ventral axis, which we consider next.

5.5 The dorso-ventral polarity of the egg is specified by localization of maternal proteins in the vitelline envelope

The dorso-ventral axis is specified by a different set of maternal genes from those that specify the anterior-posterior axis. But, like the antero-posterior axis, it is initially established in the follicle cells surrounding the unfertilized egg in the ovary. The ventral end of the axis is determined by the localized processing of a maternal protein, the spätzle protein, which is uniformly secreted into the extra-embryonic vitelline space by the follicle cells.

The key gene that defines dorso-ventral polarity in the follicle cells is *pipe*, which is transcribed into mRNA only in the follicle cells that surround the

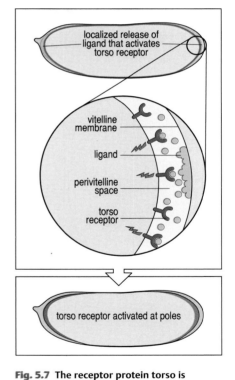

Fig. 5.7 The receptor protein torso is involved in specifying the terminal regions of the embryo. The receptor protein encoded by the gene *torso* is present throughout the egg plasma membrane. Its ligand is laid down in the vitelline membrane at each end of the egg during oogenesis. After fertilization, the ligand is released and diffuses across the perivitelline space to activate the torso protein at the ends of the embryo only.

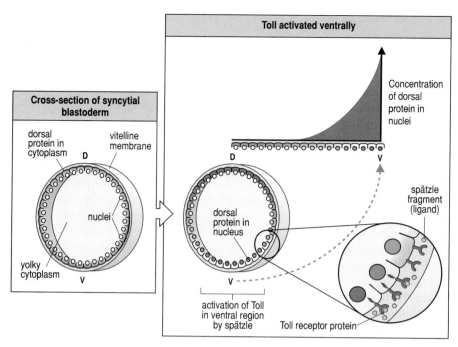

Fig. 5.8 Toll protein activation results in a gradient of intranuclear dorsal protein along the dorso-ventral axis. Before Toll protein is activated, the dorsal protein (red) is distributed throughout the peripheral band of cytoplasm. The Toll protein is a receptor that is only activated in the ventral region, by a maternally derived ligand (the spätzle fragment), which is processed in the perivitelline space after fertilization. The localized activation of Toll results in the entry of dorsal protein into nearby nuclei. The intranuclear concentration of dorsal protein is greatest in ventral nuclei, resulting in a ventral to dorsal gradient.

future ventral region—about one third—of the developing egg. *pipe* codes for a secreted enzyme that is part of the chain of interactions during egg development that includes proteins coded for by *easter, nudel, windbeutel* and at least seven other genes. This pathway leads eventually to the processing of the spätzle protein, which has been secreted uniformly by the follicle cells into the perivitelline space, to produce a 'spätzle fragment' which is localized in the perivitelline space on the ventral side of the egg only.

Like the ligand for the torso protein, the spätzle fragment is the localized ligand for a receptor protein that is distributed throughout the egg plasma membrane. In this case the receptor is the product of the maternal gene *Toll*. Because of the localized nature of the processing reactions, the spätzle protein fragment is only present in the ventral perivitelline space. Thus, it activates the receptor protein Toll only in the future ventral region of the embryo. Toll activation is greatest where the concentration of its ligand is highest, and falls off rapidly, probably due to the limited amount of ligand being mopped up by the receptors. Activation of Toll sends a signal to the adjacent cytoplasm of the embryo. At this stage, the embryo is still a syncytial blastoderm and this signal causes a maternal gene product in the adjoining cytoplasm—the dorsal protein—to enter nearby nuclei (Fig. 5.8). This protein, encoded by the *dorsal* gene, is a transcription factor with a vital role in organizing the dorso-ventral axis.

5.6 Positional information along the dorso-ventral axis is provided by the dorsal protein

The initial dorso-ventral organization of the embryo is established at right angles to the antero-posterior axis at about the same time that this axis is being divided into terminal, anterior, and posterior regions. The embryo initially becomes divided into four regions along the dorso-ventral axis (see

Fig. 5.1). Patterning along the dorso-ventral axis is controlled by the distribution of the maternal protein, dorsal.

Unlike the bicoid protein, the dorsal protein is uniformly distributed in the egg. Initially it is restricted to the cytoplasm, but under the influence of signals from the ventrally activated Toll proteins it enters nuclei in a graded fashion, with the highest concentration in ventral nuclei and the concentration progressively decreasing in a ventral to dorsal direction, as the Toll signal becomes weaker (see Fig. 5.8). The greater the activation of Toll by the spätzle protein, the more dorsal protein enters the nuclei. There is little or no dorsal protein in the nuclei in the dorsal regions of the embryo. The role of Toll was first established by the observation that embryos lacking it are strongly 'dorsalized'—that is, no ventral structures develop. In these embryos, the dorsal protein does not enter the nuclei but remains uniformly distributed in the cytoplasm. Transfer of wild-type cytoplasm into *Toll* mutant embryos results in specification of a new dorso-ventral axis, the ventral region always corresponding to the site of injection. The Toll present in the wild-type cytoplasm enters the membrane at the site of the injection of cytoplasm. In the absence of *Toll* expression in the mutant embryos, the spätzle fragments, initially produced only on the ventral side, diffuse throughout the perivitelline space because there is no Toll protein to bind them. They bind to and activate the injected Toll proteins and set in motion the chain of events that leads to dorsal protein entering nearby nuclei, thus defining the ventral region at the site of cytoplasm injection.

In the absence of a signal from Toll protein, dorsal protein is prevented from entering nuclei by being bound in the cytoplasm to another maternal gene product, the cactus protein. As a result of Toll activation, cactus protein is degraded and no longer binds the dorsal protein, which is then free to enter the nuclei (Fig. 5.9). In embryos lacking cactus protein, almost all of the dorsal protein is found in the nuclei; there is a very poor concentration gradient and the embryos are 'ventralized'—that is, no dorsal structures develop.

The interaction between the dorsal and cactus proteins is of more than local interest: dorsal protein is a transcription factor with considerable homology to the vertebrate transcription factor NF-κB, which is involved in regulation of gene expression in the B cells of the immune system. NF-κB is found in B-cell cytoplasm in a complex with another protein, I-κB, which prevents it from entering the nucleus before the cell has received the appropriate signal that dissociates the complex. I-κB has homology with the *Drosophila* cactus protein. So what might seem at first sight a rather specialized mechanism for confining transcription factors to the cytoplasm until it is time for them to enter the nucleus, is likely to be widely used for controlling cell differentiation.

Having considered the importance of localized maternal gene products in the egg in setting the basic framework for future development, we now look at how they come to be localized so precisely.

Summary

Maternal genes act in the ovary of the mother fly to set up differences in the egg in the form of localized deposits of mRNAs and proteins. After fertilization, maternal mRNAs are translated and provide the embryonic nuclei with positional information in the form of protein gradients or localized

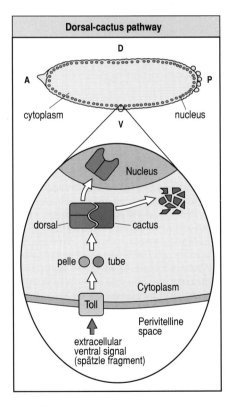

Fig. 5.9 The mechanism of localization of dorsal protein to the nucleus. In unfertilized eggs, dorsal protein is present in the cytoplasm bound to the cactus protein, which prevents it entering nuclei. The signal delivered by Toll activation is transmitted along an intracellular signaling pathway involving other maternal gene products (e.g. those of *tube* and *pelle*), with the end result that cactus protein is degraded and so no longer binds to dorsal protein, which can then enter the nucleus.

protein. Along the antero-posterior axis there is an anterior to posterior gradient of maternal bicoid protein, which controls patterning of the anterior region. For normal development it is essential that maternal hunchback protein is absent from the posterior region and its suppression is the function of the posterior to anterior gradient of nanos protein. The extremities of the embryo are specified by localized activation of the receptor protein torso at the poles. The dorso-ventral axis is established by intranuclear localization of the dorsal protein in a graded manner (ventral to dorsal), as a result of ventrally localized activation of the receptor protein Toll by a fragment of the protein spätzle.

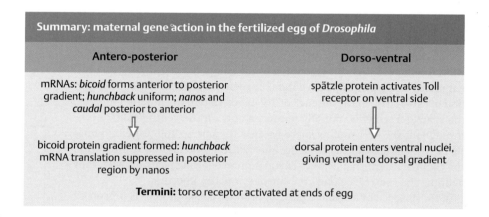

Summary: maternal gene action in the fertilized egg of *Drosophila*

Antero-posterior	Dorso-ventral
mRNAs: *bicoid* forms anterior to posterior gradient; *hunchback* uniform; *nanos* and *caudal* posterior to anterior	spätzle protein activates Toll receptor on ventral side
⇩	⇩
bicoid protein gradient formed: *hunchback* mRNA translation suppressed in posterior region by nanos	dorsal protein enters ventral nuclei, giving ventral to dorsal gradient

Termini: torso receptor activated at ends of egg

Polarization of the body axes during oogenesis

Fig. 5.10 Egg development in *Drosophila*. Oocyte development begins in a germarium, with stem cells at one end. One stem cell will divide four times to give 16 cells with cytoplasmic connections between each other. One of the cells that is connected to four others will become the oocyte, the others will become nurse cells. The nurse cells and oocyte become surrounded by follicle cells and the resulting structure buds off from the germarium as an egg chamber. Successively produced egg chambers are still attached to each other at the poles. The oocyte grows as the nurse cells provide material through the cytoplasmic bridges. The follicle cells have a key role in patterning the oocyte.

When the *Drosophila* egg is released from the ovary it already has a well-defined organization. *bicoid* mRNA is located at the anterior end and *nanos* and *caudal* mRNAs at the opposite end. The ligand for the torso protein is present in the vitelline envelope at both poles, and other maternal proteins are localized in the ventral vitelline envelope. Numerous other mRNAs and proteins, such as Toll, torso, dorsal, and cactus proteins, are distributed uniformly. How do these maternal mRNAs and proteins get into the egg during its period of development in the ovary—oogenesis—and how are they localized in the correct places?

The development of an egg in the *Drosophila* ovary is shown in Fig. 5.10. A stem cell undergoes four mitotic divisions to give 16 cells with cytoplasmic bridges between each other. One of these 16 cells will become the **oocyte**; the other 15 will develop into **nurse cells**, which produce large quantities

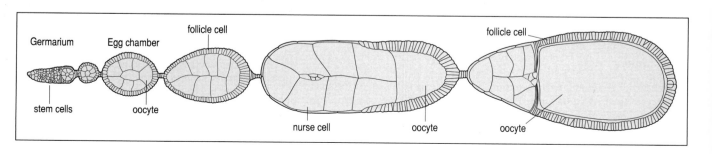

Germarium   Egg chamber   follicle cell   follicle cell

stem cells   oocyte   nurse cell   oocyte   oocyte

of proteins and RNAs that are exported into the egg through the cytoplasmic bridges. Somatic ovarian cells form a sheath of **follicle cells** around the nurse cells and oocyte to form the egg chamber, and they have a key role in patterning the egg's axes. There are various types of follicle cells, which express different genes and have different effects on the oocyte (Fig. 5.11). Follicle cells also secrete the materials of the vitelline envelope and eggshell that surround the mature egg.

5.7 Antero-posterior and dorso-ventral axes of the oocyte are specified by interactions with follicle cells

The first visible sign of antero-posterior polarization during oogenesis is the movement of the oocyte towards one end of the egg chamber, where it comes into contact with the follicle cells (Fig. 5.12). The localization of the oocyte requires cadherin-dependent adhesion (see Box 8A, p. 255) between the two cell types. Both the oocyte and the posterior follicle cells express higher levels of the adhesion molecule E-cadherin than other follicle cells and the nurse cells, and this leads to the ooctye's posterior localization. If E-cadherin is removed, the oocyte is randomly positioned. The oocyte then induces these follicle cells to adopt a posterior fate, while the follicle cells at the other end of the egg chamber, which are not in contact with the oocyte, remain unaffected and become the anterior follicle cells. The inductive signal from the oocyte is transmitted by the gurken protein, which belongs to the transforming growth factor-α (TGF-α) family. The gurken protein is

Fig. 5.11 ***Drosophila* oocyte development.** A developing *Drosophila* oocyte (right) is shown attached to its 15 nurse cells (left) and surrounded by a monolayer of 700 follicle cells. The oocyte and follicle layer are cooperating at this time to define the future dorso-ventral axis of the egg and embryo, as indicated by the expression of a gene only in the follicle cells overlying the dorsal anterior region of the oocyte (blue staining).

Photograph courtesy of A. Spradling.

Fig. 5.12 Specification of the antero-posterior and dorso-ventral axes during *Drosophila* oogenesis. The oocyte moves to the posterior end of the egg chamber and comes into contact with the polar follicle cells. It is separated from follicle cells at the anterior end (blue) by the nurse cells. *gurken* mRNA is synthesized in the oocyte and the gurken protein secreted locally. The binding of this protein to the receptor protein torpedo on the adjacent follicle cell initiates their specification as posterior polar follicle cells (yellow). They send a signal back to the oocyte that reorganizes the oocyte microtubule cytoskeleton (green). This directs the localization of the bicoid and oskar proteins to the anterior and posterior ends of the oocyte, respectively, thus defining the antero-posterior axis. Subsequent movement of the nucleus toward the future dorsal side and the local release of gurken protein then specifies the adjacent follicle cells as dorsal follicle cells and that side of the oocyte as the future dorsal side. After Gonzáles-Reyes, A., *et al.*: 1995.

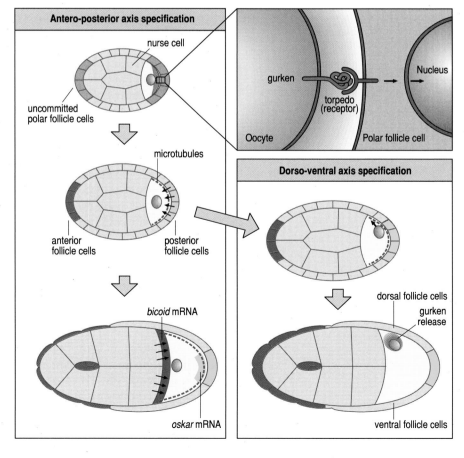

synthesized and secreted by the oocyte at the posterior end, where the oocyte nucleus is located at the time. It binds to a receptor protein in the follicle cell plasma membrane encoded by the *torpedo* gene. The torpedo protein is a transmembrane receptor tyrosine kinase similar to the epidermal growth factor receptor.

The posterior follicle cells send a signal back to the oocyte that results in a reorganization of the oocyte's microtubule cytoskeleton into an array of microtubules stretching from the anterior end towards the posterior end. This microtubule organization is disrupted in *par-1* mutants. Microtubule reorganization is essential for the localization of *bicoid* mRNA at the anterior end of the egg. *bicoid* mRNA is made by nurse cells located next to the anterior end of the developing oocyte, and is transferred from them to the egg. The *bicoid* mRNA interacts with the microtubule array in such a way that it is moved toward the anterior end and retained there. Similarly, *oskar* mRNA, which specifies the egg posterior germ plasm that gives rise to the germ cells, is delivered into the oocyte by nurse cells and moved to the posterior end through its interaction with the microtubule array (see Chapter 12). *nanos* mRNA is also localized to the posterior end. Several maternal genes are necessary for the localization of *bicoid* mRNA. Mother flies mutant for the gene *exuperantia*, for example, have eggs in which the *bicoid* mRNA is distributed throughout the egg and not restricted to the anterior. Therefore, *exuperantia* must be involved in the anterior localization process.

The setting-up of the egg's dorso-ventral axis involves a later set of oocyte–follicle cell interactions, which occur after the posterior end of the oocyte has been specified and which depend on the previous reorganization of the microtubule array. The oocyte nucleus moves along the microtubules from the posterior of the oocyte to a site on the anterior margin. In this new position, the *gurken* gene is expressed in the oocyte nucleus again. This time, the locally secreted gurken protein acts as a signal to adjacent follicle cells on one side of the oocyte, specifying them as dorsal follicle cells; the side away from the nucleus thus becomes the ventral region by default. The ventral follicle cells produce proteins, like pipe (see Section 5.5), that are only deposited in the ventral vitelline envelope.

The gurken protein can polarize both axes by its interactions with different sets of follicle cells, which implies that an earlier mechanism has already made some follicle cells, such as the polar follicle cells at each end of the egg chamber, different from the others. Such a difference would ensure that only polar follicle cells could respond to the gurken protein signal to become posterior cells.

The ligand for the torso protein, which distinguishes the termini, is synthesized and secreted by both posterior and anterior follicle cells, but not by the other follicle cells. It is thus only deposited in the vitelline envelope at both ends of the egg during oogenesis.

Summary

Nurse cells surrounding the *Drosophila* oocyte in the ovarian follicle provide it with large amounts of mRNAs and proteins, some of which become localized in particular sites. The oocyte produces a local signal which induces follicle cells at one end to become posterior follicle cells. The posterior follicle cells cause a reorganization of the oocyte cytoskeleton that localizes *bicoid* mRNA to the anterior end and other mRNAs to the posterior end of the oocyte. The dorso-ventral axis of the oocyte is also initiated by a local signal from the oocyte to certain follicle cells, which then become dorsal follicle cells. Follicle cells on the opposite side of the oocyte specify the ventral side of the oocyte by deposition of maternal proteins in the ventral vitelline envelope. Follicle cells at either end of the oocyte similarly specify the termini by localized deposition of maternal protein in the vitelline envelope.

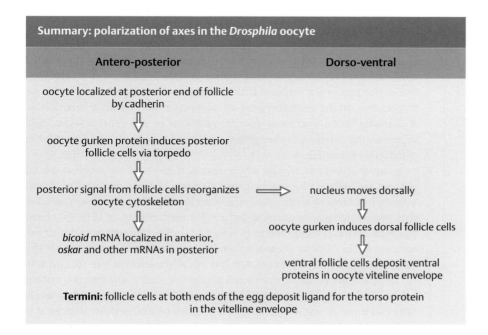

Zygotic genes pattern the early embryo

Understanding in such detail how the main body axes of *Drosophila* are specified is a major achievement, and those who work on other animals like the frog and chick, where very little is known about their developmental genetics, are justifiably somewhat envious. We have seen how gradients of bicoid, hunchback, and caudal proteins are established along the antero-posterior axis, and how intranuclear dorsal protein is graded along the ventral to dorsal axis. This maternally derived framework of positional information is interpreted and elaborated on by zygotic genes to give each region of the embryo an identity. Most of the zygotic genes first activated along the antero-posterior and dorso-ventral axes encode transcription factors, which are thus localized along the axes and activate yet more zygotic genes. We first consider the patterning along the dorso-ventral axes, which is somewhat simpler than that along the antero-posterior axis.

5.8 The expression of zygotic genes along the dorso-ventral axis is controlled by dorsal protein

After dorsal protein has entered the nuclei, its effects on gene expression divide the dorso-ventral axis into well-defined regions and also specify the ventral-most cells as prospective mesoderm. Going from ventral to dorsal,

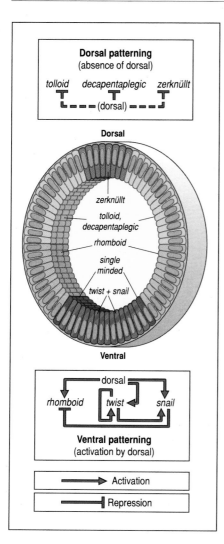

Fig. 5.13 Model for the subdivision of the dorso-ventral axis into different regions by the gradient in nuclear dorsal protein. In the dorsal region, where nuclear dorsal protein is absent, *tolloid*, *zerknüllt*, and *decapentaplegic* are not repressed. In the ventral region, the dorsal protein activates the genes *twist*, *snail*, and *rhomboid*. *twist* is autoregulatory, maintaining its own expression, and also activates *snail*; snail protein inhibits *rhomboid* expression. The gene *single-minded* is activated by the dorsal protein but its localization is dependent on the action of other genes.

the main regions are mesoderm, ventral ectoderm (prospective neurectoderm), dorsal ectoderm (prospective dorsal epidermis), and prospective amnioserosa (which is an extra-embryonic membrane on the dorsal side of the embryo that is sloughed off as embryonic development is completed). The mesoderm gives rise to internal soft tissues such as muscle and connective tissue; the neurectoderm gives rise to epidermis as well as all the nervous tissue; the dorsal ectoderm gives rise only to epidermis. The third germ layer, the endoderm, which is located at either end of the embryo, and which we do not consider here, gives rise to the mid-gut.

Patterning along the axes poses a problem like that of patterning the French flag (see Section 1.13). Expression of zygotic genes in localized regions along the dorso-ventral axis is initially controlled by the graded concentration of the intranuclear dorsal protein, which falls off rapidly in the dorsal half of the embryo; little dorsal protein is found in nuclei above the equator. In the ventral region, dorsal protein has two main functions— it activates certain genes at specific positions in the ventral region and represses the activity of other genes, which are therefore only expressed in the dorsal region (Fig. 5.13).

In the ventral-most region, where concentrations of intranuclear dorsal protein are highest, the zygotic genes *twist* and *snail* are activated by dorsal protein in a strip of nuclei along the ventral side of the embryo; soon after this the blastoderm becomes cellular. This ventral strip of cells will form the mesoderm. The expression of *twist* and *snail* is required both for development of the cells as mesoderm and for gastrulation, during which the ventral band of cells moves into the interior of the embryo (see Section 8.8). In the future neurectoderm, which will give rise both to the nervous system and to larval ventral epidermis, the gene *rhomboid* is activated at low levels of dorsal protein, but is not expressed in more ventral regions because it is repressed by the snail protein.

The genes *decapentaplegic*, *tolloid*, and *zerknüllt* are repressed by dorsal protein and so their activity is confined to the more dorsal regions of the embryo, where there is virtually no dorsal protein in the nuclei. *zerknüllt* is expressed most dorsally and appears to specify the amnioserosa. *decapentaplegic* is a key gene in the specification of pattern in the dorsal part of the dorso-ventral axis and its role is considered in detail in the next section.

Mutations in the maternal dorso-ventral genes can cause dorsalization or ventralization of the embryo. In dorsalized embryos, dorsal protein is excluded uniformly from the nuclei. This has a number of effects, one of which is that the *decapentaplegic* gene is expressed everywhere, in line with the observation that it is normally repressed in the ventral region by high intranuclear concentrations of the dorsal protein. By contrast, *twist* and *snail* are not expressed at all in dorsalized embryos, as they are activated by high intranuclear concentrations of dorsal protein. Just the opposite result is obtained in ventralized embryos, where the dorsal protein is present at high concentration in all the nuclei; *twist* and *snail* are expressed throughout and *decapentaplegic* is not expressed at all (Fig. 5.14).

Genes whose expression is regulated by the dorsal protein, such as *twist*, *snail*, and *decapentaplegic*, contain binding sites for dorsal protein in their regulatory regions that activate or repress gene expression at particular concentrations of the protein. (These binding sites may not be determined by the DNA sequence but by proteins bound to adjacent sites.) This **threshold effect** on gene expression is the result of the integrating function of

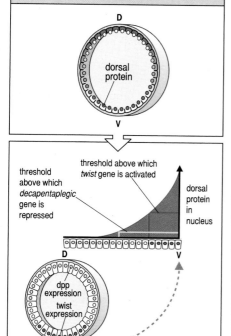

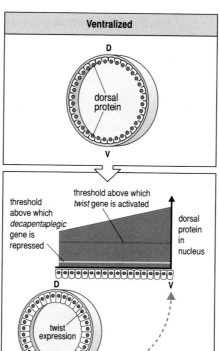

Fig. 5.14 The nuclear gradient in dorsal protein is interpreted by the activation of other genes, such as *twist* and *decapentaplegic*. Left panels: in normal larvae, the *twist* gene is activated above a certain threshold concentration (green line) of dorsal protein, whereas above a lower threshold (yellow line), the *decapentaplegic* gene is repressed. Right panels: in ventralized larvae, the dorsal protein is present in all nuclei; *twist* is now also expressed everywhere, whereas *decapentaplegic* is not expressed at all, because dorsal protein is above the threshold level required to repress it everywhere.

these regulatory binding sites. The ability of genes to respond in a threshold-like manner to varying concentrations of dorsal protein is due to the presence of both high- and low-affinity binding sites for dorsal protein in their regulatory regions. In the most ventral regions (a strip 12–14 cells wide), where the concentration of dorsal protein is high, low-affinity sites delimit gene expression, whereas high-affinity sites control expression in slightly more dorsal regions (up to 20 cells from the ventral midline). It is very likely that the threshold response involves cooperativity between the different binding sites; that is, binding at one site makes binding at a nearby site easier and so facilitates further binding. In addition, inhibitory interactions with other gene products are involved. For example, the snail protein represses the expression of certain genes in the ventral regions, and thus helps to confine the expression of genes such as *rhomboid* to the neurectoderm.

The gradient of dorsal protein is therefore effectively acting as a morphogen gradient along the dorso-ventral axis, activating specific genes at different threshold concentrations, and so defining the dorso-ventral pattern. The regulatory sequences in these genes can be thought of as developmental switches, which when thrown by the binding of transcription factors, activate genes and set cells off along new developmental pathways. The dorsal protein gradient is one solution to the French flag problem. But it is not the whole story—yet another gradient is also involved.

5.9 The decapentaplegic protein acts as a morphogen to pattern the dorsal region

As with the antero-posterior axis, each end of the dorso-ventral axis is specified by different proteins. The gradient of dorsal protein, with its high point in the ventral-most region, specifies the initial pattern of zygotic gene activity, and patterns the ventral mesoderm. But the dorsal region is not similarly specified by a low-level gradient in dorsal protein. Indeed, there is little or no dorsal protein in the nuclei of the dorsal half of the embryo. The more dorsal part of the dorso-ventral pattern is thought to be determined by a gradient in the activity of the decapentaplegic protein.

Soon after the gradient of intranuclear dorsal protein has become established, the embryo becomes cellular, and transcription factors can no longer diffuse between nuclei. Secreted or transmembrane proteins and their corresponding receptors must now be used to transmit signals between cells. The decapentaplegic protein is one such secreted signaling protein. It is a homolog of BMP-4, a member of the transforming growth factor-β (TGF-β) family of vertebrate growth factors (see Box 3A, p. 68) and, as we shall see, is involved in a variety of signaling processes throughout *Drosophila* development.

The *decapentaplegic* gene is expressed throughout the dorsal region, where dorsal protein is not present in the nuclei. Evidence for a gradient in decapentaplegic protein activity comes from experiments in which *decapentaplegic* mRNA is introduced into an early wild-type embryo. As more mRNA is introduced and the concentration of decapentaplegic protein increases above the normal level, the cells along the dorso-ventral axis adopt a more dorsal fate than they would normally. Ventral ectoderm becomes dorsal ectoderm and, at very high concentrations of *decapentaplegic* mRNA, all the ectoderm develops as the dorsal-most region—the amnioserosa.

The graded activity of decapentaplegic protein along the dorso-ventral axis at the cellular blastoderm stage is similar to that of BMP-4 in the *Xenopus* gastrula (see Section 3.19). The decapentaplegic gradient is formed as a result of its interaction with various proteins. One is a secreted protein, short gastrulation (sog), which is expressed in the middle region (the neurectoderm, which gives rise to neural tissue). By binding decapentaplegic and preventing it spreading into the neurectoderm, sog protein helps to confine decapentaplegic activity to the dorsal region. It is thought that sog also diffuses dorsally and is degraded by the tolloid protein, which is expressed in the dorsal region, thus setting up a gradient in sog with the

Fig. 5.15 Decapentaplegic protein activity is restricted to the dorsal-most region of the embryo by the antagonistic activity of the short gastrulation protein. The gene *short gastrulation* (sog) is expressed in the non-neural ectoderm of the early blastoderm (left panel). The sog protein prevents decapentaplegic (dpp) signaling from spreading into the neurectoderm by interfering with the positive feedback loop created by decapentaplegic diffusing and activating expression of its own gene. The right panel shows the situation in the mid-blastoderm embryo. The dorsal region has been subdivided by sog activity into a region of high dpp signaling, which will become the aminoserosa, and a zone of lower dpp signaling, which is the dorsal ectoderm. After Bier, E.: 1999.

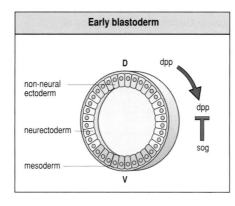

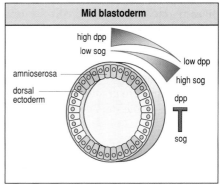

high point in the neurectoderm and the low point dorsally. As sog blocks decapentaplegic signaling, this in turn generates a gradient of decapentaplegic activity with its high point in the dorsal-most region (Fig. 5.15). But there is still an unsolved puzzle, as sog actually appears to be required in order to maintain the high levels of decapentaplegic in that dorsal-most region which specify the amnioserosa. It is likely that interactions of other proteins with the complex of sog and decapentaplegic also determine the exact shape of the decapentaplegic gradient. This example well illustrates the point that morphogen gradients are not necessarily formed by simple diffusion.

The patterning of the dorso-ventral axis therefore involves two gradients, one of dorsal protein and one of decapentaplegic protein, with high points at different ends. Together they result in the dorso-ventral axis becoming divided up into several regions of unique gene activity and developmental fate. We now return to the antero-posterior axis at an earlier stage, while the embryo is still acellular.

5.10 The antero-posterior axis is divided up into broad regions by gap gene expression

The **gap genes** are the first zygotic genes to be expressed along the antero-posterior axis, and all code for transcription factors. Their expression is initiated by the antero-posterior gradient of bicoid protein while the embryo is still a syncytial blastoderm. Bicoid protein primarily activates anterior expression of the gap gene *hunchback*, which in turn is instrumental in switching on the expression of the other gap genes, amongst which are *giant*, *Krüppel*, and *knirps*, which are expressed in this sequence along the antero-posterior axis (Fig. 5.16). (*giant* is in fact expressed in two bands, one anterior and one posterior, but its posterior expression does not concern us here.)

Gap genes were initially recognized by their mutant phenotypes, in which quite large sections of the body pattern along the antero-posterior axis are missing. Although the mutant phenotype of a gap gene usually shows a gap in the antero-posterior pattern in more-or-less the region in which the gene is normally expressed, there are also more wide-ranging effects. This is because gap gene expression is also essential for later development along the axis.

As the blastoderm is still acellular at the stage at which the gap genes are expressed, the gap gene proteins can diffuse away from their site of synthesis. They are short-lived proteins with a half-life of minutes. Their distribution therefore extends only slightly beyond the region in which the gene is expressed, and this typically gives a bell-shaped protein concentration profile. *hunchback* is exceptional as it is expressed over a broad anterior region and so sets up a steep antero-posterior gradient. The control of zygotic *hunchback* expression by bicoid protein is best understood and will be considered first.

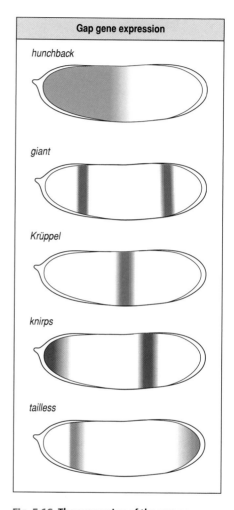

Fig. 5.16 The expression of the gap genes *hunchback*, *Krüppel*, *giant*, *knirps*, and *tailless* in the early *Drosophila* embryo. Gap gene expression at different points along the antero-posterior axis is controlled by the concentration of bicoid and hunchback proteins, together with interactions between the gap genes themselves. The expression pattern of the gap genes provides an aperiodic pattern of transcription factors along the antero-posterior axis, which delimits broad body regions.

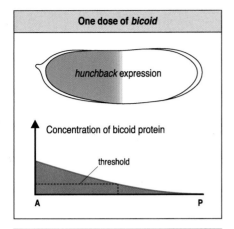

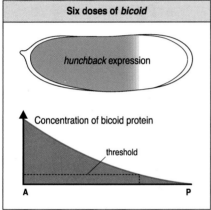

Fig. 5.17 Maternal bicoid protein controls zygotic *hunchback* expression. If the dose of maternal *bicoid* is increased sixfold, the extent of the bicoid protein gradient also increases. The activity of the *hunchback* gene is determined by the threshold concentration of bicoid protein, so at the higher dose, its region of expression is extended toward the posterior end because the region in which bicoid protein concentration exceeds the threshold level also extends more posteriorly (see graph, bottom panel).

5.11 Bicoid protein provides a positional signal for the anterior expression of *hunchback*

Zygotic expression of the *hunchback* gene in normal embryos occurs over most of the anterior half of the embryo. This zygotic expression is superimposed on a low level of maternal *hunchback* mRNA, whose translation is suppressed posteriorly by nanos protein (see Section 5.3). This results in a gradient of hunchback protein in the posterior half of the embryo, running anterior to posterior.

The localized anterior expression of *hunchback* is an interpretation of the positional information provided by the bicoid protein gradient. The *hunchback* gene is switched on only when bicoid protein, a transcription factor, is present at a certain threshold concentration. This level is attained only in the anterior third of the embryo, close to the site of bicoid protein synthesis, which restricts *hunchback* expression to this region. There is some evidence that maternal hunchback protein is also necessary for spatial control of zygotic *hunchback* expression.

The relationship between bicoid protein concentration and *hunchback* gene expression can be illustrated by looking at how *hunchback* expression changes when the bicoid protein concentration gradient is changed by increasing the maternal dosage of the *bicoid* gene (Fig. 5.17). The result is that expression of *hunchback* extends more posteriorly, because the region in which the concentration of the bicoid protein is above the threshold for *hunchback* gene activation is also extended posteriorly. By calibrating the extension of *hunchback* expression with the maternal dosage of the *bicoid* gene, it can be calculated that a twofold increase in the concentration of bicoid protein can switch *hunchback* gene expression from off to on.

The bicoid protein is a member of the homeodomain family of transcriptional activators and activates the *hunchback* gene by binding to regulatory sites within the promoter region. Direct evidence of *hunchback* activation by bicoid protein has been obtained by gene transfer experiments (Box 5A) using a fusion gene constructed from *hunchback* promoter regions and a bacterial reporter gene, *lacZ*, and then introduced into the fly genome. *lacZ* codes for the enzyme β-galactosidase, which is easily made visible by histochemical staining. The extent of *lacZ* expression in embryos carrying this transgene exactly parallels normal *hunchback* expression if the promoter region of the transgene is complete, but not if a large part of it is deleted (Fig. 5.18). The large promoter region required for completely normal gene expression can be whittled down to an essential sequence of 263 base pairs that will still give almost normal activity in this situation. This sequence has several sites at which bicoid protein can bind, and it seems that cooperative binding of bicoid proteins is involved in establishing the threshold response.

The regulatory sequences of genes such as these are yet further examples of developmental switches that direct nuclei along a new developmental pathway. We will encounter many more examples of these transcriptional switches in *Drosophila* early development.

Box 5A Transgenic flies

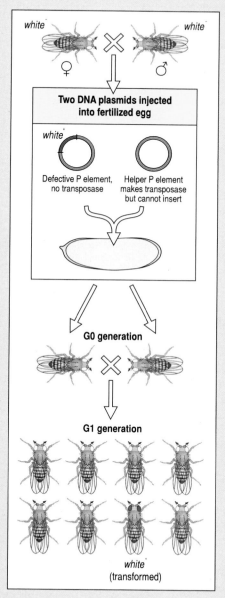

As discussed in Boxes 3C (p. 89) and 4B (p. 122), transgenic mice can be made in which the function of developmental genes can be studied. Transgenic fruit flies can also be created, and have contributed greatly to *Drosophila* developmental genetics. They are made by inserting a known sequence of DNA into the *Drosophila*'s chromosomal DNA, using as a carrier a **transposon** that occurs naturally in some strains of *Drosophila*. This transposon is known as a **P element** and the technique as **P-element mediated transformation** (see figure).

P elements can insert at almost any site on a chromosome, and can also hop from one site on a chromosome to another within the germ cells, an action that requires an enzyme called transposase. As hopping can cause genomic instability, carrier P elements have had their own transposase gene removed. The transposase required to insert the P element initially is instead provided by a helper P element, which cannot itself insert into the host chromosomes and is thus quickly lost from cells. The two elements are injected together into the posterior end of the egg where the germ cells are made.

Usually a marker gene, such as the wild-type *white*⁺ gene, is added to the P element. When *white*⁺ is the marker, the P element is inserted into host flies homozygous for the mutant *white*⁻ gene (which have white eyes rather than the red eyes of the wild-type *Drosophila*). Red eyes are dominant over white and so flies in which the P element has become integrated into the chromosome, and is being expressed, can be detected by their red eyes.

In the first generation, all flies have white eyes, as any P element that has integrated is still restricted to the germ cells. But in the second generation, a few flies will have wild-type red eyes, showing that they carry the inserted P element in their somatic cells.

This technique can be used to increase the number of copies of a particular gene, to introduce a mutated gene that, for example, has its control or coding regions altered in a known way, or to introduce new genes. It is also possible to introduce genes that carry a marker coding sequence such as *lacZ* (encoding the bacterial enzyme β-galactosidase), whose expression is detectable by histochemical staining.

The ability to turn on the expression of a gene in a particular place and time during development is very useful for analyzing the gene's role in development. One approach is by targeted gene expression using GAL4. GAL4 is a transcription factor from yeast which can activate transcription of any gene whose promoter has a GAL4-binding site. *Drosophila* genes with GAL4-responsive promoters can be created by inserting the appropriate GAL4-binding site by standard genetic techniques.

To turn on the target gene, GAL4 itself has to be produced in the embryo. In one approach, GAL4 can be produced in a designated region or at a particular time in development by introducing a transgene in which the yeast *GAL4* coding region is attached to a *Drosophila* regulatory region known to be activated in that situation. A second, more versatile, approach is based on the so-called enhancer-trap technique. The *GAL4* coding sequence is attached to a vector that integrates randomly into the *Drosophila* genome. The *GAL4* gene will come under the control of the promoter and enhancer region adjacent to its site of integration, and so GAL4 will be produced where or when that gene is normally expressed. Thus a variety of *Drosophila* lines with differing patterns of GAL4 expression can be produced.

The selected target gene will be silent in the absence of GAL4. To activate it in one of the chosen patterns, those flies that express GAL4 in that pattern are crossed with flies in which the target gene has GAL4-binding sites in its regulatory region. The effect of the novel pattern of target gene expression on development can then be observed. Using the GAL4 system, the pair-rule gene *even-skipped* was expressed in even-numbered parasegments and this led to changes in the pattern of denticles on the cuticle.

Another approach to targeted gene expression is to give the selected gene a heat-shock promoter. This enables the gene to be switched on by a sudden rise in the temperature at which the embryos are being kept. By adjusting the temperature, the timing of expression of genes attached to this promoter can be controlled; the effects of expressing a gene at different stages of development can be studied in this way.

Normal promoter	Partial promoter	Promoter deleted

Fig. 5.18 Zygotic *hunchback* expression is controlled by bicoid protein. *hunchback* expression is visualized by joining the bacterial *lacZ* gene to the control region of the *Drosophila hunchback* gene and inserting this construct into the fly's genome. With the normal *hunchback* control region, *lacZ* is expressed in the anterior half of the embryo (left); with only a partial control region its expression is more restricted (center); and when the construct lacks a bicoid-binding site, *lacZ* expression is absent (right). *lacZ* expression is visualized by histochemical staining for the *lacZ* gene product, the enzyme β-galactosidase.

Photographs courtesy of D. Tautz.

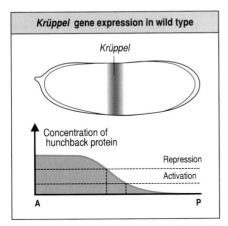

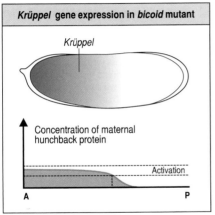

5.12 The gradient in hunchback protein activates and represses other gap genes

Expression of the other gap genes is localized in bands across the antero-posterior axis (see Fig. 5.16). The hunchback protein is itself a transcription factor and acts as a morphogen to which the other gap genes respond. The bands of gap gene expression are delimited by mechanisms that depend on the gene control regions being sensitive to different concentrations of hunchback protein, and also to other proteins, including bicoid protein. Expression of the *Krüppel* gene, for example, is activated by a combination of bicoid protein and low levels of hunchback, but is repressed at high concentrations of hunchback protein. Within this concentration 'window' *Krüppel* remains activated (Fig. 5.19, top panel). But below the lower threshold concentration of hunchback, it is not activated. In this way, the gradient in hunchback protein precisely locates a band of *Krüppel* gene activity near the center of the embryo. Refined spatial localization is brought about by repression of *Krüppel* by other gap gene proteins.

Such relationships were worked out by altering the concentration profile of hunchback protein systematically, while all other known influences were eliminated or held constant. Increasing the dose of hunchback protein, for example, results in a posterior shift in its concentration profile, and this results in a posterior shift in the posterior boundary of *Krüppel* expression. In another set of experiments on embryos lacking bicoid protein, so that only the maternal hunchback protein gradient is present, the level of hunchback protein is such that *Krüppel* is even activated at the anterior end of the embryo (Fig. 5.19, bottom panel).

The hunchback protein is also involved in specifying the anterior border of the bands of expression of the gap genes *knirps* and *giant*, again by a mechanism involving thresholds for repression and activation of these genes. At high concentrations of hunchback protein, *knirps* is repressed, and this specifies its anterior margin of expression. The posterior margin of the *knirps* band is specified by a similar type of interaction with the product of another gap gene, *tailless*. Where the regions of expression of the gap genes overlap, there is extensive cross-inhibition between them, their proteins all being transcription factors. These interactions are essential to sharpen and stabilize the pattern of gap gene expression. For example, the

Fig. 5.19 *Krüppel* gene activity is specified by *hunchback* protein. Top panel: above a threshold concentration of hunchback protein, the *Krüppel* gene is repressed; at a lower concentration, above another threshold value, it is activated. Bottom panel: in mutants lacking the *bicoid* gene, and thus also lacking zygotic *hunchback* gene expression, only maternal hunchback protein is present, which is located at the anterior end of the embryo at a relatively low level. In these mutants, *Krüppel* is activated at the anterior end of the embryo, giving an abnormal pattern.

anterior border of *Krüppel* expression lies four to five nuclei posterior to nuclei that express *giant*, and that anterior border is set by low levels of giant protein.

The antero-posterior axis becomes divided into a number of unique regions on the basis of the overlapping and graded distributions of different transcription factors. This beautifully elegant method of delimiting regions can, however, only work in an embryo such as the acellular blastoderm of *Drosophila*, where the transcription factors are able to diffuse throughout the embryo. This regional distribution of the gap gene products provides the starting point for the next stage in development—the activation of the pair-rule genes and the beginning of segmentation.

Summary

Gradients of maternally derived transcription factors along the dorso-ventral and antero-posterior axes provide positional information that activates zygotic genes at specific locations along these axes. The dorso-ventral axis becomes divided into four regions: ventral mesoderm, ventral ectoderm (neurectoderm), dorsal ectoderm (dorsal epidermis), and amnioserosa. A ventral to dorsal gradient of maternal dorsal protein both specifies the ventral mesoderm and defines the dorsal region; a second gradient, of the decapentaplegic protein, specifies the dorsal ectoderm. Along the antero-posterior axis the zygotic gap genes are activated by the bicoid protein gradient to specify general body regions. Interactions between the gap genes, all of which code for transcription factors, help to define their borders of expression. Patterning along the dorso-ventral and antero-posterior axes divides the embryo into a number of discrete regions, each characterized by a unique pattern of zygotic gene activity.

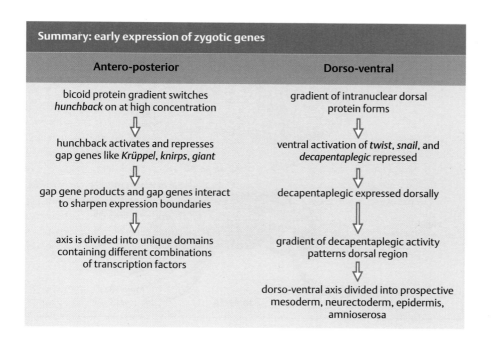

Summary: early expression of zygotic genes

Antero-posterior	Dorso-ventral
bicoid protein gradient switches *hunchback* on at high concentration	gradient of intranuclear dorsal protein forms
⇩	⇩
hunchback activates and represses gap genes like *Krüppel, knirps, giant*	ventral activation of *twist, snail*, and *decapentaplegic* repressed
⇩	⇩
gap gene products and gap genes interact to sharpen expression boundaries	decapentaplegic expressed dorsally
⇩	⇩
axis is divided into unique domains containing different combinations of transcription factors	gradient of decapentaplegic activity patterns dorsal region
	⇩
	dorso-ventral axis divided into prospective mesoderm, neurectoderm, epidermis, amnioserosa

Segmentation: activation of the pair-rule genes

The most obvious feature of a *Drosophila* larva is the regular segmentation of the larval cuticle along the antero-posterior axis, each segment carrying cuticular structures that define it as, for example, thorax or abdomen. This pattern of segmentation is carried over into the adult, in which each segment has its own identity. Adult appendages such as wings, halteres, and legs are attached to particular segments (Fig. 5.20). But the discernible segments of the mature larva are not in fact the first units of segmentation along this axis. The basic developmental modules, whose definition we will follow in some detail, are the **parasegments**, which are specified first and from which the segments derive.

Fig. 5.20 The relationship between parasegments and segments in the early embryo, late embryo, and adult fly. Initially, pair-rule genes are expressed in the embryo as stripes in every second parasegment. *even-skipped* (yellow), for example, is expressed in odd-numbered parasegments. The segment polarity selector gene *engrailed* (blue) is expressed in the anterior region of every parasegment, and delimits the anterior margin of each parasegment. Each larval segment is composed of the posterior region of one parasegment and the anterior region of the next. The anterior region of a parasegment becomes the posterior portion of a segment. Segments are thus offset from the original parasegments by about half a segment, and *engrailed* is expressed in the posterior region of each segment. In this figure, a and p refer to the anterior and posterior compartments of segments or parasegments. The segment specification is carried over into the adult and results in particular appendages, such as legs and wings, developing on specific segments only. C1, C2, and C3 represent segments which become fused to form the head region. T, thoracic segments; A, abdominal segments. After Lawrence, P.: 1992.

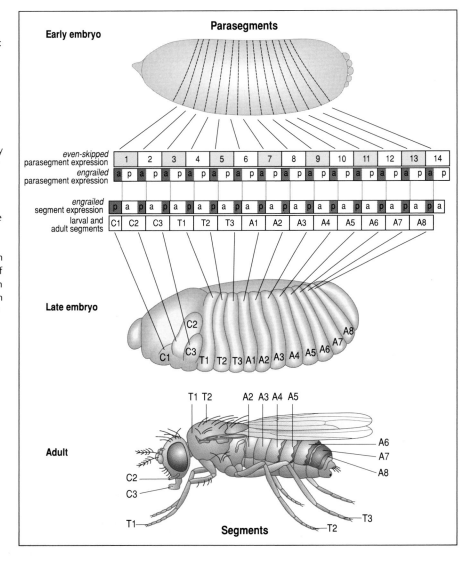

5.13 Parasegments are delimited by expression of pair-rule genes in a periodic pattern

The first visible signs of segmentation in the embryo are transient grooves that appear on the surface of the embryo after gastrulation. These grooves define the parasegments. There are 14 parasegments, which are the fundamental units in the segmentation of the *Drosophila* embryo. Once each parasegment is delimited, it behaves as an independent developmental unit, under the control of a particular set of genes, and in this sense at least, the embryo can be thought of as being built up piecemeal. The parasegments are initially similar but each will eventually acquire its own unique identity. The parasegments are out of register with the final segments by about half a segment; each segment is therefore made up of the posterior region of one parasegment and the anterior region of the next (see Fig. 5.20). In the head region the segmental arrangement is lost when the anterior parasegments fuse.

The parasegments are delimited by the action of the **pair-rule genes**, each of which is expressed in a series of seven transverse stripes along the embryo, each stripe corresponding to every second parasegment. When pair-rule gene expression is visualized by staining for the pair-rule proteins, a striking zebra-striped embryo is revealed (Fig. 5.21).

The positions of the stripes of pair-rule gene expression are determined by the pattern of gap gene expression—a non-repeating pattern of gap gene activity is converted into repeating stripes of pair-rule gene expression. We now consider how this is achieved.

5.14 Gap gene activity positions stripes of pair-rule gene expression

Pair-rule genes are expressed in stripes, with a periodicity corresponding to alternate parasegments. Mutations in these genes thus affect alternate segments. Some pair-rule genes (e.g. *even-skipped*) define odd-numbered parasegments, whereas others (e.g. *fushi tarazu*) define even-numbered parasegments. The striped pattern of expression of pair-rule genes is present even before cells are formed, while the embryo is still a syncytium, although cellularization occurs soon after expression begins. Each pair-rule gene is expressed in seven stripes, each of which is only a few cells wide. In some genes, such as *even-skipped*, the anterior margin of the stripe corresponds to the anterior boundary of a parasegment; the domains of

Anterior / Posterior

Fig. 5.21 The striped patterns of activity of pair-rule genes in the *Drosophila* embryo just before cellularization. Parasegments are delimited by pair-rule gene expression, each pair-rule gene being expressed in alternate parasegments. Expression of the pair-rule genes *even-skipped* (blue) and *fushi tarazu* (brown) is visualized by staining with antibody for their protein products. *even-skipped* is expressed in odd-numbered parasegments, *fushi tarazu* in even-numbered parasegments. Scale bar = 0.1 mm.
Photograph from Lawrence, P.: 1992.

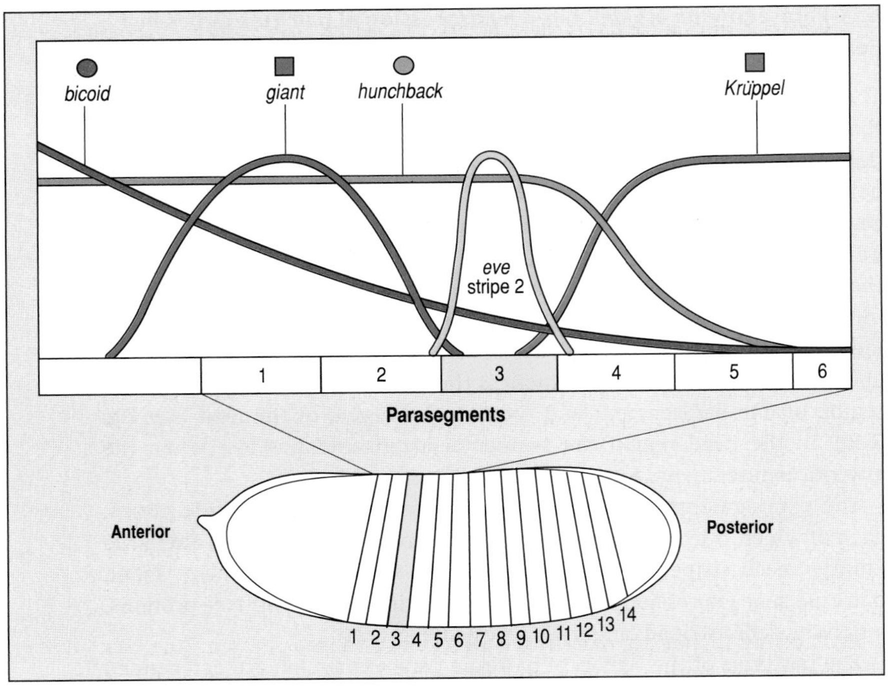

Fig. 5.22 The specification of the second *even-skipped* (*eve*) stripe by gap gene proteins. The different concentrations of transcription factors encoded by the gap genes *hunchback, giant,* and *Krüppel* localize *even-skipped*—expressed in a narrow stripe at a particular point along their gradients—in parasegment 3. Bicoid and hunchback proteins activate the gene in a broad domain, and the anterior and posterior borders are formed through repression by giant and Krüppel proteins, respectively.

expression of other pair-rule genes, however, cross parasegment boundaries.

The striped expression pattern appears gradually; the *even-skipped* gene, for example, is initially expressed at a low level in all nuclei. A single broad stripe of gene expression then appears anteriorly and narrows as the other stripes develop. Each stripe is initially fuzzy, but eventually acquires a sharp anterior margin. At first sight this type of patterning would seem to require some underlying periodic process, such as the setting up of a wave-like concentration of a morphogen, with each stripe forming at the crest of a wave. It was surprising, therefore, to discover that each stripe is specified independently.

As an example of how the pair-rule stripes are generated we will look in detail at the expression of the second *even-skipped* stripe (Fig. 5.22). The appearance of this stripe depends on the normal expression of *bicoid* and of the three gap genes *hunchback, Krüppel,* and *giant* (only the anterior band of *giant* expression is involved in specifying the second *even-skipped* stripe, see Fig. 5.16). Bicoid and hunchback proteins are required to activate the *even-skipped* gene, but they do not define the boundaries of the stripe. These are defined by Krüppel and giant proteins, by a mechanism based on repression of *even-skipped*. When concentrations of Krüppel and giant proteins are above certain threshold levels, *even-skipped* is repressed, even if bicoid and hunchback proteins are present. The anterior edge of the stripe is localized at the point of threshold concentration of giant protein, whereas the posterior border is similarly specified by Krüppel protein.

The independent localization of each of the stripes by the gap gene transcription factors requires that, in each stripe, pair-rule genes respond to different concentrations and combinations of the gap gene transcription factors. The pair-rule genes thus require complex control regions with multiple binding sites for each of the different factors. Examination of the

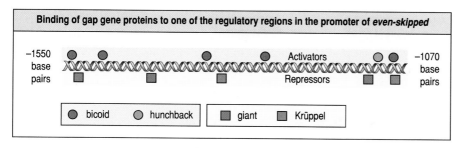

Fig. 5.23 Sites of action of activating and repressing transcription factors in the region of the *even-skipped* promoter involved in expression of the second *even-skipped* stripe. A promoter region of around 500 base pairs, located between 1070 and 1550 base pairs upstream of the transcription start site, directs formation of the second *even-skipped* stripe. Gene expression occurs when the bicoid and hunchback transcription factors are present above a threshold concentration, with the giant and Krüppel proteins acting as repressors where they are above threshold levels. The repressors may act by preventing binding of activators.

regulatory regions of the *even-skipped* gene reveals a number of separate regions, each controlling the localization of a different stripe. Using the *lacZ* reporter gene technique described in Box 5A (p. 163), regulatory regions of around 500 base pairs have been isolated, and each determines the expression of a single stripe (Fig. 5.23).

The presence of control regions which, when activated, can lead to gene expression in a specific position in the embryo, is an important principle controlling gene action in development. Other examples are provided by the localized expression of the gap genes, and of genes along the dorso-ventral axis.

Each of the regulatory regions on such genes contains binding sites for different transcription factors, some of which activate the gene, while others repress it. In this way, the gap genes regulate pair-rule gene expression in each parasegment. Some pair-rule genes, such as *fushi tarazu*, may not be regulated by the gap genes directly, but may depend on the prior expression of primary pair-rule genes such as *even-skipped* and *hairy*. With the initiation of pair-rule gene expression the embryo becomes segmented; it is now divided into a number of unique regions, which are characterized mainly by the combinations of transcription factors being expressed in each. These include proteins encoded by the gap genes, the pair-rule genes, and the genes expressed along the dorso-ventral axis.

The transcription factors encoded by the pair-rule genes set up the spatial framework for the next round of patterning by transcriptional activation. This involves the further patterning of the parasegments, the development of the final segmentation and the acquisition of segment identity, which is considered in the following sections.

Summary

The activation of the pair-rule genes by the gap genes results in the transformation of the embryonic pattern along the antero-posterior axis from an aperiodic regionalization to a periodic one. The pair-rule genes define 14 parasegments. Each parasegment is defined by narrow stripes of pair-rule gene activity. These stripes are uniquely defined by the local concentration of gap gene transcription factors acting on the regulatory regions of the pair-rule genes. Each pair-rule gene is expressed in alternate parasegments—some in odd-numbered, others in even-numbered. Most pair-rule genes code for transcription factors.

Summary: pair-rule genes and segmentation

production of local combinations of gap gene transcription factors

activation of each pair-rule gene in seven transverse stripes along the antero-posterior axis

pair-rule gene expression defines 14 parasegments, each pair-rule gene
being expressed in alternate parasegments

Segment polarity genes and compartments

The expression of the pair-rule genes defines the anterior boundaries of all
14 parasegments but, like the gap genes, their activity is only temporary.
Moreover, at this time the blastoderm becomes cellularized. How, therefore, are the positions of parasegment boundaries fixed, and how do the
final segment boundaries of the larval epidermis become established? This
is the role of the **segment polarity genes**. Unlike the gap genes and pair-
rule genes, which encode transcription factors, the segment polarity genes
are a diverse group of genes that bear no obvious relation to each other in
their protein products or mechanism of operation. They are known as segment polarity genes because the effect of their mutant forms is generally
to upset the antero-posterior polarity of the segments, producing
mirror-image or tandem duplications of either anterior or posterior parts.

Segment polarity genes are activated in response to pair-rule gene
expression. They are each expressed in 14 transverse stripes, one stripe
corresponding to each parasegment. During pair-rule gene expression, the
blastoderm becomes cellularized, so the segment polarity genes are acting
in a cellular rather than a syncytial environment. One of the segment polarity genes to be activated by the pair-rule genes is *engrailed*, which is
expressed in the anterior region of every parasegment. *engrailed* is of
particular interest as its expression delimits a boundary of cell lineage
restriction. Moreover, *engrailed* is also a **selector gene**, a gene that confers a
particular identity on a region or regions by controlling the activity of other
genes, and which continues to act for an extended period.

5.15 Expression of the *engrailed* gene delimits a cell-lineage boundary and defines a compartment

The *engrailed* gene has a key role in segmentation and, unlike the pair-rule
and gap genes, whose activity is transitory, it is expressed throughout the
life of the fly. *engrailed* activity first appears at the time of cellularization as
a series of 14 transverse stripes. Fig. 5.24 shows expression at the later
extended germ-band stage, when part of the ventral blastoderm (the germ
band) has extended over the dorsal side of the embryo. *engrailed* is initially
expressed in a single line of cells at the anterior margin of each parasegment, which is itself only about three cells wide (Fig. 5.25). It is likely that
this periodic pattern of *engrailed* activity is the result of the combinatorial
action of transcription factors encoded by pair-rule genes, including *fushi
tarazu* and *even-skipped*. Evidence that the pair-rule genes do control *engrailed*
expression is provided, for example, by embryos carrying mutations in *fushi*

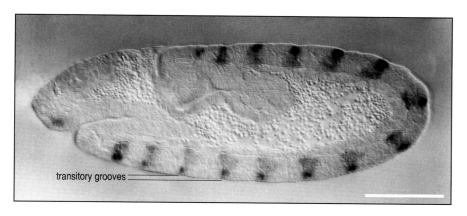

transitory grooves

Fig. 5.24 The expression of the *engrailed* gene in a late (stage 11) *Drosophila* embryo. The gene is expressed in the anterior region of each parasegment and the transitory grooves between parasegments can be seen. At this stage of development, the germ band has temporarily extended and is curved over the back of the embryo. Scale bar = 0.1 mm.

Photograph from Lawrence, P.: 1992.

tarazu, in which *engrailed* expression is absent only in even-numbered parasegments (in which *fushi tarazu* is normally expressed).

The anterior margin of the parasegment has a very important property—it is a boundary of **cell-lineage restriction** (such as the boundaries between the rhombomeres of the vertebrate hindbrain discussed in Section 4.10). Cells and their descendants from one parasegment never move into adjacent ones. This implies that the cells in a parasegment are under some common genetic control which both prevents them mixing with their

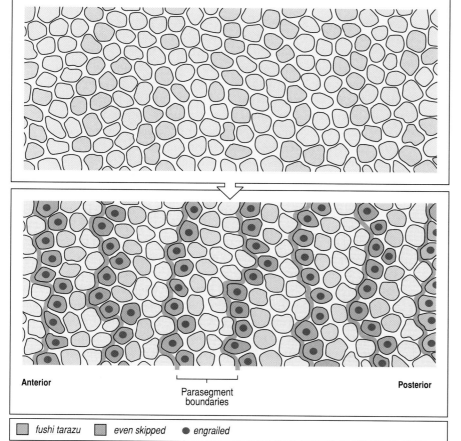

Anterior

Parasegment boundaries

Posterior

☐ *fushi tarazu*   ☐ *even skipped*   ● *engrailed*

Fig. 5.25 The expression of the pair-rule genes *fushi tarazu* (blue), *even-skipped* (pink), and *engrailed* (purple dots) in parasegments. *engrailed* is expressed at the anterior margin of each stripe and delimits the anterior border of each parasegment. The boundaries of the parasegments become sharper and straighter later on. After Lawrence, P.: 1992.

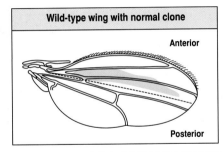

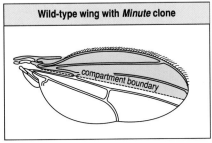

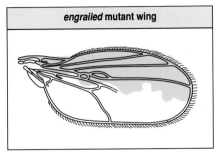

Fig. 5.26 The boundary between anterior and posterior compartments in the wing can be demonstrated by marked cell clones. Top panel: in the wild-type wing, clones marked by mitotic recombination in the embryo, which gives marked cells a phenotype different from other cells of the wing, are too small to demonstrate the compartment boundary. Middle panel: the use of the *Minute* technique (see Box 5B, p. 173) produces an increased rate of cell division in the marked cell and gives clones large enough to determine that cells from one compartment do not cross the boundary into an adjoining compartment. Bottom panel: in the wing of an *engrailed* mutant in which the engrailed protein is not produced, there is no posterior compartment or boundary. Clones in the anterior part of the wing cross over into the posterior region and the posterior region is transformed into a more anterior-like structure, bearing anterior-type hairs on its margin. The *engrailed* gene is required for the maintenance of the character of the posterior compartment, and for the formation of the boundary.

neighbors and controls their later development. Such domains of lineage restriction are known as **compartments**.

The existence of compartments can be detected by cell lineage studies. Cell lineage can be followed by marking single cells in the early embryo so that all their descendants (clones) can be identified at later stages of development. One technique is to inject the egg with a harmless fluorescent compound which is incorporated into all the cells of the embryo. In the early embryo, a fine beam of ultraviolet light directed onto a single cell activates the fluorescent compound. All the descendants of that cell will be fluorescent and can therefore be identified. Examination of these clones, together with an analysis of *engrailed* expression, shows that cells at the anterior margin of the parasegment have no descendants that lie on the other side of that margin. The anterior margin is therefore a boundary of lineage restriction; that is, cells and their descendants that are on one or other side of the boundary when it is formed, never cross it.

The lineage restriction of the anterior margin of the parasegment is carried over into the segments of the larva and the adult. But because the anterior part of a parasegment becomes the posterior part of a segment, the lineage restriction within the segment is between anterior and posterior regions. The segment is thus divided into anterior and posterior compartments, with *engrailed* expression defining the posterior compartment. A compartment can be defined as a region in the embryo that contains all the descendants of the cells present when the compartment is set up, and no others.

In a compartment, the cells may also all be under a common genetic control. Imagine a group of cells becoming divided into two separate regional populations by means of signals that turn on different genes in each region. The regions will be compartments if there is never any mixing of the descendants of the original cells across the boundary between them. This is illustrated by the behavior of cells in the wing of the adult fly. The adult wing is chosen as it is easier to distinguish the lineage restriction in the adult structures, in which there has been considerable cell division, compared to embryonic and early larval structures where there has been little cell division since the initial determining events.

Compartments are made visible by making genetically mosaic flies composed of two distinguishable kinds of cells (Box 5B). Single cells in the embryonic blastoderm, or in the epidermis of the larva, can be given a distinctive phenotype by X-ray- or laser-induced mitotic recombination. The fate of all the descendants of this marked cell can then be followed. Their behavior depends on the stage of development at which the founder nucleus was marked. Descendants of nuclei marked at early stages during cleavage become part of many tissues and organs, but those of nuclei marked at the blastoderm stage or later have a more restricted fate. They are found only in the anterior or the posterior part of each segment (or of an appendage like a wing), never throughout the whole segment.

Because the cells of imaginal discs divide only about ten times after the cellular blastoderm stage, the clones of marked cells are small even in the adult, and it is not easy to detect a boundary of lineage restriction (Fig. 5.26, top panel). Clone size can be increased by the *Minute* technique, which results in the marked cell dividing many more times than the other cells (see Box 5B). A single clone of such cells can almost fill either the anterior or posterior part of the wing, meaning that the boundary that the cells never

Box 5B Genetic mosaics and mitotic recombination

Genetic mosaics are embryos derived from a single genome but in which there is a mixture of cells with rearranged or inactivated genes. In flies, genetic mosaics can be generated by inducing rare mitotic recombination events in the embryo or larva. Chromosome breaks induced with X-rays result in the exchange of material between homologous chromosomes, just after the chromosomes have replicated into chromatids. Such an event can generate a single cell with a unique genetic construction that will be inherited by all the cell's descendants, which in flies usually form a coherent patch of tissue.

Easily distinguishable mutations like the recessive *multiple wing hairs* can be used to identify the marked clone; if a cell homozygous for this mutation is generated by mitotic recombination in a heterozygous larva, all of its descendant cells will have multiple hairs (see figure).

Marked epidermal clones made by this method are usually small as there is little cell proliferation after the recombination event. Larger clones can be made using the *Minute* technique. The cells in flies carrying a mutation in the *Minute* gene grow more slowly than those in the wild type. By using flies heterozygous for the *Minute* mutation, clones can be made in which the mitotic recombination event has generated a marked cell that is normal because it has lost the *Minute* mutation, and so is wild type. This normal cell proliferates faster than the slower-growing background *Minute* heterozygous cells and thus large clones of the marked cells are produced (see Fig. 5.26). The mitotic recombination technique has many applications. If clones of marked cells are generated at different stages of development, one can trace the fate of the altered cells and thus see what structures they are able to contribute to. This can provide information on their state of determination or specification at different developmental stages. The technique can also be used to study the localized effects of homozygous recessive mutations that are lethal if present in the homozygous state throughout the whole animal. Illustration after Lawrence, P.: 1992.

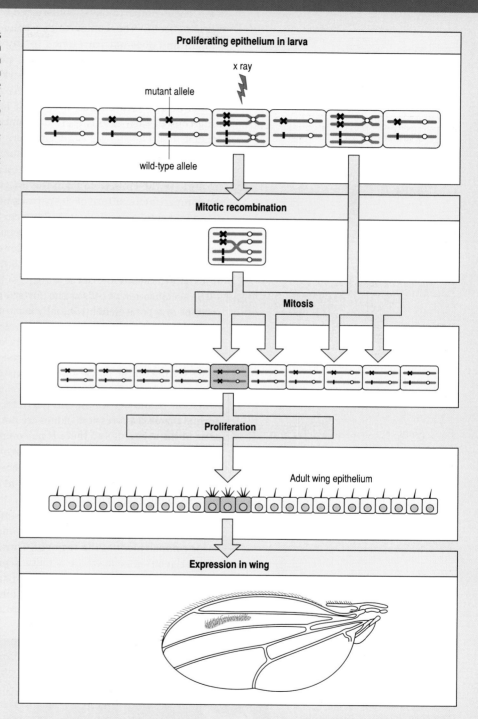

transgress becomes much more evident (see Fig. 5.26, middle panel). This boundary separates the anterior and posterior compartments. In a normal wing, each compartment is constructed by all the descendants of a set of founding cells. The compartment boundary is remarkably sharp and straight and does not correspond to any structural features in the wing. These experiments also show that the pattern of the wing is in no way dependent on cell lineage. A single marked embryonic cell can give rise to about a twentieth of the cells of the adult wing or, using the *Minute* technique to increase clone size, about half the wing. The lineage of the wing cells in each case is quite different, yet the wing's pattern is quite normal.

The compartment pattern in the adult *Drosophila* wing is carried over from its initial specification in the imaginal discs. When epidermal cells are set aside in the embryo to form the imaginal discs, each disc carries over the compartment pattern of the parasegments from which it arises. A wing disc, for example, develops at the boundary between the two parasegments that contribute to the second thoracic segment. Thus, a wing is divided into an anterior compartment and a posterior compartment, the compartment boundary (the old parasegment boundary) running in a straight line more or less down the middle of the wing.

The specification of cells as the posterior compartment of a segment (the anterior of a parasegment) initially occurs when the parasegments are set up, and is due to the *engrailed* gene. Expression of *engrailed* is required both to confer a 'posterior segment' identity on the cells and to change their surface properties so that they cannot mix with the cells adjacent to them, hence setting up the parasegment (compartment) boundary.

Direct evidence for this comes from examining the behavior of clones of wing cells in an *engrailed* mutant (Fig. 5.26, bottom panel). In the absence of normal *engrailed* expression, clones are not confined to anterior or posterior parts of the segment and there is no compartment boundary. Moreover, in *engrailed* mutants, the posterior compartment is partly transformed so that it comes to resemble the pattern of the anterior part of the wing. For example, bristles normally found only at the anterior margin of the wing are also found at the posterior margin.

The *engrailed* gene needs to be expressed continuously throughout larval and pupal stages and into the adult to maintain the character of the posterior compartment of the segment. Thus, *engrailed* is also an example of a selector gene—a gene whose activity is sufficient to cause cells to adopt a particular fate. Selector genes can control the development of a region such as a compartment and, by controlling the activity of other genes, give the region a particular identity.

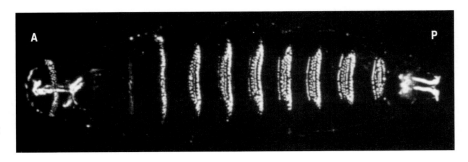

Fig. 5.27 Each segment of the *Drosophila* larva has a characteristic pattern of denticles on its ventral surface. Denticles are confined to the anterior regions of segments, with each segment having its own pattern.

5.16 Segment polarity genes pattern the segments and stabilize parasegment and segment boundaries

Each larval segment has a well-defined antero-posterior pattern, which is easily seen on the ventral epidermis of the abdomen: the anterior region of each segment bears denticles (outgrowths of the chitinous cuticle), and the posterior region is naked (Fig. 5.27). The rows of denticles, of which there are six types, make a distinct pattern, and this is thought to reflect an antero-posterior polarity gradient in each segment. Mutations in segment polarity genes often alter the denticle pattern, and this is how they were first discovered. For example, a mutation in the segment polarity gene *wingless* results in the whole of the ventral abdomen being covered in denticles, but in the posterior half of each segment the denticle pattern is reversed. In this mutant, the anterior region of each segment is duplicated in mirror-image polarity, and the posterior region is lost. Mutations in the segment polarity gene *hedgehog* give a similar phenotype. *wingless* encodes a secreted protein and is related to the *Wnt* genes, which are key players in signaling during pattern formation in vertebrates (see Chapters 3 and 4). Hedgehog is a secreted or membrane-associated protein which binds to the membrane protein patched and lifts the inhibition by patched of signaling by another membrane protein, smoothened (Fig. 5.28). We shall meet the hedgehog and wingless proteins again as, like many other signaling proteins, they are involved in more than one patterning process during development.

The larval denticle patterns depend on the correct establishment and maintenance of parasegment boundaries. Segment polarity genes are expressed in restricted domains within the parasegment (Fig. 5.29). Maintenance of a parasegment boundary depends on an intercellular signaling circuit being set up between adjacent cells on either side of the boundary. This circuit involves interactions between *engrailed*, *wingless*, *hedgehog* and other segment polarity genes. On one side of the boundary, cells expressing *engrailed* also express and secrete the hedgehog protein, which acts on the adjacent cells on the other side of the boundary to maintain *wingless* expression in those cells. The secreted wingless protein in turn provides a signal that feeds back over the boundary to maintain *hedgehog* and *engrailed* expression, thus stabilizing and maintaining the compartment boundary (Fig. 5.30). A computer model of this circuitry shows that it is remarkably robust and resistant to variations in the genetic constants governing

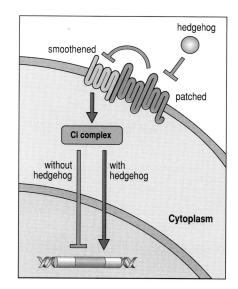

Fig. 5.28 The hedgehog signaling pathway. When hedgehog protein binds to the membrane receptor patched, this receptor no longer inhibits the membrane protein smoothened, which results in the initiation of a signal transduction pathway. This acts on the Cubitus interruptus (Ci) protein complex, releasing the Ci protein which then enters the cell nucleus and acts as a gene activator. In the absence of this pathway, the Ci complex releases a part of the Ci protein which acts as a repressor in the nucleus. Other proteins can act as inhibitors of the complex, preventing gene activation.

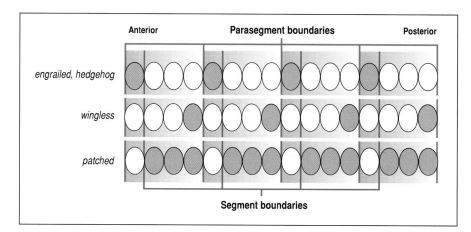

Fig. 5.29 The domains of expression of the segment polarity genes. At the cellular blastoderm stage, the *engrailed* gene is expressed at the anterior margin of each parasegment together with *hedgehog*, while *wingless* is expressed at the posterior margin. After gastrulation, *patched* is expressed in all of the cells that express neither *engrailed* nor *hedgehog*. About two cell divisions occur between the time that the parasegments are delimited and the time the embryo hatches.

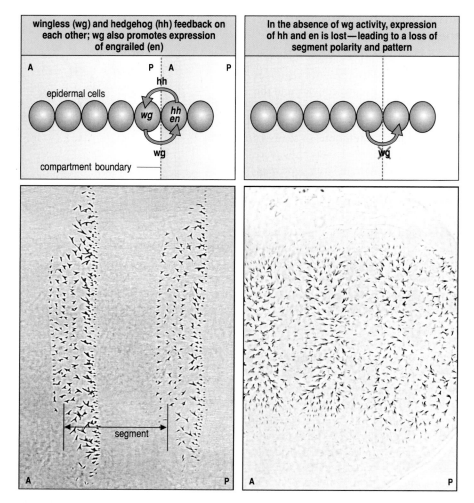

Fig. 5.30 Interactions between *hedgehog*, *wingless*, and *engrailed* genes and proteins at the compartment boundary control denticle pattern. Top left panel: the *engrailed* gene, which encodes a homeodomain transcription factor, is expressed in cells along the anterior margin of the parasegment; these cells also express the segment polarity gene *hedgehog* and secrete the hedgehog protein. hedgehog protein activates and maintains expression of the segment polarity gene *wingless* in adjacent cells across the compartment boundary. wingless protein feeds back on *engrailed*-expressing cells to maintain expression of the *engrailed* gene and *hedgehog*. These interactions stabilize and maintain the compartment boundary. Top right panel: in mutants where the *wingless* gene is inactivated and the wingless protein is absent, neither *hedgehog* nor *engrailed* genes are expressed. This leads to loss of the compartment boundary and the normally well-defined pattern of denticles within each abdominal segment. Bottom left panel: in the wild-type larva the denticle bands in the ventral cuticle are confined to the anterior part of the segment and are dependent on the activity of *hedgehog* and *wingless* genes. Bottom right panel: in the *wingless* mutant, denticles are present across the whole ventral surface of the segment in what looks like a mirror-image repeat of the anterior segment pattern.

Photographs from Lawrence, P.: 1992.

the circuit's behavior. In the absence of *wingless* activity the pattern is disrupted.

The pattern of denticle belts is specified in response to the signals provided by the hedgehog and wingless proteins see Fig. 5.30. Hedgehog signaling coming from cells at the boundary patterns the cuticle both in the compartment anterior to it and in the compartment in which it is expressed. These signals lead, by a complex set of interactions, to the expression of a number of other genes in non-overlapping narrow stripes, which specify the cuticle pattern.

Other segment polarity genes encode the intracellular components of the signaling pathways and receptors for secreted signaling molecules such as hedgehog and wingless. Once compartment boundaries are consolidated, gradients of signals that pattern the segment are then thought to be set up within these boundaries. The wingless protein, for example, becomes distributed asymmetrically; its movement is less in cells expressing *engrailed*, and so its movement posterior to the boundary is less than that in an anterior direction.

Patterning of the cuticle within compartment boundaries is reflected in the very well-defined cuticle pattern on the abdominal segments of the adult fly (Fig. 5.31). The cuticle of each segment comprises a number of cuticle types—at least five in the anterior compartment and three in the posterior compartment. *Engrailed* (*en*) is 'on' in the posterior compartment, and determines it as a posterior compartment. Clones of cells in which *engrailed* is not expressed, *en⁻* clones, transform into anterior-type cuticle.

5.17 Compartment boundaries are involved in patterning and polarizing segments

Good evidence for the role of a compartment boundary, in this case a segment boundary, in controlling pattern and polarization of the segment comes from structures on the epidermis of two other insects—the plant-feeding bug *Oncopeltus*, and the wax moth *Galleria*. *Oncopeltus* adults have a large number of hairs covering each segment and all the hairs point in an anterior to posterior direction, just as if they were little arrows (Fig. 5.32, left panel). Some individuals have a gap in the segment boundary and this results in a rather precise pattern of hairs with altered orientation: many of the hairs near the gap point in the reverse direction (Fig. 5.32, center and right panels). One can explain this by assuming that there is a gradient—possibly of a morphogen—running from the anterior boundary to the posterior boundary of each segment. If the slope of the gradient gives the hairs their polarity, they will always point down the gradient. A gap in a segment boundary would result in a local change in the gradient,

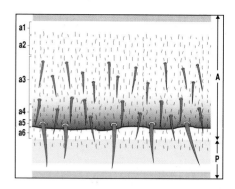

Fig. 5.31 The cuticle of each segment in the abdomen of the adult *Drosophila* is divided into an anterior and a posterior compartment. A number of different regions, a1 to a6, can be distinguished in the anterior compartment, characterized by different bristles, pigmentation and gene expression.

Fig. 5.32 Gradients could specify polarity in the segments of *Oncopeltus*. The hairs on the cuticle point in a posterior direction and this may reflect an underlying gradient in a morphogen that is partly maintained by the segment boundary (left panels). When there is a gap in the boundary (center panels) the sharp discontinuity in morphogen concentration smoothes out, with a resulting local reversal of the gradient and of the direction in which the hairs point. This is illustrated in the photograph (right panel).

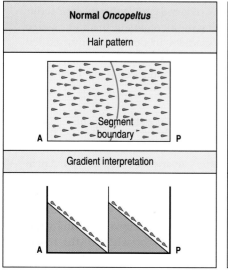
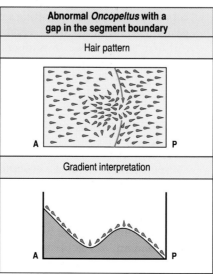
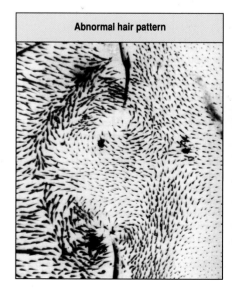

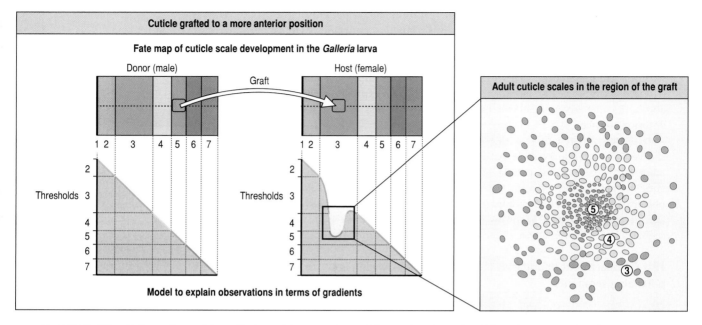

Fig. 5.33 Positional information could specify the cuticle pattern in *Galleria*. There are seven types of cuticle scale in each segment of the adult moth, arranged in consecutive bands. If a small piece of cuticle from a male larva is grafted to a more anterior position in the segment of a female, there is a local alteration in both the pattern and orientation of the cuticle scales in the adult. Grafts are carried out between males and females as female cells have distinctive nuclei, so the transplanted cells and their descendants can be identified. The new pattern can be understood in terms of changes in a gradient in the segment that provides positional information for scale development.

the sharp change in concentration at the normal boundary being smoothed out and forming a local gradient running in the opposite direction. This would result in the hairs in this region pointing in the opposite direction.

Evidence for such a gradient, providing not only polarity put positional information, is provided by grafting experiments on the larval cuticle of *Galleria* (Fig. 5.33). In the adult moth there are seven different types of cuticle in each segment. Grafting a piece of larval cuticle to a more anterior position results in a striking repatterning in the region of the graft, which can best be explained in terms of a gradient in positional information determining the character and polarity of the cuticle in each region. While both the above observations are open to other interpretations, even these involve gradients.

5.18 Some insects use different mechanisms for patterning the body plan

Drosophila belongs to an evolutionarily advanced group of insects, in which all the segments are specified more or less at the same time in development, as shown both by the striped patterns of pair-rule gene expression in the acellular blastoderm and the appearance of all the segments shortly after gastrulation. This type of development is known as **long-germ development**, as the blastoderm corresponds to the whole of the future embryo. All of the segments form at about the same time. Many other insects, such as the flour beetle *Tribolium*, have a **short-germ development**. In short-germ development the blastoderm is short and forms only anterior segments. The posterior segments are formed after completion of the blastoderm stage and gastrulation. Thus, most segments are formed from a cellular blastoderm and posterior segments are added by growth in the posterior region (Fig. 5.34). In spite of early differences between them, the mature germ band stages of both long- and short-germ insect embryos look similar. This, therefore, is a common stage—the **phylotypic stage** (see Fig. 2.2)—through which all insect embryos develop.

The question clearly arises as to what processes are common to the specification of the body plan in long- and short-germ insects. One clear difference is that whereas the patterning of the body plan in the long-germ band of *Drosophila* takes place before cell boundaries form, much of the body plan in short-germ insects is laid down at a later stage, when the posterior segments are generated during growth. At this stage the embryo is multicellular. So are the same genes involved?

There is very good evidence that the same genes and development processes are involved in the patterning of *Tribolium* and *Drosophila*. For example, the gap gene *Krüppel* is expressed at the posterior end of the *Tribolium* embryo at the blastoderm stage and not in the middle region, as in *Drosophila* (Fig. 5.35). It therefore seems to be specifying the same part of the body in the two insects. Similarly, only two repeats of the pair-rule stripes are present at the blastoderm stage, together with a posterior cap of pair-rule gene expression, in contrast to the seven in *Drosophila*. The genes *wingless* and *engrailed* are also expressed in a similar relation to that in *Drosophila*.

Although not many genes have been studied in detail in other insects, at least one, the segment polarity gene *engrailed*, is known to be expressed in the posterior region of segments in a variety of insects. The pair-rule gene *even-skipped* (see Section 5.14), while present in the grasshopper (a short-germ insect), may not have a similar role in segmentation. It is, however, involved later in the development of the nervous system in the grasshopper, and is also expressed at the posterior end of the growing germ band.

Experiments with another insect, the leaf-hopper *Euscelis*, show that it has a mechanism of antero-posterior axis determination with a strong resemblance to the mechanism involving the *bicoid* gradient (see Section 5.2). *Euscelis* has a mode of development intermediate between long- and short-germ insects. It has a rather long egg which contains a ball of symbiotic bacteria at the posterior end. This ball can be moved with a microneedle to more anterior regions. Some posterior cytoplasm is moved along with it, thus providing a serendipitous cytoplasmic marker, which is useful in transplantation experiments.

Two experiments provide evidence for a morphogen gradient in the

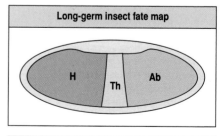

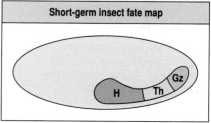

Fig. 5.34 Differences in the development of long-germ and short-germ insects. Top panel: the general fate map of long-germ insects such as *Drosophila* shows that the whole of the body plan—head (H), thorax (Th), and abdomen (Ab)—is present at the time of initial germ-band formation. Bottom panel: in short-germ insects, only the anterior regions of the body plan are present at this embryonic stage. Most of the abdominal segments develop later, after gastrulation, from a posterior growth zone (Gz).

Fig. 5.35 Gap and pair-rule gene expression in long-germ and short-germ insects at the time of germ band formation. *Krüppel* (red) is a gap gene and *hairy* (green) is a pair-rule gene. The position of the *Krüppel* stripe in the short-germ embryo of *Tribolium* indicates that posterior regions of the body are not yet present at this stage. Similarly, only three *hairy* stripes, corresponding to the first three *hairy* stripes in a long-germ embryo, are present in *Tribolium*.

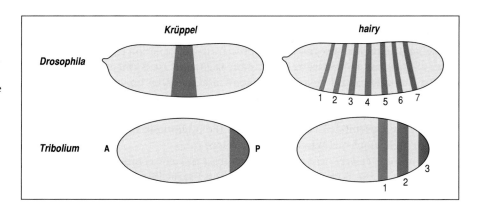

Euscelis egg, with the high point at the posterior end. First, if a ligature is tied around the fertilized egg to prevent communication between the anterior and posterior regions, the result is a gap in the body plan, with some regions failing to develop. Second, if the ball of bacteria with posterior cytoplasm is moved anteriorly, and then a ligature is tied behind it (Fig. 5.36), a typical result is that a complete set of structures forms anterior to the ligature, while a mirror-image set of incomplete structures forms

Fig. 5.36 An antero-posterior gradient of morphogen in the egg of the leaf-hopper *Euscelis* appears to direct development along the antero-posterior axis. Top panels: *Euscelis* eggs carry a ball of symbiotic bacteria at the posterior pole, and some posterior cytoplasm can be moved toward the middle of the embryo by pushing the ball forward. The egg can then be ligatured so that postulated diffusible factors emanating from the transplanted cytoplasm and the posterior pole cannot cross the ligature. This manipulation results in alteration of the pattern of segments in line with the presence of a morphogen gradient in the normal egg, with a high point in the posterior cytoplasm. Bottom panel: isolated germ-band stage embryo of normal *Euscelis*. Capital letters indicate body regions identified in the experiments illustrated in the top panels.

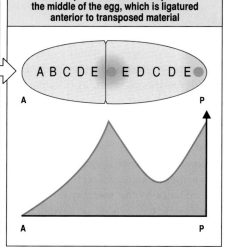

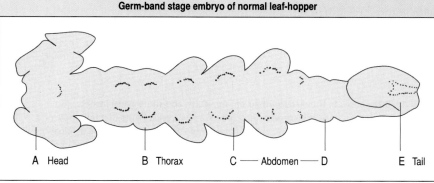

posteriorly. Both these results can be accounted for in terms of a diffusible morphogen gradient, with its high point and source at the posterior end.

Looking at some other insects, differences in early development are much more dramatic. In certain parasitic wasps, the egg is small and undergoes cleavage to form a ball of cells, which then falls apart. Each of the resulting small clusters of cells—there can be as many as 400—can develop into a separate embryo. This wasp's development apparently does not depend on maternal information to specify the body axes, and in this respect resembles that of the early mammalian embryo.

Summary

The segment polarity genes are involved in patterning the parasegments. One of the first genes to be activated is *engrailed*, which is expressed at the anterior margin of each parasegment, delineating a group of cells which becomes the posterior compartment of a segment. *engrailed* is also a selector gene in that it provides a group of cells with a long-term regional identity. *engrailed* expression shows lineage restriction: cells expressing *engrailed* define the posterior compartment of a segment, and cells never cross from the posterior into the anterior compartment. *engrailed* is turned on by the pair-rule genes, and its expression is maintained by the segment polarity genes *wingless* and *hedgehog*, which stabilize the compartment boundary. Studies in other insects suggest that a discrete gradient of positional information is set up in each segment, delimited by the boundaries, and this patterns the segment. In contrast to *Drosophila*, some insects have a short-germ pattern of development in which posterior segments are added by growth after the cellular blastoderm stage.

Summary: segment polarity gene expression defines segment compartments

pair-rule gene expression

segment polarity and selector gene *engrailed* activated in anterior of each parasegment, defining anterior compartment of parasegment and posterior compartment of segment

engrailed-expressing cells also express segment polarity gene *hedgehog*

cells on other side of compartment boundary express segment polarity gene *wingless*

engrailed expression maintained and compartment boundary stabilized by wingless and hedgehog proteins

compartment boundary provides signaling center from which segment is patterned

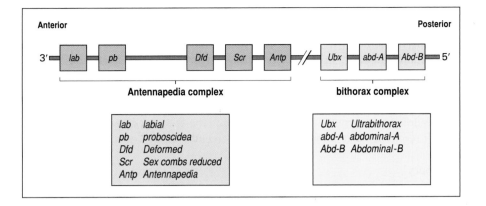

Fig. 5.37 The Antennapedia and bithorax homeotic selector gene complexes. The order of the genes from 3′ to 5′ in each complex reflects both the order of their spatial expression (anterior to posterior) and the timing of expression (3′ earliest).

Segmentation: selector and homeotic genes

Each segment has a unique identity, most easily seen in the characteristic pattern of denticles on the ventral surface of the larva (see Fig. 5.27). Since the same segment polarity genes are turned on in each segment, what makes the segments different from each other? Specification of segment identity is carried out by a class of master regulatory genes, known as **homeotic selector genes**, which set the future developmental pathway of each segment. A selector gene controls the activity of other genes and is required throughout development to maintain this pattern of gene expression. The *Drosophila* selector genes that control segment identity are organized into two gene complexes (Fig. 5.37), which together are broadly homologous to a single Hox gene complex in vertebrates (see Box 4A, p. 117). As we saw in Chapter 4, such genes also control patterning in vertebrates; however, they were first identified in *Drosophila*, where they are also known as HOM genes, and it is in *Drosophila* that their actions are best understood.

The two homeotic complexes in *Drosophila*—the bithorax and Antennapedia gene complexes—are named after the unusual and striking mutations that first revealed their existence. Flies with the *bithorax* mutation have part of the haltere (the balancing organ on the third thoracic segment) transformed into part of a wing (Fig. 5.38), whereas flies with the dominant *Antennapedia* mutation have their antennae transformed into legs. Genes identified by such mutations are called **homeotic genes** because when mutated they result in **homeosis**—the transformation of a whole segment or structure into another related one, as in the transformation of antenna to leg. These bizarre transformations arise out of the homeotic selector genes' key role as positional identity specifiers. They control the activity of other genes in the segments, thus determining, for example, that a particular imaginal disc will develop as wing or haltere. The bithorax complex controls the development of parasegments 5–14, while the Antennapedia complex controls the identity of the more anterior parasegments. The action of the bithorax complex of selector genes is the best understood and will therefore be discussed first.

Neither pair-rule nor Hox genes are involved in the specification and formation of the anterior head segments of *Drosophila*. Instead, the genes *orthodenticle*, *empty spiracles* and *buttonhead* have key roles. Only *empty*

Fig. 5.38 Homeotic transformation of the wing and haltere by mutations in the bithorax complex. Top panel: in the normal adult, both wing and haltere are divided into an anterior (A) and a posterior (P) compartment. Middle panel: the *bithorax* mutation transforms the anterior compartment of the haltere into an anterior wing region. The mutation *postbithorax* acts similarly on the posterior compartment, converting it into posterior wing (not illustrated). Bottom panel: if both mutations are present together, the effect is additive and the haltere is transformed into a complete wing, producing a four-winged fly.

Photograph courtesy of E. Lewis, from Bender, W., et al.: 1983. Illustration after Lawrence, P.: 1992.

spiracles has a Hox-like homeobox. Like the Hox genes, however, these genes specify head segments anterior to the mandibular by their combinatorial pattern of expression.

5.19 Homeotic selector genes of the bithorax complex are responsible for diversification of the posterior segments

The bithorax complex of *Drosophila* comprises three homeobox genes: *Ultrabithorax*, *adbominal-A*, and *Abdominal-B*. These genes are expressed in the parasegments in a combinatorial manner (Fig. 5.39, top panel). *Ultrabithorax* is expressed in all parasegments from 5–12, *abdominal-A* is expressed more posteriorly in parasegments 7–13, and *Abdominal-B* more posteriorly still from parasegment 10 onwards. Because the genes are also active to varying extents in different parasegments, their combined activities define the character of each parasegment. *Abdominal-B* also suppresses *Ultrabithorax*, such that *Ultrabithorax* expression is very low by parasegment 14, as *Abdominal-B* expression increases. The pattern of activity of the bithorax complex genes is determined by gap and pair-rule genes.

The role of the bithorax complex was first indicated by classical genetic experiments. In larvae lacking the whole of the bithorax complex (see Fig. 5.39, second panel), every parasegment from 5–13 develops in the same way and resembles parasegment 4. The bithorax complex is therefore esential for the diversification of these segments, whose basic pattern is represented by parasegment 4. This parasegment can be considered as being a type of 'default' state, which is modified in all the parasegments posterior to it by the proteins encoded by the bithorax complex. It is because the genes of the bithorax complex are able to superimpose a new identity on the default state that they are called selector genes.

An indication of the role of each gene of the bithorax complex can be deduced by looking at embryos constructed so that the genes are put one at a time into embryos lacking the whole complex (see Fig. 5.39, bottom three panels). If only the *Ultrabithorax* gene is present, the resulting larva has one parasegment 4, one parasegment 5, and eight parasegments 6. Clearly, *Ultrabithorax* has some effect on all parasegments from 5 backward and can specify parasegments 5 and 6. If *abdominal-A* and *Ultrabithorax* are put into the embryo, then the larva has parasegments 4, 5, 6, 7 and 8, followed by five parasegments 9. So *abdominal-A* affects parasegments from 7 backward, and in combination, *Ultrabithorax* and *abdominal-A* can specify the character of parasegments 7, 8 and 9. Similar principles apply to *Abdominal-B*, whose domain of influence extends from parasegment 10 backward, and which is expressed most strongly in parasegment 14. Differences between segments may reflect differences in the spatial and temporal pattern of HOM gene expression.

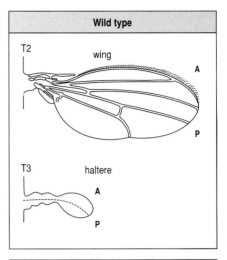

Wild type

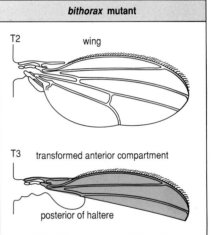

bithorax mutant

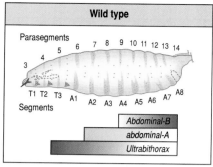

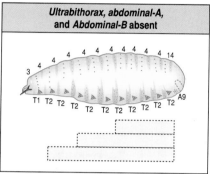

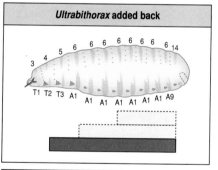

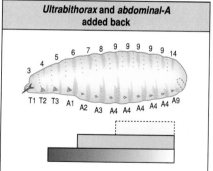

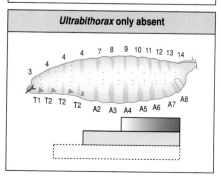

Fig. 5.39 The spatial pattern of expression of genes of the bithorax complex characterizes each parasegment. In the wild-type embryo (top panel) the expression of the genes *Ultrabithorax*, *abdominal-A*, and *Abdominal-B* are required to confer an identity on each parasegment. Mutations in the bithorax complex result in homeotic transformations in the character of the parasegments and the segments derived from them. When the bithorax complex is completely absent (second panel), parasegments 5–13 are converted into nine parasegments 4 (corresponding to segment T2 in the larva), as shown by the denticle and bristle patterns on the cuticle. The three lower panels show the transformations caused by the absence of different combinations of genes. When the *Ultrabithorax* gene alone is absent (bottom panel), parasegments 5 and 6 are converted into 4. In each case the spatial extent of gene expression is detected by *in situ* hybridization (see Box 3B, p. 70). Note that the specification of parasegment 14 is relatively unaffected by the bithorax complex.

These results illustrate an important principle, namely that the character of the parasegments is specified by the genes of the bithorax complex acting in a combinatorial manner. Their combinatorial effect can also been seen by taking the genes away, one at a time, from the wild type. In the absence of *Ultrabithorax*, for example, parasegments 5 and 6 become converted to parasegment 4 (see Fig. 5.39, bottom panel). There is a further effect on the cuticle pattern in parasegments 7–14 in that structures characteristic of the thorax are now present in the abdomen, showing that *Ultrabithorax* exerts an effect in all these segments. Such abnormalities may result from expression of 'nonsense' combinations of the bithorax genes. For example, in such a mutant, *abdominal-A* protein is present in parasegments 7–9 without *Ultrabithorax* protein, and this is a combination never found normally.

While the gap and pair-rule proteins control the pattern of HOM gene expression, these proteins disappear after about 4 hours. The continued correct expression of the homeotic genes involves two groups of genes— the Polycomb and Trithorax groups. The proteins of the Polycomb group maintain transcriptional repression of homeotic genes where they are initially off, while the Trithorax group maintain expression in those cells where HOM genes have been turned on. These genes thus maintain expression of the HOM complex.

The downstream targets of the proteins coded for by the HOM genes are largely unknown, but there may be a very large number of genes whose expression is affected.

5.20 The Antennapedia complex controls specification of anterior regions

The Antennapedia complex comprises five homeobox genes (see Fig. 5.37), which control the behavior of the parasegments anterior to parasegment 5 in a manner similar to the bithorax complex. Since there are no novel principles involved we just mention its role briefly. Several genes within the complex are critically involved in specifying particular parasegments. Mutations in the gene *Deformed* affect the ectodermally derived structures of parasegments 0 and 1, those in *Sex combs reduced* affect parasegments 2 and 3, and those in the *Antennapedia* gene affect parasegments 4 and 5.

5.21 **The order of HOM gene expression corresponds to the order of genes along the chromosome**

The bithorax and Antennapedia complexes possess some striking features of gene organization. In both, the order of the genes in the complex is the same as the spatial and temporal order in which they are expressed along the antero-posterior axis during development. *Ultrabithorax*, for example, is on the 3′ side of *abdominal-A* on the chromosome, is more anterior in its pattern of expression, and is activated earlier. As we have already seen, the related Hox gene complexes of vertebrates (see Box 4A, p. 117), whose ancestors diverged from those of arthropods hundreds of millions of years ago, show the same correspondence between gene order and order of expression. This highly conserved temporal and spatial co-linearity of gene order and expression must be related to the mechanisms that control the expression of these genes.

The complex yet subtle control to which the genes of the bithorax complex are subject is seen in experiments in which production of Ultrabithorax protein is forced in all segments. This is achieved by linking the Ultrabithorax protein-coding sequence to a heat-shock promoter—a promoter that is activated at 29°C—and introducing the novel DNA construct into the fly genome using a P element (see Box 5A, p. 163). When this transgenic embryo is given a heat shock for a few minutes, the extra *Ultrabithorax* gene is transcribed and the protein is made at high levels in all cells. This has no effect on posterior parasegments (in which Ultrabithorax protein is normally present), with the exception of parasegment 5 which, for unknown reasons, is transformed into parasegment 6 (this may reflect a quantitative effect of the Ultrabithorax protein). However, all parasegments anterior to 5 are also transformed into parasegment 6. While this seems a reasonably simple and expected result, consider what happens to parasegment 13.

Transcription of *Ultrabithorax* is normally suppressed in parasegment 13 in wild-type embryos, yet when the protein itself is produced as a result of heat shock, it has no effect. By some mechanism, the Ultrabithorax protein is rendered inactive in this parasegment. This phenomenon is quite common in the specification of the parasegments and is known as phenotypic suppression or posterior prevalence (see Section 4.4): HOM gene products normally expressed in anterior regions are suppressed by more posterior products.

While the role of the bithorax and Antennapedia complexes in controlling segment identity is well established, we do not yet know very much about their interactions with the downstream target genes that specify the structures that give the segments their unique identities. What, for example, is the pathway that leads to a thoracic rather than an abdominal segment structure? Are just a few or many target genes involved? It may be that there are many targets with different functions which act in concert to form a particular structure. This may help to understand how changes in a single gene could cause a homeotic transformation like antenna into a leg.

The role of the Hox genes in the evolution of different body patterns will be considered in Chapter 15. We finish this chapter by considering one downstream pathway that has been partly worked out—the influence of HOM gene expression in the mesoderm on the segment-specific development of the endoderm.

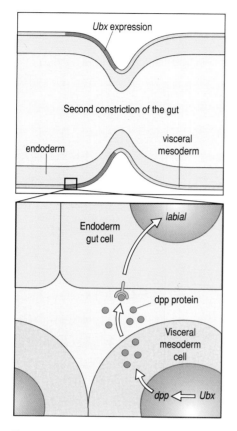

Fig. 5.40 Gene expression in the visceral mesoderm patterns the underlying gut endoderm. Top panel: *Ultrabithorax* (*Ubx*) is expressed in the visceral mesoderm near the second constriction of the gut. Bottom panel: it activates the gene *decapentaplegic* (*dpp*) in visceral mesoderm cells. This results in secretion of decapentaplegic protein from the visceral mesoderm, which induces expression of the gene *labial* in the adjacent cells of the gut endoderm. Patterning laid down in the visceral mesoderm is thus transferred to the gut endoderm.

5.22 HOM gene expression in visceral mesoderm controls the structure of the adjacent gut

Discussion of the actions of the gene *engrailed* and of the bithorax and Antennapedia complexes has so far referred only to their effects on ecto-dermal structures, especially the epidermis or cuticle. These genes are also expressed in internal tissues such as the somatic and visceral mesoderm of the embryo. The somatic mesoderm gives rise to the main body muscles, while the visceral mesoderm generates the smooth muscle surrounding the gut. In the somatic mesoderm, the pattern of expression of the bithorax complex is simpler than in the ectoderm, but in general corresponds with it. The visceral mesoderm, however, deserves special attention as HOM gene expression there appears to induce pattern in the gut endoderm.

The developing midgut has three constrictions, the second of which occurs in parasegment 7. Most of the HOM selector genes are not expressed in the endoderm, and its segment-specific character is conferred by induc-tion as a result of HOM gene expression in the visceral mesoderm surround-ing it. The detailed pattern of expression of the bithorax complex in the visceral mesoderm is somewhat different from that in both ectoderm and somatic mesoderm. The *Ultrabithorax* gene, for example, is expressed only in the parasegment of the visceral mesoderm adjacent to the second of the three gut constrictions (Fig. 5.40, top panel). If *Ultrabithorax* expression is absent, this constriction does not develop and the gut is abnormal.

Ultrabithorax itself is not expressed in the endoderm. Rather, it appears to exert its effect on the gut by its action on two other genes—*decapentaplegic* and *labial*, which is part of the Antennapedia complex. In normal embryos, both these genes are expressed in the region of the gut constriction, *decapentaplegic* in the visceral mesoderm and *labial* in the endoderm, but in the absence of *Ultrabithorax* both are hardly expressed at all. Thus, the Ultrabithorax protein is necessary for the activation of the *decapentaplegic* gene in the visceral mesoderm. The decapentaplegic protein then diffuses from the visceral mesoderm to the adjacent endoderm, where it stimulates a signaling pathway that results in the activation of the *labial* gene, which is involved in gut morphogenesis (see Fig. 5.40, bottom panel). Thus, the visceral mesoderm may pattern the endoderm by transfer of pos-itional information from one germ layer to another by means of extracel-lular signals, a process reminiscent of the induction of the nervous system in vertebrates (see Chapter 4).

Summary

Segment identity is conferred by action of selector or homeotic genes, which direct the development of different parasegments so that each acquires a unique identity. Two clusters of selector genes are involved in specification of segment identity in *Drosophila*: the Antennapedia compex, which controls parasegment identity in the head and first thoracic seg-ment, and the bithorax complex, which acts on the remaining paraseg-ments. Segment identity seems to be determined by the combination of genes active in a particular region. Selector genes need to be switched on continuously throughout development to maintain the required pheno-type. The spatial pattern of expression of selector genes along the embryo body is largely determined by the preceding gap gene activity. Mutations in genes of the Antennapedia and bithorax complexes can bring about homeotic transformations, in which one segment or structure is changed into a related one, for example an antenna into a leg. The Antennapedia and bithorax complexes are remarkable in that gene order on the chromosome corresponds to the spatial and temporal order of gene expression along the body. Selector genes of the two complexes are expressed in the ectoderm and in the underlying somatic and visceral mesoderm. They are not expressed in the endoderm from which the gut develops, but induction of the endoderm by visceral mesoderm seems to transfer to the endoderm some of the segment-specific pattern conferred on the mesoderm by selector gene activity.

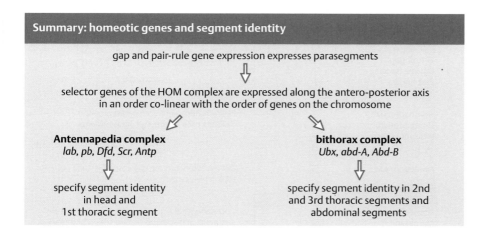

Summary: homeotic genes and segment identity

gap and pair-rule gene expression expresses parasegments

selector genes of the HOM complex are expressed along the antero-posterior axis in an order co-linear with the order of genes on the chromosome

Antennapedia complex
lab, pb, Dfd, Scr, Antp

bithorax complex
Ubx, abd-A, Abd-B

specify segment identity in head and 1st thoracic segment

specify segment identity in 2nd and 3rd thoracic segments and abdominal segments

SUMMARY TO CHAPTER 5

How genes control early development is better understood in the fruit fly *Drosophila* than in any other organism. The earliest stages of development in *Drosophila* take place when the embryo is a multinucleate syncytium. Maternal gene products deposited in the egg in a particular spatial pattern during oogenesis determine the main body axes and lay down a framework of positional information which activates a cascade of zygotic gene activity that further patterns the body. The first zygotic genes to be activated along the antero-posterior axis are the gap genes, all encoding transcription factors, whose pattern of expression divides the embryo into a number of regions. The transition to a segmented body organization then starts with the activity of the pair-rule genes, whose sites of expression are specified by the gap gene proteins and which divide the axis into 14 parasegments. Along the dorso-ventral axis, zygotic gene expression also defines several regions, including the future mesoderm and future neural tissue. At the time of pair-rule gene expression, the embryo becomes cellularized and is no longer a syncytium. Segment polarity genes pattern the segment. The identity of the segments is determined by two homeotic gene complexes which contain selector genes. The spatial pattern of expression of these genes is largely determined by gap gene activity. The order of genes along the complexes correlates with their spatial order of expression. Patterning of the gut endoderm involves transfer of patterning from the visceral endoderm.

The main genes involved in specifying pattern in the early *Drosophila* embryo				
Gene	**Maternal/ zygotic**	**Nature of protein**	**Transcription factor (T), receptor (R), or signal protein (S)**	**Function (where known)**

Antero-posterior System

Gene	M/Z	Nature of protein	T/R/S	Function
bicoid	M	Homeodomain	T	Morphogen, provides positional information along AP axis
hunchback	M/Z	Zinc fingers	T	Morphogen, provides positional information along AP axis
nanos	M	RNA-binding protein		Helps to establish AP gradient of hunchback protein
caudal	M	Homeodomain	T	Involved in specifying posterior region
gurken	M	Secreted protein of TGF-α family	S	Posterior oocyte–follicle cell signaling
oskar	M			Pole-cell determination

Terminal system

Gene	M/Z	Nature of protein	T/R/S	Function
torso	M	Receptor tyrosine kinase	R	Terminal specification
trunk	M		S	Ligand for torso

Gap genes

Gene	M/Z	Nature of protein	T/R/S	Function
hunchback	Z	Zinc fingers	T	
Krüppel	Z	Zinc fingers	T	
knirps	Z	Zinc fingers	T	Localize pair-rule gene expression
giant	Z	Leucine fingers	T	
tailless	Z	Zinc fingers	T	

Pair-rule genes

Gene	M/Z	Nature of protein	T/R/S	Function
even-skipped	Z	Homeodomain	T	Delimits odd-numbered parasegments
fushi tarazu	Z	Homeodomain	T	Delimits even-numbered parasegments
hairy	Z	Helix-loop-helix	T	

Segment polarity genes

Gene	M/Z	Nature of protein	T/R/S	Function
engrailed	Z	Homeodomain	T	Defines anterior region of parasegment and posterior region of segment
hedgehog	Z	Membrane or secreted	S	
wingless	Z	Secreted	S	
gooseberry	Z	Homeodomain	T	Components of signaling pathways that pattern segments and stabilize compartment boundaries
patched	Z	Membrane	R	
smoothened	Z	G-protein coupled	R	

Selector genes bithorax complex

Gene	M/Z	Nature of protein	T/R/S	Function
Ultrabithorax	Z	Homeodomain	T	
abdominal-A	Z	Homeodomain	T	Combinatorial activity confers identity on parasegments 5–13
Abdominal-B	Z	Homeodomain	T	

Antennapedia complex

Gene	M/Z	Nature of protein	T/R/S	Function
Deformed	Z	Homeodomain	T	
Sex combs reduced	Z	Homeodomain	T	Combinatorial activity confers identity on parasegments anterior to 5
Antennapedia	Z	Homeodomain	T	
labial	Z	Homeodomain	T	

Maintenance genes

Gene	M/Z	Nature of protein	T/R/S	Function
Polycomb group	Z		T	Maintain state of homeotic genes
Trithorax	Z		T	

Dorso-ventral System

Maternal genes

Gene	M/Z	Nature of protein	T/R/S	Function
Toll	M	Membrane	R	Activation results in dorsal protein entering nucleus
spätzle	M	Extracellular	S	Ligand for Toll protein
dorsal	M		T	Morphogen, sets dorso-ventral polarity
cactus	M			Binds dorsal protein and prevents it entering nucleus
gurken	M	Secreted protein of TGF-α family	S	Specifies oocyte axis
pipe	M	Sulfotransferase	Enzyme	Part of pathway leading to spätzle processing

Zygotic genes

Gene	M/Z	Nature of protein	T/R/S	Function
twist	Z	Helix-loop-helix	T	Define mesoderm
snail	Z	Zinc finger	T	
rhomboid	Z	Membrane protein	S	
single-minded	Z			
zerknüllt	Z	Homeodomain	T	Confer regional identity on dorso-ventral axis
decapenta-plegic	Z	Secreted protein of TGF-β family	S	
tolloid	Z	BMP-2 family	S	
short gastrulation	Z		S	

GENERAL REFERENCES

Adams, M.D., *et al.*: **The genome sequence of *Drosophila melanogaster*.** *Science* 2000, **267**: 2185–2195.

Lawrence, P.A.: *The Making of a Fly.* Oxford: Blackwell Scientific Publications, 1992.

SECTION REFERENCES

5.1 Three classes of maternal genes specify the antero-posterior axis

St Johnston, D., Nüsslein-Volhard, C.: **The origin of pattern and polarity in the *Drosophila* embryo.** *Cell* 1992, **68**: 201–219.

5.2 The *bicoid* gene provides an antero-posterior morphogen gradient

Driever, W., Nüsslein-Volhard, C.: **The bicoid protein determines position in the *Drosophila* embryo in a concentration dependent manner.** *Cell* 1988, **54**: 95–104.

5.3 The posterior pattern is controlled by the gradients of nanos and caudal proteins

Irish, V., Lehmann, R., Akam, M.: **The *Drosophila* posterior-group gene nanos functions by repressing *hunchback* activity.** *Nature* 1989, **338**: 646–648.

Murafta, Y., Wharton, R.P.: **Binding of pumilio to maternal *hunchback* mRNA is required for posterior patterning in *Drosophila* embryos.** *Cell* 1995, **80**: 747–756.

Rivera-Pomar, R., Lu, X., Perrimon, N., Taubert, H., Jackle, H.: **Activation of posterior gap gene expression in the *Drosophila* blastoderm.** *Nature* 1995, **376**: 253–256.

Struhl, G.: **Differing strategies for organizing anterior and posterior body pattern in *Drosophila* embryos.** *Nature* 1989, **338**: 741–744.

5.4 The anterior and posterior extremities of the embryo are specified by cell-surface receptor activation

Casanova, J., Struhl, G.: **Localized surface activity of *torso*, a receptor tyrosine kinase, specifies terminal body pattern in *Drosophila*.** *Genes Dev.* 1989, **3**: 2025–2038.

5.5 The dorso-ventral polarity of the egg is specified by localization of maternal proteins in the vitelline envelope

Anderson, K.V.: **Pinning down positional information: dorsal-ventral polarity in the *Drosophila* embryo.** *Cell* 1998, **95**: 439–442.

5.6 Positional information along the dorso-ventral axis is provided by the dorsal protein

Belvin, M.P., Anderson, K.V.: **A conserved signaling pathway: the *Drosophila* Toll-dorsal pathway.** *Annu. Rev. Cell Dev. Biol.* 1996, **12**: 393–416.

Roth, S., Stein, D., Nüsslein-Volhard, C.: **A gradient of nuclear localization of the dorsal protein determines dorso-ventral pattern in the *Drosophila* embryo.** *Cell* 1989, **59**: 1189–1202.

Steward, R., Govind, R.: **Dorsal-ventral polarity in the *Drosophila* embryo.** *Curr. Opin. Genet. Dev.* 1993, **3**: 556–561.

5.7 Antero-posterior and dorso-ventral axes of the oocyte are specified by interactions with follicle cells

Gonzalez-Reyes, A., St Johnston, D.: **The *Drosophila* AP axis is polarised by the cadherin-mediated positioning of the oocyte.** *Development* 1998, **125**: 3635–3644.

Gonzalez-Reyes, A., Elliott, H., St Johnston, D.: **Polarization of both major body axes in *Drosophila* by *gurken-torpedo* signaling.** *Nature* 1995, **375**: 654–658.

Riechmann, V., Ephrussi, A.: **Axis formation during *Drosophila* oogenesis.** *Curr. Opin. Genet. Dev.* 2001, **11**: 374–383.

Roth, S., Neuman-Silberberg, F.S., Barcelo, G., Schupbach, T.: **Cornichon and the EGF receptor signaling process are necessary for both anterior-posterior and dorsal-ventral pattern formation in *Drosophila*.** *Cell* 1995, **81**: 967–978.

Shulman, J.M., Benton, R., St Johnston, D.: **The *Drosophila* homolog of *C. elegans* PAR-1 organizes the oocyte cytoskeleton and directs *oskar* mRNA localization to the posterior pole.** *Cell* 2000, **101**: 377–388.

van Eeden, F., St Johnston, D.: **The polarisation of the anterior-posterior and dorsal–ventral axes during *Drosophila* oogenesis.** *Curr. Opin. Genet. Dev.* 1999, **9**: 396–404.

5.8 The expression of zygotic genes along the dorso-ventral axis is controlled by dorsal protein

Jiang, G., Levine, M.: **Binding affinities and cooperative interactions with HLH activators delimit threshold responses to the dorsal gradient morphogen.** *Cell* 1993, **72**: 741–752.

5.9 The decapentaplegic protein acts as a morphogen to pattern the dorsal region

Bier, E.: **A unity of opposites.** *Nature* 1999, **398**: 375–376.

Harland, R.M.: **A twist on embryonic signalling.** *Nature* 2001, **410**: 423–424.

Rusch, J., Levine, M.: **Threshold responses to the dorsal regulatory gradient and the subdivision of primary tissue territories in the *Drosophila* embryo.** *Curr. Opin. Genet. Dev.* 1996, **6**: 416–423.

Wharton, K.A., Ray, R.P., Gelbart, W.M.: **An activity gradient of decapentaplegic is necessary for the specification of dorsal pattern elements in the *Drosophila* embryo.** *Development* 1993, **117**: 807–822.

5.10 The antero-posterior axis is divided up into broad regions by gap gene expression

Hülskamp, M., Tautz, D.: **Gap genes and gradients—the logic behind the gaps.** *BioEssays* 1991, **13**: 261–268.

5.11 Bicoid protein provides a positional signal for the anterior expression of *hunchback*

Brand, A.H., Perrimon, N.: **Targeted gene expression as a means of altering cell fates and generating dominant phenotypes.** *Development* 1993, **118**: 401–415.

Simpson-Brose, M., Treisman, J., Desplan, C.: **Synergy between the hunchback and bicoid morphogens is required for anterior patterning in *Drosophila*.** *Cell* 1994, **78**: 855–865.

Struhl, G., Struhl, K., Macdonald, P.M.: **The gradient morphogen bicoid is a concentration-dependent transcriptional activator.** *Cell* 1989, **57**: 1259–1273.

5.12 The gradient in hunchback protein activates and represses other gap genes

Rivera-Pomar, R., Jäckle, H.: **From gradients to stripes in** *Drosophila* **embryogenesis: filling in the gaps.** *Trends Genet.* 1996, **12**: 478–483.

Struhl, G., Johnston, P., Lawrence, P.A.: **Control of** *Drosophila* **body pattern by the hunchback morphogen gradient.** *Cell* 1992, **69**: 237–249.

Wu, X., Vakani, R., Small, S.: **Two distinct mechanisms for differential positioning of gene expression borders involving the** *Drosophila* **gap protein giant.** *Development* 1998, **125**: 3765–3774.

5.14 Gap gene activity positions stripes of pair-rule gene expression

Small, S., Levine, M.: **The initiation of pair-rule stripes in the** *Drosophila* **blastoderm.** *Curr. Opin. Genet. Dev.* 1991, **1**: 255–260.

5.15 Expression of the *engrailed* **gene delimits a cell lineage boundary and defines a compartment**

Dahmann, C., Basler, K.: **Compartment boundaries: at the edge of development.** *Trends Genet.* 1999, **15**: 320–326.

Gray, S., Cai, H., Barolo, S., Levine, M.: **Transcriptional repression in the** *Drosophila* **embryo.** *Phil. Trans. R. Soc. Lond.* 1995, **349**: 257–262.

Harrison, D.A., Perrimon, N.: **Simple and efficient generation of marked clones in** *Drosophila.* *Curr. Biol.* 1993, **3**: 424–433.

Vincent, J.P., O'Farrell, P.H.: **The state of engrailed expression is not clonally transmitted during early** *Drosophila* **development.** *Cell* 1992, **68**: 923–931.

5.16 Segment polarity genes pattern the segments and stabilize parasegment and segment boundaries

Alexandre, C., Lecourtois, M., Vincent, J-P.: **Wingless and hedgehog pattern** *Drosophila* **denticle belts by regulating the production of short-range signals.** *Development* 1999, **126**: 5689–5698.

Johnson, R.L., Scott, M.P.: **New players and puzzles in the Hedgehog signaling pathway.** *Curr. Opin. Genet. Dev.* 1998, **8**: 450–456.

Lawrence, P.A., Casal, J., Struhl, G.: **Hedgehog and engrailed: pattern formation and polarity in the** *Drosophila* **abdomen.** *Development* 1999, **126**: 2431–2439.

von Dassow, G., Meir, E., Munro, E.M., Odell, G.M.: **The segment polarity network is a robust developmental module.** *Nature* 2000, **406**: 188–192.

5.18 Some insects use different mechanisms for patterning the body plan

Akam, M., Dawes, R.: **More than one way to slice an egg.** *Curr. Biol.* 1992, **8**: 395–398.

Brown, S.J., Parrish, J.K., Beeman, R.W., Denell, R.E.: **Molecular characterization and embryonic expression of the even-skipped ortholog of** *Tribolium castaneum.* *Mech. Dev.* 1997, **61**: 165–173.

French, V.: **Segmentation (and** *eve***) in very odd insect embryos.** *BioEssays* 1996, **18**: 435–438.

Nagy, L.M.: **A glance posterior.** *Curr. Biol.* 1994, **4**: 811–814.

Sander, K.: **Pattern formation in the insect embryo.** In *Cell Patterning, Ciba Found. Symp. 29.* London: Ciba Foundation, 1975: 241–263.

Tautz, D., Sommer, R.J.: **Evolution of segmentation genes in insects.** *Trends Genet.* 1995, **11**: 23–27.

5.19 Homeotic selector genes of the bithorax complex are responsible for diversification of the posterior segments

Castelli-Gair, J., Akam, M.: **How the Hox gene** *Ultrabithorax* **specifies two different segments: the significance of spatial and temporal regulation within metameres.** *Development* 1995, **121**: 2973–2982.

Duncan, I.: **How do single homeotic genes control multiple segment identities?** *BioEssays* 1996, **18**: 91–94.

Lawrence, P.A., Morata, G.: **Homeobox genes: their function in** *Drosophila* **segmentation and pattern formation.** *Cell* 1994, **78**: 181–189.

Liang, Z., Biggin, M.D.: **Eve and ftz regulate a wide array of genes in blastoderm embryos: the selector homeoproteins directly or indirectly regulate most genes in** *Drosophila.* *Development* 1998, **125**: 4471–4482.

Mann, R.S., Morata, G.: **The developmental and molecular biology of genes that subdivide the body of** *Drosophila.* *Annu. Rev. Cell Dev. Biol.* 2000, **16**: 143–271.

Mannervik, M.: **Target genes of homeodomain proteins.** *BioEssays* 1999, **4**: 267–270.

Simon, J.: **Locking in stable states of gene expression: transcriptional control during** *Drosophila* **development.** *Curr. Opin. Cell Biol.* 1995, **7**: 376–385.

5.21 The order of HOM gene expression corresponds to the order of genes along the chromosome

Morata, G.: **Homeotic genes of** *Drosophila.* *Curr. Opin. Genet. Dev.* 1993, **3**: 606–614.

5.22 HOM gene expression in visceral mesoderm controls the structure of the adjacent gut

Bienz, M.: **Endoderm induction in** *Drosophila***: the nuclear targets of the inducing signals.** *Curr. Opin. Genet. Dev.* 1997, **7**: 683–685.

Immergluck, K., Lawrence, P.A., Bienz, M.: **Induction across germ layers in** *Drosophila* **mediated by a genetic cascade.** *Cell* 1990, **62**: 261–268.

Development of nematodes, sea urchins, ascidians, and slime molds

6

- Nematodes
- Echinoderms
- Ascidians
- Cellular slime molds

"Sometimes we did things in a different way, were given our identity one by one, and spoke only to our neighbors."

This chapter considers aspects of body plan development in three model invertebrate organisms—nematodes, echinoderms, and ascidians—and ends with a short discussion on the cellular slime molds, which represent a very simple developmental system. These organisms will highlight the similarities and differences in developmental mechanisms compared with those organisms we have already considered. The evolutionary relationships between the organisms discussed in this chapter are shown in Fig. 6.1. All, with the exception of the cellular slime mold, conform to the general

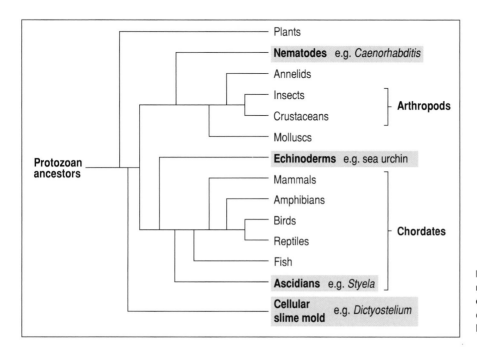

Fig. 6.1 Phylogenetic tree showing relationships between the organisms considered in this book. Those organisms discussed in this chapter are highlighted in blue.

plan of animal development: cleavage leads to a blastula, which undergoes gastrulation with the emergence of a body plan.

There is an old, and now less fashionable, distinction sometimes made between so-called regulative and mosaic development—the former involving mainly cell–cell interactions, while the latter is based on localized cytoplasmic factors and their distribution through asymmetric cell divisions (see Section 1.10). The organisms discussed in earlier chapters, such as the vertebrates, are examples of mainly regulative development, whereas some of those discussed in this chapter have a mosaic-like development, although there is an element of both processes in most organisms.

An important feature of the development of some invertebrates, such as nematodes and the chordate ascidians, is that cell fate is often specified on a cell-by-cell basis, rather than in groups of cells as in flies and vertebrates, and in general does not rely on positional information established by gradients of morphogens. The early embryos of many invertebrates contain far fewer cells than those of vertebrates or flies, with each cell acquiring a unique identity at an early stage of development. For example, in the nematode there are only 26 cells when gastrulation starts, compared to thousands in vertebrates. Specification on a cell-by-cell basis often seems to make use of a developmental mechanism that is less commonly found in early insect and vertebrate development, namely **asymmetric cell division** and the unequal distribution of cytoplasmic factors as a means of determining cell fate (see Section 1.15). Daughter cells resulting from asymmetric cell division often adopt different fates, not as a result of extracellular signals but autonomously, as a result of the unequal distribution of some factor between them. However, asymmetric cell division in the early stages of development does not mean that cell–cell interactions are absent or unimportant in these organisms.

We begin with the nematode *Caenorhabditis elegans*, which has been studied intensively and in which many key developmental genes have been identified. In this animal, specification is largely on a cell-by-cell basis. We then consider echinoderms, which develop much more like vertebrates, in that their embryos rely heavily on intercellular interactions and are highly regulative, and patterning involves groups of cells. Next we consider ascidians, with particular emphasis on the role of localization of cytoplasmic determinants in their early development. Finally, we look at patterning in the cellular slime mold, which represents a primitive and very different developmental system.

Nematodes

It is a triumph of direct observation that, with the aid of Nomarski interference microscopy, the complete lineage of every cell in the nematode *C. elegans* has been worked out (see Fig. 2.37). The pattern of cell division is invariant—it is the same in every embryo. The larva, when it hatches, is made up of 558 cells, and after four further molts this number has increased to 959, excluding the germ cells, which vary in number. This is not the total number of cells derived from the egg, as 131 cells die during development due to programmed cell death (apoptosis). As the fate of every cell at each stage is known, a fate map can be accurately drawn at any stage and thus has a precision not found in any vertebrate. However, as with any fate map, even where there is an invariant cell lineage this precision in no way implies that the lineage determines the fate or that the fate of the cells cannot be altered. As we shall see, cell interactions have a major role in determining cell fate in the nematode.

The complete genome of *C. elegans* has been sequenced, but only about 200 of the nearly 19,000 genes have so far been found to affect developmental patterning and differentiation. Of these 200, many are related to the genes that control development in vertebrates and *Drosophila*; they include Hox genes (see Box 4A, p. 117), and genes for signaling molecules of the TGF-β, Wnt, and Notch pathways (see Box 3A, p. 68). A notable exception is the hedgehog signaling pathway (see Section 5.16), which is not present in *Caenorhabditis*. Although zygotic gene expression begins at the four-cell stage, maternal components control almost all of the development up to gastrulation at the 28-cell stage.

6.1 The developmental axes in the nematode are determined by asymmetric cell division and cell–cell interactions

The first cleavage of the nematode egg is unequal, dividing the egg into a large anterior AB cell and a smaller posterior P_1 cell. This asymmetry defines the antero-posterior axis. The P_1 cell behaves rather like a stem cell; at each further division it produces one P-type cell, and one daughter cell that will embark on another developmental pathway. For the first three divisions, the P cell daughters give rise to body cells but after the fourth cleavage they only give rise to germ cells (Fig. 6.2). Division of the AB cell

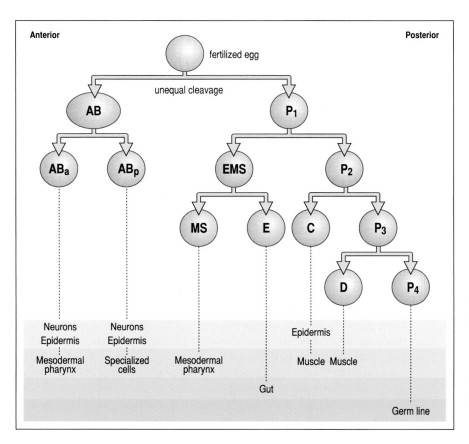

Fig. 6.2 Early cell lineage in the nematode *Caenorhabditis elegans*. The cleavage pattern is invariant. The first cleavage divides the egg into a large AB cell and a smaller P_1 cell. The descendants of these cells have constant lineages and fates. For example, germ cells all come from the P_4 cell, a descendant of P_1, and the gut from the E cell. After Sulston, J.E., *et al*.: 1983.

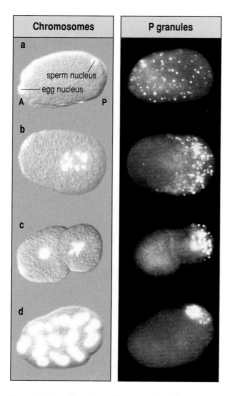

| Chromosomes | P granules |

a

sperm nucleus

egg nucleus

A   P

b

c

d

Fig. 6.3 Localization of P granules after fertilization. The movement of P granules is shown during the development of a fertilized egg of *C. elegans* from sperm entry up to the 26-cell stage. In the left panel, the embryo is stained for DNA, to visualize the chromosomes, and in the right panel for P granules. (a) Fertilized egg with egg nucleus at anterior end and sperm nucleus at posterior end. At this stage the P granules are distributed throughout the egg. (b) Fusion of sperm and egg nucleus. The P granules have already moved to the posterior end. (c) Two-cell stage. The P granules are localized in the posterior cell. (d) 26-cell stage. All the P granules are in the P_4 cell, which will give rise to the germline only.

Photographs courtesy of W. Wood, from Strome, S., et al.: 1983.

gives rise to anterior and posterior AB daughter cells. The anterior cell AB_a gives rise to typically ectodermal tissues, such as the epidermis (hypodermis) and nervous system, but also to a portion of the mesoderm of the pharynx. The posterior daughter AB_p also makes neurons and epidermis as well as some specialized cells. At second cleavage, P_1 divides asymmetrically, to give P_2 and EMS, which will subsequently divide into MS and E. MS gives rise to mesodermal pharynx, while E is the sole precursor of the gut. The C daughter cell, which is derived from the P_2 cell at the third cleavage, forms epidermis and muscle; D from the fourth cleavage of the P cell produces only muscle. All the cells undergo further invariant divisions, and by about 100 minutes after fertilization, gastrulation begins.

Before fertilization, there is no evidence of any asymmetry in the nematode egg. The first cleavage, which is both unequal and asymmetric, is related to the point of sperm entry, which marks the future posterior end. The first cleavage defines the future antero-posterior axis, with the large AB cell marking the anterior end and the smaller P_1 cell the posterior. Preceding this cleavage, a cap of actin microfilaments forms at the anterior end of the egg, and a set of granules, the so-called **P granules**, becomes localized at the posterior end. The P granules remain localized in the P daughters of cell division, eventually becoming localized to the P_4 cell, which gives rise to the germline (Fig. 6.3). The importance of the microfilaments in controlling the orientation of the cleavage divisions can be shown by destroying their integrity with a short pulse of cytochalasin D during the first cleavage, which results in aberrant early divisions and development. More than 100 maternal gene products are involved in controlling the first cleavage division.

Clearly, some sort of structural inhomogeneity in the egg after fertilization controls this early manifestation of antero-posterior polarity. It cannot, however, be the P granules themselves, whose distribution is a reflection rather than the cause of the polarity. Experiments in which egg cytoplasm is removed suggest the presence of some component in the posterior cytoplasm that determines the polarity. If more than 25% of the egg cytoplasm is extruded from the posterior end after fertilization, the unequal pattern of early cleavage is lost, but up to 40% of cytoplasm can be extruded from the anterior end without altering the early pattern.

A very early marker of the antero-posterior axis is the protein PAR-1, encoded by the maternal gene *par-1*, which is localized in the future posterior region of the egg after fertilization. Localization of the P granules and PAR-1, and the subsequent establishment of antero-posterior polarity, requires microtubules, which emanate from a microtubule-organizing center at the posterior of the fertilized egg. This center arises from the aster of microtubules contributed by the sperm nucleus. Maternal mutations in *par-1* disrupt the asymmetry of the first cleavage. In such mutants, P granules fail to localize, and further cleavage and development is abnormal. The protein PAR-1 probably acts by interacting with the egg cell's microfilament cytoskeleton. After first cleavage, PAR-1 is localized in the P_1 cell. In this way, as in *Drosophila* (see Section 5.7), the *par-1* gene is involved in specifying the antero-posterior axis.

In spite of the highly determinate cell lineage in the nematode, cell–cell interactions are involved in specifying the dorso-ventral axis. At the time of the second cleavage, if the future anterior AB_a cell is pushed and rotated with a glass needle, not only is the antero-posterior order of the daughter

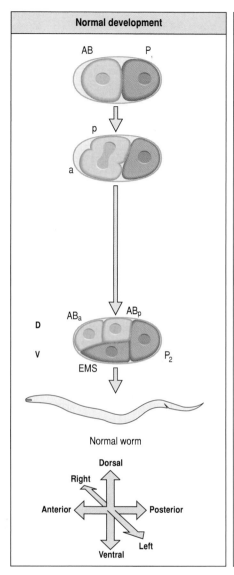

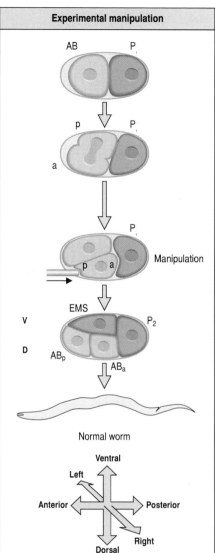

Fig. 6.4 Reversal of dorso-ventral polarity at the four-cell stage of the nematode. In normal embryos, the AB cell rotates at cleavage so that AB_a is anterior. If the AB cell is mechanically rotated in the opposite direction at second cleavage, the AB_p cell is now anterior. This manipulation also displaces the P_1 cell so that when it divides, the position of the EMS daughter cell is reversed with respect to the AB cells. Development is still normal, but the dorso-ventral axis is inverted and so too is left–right asymmetry. After Sulston, J.E., *et al.*: 1983.

AB cells reversed, but the cleavage of P_1 is also affected, so that the position of the P_1 daughter cell, EMS, relative to the AB cells is inverted (Fig. 6.4). The manipulated embryo develops completely normally, but the dorso-ventral axis is reversed, showing firstly that the axis is not yet determined at this stage, for by inverting the position of the EMS cell the dorso-ventral axis is also inverted, and secondly that cell interactions are involved in specifying the cells of the early embryo. By inverting the position of the EMS, the dorso-ventral behavior of P_1 is also inverted. This has the further implication that the left–right axis is also not yet determined; indeed this axis can also be inverted, as we shall now see.

Adult worms have a well-defined left–right asymmetry in their internal structures, and this axis is determined after the dorso-ventral axis is established. The pattern of development on the left and right sides of the nematode embryo shows striking differences; indeed, the differences during early development are more marked than in many other animals. Not only

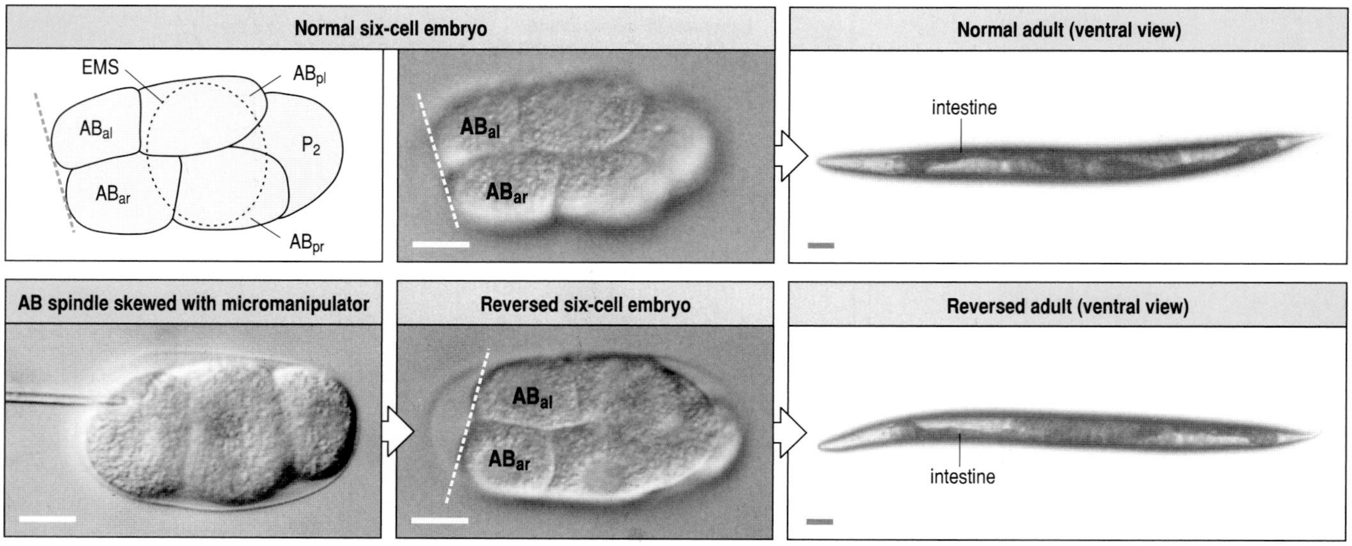

Fig. 6.5 Reversal of handedness in *C. elegans*. At the six-cell stage in a normal embryo, the AB_a cell on the left side (Ab_{al}) is slightly anterior to that on the right (top left panels, scale bar = 10 μm). Manipulation that makes the AB_a cell on the right side more anterior results in an animal with reversed handedness (bottom left panels, scale bars = 10 μm). The right panels show the resulting normal and reversed adults; scale bars = 50 μm.

Photographs courtesy of W. Wood, from Wood, W.B.: 1991.

do cell lineages on left and right differ, but some cells even cross from one side to the other. The concept of left and right, as discussed previously (see Section 3.7), only has meaning once the antero-posterior and dorso-ventral axes are defined. Since the dorso-ventral axis in the nematode can still be reversed at the two-cell to four-cell stage, left and right are also not yet specified.

Specification of left and right occurs at the third cleavage, and handedness can be reversed by experimental manipulation at this stage. Following the division of the AB cell into anterior and posterior blastomeres AB_a and AB_p, each of these divides at the third cleavage to produce laterally disposed right and left daughter cells. The plane of cleavage is, however, slightly asymmetric so that the left daughter cell lies just a little anterior to its right-hand sister. If the cells are manipulated with a glass rod during this cleavage, their positioning can be reversed so that the right-hand cell lies slightly anterior (Fig. 6.5). This small manipulation is sufficient to reverse the handedness of the animal.

6.2 Asymmetric divisions and cell–cell interactions specify cell fate in the early nematode embryo

Although cell lineage in the nematode is invariant, experimental evidence such as that described above shows that cell–cell interactions are of crucial importance in specifying cell fate in the early embryo. Otherwise, the reversal of AB_a and AB_p (which normally have different cell fates) by micromanipulation, just as they are being formed (see Fig. 6.4), would not give rise to a normal worm. AB_a and AB_p must initially be equivalent, and their fate must be specified by interactions with adjacent cells. Evidence for such an interaction comes from removing P_1 at the first cleavage, as pharyngeal cells, a normal product of AB_a, are then not made.

What then are the interactions that specify the non-equivalence of the two AB descendants? If AB_p is prevented from contacting P_2 it develops as an AB_a cell. Thus, the P_2 blastomere is responsible for specifying AB_p. The induction of AB_p by P_2 involves proteins encoded by the maternal genes *glp-1* and *apx-1*, which are similar, respectively, to the vertebrate and insect proteins Notch and Delta, which are involved in interactions between adjacent cells in many developmental situations (see, for example, Section 11.2).

The protein GLP-1 is a transmembrane receptor and is one of the earliest proteins to be spatially localized during embryogenesis in the nematode. Although *glp-1* mRNA is uniformly present throughout the embryo, its translation is repressed in the posterior P cell; at the two-cell stage the GLP-1 protein is thus only expressed in the anterior AB cell. This strategy for protein localization is highly reminiscent of the localization of maternal hunchback protein in *Drosophila* (see Section 5.3); the 3' untranslated region of the mRNA is involved in the suppression of translation. The P_2 cell also signals to the EMS cell via a Wnt-like pathway.

After second cleavage the AB_a and AB_p cells both contain the receptor protein GLP-1. The two cells are then directed to become different by a local inductive signal, sent to the AB_p cellfrom the P_2 cell at the four-cell stage (Fig. 6.6). This signal is thought to be the protein APX-1 (produced by the P_2 cell) acting as an activating ligand for the GLP-1 receptor. As a result of this induction, the descendants of AB_a and AB_p respond differently to later signals from the adjacent MS cell (a daughter of EMS).

Other maternal-effect genes involved in these early cell–cell interactions have been identified. Specification of the EMS cell involves the expression of the *skin excess* (*skn*) gene. The two cells which EMS gives rise to, E and MS, produce intestinal and muscle cell types, respectively. Mutations in *skn-1* result in E making mainly muscle instead of intestinal cells. *skn-1* mRNA is uniform at the two-cell stage but the protein is at much higher levels in the nucleus of P_1 than in AB.

Gut development is also dependent on induction. The nematode gut develops from a single blastomere, the E cell, which is formed at the eight-cell stage as one of the daughters of EMS, which divides at third cleavage to give an anterior MS cell and a posterior E cell (see Fig. 6.2). If isolated from the influence of its neighbors at the four-cell stage, an EMS cell can develop gut structures. However, this property of EMS depends crucially on just when it is isolated. If isolated at the beginning of the four-cell stage, EMS cannot develop gut structures, suggesting that its ability to form gut requires an interaction with other cells at the beginning of this stage. The cell necessary to induce gut formation from EMS is the P_2 cell, as removal of P_2 at the early four-cell stage results in no gut being formed. Recombining isolated EMS and P_2 cells restores gut development, whereas recombining EMS with other cells from the four-cell stage has no effect.

We thus begin to see how cell–cell interactions and localized cytoplasmic determinants together specify cell fate in the early nematode embryo. The results of killing individual cells by laser ablation at the 32-cell stage, when gastrulation begins, suggest that many of the cell lineages are now determined, as when a cell is destroyed at this stage there is no regulation and its normal descendants are absent. Cell–cell interactions are, however, required later to effect cells' final differentiation.

Cell differentiation is closely linked to cell division. Each blastomere undergoes a unique and nearly invariant series of cleavages that successively divide cells into anterior and posterior daughter cells. The pattern of cell division, that is, whether the final differentiated cell is descended through the anterior (a) or posterior (p) cell at each division, appears to

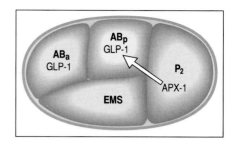

Fig. 6.6 An early inductive event specifies AB_p. AB_p becomes different from AB_a as a result of a signal emanating from the adjacent P_2 cell, thought to be the protein APX-1. This signal is received by the receptor GLP-1 on the AB_p cell. After Mello, C.C., *et al.*: 1994.

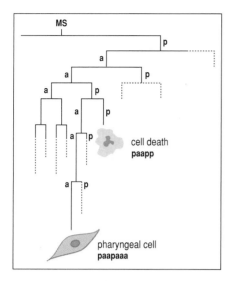

Fig. 6.7 Cell fate is linked to the pattern of cell divisions. The figure illustrates part of the cell lineages generated by the MS blastomere, which give rise to body muscle and the mesodermal cells of the pharynx. Each division produces an anterior (a) and a posterior (p) cell. The lineage paapp, for example, always gives rise to a cell that undergoes programmed cell death and dies by apoptosis. The lineage paapaaa, on the other hand, always results in a particular pharyngeal cell.

specify cell fate. In the lineage generated from the MS blastomere, for example, the cell resulting from the sequence p-a-a-p-p undergoes programmed cell death (Fig. 6.7), whereas that resulting from p-a-a-p-a-a-a gives rise to a particular pharyngeal cell. What is the causal relation between the pattern of division and cell differentiation? An answer appears to lie in the anterior daughter cell having a higher level of a nuclear protein, POP-1 (encoded by the maternal gene *pop-1*), compared with its posterior sister. In the case of the division of the EMS blastomere, for example, the anterior cell, which becomes the MS blastomere, has a higher level of POP-1 in the nucleus compared with the posterior cell, which becomes E. In a *pop-1* mutant that does not make any POP-1 protein, MS adopts an E-like fate because the antero-posterior distinction is lost. How the antero-posterior difference is maintained at each cleavage is not known, but it involves the Wnt signaling pathway.

At the 80-cell gastrula stage a fate map can be made for the nematode embryo (Fig. 6.8). At this stage, cells from different lineages that will contribute to the same organ, such as the pharynx or gut, cluster together. Genes that appear to act as organ identity genes may now be expressed by the cells in the cluster. The *pha-4* gene, for example, seems to be an organ identity gene for the pharynx. Mutations in *pha-4* result in loss of the pharynx, whereas ectopic expression of normal *pha-4* in all the cells of the embryo leads to all cells expressing pharyngeal cell markers.

6.3 A small cluster of Hox genes specifies cell fate along the antero-posterior axis

Although the body plan of the nematode differs greatly from that of vertebrates or *Drosophila*, and there is no segmental pattern along the antero-posterior axis, homeobox-containing genes (see Box 4A, p. 117) are involved in specifying cell fate along this axis in the nematode. The nematode contains a large number of homeobox genes, of which only four are similar to the Antennapedia class of *Drosophila* Hox genes and the Hox genes of vertebrates. These four genes—*lin-39*, *ceh-13*, *mab-5*, and *egl-5*—are arranged in a cluster, the Hox cluster, and in a similar order on the chromosome to their *Drosophila* homologs. A fifth homeobox gene of the nematode cluster, *ceh-23*, is less related to the Antennapedia class. *lin-39*, *mab-5*, and *egl-5* are expressed as zygotic genes in different positions along the antero-posterior axis during embryonic development; their spatial order of expression corresponds to their order along the chromosome, as in the other organisms (Fig. 6.9). For example, *egl-5* provides positional identity for posterior structures. *ceh-13* is required for the anterior organization of the embryo. Although the Hox genes are expressed in the embryo, it seems that their primary function in the nematode is in post-embryonic larval development, as mutations in these genes mainly affect larval development. However, two recently discovered Hox genes, which lie outside the Hox cluster, are required for embryonic development, in this case for posterior structures. Mutations in these genes can cause cells in one region of the body to adopt fates characteristic of other body regions. For example, a mutation in *lin-39* can result in mid-body cells expressing fates characteristic of more anterior or posterior body regions.

Despite the fact that nematode Hox gene expression occurs in a regional pattern, the pattern may not be position dependent. For example, in the larva, cells expressing the Hox gene *mab-5* all occur in the same region (see Fig. 6.9) but are quite unrelated by lineage. Yet, despite appearances, the expression of *mab-5* is not due to extracellular positional signals, but is determined autonomously and independently in each of the lineages.

6.4 Genes control graded temporal information in nematode development

Embryonic development gives rise to a larva of 550 cells and there are then four larval stages which produce the adult. Because each cell in the developing nematode can be identified by its lineage and position, genes that control the fates of individual cells at specific times in development can also be identified. This enables the genetic control of timing during development to be studied. The order of developmental processes is of central importance, as well as the time at which they occur. Genes must be expressed both in the right place and at the right time. One well-studied example of timing in nematode development is the generation of different patterns of cell division and differentiation in the four larval stages of *C. elegans*, which are easily distinguished in the developing cuticle.

Mutations in two genes, *lin-4* and *lin-14*, change the timing of cell divisions in many tissues and cell types. Mutations that alter the timing of developmental events are called **heterochronic**. Mutations in *lin-4* and *lin-14* can result in both 'retarded' and 'precocious' development so that, for example, some stage-specific events such as molting and larval cuticle synthesis are repeated at abnormally late stages, leading to retardation of normal events such as adult cuticle synthesis.

Examples of changes in developmental timing that result from mutations in *lin-14* are provided by the lateral hypodermal T-cell lineage called T.ap (Fig. 6.10). In wild-type embryos, the T cell generates a lineage that gives rise to epidermal cells, neurons, and their support cells in both the first (L1) and second (L2) larval stages. During the later larval stages, L3 and L4, some of the T cell descendants divide to give rise to other structures. Gain-of-function (gf) mutations in *lin-14* result in retarded development. Post-embryonic development begins normally, but the developmental patterns of the first or second larval stages are repeated. Loss-of-function (lf) mutations in *lin-14* result in a precocious phenotype—the pattern of cell divisions seen in early larval stages is lost and post-embryonic development starts with cell divisions normally seen in the second larval stage.

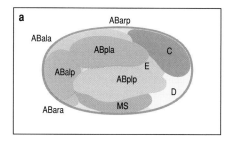

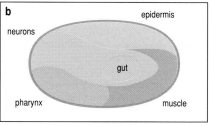

Fig. 6.8 Fate map of the 80-cell gastrula of *C. elegans*. In panel a, the regions of the embryo are color-coded according to their blastomere origin. In panel b they are colored according to the organs or tissues they will ultimately give rise to. At this stage in development, cells from different lineages that contribute to the same tissue or organ have been brought together in the embryo. For clarity, the domains occupied by ABpra and ABprp are not shown in panel a. Anterior is to the left, ventral down. After Labouesse, M., Mango, S.E.: 1999.

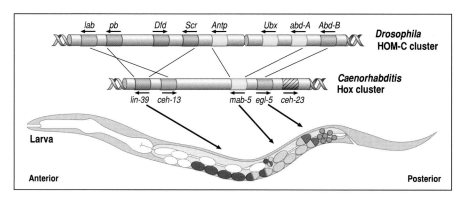

Fig. 6.9 The *C. elegans* Hox gene cluster and its relation to the HOM-C gene cluster of *Drosophila*. The nematode contains a cluster of five Hox genes, four of which show homologies with those of the Antennapedia complex of the fly. The pattern of expression of three of the genes in the larva is shown. After Bürglin, T.R., *et al.*: 1993.

Fig. 6.10 Cell lineage patterns in wild-type and heterochronic mutants of C. elegans. The lineage of the T blast cell (T.ap) continues through four larval stages (left panel). Mutants in the gene *lin-14* show disturbance in the timing of cell division, resulting in changes in the patterns of cell lineage. Loss-of-function mutations result in a precocious lineage pattern, with the pattern of development of early stages being lost (center panel). Gain-of-function mutations result in retarded lineage patterns, with the patterns of the early larval stage being repeated (right panel).

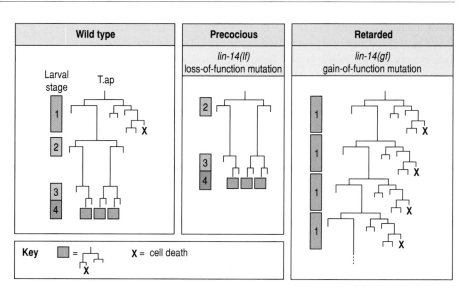

It has been suggested that the genes that control timing of developmental events may do so by controlling the concentration of some substance, causing it to decrease with time (Fig. 6.11). This temporal gradient could control development in much the same way that a spatial gradient can control patterning. This type of timing mechanism seems to operate in *C. elegans* as the concentration of LIN-14 protein drops tenfold between the first and later larval stages. Differences in the concentration of LIN-14 at different stages of development may specify the fates of cells, with high concentrations specifying early fates and low concentrations later fates. Thus, the decrease of LIN-14 during development could provide the basis of a precisely ordered temporal sequence of cell activities. Dominant gain-of-function mutations of *lin-14* keep LIN-14 protein levels high, and so in these mutants the cells continue to behave as if they are at an earlier larval stage. In contrast, loss-of-function mutations result in abnormally low concentrations of LIN-14, and the larvae therefore behave as if they were at a later larval stage.

The concentration of LIN-14 protein is post-transcriptionally regulated in an interesting and unusual way. The translation of *lin-14* mRNA can be repressed by *lin-4* RNA, which complexes with the *lin-14* mRNA. Increasing synthesis of *lin-4* RNA during the later larval stages could thus generate the temporal gradient in LIN-14. Support for this idea comes from mutations in *lin-4*. Loss-of-function mutations in *lin-4* have the same effect as gain-of-function mutations in *lin-14*.

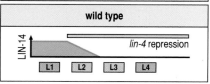

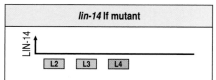

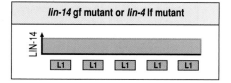

Fig. 6.11 A model for the control of the temporal pattern of C. elegans larval development. Top panel: the stage-specific pattern of larval development is determined by a temporal gradient of the protein LIN-14, which decreases during larval development. A high concentration in the early stages specifies the pattern of early-stage development as shown in Fig. 6.10 (left panel). The reduction in LIN-14 at later stages is due to the inhibition of *lin-14* mRNA translation by *lin-4* mRNA. Bottom panels: loss-of-function (lf) mutations in *lin-14* result in the absence of the first larval stage (L1) lineage, whereas gain-of-function (gf) mutations, which maintain a high level of LIN-14 throughout development, keep the lineage in an L1 phase. Loss-of-function mutations in *lin-4* result in a lifting of repression of *lin-14*, a continued high activity of LIN-14, and a repetition of the L1 lineage.

Summary

Specification of cell fate in the nematode embryo is intimately linked to the pattern of cleavage, and provides an excellent example of the subtle relationships between maternally specified cytoplasmic differences and very local and immediate cell–cell interactions. The antero-posterior axis is specified at the first cleavage and specification of the dorso-ventral and left–right axes involves local cell–cell interactions. Gene products are asymmetrically distributed during the early cleavage stages, but specification of cell fates in the early embryo is crucially dependent on local cell–cell interactions. The development of the gut, which is derived from a single cell, requires an inductive signal by an adjacent cell. A small cluster of homeobox genes provides positional identity along the antero-posterior axis of the larva. The timing of developmental events in the larva may be based on the concentration of a substance that decreases with time.

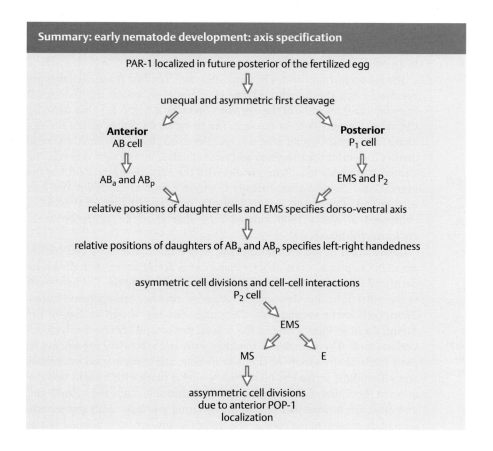

Summary: early nematode development: axis specification

PAR-1 localized in future posterior of the fertilized egg

unequal and asymmetric first cleavage

Anterior
AB cell

Posterior
P_1 cell

AB_a and AB_p

EMS and P_2

relative positions of daughter cells and EMS specifies dorso-ventral axis

relative positions of daughters of AB_a and AB_p specifies left-right handedness

asymmetric cell divisions and cell-cell interactions
P_2 cell

EMS

MS   E

assymmetric cell divisions
due to anterior POP-1
localization

Echinoderms

Echinoderms include the sea urchins and starfish. Because of their transparency and ease of handling, sea urchin embryos have long been used as a model developmental system. Developmental studies are confined to the development of the larva, as metamorphosis into the adult is a complex and poorly understood process. The sea urchin embryo is classically regarded as a model of regulative development. It formed the basis for

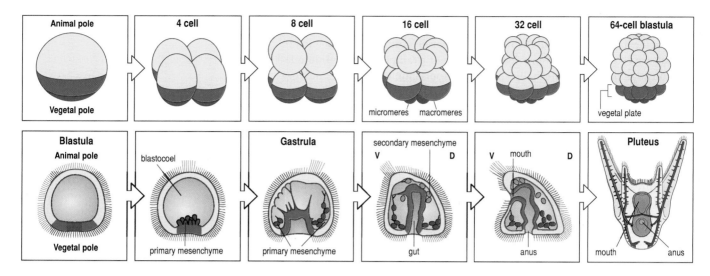

Fig. 6.12 Development of the sea urchin embryo. Top panels: external view of cleavage to the 64-cell stage. The first two cleavages divide the egg along the animal–vegetal axis. The third cleavage divides the embryo into animal and vegetal halves. At the fourth cleavage, which is unequal, four small micromeres (red) are formed at the vegetal pole, only two of which can be seen here. Further cleavage results in a hollow blastula. Bottom panels: gastrulation and development of the pluteus larva. Sections of the developing embryo through the plane of the animal–vegetal axis are shown, starting from a hollow blastula. Gastrulation begins at the vegetal pole, with the entry of about 40 primary mesenchyme cells into the interior of the blastula. The gut invaginates from this site and fuses with the mouth, which invaginates from the opposite side of the embryo. The secondary mesenchyme comes from the tip of the invagination. During further development, growth of skeletal rods laid down by the primary mesenchyme results in the extension of the four 'arms' of the pluteus larva. The pluteus is depicted from its outside with the mouth uppermost.

Driesch's ideas at the beginning of the century, that the position of the cells in an embryo determined their fate (see Section 1.3).

The sea urchin egg divides by radial cleavage. The first three cleavages are symmetric but the fourth cleavage is asymmetric, producing four small micromeres at one pole of the egg, the vegetal pole (Fig. 6.12), thus defining the animal–vegetal axis of the egg. The first two cleavages divide the egg along the animal–vegetal axis, whereas the third cleavage is equatorial and divides the embryo into animal and vegetal halves. At the next cleavage the animal cells divide in a plane parallel with the animal–vegetal axis, but the vegetal cells divide asymmetrically to produce four macromeres and four micromeres. Continued cleavage results in a hollow spherical blastula composed of about 1000 ciliated cells that form an epithelial sheet enclosing the blastocoel.

Gastrulation in sea urchin embryos, whose mechanisms we consider in detail in Chapter 8, starts about 10 hours after fertilization, with the mesoderm and endoderm moving inside from the vegetal region. The first event is the entry into the blastocoel of about 40 primary mesenchyme (mesoderm) cells at the vegetal pole. They migrate along the inner face of the blastula wall to form a ring in the vegetal region and lay down calcareous skeletal rods. The endoderm, together with the secondary mesenchyme, then starts to invaginate at the vegetal pole, the invagination eventually stretching right across the blastocoel, where it fuses with a small invagination in the future mouth region on the ventral side. Thus the mouth, gut, and anus are formed. Before the invaginating gut fuses with the mouth, secondary mesenchyme cells at the tip of the invagination migrate out as single cells and give rise to mesoderm, such as muscle and pigment cells. The embryo is then a feeding **pluteus** larva (see Fig. 6.12).

6.5 The sea urchin egg is polarized along the animal–vegetal axis

The sea urchin egg has a well-defined animal–vegetal polarity that appears to be related to the site of attachment of the egg in the ovary. Polarity is marked in some species by a fine canal at the animal pole, whereas in other species there is a band of pigment granules in the vegetal region. Early development is intimately linked to this egg axis, and it can be considered

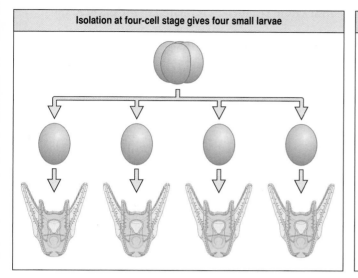

Isolation at four-cell stage gives four small larvae

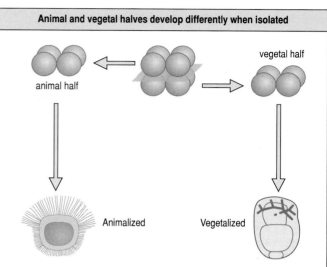

Animal and vegetal halves develop differently when isolated

as corresponding to the antero-posterior axis of the larva. The first two planes of cleavage are always parallel to the animal–vegetal axis and at the fourth cleavage, which is unequal (see Fig. 6.12), the micromeres are formed at the vegetal pole. The micromeres give rise to the primary mesenchyme, and both the micromeres and the primary mesenchyme may initially be specified by cytoplasmic factors localized at the vegetal pole of the egg.

The animal–vegetal axis is stable and cannot be altered by centrifugation, which redistributes larger organelles such as mitochondria and yolk platelets. Isolated fragments of egg also retain their original polarity. Blastomeres isolated at the two-cell and four-cell stages, each of which contains a complete animal–vegetal axis, give rise to mostly normal but small pluteus larvae (Fig. 6.13, left panel). Similarly, eggs that are fused together with their animal–vegetal axes parallel to each other form giant, but otherwise normal, larvae.

There is, by contrast, a striking difference in the development of animal and vegetal halves if they are isolated at the eight-cell stage. An isolated animal half merely forms a hollow sphere of ciliated ectoderm, whereas a vegetal half develops into a larva that is variable in form but is usually **vegetalized**, that is, it has a large gut and skeletal rods, and a reduced ectoderm lacking the mouth region (see Fig. 6.13, right panel). On occasion, however, vegetal halves from the eight-cell stage can form normal pluteus larvae if the third cleavage is slightly displaced toward the animal pole. These observations show that, as in amphibians, maternal cytoplasmic differences present in the egg can specify all three germ layers. Some spatially restricted mRNAs that may encode proteins involved in such specification along the animal–vegetal axis have already been identified.

Even though there are cytoplasmic differences along the animal–vegetal axis it is clear that the sea urchin embryo has considerable capacity for regulation, implying the occurrence of cell–cell interactions. As we see below, important developmental signals are produced by the micromeres in the vegetal region.

Fig. 6.13 Development of isolated sea urchin blastomeres. Left panel: if isolated at the four-cell stage, each blastomere develops into a small but normal larva. Right panel: an isolated animal half from the eight-cell stage forms a hollow sphere of ciliated ectoderm, whereas an isolated vegetal half usually develops into a highly abnormal embryo with a large gut, and some skeletal structures, but with a reduced ectoderm.

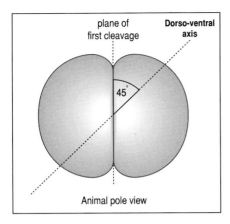

Fig. 6.14 Position of the dorso-ventral axis of the sea urchin S. purpuratus embryo in relation to first cleavage. When viewed from the animal pole, the dorso-ventral axis is usually 45° clockwise from the plane of first cleavage. Note that its polarity is not determined at this stage, only the axis.

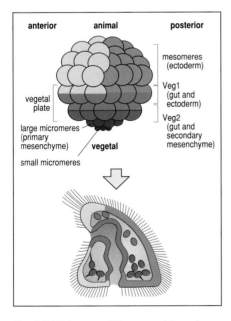

Fig. 6.15 Fate map of the sea urchin embryo. Lineage analysis has shown four main regions at the 60-cell stage. The embryo is divided along the animal–vegetal axis into three bands: the micromeres, which give rise to the primary mesenchyme; the vegetal plate, which gives rise to the gut, secondary mesenchyme, and some ectoderm; and the mesomeres, which give rise to ectoderm. The ectoderm is divided into anterior and posterior regions. After Logan, C.Y., McClay, D.R.: 1997.

6.6 The dorso-ventral axis in sea urchins is related to the plane of the first cleavage

The dorso-ventral axis of a sea urchin pluteus larva is defined with respect to the mouth, which develops on the ventral side (see Fig. 6.12), but there are also clear differences in skeletal patterning in the ventral and dorsal regions. The ventral side can be recognized well before mouth invagination, because the primary mesenchyme migrates to form two columns of cells on the future ventral face. The migration of these cells is considered in Chapter 8.

Unlike the animal–vegetal axis, the dorso-ventral axis cannot be identified in the egg and appears to be labile up to as late as the 16-cell stage. However, in normal embryos this axis is related to the plane of the first cleavage. In the sea urchin *Strongylocentrotus purpuratus*, a good correlation was found between the dorso-ventral axis and the plane of cleavage: the future dorso-ventral axis lies 45° clockwise from the first cleavage plane as viewed from the animal pole (Fig. 6.14). In other species, however, the relationship between the dorso-ventral axis and plane of cleavage is different; the axis may coincide with the plane of cleavage or be at right angles to it. It is worth recalling that in *Xenopus* the first cleavage usually corresponds with the plane of bilateral symmetry. In sea urchins, the cytoskeletal actin gene *CY111a* is transcribed during cleavage on the dorsal side of the embryo, suggesting an involvement of the cytoskeleton in determining the plane of cleavage.

6.7 The sea urchin fate map is finely specified, yet considerable regulation is possible

At the 60-cell stage, four regions of cells can be distinguished in the sea urchin embryo along the animal–vegetal axis and a simplified fate map can be constructed (Fig. 6.15). This is composed of three bands of cells: the large and small micromeres at the vegetal pole, which will give rise to mesoderm (the primary mesenchyme that forms the skeleton); the vegetal plate, which comprises the next two tiers of cells (Veg1 and Veg2) and gives rise to endoderm, mesoderm (the secondary mesenchyme that forms muscle and connective tissues), and some ectoderm; and the remainder of the embryo, which gives rise to ectoderm and is divided into anterior and posterior regions.

The small and large micromeres are generated from the four micromeres at the fifth cleavage division. The small micromeres contribute to the coelomic sacs, which form from the endoderm during late gastrulation, whereas the large micromeres give rise to the skeletal rods. The fate of the micromeres is maternally determined. Isolated micromeres give rise to skeletogenic cells when isolated and no other fate has been observed, no matter to what site they are grafted. The vegetal region of vegetal plate layer Veg2 (see Fig. 6.15) gives rise to the secondary mesenchyme cells, whereas the endoderm, from which the gut is formed, comes from the remainder of Veg2 and part of Veg1. The position of the endoderm–ectoderm boundary in Veg1 depends on signals from Veg2. Veg1 gives rise to the hindgut and portions of the midgut. The gene *Endo-16*, a marker of endoderm, is expressed in Veg1, and is later confined to the midgut. The endoderm cells derived from Veg1 are already expressing *Endo-16* before they invaginate at gastrulation.

Using vital dyes as lineage tracers, the pattern of cleavage has been shown to be invariant but complex, particularly for the ectoderm. However, unlike the nematode, the invariance of cleavage in the sea urchin seems irrelevant to normal development. Compressing an early sea urchin embryo by squashing it with a cover slip, thus altering the pattern of cleavage, still results in a normal embryo after further development. Moreover, as the sea urchin embryo is capable of considerable regulation, specification of fate cannot be closely linked to a cleavage pattern.

When isolated and cultured, some regions of the embryo develop more or less in line with their normal fate: for example, micromeres isolated at the 16-cell stage will form mesenchyme-like cells and may even form skeletal rods. An isolated animal half of presumptive ectoderm forms an animalized ciliated epithelial ball, but there is no indication of mouth formation. By contrast, a vegetal half at the 16-cell stage will develop into a vegetalized embryo with not only a large gut, but also ciliated ectoderm and some skeletogenic mesenchyme. In this tissue, therefore, the fate of some of the cells has been changed from prospective gut endoderm to mesoderm and ectoderm. This capacity for regulation is consistent with the observation that whole vegetal halves isolated at the third cleavage can, on occasion, regulate to give quite normal larvae but only if the third cleavage is displaced toward the animal pole.

These isolation experiments suggest that localized cytoplasmic factors are involved in specifying cell fate, but do not tell the full story as they do not reveal the role of inductive cell–cell interactions in the normal process of development, and the enormous capacity of the early embryo for regulation. A particularly dramatic example of regulation is seen when a meridional half embryo is combined with an animal half; this unlikely combination can develop into a normal embryo (Fig. 6.16), owing to the organizing and inducing properties of the vegetal region. Regulation can occur quite late in development; if the skeleton-forming primary mesenchyme cells are removed during gastrulation by flushing them out with a fine pipette, some of the secondary mesenchyme takes over their role and becomes skeletogenic.

The inductive and organizing role of the vegetal region is crucial for understanding both normal development and regulation, and we now consider this in more detail.

6.8 The vegetal region of the sea urchin embryo acts as an organizer

Striking evidence for an organizer-like region in the vegetal region of the sea urchin embryo comes from combining isolated micromeres (which normally give rise to skeletogenic mesenchyme) with an isolated animal half from a 32-cell embryo. This combination of cells can give rise to an almost normal larva, showing that, as in amphibians, the vegetal region contains an organizer that can induce the formation of an almost complete body axis (Fig. 6.17). The micromeres clearly induce some of the presumptive ectodermal cells in the animal half to form a gut, and the ectoderm becomes correctly patterned. Further evidence for the organizer-like properties of the micromeres comes from grafting them to the side of an intact embryo. The grafted micromeres induce endoderm in the presumptive ectoderm at that site, which then invaginates to form a second gut. There is some evidence that the closer the graft is to the vegetal region, the greater

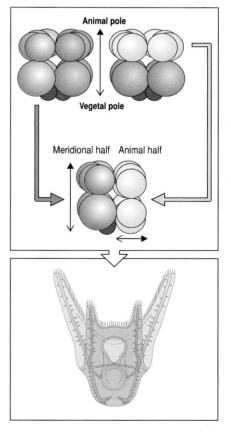

Fig. 6.16 Regulation in sea urchin development. Combining a meridional half of an eight-cell embryo with an animal half can result in a normal larva, even though the relationship between the cells in the recombinant embryo is abnormal.

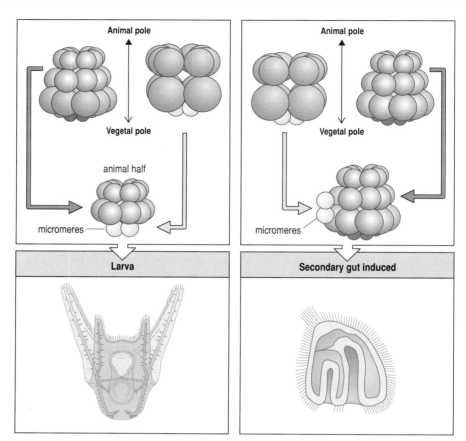

Fig. 6.17 The inductive action of micromeres. Left panels: combining the four micromeres from a 16-cell sea urchin embryo with an animal half from a 32-cell embryo results in a normal larva. (An animal half cultured on its own merely forms ectoderm.) Right panels: micromeres implanted into the side of another 32-cell embryo induce the formation of another gut at the implantation site.

the size of the invagination, implying a graded ability to respond to a signal from micromeres along the animal–vegetal axis. If the micromeres are removed when they are formed, regulation occurs and a normal larva will still develop. The most vegetal region takes on the property of micromeres, giving rise to skeleton-forming cells and acquiring organizing properties.

The normal fate of Veg2 is dependent on signals from the adjacent micromeres. Ablation of the mesomeres at the fourth cleavage results in abnormal development, but if they are removed at the sixth cleavage, after 2 to 3 hours of contact, gastrulation is delayed but normal larvae can develop. The micromeres are also responsible for inducing secondary mesenchyme, and this induction involves the Notch signaling pathway.

As in amphibian embryos, the fate of cells in the early sea urchin embryo can be altered by the action of lithium. In the sea urchin, treatment of cleavage stage embryos with lithium chloride causes vegetalization, resulting in embryos with a reduced ectoderm and an enlarged gut. The main effect of lithium on whole embryos is to shift the border between endoderm and ectoderm toward the animal pole. Thus, animal halves clearly have the capacity to develop endoderm and mesoderm but this capacity is suppressed in normal embryos. Moreover, in lithium-treated animal halves, the most vegetal region of the animal half acquires the organizer-like properties of the micromeres and can be used to reconstitute a complete embryo with an isolated, untreated animal half.

The molecular pathway initiating vegetal development is maternally determined and involves β-catenin, which is part of the Wnt signaling pathway and has a key role in early *Xenopus* development (see Section 3.4). It promotes vegetal fates in early sea urchin embryos. β-Catenin accumulates in the nuclei of the vegetal cells at the time they are acquiring their identity. Glycogen synthase kinase-3 (GSK-3β; see Section 3.4) is part of this pathway and targets β-catenin for degradation. Lithium chloride, which inhibits GSK-3β, vegetalizes the embryo and causes an enhancement and expansion of β-catenin accumulation in nuclei in the vegetal region. Injection of mRNA encoding a form of cadherin, an adhesion molecule that binds to β-catenin, completely prevents its nuclear accumulation and completely inhibits the development of both endoderm and mesoderm. Micromeres depleted of β-catenin are unable to induce endoderm when transplanted to animal regions. Overexpression of an 'activated' form of *Xenopus* β-catenin, on the other hand, causes vegetalization in the sea urchin embryo and can even induce endoderm in animal caps. Further evidence for the role of β-catenin in vegetal development comes from overexpression of GSK-3β, which promotes β-catenin degradation. This inhibits vegetal development, whereas a dominant-negative form of GSK-3β, which is unable to act, has the opposite effect—vegetalization.

There are similarities in the signals patterning the animal–vegetal axis of the sea urchin embryo and the dorso-ventral axis of amphibians. BMP-4 (see Section 3.19) is a strong ventralizing signal in *Xenopus*. Injection of either *Xenopus BMP-4* mRNA (or its sea urchin equivalent) into a sea urchin egg results in animalization. Noggin protein interacts with BMP-4 in *Xenopus*, and injection of its mRNA into sea urchin eggs results in vegetalization. These results suggest that patterning along the sea urchin animal–vegetal axis may be based on mechanisms similar to those patterning the dorso-ventral axis of amphibians.

One can now begin to make a model for patterning along the animal axis in terms of gradients that were proposed many years ago, but whose molecular basis was quite unknown. A maternally specified gradient from the vegetal to the animal pole would be based on β-catenin. This would be opposed by BMP-4, which is expressed more or less uniformly in the animal half. Interaction between these two systems could specify the endoderm–ectoderm boundary.

The Hox cluster in sea urchins contains ten genes but only two of these are expressed during development of the embryo and their function is not known. *SpHox7* is detected at the blastula stage in the ectoderm on the side away from the mouth, and *SpHox11/13b* is widely expressed in the blastula stage. The other Hox genes are expressed in a complex pattern during the development of the adult sea urchin at metamorphosis.

6.9 The regulatory regions of sea urchin developmental genes are complex and modular

In previous chapters, we saw how complex the regulatory regions of developmental genes can be. For example, the regulatory region of a *Drosophila* pair-rule gene such as *even-skipped* contains many binding sites for transcription factors that can both activate and repress gene activity (see Fig. 5.23). These target sites are grouped into discrete subregions—regulatory modules—which each control expression of the gene at a particular site in the embryo.

Although many of the genes controlling sea urchin development have yet

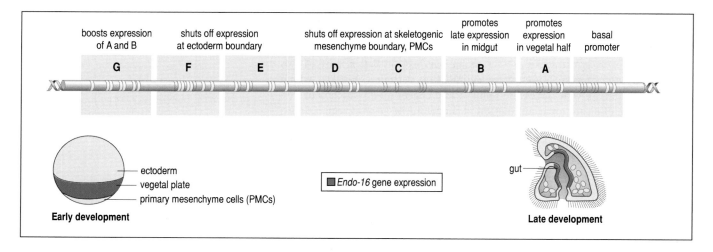

Fig. 6.18 Modular organization of the *Endo-16* gene regulatory region. The modular subregions are indicated A to G. Thirteen different transcription factors bind to the regulatory sites in these modules. The spatial domains of the embryo in which each module regulates gene expression are indicated, together with other regulatory modules. Early in development, gene regulation restricts *Endo-16* gene activity to the vegetal plate and the endoderm of the future gut; later in development it is restricted to the midgut.

to be identified, several whose expression varies with both space and time have been analyzed in detail. Their regulatory regions prove to be complex and modular, as illustrated by the regulatory region of the sea urchin *Endo-16* gene, a gene coding for a secreted glycoprotein of unknown function. *Endo-16* is initially expressed in the vegetal region of the blastula, in the presumptive endoderm (Fig. 6.18). After gastrulation, *Endo-16* expression increases and becomes confined to the middle region of the gut. The regulatory domain controlling *Endo-16* expression is about 2200 base pairs long and contains at least 30 target sites, to which 13 different transcription factors can bind. These regulatory sites seem to fall into several subregions, or modules. The function of each module has been determined by attaching the prospective module to a reporter gene and injecting the recombinant DNA construct into a sea urchin egg. In such experiments, module A, for example, is found to promote expression of its attached gene in the presumptive endoderm, whereas modules D and C prevent expression in the primary mesenchyme. Midgut expression is under control of module B. Modules D, C, E, and F are activated by the action of lithium and are involved in the vegetalizing action of *Endo-16*.

The modular nature of the *Endo-16* control region is similar to that of the pair-rule genes in *Drosophila*, where different modules are responsible for expression in each of the seven stripes. The sea urchin example illustrates yet again the importance of gene control regions in integrating and interpreting developmental signals.

Summary

The sea urchin egg has a stable animal–vegetal axis, which corresponds to the future antero-posterior axis and appears to be specified by the localization of maternal cytoplasmic factors in the egg. Maternal factors are also probably involved in specifying an organizer-like region in the most vegetal region of the egg, where the micromeres develop. The dorso-ventral axis is more labile, but is related to the plane of first cleavage. Sea urchin embryos have a remarkable capacity for regulation even though they also have an invariant cell lineage. Cell–cell interactions are clearly involved in specifying pattern along the animal–vegetal axis, and the micromeres at the vegetal pole can act as an organizer region. Patterning along the animal–vegetal axis involves signals similar to those along the amphibian dorso-ventral axis, namely the Wnt pathway and β-catenin. Lithium has a powerful vegetalizing effect, shifting the endoderm–ectoderm border toward the animal pole, and inducing endoderm and mesoderm differentiation in isolated animal halves. The site and time of expression of developmental genes are controlled by complex gene regulatory regions.

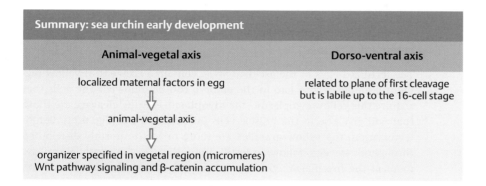

Summary: sea urchin early development	
Animal-vegetal axis	**Dorso-ventral axis**
localized maternal factors in egg ⇩ animal-vegetal axis ⇩ organizer specified in vegetal region (micromeres) Wnt pathway signaling and β-catenin accumulation	related to plane of first cleavage but is labile up to the 16-cell stage

Ascidians

Adult ascidians, also known as tunicates, are sessile marine animals. They are urochordates, which are included in the same phylum—the Chordata—as the vertebrates, because their free-living, tadpole-like larvae (Fig. 6.19) possess a notochord, neural tube, and muscles, and are rather similar to vertebrate neurulas. Unlike vertebrate embryos, however, ascidian embryos have an invariant cleavage pattern (Fig. 6.20), and localized cytoplasmic factors appear to have a much more important role in specifying cell fate. Ascidian embryogenesis has long been regarded as a typical example of mosaic development, with cytoplasmic factors specifying cell fate during cleavage, and cell interactions playing only a relatively minor part. It is now clear, however, that cell interactions are more important in ascidian development than was previously thought.

The developmental axes in ascidians are related to the early pattern of cleavage. The first cleavage passes through the site of polar body formation (see Fig. 6.20) and defines the future antero-posterior axis, usually dividing the embryo bilaterally. The unfertilized egg itself is highly regulative; if it is divided by a plane parallel to the future antero-posterior axis, each half, when fertilized, will develop into a complete larva. After a few cleavages, however, the cells become more committed to their prospective fate and there appears to be little capacity for regulation.

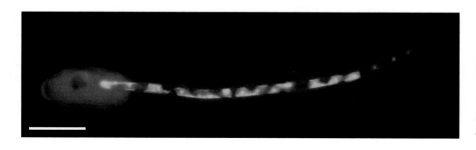

Fig. 6.19 Ascidian larva (*Ciona intestinalis*). Some of the notochord cells are labeled green. Scale bar = 0.1 mm.

Photograph courtesy of J. Corbo, from Corbo, J.C., et al.: 1997.

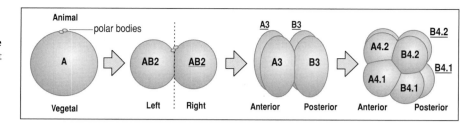

Fig. 6.20 Cleavages in the egg of the ascidian *Halocynthia rosetryi*. The first cleavage divides the embryo into left and right halves. The second cleavage divides the embryo along the antero-posterior axis.

6.10 In ascidians, muscle may be specified by localized cytoplasmic factors

The fertilized eggs of the ascidian *Styela* contain a crescent of yellow pigment granules whose fate in the embryo can be followed. The cells that acquire this yellow cytoplasm—the **myoplasm**—during cleavage are those that will give rise to the muscle cells of the larval tail (Fig. 6.21). Before fertilization, the yellow granules are more or less uniformly distributed throughout the egg; following fertilization, there is a dramatic rearrangement of the myoplasm, which occurs in two stages. The myoplasm first moves toward the vegetal pole and then moves laterally and toward the equator of the egg to form the yellow crescent. The crescent marks the future posterior end of the embryo, where gastrulation is initiated. The movements of the myoplasm are associated with cytoskeletal components, principally actin filaments lying beneath the plasma membrane and a deep network of intermediate filaments. These move with the myoplasm and end up in the yellow crescent; microtubules are also implicated in the second stage of localization.

During cleavage, the myoplasm becomes confined to particular cells, and by the eight-cell stage is largely confined to the two dorsal posterior cells, with a small amount in adjacent cells (see Fig. 6.21). In the species *Halocynthia*, the cell lineage has been worked out in detail by the injection of a tracer into the early cells. The two posterior B4.1 cells (see Fig. 6.20) that contain myoplasm contribute to the primary muscle cells, 28 of the 42 muscle cells that lie on each side of the tail. The secondary muscle cells are derived from blastomeres adjacent to B4.1 at the eight-cell stage and end up at the top of the tail. The lineage is complex; for example, at the 128-cell stage one of the descendants of B4.1 will still give rise to both muscle and endoderm. So, while there is a good correlation between muscle development and yellow cytoplasm, these observations alone do not establish that something in the yellow cytoplasm causes differentiation of the cells into muscle.

Experiments that alter the distribution of the myoplasm have provided

Fig. 6.21 Muscle development and cytoplasmic determinants in the ascidian *Styela*. Following fertilization, the myoplasm, which is colored with yellow granules, moves laterally and toward the equator. This movement forms a yellow crescent at the future posterior end of the embryo. Gastrulation starts at this site. The muscle of the ascidian tadpole's tail comes both from cells that contain the yellow myoplasm and the cells adjacent to them. After Conklin, E.G.: 1905.

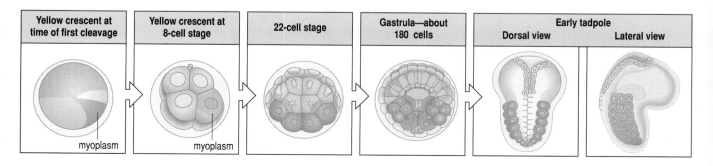

evidence for the involvement of cytoplasmic determinants in muscle differentiation and also a requirement for interactions with other cells. Mechanical deformation of the embryo can cause the myoplasm to be incorporated into more cells than usual; at least some of these 'new' myoplasm-containing cells then develop into muscle. But injection of myoplasm into cells that would not normally form muscle results in only a small number of cases of differentiation into muscle cells. Also, when the volume of the egg is reduced before cleavage, by removal of cytoplasm but not myoplasm, so that more cells than normal acquire myoplasm, there is no increase in the number of muscle cells that develop. The evidence for the ability of myoplasm on its own to specify muscle cells is thus suggestive, but not conclusive.

Persuasive evidence that a factor in myoplasm can indeed specify muscle cells is, however, provided by experiments with the *macho-1* gene. *macho-1* mRNA is localized in the myoplasm and its depletion results in loss of the primary muscle cells in the tail. Moreover, injection of *macho-1* mRNA into non-muscle cell lineages causes ectopic muscle differentiation.

Isolation of blastomeres at the 32- and 64-cell stage show that a signal, which is thought to be a form of fibroblast growth factor (FGF), from the endoderm is required to suppress muscle formation in blastomeres whose normal fate is mesenchyme. In the absence of such signals they develop muscle, as a result of the presence of maternal muscle determinants.

6.11 Notochord development in ascidians requires induction

The notochord of the *Halocynthia* larva consists of a single row of 40 cells aligned along the center of the tail. Cells of the A lineage give rise to 32 of the notochord cells, the remaining ones being derived from the B lineage. A-lineage cells that give rise only to notochord can be identified at the 64-cell stage, whereas B-lineage cells that give rise only to notochord cannot be identified until the 110-cell stage (Fig. 6.22). Blastomeres that would normally give rise to notochord do not do so if they are isolated at the 32-cell stage, unless they are induced by being combined with vegetal blastomeres. However, prospective notochord cells of the B lineage isolated at the 110-cell stage do develop into notochord. So, for notochord to develop, induction by vegetal cells is required somewhere between the 32-cell and 110-cell stage. FGF can induce notochord and is probably the inducer *in vivo*. Although the early development of ascidians is very different from that of vertebrates, the presence of a notochord and its induction by vegetal cells present striking parallels between the two groups. Moreover, the same genes are involved in notochord specification in both ascidians and vertebrates.

The gene *Brachyury* is expressed in early mesoderm in vertebrates (see Section 3.20) and then becomes confined to the notochord. Expression of the ascidian homolog of *Brachyury* is first detected at the 64-cell stage in the A-lineage precursors of the notochord, and this stage appears to correspond with the time at which induction is complete. Ectopic expression of the *Brachyury* gene of the ascidian *Ciona* can transform endoderm to notochord. Thus, there appears to be a similar mechanism involved in notochord formation in all chordates.

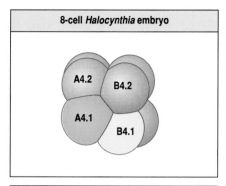

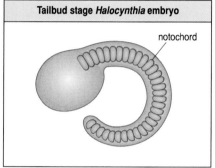

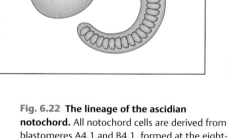

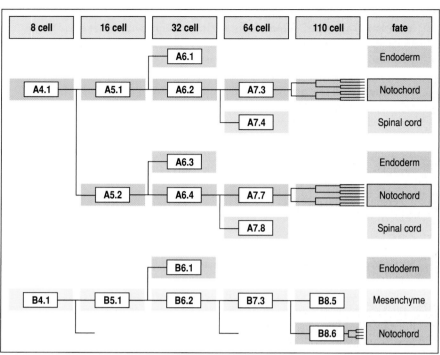

Fig. 6.22 The lineage of the ascidian notochord. All notochord cells are derived from blastomeres A4.1 and B4.1, formed at the eight-cell stage (see Fig. 6.20). Only the lineage on the left side is illustrated, as that on the right side is the same. After Nakatani, Y., *et al.*: 1996.

Summary

In ascidians, there is evidence that localized cytoplasmic factors are involved in specifying cell fate, particularly of muscle, but that cell interactions are also involved. The notochord develops through a well-defined lineage but requires induction. The ascidian homolog of the vertebrate gene *Brachyury*, which is involved in specifying the notochord in vertebrates, is expressed in the presumptive ascidian notochord after induction.

Cellular slime molds

The cellular slime molds, such as *Dictyostelium discoideum*, have some properties similar to both animals and plants. The cellulose cell walls formed during their development, and the presence of reproductive spores are plant-like, whereas the cell movements involved in their morphogenesis are animal-like. Analysis of protein sequences suggests that slime molds diverged from the eukaryotic lineage after the ancestors of plants but before those of animals (see Fig. 6.1). The slime molds have a developmental strategy with similarities to those of both animals and plants. The slime mold's simple life cycle (Fig. 6.23) and the ability to grow large numbers of cells easily, together with the possibility of genetic manipulation, make it an attractive system for developmental study.

Slime molds alternate between a unicellular and a multicellular phase of life cycle. The multicellular phase is not generated through the division of a single large cell to form an embryo, but by the aggregation of hitherto free-living single cells. Cellular slime molds grow and multiply as unicellular

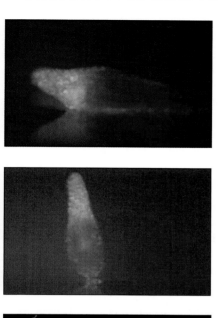

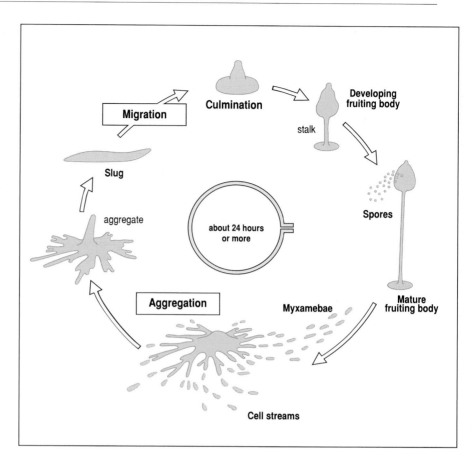

ameba-like myxamebae, which feed on bacteria. When their food supply runs out, the cells aggregate to form a migrating multicellular 'slug', typically containing some 100,000 cells. The slug develops into a fruiting body, consisting of a stalk composed of dead vacuolated cells supporting a mass of spores. Each spore develops into a new myxameba, which multiplies by division, and eventually gives rise to a new aggregate and slug. Thus, the simple life cycle involves differentiation into two main cell types—stalk cells and spore cells—together with major cell movements and rearrangements, and so provides an interesting example of the diversity of developmental strategies used by multicellular organisms.

Although many developmental mutants have been isolated from the slime mold *D. discoideum*, until recently it has proved very difficult to isolate the corresponding genes. However, a technique rather similar to transposon tagging (see Box 5A, p. 163) is now available. This is based on 'gene tagging' in which insertional mutations are created in a gene and the mutation recovered by recognition of the flanking sequences. This will make cloning of important developmental genes much easier. Some 300 genes are estimated to be essential for the development of *Dictyostelium*, as distinct from those required for its growth.

There appears to be a mechanism for sensing aggregate size. Normal aggregaates of *Dictyostelium* cells tend to be of similar size. A gene controlling aggregate size is *smlA*, which when deleted results in aggregates containing less than a tenth the usual number of cells. The presence of just 5%

Fig. 6.23 Life cycle of the cellular slime mold *Dictyostelium discoideum*. Unicellular myxamebae aggregate to form a multicellular 'slug' which, after migration, undergoes a process of culmination to form a differentiated, multicellular fruiting body. This is composed of a stalk with spores at the tip. The photographs show: top panel, a living migrating slug in which the prestalk region can be visualized (prestalk A cells (green); prestalk O cells (red)) because the prestalk A and O cells are expressing different spectral variants of green fluorescent protein. The prespore region is unstained. Yellow color indicates regions where prestalk A and prestalk O cells physically overlap, and the few cells that coexpress the two genes used to distinguish these cell types. Middle panel: a slug that has stopped migrating and stands on end. Cells in the prestalk region move first to the apex; then they reverse their direction of movement and travel towards the base, as indicated by the green mass of prestalk A cells visible in the prespore region. Bottom panel; mature *Dictyostelium* fruiting bodies.

Photographs in the top and center panels courtesy of P. Dorman, T. Abe, J.G. Williams and C.J. Weijer. The photograph in the bottom panel courtesy L. Blanton.

smlA mutant cells in a mixture with wild-type cells causes small aggregates to develop. The mutant *smlA* cells secrete an excessive amount of a large protein complex which seems to act as a counting factor, although the mechanism is unknown. Mutants that reduce the amount of this counting factor make excessively large aggregates. The aggregation process is of considerable interest in its own right, involving as it does a chemotactic mechanism, which will be considered in detail in Chapter 8. Here, we focus on the patterning of the migrating slug and the fruiting body, and particularly the patterning of the prespore and prestalk cells.

6.12 Patterning of the slime mold slug involves cell sorting and positional signaling

In the migrating slug, prestalk cells are at the anterior end, and prespore cells are at the posterior end. When it is ready to form a fruiting body, the slug rests on its hind end and the prestalk cells form a stalk by migrating down through the prespore region, pushing the prespore region upward (Fig. 6.24).

What specifies the prespore cells posteriorly and the prestalk cells at the front? This problem is particularly well exemplified by the ability of the slugs to regulate and so maintain the basic pattern. The proportion of stalk to spore cells is more or less constant over a 1000-fold range of slug sizes. The capacity for regulation is also shown by the fact that isolated anterior and posterior regions are capable of forming properly proportioned fruiting bodies. Two classes of mechanism have been proposed to account for this patterning and regulation. The first suggests that there is some mechanism for providing positional information to the cells in the slug so that their relative position is specified. This information could be in the form of a morphogen gradient with a single threshold for specifying prestalk and prespore cells. The second suggests that regulation involves the differentiation of cells at random locations, which then sort out to give the normal pattern. It appears that both mechanisms are involved in patterning the slug.

Patterning does not simply involve specifying cells as prespore and prestalk. Within the prestalk region there is a diversity of cell types. A, O, B,

Fig. 6.24 Formation of the fruiting body of *D. discoideum.* In the migrating slug the prestalk cells (blue) are at the front (left panel). At culmination the slug becomes upright and the prestalk cells move up to the apex and down a stalk tube (center panel). The growth of the stalk tube lifts up the prespore region (gray) to the tip. A basal disc (red) forms at the base (right panel).

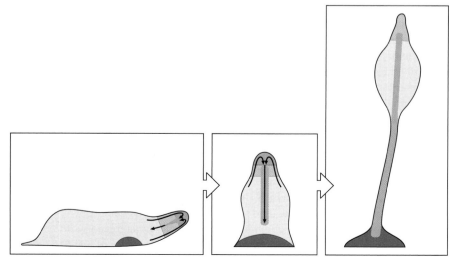

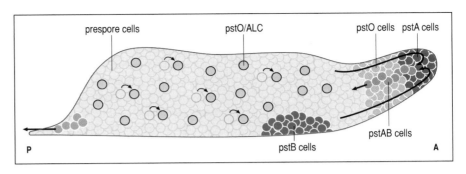

Fig. 6.25 Cell flow and differentiation in the migrating slime mold slug. At the tip are prestalk A (pstA) and prestalk AB (pstAB) cells. Some pstAB cells fall back and are lost at the rear of the slug. The lost pstAB cells are replaced by pstO/ALC cells which are derived from prespore cells. All of the rear of the slug contains prespore cells.

and AB prestalk cells are distinguished by the differential expression of the genes *ecmA* and *ecmB*, which code for extracellular matrix proteins. Prestalk A cells occupy the anterior half of the prestalk region and express *ecmA* strongly, while pstO cells occupy the rear half and express the *ecmA* gene less strongly. Prestalk AB cells express both *ecmA* and *ecmB* genes and are arranged in a funnel-shaped region near the tip of the slug (Fig. 6.25). Prestalk B cells express only the *ecmB* gene near the slug's prestalk and prespore boundary, and move downward at the culmination stage (see Fig. 6.24) to form the basal disc. This arrangement of cell types seems to be due to some form of cell sorting, rather than to positional signaling. Differentiation of prestalk A cells occurs at the periphery of the aggregate; only after differentiation do they accumulate at the tip. Also, when cells are removed from the anterior region of a slug and injected into another slug they return to their original position, which suggests that cell sorting is operating.

The migrating slug has a surprisingly dynamic structure, with respect both to cell differentiation and cell movement. At the posterior end of the slug, prestalk AB cells are lost, apparently having drifted backward from the tip region. They are replaced by pstA cells, which in turn are replaced by pstO cells. In the prespore region, posterior to the tip, are another class of cells, pstO/ALC (anterior-like cells), that appear to be very similar to the pstO cells, which express *ecmA*. During slug migration, some of these pstO/ALC cells move to the anterior end and become, at first, pstO cells, and then pstA cells. To maintain the correct proportion of the different types of cells, some prespore cells convert to pstO/ALC. Thus, the correct ratio of prespore to prestalk cells is maintained. The pattern of cell flow during slug migration also foreshadows the behavior of prestalk cells during fruiting body formation, when stalk cells form a stalk tube, which is pushed down through the prespore region.

It seems that the initial specification of prestalk and prespore cells occurs at the time of aggregation and may depend on a positional signal. As the cells come together to form a mound, prestalk A cells are confined to an outer ring, suggesting that this specification could result from a positional signal.

6.13 Chemical signals direct cell differentiation in the slime mold

Studies on isolated cells have led to the identification of possible morphogens which can direct stalk and spore cell differentiation: extracellular cyclic AMP is required for prespore cell differentiation, and prespore-specific genes can be induced by adding cyclic AMP. (Cyclic AMP is also involved in cell signaling and chemotaxis in cellular slime molds, as discussed in Chapter 8.) Differentiation-inducing factor (DIF) is a chlorinated hexaphenone released by developing cells, which induces prestalk and stalk cell differentiation. It seems that pstA and pstO prestalk cells are stimulated to differentiate by cyclic AMP in the presence of DIF, whereas differentiation of B-type prestalk cells is inhibited in these conditions. Mutants lacking glycogen synthase kinase-3 (GSK-3), an enzyme that we have already seen playing an important part in the Wnt pathway in early animal development (see Section 6.8), have fruiting bodies with large basal discs and few spores. Therefore, GSK-3 seems to mediate the initial decision to become a spore cell. In the absence of GSK-3 activity, differentiation of prestalk B cells occurs earlier and with a higher proportion of cells being differentiated. Thus, cyclic AMP may act to increase GSK-3, which prevents differentiation into prestalk B cells and favors prespore differentiation.

During development, DIF is extensively metabolized and thus destroyed by a cytoplasmic enzyme whose activity is induced by DIF itself. This induction by DIF of its own destructive agent seems to provide a negative feedback loop that controls the level of DIF, and possibly the number of stalk cells. While this feedback may operate during the aggregation phase, it poses a puzzle with respect to the slug: the apparent concentration of DIF in the prespore zone is 10 times higher than that necessary to inhibit prespore differentiation in isolated cells. Moreover, DIF is graded, with its highest concentration point in the posterior region, whereas the anterior region appears to be a localized sink for DIF. One attractive possibility is that, as part of their program of differentiation, prespore cells become insensitive to DIF.

Summary

In spite of the cellular slime mold appearing to provide a simple model for pattern formation and differentiation, giving rise to just two main cell types, patterning has turned out to be quite complex. Both positional signaling and cell sorting are involved, and in the migrating slug there is a dynamic balance of cell types as prestalk cells are lost and cells differentiate to replace them. This involves changes in both cell state and cell movement. The identification of cyclic AMP and differentiation-inducing factor (DIF) as factors controlling differentiation, together with the possibility of genetic manipulation, make the cellular slime mold a most promising system for developmental studies.

Summary: slime mold aggregation and differentiation

free-living myxamebae aggregate to form a multicellular 'slug'

⇩ cyclic AMP and DIF

specification of prestalk and prespore cells within the slug

⇩

slug migrates with prestalk cells at anterior end

⇩

slug stops moving and develops into a fruiting body with a cellular stalk and a head of spores

SUMMARY TO CHAPTER 6

In many invertebrates such as the nematode and the ascidian, cytoplasmic localization and cell lineage, together with local intercellular interactions, seem to be very important in specifying cell fate. However, there is very little evidence in any of these animals for key signaling regions corresponding to the vertebrate organizer, or for morphogen gradients as in *Drosophila*. This may reflect the predominance of mechanisms that specify cell fate on a cell-by-cell basis, rather than by groups of cells as in vertebrates and insects.

Although genetic evidence of the kind available in *Drosophila* is lacking in other invertebrates, except in the case of the nematode *Caenorhabditis elegans*, it is clear that maternal genes are crucial in specifying early patterning. There are numerous examples of the first cleavage in the embryo being related to the establishment of the embryo's axes and evidence that this plane of cleavage is maternally determined. Asymmetric cleavage associated with the unequal distribution of cytoplasmic determinants seems to be very important at early stages. Cytoplasmic movements that are involved in this cytoplasmic localization often occur after fertilization, and in this respect they may resemble the cortical rotation of the amphibian egg.

There can be no doubt that there are cell interactions, even at early stages, in all of the organisms in this chapter; an example is the specification of left–right asymmetry in nematodes, where a small alteration in cell relationships can reverse their handedness. But in most of these embryos the interactions are highly local, only affecting an adjacent cell. This is unlike vertebrates, in which interactions may occur over several cell diameters and involve groups of cells.

By contrast, in sea urchin development, which is highly regulative, cell interactions play a major role, and thus echinoderms resemble the vertebrates in this respect. The micromeres at the vegetal pole have some properties similar to those of the vertebrate organizer, and cells are specified in groups rather than on a cell-by-cell basis.

Cellular slime molds are organisms of ancient origin, diverging at around the time of the plant–animal split. Patterning of their multicellular phase involves cell sorting and positional signaling.

REFERENCES

6.1 **The developmental axes in the nematode are determined by asymmetric cell division and cell–cell interactions**

Bowerman, B.: Embryonic polarity: Protein stability in asymmetric cell division. *Curr. Biol.* 2000, **10**: R637–R641.

Guo, S., Kemphues, K.J.: *par-1*, a gene required for establishing polarity in C. elegans embryos, encodes a putative Ser/Thr kinase that is asymmetrically distributed. *Cell* 1995, **81**: 611–620.

Kemphues, K.: PARsing embryonic polarity. *Cell* 2000, **101**: 345–348.

Ruvkun, G., Habert, O: The taxonomy of developmental control in *Caenorhabditis elegans*. *Science* 1998, **282**: 2033–2040.

Wallenfang, M.W., Seydoux, G.: Polarization of the anterior–posterior axis of C. elegans is a microtubule-directed process. *Nature* 2000, **408**: 89–92.

Wood, W.B.: Evidence from reversal of handedness in *C. elegans* embryos for early cell interactions determining cell fates. *Nature* 1991, **349**: 536–538.

6.2 **Asymmetric divisions and cell–cell interactions specify cell fate in the early nematode embryo**

Evans, T.C., Crittenden, S.L., Kodoyianni, V., Kimble, J.: Translational control of maternal glp-1 mRNA establishes an asymmetry in the C. elegans embryo. *Cell* 1994, **77**: 183–194.

Goldstein, B.: Induction of gut in *Caenorhabditis elegans* embryos. *Nature* 1992, **357**: 255–257.

Kirby, C., Kusch, M., Kemphues, K.: Mutations in the *par* genes of *Caenorhabditis elegans* affect cytoplasmic reorganization during the first cell cycle. *Dev. Biol.* 1990, **142**: 203–215.

Labouesse, M., Mango, S.E.: Patterning the C. elegans embryo. *Trends Genet.* 1999, **15**: 307–313.

Lin, R., Hill, R.J., Priess, J.R.: POP-1 and anterior-posterior fate decisions in C. elegans embryos. *Cell* 1998, **92**: 229–239.

Mello, G.C., Draper, B.W., Priess, J.R.: **The maternal genes** *apx-1* **and** *glp-1* **and the establishment of dorsal-ventral polarity in the early** *C. elegans* **embryo.** *Cell* 1994, **77**: 95–106.

Priess, J.R., Thomson, J.N.: **Cellular interactions in early** *C. elegans* **embryos.** *Cell* 1987, **48**: 241–250.

Schnabel, R.: **Early determinative events in** *Caenorhabditis elegans.* *Curr. Opin. Genet. Dev.* 1991, **1**: 179–184.

Strome, S., Wood, W.B.: **Generation of asymmetry and segregation of germ-line granules in early** *C. elegans* **embryos.** *Cell* 1983, **35**: 15–25.

Sulston, J.E., Schierenberg, E., White, J.G., Thomson, J.N.: **The embryonic cell lineage of the nematode** *Caenorhabditis elegans.* *Dev. Biol.* 1983, **100**: 64–119.

Tax, F.E., Thomas, J.H.: **Cell–cell interactions. Receiving signals in the nematode embryo.** *Curr. Biol.* 1994, **4**: 914–916.

6.3 A small cluster of Hox genes specifies cell fate along the antero-posterior axis

Braunschwig, K., Wittmann, C., Schnabel, R., Bürglin, T.R.: **Anterior organization of the** *Caenorhabditis* **embryo by the labial-like Hox gene** *ceh-13.* *Development* 1999, **126**: 1537–1546.

Bürglin, T.R., Ruvkun, G.: **The** *Caenorhabditis elegans* **homeobox gene cluster.** *Curr. Opin. Genet. Dev.* 1993, 3: 615–620.

Clark, S.G., Chisholm, A.D., Horvitz, H.R.: **Control of cell fates in the central body region of** *C. elegans* **by the homeobox gene** *lin-39.* *Cell* 1993, **74**: 43–55.

Cowing, D., Kenyon, C.: **Correct Hox gene expression established independently of position in** *Caenorhabditis elegans.* *Nature* 1996, **382**: 353–356.

Salser, S.J., Kenyon, C.: **Patterning** *C. elegans*: **homeotic cluster genes, cell fates and cell migrations.** *Trends Genet.* 1994, **10**: 159–164.

Salser, S.J., Kenyon, C.: **A** *C. elegans* **Hox gene switches on, off, on and off again to regulate proliferation, differentiation and morphogenesis.** *Development* 1996, **122**: 1651–1661.

Van Auken, K., Weaver, D.C., Edgar, L.G., Wood, W.B.: *Caenorhabditis elegans* **embryonic axial patterning requires two recently discovered posterior-group Hox genes.** *Proc. Natl Acad. Sci. USA* 2000, **97**: 4499–4503.

6.4 Genes control graded temporal information in nematode development

Ambros, V.: **Control of developmental timing in** *Caenorhabditis elegans.* *Curr. Opin. Genet. Dev.* 2000, **10**: 428–433.

Ambros, V., Horvitz, H.R: **Heterochronic mutants of the nematode** *Caenorhabditis elegans.* *Science* 1984, **226**: 409–416.

Austin, J., Kenyon, C.: **Developmental timekeeping: marking time with antisense.** *Curr. Biol.* 1994, **4**: 366–396.

Ferreira, H.B., Zhang, Y., Zhao, C., Emmons, S.W.: **Patterning of** *Caenorhabditis elegans* **posterior structures by the Abdominal-B homolog, egl-5.** *Dev. Biol.* 1999, **207**: 215–228.

6.5 The sea urchin egg is polarized along the animal–vegetal axis

Horstadius, S.: *Experimental Embryology of Echinoderms.* Oxford: Clarendon Press, 1973.

Wilt, F.H.: **Determination and morphogenesis in the sea urchin embryo.** *Development* 1987, **100**: 559–575.

6.6 The dorso-ventral axis in sea urchins is related to the plane of the first cleavage

Cameron, R.A., Fraser, S.E., Britten, R.J., Davidson, E.H.: **The oral–aboral axis of a sea urchin embryo is specified by first cleavage.** *Development* 1989, **106**: 641–647.

Jeffrey, W.R.: **Axis determination in sea urchin embryos: from confusion to evolution.** *Trends Genet.* 1992, **8**: 223–225.

6.7 The sea urchin fate map is finely specified, yet considerable regulation is possible

Davidson, E.H., Cameron, R.S., Ransick, A.: **Specification of cell fate in the sea urchin embryo: summary and some proposed mechanisms.** *Development* 1998, **125**: 3269–3290.

Ettensohn, C.A.: **Cell interactions and mesodermal cell fates in the sea urchin embryo.** *Development* (Suppl.) 1992, 43–51.

Livingston, B.T., Wilt, F.H.: **Lithium evokes expression of vegetal-specific molecules in the animal blastomeres of sea-urchin embryos.** *Proc. Natl Acad. Sci. USA* 1989, **86**: 3669–3673.

Logan, C.Y., McClay, D.R.: **The allocation of early blastomeres to the ectoderm and endoderm is variable in the sea urchin embryo.** *Development* 1997, **124**: 2213–2223.

6.8 The vegetal region of the sea urchin embryo acts as an organizer

Angerer, L.M., Angerer, R.C.: **Animal–vegetal axis patterning mechanisms in the early sea urchin embryo.** *Dev. Biol.* 2000, **218**: 1–12.

Angerer, L.M., Oleksyn, D.W., Logan, C.Y., McClay, D.R., Dale, L., Angerer, R.C.: **A BMP pathway regulates cell fate allocation along the sea urchin animal–vegetal embryonic axis.** *Development* 2000, **127**: 1105–1114.

Arenas-Mena, A., Cameron, A.R., Davidson, E.H.: **Spatial expression of Hox cluster genes in the ontogeny of a sea urchin.** *Development* 2000, **127**: 4631–4643.

Logan, C.Y., Miller, J.R., Ferkowicz, M.J., McClay, D.R.: **Nuclear β-catenin is required to specify vegetal cell fates in the sea urchin embryo.** *Development* 1999, **126**: 345–357.

Ransick, A., Davidson, E.H.: **A complete second gut induced by transplanted micromeres in the sea urchin embryo.** *Science* 1993, **259**: 1134–1138.

Sweet, H.C., Hodor, P.G., Ettensohn, C.A.: **The role of micromere signaling in Notch activation and mesoderm specification during sea urchin embryogenesis.** *Development* 1999, **126**: 5255–5265.

Wessel, G.M., Wikramanayake, A.: **How to grow a gut: ontogeny of the endoderm in the sea urchin embryo.** *BioEssays* 1999, **21**: 459–471.

6.9 The regulatory regions of sea urchin developmental genes are complex and modular

Davidson, E.H.: **A view from the genome: spatial control of transcription in sea urchin development.** *Curr. Opin. Genet. Dev.* 1999, **9**: 530–541.

Kirchnamer, C.V., Yuh, C.V., Davidson, E.H.: **Modular** *cis-**regulatory organization of developmentally expressed genes: two genes transcribed territorially in the sea urchin embryo, and additional examples.** *Proc. Natl Acad. Sci. USA* 1996, **93**: 9322–9328.

6.10 In ascidians, muscle may be specified by localized cytoplasmic factors

Bates, W.R.: Development of myoplasm-enriched ascidian embryos. *Dev. Biol.* 1988, **129**: 241–252.

di Gregorio, A., Levine, M.: Ascidian embryogenesis and the origins of the chordate body plan. *Curr. Opin. Genet. Dev.* 1998, **7**: 457–463.

Kim, G.J., Nishida, H.: Suppression of muscle fate by cellular interaction as required for mesenchyme formation during ascidian embryogenesis. *Dev. Biol.* 1999, **214**: 9–22.

Nishida, H.: Determinative mechanisms in secondary muscle lineages of ascidian embryos: development of muscle-specific features in isolated muscle progenitor cells. *Development* 1990, **108**: 559–568.

Nishida, H., Sawada, K.: *macho-1* encodes a localized mRNA in ascidian eggs that specifies muscle fate during embryogenesis. *Nature* 2001, **409**: 724–728.

Satoh, N.: *The Developmental Biology of Ascidians*. Cambridge: Cambridge University Press, 1994.

6.11 Notochord development in ascidians requires induction

Corbo, J.C., Levine, M., Zeller, R.W.: Characterization of a notochord-specific enhancer from the *Brachyury* promoter region of the ascidian, *Ciona intestinalis*. *Development* 1997, **124**: 589–602.

Kim, G.J., Yamada, A., Nishidda, H.: An FGF signal from endoderm and localized factors in the posterior–vegetal egg cytoplasm pattern the mesodermal tissues in the ascidian embryo. *Development* 2000, **127**: 2853–2862.

Nakatani, Y., Nishida, H.: Induction of notochord during ascidian embryogenesis. *Dev. Biol.* 1994, **166**: 289–299.

Nakatani, Y., Yasuo, H., Satoh, N., Nishida, H.: Basic fibroblast growth factor induces notochord formation and the expression of As-T, a Brachyury homolog, during ascidian embryogenesis. *Development* 1996, **122**: 2023–2031.

6.12 Patterning of the slime mold slug involves cell sorting and positional signaling

Abe, T., Early, A., Siegert, F., Weijer, C., Williams, J.: Patterns of cell movement within the *Dictyostelium* slug revealed by cell type-specific, surface labelling of cells. *Cell* 1994, **77**: 687–699.

Brown, J.M., Firtel, R.A. Regulation of cell-fate determination in *Dictyostelium*. *Dev. Biol.* 1999, **216**: 426–441.

Brown, J.M., Firtel, R.A.: Just the right size: cell counting in *Dictyostelium*. *Trends Genet.* 2000, **16**: 191–193.

Early, A., Abe, T., Williams, J.: Evidence for positional differentiation of prestalk cells and for a morphogenetic gradient in *Dictyostelium*. *Cell* 1995, **83**: 91–99.

Jermyn, K., Traynor, D., Williams, J.: The initiation of basal disc formation in *Dictyostelium discoideum* is an early event in culmination. *Development* 1996, **122**: 753–760.

6.13 Chemical signals direct cell differentiation in the slime mold

Brisco, F., Firtel, R.A.: A kinase for cell fate determination? *Curr. Biol.* 1995, **5**: 228–231.

Gross, J.D.: Developmental decisions in *Dictyostelium discoideum*. *Microbiol. Rev.* 1994, **58**: 330–351.

Kay, R.R.: Differentiation and patterning in *Dictyostelium*. *Curr. Opin. Genet. Dev.* 1994, **4**: 637–641.

Plant development

7

- Embryonic development
- Meristems
- Flower development

The plant kingdom is very large, ranging from the algae, many of which are unicellular, to the multicellular land plants which exist in a prodigious variety of forms. Plants and animals probably evolved the process of multicellular development separately, their last common ancestor being a unicellular eukaryote. This raises the interesting question as to whether land plants and animals use the same fundamental mechanisms. As we shall see in this chapter, the regulation of spatial patterns of gene expression in, for example, flowers, is similar to the logic we have seen in relation to homeotic genes in animals. By contrast, the means by which cells interact and recognize each other seems quite different and in plants is as yet poorly understood. Environment also has a much greater impact on plant development compared to that of animals. Plant and animal cells share many common internal features and much basic biochemistry, but there are some crucial differences that have a bearing on plant development. One of the most important is that plant cells are surrounded by a framework of relatively rigid cell walls. There is therefore virtually no cell migration in plants, and major changes in the shape of the developing plant cannot be achieved by the movement and folding of sheets of cells, as they are in early animal embryogenesis. In plant development, form is largely generated by differences in rates of cell division and by division in different planes, followed by directed enlargement of the cells.

A fundamental difference between plants and animals is that all post-embryonic growth in higher plants occurs from localized regions called **meristems**, which give rise to all the 'adult' structures of the plant—shoots, roots, stems, leaves, and flowers. Meristems contain cells that have the capacity to divide repeatedly in the same way as animal stem cells, and the progeny of these cells can give rise to a variety of tissues. Two meristems are established in the embryo, one located at the tip of the root and the other at the tip of the shoot, and almost all the other meristems of an adult plant derive from these. Because of this mode of growth, the developmental patterning of organs in a plant does not occur only in the embryo, but continues in the meristems throughout its life.

One outstanding question in plant development is how a cell's fate is determined. Many structures in plants develop by stereotyped patterns of cell division, but it is not yet clear to what extent a cell's behavior and subsequent differentiation is determined by these patterned divisions, and whether they are influenced by factors intrinsic to the cell, or whether interactions between cells are more important. A cell's fate can be altered by a change in its position within the meristem, which implies that signals from other cells must have some influence. Far less is known about how plant cells communicate with each other and transduce hormonal and other signals than is the case in animal cells. The cell wall would seem to impose a barrier to the passage of large molecules, such as proteins, although it is very thin in some regions such as the meristems, but all known plant growth hormones, such as auxins, gibberellins, cytokinins, and ethylene, are small molecules that easily penetrate cell walls. It may be that plant cells also communicate with each other through the fine cytoplasmic channels—plasmodesmata—which link neighboring plant cells through the cell wall and there is evidence that both proteins and RNA can move between plant cells through the plasmodesmata.

Another important difference between plant and animal cells is that a complete, fertile plant can develop from a single somatic cell and not just from a fertilized egg. This suggests that, unlike the differentiated cells of adult animals, some differentiated cells of the adult plant may retain totipotency. Perhaps they do not become fully determined in the sense that adult animal cells do, or perhaps they are able to escape from the determined state, although how this could be achieved is as yet unknown. In any case, this difference between plant and animal cells illustrates the dangers of the wholesale application to plant development of concepts derived from animal development. Nevertheless, the genetic analysis of development in plants is turning up instances of genetic strategies for developmental patterning rather similar to those of animals (see Section 7.12).

In this chapter, we first consider the embryogenesis and further development of higher plants, with particular focus on the genetic control of development in flowering plants (the angiosperms), and on how cell fate may be determined. The small annual thale-cress, *Arabidopsis thaliana*, has become the *Drosophila* of the plant world (see Section 2.7 for the developmental stages and life-history of *Arabidopsis*), as it is easily grown and is more amenable to genetic analysis than many other plants, and we use it as an example wherever possible. We then look at the very early development of the zygote in multicellular algae, where it is more easily studied than in higher plants.

Embryonic development

In the same way as an animal zygote, the fertilized plant egg cell undergoes repeated cell division, cell growth, and differentiation to form a multicellular embryo. There is enormous diversity in the pattern of early cell divisions in different types of plants, but the significance of any variation, in terms of developmental strategy, is not known. A very common early pattern of cell division in higher plants is seen with the first division of the zygote. This occurs at right angles to the long axis of the embryo and so divides the zygote into apical and basal regions. In many species this first division is unequal, but in most of these cases it is not known to what extent this reflects a developmentally significant asymmetric cell division that results in daughter cells with different identities.

7.1 Both asymmetric cell divisions and cell position pattern the early embryos of flowering plants

There is great diversity in the pattern of cell division in the embryos of different species of angiosperms. One common principle is that the first division in angiosperm embryos is transverse with respect to the long axis of the zygote, dividing the zygote into an apical cell and a basal cell. In some species the basal cell contributes little to further development of the embryo but divides to give rise to the suspensor, which may be several cells long. The apical cell undergoes a complex pattern of divisions which in dicotyledons such as *Capsella bursa-pastoris* (shepherd's purse) and *Arabidopsis* leads in about 5 days to a globular-stage and then to a heart-stage embryo (Box 7A, p. 224).

A rough fate map for *Arabidopsis* can be constructed for the very early embryo on the basis of the pattern of cell divisions, by following the fate of the embryonic cells and finding what region(s) each gives rise to (Fig. 7.1). When there are just eight cells at the apical end—the so-called octant stage—one can begin to map the future main regions of the heart-shaped stage, at which stage a clear fate map can be drawn. The heart-stage embryo can be divided into three main regions: the apical region, which gives rise to the shoot meristem and cotyledons; the central region, which forms the

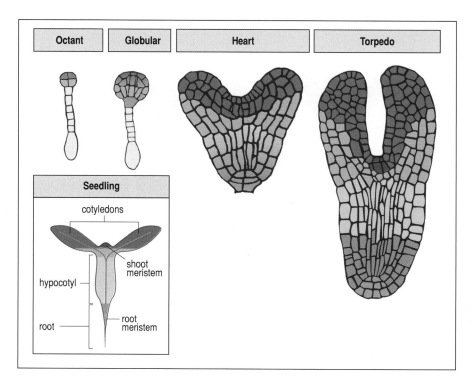

Fig. 7.1 Fate map of the *Arabidopsis* embryo. The stereotyped pattern of cell division in dicotyledon embryos means that at the globular stage it is already possible to map the three regions that will give rise to the cotyledons (dark green) and shoot meristem (red), the hypocotyl (yellow), and the root meristem (purple), in the seedling. After Scheres, B., *et al.*: 1994.

Box 7A Angiosperm embryogenesis

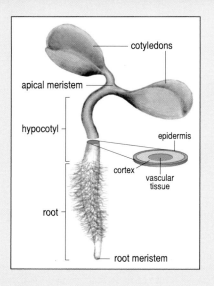

In flowering plants (angiosperms), the egg cell is contained within an ovule inside the ovary in the flower (right inset in the bottom panel). At fertilization, a pollen grain deposited on the surface of the stigma puts out a pollen tube, down which two male gametes migrate into the ovule (see Section 2.7). One male gamete fertilizes the egg cell while the other com-bines with another cell inside the ovule to form a specialized nutritive tissue, the endosperm, which surrounds, and provides the food source for, the development of the embryo.

Capsella bursa-pastoris (shepherd's purse) is a typical dicotyledon. The first, asymmetric, cleavage divides the zygote transversely into an apical and a basal cell (bottom panel). The basal cell then divides several times to form a single row of cells-the suspensor-which in many angiosperm embryos takes no further part in embryonic development, but may have an absorptive function; in *Capsella*, however, it contributes to the root meristem. Most of the embryo is derived from the terminal cell. This undergoes a series of stereotyped divisions, in which a precise pattern of cleavages in different planes gives rise to the heart-shaped embryonic stage typical of dicotyledons. This develops into a mature embryo which consists of a main axis, with a meristem at either end, and two cotyledons, which are storage organs.

The early embryo becomes differentiated into three main tissues-the outer epidermis, the prospective vascular tissue, which runs through the center of the main axis and cotyledons, and the ground tissue (prospective cortex) that surrounds it. Monocotyledons, such as maize, have a single cotyledon; dicotyledons, such as *Arabidopsis*, have two.

The ovule containing the embryo matures into a seed (left inset in the bottom panel), which remains dormant until suitable external conditions trigger germination and growth of the seedling. A typical dicotyledon seedling (left panel) comprises the shoot apical meristem, two cotyledons, the trunk of the seedling (the hypocotyl) and the root apical meristem. The seedling may be thought of as the phylotypic stage of flowering plants. The seedling body plan is simple. There is one main axis—the apical-basal axis—which defines the polarity of the plant. The shoot forms at the apical pole and the root at the basal pole. The plant axis has radial symmetry in cross-section, which is evident in the hypocotyl and is continued in the root and shoot. In the center is the vascular tissue, which is surrounded by cortex, and an outer covering of epidermis.

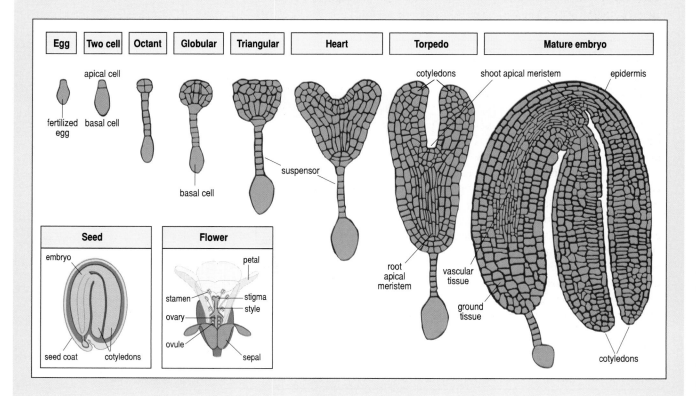

future hypocotyl (embryonic stem); and the basal region, which gives rise to the root meristem. There is a radial pattern made up of concentric rings of tissue layers: epidermis, ground tissue (cortex and endodermis), and vascular tissue. It is not known to what extent the fate of the cells is determined at this stage, or whether their fate is in fact dependent on the early pattern of asymmetric cleavages. It seems that cell lineage is not crucial, as clonal analysis reveals considerable variability. Mutations in the *FASS* gene totally alter the pattern of cell division and disorganize the cells, yet although the resultant seedling is misshapen, it has roots, shoots, and flowers in the correct places.

A very different pattern of cleavage occurs in the embryo of the monocotyledon *Zea mays* (maize). Because later cell divisions are irregular, both in orientation and sequence, it is much more difficult to make a fate map at an early stage. The position of the future shoot apical meristem can only be identified after about 9 days, when the embryo already consists of many cells. The shoot meristem can then be detected lying at the base of the single cotyledon. The root meristem can be detected several days later in the region near the suspensor. The inability to make a fate map of the early embryo reflects the fact that cell lineages in maize embryos are variable. This suggests that it is relative cell position that determines cell fate in maize embryos.

7.2 The patterning of particular regions of the *Arabidopsis* embryo can be altered by mutation

As an angiosperm embryo is enclosed within a seed, it is not easily accessible to experimental microsurgical manipulation. The course of development can, however, be altered by mutation. The effects of mutations that alter the basic body plan in *Arabidopsis* can be recognized at the heart-shaped embryo stage. Mutations in zygotic genes, affecting different parts of the pattern along the apical–basal axis, have been found in *Arabidopsis*. In apical mutants the cotyledons and shoot meristem are missing; in central mutants there is no hypocotyl and the cotyledons appear to be attached to the root; in basal mutants there is no hypocotyl or root (Fig. 7.2).

The phenotypes of these mutants suggest that the patterning of particular regions of the embryo along the apical–basal axis is controlled by specific genes. The control mechanisms are not yet known, but several of the genes involved have been identified (see Fig. 7.2). The basal mutation *monopterous* eliminates the hypocotyl and the root meristem, and abnormal cell divisions can be observed at the octant stage in the central and basal regions. *fäckel* mutations affect the central region and *gurke* mutants are apical mutants, lacking shoot meristems and cotyledons. Mutations in the gene *SHOOT MERISTEMLESS* (*STM*) completely block the formation of the shoot apical meristem, but have no effect on the root meristem or other parts of the embryo. *SHOOT MERISTEMLESS* probably codes for a transcription factor. Its pattern of expression develops gradually, which is also typical of several other genes that characterize the shoot apical meristem. Expression is first detected in the globular stage in one or two cells and only later in the central region between the two cotyledons (Fig. 7.3).

Radial pattern formation commences at the eight-cell stage and involves oriented cell division. **Periclinal** divisions, in which the plane of division is parallel to the outer surface, give rise to a new tissue layer, and **anticlinal**

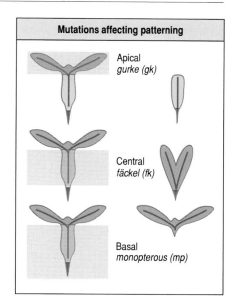

Fig. 7.2 Deletion of pattern by mutation along the apical–basal axis in *Arabidopsis* embryos. Three main classes of mutants have been discovered—apical, central, and basal. The regions deleted in each class of mutants are shown highlighted on the wild-type seedlings on the left, with the resulting mutant seedlings on the right. Examples of genes in which mutations give rise to these phenotypes are given. After Mayer, U., *et al*.: 1991.

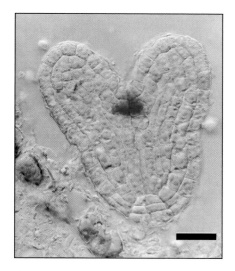

Fig. 7.3 Section through a late heart-stage *Arabidopsis* embryo showing expression of *SHOOT MERISTEMLESS* (*STM*). At this stage, the *STM* RNA (stained red) is expressed in cells located between the cotyledons. Scale bar = 25 mm.

Photograph courtesy of K. Barton. From Long, J.A., Barton, K.B.: 1998.

divisions, in which the plane of cell division is at right angles to the outer surface, increase the number of cells in a layer. It seems that tissue fate is determined early in development, because in mutant embryos the epidermal cells are enlarged from an early stage whereas the endodermis is absent when the *short root* mutation is present.

Other mutations have been discovered that affect patterning of the *Arabidopsis* embryo. A mutation in the *LEAFY COTYLEDON* gene, for example, results in what appears to be a homeotic transformation of cotyledons into leaves. A feature of this transformation is the appearance of leaf hairs (trichomes) on the cotyledons; these are normally only found on leaves and stems. In addition, the vascular tissue in the transformed cotyledons is intermediate in complexity between that of leaves and cotyledons.

Distinct from the pattern mutations are those that affect morphogenesis—the shape of the embryo and seedling. One such mutation, *fass*, mentioned above, randomizes patterns of cell division and results in fat, short seedlings in which the cells are more rounded and irregularly spaced than the orderly stacks of elongated cells in a wild-type embryo. But although *fass* seedlings are squat, because of an enlarged vascular system and multiple cortical layers, the basic radial pattern is normal. This again suggests that the initial embryonic patterning is not dependent on cell shape or size, but on the relative positions of the cells. A somewhat surprising feature of the early development of the embryo in *Arabidopsis* is that the expression of many paternal genes is delayed.

7.3 Plant somatic cells can give rise to embryos and seedlings

As gardeners well know, plants have amazing powers of regeneration. A complete new plant can develop from a small piece of stem or root, or even from the cut edge of a leaf. This reflects an important difference between the developmental potential of plant and animal cells. In animals, with few exceptions, cell determination and differentiation are irreversible. By contrast, many somatic plant cells remain **totipotent**. Somatic cells from roots, leaves and stems, and even, for some species, a single isolated protoplast, can be grown in culture and induced by treatment with the appropriate growth hormones to give rise to a new plant (Fig. 7.4). Careful observation of plant cells proliferating in culture has revealed that some of the dividing cells give rise to cell clusters that pass through a stage strongly resembling normal embryonic development, although the pattern of cell divisions are not the same as those in the embryo. These 'embryos' can then develop into seedlings. The initial work on plant regeneration in culture was done with carrot cells, but a wide variety of plants can form embryos from single cells in this way. The phenomenon has been studied most in dicotyledons, such as carrot, potato, petunia, and tobacco, which are easy to propagate by these means.

The ability of single somatic cells to give rise to whole plants has two important implications for plant development. The first is that maternal determinants may be of little or no importance in plant embryogenesis, as it is unlikely that every somatic cell would still be carrying such determinants. Second, it suggests that many cells in the 'adult' plant body are not fully determined with respect to their fate, but remain totipotent. Of course, this totipotency seems only to be expressed under special conditions, but it is, nevertheless, quite unlike the behavior of animal cells. It is

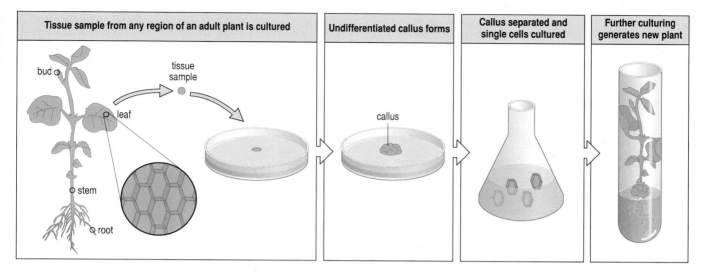

| Tissue sample from any region of an adult plant is cultured | Undifferentiated callus forms | Callus separated and single cells cultured | Further culturing generates new plant |

Fig. 7.4 Cultured somatic cells from a mature plant can form an embryo and regenerate a new plant. The illustration shows the generation of a plant from single cells. If a small piece of tissue from a plant stem or leaf is placed on a solid agar medium containing the appropriate nutrients and growth hormones, the cells will start to divide to form a disorganized mass of undifferentiated cells—a callus. The callus cells are then separated and grown in liquid culture, again containing the appropriate growth hormones. In suspension culture, some of the callus cells divide to form small cell clusters. These cell clusters resemble the globular stage of a dicotyledon embryo, and with further culture on solid medium, develop through the heart-shaped and later stages to regenerate a complete new plant.

as if such plant cells have no long-term developmental memory, or that such memory is easily reset.

7.4 Electrical currents are involved in polarizing the *Fucus* zygote

Embryonic development has also been studied in algae. In the brown alga *Fucus*, the body axis (apical–basal) is determined by environmental signals while the zygote is still a single cell. Determination of the axis leads to asymmetric cleavage of the zygote transverse to the axis, which delineates the two main regions of the algal body. Studies of *Fucus* illustrate the importance of the cell wall as a potential source of developmental signals. The eggs of *Fucus* are fertilized externally, and the zygote drifts through the water until it encounters a suitable surface on which to settle. Once settled, it begins to develop. At first cleavage, the zygote is divided into two cells of different size, the smaller of which gives rise to the root-like rhizoid, and the other to the thallus, or 'leafy' part of the alga (Fig. 7.5). This asymmetric cell division reflects an apical–basal polarity that is set up in the zygote before cleavage begins, and which can be determined by the site of sperm entry. A similar transverse asymmetric division indicates the direction of apical–basal polarity in the cleaving zygote of many higher plants. The first sign of asymmetry in the fertilized *Fucus* egg appears before the first cleavage as a protrusion in the region of the future rhizoid. Even though the fertilized egg may be polarized, experimental manipulations have shown that this polarity can be easily overriden by environmental signals. In this, *Fucus* resembles those animal embryos, such as *Xenopus*, which have their axes specified by external signals.

A wide range of environmental stimuli, including illumination, pH gradients, and water flow can set the polarity. (In default of such signals, there is an intrinsic mechanism for determining polarity by the point of sperm entry, which corresponds to the future rhizoid region.) The effects of light are dramatically demonstrated by an experiment in which fertilized *Fucus* eggs are placed in a narrow capillary tube that prevents their rotation and a light is shone at one end of the tube. The rhizoids all grow out of the shaded sides of the eggs, irrespective of the site of sperm entry.

Associated with polarization of the egg is the development of an

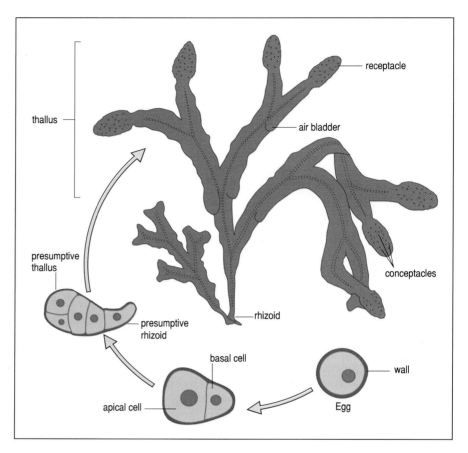

Fig. 7.5 The life history of the multicellular brown alga *Fucus spiralis*. The algal body consists of a flattened thallus divided into fronds, which is anchored to a rock or other hard substrate by the rhizoid or holdfast. Swollen fruiting bodies, the receptacles, develop at the tips of the fronds. They are pitted with small, jelly-filled cavities, the conceptacles, in which the gametes are produced. Male and female gametes develop in different receptacles. Mature gametes are released from the conceptacles and the eggs are fertilized externally by the motile sperms. The fertilized egg drifts through the water until it settles on a suitable substrate and begins to develop. The first cleavage of the zygote is unequal, producing a smaller basal cell that gives rise to the rhizoid, and a larger apical cell that gives rise to the thallus.

electrical current that flows in the direction of the axis (Fig. 7.6). The current must be partly carried by calcium ions, as local application of a calcium ionophore, which causes a local flow of calcium ions into the cell, causes rhizoid outgrowth at the site of application. All the environmental signals may act by causing calcium channels to open locally, with the resulting current causing further calcium channels to cluster at the site, and calcium pumps (which pump calcium ions out of the cell) to cluster at the other end of the axis. This positive feedback mechanism would act to maintain the direction of flow. The ion flow affects internal components of the egg, which now become localized in the region where calcium ions are entering the cell. In the absence of environmental influences, the site of sperm entry can determine where the rhizoid will form.

Although ion currents establish the axis, other processes are required to stabilize it. By 12 hours after fertilization the axis is fixed, and it only remains labile—that is, its position can be altered—for up to 7 hours after fertilization. Intracellular components are transported to, and accumulate in, the cytoplasm at the site of rhizoid formation. A particular polysaccharide is transported to the rhizoid site and inserted into the rhizoid wall. It seems that both the cytoskeleton and cell wall are involved in fixation of the axis. Actin filaments soon become localized at the rhizoid pole and the introduction of agents that disrupt them, such as cytochalasin B, prevent axis formation. Removal of the cell wall by enzyme treatment also prevents axis formation. Thus, an attractive hypothesis for polarization of the egg

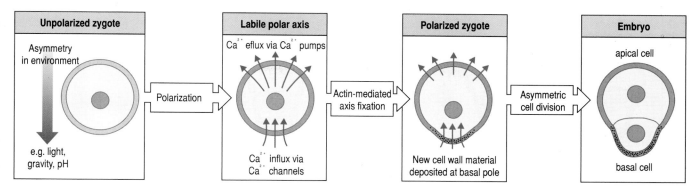

Fig. 7.6 Polarization of the *Fucus* zygote. The fertilized egg is initially unpolarized. Environmental factors, such as illumination from one side, gravity, or pH gradients, impose an asymmetry on the zygote that results in the development of an electrical current through it. This confers a polarity on the zygote. The current is partly carried by calcium ions. The environmental influences result in an increased flow of calcium ions into the cell in one region and an increased outflow on the opposite side. In the case of directional illumination, the shaded side becomes the site of increased calcium influx, which becomes the basal pole of the zygote and the site of future rhizoid formation. The apical–basal axis is fixed by the accumulation of cell vesicles, mitochondria, and Golgi bodies at the site of calcium influx. Cell wall material is deposited at this site and actin filaments become localized there. The first cleavage occurs transverse to the apical–basal axis, and is asymmetric, resulting in a smaller basal cell and a larger apical cell. After Alberts, B., *et al.*: 1989.

suggests that the calcium currents localize an actin network in the region of the future rhizoid, and set in motion a series of events leading to localized polysaccharide insertion into the cell wall.

7.5 Cell fate in early *Fucus* development is determined by the cell wall

The pattern of early cleavage in *Fucus* is similar to that in some higher plant embryos. Because the *Fucus* embryo is accessible to microsurgical manipulation, it is possible to gain insight into mechanisms specifying early cell fate, in particular the development of the thallus from apical cells and the rhizoid from the basal cell. The fates of both presumptive rhizoid and thallus cells can be changed by experimental manipulation of the two-cell embryo. If the wall of the basal cell is destroyed with a laser microbeam, the presumptive rhizoid cell is extruded as a spherical protoplast. Within an hour, the isolated protoplast regenerates a cell wall and then goes on to develop as a normal complete embryo, giving rise to both rhizoid and thallus cells. This shows quite clearly that the fate of the basal cell is not fixed at this stage and can be altered. It also implies that components of the cell wall may be involved in directing the basal cell's future development.

More evidence for the role of the cell wall in directing cell fate comes from experiments with apical cells (Fig. 7.7). The thallus and rhizoid can be distinguished by their different patterns of cell division. An intact apical cell, complete with its cell wall, isolated from the two-cell embryo, continues to divide in a normal 'thallus' pattern until at least the eight-cell stage. But if the wall of the basal cell is still attached, those thallus cells that come into contact with it develop into rhizoid cells, while the rest of the cells develop as thallus cells. So development as rhizoid cells is dependent on contact with the basal cell wall. This provides strong evidence that the position-dependent differentiation of the two cells of the early *Fucus* embryo may be linked to factors on the inner face of the cell wall. Further evidence comes from laser ablation, which shows that cell walls derived

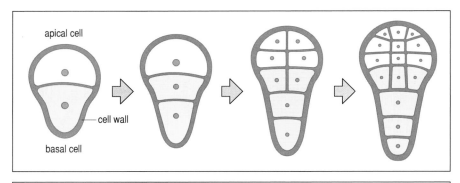

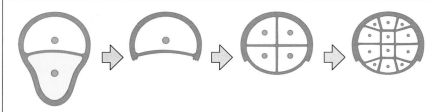

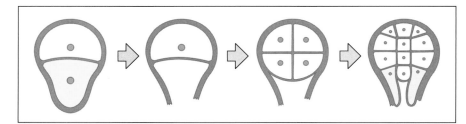

Fig. 7.7 Cell walls can determine cell fate in *Fucus*. Top panel: normal embryonic development. Presumptive rhizoid cells are shown in yellow. Middle panel: development of a complete apical cell when isolated from a two-cell embryo. The isolated apical cell develops by a normal series of cleavages to produce thallus cells only. Bottom panel: development of an isolated apical cell with part of the basal cell wall left attached. As the presumptive thallus cells divide, those that come into contact with the former basal cell wall develop differently, becoming rhizoid rather than thallus cells.

from rhizoid and thallus can confer cell-specific properties on adjacent cells. Thus, if a rhizoid protoplast contacts a wall different from its original wall it develops characteristics specified by the contacted wall. At later stages there is evidence for signaling between the cells.

7.6 Differences in cell size resulting from unequal divisions could specify cell type in the *Volvox* embryo

Many cell divisions in early plant development are unequal. Studies in the green alga *Volvox* show that these differences in size could be important in specifying cell fate. *Volvox* is an extremely simple multicellular organism. An asexual adult colony is made up of just two cell types arranged in an orderly pattern. Around 2000 biflagellate somatic cells are uniformly distributed over the surface of a transparent gelatinous sphere, with 16 larger asexual reproductive cells, known as gonidia, lying beneath the surface (Fig. 7.8). When mature, each gonidium undergoes repeated cleavage to form a juvenile *Volvox* colony *in situ*. These are eventually released from the parent colony, which then dies.

The pattern of 2000 small somatic cells and just 16 large reproductive ones has its origin in the asymmetric cleavage of the gonidium. The first five cleavages are symmetric and result in a hollow spherical embryo of 32 cells with a clear asymmetry in cell packing. At the sixth cleavage, the 16 anterior cells divide asymmetrically to give 16 small and 16 larger cells. The larger cells lie posteriorly and give rise to the gonidia; they divide

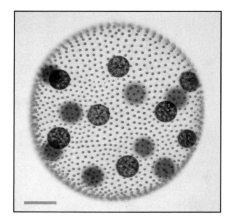

Fig. 7.8 Asexual colony of *Volvox carteri*. This colonial green alga consists of 2000 biflagellate somatic cells surrounding 16 larger gonidia. Each gonidium can give rise to a new colony by asymmetric cell division. Scale bar = 10 μm.

Photograph courtesy of D. Kirk.

asymmetrically twice more and then stop. The smaller cells continue to divide for another three or four cycles and differentiate as somatic cells. The 16 small cells that were originally at the posterior pole also divide symmetrically and produce somatic cells.

There could be several possible causes for the difference in developmental fate between the large gonidia and the smaller somatic cells. It might, for example, be the result of asymmetric distribution of a localized cytoplasmic determinant. Experimental evidence suggests, however, that the difference in size is the determining factor. In mutants where the number of divisions is reduced, and so all the cells are larger, the number of gonidia is increased. And when heat shock is used to interrupt cleavage in otherwise normal embryos, the presumptive somatic cells remain large and develop as gonidia. Even more striking is the demonstration that when somatic cells are connected by microsurgery to make one larger cell, this cell will develop into a gonidium. How the differences in size are transduced into differences in gene expression so that somatic cells or gonidia develop is not yet known.

Summary

Early embryonic development in many plants is characterized by asymmetric cell division of the zygote, which specifies apical and basal regions. Embryonic development in flowering plants establishes the shoot and root meristems from which the adult plant develops. Patterning of the embryo of flowering plants seems to rely less on stereotyped patterns of cell divisions than on cell–cell interactions. Particular genes control radial differentiation and patterning of specific regions along the apical–basal axis of the *Arabidopsis* embryo. One major difference between plants and animals is that, in culture, a single somatic plant cell can develop through an embryo-like stage and regenerate a complete new plant, indicating that some differentiated plant cells retain totipotency. The free-living 'embryos' of the algae *Fucus* and *Volvox* provide useful and accessible models for studying early pattern formation. The apical–basal axis of the *Fucus* zygote, which defines the first plane of cleavage and the future rhizoid and thallus, is specified by environmental signals. It is fixed by electrical currents and stabilized by interactions between the cytoskeleton and cell wall. In *Fucus*, cell wall components seem to be a major determinant of cell fate. In the colonial green alga *Volvox*, cell fate is determined by differences in cell size arising from asymmetric divisions.

Summary: early development in flowering plants

first asymmetric cell division in embryo establishes apical–basal axis

⇩

embryonic cell fate is determined by position

⇩

shoot and root meristems of seedling give rise to all adult plant structures

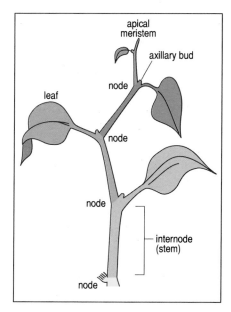

Fig. 7.9 Plant shoots grow in a modular fashion. The shoot apical meristem produces a repeated basic unit of structure called a module. The vegetative shoot module typically consists of internode, node, leaf, and axillary bud (from which a side branch may develop). Successive modules are shown here in different shades of green. As the plant grows, the internodes behind the meristem lengthen and the leaves expand. After Alberts, B., *et al.*: 1989.

Meristems

The embryonic root and shoot meristems give rise to all the structures of the adult plant. Unlike animals, where the late embryo can usually be regarded as a miniature version of the adult, in plants the adult structures are derived from just two regions of the embryo, the shoot and root meristems. The shoot meristem, for example, gives rise to leaves, internodes, and flowers. As the shoot meristem develops, lateral outgrowths from the meristem give rise to leaves and additional meristems. Outgrowths from a meristem that will give rise to an organ such as a leaf or flower are known as **primordia**. These outgrowths are initially small but increase in size due to cell proliferation and enlargement. There is usually a time delay between the initiation of two successive leaves, and this results in a plant shoot being formed from repeated modules (Fig. 7.9). Each module consists of an **internode** (the cells produced by the meristem between successive leaf initiations), a **node** and its associated leaf, and an axillary bud. The axillary bud itself contains a meristem, and can form a side shoot when the inhibitory influence of the shoot tip is removed. Root growth is not so obviously modular, but similar considerations apply, as new lateral meristems initiated behind the root apical meristem give rise to lateral roots.

Meristems are small, rarely more than 250 µm in diameter in angiosperms, and contain relatively small, undifferentiated, cells. Many of the cell divisions in normal plant development occur within the meristems, or soon after a cell leaves a meristem, and much of the subsequent growth is due to cell enlargement. Since a meristem remains relatively constant in size during growth, cells are continually leaving it, and as they do so they begin to differentiate. The central region of the shoot meristem or root meristem is commonly referred to as the **promeristem**. This central region contains cells known as **initials**, which behave in the same way as animal stem cells. They are self-renewing and give rise to the cells of the meristem. These initials generally divide rather slowly; their progeny divide more rapidly as they move toward the peripheral parts of the meristem. In the shoot, individual initials eventually get displaced from the meristem and their place is taken by another cell, which then behaves as an initial.

7.7 The fate of a cell in the shoot meristem is dependent on its position

The shoot meristem of dicotyledons gives rise to the stem and leaves. It is made up of three layers (Fig. 7.10). L1 is the outermost layer and is just one cell thick. L2 is also one cell thick and lies beneath L1. In both L1 and L2, cell divisions are anticlinal—that is, the new wall is in a plane perpendicular to the layer—thus maintaining the organization of these layers. The innermost layer is L3, in which the cells can divide in any plane. L1 and L2 are often known as the tunica, and L3 as the corpus.

The fate of the cells in each layer has been determined using plant chimeras. Chimeric tissues are composed of cells of two different genotypes which can be distinguished from each other by distinctive features such as polyploid nuclei (nuclei containing extra sets of chromosomes) or pigmentation. Chimeras can be made by treatment of the meristem with radiation or with chemicals such as colchicine, which induce polyploidy by blocking

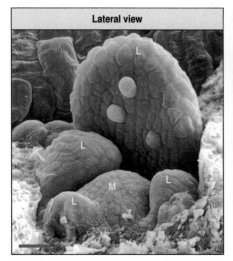
Lateral view

Apical view

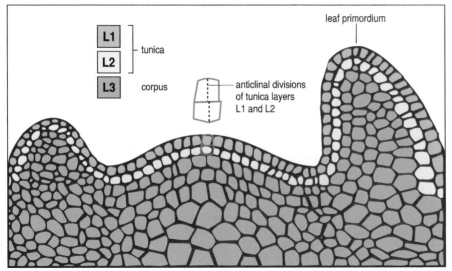

Fig. 7.10 Apical meristem of Arabidopsis. Top panels: scanning electron micrographs showing the organization of the meristem at the young vegetative apex of *Arabidopsis*. The plant is a *clavata1–1* mutant, which has a broadened apex, allowing for a clearer visualization of the leaf primordia (L) and the meristem (M). Scale bar = 10 μm. Bottom panel: diagram of a vertical section through the apex of a shoot. The three-layered structure of the meristem is apparent in the most apical region. In layer 1 (L1) and layer 2 (L2), the plane of cell division is anticlinal, that is at right angles to the surface of the shoot. Cells in layer 3 (L3, the corpus) can divide in any plane. A leaf primordium is shown forming at one side of the meristem.

Photographs courtesy of M. Griffiths.

nuclear division but not chromosomal division. So-called **periclinal chimeras** can be produced, in which one of the three layers of the meristem has a genetic marker that distinguishes its cells and their progeny from those of the other two layers (Fig. 7.11).

By following the fate of the marked cells one can see what structures each layer gives rise to. In angiosperms, layer L1 gives rise to the epidermis of all the plant structures, while L2 and L3 can both contribute to cortex and vascular structures. Leaves and floral organs are produced mainly from L2, and L3 contributes mainly to the stem. Although the three layers maintain their identity in the central region of the meristem over long periods of growth, cells in either L1 or L2 occasionally divide periclinally so that new cell walls are formed parallel to the surface of the meristem, and thus one of the new cells invades an adjacent layer. This migrant cell now develops according to its new position, showing that cell fate is not necessarily determined by the meristem layer in which the cell originated, and that some sort of intercellular signaling is involved. The L2 layer also becomes disrupted by periclinal divisions as soon as leaves begin to form.

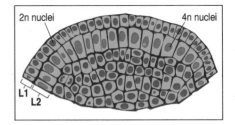

Fig. 7.11 Chimeric meristem (periclinal chimera) composed of cells of two different genotypes. In L1 the cells are diploid whereas the cells of L2 are tetraploid—they have double the normal chromosome number, and are larger and easily recognized. After Steeves, T.A., *et al.*: 1989.

Fig. 7.12 Tobacco plant mericlinal chimera. This plant has grown from an embryonic shoot meristem in which an albino mutation has occurred in a cell of the L2 layer. The affected area occupies about one third of the circumference of the shoot, suggesting that there are three apical initial cells in the embryonic shoot meristem.

Photograph courtesy of S. Poethig.

It is important to know what structures in the plant, such as particular internodes and leaves, arise from which regions of the meristem. This can be determined using clonal analysis in a manner similar to that used in *Drosophila* research (see Box 5B, p. 173). Individual mutant, or otherwise distinguishable, meristematic cells can be generated by X-irradiation or by the activation of a transposable element, which can result, for example, in marked cells of a different color from the rest of the plant. Because there is no cell migration during plant development, the clone of cells deriving from the marked meristem cell forms a coherent sector of marked cells as the plant grows. The resulting chimeric plants are known as **mericlinal chimeras**, as only one sector of the plant or of an organ is marked by the clone (Fig. 7.12).

Analysis of maize mericlinal chimeras grown from irradiated seeds shows that the most common pattern of development is of a marked sector

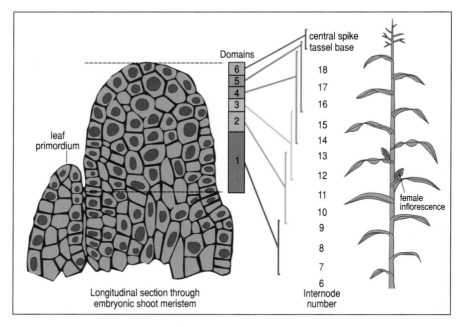

Fig. 7.13 Probabilistic fate map of the shoot apical meristem in the mature maize embryo, derived from clonal analysis. At the stage of development when marked clones were generated, the first six nodes had already been specified and their leaf primordia initiated. They are therefore already outside the meristem and are excluded from the analysis. A longitudinal section through the embryonic apical dome is shown on the left. Clonal analysis shows that the apical dome can be divided into six vertically stacked domains, each of which contains a set of initials that can give rise to a particular part of the plant. The number of initials in layers L1 and L2 of each domain at the embryonic stage can be estimated from the final extent of the marked sector in the mature plant. The fate of each domain in the maize plant, as estimated from the collective results of clonal analysis of many different plants, is shown on the right. Thus domain 6, comprising the three most apical L1 cells of the meristem, will give rise to the terminal male inflorescence. The fate of cells in the other domains is less circumscribed. Domain 5, for example, which consists of a region of around eight L1 cells surrounding domain 6 and the underlying four or so L2 cells, can contribute to nodes 16, 17, or 18; domain 4 to nodes 14–18; and domain 3 to nodes 12–15. The female inflorescences, from which the cobs develop, grow at intervals up the stem, adjacent to the leaves, and are also derived from the corresponding domains shown. After McDaniel, C.N., *et al.*: 1988.

that appears at the base of an internode and extends apically, terminating in a leaf. Some marked clones populate just a single internode whereas others populate sectors extending through as many as 13 internodes. This tells us that initials eventually become displaced from the meristem, although some remain as initials for a long time, contributing to several node–internode modules in succession. In similar experiments in sunflowers, marked clones extending through several internodes and up into the flower are found, showing that the same initial can contribute to both vegetative and floral structures.

When the varied patterns of the sectors that arise from individual cells are examined in a number of plants, clonal sectors can often be found populating similar, but not identical, regions of the plant body, showing that the fate of a particular cell is not fixed. Nevertheless, the similarity suggests a predictable pattern of cell division. The clonal analysis experiments indicate that initial cells are simply those that happen to be in the central region of the meristem at any given time. Most initials are displaced from the meristem after a time, and are replaced by other cells. Their self-renewing behavior as initials is a consequence of their position, rather than of some intrinsic difference from other meristem cells.

Using the results of clonal analysis, a rough fate map has been constructed for the shoot apical meristem of the mature maize embryo in the seed. The meristem fate map can only be approximate because it is not possible to mark a cell at a particular location in the embryonic meristem and then follow its development, as it is inaccessible inside the seed. Plant fate maps are thus more probabilistic and are based on indirect methods. The frequency of sector occurrence in a given leaf, for example, reflects the number of cells in the meristem that give rise to that leaf. The final size of the clone in relation to the total size of the organ, such as a leaf, through which the clone extends, gives the apparent cell number in the meristem that gave rise to that leaf. If, for example, a marked clone occupies about a quarter of a leaf, then, assuming the marked clone arises from a single marked cell, the number of meristem initials that contributed to that leaf will be around four.

After adjustments to the apparent cell number are made in the light of the actual number and position of cells in the meristem, a probabilistic fate map can be constructed. These maps show the domains in the meristem that are likely to give rise to a particular region of the plant body (Fig. 7.13 opposite). In the mature embryonic apical meristem of maize, the primordia of the first six leaves are already present and are excluded from the fate map shown here. At the time the clones are induced, the meristem contains about 335 cells, which will give rise to 12 more leaves, the female inflorescence, and the terminal male inflorescence (the tassel and spike).

A probabilistic fate map has also been constructed using clonal analysis for the embryonic shoot meristem of *Arabidopsis* (Fig. 7.14), which shows that most of the cells of the meristem give rise to the first six leaves, whereas the remainder of the shoot is derived from a very few cells. The number of leaves in *Arabidopsis* is not fixed—growth is said to be indeterminate. There is no relationship between particular lineages and particular structures, which suggests that position is crucial in determining cell fate.

The signals involved in specifying position are not known, but clues emerge from studies on genes that control the behavior of the cells in the

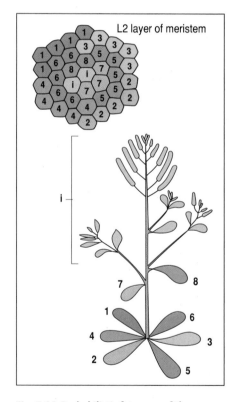

Fig. 7.14 Probabilistic fate map of the embryonic shoot meristem of *Arabidopsis*. The L2 layer of the meristem is depicted as if flattened out and viewed from above. The numbers indicate the leaf, as shown on the plant below, to which each group of the meristem cells contributes, and indicate the sequence in which the leaves are formed. The inflorescence shoot (i) is derived from a small number of cells in the center of the layer. After Irish, V.E.: 1991.

meristem. One gene that has been implicated is *SHOOT MERISTEMLESS*, which is involved in specifying the shoot meristem in the development of the embryo and appears to maintain some of the cells in an uncommitted pluripotent state so that they can behave as initials (Section 7.2). Mutation in *SHOOT MERISTEMLESS* results in all the meristem cells being incorporated into organ primordia. By contrast, another set of genes, the *CLAVATA* genes, regulate the transition of meristem initials to organ differentiation. Mutations in these genes result in an increase in the size of the meristem owing to an increase in the number of central undifferentiated cells and a relatively decreased rate of organ formation. *CLAVATA1* codes for a receptor kinase and *CLAVATA3* for its likely ligand. One model suggests that *CLAVATA3* confers a prospective identity on the initials, and that its activity is maintained by cells expressing the homeobox gene *WUSCHEL*; cells expressing *WUSCHEL* lie beneath the initials and the initials in turn inhibit, via *CLAVATA3*, the expansion of the *WUSCHEL* expression domain, forming a self-regulatory loop.

Mutations in the *knotted-1* gene in maize, which is homologous to *SHOOT MERISTEMLESS*, result in outgrowths—knots—of tissue around the lateral veins in the leaf blade. The transcription factor encoded by *knotted-1* is normally only present in meristems and vascular tissue. The idea that it can maintain cells in an undifferentiated and pluripotent state comes from experiments in which the maize gene is overexpressed as a transgene in tobacco plants using the techniques described in Box 7B. Transgenic plants expressing high levels of the protein form small ectopic shoots on leaf surfaces. The dominant maize KNOTTED phenotype also results from ectopic expression of *knotted-1* in leaves, causing excess cell division. The knotted protein, which is normally synthesized in all layers except L1, is found in L1, showing that it can move between cells, perhaps via plasmodesmata. Evidence that transcription factors can move between plant cells is considered in Section 7.13.

7.8 Meristem development is dependent on signals from other parts of the plant

To what extent does the behavior of a meristem depend on other parts of the plant? It seems to have some autonomy, as if a meristem is isolated from adjacent tissues by excision, it will continue to develop, although often at a much slower rate. Excised shoot apical meristems of a variety of plants can be grown in culture, where they will develop into shoots complete with leaves if the growth hormones auxin and cytokinin are added. The behavior of the meristem *in situ*, is, however, influenced by interactions with the rest of the plant.

The apical meristem of maize gives rise to a succession of nodes, and terminates in a male flower, the tassel. The number of nodes before flowering is usually between 16 and 22. This number is not controlled by the meristem alone, but by its interaction with the rest of the plant. The evidence for this comes from culturing shoot tips of maize consisting of the apical meristem and one or two leaf primordia. Such meristems taken from plants that have already formed as many as 10 nodes develop into normal plants with the correct number of nodes. Thus, the isolated meristem has no memory of the number of nodes it has already formed, and repeats the process from the beginning. The control over the number of nodes must

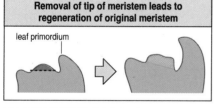

Removal of tip of meristem leads to regeneration of original meristem

leaf primordium

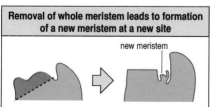

Removal of whole meristem leads to formation of a new meristem at a new site

new meristem

Fig. 7.15 Regulation of a shoot meristem. Top panel: removal of the tip of the meristem leaves most of the promeristem *in situ* and a more or less normal meristem regenerates. Bottom panel: removal of most of the meristem also removes virtually all of the promeristem. A small new meristem forms at the site where a small amount of promeristem remains. After Sachs, T.: 1994.

Box 7B Transgenic plants

One of the most common ways of generating transgenic plants containing new and modified genes is through infection of plant tissue in culture with the bacterium *Agrobacterium tumefaciens*, the causal agent of crown gall tumors. *Agrobacterium* is a natural genetic engineer. It contains a **plasmid**—the Ti plasmid—that contains the genes required for the transformation and proliferation of infected cells to form a callus. During infection, a portion of this plasmid-the T-DNA (shown in red below)-is transferred into the genome of the plant cell, where it becomes stably integrated. Genes experimentally inserted into the T-DNA will therefore also be transferred into the plant cell chromosomes. Ti plasmids, modified so that they do not cause tumors but still retain the ability to transfer T-DNA, are widely used as vectors for gene transfer in dicotyledonous plants. The genetically modified plant cells of the callus can then be grown into a complete new transgenic plant that carries the introduced gene in all its cells and can transmit it to the next generation.

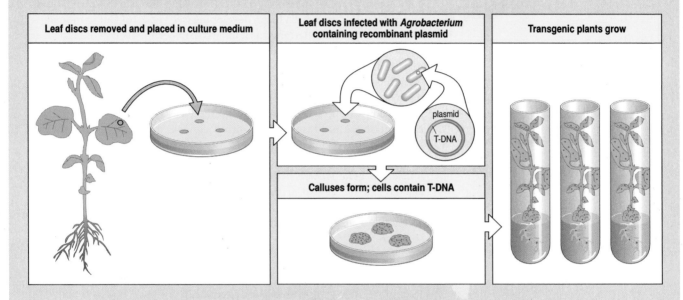

therefore involve signals from the developing plant to the meristem, directing that region to terminate node formation and form a tassel. The early meristem is thus not determined with respect to the number of nodes it will form.

Meristems are capable of regulation. If, for example, a pea seedling meristem is divided in two by a vertical incision, each half becomes reorganized into a complete meristem, which gives rise to a normal shoot. In lupins, division of the shoot meristem into four quadrants gives four new meristems that each develop a shoot. Provided some subpopulation of promeristem cells is present, a normal meristem can regenerate. So, if only a small part of the most apical region is removed, a normal meristem may still regenerate. This regulative behavior is in line with cell–cell interactions being a major determinant of cell fate in the meristem. Removal of all of the meristem does not lead to a new meristem forming, but it allows the incipient meristem at the base of the leaf to develop (Fig. 7.15), opposite. In the presence of the original meristem, this prospective meristem would have remained inactive, as growing meristems inhibit the growth of other meristems nearby.

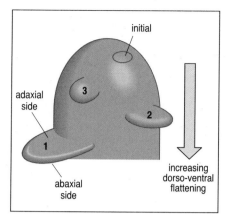

Fig. 7.16 Development of the dorso-ventral axis in leaf primordia. In this diagram, the leaf primordia developing from a shoot apical meristem are labeled 1 (oldest) to 3 (youngest). As they mature, the primordia elongate and begin to flatten, acquiring a clear dorso-ventral axis, as in 1 and 2. The initial predicts the site of emergence of the next leaf primordium.

7.9 Gene activity patterns the proximo-distal and adaxial–abaxial axes of leaves developing from the meristem

Leaves develop from groups of initial cells within the peripheral zone of the shoot apical meristem. These initials form two new axes related to the leaf—proximo-distal (leaf base to leaf tip) and **adaxial–abaxial** (upper surface to lower surface, or dorsal to ventral). The first indication of leaf initiation in the meristem is usually a swelling of a region to the side of the apex to form a leaf primordium (Fig. 7.16). This small protrusion is the result of increased localized cell multiplication and altered patterns of cell division. It also reflects changes in polarized cell expansion.

In *Arabidopsis*, the flattening of the leaf along the adaxial–abaxial axis occurs after leaf primordia begin developing (see Fig. 7.16), but in monocots like maize, the leaf is flattened as it emerges. Experiments in which the leaf primordia were separated from the meristem suggest that signals from the meristem promote adaxial cell fate.

Arabidopsis leaf primordia emerge from the shoot meristem with distinct programs of development in the upper and lower halves. This asymmetry can be seen from the beginning as the leaf primordium has a crescent shape in cross-section, with a convex 'outer' (abaxial) side and a concave 'inner' (adaxial) side (Fig. 7.17). Different genes are expressed in the future adaxial side and in the future abaxial side. For example, the *Arabidopsis* gene *FILAMENTOUS FLOWER* (*FIL*) is normally expressed in the abaxial side of the leaf primordium, and specifies an abaxial cell fate. Its ectopic expression throughout the leaf primordium can cause all cells to adopt an abaxial cell fate, and the leaf develops as an arrested cylindrical structure.

In *Antirrhinum*, the *PHANTASTICA* gene is required for adaxial cell identity;

Fig. 7.17 Leaf phyllotaxis. In shoots where single leaves are arranged spirally up the stem, the leaf primordia arise sequentially in a mathematically regular pattern in the meristem. Leaf primordia arise around the sides of the apical dome, just outside the promeristem region. A new leaf primordium is formed slightly above and at a fixed radial angle from the previous leaf, often generating a helical arrangement of primordia visible at the apex. Top panel: lateral views of the shoot apex. Bottom panel: view looking down on cross-sections through the apex near the tip, at successive stages from the top panel. After Poethig, R.S., *et al.*: 1985 (top panel); and Sachs, T.: 1994 (bottom panel).

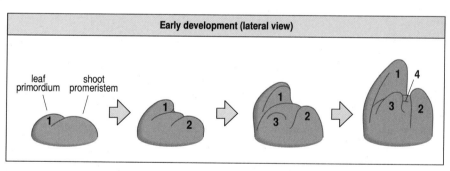

mutations in this gene result in radially symmetrical leaves with only abaxial cell types characteristic of the underside of the leaf. In *Arabidopsis*, a dominant mutation of the *PHABULOSA-1D* gene has a similar but reverse effect. It causes all the prospective leaf cells to become adaxial and a cylindrical leaf again develops. It has been suggested that, in normal plants, the interaction between adaxial and abaxial initial cells at the boundary between them initiates lateral growth, resulting in the formation of the leaf blade and the flattening of the leaf. The importance of boundaries in controlling pattern and form has already been seen in the parasegments of *Drosophila* (Section 5.19) and further examples will be found in Chapter 10.

Development along the proximo-distal axis also appears to be under genetic control. Like other grasses, a maize leaf primordium is composed of prospective leaf-sheath tissue proximal to the stem and prospective leaf-blade tissue distally. Mutations in certain genes result in distal cells taking on more proximal identities, for example, making sheath in place of blade. Similar proximo-distal shifts in pattern occur in *Arabidopsis* as a result of mutation. Positional identity along the proximo-distal axis may reflect the developmental age of the cells, distal cells adopting a different fate from proximal cells because they mature later.

7.10 · The regular arrangement of leaves on a stem and trichomes on leaves is generated by lateral inhibition

As the shoot grows, leaves are generated within the meristem at regular intervals and with a particular spacing. Leaves are arranged along a shoot in a variety of ways in different plants, and the particular arrangement—or **phyllotaxy**—is reflected in the arrangement of leaf primordia in the meristem. Leaves can occur singly at each node, in pairs, or in whorls of three or more. A common arrangement is the positioning of single leaves spirally up the stem, which can sometimes form a striking helical pattern in the shoot apex.

In plants in which leaves are borne spirally, a new leaf primordium forms at the center of the first available space outside the promeristem and above the previous primordium (see Fig. 7.17). This pattern suggests a mechanism for leaf arrangement based on lateral inhibition (see Section 1.14), in which each leaf primordium inhibits the formation of a new leaf within a given distance. In this model, inhibitory signals emanating from recently initiated primordia prevent leaves from forming close to each other. There is some experimental evidence for this. In ferns, leaf primordia are widely spaced, allowing experimental microsurgical interference. Destruction of the site of the next primordium to be formed results in a shift toward that site by the future primordium whose position is closest to it (Fig. 7.18).

Lateral inhibition is also involved in the regular spacing of the hair-bearing cells, or **trichomes**, in the leaf epidermis, which are always separated from each other by three or four epidermal cells. The pattern is not related to cell lineage, but arises by lateral inhibition, in which all cells have the potential to develop as trichomes, but the trichome fate is suppressed in cells immediately surrounding any cell that has begun to develop as a trichome. Studies on mutants affecting this pattern suggest a model in which the product of the gene *TRIPTYCHON* (*TRY*) inhibits trichome development by suppressing the transcription factor glabrous1 (GL1), which is required for trichome differentiation. In those cells that have a

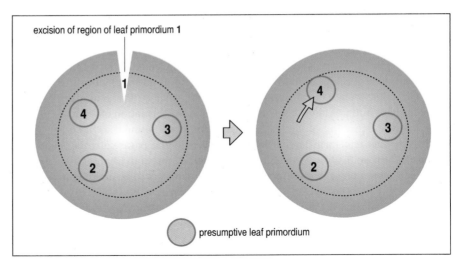

excision of region of leaf primordium **1**

presumptive leaf primordium

Fig. 7.18 Lateral inhibition may position leaf primordia. Leaf primordia on a fern shoot tip form in a regular order in positions 1 to 4. Primordia appear to form as far as possible from existing primordia, so 2 forms almost opposite 1. Normally, 4 will develop between 1 and 2, but if 1 is excised, 4 forms much further from 2 because the inhibitory action of 1 has been removed.

slight start in trichome development, there is a positive feedback loop involving GL1 that leads to the production of TRY and thus inhibition of trichome development in the adjacent cells.

7.11 Root tissues are produced from root apical meristems by a highly stereotyped pattern of cell divisions

The organization of tissues in the *Arabidopsis* root is shown in Fig. 7.19. The radial pattern comprises single layers of epidermal, cortical, endodermal, and pericycle cells. Root apical meristems resemble shoot apical meristems in many ways and give rise to the root in a similar manner to shoot generation. But there are some important differences between the root and shoot meristems. The shoot meristem is at the extreme tip of the shoot whereas the root meristem is covered by a root cap (which is itself derived from one of the layers of the meristem); also, there is no obvious segmental arrangement at the root tip resembling the node–internode–leaf module.

The root is set up early and its origin can be identified in the late heart-stage embryo (Fig. 7.20). Clonal analysis has shown that the root meristem can be traced back to a set of initial cells that come from a single tier of cells in the heart-stage embryo. Each column, or file, of cells in the root has its origin in a particular initial cell in the meristem, and each initial cell has a stereotyped pattern of cell divisions that gives rise to each column. The initials surrounding the quiescent center give rise to all the cell types, including endodermis, cortex and epidermis. The gene *SCARECROW*, thought to code for a transcription factor, is necessary for asymmetric cell divisions and mutations in this gene give a single layer with characteristics of both endodermis and cortex. The normal pattern of cell divisions is not obligatory, however. Destruction of some meristem cells with a laser beam alters the pattern of cell divisions, but the tissues are normal because novel cell divisions replace the cells that have been destroyed. As discussed earlier, *fass* mutants, which have disrupted cell divisions, still have a relatively normal patterning in the root. Such observations show the importance of cell–cell interactions in patterning the root.

The quiescent center has a greatly reduced rate of cell division compared to other parts of the root. Microsurgical removal of parts of the root apical

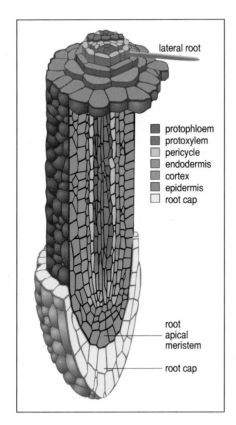

lateral root

- ■ protophloem
- ■ protoxylem
- □ pericycle
- ■ endodermis
- ■ cortex
- ■ epidermis
- □ root cap

root apical meristem

root cap

Fig. 7.19 The structure of the root tip in *Arabidopsis*. Roots have a radial organization. In the center of the growing root tip is the future vascular tissue (protoxylem and protophloem). This is surrounded by further tissue layers.

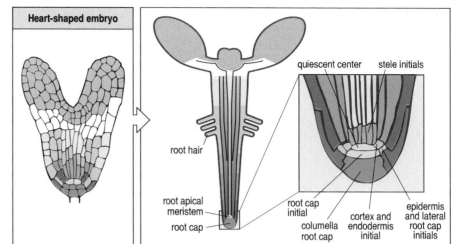

Fig. 7.20 Fate map of root regions in the heart-stage *Arabidopsis* embryo. The root grows by the division of a set of initial cells. The root meristem comes from a small number of cells in the heart-shaped embryo. Each tissue in the root is derived from the division of a particular initial cell. At the center of the root meristem is a quiescent center, which does not divide. After Scheres, B., *et al.*: 1994.

meristem showed that regeneration of a meristem can occur, but is always preceded by the formation of a new quiescent center. Laser ablation of individual quiescent cells showed that a key function of the quiescent center is to prevent the immediately adjacent initials from differentiating and thus to maintain their stem-cell-like character.

The plant hormone auxin (indoleacetic acid) plays a role in specifying the pattern of root cells, and mutations affecting its localization lead to root defects. Auxin is synthesized elsewhere and moves into root cells by means of a special set of membrane transport proteins. There is evidence that cells with raised auxin levels transport auxin better, and so there is a positive feedback loop that raises the local concentration. The gene *PIN1* codes for a protein involved in auxin transport. At the globular stage of embryonic development (see Fig. 7.1) PIN1 protein is localized in cells in the future root region and thus the highest level of auxin is found adjacent to where the quiescent center will develop.

Auxin is also involved in the ability of plants to regenerate from a small piece of stem. In general, roots form from the end of the stem that was originally closest to the root, whereas shoots tend to develop from dormant buds at the end that was nearest to the shoot. This polarized regeneration is related to vascular differentiation and to the polarized transport of auxin. Transport of auxin from its source in the shoot tip toward the root leads to an accumulation of auxin at the 'root' end of the stem cutting, where it induces the formation of roots. One hypothesis suggests that polarity is both induced and expressed by the oriented flow of auxin.

Further examples of cell–cell interactions relate to the two cell types in the root epidermis—the hair-forming trichoblasts and hairless atrichoblasts (Fig. 7.21). These form alternating files of cells on the surface of the developing root and their character appears to depend on positional signals from the underlying cells. Whereas most cell divisions in the future epidermis are anticlinal (at right angles to the root surface), thus increasing the number of cells per file, occasionally a longitudinal periclinal division occurs, pushing one of the daughter cells into an adjacent file. The daughter cell then assumes a fate corresponding to its new position. The subsequent spacing of trichoblasts within the file is due to lateral inhibition (see Section 7.10).

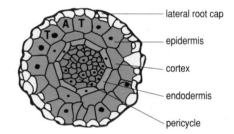

Fig. 7.21 Organization of cell types in the root epidermis. The epidermis is composed of two types of cells—trichoblasts (T) which will form root hairs, and atrichoblasts (A) which will not. Trichoblasts overlie the junction between two cortical cells and atrichoblasts are located over the outer tangential wall of cortical cells. After Dolan, L., Scheres, B.: 1998.

Summary

Meristems are the growing points of a plant. They consist of small regions of undifferentiated cells that are capable of repeated division. The apical meristems, found at the tip of shoots and roots, give rise to all the plant organs—roots, stem, leaves, and flowers. The fate of a cell in the shoot meristem clearly depends upon its position rather than its lineage, because when a cell is displaced from one layer and becomes part of another it adopts the fate of its new layer. Fate maps of the embryonic shoot meristem in maize show that it can be divided into domains, each of which can contribute to the tissues of a particular region of the plant. Cell–cell inter-actions have a role in determining cell fate and, in line with this, meristems are capable of regulation when parts of them are removed. The shoot meri-stem gives rise to leaves in species-specific patterns—phyllotaxy—which seem best accounted for in terms of lateral inhibition. Lateral inhibition is also involved in the regular spacing of hair cells or trichomes on root and leaf surfaces. In the root meristem, the cells are organized rather differ-ently from those in the shoot meristem, and there is a much more stereo-typed pattern of cell division. A set of initial cells maintain root structure by dividing along different planes.

Summary: meristems give rise to all adult tissues	
Shoot meristem	**Root meristem**
cell fate is determined by position	stereotyped division of initials along different planes
⇩	⇩
shoot meristem gives rise to stem internodes, leaves and flowers	root meristem gives rise to all root structures
⇩	
arrangement of leaves on stem may be due to a lateral inhibition mechanism at shoot apex	

Fig. 7.22 Scanning electron micrograph of an *Arabidopsis* inflorescence meristem. The central inflorescence meristem (shoot apical meristem, SAM) is surrounded by a series of floral meristems (FM) of varying developmental ages. The inflorescence meristem grows indeterminately, with cell divisions providing new cells for the stem below, and new floral meristems on its flanks. The floral meristems (or floral primordia) arise one at a time in a spiral pattern. The most mature of the developing flowers is on the right (FM1), showing the initiation of sepal primordia surrounding a still-undifferentiated floral meristem. Eventually, such a floral meristem will also form petal, stamen, and carpel primordia.

Photograph from Meyerowitz, E.M., et al.: 1991.

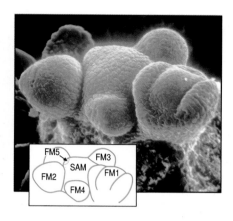

Flower development

Flowers contain the reproductive cells of higher plants and develop from the shoot meristem. In most plants, the transition from a vegetative shoot meristem to a **floral meristem** that produces a flower is largely or absolutely under environmental control, with day length and temperature being important determining factors. Flowers, with their arrangement of floral organs (sepals, petals, stamens, and carpels), are rather complex, and it is a major challenge to understand how they arise from the floral meristem. Here, we consider the mechanisms that pattern the flower, particularly those genes that specify the identity of the floral organs.

Three classes of genes seem to control the basic patterning of a flower. Organ identity genes specify the identity of the different floral organs, and have a function equivalent to homeotic selector genes in animals (see Section 5.15). Cadastral genes set the boundaries for expression of the organ identity genes and prevent their ectopic expression. Expression of the meristem identity genes converts a vegetative shoot meristem into one that makes flowers.

7.12 Homeotic genes control organ identity in the flower

In a shoot with multiple flowers, like those of *Arabidopsis*, the shoot meristem becomes converted into an **inflorescence meristem**, which can form one or more floral meristems, each of which develops into a single flower (Fig. 7.22). The **floral organ primordia**, from which the individual parts of the flower develop, arise in the floral meristem by patterned cell division followed by cell differentiation and enlargement. A flower is composed of four concentric whorls of structures (Fig. 7.23), which reflect the arrangement of the floral organ primordia in the meristem. The sepals (whorl 1) arise from the outermost ring of meristem tissue, and the petals (whorl 2) from a ring of tissue lying immediately inside it. An inner ring of tissue gives rise to the male reproductive organs—the stamens (whorl 3). The female reproductive organs—the carpels (whorl 4)—develop from the center of the meristem. In a floral meristem of *Arabidopsis*, there are 16 separate

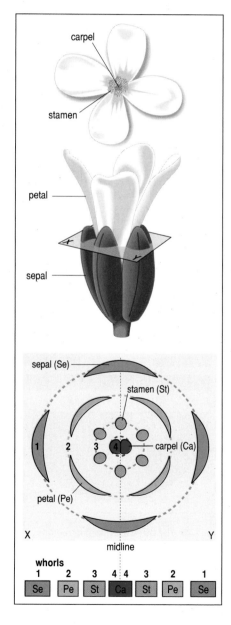

Fig. 7.23 Structure of an *Arabidopsis* flower. Top: *Arabidopsis* flowers are radially symmetrical and have an outer ring of four identical green sepals, enclosing four identical white petals, within which is a ring of six stamens, with two carpels in the center. Bottom: floral diagram of the *Arabidopsis* flower representing a cross-section taken in the plane indicated in the top diagram. This is a conventional representation of the arrangement of the parts of the flower, showing the number of flower parts in each whorl and their arrangement relative to each other. After Coen, E.S., *et al.*: 1991.

Fig. 7.24 Homeotic floral mutations in *Arabidopsis*. Left panel: an *apetala2* mutant has whorls of carpels and stamens in place of sepals and petals. Center panel: an *apetala3* mutant has two whorls of sepals and two of carpels. Right panel: *agamous* mutants have a whorl of petals and sepals in place of stamens and carpels. Transformations of whorls are shown inset, and can be compared to the wild-type arrangement, as shown in Fig. 7.23.

Photographs from Meyerowitz, E.M., et al.: 1991 (left panel), and Bowman, J.L., et al.: 1989 (center panel).

apetala2 mutant

| Ca | St | St | Ca | Ca | St | St | Ca |

apetala3 mutant

| Se | Se | Ca | Ca | Ca | Ca | Se | Se |

agamous mutant

| Se | Pe | Pe | Se | Se | Pe | Pe | Se |

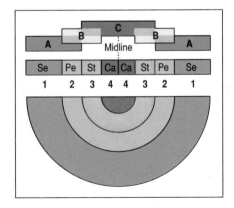

Fig. 7.25 The three overlapping regions of the *Arabidopsis* floral meristem that have been identified by the homeotic floral identity mutations. Region A corresponds to whorls 1 and 2, B to whorls 2 and 3, and C to whorls 3 and 4.

primordia, giving rise to a flower with four sepals, four petals, six stamens and a pistil made up of two carpels (see Fig. 7.23).

The primordia arise at specific positions within the meristem, where they develop into their characteristic structures. During flower development in the snapdragon (*Antirrhinum*), there is lineage restriction to particular whorls, rather like the lineage restriction to compartments in *Drosophila* (see Section 5.15). A key question is how the cells of the developing flower acquire and record their positional identity. In *Arabidopsis* and other plants, **homeotic mutations** have been discovered that result in the development of abnormal flowers in which one type of flower part is replaced by another. In the mutant *apetala2*, for example, the sepals are replaced by carpels and the petals by stamens; in the *pistillata* mutant, petals are replaced by sepals and stamens by carpels. These mutations identify the floral organ identity genes, and have enabled their mode of action to be determined.

The homeotic floral mutations in *Arabidopsis* fall into three classes, each of which affects the organs of two adjacent whorls (Fig. 7.24). The first class of mutations, of which *apetala2* is an example, affect whorls 1 and 2, giving carpels instead of sepals in whorl 1, and stamens instead of petals in whorl 2. The phenotype of the flower, going from the outside to the center, is therefore carpel, stamen, stamen, carpel. The second class of homeotic floral mutations affect whorls 2 and 3. In this class, *apetala3* and *pistillata* give sepals instead of petals in whorl 2 and carpels instead of stamens in whorl 3, with a phenotype sepal, sepal, carpel, carpel. The third class of mutations affects whorls 3 and 4 and gives petals instead of stamens in whorl 3 and sepals or variable structures in whorl 4. The mutant *agamous*, which belongs to this class, has an extra set of sepals and petals in the center instead of the reproductive organs.

These mutant phenotypes can be accounted for by quite a simple model of gene activity. Let us assume that the floral meristem is divided into three overlapping regions, A, B, and C, each region corresponding to the site of action of one of the three classes of homeotic mutations (Fig. 7.25). Region A thus covers whorl 1 and 2; B, whorls 2 and 3; and C, whorls 3 and 4. We next assume that there are three regulatory functions—*a*, *b*, and *c*—which function in regions A, B, and C, respectively, and which, combinatorially, can give each whorl a unique identity and so specify organ identity (Fig. 7.26, left panel). *a* is expressed in whorls 1 and 2, *b* in 2 and 3, and *c* in whorls 3 and 4. In addition, *a* function inhibits *c* function in whorls 1 and 2

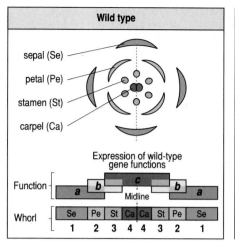 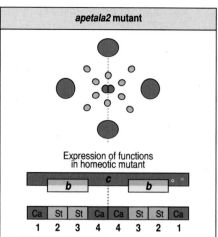 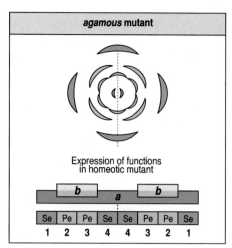

Fig. 7.26 Model for the patterning of the *Arabidopsis* flower. In the wild-type flower (left panel), it is assumed that three regulatory functions, *a*, *b*, and *c*, are expressed in whorls 1 and 2, 2 and 3, and 3 and 4, respectively. *a* alone specifies sepals, *a* and *b* together specify petals, *b* and *c* stamens, and *c* alone carpels. Mutations alter the regions within the meristem where these functions are expressed. In the *apetala2* mutant (center), function *a* is absent and *c* spreads throughout the meristem, resulting in the half-flower pattern of carpel, stamen, stamen, carpel. In the *agamous* mutant (right) there is no *c* function and expression of *a* spreads throughout the meristem, resulting in the half-flower pattern of sepal, petal, petal, sepal. After Dennis, E., *et al.*: 1993.

and *c* function inhibits *a* function in whorls 3 and 4—that is, *a* and *c* functions are mutually exclusive. Any floral meristem in which *a* activity alone is present (that is, whorl 1) will develop sepals; *a* and *b* acting together direct petal development in whorl 2; *b* and *c* acting together specify stamens in whorl 3; and *c* alone specifies carpel formation in whorl 4. What the homeotic mutations do is eliminate the functions of *a*, *b*, or *c*. So mutations in the first class (such as *apetala2* mutations) disrupt function *a*; expression of *c* is then not prevented in whorls 1 and 2, and so *c* is expressed in all whorls, giving the phenotype carpel, stamen, stamen, carpel (see Fig. 7.26, center panel). Mutations in *b*, the second class of mutations (such as *apetala3*), result in only *a* functioning in whorls 1 and 2, and *c* in whorls 3 and 4, giving sepal, sepal, carpel, carpel. Mutations in *c* function genes (such as *agamous*), result in *a* activity in all whorls, giving the phenotype sepal, petal, petal, sepal (see Fig. 7.26, right panel).

All the floral homeotic mutants discovered so far can be quite satisfactorily accounted for by this model, and particular genes can be assigned to each of the controlling functions. Function *a* corresponds to the activity of *a* function genes such as *APETALA2*, *b* to *APETALA3* and *PISTILLATA*, and *c* to *AGAMOUS*. The model also accounts for the phenotype of double mutants, such as *apetala2* with *apetala3*, and *apetala3* with *pistillata* (Fig. 7.27).

Another way of examining the action of these homeotic genes is to look at the phenotype of the flower when the action of all of them is lacking, and then study the effects when they are added back one by one (Fig. 7.28). A similar approach was taken with the homeotic genes that specify segment identity in *Drosophila* (see Fig. 5.39). In the absence of all three classes of gene, the flower consists of whorls of identical leaf-like organs; this can be regarded as a developmental ground state. Addition of *a* function genes to the ground state results in all sepals, while addition of *c* function to the ground state gives all carpels. Addition of both *a* and *c* results in flowers

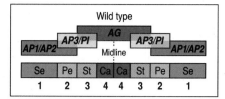

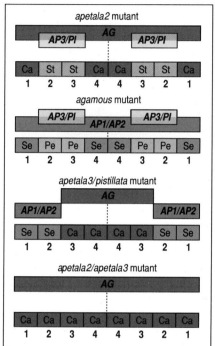

Fig. 7.27 Model of gene action controlling flower pattern in *Arabidopsis*. The gene *APETALA1* (*AP1*) is expressed in whorls 1 and 2; *APETALA3* (*AP3*) and *PISTILLATA* (*PI*) in whorls 2 and 3; and *AGAMOUS* (*AG*) in whorls 3 and 4. Sepals require *APETALA1* and *APETALA2*, (*APETALA2* is expressed in all whorls, but acts in organ identity specification only along with the whorl 1- and 2-limited *APETALA1* gene); petals require a combination of *APETALA1* and *APETALA2* together with *APETALA3* and *PISTILLATA*; stamens require *APETALA3* and *PISTILLATA* with *AGAMOUS*; and carpels require *AGAMOUS* alone. Mutations that alter the pattern of expression of one or more of these genes result in differing patterns of expression of the others and homeotic transformations of the floral parts. After Meyerowitz, E.M., *et al.*: 1991.

with half-flower phenotype sepal, sepal, carpel, carpel, showing that, in accord with the above model, expression of *a* and *c* are now restricted to regions A and C, respectively, as a result of their mutually inhibitory actions. However, other genes may also be required.

The model clearly predicts particular spatial patterns of gene activity in the meristem. For example, *b* class genes should be expressed only in the B region, that is, in whorls 2 and 3. Expression of the *b* class gene *APETALA3* is indeed first detected by *in situ* hybridization (see Box 3B, p. 70) in the floral meristem at the time the sepal primordia begin to form, and is restricted to whorls of cells that will give rise to petals and stamens. By contrast, the *c* class gene *AGAMOUS* is expressed in the meristem region that gives rise to whorls 3 and 4 (Fig. 7.29).

From the DNA sequence, one can deduce that the homeotic proteins encoded by floral identity genes such as *APETALA1* and *AGAMOUS* contain a conserved sequence of 57 amino acids, known as the MADS box, which can bind to DNA. The MADS box is also present in some transcription factors from yeast and animals. This makes it highly likely that, as with the homeotic selector genes of *Drosophila*, the floral identity genes are transcription factors. The MADS box transcription factors are expressed in regions of the flower that exhibit homeotic transformations when one gene is absent. In line with the model of expression described above, organ identity genes have been found to be sufficient to specify new organ identity. Thus, expression of *AGAMOUS*—a *c* function gene—with a constitutive promoter represses *a* function in whorls 1 and 2; the result is phenotypically the same as loss-of-function *a* mutants. The flower has carpels in place of sepals and stamens in place of petals.

Further evidence for this model comes from expressing *APETALA3* and *PISTILLATA* throughout the flower. The result is that the two outer whorls are petals and the two inner ones are stamens. This shows that expression of these two genes is sufficient to provide *b* organ identity—petals and stamens—in combination with class *a* and *c* genes. All that these results emphasize the similarity in function between the homeotic genes in animals and those controlling organ identity in flowers. The similarity with the HOM complex of *Drosophila* is further illustrated by the role of the *CURLY LEAF* gene of *Arabidopsis*, which is necessary for the stable maintenance of homeotic gene activity. *CURLY LEAF* is related to the Polycomb family of *Drosophila* and is similarly required for stable repression of homeotic genes.

Despite the enormous variation in the flowers of different species, the mechanisms underlying flower development seem to be very similar. For example, there are striking similarities between the floral control genes of *Arabidopsis* and the snapdragon, *Antirrhinum*. In developing *Antirrhinum*

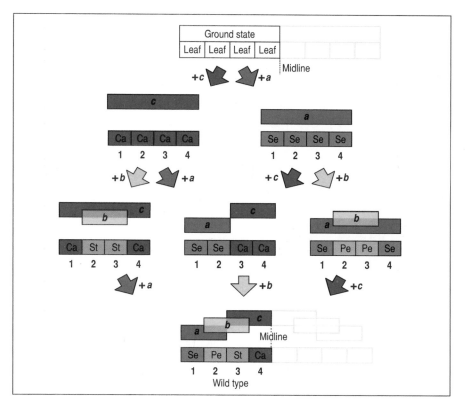

Fig. 7.28 The combinatorial action of the *a*, *b*, and *c* functions in patterning the flower. The floral organs can be considered to be a modification of a ground state in which only leaves are present. Leaves are modified into sepals, petals, stamens, and carpels by the expression of the floral identity functions *a*, *b*, and *c*. Addition of function *a* alone transforms the ground state of leaf into all sepals (Se), while addition of function *c* alone results in all organs becoming carpels (Ca). Addition of different combinations of functions to specific whorls results in the formation of different structures. This model assumes that *a* and *c* restrict each other's expression. After Coen, E.S., *et al.*: 1991.

flowers, the patterns of activity of the corresponding genes fit well with the cell lineage restriction to whorls, seen above for *Arabidopsis*. Before the emergence of organ primordia, cells have not acquired organ identity and there is no lineage restriction between whorls. This only occurs at the time when the pentagonal symmetry of the flower becomes visible and organ identity genes are expressed.

How are the spatial patterns of expression of these homeotic genes controlled? We do not yet know, but genes that restrict the expression of the homeotic genes have been discovered in *Arabidopsis*. These are the cadastral genes, of which *SUPERMAN* is an example. Plants with a mutation in this gene have stamens instead of carpels in the fourth whorl. The action of *SUPERMAN* seems to prevent the expression of *APETALA3* and *PISTILLATA* in the fourth whorl. *SUPERMAN* is expressed in the third whorl, and maintains the boundary between the third and fourth whorls.

Fig. 7.29 Expression of *APETALA3* and *AGAMOUS* during flower development. *In situ* hybridization shows that *AGAMOUS* is expressed in the central whorls (left panel), whereas *APETALA3* is expressed in the outer whorls that give rise to petals and stamens (right panel).

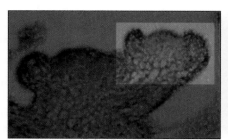

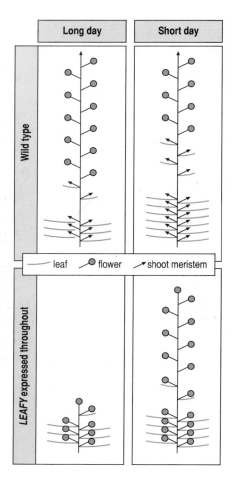

Fig. 7.30 Day length and *LEAFY* expression can control flowering. If *Arabidopsis* is grown under long daylength conditions, few lateral shoots are formed before the apical shoot meristem begins to form floral meristems. Lateral shoots can also give rise to flowers. When grown under short daylength conditions, flowering is delayed and there are more lateral shoots and flowers. Expression of the *LEAFY* gene throughout the plant results in the lateral shoot meristems that would normally form being converted to floral meristems.

7.13 The transition of a shoot meristem to a floral meristem is under environmental and genetic control

Flowering in *Arabidopsis* is controlled by internal and external factors, with day length having an important role (Fig. 7.30, top panels). Most flowering plants have a vegetative phase during which the apical meristem generates leaves. Then, triggered by environmental signals, they switch to a reproductive phase and the apical meristem gives rise to flowers. An obvious environmental signal for triggering flowering is the length of the day. Grafting experiments have shown that the duration of the light period is sensed by the leaves and a diffusible flower-inducing signal is transmitted to the shoot meristem. The nature of this signal is not known, but several of the genes involved have been identified. One is *CONSTANS*, which codes for a transcription factor; its expression is increased in response to light.

There are two types of transition from vegetative growth to flowering. In the determinate type, the inflorescence meristem becomes a terminal flower, whereas in the indeterminate type the inflorescence meristem gives rise to a number of floral meristems. The transition in *Arabidopsis* is of the indeterminate type. A primary response to floral inductive signals in *Arabidopsis* is the transcription of floral meristem identity genes such as *LEAFY* and *APETALA1*, which are necessary and sufficient for this transition. *LEAFY* activates *AGAMOUS* in the center of the flower. Mutations in these genes partly transform flowers into shoots. In a *leafy* mutant, the flowers are transformed into spirally arranged sepal-like organs along the stem, whereas expression of *LEAFY* throughout the plant is sufficient to determine floral fate in lateral shoot meristems (Fig. 7.30, bottom panels). These are the potential meristems that are formed at the base of each leaf during vegetative growth. The gene *FLORICAULA* in *Antirrhinum* functions in a similar way to *LEAFY*. The abnormal expression of *FLORICAULA* in just one layer of a shoot meristem can allow flower development and induce the genes required for flower development in layers in which *FLORICAULA* is normally not expressed. This clearly shows that one layer of the meristem can induce development in adjacent layers. Unlike animals, where transcription factors produced in one cell appear to be confined to that cell, it is possible that the induction of one meristem layer by another could be due to the movement of transcription factors, such as floricaula protein, from cell to cell.

Direct evidence that at least one transcription factor, the leafy protein, can move between cells has recently been obtained using chimeras. *leafy* mutants never form flowers, but normal flowers developed from meristems in which half the cells were expressing the *LEAFY* gene, and half were not, showing that leafy protein must have moved from the cells in which it was being produced. Leafy protein was also detected directly in the mutant cells not expressing the gene.

7.14 The *Antirrhinum* flower is patterned dorso-ventrally as well as radially

Like *Arabidopsis* flowers, those of *Antirrhinum* consist of four whorls, but unlike *Arabidopsis* flowers, they have five sepals, five petals, four stamens, and two united carpels (Fig. 7.31, left). Floral homeotic mutations similar to those in *Arabidopsis* occur in *Antirrhinum*, and floral organ identity is almost

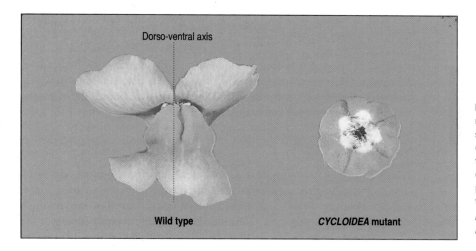

Fig. 7.31 Mutations in *CYCLOIDEA* make the *Antirrhinum* flower symmetrical. In the wild-type flower (left) the petal pattern is different along the dorso-ventral axis. In the mutant (right) the flower is symmetrical. All the petals are like the most ventral one in the wild type and are folded back. *Photograph courtesy of E. Coen, from Coen, E.S., et al.: 1991.*

certainly specified in the same way. Several of the *Antirrhinum* homeotic genes have extensive homology with those of *Arabidopsis*, the MADS box in particular being well conserved.

However, an extra element of patterning is required in the *Antirrhinum* flower, which has a bilateral symmetry imposed on the basic radial pattern common to all flowers. In whorl 2, the upper two petal lobes have a shape quite distinct from the lower three, giving the flower its characteristic snapdragon appearance. In whorl 3, the uppermost stamen is absent, as its development is aborted early on. The *Antirrhinum* flower therefore has a distinct dorso-ventral axis. Another group of homeotic genes, different from those that govern floral organ identity, appear to act in this dorso-ventral patterning. For example, mutations in the gene *CYCLOIDEA*, which is expressed in the dorsal region, abolish dorso-ventral polarity and produce flowers that are more radially symmetrical (see Fig. 7.31, above).

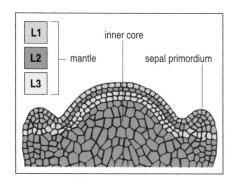

Fig. 7.32 **Floral meristem.** The meristem is composed of layers L1, L2, and L3. The inner core cells are derived from L3. The sepal primordia are just beginning to develop. After Drews, G.N., *et al.*: 1989.

7.15 The internal meristem layer can specify floral meristem patterning

Although all three layers of a floral meristem (Fig. 7.32) are involved in organogenesis, the contribution of cells from each layer to a particular structure may be variable. Cells from one layer can become part of another layer without disrupting normal morphology, suggesting that a cell's position in the meristem is the main determinant of its future behavior. Some insight into positional signaling and patterning in the floral meristem can be obtained by making periclinal chimeras (see Section 7.7) from cells that have different genotypes and that give rise to different types of flower. From such chimeras, one can find out whether the cells develop autonomously according to their own genotype, or whether their behavior is controlled by signals from other cells.

Chimeras can be generated by grafting between two plants of different genotypes. A new shoot meristem forms at the junction of the graft, and sometimes contains cells from both genotypes. Such chimeras can be made between wild-type tomato plants and tomato plants carrying the mutation *fasciated*, in which the flower has an increased number of floral organs per whorl. Similar flowers are found in chimeras in which only layer L3

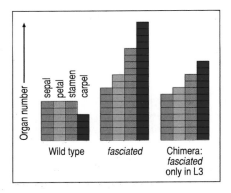

Fig. 7.33 Floral organ number in chimeras of wild-type and *fasciated* tomato plants. In the *fasciated* mutant there are more organs in the flower than in wild-type plants. In chimeras in which only layer L3 of the floral meristem contains *fasciated* mutant cells, the number of organs per flower is still increased, showing that L3 can control cell behavior in the outer layers of the meristem.

contains *fasciated* cells (Fig. 7.33). The increased number of floral organs is associated with an overall increase in the size of the floral meristem, and this cannot be achieved unless the *fasciated* cells of layer L3 induce the wild-type L1 cells to divide more frequently than normal. The mechanism of intercellular signaling between L3 and L1 is not yet known. These results, together with those described for *FLORICAULA* above, illustrate the importance of signaling between layers in flower development.

Summary

Before flowering, the vegetative shoot apical meristem becomes converted into an inflorescence meristem which either then becomes a flower or produces a series of floral meristems, each of which develops into a single flower. Genes involved in the initiation of flowering and patterning of the flower have been identified in both *Arabidopsis* and *Antirrhinum*. Expression of meristem identity genes is required for the formation of floral meristems from the inflorescence meristem. Homeotic floral organ identity genes, which specify the organ types found in the flowers have been identified from mutations that transform one flower part into another. On the basis of these mutations, a model has been proposed in which the floral meristem is divided into three concentric overlapping regions, in each of which certain floral identity genes act in a combinatorial manner to specify the organ type appropriate to each whorl. Studies with chimeric plants have shown that different meristem layers communicate with each other during flower development and that transcription factors can move between cells.

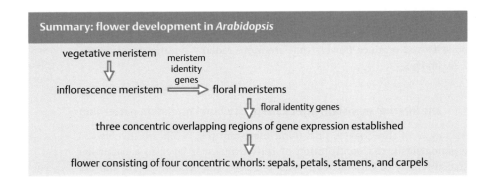

SUMMARY TO CHAPTER 7

A distinctive feature of plant development is the presence of relatively rigid walls and the absence of any cell migration. Another is that a single isolated somatic cell from a plant can regenerate into a complete new plant. Early embryonic development is characterized by asymmetric cell division of the fertilized egg, which specifies the future apical and basal regions. In the alga *Fucus*, this plane of cleavage is specified by environmental signals. During early development of flowering plants, both asymmetric cell division and cell–cell interactions are involved in patterning the body plan. During this process, the shoot and root meristems are specified and these meristems give rise to all the organs of the plant—stems, leaves, flowers, and roots. The shoot meristem gives rise to leaves in well-defined positions, a process involving lateral inhibition. The shoot meristem eventually becomes converted to an inflorescence meristem, which either becomes a floral meristem (in determinate inflorescences) or gives rise to a series of floral meristems, retaining its shoot meristem identity indefinitely (in indeterminate inflorescences). In floral meristems, each of which develops into a flower, homeotic floral organ identity genes act in combination to specify the floral organ types.

GENERAL REFERENCES

Meyerowitz, E.M.: **Plants and the logic of development.** *Genetics* 1997, **145**: 5–9.

Pickard, B.G., Beachy, R.N.: **Intercellular connections are developmentally controlled to help move molecules through the plant.** *Cell* 1999, **98**: 5–8.

SECTION REFERENCES

7.1 Both asymmetric cell divisions and cell position pattern the early embryos of flowering plants

Mayer, U., Jürgens, G.: **Pattern formation in plant embryogenesis: a reassessment.** *Semin. Cell Dev. Biol.* 1998, **9**: 187–193.

Meyerowitz, E.M.: **Plant development: local control, global patterning.** *Curr. Opin. Genet. Dev.* 1996, **6**: 475–479.

Meyerowitz, E.M.: **Genetic control of cell division patterns in developing plants.** *Cell* 1997, **88**: 299–308.

Torres-Ruiz, R.A., Jürgens, G.: **Mutations in the FASS gene uncouple pattern formation and morphogenesis in *Arabidopsis* development.** *Development* 1994, **120**: 2967–2978.

7.2 The patterning of particular regions of the *Arabidopsis* embryo can be altered by mutation

Brand, U., Hobe, M., Simon, R.: **Functional domains in plant shoot meristems.** *BioEssays* 2001, **23**: 134–141.

Jürgens, G.: **Axis formation in plant embryogenesis: cues and clues.** *Cell* 1995, **81**: 467–470.

Jürgens, G., Torres-Ruiz, R.A., Berleth, T.: **Embryonic pattern formation in flowering plants.** *Annu. Rev. Genet.* 1994, **28**: 351–371.

Lloyd, C.: **Plant morphogenesis: life on a different plane.** *Curr. Biol.* 1995, **5**: 1085–1087.

Long, J.A., Moan, E.I., Medford, J.I., Barton, M.K.: **A member of the knotted class of homeodomain proteins encoded by the STM gene of *Arabidopsis*.** *Nature* 1995, **379**: 66–69.

Ma, H.: **Gene regulation: Better late than never?** *Curr. Biol.* 2000, **10**: R365–368.

7.3 Plant somatic cells can give rise to embryos and seedlings

Zimmerman, J.L.: **Somatic embryogenesis: a model for early development in higher plants.** *Plant Cell* 1993, **5**: 1411–1423.

7.4 Electrical currents are involved in polarizing the *Fucus* zygote

Brownlee, C., Bouget, F-Y.: **Polarity determination in *Fucus*: from zygote to multicellular embryo.** *Semin. Cell Dev. Biol.* 1998, **9**: 179–185.

Goodner, B., Quatrano, R.S.: ***Fucus* embryogenesis: A model to study the establishment of polarity.** *Plant Cell* 1993, **5**: 1471–1481.

Hable, W.E., Kropf, D.L.: **Sperm entry induces polarity in fucoid zygotes.** *Development* 2000, **127**: 493–501.

Robinson, K.R., Cone, R.: **Polarization of fucoid eggs by a calcium ionophore gradient.** *Science* 1980, **207**: 77–78.

7.5 Cell fate in early *Fucus* development is determined by the cell wall

Berger, F., Taylor, A., Brownlee, C.: **Cell fate determination by the cell wall in early *Fucus* development.** *Science* 1994, **263**: 1421–1423.

Bouget, F.-Y., Berger, F., Brownlee, C.: **Position-dependent control of cell fate in the *Fucus* embryo: role of intercellular communication.** *Development* 1998, **125**: 1999–2008.

Shaw, S.L., Quatrano, R.S.: **The role of targeted secretion in the establishment of cell polarity and the orientation of the division plane in *Fucus* zygotes.** *Development* 1996, **122**: 2623–2630.

7.6 Differences in cell size resulting from unequal divisions could specify cell type in the *Volvox* embryo

Kirk, M.M., Ransick, A., McRae, S.E., Kirk, D.L.: **The relationship between cell size and cell fate in *Volvox carteri*.** *J. Cell Biol.* 1993, **123**: 191–208.

7.7 The fate of a cell in the shoot meristem is dependent on its position

Clark, S.E.: **Cell signalling at the shoot meristem.** *Nat. Rev. Mol. Cell Biol.* 2001, **2**: 277–284.

Irish, V.F.: **Cell lineage in plant development.** *Curr. Opin. Genet. Dev.* 1991, **1**: 169–173.

Langdale, J.A.: **Plant morphogenesis: more knots untied.** *Curr. Biol.* 1994, **4**: 529–531.

Laux, T., Mayer, K.F.X.: **Cell fate regulation in the shoot meristem.** *Semin. Cell Dev. Biol.* 1998, **9**: 195–200.

Schoof, H., Lenhard, M., Haecker, A., Mayer, K.F.X., Jurgens, G., Laux, T.: **The stem cell population of arabidopsis shoot meristems is maintained by a regulatory loop between the CLAVATA and WUSCHEL genes.** *Cell* 2000, **100**: 635–644.

Sinha, N.R., Williams, R.E., Hake, S.: **Overexpression of the maize homeobox gene *knotted-1*, causes a switch from determinate to indeterminate cell fates.** *Genes Dev.* 1993, **7**: 787–795.

Turner, I.J., Pumfrey, J.E.: **Cell fate in the shoot apical meristem of *Arabidopsis thaliana*.** *Development* 1992, **115**: 755–764.

Waites R., Simon, R.: **Signaling cell fate in plant meristems: three clubs on one tousle.** *Cell* 2000, **103**: 835–838.

7.8 Meristem development is dependent on signals from other parts of the plant

Doerner, P.: Shoot meristems: intercellular signals keep the balance. *Curr. Biol.* 1999, **9**: R377–R380.

Irish, E.E., Nelson, T.M.: Development of maize plants from cultured shoot apices. *Planta* 1988, **175**: 9–12.

Sachs, T.: *Pattern Formation in Plant Tissues.* Cambridge: Cambridge University Press, 1994.

7.9 Gene activity patterns the proximo-distal and adaxial–abaxial axes of leaves developing from the meristem

Bowman, J.L.: Axial patterning in leaves and other lateral organs. *Curr. Opin. Genet. Dev.* 2000, **10**: 399–404.

Waites, R., Selvadurai, H.R.N., Oliver, I.R., Hudson, A.: The *PHANTASTICA* gene encodes a MYB transcription factor involved in growth and dorsoventrality of lateral organs in *Antirrhinum*. *Cell* 1998, **93**: 779–789.

7.10 The regular arrangement of leaves on a stem and trichomes on leaves is generated by lateral inhibition

Mitchison, G.J.: Phyllotaxis and the Fibonacci series. *Science* 1977, **196**: 270–275.

Scheres, B.: Non-linear signaling for pattern formation. *Curr. Opin. Plant Biol.* 2000, **3**: 412–417.

Smith, L.G., Hake, S.: The initiation and determination of leaves. *Plant Cell* 1992, **4**: 1017–1027.

7.11 Root tissues are produced from root apical meristems by a highly stereotyped pattern of cell divisions

Benfey, P.N., Schiefelbein, J.W.: Getting to the root of plant development: genetics of *Arabidopsis* root formation. *Trends Genet.* 1994, **10**: 84–88.

Berger, F., Haseloff, J., Schiefelbein, J., Dolan, L.: Positional information in root epidermis is defined during embryogenesis and acts in domains with strict boundaries. *Curr. Biol.* 1998, **8**: 421–430.

Costa, S., Dolan, L.: Development of the root pole and patterning in *Arabidopsis* roots. *Curr. Opin. Genet. Dev.* 2000, **10**: 405–409.

Doerner, P.: Root development: quiescent center not so mute after all. *Curr. Biol.* 1998, **8**: R42–R44.

Dolan, L., Roberts, K.: Plant development: Pulled up by the roots. *Curr. Opin. Genet. Dev.* 1995, **5**: 432–438.

Dolan, L., Scheres, B.: Root pattern: shooting in the dark? *Cell Dev. Biol.* 1998, **9**: 201–206.

Sabatini, S., Beis, D., Wolkenfeldt, H., Murfett, J., Guilfoyle, T., Malamy, J., Benfey, P., Leyser, O., Bechtold, N., Weisbeek, P., Scheres, B.: An auxin-dependent distal organizer of pattern and polarity in the *Arabidopsis* root. *Cell* 1999, **99**: 463–472.

Scheres, B., McKhann, H.I., van den Berg, C.: Roots redefined: anatomical and genetic analysis of root development. *Plant Physiol.* 1996, **111**: 959–964.

van den Berg, C., Willemsen, V., Hendriks, G., Weisbeek, P., Scheres, B.: Short-range control of cell differentiation in the *Arabidopsis* root meristem. *Nature* 1997, **390**: 287–289.

7.12 Homeotic genes control organ identity in the flower

Bowman, J.L., Sakai, H., Jack, T., Weigel, D., Mayer, U., Meyerowitz, E.M.: *SUPERMAN,* a regulator of floral homeotic genes in *Arabidopsis*. *Development* 1992, **114**: 599–615.

Coen, E.S., Meyerowitz, E.M.: The war of the whorls: genetic interactions controlling flower development. *Nature* 1991, **353**: 31–37.

Goodrich, J., Puangsomlee, P., Martin, M., Meyerowitz, E.M., Coupland, G.: A Polycomb-group gene regulates homeotic gene expression in *Arabidopsis*. *Nature* 1997, **386**: 44–51.

Irish, V.F.: Patterning the flower. *Dev. Biol.* 1999, **209**: 211–220.

Krizek, B.A., Meyerowitz, E.M.: The *Arabidopsis* homeotic genes *APETALA3* and *PISTILLATA* are sufficient to provide the B class organ identity function. *Development* 1996, **122**: 11–22.

Ma, H., dePamphilis, C.: The ABCs of floral evolution. *Cell* 2000, **101**: 5–8.

Meyerowitz, E.M.: The genetics of flower development. *Sci. Am.* 1994, **271**: 40–47.

Meyerowitz, E.M., Bowman, J.L., Brockman, L.L., Drews, G.M., Jack, T., Sieburth, L.E., Weigel, D.: A genetic and molecular model for flower development in *Arabidopsis thaliana*. *Development Suppl.* 1991, **1**: 157–167.

Sakai, H., Medrano, L.J., Meyerowitz, E.M.: Role of *SUPERMAN* in maintaining Arabidopsis floral whorl boundaries. *Nature* 1994, **378**: 199–203.

Vincent, C.A., Carpenter, R., Coen, E.S.: Cell lineage patterns and homeotic gene activity during *Antirrhinum* flower development. *Curr. Biol.* 1995, **5**: 1449–1458.

Wagner, D., Sablowski, R.W.M., Meyerowitz, E.M.: Transcriptional activation of *APETALA 1* by *LEAFY*. *Science* 1999, **285**: 582–584.

7.13 The transition of a shoot meristem to a floral meristem is under environmental and genetic control

Becroft, P.W.: Intercellular induction of homeotic gene expression in flower development. *Trends Genet.* 1995, **11**: 253–255.

Hake, S.: Transcription factors on the move. *Trends Genet.* 2001, **17**: 2–3.

Lohmann, J.U., Hong, R.L., Hobe, M., Busch, M.A., Parcy, F., Simon, R., Weigel, D.: A molecular link between stem cell regulation and floral patterning in *Arabidopsis*. *Cell* 2001, **105**: 793–803.

Weigel, D., Nilsson, O.: A developmental switch sufficient for flower initiation in diverse plants. *Nature* 1995, **327**: 495–500.

7.14 The *Antirrhinum* flower is patterned dorso-ventrally as well as radially

Coen, E.S.: Floral symmetry. *EMBO J.* 1996, **15**: 6777–6788.

Luo, D., Carpenter, R., Vincent, C., Copsey, L., Coen, E.: Origin of floral asymmetry in *Antirrhinum*. *Nature* 1996, **383**: 794–799.

7.15 The internal meristem layer can specify floral meristem patterning

Szymkowiak, E.J., Sussex, I.M.: The internal meristem layer (L3) determines floral meristem size and carpel number in tomato periclinal chimeras. *Plant Cell* 1992, **4**: 1089–1100.

Morphogenesis: change in form in the early embryo

8

- Cell adhesion
- Cleavage and formation of the blastula
- Gastrulation
- Neural tube formation
- Cell migration
- Directed dilation

"We tend to stick together but can change neighbors to form interesting shapes and sometimes move away to sites where we can get the best grip."

In previous chapters, we have discussed early development mainly from the viewpoint of developmental patterning and the assignment of cell fate. In this chapter, we look at embryonic development from a different perspective. Generation of form in early animal development involves rearrangement of the cell layers and movement of cells from one location to another. All animal embryos, for example, undergo a dramatic change in shape during their early development. This occurs principally during **gastrulation**, when the gut is formed and the basic animal body plan is laid down. Gastrulation involves extensive rearrangement of cell layers and the directed movement of cells from one location to another. In plants there is no cell movement, and changes in form are generated by cell division and expansion only.

To understand morphogenesis we need to look at the mechanical forces that cells can exert and the underlying cellular properties that generate these forces. If pattern formation is likened to painting, the developmental phenomena described in this chapter are more akin to modeling a formless lump of clay into a recognizable shape. Change in form is largely a problem in cell mechanics: it requires an understanding of the forces involved in bringing about changes in cell shape and cell migration. It is not only the mechanical forces underlying generation of form that need to be identified, but also how they are controlled, and which molecules are involved in these processes.

Two key cellular properties involved in changes in animal embryonic form are **cell adhesiveness** and **cell motility**. Animal cells stick to one another, and to the extracellular matrix, through interactions involving cell-surface proteins. Changes in these proteins can therefore determine both the strength of cell adhesion and its specificity. The second key cellular property, cell motility, encompasses the ability of cells to migrate to new locations and to change their shape when, for example, a sheet of cells folds—a very common feature in animal embryonic development. The ability of cells to move and change shape is determined by rearrangements in their internal cytoskeletal structures. An additional force that operates during morphogenesis, particularly in plants, is hydrostatic pressure, which is generated by osmosis and fluid accumulation. Other morphogenetic mechanisms such as cell growth, cell proliferation, and cell death will be considered in relation to the development of particular organs (see Chapter 10) and of the nervous system (see Chapter 11).

At the molecular level, changes in embryonic form can be ultimately considered as the consequence of the precise spatio-temporal expression of the molecules that control cell adhesion, cell motility, oriented cell division, and the generation of hydrostatic pressure. An attractive hypothesis is that pattern-determining genes such as the Hox genes activate other genes that control the expression of such molecules, and there is some evidence for this. It is likely that changes in embryonic form are brought about by an earlier patterning process that determines which cells will express those gene products required to generate and harness the appropriate forces.

The changes in form considered in this chapter are mainly those involved in laying down the animal body plan. First, we look at the mechanisms by which cleavage of the zygote gives rise to the simple shape of the early embryo, of which the spherical blastulas of the mouse, sea urchin, and amphibian are good examples. We then consider gastrulation, which transforms this essentially spherical sheet of cells into a complex three-dimensional animal body, and neurulation—the formation of the neural tube in vertebrates—which also involves folding of cell sheets and rearrangement of cell layers. In vertebrates, migration of cells from the neural crest after neurulation generates a variety of structures in the trunk and head, and we consider how these cells migrate to their correct sites. We also consider the role of chemotaxis and signal propagation in the aggregation of unicellular amebae into the fruiting body in the cellular slime mold *Dictyostelium discoideum*. Finally we look at directed dilation, in which hydrostatic pressure is the force driving changes in shape, particularly in plant cells.

We begin by considering how cells adhere to each other, and how differences in adhesiveness and specificity of adhesion may be involved in maintaining boundaries between tissues.

Cell adhesion

The late embryo and the adult are composed of a variety of differentiated cell types, which are grouped together in tissues such as skin and cartilage. The integrity of tissues is maintained by adhesive interactions both between cells and between cells and the extracellular matrix; differences in cell adhesiveness also play a part in maintaining the boundaries between different tissues and structures. Cells are stuck together by **adhesion molecules**, which are proteins carried on the cell surface that can bind to other molecules on cell surfaces or in the extracellular matrix (Box 8A). The particular adhesion molecules expressed by a cell determine which cells it can adhere to; changes in expression of adhesion molecules are known to be involved in many developmental phenomena, such as neurulation in vertebrates.

Box 8A Cell adhesion molecules

Three classes of adhesion molecules are important in development. The **cadherins** are transmembrane proteins which, in the presence of calcium ions (Ca^{2+}), adhere to cadherins on the surface of another cell. They are involved in cell-cell adhesion only. Calcium-independent cell-cell adhesion involves a different structural class of proteins—members of the large **immunoglobulin superfamily**. The neural cell adhesion molecule (N-CAM), which was first isolated from neural tissue, is a typical member of this family. Some immunoglobulin superfamily members, such as N-CAM, bind to similar molecules on other cells; others bind to a different class of adhesion molecule, the **integrins**. These are the third class of adhesion molecule involved in development. Integrins act as receptors for molecules of the extracellular matrix, which they can use to mediate adhesion between a cell and its substratum.

About 30 different types of cadherins have been identified in vertebrates. Cadherins bind to each other through one or more binding sites located within the extracellular amino-terminal 100 amino acids. In general, a cadherin binds only to another cadherin of the same type, but they can also bind to some other molecules. A typical cadherin is E-cadherin (also known as uvomorulin), which is involved in the generation of a polarized epithelial sheet in the early mammalian embryo. Cadherins are the adhesive components in intermediate junctions (belt desmosomes), adhesive cell junctions that are present in many tissues, notably in epithelia, where they are part of a special region known as the junctional complex.

As cells approach and touch one another, the cadherins cluster at the site of contact. They also interact with the cell's cytoskeleton through the connection of their cytoplasmic tails with intracellular **catenins**, and thus can be involved in

transmitting signals to the cell's interior. The interaction with the cytoskeleton is required for normal cell adhesion.

Adhesion to the extracellular matrix, which contains proteins such as collagen, fibronectin, laminin, and tenascin, as well as proteoglycans, is mediated by integrins, which bind to these matrix molecules. Each integrin is made up of two different subunits, an α subunit and a β subunit. At least 20 different integrins have been found in vertebrates, made up from nine known β subunits and 15 known α subunits. Many extracellular matrix molecules are recognized by more than one integrin.

Integrins not only bind to other molecules through their extracellular face, but they also associate with the actin filaments of the cell's cytoskeleton through complexes of proteins that are in contact with their cytoplasmic region. This association may enable integrins to transmit information about the extracellular environment, such as extracellular matrix composition or the type of intercellular contact. Integrins can thus mediate signals from the matrix that affect cell shape, motility, metabolism, and cell differentiation.

8.1 Sorting out of dissociated cells demonstrates differences in cell adhesiveness in different tissues

Differences in adhesiveness are illustrated by an experiment in which different tissues are confronted with one another in an artificial setting. Two pieces of early endoderm from an amphibian blastula will fuse to form a smooth sphere, but when a piece of early endoderm and a piece of early ectoderm are combined they initially fuse, but the endodermal and ectodermal cells eventually separate, until only a narrow bridge connects the two types of tissue (Fig. 8.1).

Cellular affinities are further illustrated by disaggregating cells of the presumptive epidermis and presumptive neural plate of an amphibian neurula by treatment with an alkaline solution, then mixing them together

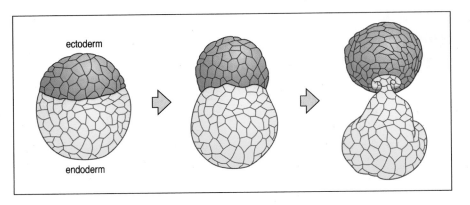

Fig. 8.1 Separation of embryonic tissues with different adhesive properties. When a piece of early ectoderm (blue) and early endoderm (yellow) from an amphibian blastula are placed together, they initially fuse but then separate until only a narrow strip of tissue joins the two types.

and allowing them to reaggregate (Fig. 8.2). The cells of the mixed cell mass exchange neighbors and move in such a way that the epidermal cells are eventually found on the outer face of the aggregate, surrounding a mass of neural cells in the interior. The same types of cell are now in contact with each other. When ectodermal and mesodermal cells are mixed, they similarly sort themselves out to form an aggregate with ectoderm on the outside and mesoderm on the inside. The sorting out of cells of different types is the result of cell movement and differential adhesiveness. Initially, cells move about randomly in the aggregate, exchanging weaker for stronger adhesions. In their final distribution, intercellular binding strengths in the system as a whole are maximized. In general, if the adhesion between unlike cells is weaker than the average of the adhesions between like cells, the cells will segregate type-specifically, with the more cohesive tissue tending to be enveloped by the less cohesive one. These experiments show how differential cell adhesion can stabilize the boundaries between tissues. Ephrins can also have a role in cell sorting. As we have seen (Section 4.10), they are involved in maintaining rhombomere boundaries in the hindbrain by preventing cells from different rhombomeres intermingling.

8.2 Cadherins can provide adhesive specificity

Differential adhesiveness between cells is the result of the presence of differences in the kinds and numbers of adhesion molecules on cell surfaces. To study the role of these molecules, genes encoding them are introduced and expressed in cells that do not normally produce them, a procedure known as **transfection**. Evidence that the cadherin class of adhesion molecules can provide adhesive specificity comes from studies in which cells with different cadherins on their surface are mixed together.

Cultured fibroblast cells of the L-cell line do not normally adhere strongly to each other, nor do they express cadherins on their surface. L cells transfected with E-cadherin DNA express that cadherin on their surface. They adhere to one another, forming a structure resembling a compact epithelium. The adherence is both calcium dependent, indicating that it is due

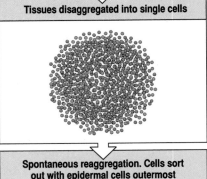

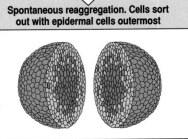

Fig. 8.2 Sorting out of different cell types. Ectoderm from the presumptive neural plate (blue) and from the presumptive epidermis (gray) of early amphibian neurulas are disaggregated into single cells through treatment with an alkaline solution. The cells, when mixed together, sort out with the epidermal cells on the outside.

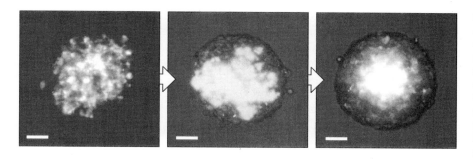

Fig. 8.3 Sorting out of cells with quantitative differences in cell adhesion molecules. Two cell lines with different amounts of P-cadherin on their surface sort out, with the cells containing the most P-cadherin on the inside. The fluorescently stained cells are those expressing the greater amount of P-cadherin. Scale bars = 0.1 mm.

Photographs courtesy of M. Steinberg, from Steinberg, M.S., et al.: 1994.

to the cadherin, and specific, as the transfected cells do not adhere to untransfected L cells, which lack surface cadherins. When groups of cells are transfected with different types of cadherin and mixed together in suspension, only those cells expressing the same cadherin adhere strongly to each other: cells expressing E-cadherin adhere strongly to other cells expressing E-cadherin, but only weakly to cells expressing P- or N-cadherin. The amount of cadherin on the cell surface can also have an effect on adhesion. If cells expressing different amounts of the same cadherin are mixed together, those cells expressing more cadherin on their surface form an inner ball, surrounded by the cells expressing less cadherin on their surface (Fig. 8.3). Thus, quantitative differences in cell adhesion molecules could maintain differential cell adhesion.

Cadherin molecules bind to each other through their extracellular domains, but adhesion is not exclusively controlled by these binding domains: the cytoplasmic domain is also important (see Box 8A, p. 255). This domain can associate with the actin filaments of the cytoskeleton by means of a protein complex containing catenins. Failure of the cytoplasmic domain to associate with the cytoskeleton results in weak adhesion only. In early *Xenopus* blastulas, E-cadherin is expressed in the ectoderm just before gastrulation and N-cadherin appears in the prospective neural plate. If an E-cadherin lacking an extracellular domain is produced in these blastulas from injected mutated mRNA, the mutant cadherin will compete with the embryo's own intact cadherin molecules for the cytoskeletal association sites. The defective cadherin cannot influence cell–cell adhesion as it has no extracellular domain, but it blocks the intact cadherin's access to the cytoskeleton. The result is disruption of the ectoderm during gastrulation, indicating that a cadherin molecule must bind both to its partner on the opposing cell and to the cytoskeleton of its own cell to create a stable adhesion. The initial binding of the extracellular cadherin domains transmits a signal to the cytoskeleton, which then stabilizes the interaction.

Summary

The adhesion of cells to each other and to the extracellular matrix maintains the integrity of tissues and the boundaries between them. The associations of cells with each other are determined by the cell adhesion molecules they express on their surface: cells bearing different adhesion molecules, or different quantities of the same molecule, sort out into separate tissues. Cell–cell adhesion is due mainly to two classes of surface proteins: the cadherins, which bind in a calcium-dependent manner to identical cadherins on another cell surface, and members of the immunoglobulin superfamily, some of which bind to similar molecules on other cells while others bind to different molecules, such as integrins. The binding of this second class of surface proteins is calcium independent.

Adhesion to the extracellular matrix is mediated by a third class of adhesion molecule—the integrins. Stable cell adhesion by cadherins involves the external binding domains and interactions with the cytoskeleton, with the cytoplasmic domain of the cadherin molecules binding to a protein complex that contains catenins.

Cleavage and formation of the blastula

In animal eggs, the first step in embryonic development is the division of the fertilized egg by cleavage into a number of smaller cells (blastomeres), leading in many animals to the formation of a hollow sphere of cells—the blastula (see Chapter 2). Cleavage involves short cell cycles, in which cell division and mitosis succeed each other repeatedly without any intervening periods of cell growth. During cleavage therefore, the mass of the embryo does not increase. Early cleavage patterns can vary widely between different groups of animals (Fig. 8.4). The simplest pattern of division is radial, in which successive symmetric cleavages divide the embryo into equal-sized cells. This pattern is seen in the first three cleavages in the sea urchin. Later cleavages, such as those that form the micromeres, are unequal, one daughter cell being smaller than the other. The first cleavage of the nematode egg produces two cells of unequal size. The spirally cleaving eggs of molluscs and annelids illustrate yet another pattern of cleavage, in which successive divisions of each blastomere, in different planes, produce a spiral arrangement of cells; many of these divisions are unequal.

The amount of yolk in the egg can influence the pattern of cleavage. In

Fig. 8.4 Different patterns of early cleavage are found in different animal groups. Radial cleavages are equal and symmetric (sea urchin). Unequal cleavage (nematode) results in one daughter cell being larger than the other. In spiral cleavage, the mitotic apparatus is oriented at an oblique angle to the long axis of the cell.

Radial cleavage—sea urchin	Unequal cleavage—nematode	Spiral cleavage—annelids and molluscs

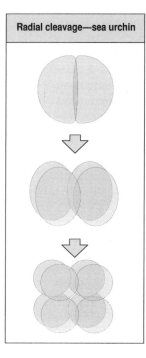 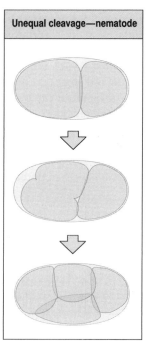 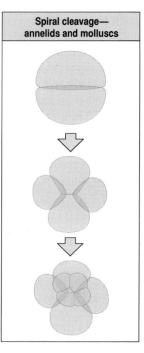

yolky eggs undergoing symmetric cleavage, a cleavage furrow develops in the least yolky region and gradually spreads across the egg, but its progress is slowed or even halted by the presence of the yolk. Cleavage may thus be incomplete for some time. This effect is most pronounced in the heavily yolked eggs of birds and zebrafish, where complete divisions are restricted to a region at one end of the egg, and the embryo is formed as a cap of cells sitting on top of the yolk (see Fig. 2.10). Even in moderately yolky eggs, such as those of amphibians, the presence of the yolk can influence cleavage patterns. In frogs, for example, the later cleavages are unequal and asynchronous, resulting in an animal half composed of a mass of small cells and a yolky vegetal region composed of fewer and larger cells.

Two key questions arise relating to early cleavage: how are the positions of the cleavage planes determined; and how can cleavage lead to a hollow blastula (or its equivalent), which has a clear inside–outside polarity?

8.3 The asters of the mitotic apparatus determine the plane of cleavage at cell division

The orientation of the plane of cleavage at cell division is important in a variety of situations. We have already seen its role in relation to unequal cleavages in early invertebrate development, where unequal divisions produce cells with differently specified fates. This is because of the asymmetric distribution of cytoplasmic determinants (see Chapter 6). The plane of cleavage can also be of great importance in later morphogenesis and growth. It determines, for example, whether an epithelial sheet remains as a single cell layer or becomes multilayered.

Experiments show that the plane of cleavage in animal cells is not specified by the mitotic spindle itself, but by the asters at each pole. To disrupt its normal cleavage, a sea urchin egg is deformed by a glass bead. This displaces the mitotic spindle and interrupts the first cleavage furrow, creating a horseshoe-shaped cell in which the two new nuclei are segregated into the separate arms of the horseshoe (Fig. 8.5). At the next division, two cleavage furrows bisect the spindles formed in the arms of the horseshoe, but a third furrow also forms between the two adjacent asters not connected by a spindle. How the asters specify a furrow between them is not known.

Fig. 8.5 The plane of cleavage in animal cells is determined by the asters and not the mitotic spindle. If the mitotic apparatus of a fertilized sea urchin egg is displaced at the first cleavage by a glass bead, the cleavage furrow forms only on the side of the egg to which the mitotic apparatus has been moved. At the next cleavage, furrows bisect each mitotic apparatus, and a furrow also forms between the two adjacent asters, even though there is no spindle between them.

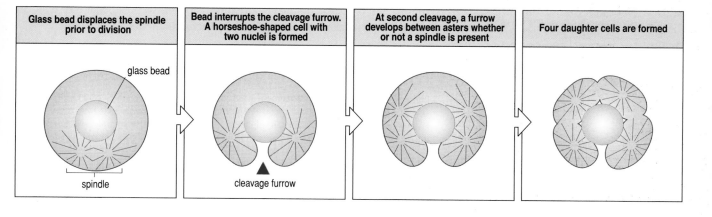

| Glass bead displaces the spindle prior to division | Bead interrupts the cleavage furrow. A horseshoe-shaped cell with two nuclei is formed | At second cleavage, a furrow develops between asters whether or not a spindle is present | Four daughter cells are formed |

glass bead

spindle

cleavage furrow

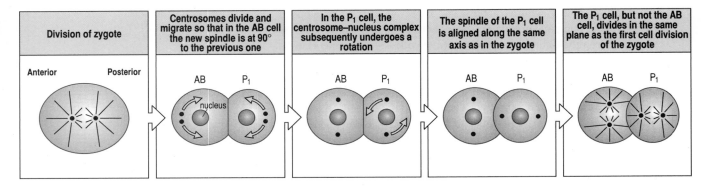

Fig. 8.6 Different cells have different planes of cleavage because of the behavior of their centrosomes. In the nematode, the first cleavage of the zygote divides the cells into an anterior AB cell and a posterior P_1 cell. At the next division, the centrosomes of the two cells move in different directions. In the AB cell, the duplicated centrosomes move so that the next cleavage is at right angles to the first cleavage. In the P_1 cell, the nucleus and duplicated centrosomes rotate so that the cleavage of P_1 is in the same plane as the first cleavage. After Strome, S.: 1993.

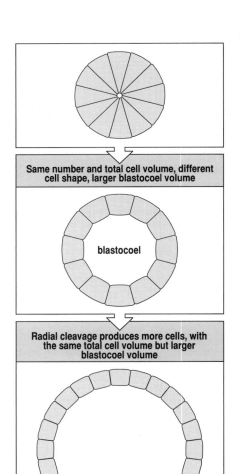

Same number and total cell volume, different cell shape, larger blastocoel volume

blastocoel

Radial cleavage produces more cells, with the same total cell volume but larger blastocoel volume

The asters in a dividing cell are composed of microtubules radiating out from a centrosome, which acts as an organizing center for microtubule growth. In most animal cells, the centrosome contains a pair of centrioles, which consist of an array of microtubules. Before mitosis, the centrosome becomes duplicated and the daughter centrosomes move to opposite sides of the nucleus and form asters. In the cells of the early embryo, they usually take up positions that cause the plane of cleavage of the cell to be at right angles to the plane of the previous cleavage. An example of this is cleavage of the AB cell, one of the pair of cells formed by cleavage of the nematode zygote (Fig. 8.6). By contrast, the other member of the pair, the P_1 cell, exemplifies the fact that there are many cases where successive cleavages are not at right angles to each other. This is a result of local cytoplasmic factors or cytoskeletal components of the cell causing some centrosome pairs to take up a different orientation with respect to the previous plane of cleavage.

Planes of cleavage can play a very important part in determining the form of the embryo, particularly in early animal and plant development. Higher plant cells lack obvious centrosomes and their spindles lack asters, and instead of a contractile mechanism pinching the cell in two, a new cell wall forms in the plane of division. The plane in which the new cell wall forms is not determined, it seems, by the orientation of the mitotic spindle, but is instead defined before mitosis begins by the appearance of a circumferential band of microtubules and actin filaments.

Fig. 8.7 Cell packing can determine the volume of a blastula. Top panel: when adjacent cells in a spherical sheet, such as those of a blastula, make contact with each other over large areas of cell surface, the overall volume of the cell sheet is relatively small as there is little space at its center. Middle panel: with a decrease in the area of cell contact, the size of the internal space (the blastocoel), and thus the overall volume of the blastula, is greatly increased without any corresponding increase in cell numbers or total cell volume. Bottom panel: if the number of cells is increased by radial cleavage, and the packing remains the same, the blastocoel volume will increase further, again without any increase in total cell volume.

8.4 Cells become polarized in early mouse and sea urchin blastulas

Successive cleavages usually result in the formation of a sheet of cells enclosing a hollow fluid-filled blastula. Cell packing can be important in determining the form of the early embryo, as can the planes of cleavage (Fig. 8.7 opposite). If all the planes of cleavage are radial, that is at right angles to the surface, then the cells remain in a single layer and the volume inside increases at each division. This occurs in the sea urchin blastula. In both the sea urchin blastula and the mammalian morula, the cells of the epithelial sheet become radially polarized, with distinct differences between the outer (apical) face and the inner (basal) surface facing into the blastocoel.

In the sea urchin, cleavage eventually produces a hollow spherical blastula composed of a polarized epithelium one cell thick (Fig. 8.8). All planes of cleavage are radial to the surface, thus maintaining a single layer of cells. The surface of the egg, which is covered in microvilli, becomes the outer surface of the blastula and develops an external layer of extracellular matrix, the hyaline layer, to which the cells are attached. Specialized junctions that include desmosomes (see Box 8A, p. 255) develop between adjacent cells and the cell contents become polarized, with the Golgi apparatus oriented toward the apical (outer) surface. On the inner surface of the cell layer, a basal lamina (an organized layer of extracellular matrix) is laid down.

The volume of the hollow interior of the blastula, the blastocoel, increases with each cleavage. While this may partly be due to fluid entering the blastocoel (see Section 8.5), it is also a result of the cells becoming smaller at each division while the total cell mass remains the same. At each cleavage the number of cells increases, the cells become smaller, and the cell layer thinner. Thus, both the surface area of the blastula and the size of the blastocoel increase (see Fig. 8.7).

In the mouse embryo, the earliest sign of structural differentiation takes place at the eight-cell stage when the embryo undergoes **compaction** (Fig. 8.9). The blastomeres flatten against each other, thus maximizing cell–cell contacts, and the microvilli, which hitherto were uniformly distributed on cell surfaces, become confined to the apical surface of the cells (Fig. 8.10). Subsequently, some cleavages in the outer cell layer occur tangentially (parallel to the surface), each producing one polarized and one

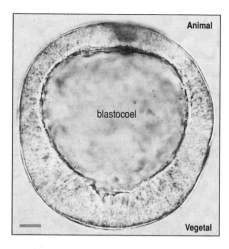

Fig. 8.8 Blastula of a sea urchin embryo. A single layer of cells surrounds the hollow blastocoel. Scale bar = 10 μm.

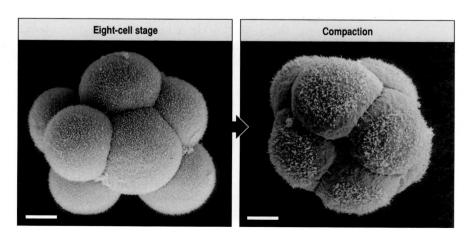

Fig. 8.9 Compaction of the mouse embryo. At the eight-cell stage the cells have relatively smooth surfaces and microvilli are distributed uniformly over the surface. At compaction, microvilli are confined to the outer surface and cells increase their contact with one another. Scale bar = 10 μm.
Photographs courtesy of T. Bloom, from Bloom, T.L.: 1989.

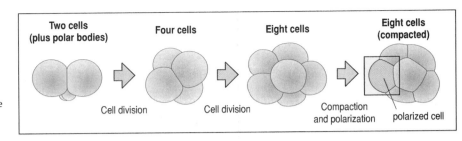

Fig. 8.10 Polarization of cells during cleavage of the mouse embryo. At the eight-cell stage, compaction takes place, with the cells forming extensive contacts (top panel). The cells also become polarized; for example, the microvilli, which were initially uniformly distributed over the cell surface, become confined to the outward-facing cell surface. Future cleavages can thus divide the cell either into two polarized cells (by a radial cleavage, bottom left panel) or into a polarized and a nonpolarized cell (by a tangential cleavage, bottom right panel). The cleavages give rise to the inner cell mass.

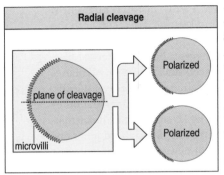

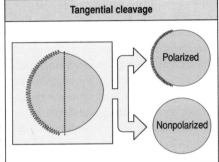

nonpolarized daughter cell. Radial cleavages give two polarized cells. The nonpolarized cells become the inner cell mass, which will give rise to the embryo proper, whereas the outer cells become the trophectoderm and form extra-embryonic structures.

The changes in intercellular contacts that bring about compaction are probably caused by changes in the association between the cell adhesion molecule E-cadherin (uvomorulin) and the underlying cell cortex. At the two-cell and four-cell stages, E-cadherin is uniformly distributed over blastomere surfaces and contact between cells is not extensive. Only at the eight-cell stage does it become restricted to regions of intercellular contact, where it presumably now acts for the first time as an adhesion molecule. This change in the adhesive properties of E-cadherin may involve an association between the E-cadherin and cytoskeletal elements in the cell cortex (see Box 8A, p. 255). Activation of E-cadherin may involve transduction of a signal and the action of protein kinase C. If this kinase is activated before the eight-cell stage, compaction occurs prematurely.

A radical remodeling of the cell cortex is associated with compaction and the localization of E-cadherin and the microvilli. The cytoskeletal proteins actin, spectrin, and myosin are cleared from the region of intercellular contact and become concentrated in a band around the apical region of the cell. It has been proposed that both the cell flattening and the redistribution of the cortical elements may be brought about by the contraction of actin filaments, which draws the cortical elements to the apical pole.

8.5 Ion transport is involved in fluid accumulation in the frog blastocoel

Accumulation of fluid in the interior of the blastula exerts an outward pressure on the blastocoel wall, and this hydrostatic pressure is one of the forces involved in forming and maintaining the spherical shape of

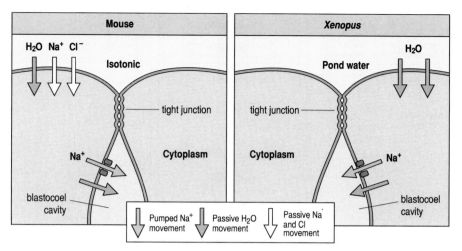

Fig. 8.11 Hydrostatic pressure can inflate the blastula. Active transport of sodium ions (Na$^+$) into the blastocoel cavity of the mouse embryo results in water and salts flowing in from the surrounding isotonic solution, creating positive hydrostatic pressure. The blastocoel cavity is sealed off from the outside medium by tight junctions between the cells at their outer edge. The *Xenopus* embryo develops in pond water, so there is no influx of ions and the supply of sodium ions probably comes from the cells themselves, where sodium is stored in an ionically inactive form.

the blastula. In the development of mammalian embryos, there is evidence that the blastocoel fluid is formed by a mechanism that involves active pumping of sodium ions into the blastocoel (Fig. 8.11). Tight junctions between cells of the outer layer appear at the eight-cell stage, and by the 32-cell stage they act as a permeability barrier to the passage of materials across the epithelial cell layer. At the same time, sodium pumps become active in the cell membranes that face the blastocoel. There appears to be a net inward flux of sodium ions: possibly, as in the frog skin, sodium enters passively at the outer face of the cell and is then pumped into the blastocoel across the inner face of the cell. As the ion concentration in the blastocoel fluid increases, water flows into the blastocoel by osmosis and the increase in hydrostatic pressure stretches the surrounding tissue layers. A similar mechanism seems to operate in the *Xenopus* blastula. In this case, as the embryo develops in pond water, which contains very little sodium, the sodium that is pumped into the blastocoel probably comes from the cells themselves, where it is stored in an ionically inactive form.

8.6 Internal cavities can be created by cell death

There are several ways of creating a hollow structure in early development. In the case of the neural tube, a tubular structure can result from the rolling up of an epithelial sheet. Another way of forming an internal space is by creating a cavity in a solid structure. One example of this is the formation of the epithelium of the epiblast in the early mouse embryo.

In the early mouse embryo, the inner cell mass gives rise to the epiblast, which is pushed across the blastocoel (see Section 2.3). Initially the epiblast is a solid mass of cells, but it later turns into an epithelial layer enclosing a fluid-filled cavity. Formation of this cavity results from programmed cell death—**apoptosis**—of the cells in the center of the epiblast (Fig. 8.12). It is likely that the surrounding cell layers—the visceral endoderm—send a signal to die to all the cells of the epiblast, and that only outer cells in contact with the basement membrane, via integrins, survive. The phenomenon of programmed cell death and its role in development is discussed further in later chapters.

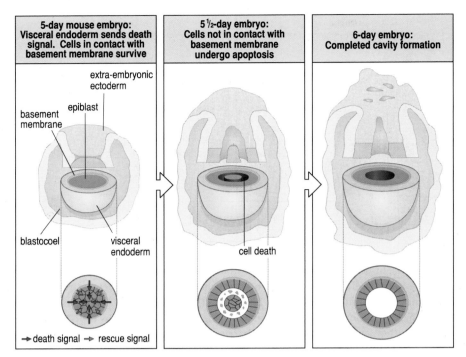

Fig. 8.12 Cavity formation in the epiblast of the mouse embryo. The inner cell mass proliferates and forms a solid mass surrounded by the visceral endoderm. Programmed cell death—apoptosis—in the epiblast results in the formation of a cavity. A signal to die may come from the visceral endoderm; only cells attached to the basement membrane receive a rescue signal and so survive. After Coucouvanis, E., *et al.*: 1995.

Summary

In many animals, the fertilized egg undergoes a cleavage stage that divides the egg into a number of small cells (blastomeres) and eventually gives rise to a hollow blastula. Various patterns of cleavage are found in different animal groups. In some animals, such as nematodes and molluscs, the plane of cleavage is of importance in determining the position of particular blastomeres in the embryo, and the distribution of cytoplasmic determinants. The plane of cleavage in animal cells is determined by the orientation

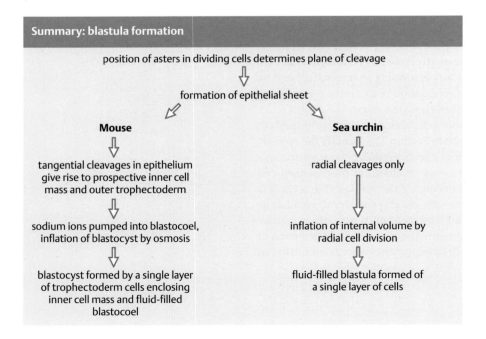

of the mitotic spindle, which in turn is due to the action of the asters, whose position is specified by the chromosomes. At the end of the cleavage period, the blastula essentially consists of a polarized epithelium surrounding a fluid-filled blastocoel. Fluid accumulation in the blastocoel may be partly due to active ion transport. Cavity formation in the early mouse epiblast is caused by cell death.

Gastrulation

Movements of cells and cell sheets at gastrulation bring most of the tissues of the blastula (or its equivalent stage) into their appropriate position in relation to the body plan. The necessity for gastrulation is clear from fate maps of the blastula in, for example, sea urchins and amphibians, where the presumptive endoderm, mesoderm, and ectoderm can be identified as adjacent regions in the epithelial sheet. After gastrulation, these tissues become completely rearranged in relation to each other: for example, the endoderm develops into an internal gut and is separated from the outer ectoderm by a layer of mesoderm. Gastrulation thus involves dramatic changes in the overall structure of the embryo, converting it into a complex three-dimensional structure. During gastrulation, a program of cell activity involving changes in cell shape and adhesiveness remodels the embryo, so that the endoderm and mesoderm move inside and only ectoderm remains on the outside. The primary force for gastrulation is provided by cell motility (Box 8B, 266). In some embryos, all this remodeling occurs with little or no accompanying increase in cell number or total cell mass.

In this section, we first consider gastrulation in sea urchins and insects, in which it is a relatively simple process. The mechanisms of gastrulation in chick and mouse are not well understood, and so *Xenopus* serves as our model for the more complex gastrulation process in vertebrates.

8.7 Gastrulation in the sea urchin involves cell migration and invagination

Just before gastrulation begins, the late sea urchin blastula consists of a single layer of cells surrounding a central fluid-filled blastocoel. The future mesoderm occupies the most vegetal region, with the future endoderm adjacent to it (Fig. 8.13). The rest of the embryo gives rise to ectoderm. On their outward-facing surface the cells are attached to an extracellular

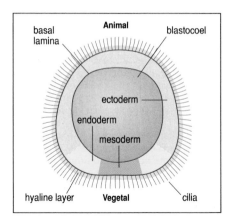

Fig. 8.13 The sea urchin blastula before gastrulation. The prospective endoderm and mesoderm are at the vegetal pole. There is an extracellular hyaline layer and a basal lamina lines the blastocoel.

Box 8B Change in cell shape and cell movement

Cleaving cell	Apical constriction	Migrating cell
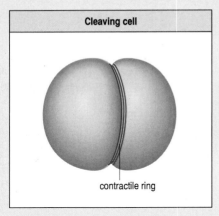	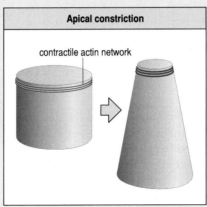	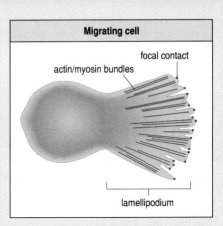

Cells actively undergo major changes in shape during development. The two main changes are associated with cell migration and with the infolding of epithelial sheets. Changes in shape are generated by the cytoskeleton, an intracellular protein framework that controls cell shape and is also involved in cell movement. There are three principal types of protein polymers in the cytoskeleton—**actin filaments (microfilaments)**, **microtubules**, and **intermediate filaments**—as well as many other proteins that interact with them. Actin filaments are primarily responsible for force generation and contractions within cells that lead to a change in shape. Microtubules play an important part in maintaining cell asymmetry and polarity, and intermediate filaments can transmit mechanical forces and provide mechanical stability.

Actin filaments are fine threads of protein about 7 nm in diameter and are polymers of the globular protein actin. They are organized into bundles and three-dimensional networks, which in most cells lie predominantly just beneath the plasma membrane, forming the gel-like cell cortex. Numerous actin-binding proteins are associated with actin filaments, and are involved in bundling them together, forming networks, and aiding the polymerization and depolymerization of the actin subunits. Actin filaments can form rapidly by polymerization of actin subunits and can be equally rapidly depolymerized. This provides the cell with a highly versatile system for assembling actin filaments in a variety of different ways and in different locations, as required. The fungal drug cytochalasin D can prevent actin polymerization and is a useful agent for investigating the role of actin networks. Actin filaments can assemble with myosin into contractile structures, which act as miniature muscles. Such contractile bundles are present in the contractile ring of a dividing animal cell (left panel), which separates the cell into two daughter cells. Localized contraction of an actin network in the cell cortex can result in apical constriction of the cell and its longitudinal elongation (center panel).

Many embryonic cells are capable of movement over a solid substratum. They move by extending a thin sheet-like layer of cytoplasm known as a lamellipodium (right panel), and long fine cytoplasmic processes called filopodia. These temporary structures contain actin filaments, which are assembled as the process extends. It is likely that contraction of the actin network elsewhere then draws the cell forward. To do this, the contractile system must be able to exert a force on the substratum and this occurs at focal contacts, points at which the advancing filopodia or lamellipodium are anchored to the substratum over which the cell is moving. *In vivo* this will often be a layer of extracellular matrix. At focal contacts, integrins (see Box 8A, p. 255) both adhere to extracellular matrix molecules through their extracellular domains, and provide an anchor point for actin filaments through their cytoplasmic domains. It is likely that integrins can mediate signal transduction across the plasma membrane at focal contacts, thus providing the cell with a means of sensing the environment and controlling cell movement.

The family of small GTP-binding proteins that includes Rho, Rac and Cdc42 has a key role in regulating the actin cytoskeleton. Rac is required for extension of lamellipodia, for example, and Cdc42 is necessary to maintain cell polarity. Signals from the environment are relayed to the cytoskeleton via these proteins and they can thus control cell movement.

hyaline layer, which contains the proteins hyalin and echinonectin. Cells are attached to adjacent cells by junctional complexes in the lateral membranes, and form a polarized epithelium. Cadherins and β-catenin are associated with lateral cell–cell contacts and accumulate at adherens junctions (belt desmosomes) from cleavage stages onwards. A basal lamina of extracellular matrix is present on the inner surface of the cells facing into the blastocoel.

Gastrulation begins with an **epithelial to mesenchymal transition**,

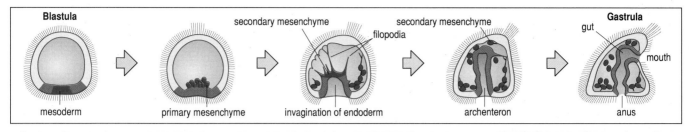

with the most vegetal mesodermal cells becoming motile and mesenchymal in form. These primary mesenchyme cells become detached from each other and from the hyaline layer, and migrate into the blastocoel as single cells (Fig. 8.14), which have lost both their epithelial polarity and their cuboid shape. The transition to mesenchyme cells and their entry into the blastocoel is foreshadowed by intense pulsatory activity on the inner face of these cells, and on occasion a small transitory infolding or invagination is seen in the surface of the blastula before migration begins properly. The cell's entry requires loss of adhesion and is thus associated with the loss of α- and β-catenins, and the endocytosis of cadherin. Having entered the blastocoel, the mesenchyme cells migrate to sites within the blastula. We consider this directed migration in Section 8.15.

The entry of the primary mesenchyme is followed by the invagination and extension of the endoderm to form the embryonic gut (the **archenteron**). The endoderm invaginates as a continuous sheet of cells (see Fig. 8.14). Formation of the gut occurs in two phases. During the initial phase, the endoderm invaginates to form a short, squat cylinder extending up to half-way across the blastocoel. There is then a short pause, after which extension continues. In this second phase, the cells at the tip of the invaginating gut, which will later detach as the secondary mesenchyme, form long filopodia, which make contact with the blastocoel wall. Filopodial extension and contraction pull the elongating gut across the blastocoel until it eventually comes in contact with and fuses with the mouth region, which forms a small invagination on the ventral side of the embryo (see Fig. 8.14). During this process the number of invaginating cells doubles, and this is in part due to cells around the site of invagination contributing to the hindgut.

How is invagination of the endoderm initiated? The simplest explanation is that a change in shape of the endodermal cells causes, and initially maintains, the change in curvature of the cell sheet (Fig. 8.15). At the site of invagination, the initially cuboid cells adopt a more elongated, wedge-shaped form, which is narrower at the outer (apical) face. This change in cell shape is the result of constriction of the cell at its apex, probably by contraction of cytoskeletal elements. The change in cell shape is initially

Fig. 8.14 Sea urchin gastrulation. Cells of the vegetal mesoderm undergo the transition to primary mesenchyme cells and enter the blastocoel at the vegetal pole. This is followed by an invagination of the endoderm, which extends inside the blastocoel toward the animal pole, forming a clear archenteron. Filopodia extending from the secondary mesenchyme cells at the tip of the invaginating endoderm contact the blastocoel wall and draw the invagination to the site of the future mouth, with which it fuses, forming the gut. Scale bar = 50 μm.

Photographs courtesy of J. Morrill.

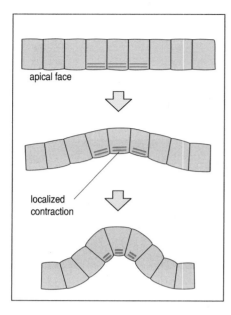

Fig. 8.15 Change in cell shape in a small number of cells can cause invagination of endodermal cells. Bundles of filaments composed of actin and other motor proteins contract at the outer edge of the cells, making them wedge-shaped. As long as the cells remain mechanically linked to adjacent cells in the sheet, this local change in cell shape draws the sheet of cells inward at that point.

Fig. 8.16 Computer simulation of the role of apical constriction in invagination. Computer simulation of the spreading of apical constriction over a region of a cell sheet shows how this can lead to an invagination. Illustration after Odell, G.M., *et al.*: 1981.

sufficient to pull the outer surface of the cell sheet inward and to maintain invagination, as shown by a computer simulation in which apical constriction, modeled to spread over the vegetal pole, results in an invagination (Fig. 8.16).

Other mechanisms have been proposed to account for the primary invagination. These include secretion of chondroitin sulfate, leading to the swelling of the apical lamina and thus buckling of the vegetal plate, and contraction of the cells along the apico-basal axis, which would cause the cells to round up and so buckle the plate inwards.

The primary invagination only takes the gut about a third of the way to its final destination. The second phase of gastrulation involves two different mechanisms. The filopodia extending from the secondary mesenchyme cells at the tip of the invaginating gut make contact with the blastocoel wall, and their contraction pulls the gut across the blastocoel. Treatments that interfere with filopodial attachment to the blastocoel wall result in failure of the gut to elongate completely, but it still reaches about two-thirds of its complete length. The filopodial-independent extension is due to active rearrangement of cells within the endodermal sheet. If a sector of cells in the vegetal mesoderm is labeled with a fluorescent dye before gastrulation, this sector is seen to become a long, narrow strip as the gut extends (Fig. 8.17). We look at this type of active cell rearrangement—known as convergent extension—in more detail later in this chapter, in relation to similar phenomena in amphibian gastrulation. In the sea urchin, this cell rearrangement requires the interaction of cells with the basal lamina. If antibodies against a basal lamina component are injected into the blastocoel, the second phase of gastrulation is blocked.

During extension the gut makes contact with the future mouth region. What guides the tip of the gut to this region? As the long filopodia at the tip of the gut initially explore the blastocoel wall, they make more stable contacts at the animal pole, where the future mouth will form. Filopodia making contact there remain attached for 20–50 times longer than when attached to other sites on the blastocoel wall. Similar differences in adhesiveness enable mesoderm cells to migrate to their correct positions. Gastrulation in the sea urchin clearly shows how changes in cell shape, changes in cell adhesiveness, and cell migration all work together to cause a major change in embryonic form.

8.8 Mesoderm invagination in *Drosophila* is due to changes in cell shape, controlled by genes that pattern the dorso-ventral axis

At the time gastrulation begins, the *Drosophila* embryo consists of a blastoderm of about 6000 cells, which forms a superficial single-celled layer (see Section 2.5). Gastrulation begins with the invagination of a longitudinal strip of future mesodermal cells (8–10 cells wide) on the ventral side of the

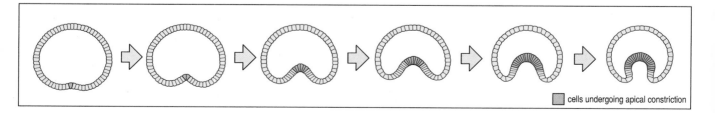

cells undergoing apical constriction

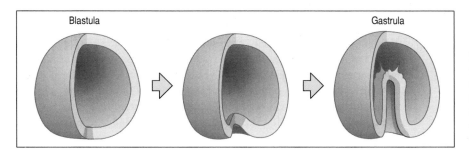

Fig. 8.17 **Gut extension during sea urchin gastrulation.** Labeling of cells in the vegetal region of the blastula shows that gut extension involves cell rearrangement within the endoderm, causing the labeled cells to become distributed in a long narrow strip.

embryo, to form a ventral furrow and then a tube. This breaks up into individual cells, which spread out to form a single layer of mesoderm on the interior face of the ectoderm (Fig. 8.18). The gut develops slightly later, by invaginations of prospective endoderm near the anterior and posterior ends of the embryo.

Invagination of the mesoderm is rapid: formation of the mesodermal tube takes about 30 minutes and cell spreading about an hour. The invagination of the mesoderm initially forms a furrow, which in cross-section looks remarkably similar to the primary invagination of the endoderm in sea urchin gastrulation. Invagination appears to occur in two phases: during the first phase, the central strip of cells develop flattened and smaller apical surfaces, possibly by apical contraction, and their nuclei move away from the periphery. This results in a ventral furrow which folds into the interior of the embryo to form a tube, the central cells forming the tube proper while the peripheral cells form a 'stem'; during the second phase, the tube dissociates into individual cells which proliferate and spread out laterally. The entry of cells into cell division is delayed during invagination and this delay is essential for the cell shape changes to occur.

Mutant *Drosophila* embryos that have been dorsalized or ventralized (see Section 5.8) show that this behavior of the mesodermal cells is autonomous, as in the sea urchin, and is not affected by adjacent tissues. In dorsalized embryos, which have no mesoderm, no cells undergo nuclear migration or apical contraction. In ventralized embryos, in which most of the cells are mesoderm, these changes occur throughout the whole of the dorso-ventral axis.

Gastrulation in *Drosophila* provides us with the possibility of linking gene action with change in cell shape during morphogenesis. Mesoderm invagination is affected by mutations in the genes *twist* and *snail*, which are expressed in the prospective mesoderm before gastrulation, and which encode transcription factors (see Section 5.8). In *twist* mutants, a small transient furrow is formed, and in *snail* mutants the prospective mesodermal

Fig. 8.18 ***Drosophila* gastrulation.** Mesodermal cells (stained), in a longitudinal strip on the ventral side, change shape and cause an invagination, which results in a ventral furrow (left panel). Further apical contraction of the cells results in the mesoderm forming a tube on the inside of the embryo (center panel). The mesodermal cells then start to migrate individually to different sites (right panel). Scale bars = 50 µm.

Photographs courtesy of M. Leptin, from Leptin, M., et al.: 1992.

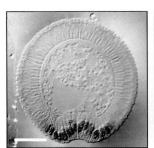
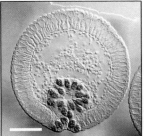
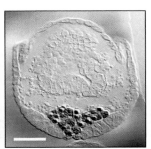

cells flatten, but no other change occurs. Double mutants of these genes show no cell shape changes and no invagination. The transcription factors encoded by these genes may therefore be controlling, either directly or indirectly, the expression of cell components such as cytoskeletal proteins that are required for the shape changes to occur. The Rho-mediated signaling pathway, which is required for changes in cell shape, is involved, as mutations in this pathway result in failure to gastrulate.

One potential downstream target for *twist* and *snail* is the gene *folded gastrulation*. It codes for a secreted protein of unknown function, which is expressed in the invaginating cells in a pattern that precisely precedes that of apical constriction. Overexpression of the gene results in ectopic constriction appearing outside the normal zone. Thus, *folded gastrulation* might be a signal for apical constriction, and its discovery at least helps bring us closer to understanding how the actions of transcription factors involved in patterning can initiate localized cell constrictions.

twist and *snail* also control all subsequent steps of mesoderm development, including expression of receptors for fibroblast growth factor (FGF). Signaling via the FGF receptors is needed for the invaginating mesodermal cells to be able to spread out and cover the inside of the ectoderm. When the central part of the mesoderm is fully internalized, its cells make contact with the ectoderm. Adherens junctions that had been established in the blastoderm are now lost, and the mesoderm dissociates into single cells which spread out to cover the ectoderm as a single-cell layer. Failure to activate the FGF signaling pathway results in the mesodermal cells remaining near the site of invagination.

One possibility is that mesodermal cell spreading is caused or facilitated by the mesodermal cells switching from expression of ectodermal E-cadherin to N-cadherin, so that they no longer adhere to the ectodermal cells. The change from E- to N-cadherin is under the control of *snail* and *twist*. The actions of *snail* suppress the production of E-cadherin in the mesoderm, while expression of *twist* induces the production of N-cadherin.

8.9 Dorsal closure in *Drosophila* and ventral closure in *C. elegans* are brought about by the action of filopodia

At about 11 hours after development of the *Drosophila* embryo, when gastrulation and germ band retraction are complete, the amnioserosa on the dorsal surface of the embryo is not yet covered by the epidermis (the amnioserosa will eventually disappear). The epidermis then moves to close this gap, zipping-up the opening from both ends (Fig. 8.19). About 2 hours after this starts, the two epithelial fronts have fused to form a seam along the dorsal midline, with the segment pattern exactly matched. The movement of the epidermis is due to the action of filopodia and lamellipodia, which extend and retract from the cells along the edges of the epithelial sheet.

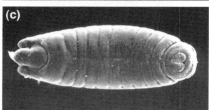

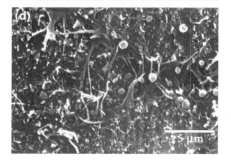

Fig. 8.19 Dorsal closure in *Drosophila*. The closure of the epidermis along the dorsal midline over a period of about 2 hours is shown in the micrographs in panels a–c. Starting at both ends of the opening, the edges of the epidermis come together over the amnioserosa (seen as the central oval area in a and b) and fuse to form the midline seam, matching the segment pattern on either side. Panel d is an electron micrograph of filopodia in the region where the edges of the epidermal cell sheet are coming together.

Photographs courtesy of A. Jacinto, from Jacinto, A., et al.: 2000.

Filopodia are active along the whole epithelial front, extending up to 10 μm from the edge, and their behavior is similar to that seen in sea urchin gastrulation (see Section 8.7). Laser ablation of filopodial cells, especially at the site of 'zippering', prevents dorsal closure. This suggests that the main force for closure is exerted at this site by the filopodia drawing the two edges together like a zip.

The *C. elegans* embryo undergoes a process similar to dorsal closure, but on the ventral surface. At the end of gastrulation, the epidermis only covers the dorsal region and the ventral region is bare. The epidermis then spreads around the embryo until its edges finally meet along the ventral midline. Time-lapse studies show that the initial ventral migration is led by four cells, two on each side, which extend filopodia towards the ventral midline. Blocking filopodial activity by laser ablation or by cytochalasin D, which blocks the formation of actin filaments, prevents ventral closure, showing that the filopodia provide the driving force. Closure probably occurs by a 'zipper' mechanism similar to that in the fly.

8.10 *Xenopus* gastrulation involves several different types of tissue movement

Gastrulation in amphibians involves a much more dramatic and complex rearrangement of the tissues than occurs in sea urchins, mainly because of the large amount of yolk present. But the outcome is the same—the transformation of a two-dimensional sheet of cells into a three-dimensional embryo, with ectoderm, mesoderm, and endoderm in the correct positions for further development of body structure. The main processes of amphibian gastrulation that we will discuss are **involution**, which is the rolling-in of the endoderm and mesoderm at the blastopore, **convergent extension** of the mesoderm, and **epiboly**, which is the spreading of the ectoderm as the endoderm and mesoderm move inside (Fig. 8.20). In the late *Xenopus* blastula, the presumptive endoderm extends from the most vegetal region to cover the presumptive mesoderm. During gastrulation, the presumptive endoderm moves inside through the blastopore to line the gut (Fig. 8.21). The blastula wall is several cells thick and the future mesoderm is present in the marginal zone as an equatorial band lying beneath presumptive endoderm. All the cells in this equatorial band move inside (see Fig. 8.20), to underlie the ectoderm extending antero-posteriorly along the dorsal midline of the embryo. In urodele amphibians, such as the newt, the presumptive mesoderm is not overlain by presumptive endoderm in the marginal zone, but is exposed on the outer surface.

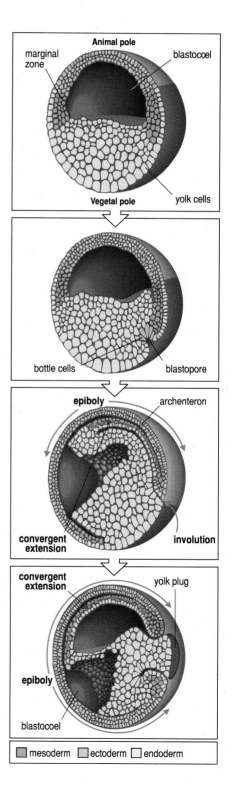

Fig. 8.20 Tissue movements during gastrulation of *Xenopus*. In the late blastula the future mesoderm (red) is in the marginal zone, overlaid by presumptive endoderm. Gastrulation is initiated by the formation of bottle cells in the blastopore region, which is followed by the involution of mesoderm over the dorsal lip of the blastopore. Marginal zone endoderm and mesoderm move inside over the dorsal lip of the blastopore (events in this region are shown in more detail in Fig. 8.21). The marginal zone endoderm, which was on the surface of the blastula, now lies ventral to the mesoderm and forms the roof of the archenteron. At the same time, the ectoderm of the animal cap spreads downward. The mesoderm converges and extends along the antero-posterior axis. The region of involution spreads ventrally to include more endoderm, and forms a circle around a plug of yolky vegetal cells. After Balinsky, B.I.: 1975.

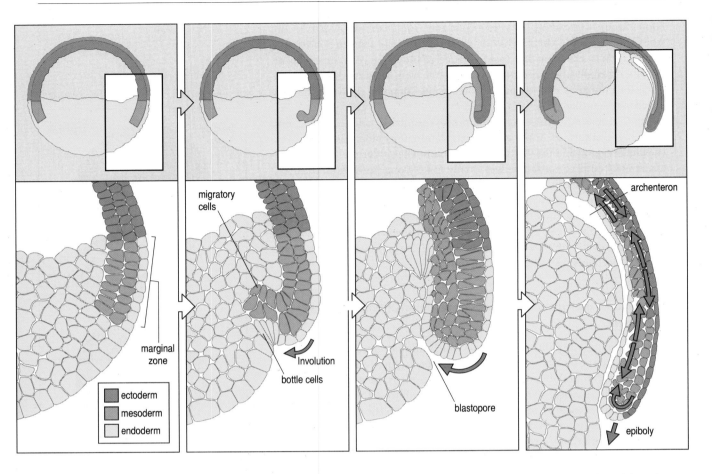

Fig. 8.21 Tissue movement in the dorsal region during formation of the blastopore and gastrulation in *Xenopus*. First panel: late blastula before gastrulation. In the marginal zone, presumptive endoderm overlies the presumptive mesoderm. Second panel: cells at the site of the blastopore undergo apical constriction and elongate, causing the involution of surrounding cells and the formation of a groove which defines the dorsal lip of the blastopore. Third panel: as gastrulation proceeds, the presumptive endoderm and mesoderm involute and then converge and extend under the ectoderm. Fourth panel: the archenteron—the future gut lined by endoderm—starts to form and the mesoderm converges and extends. At the leading edge of the mesoderm are migratory cells, which give rise to the mesoderm of the head. The ectoderm moves down by epiboly to cover the whole embryo. After Hardin, J.D., *et al.*: 1988.

Gastrulation in *Xenopus* starts at a site on the dorsal side of the blastula, toward the vegetal pole. The first visible sign of gastrulation is the formation of bottle-shaped cells (bottle cells, see Fig. 8.21, second panel) by some of the presumptive mesodermal cells. As in the sea urchin and *Drosophila*, this change in cell shape causes a small groove in the blastula surface—the **blastopore**—and defines the dorsal lip of the blastopore, which corresponds to the Spemann organizer (see Section 3.4). The layer of mesoderm and endoderm starts to roll in around the blastopore, and into the interior of the blastula. The movement starts on the dorsal side but spreads laterally and vegetally, eventually forming a circular blastopore. This rolling under of a sheet of cells against the inner surface of the sheet (see Fig. 8.21) is referred to as involution.

The first mesodermal cells to involute eventually migrate as individual cells over the blastocoel roof to give rise to very anterior mesodermal structures in the head region. Behind them is mesoderm that enters, together with the overlying endoderm, as a single multilayered sheet. For this sheet of presumptive mesodermal cells, going through the narrow blastopore is rather like going through a funnel, and the cells become rearranged by convergent extension, which is a major feature of gastrulation. The mesoderm is initially in the form of an equatorial ring, but during gastrulation it converges and extends along the antero-posterior axis, hence the term convergent extension (Fig. 8.22). Thus, cells which initially are on opposite sides of the embryo come to lie next to each other (Fig. 8.23).

In *Xenopus*, convergent extension occurs in both the mesoderm and endoderm as they involute, as well as in the future neural tissue overlying the mesoderm, which gives rise to the spinal cord. One can appreciate the dramatic nature of convergent extension of the mesoderm by viewing the early gastrula from the blastopore. The future mesoderm can be identified by the expression of the gene *Brachyury*, and can be seen as a narrow ring around the blastopore (Fig. 8.24). As this ring of tissue expressing *Brachyury* enters the gastrula it converges into a narrow band of tissue along the dorsal midline of the embryo, extending in the antero-posterior direction and only the prospective notochord continues to express *Brachyury*.

During gastrulation the future notochord begins to separate from the future somites, which lie on either side of it. The layer of endoderm which, as it rolls over the blastopore lip lies below the mesoderm, also converges and extends in a similar way. The mesoderm then lies immediately beneath the ectoderm, with the endoderm lining the roof of the archenteron, the future gut (see Fig. 8.21). While the mesoderm is in contact with the blastocoel roof the tissues do not fuse; this is due to cell repulsion mediated by cadherins. As the mesodermal and endodermal cells move inside the embryo, the ectoderm of the animal cap region increases in surface area by a rearrangement and stretching of the cells within it to form a thinner sheet, which spreads down over the vegetal region. This spreading of a tissue is referred to as epiboly.

As gastrulation proceeds, the region of involution spreads laterally and vegetally so that involution involves the vegetal endoderm and so forms a circle around a plug of yolky cells (see Fig. 8.20). With time, the blastopore contracts, forcing the yolky vegetal cells into the interior where they form the floor of the gut.

8.11 Convergent extension and epiboly are due to cell intercalation

Much has been learnt about the mechanisms of amphibian gastrulation from the use of sandwich explants taken from *Xenopus* early gastrulas (see Fig. 4.23). Two similar pieces of tissue from the dorsal sector of the marginal zone that includes the blastopore lip are placed with their inner surfaces together and cultured on an agarose-coated dish, to which the sandwich does not adhere. After healing, the explant comprises a continuous outer

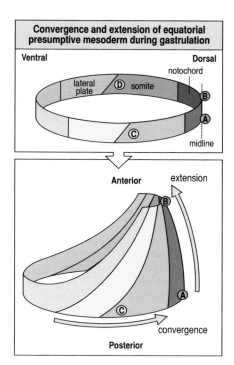

Fig. 8.22 Convergent extension of the mesoderm. Initially the mesoderm is in an equatorial ring, but during gastrulation it converges and extends along the antero-posterior axis. A–D are reference points, used to illustrate the extent of movement during convergent extension. In the bottom panel, D is hidden opposite C.

Fig. 8.23 The rearrangement of mesoderm and endoderm during gastrulation in *Xenopus*. The presumptive mesoderm is present in a ring around the blastula underlying the endoderm, which is shown peeled back. During gastrulation, both these tissues move inside the embryo through the blastopore and completely change their shape, converging and extending along the antero-posterior axis, so that points A and B move apart. Thus, the two points C and D, originally on opposite sides of the blastula, come to lie next to each other. In the neural ectoderm there is also convergent extension, as shown by the movement of points E and F. Note that no cell division or cell growth has occurred in these stages, and all these changes have occurred by rearrangement of cells within the tissues.

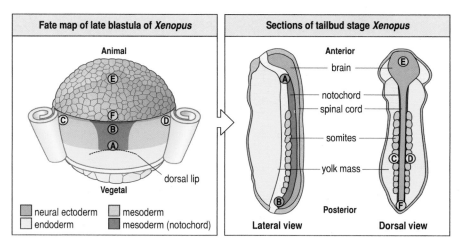

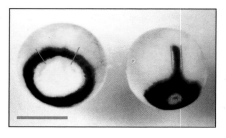

Fig. 8.24 *Brachyury* expression during *Xenopus* gastrulation illustrates convergent extension. Left: before gastrulation, the expression of *Brachyury* (dark stain) marks the future mesoderm, which is present as an equatorial ring when viewed from the vegetal pole. Right: as gastrulation proceeds, the mesoderm that will form the notochord converges and extends along the midline and only notochord cells continue to express *Brachyury*. Scale bar = 1 mm.

Photograph courtesy of J. Smith, from Smith, J.C., et al.: 1995.

epithelial sheet enclosing mesodermal cells. In principle, a single piece of tissue would suffice for these experiments, but in practice such explants curl up.

These explants undergo convergent extension in culture, and after just 4 hours the animal–vegetal dimension has increased fourfold. In these explants, convergent extension occurs, apparently independently, in two regions. One is the endoderm and mesoderm that would normally be undergoing involution; the other is the presumptive ectoderm that gives rise to the neural plate. Detailed observation of these explants during convergent extension has provided crucial information on the mechanisms involved. The key feature is cell intercalation, cells moving in between adjacent cells so that they converge toward the midline of the tissue, causing the whole tissue to narrow and to extend in an antero-posterior direction (Fig. 8.25).

Before considering in detail how cell intercalation brings about convergent extension, it is helpful to compare the two main types of intercalation that operate during gastrulation (Fig. 8.26). **Radial intercalation** occurs in the multilayered ectoderm of the animal cap, in which cells intercalate in a direction perpendicular to the surface. This leads to a thinning of the sheet of cells and an increase in its surface area. This is, in part, the cause of epiboly and the spreading of the ectoderm in the amphibian gastrula to cover the whole embryo as the endoderm and mesoderm move inside. Convergent extension involves **medio-lateral intercalation**. This results not in an increase in the surface area of the tissue, but in a change in its shape, so that it narrows along one axis and elongates along another at right angles. In amphibian gastrulation, convergent extension is along the antero-posterior axis.

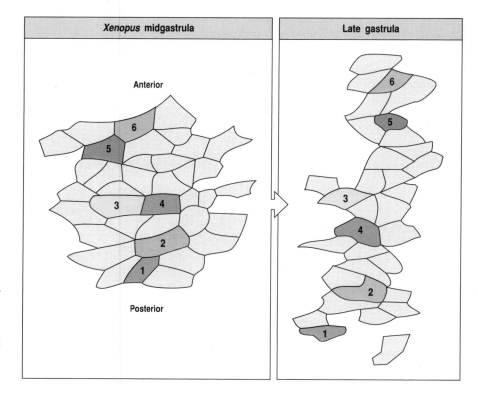

Fig. 8.25 Cell intercalation during convergent extension. In tissue explants from a *Xenopus* gastrula, cells are labeled with a fluorescent dye so that their movements can be followed. As the tissue undergoes convergent extension, the cells move in between each other (intercalate), causing the tissue to narrow and to elongate along the antero-posterior axis. After Keller, R., *et al.*: 1992.

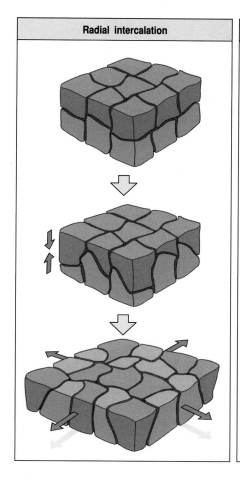

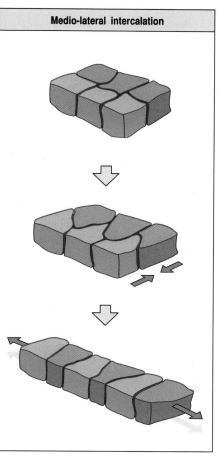

Fig. 8.26 Radial and medio-lateral intercalation. In radial intercalation, the cells intercalate in a direction perpendicular to the surface, producing an increase in surface area of the cell sheet. In medio-lateral intercalation, the sheet of cells narrows and elongates.

During convergent extension the cells take on a characteristic shape, becoming elongated in a direction at right angles to the antero-posterior axis. They also become aligned parallel to one another in a direction perpendicular to the direction of tissue extension (Fig. 8.27). Active movement is largely confined to each end of these elongated bipolar cells, enabling them to exert traction on the underlying substratum and to shuffle in between each other, always along the medio-lateral axis. There are some small protrusions on the anterior and posterior faces of the cells but these seem to hold apposed surfaces together rather than being involved in cell movement.

The boundary of a tissue undergoing convergent extension is maintained by the behavior of cells at the boundary. When an active cell tip is at the boundary of the tissue, its movement ceases and the cell becomes monopolar (see Fig. 8.27), with only the tip facing into the tissue still active. It is as if the cell tip that has reached the boundary has become fixed there. Precisely what defines a boundary is not clear, but boundaries between mesoderm, ectoderm, and endoderm seem to be specified before gastrulation commences. As gastrulation proceeds, a further boundary appears within the mesoderm, demarcating the notochord from the future somites (see Section 4.1).

Convergent extension is caused by contractile forces generated at the tips of the elongated cells. These forces both move the cells to the midline and

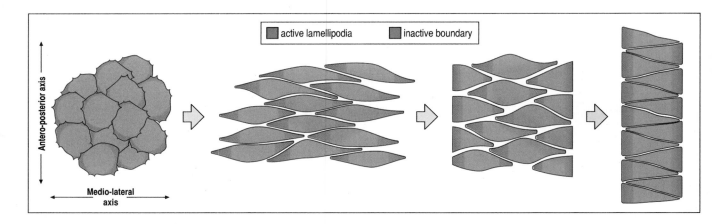

active lamellipodia   **inactive boundary**

Antero-posterior axis

Medio-lateral axis

Fig. 8.27 Cell movements during convergent extension. During convergent extension of the mesoderm in amphibians the cells become elongated in a medio-lateral direction, that is at right angles to the antero-posterior axis. Active movement is confined to the ends of these bipolar cells, which have active lamellipodia, and they shuffle past each other and intercalate. At the tissue boundary, movement ceases. After Keller, R., *et al.*: 1992.

extend the tissue. One can visualize the process of convergent extension in terms of lines of cells pulling on one another in a direction at right angles to the direction of extension. Because the cells at the boundary are trapped at one end, this tensile force causes the tissue to narrow and hence extend anteriorly. This explanation does not, of course, account for the way the cells initially become oriented in a medio-lateral direction; the mechanism by which this occurs remains completely unknown. However, the Wnt pathway, particularly the dishevelled protein, is involved in polarizing the system.

Amphibian gastrulation may seem complex, but it is, in essence, nothing more than a special pattern of cell movement and cell adhesiveness, even though we still do not fully understand how they lead to the observed changes in form. What is not yet understood is how the genes that are expressed in the Spemann organizer control these processes (see Section 3.20). One possible link is their control of a type of cadherin—paraxial protocadherin. This is expressed in the mesoderm and has a role in convergent extension, as shown by the effects of gain and loss of function mutations. This cadherin is switched off in the notochord.

8.12 Notochord elongation is caused by cell intercalation

The presumptive notochord mesoderm is among the first of the tissues to involute during gastrulation, and in the amphibian it is the first of the dorsal structures to differentiate. Initially it can be distinguished from the adjacent somitic mesoderm by the packing of its cells and by the slight gap that forms the boundary between the two tissues, possibly reflecting differences in cell adhesiveness. The notochord elongates considerably and develops into a stiff rod composed of a stack of thin flat cells shaped like pizza slices. The two main mechanisms involved in its morphogenesis are medio-lateral intercalation leading to convergent extension, and directed dilation.

After the initial elongation of the notochordal mesoderm by convergent extension, it undergoes a further dramatic narrowing in width accompanied by an increase in height (Fig. 8.28). At the same time, the cells become elongated at right angles to the main antero-posterior axis, foreshadowing the later pizza-slice arrangement of cells within the notochord (see Fig. 8.28, bottom). The cells intercalate between their neighbors again, thus causing convergent extension. The later stage in notochord elongation, involving directed dilation, is discussed later in this chapter.

Summary

At the end of cleavage the animal embryo is essentially a closed sheet of cells, which is often in the form of a sphere enclosing a fluid-filled interior. Gastrulation, strictly the formation of the gut, converts this sheet into a solid three-dimensional embryonic animal body. During gastrulation, cells move into the interior of the embryo, and the regions of endoderm and mesoderm, which were originally adjacent in the cell sheet, take up their appropriate positions in the embryo. Gastrulation results from a well-defined spatio-temporal pattern of change in cell shape, cell movement, and change in cell adhesiveness, the main forces of which are generated by localized contractions. In the sea urchin, gastrulation occurs in two phases. In the first, changes in cell shape and adhesion result in migration of meso-derm cells into the interior, and invagination of the endodermal part of the cell sheet to form the gut. In the second phase, extension of the gut to reach the mouth region on the opposite side of the embryo occurs by cell rearrangement within the endoderm, which causes the tissue to narrow and lengthen (convergent extension), and by traction of filopodia extend-ing from the gut tip against the blastocoel wall. Invagination of the mesoderm in *Drosophila* occurs by a similar mechanism to that of the inva-gination of endoderm in the sea urchin. Dorsal closure in *Drosophila* and ventral closure in *C. elegans* are driven by filopodial extension and contrac-tion. Gastrulation in *Xenopus* involves more complex movements of cell sheets and also results in elongation of the embryo along the antero-posterior axis. It involves three processes: involution, in which a double-layered sheet of endoderm and mesoderm rolls into the interior over the lip of the blastopore; convergent extension of endoderm and mesoderm in the antero-posterior direction to form the roof of the gut, and the notochord and somitic mesoderm, respectively; and epiboly—spreading of the ecto-derm from the animal cap region to cover the whole outer surface of the embryo. Both convergent extension and epiboly are due to cell intercal-ation, the cells becoming rearranged with respect to their neighbors.

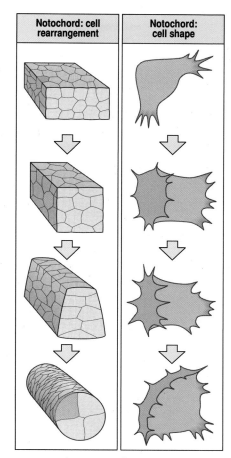

Fig. 8.28 Changes in cell arrangement and cell shape during notochord development. During convergent extension of the notochord, its height increases. At the same time, the cells become elongated. The bottom of the figure shows the eventual arrangement of cells, and their pizza slice shape. After Keller, R., *et al.*: 1989.

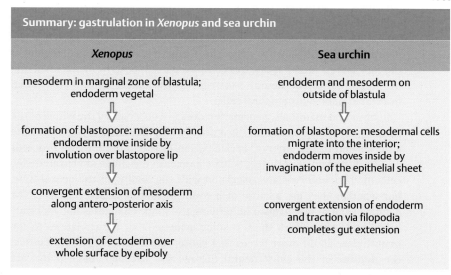

Summary: gastrulation in *Xenopus* and sea urchin	
Xenopus	**Sea urchin**
mesoderm in marginal zone of blastula; endoderm vegetal	endoderm and mesoderm on outside of blastula
⇩	⇩
formation of blastopore: mesoderm and endoderm move inside by involution over blastopore lip	formation of blastopore: mesodermal cells migrate into the interior; endoderm moves inside by invagination of the epithelial sheet
⇩	⇩
convergent extension of mesoderm along antero-posterior axis	convergent extension of endoderm and traction via filopodia completes gut extension
⇩	
extension of ectoderm over whole surface by epiboly	

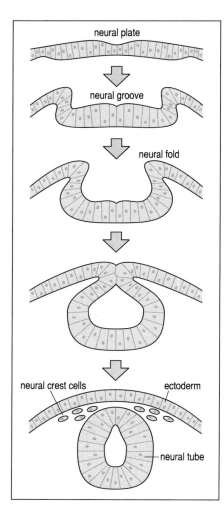

Fig. 8.29 Neural tube formation results from formation and fusion of neural folds. The tube then detaches from the ectoderm.

Neural tube formation

Neurulation in vertebrates results in the formation of the **neural tube**, an epithelial structure composed of ectoderm, which develops into the brain and spinal cord (see Section 2.2). The tissue that gives rise to the neural tube initially appears as a thickened plate of tissue—the **neural plate**—following induction by the mesoderm (see Section 4.7).

In amphibians, there is convergent extension of the future neural plate during gastrulation. There is then a change in the shape of the cells within the plate, so that the edges of the neural plate become raised above the surface, forming two parallel neural folds with a depression—the neural groove—between them (Fig. 8.29). The neural folds eventually come together along the dorsal midline of the embryo and fuse at their edges to form the neural tube, which then separates from the adjacent ectoderm. This surface layer of ectoderm becomes epidermis. Neurulation in birds and mammals proceeds in a similar manner, although the posterior region of the neural tube in both birds and mammals initially forms as a solid rod, and develops a lumen later. In fishes, the neural tube initially forms as a solid rod throughout its length.

8.13 Neural tube formation is driven by cell migration and changes in cell shape

A very early sign of neurulation in *Xenopus* is the formation of the neural groove along the midline. As with gastrulation, changes in cell shape are associated with the curvature of the neural plate and the formation of the folds, but the mechanism is not well understood. During gastrulation, the cells of the neural plate become longer and narrower than the cells of the adjacent ectoderm (Fig. 8.30). As neurulation commences and the plate begins to roll up, the cells at the edge of the plate, where it is bending, become constricted at the apical surface, making them wedge-shaped. This change in cell shape could, in principle, draw the edges of the neural plate up into folds. A local change in curvature might also be brought about by the cells at the edge of the plate crawling down the inner surface of the adjacent ectoderm (see Fig. 8.30).

The neural folds fuse in the midline, but the prospective dorsal neural tube cells are still some way from the midline and the future lumen of the neural tube is filled with cells. The formation of the dorsal region of the tube and its lumen is due to a complex pattern of movements involving the migration of the neural crest cells to the dorsal midline and radial intercalation of ventral neural cells.

In birds and mammals, neurulation does not occur along the whole of the neural plate at the same time, as it does in amphibians, but starts near the anterior end of the embryo, in the region of the midbrain, and proceeds anteriorly and posteriorly (Fig. 8.31). Changes in the shape of neural furrow cells in chick embryos are associated with the folding of the neural tube, but whether they are a cause or a result of folding is not yet clear. The detailed cellular mechanism underlying the shape changes is not yet established and in the chick there is some evidence that tissues lateral to the neural plate could exert forces that cause neural fold formation. Cells in the midline of the chick neural furrow, the so-called hinge point, are

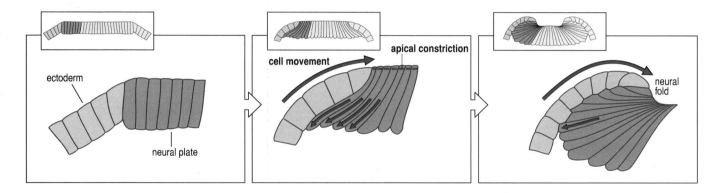

wedge-shaped. Later, wedge-shaped cells are seen at additional hinge points on the sides of the furrow, where the furrow curves round further to form a tube (see Fig. 8.31). When the folds meet they zip up into close apposition so that the lumen almost disappears; it only opens up again after the folds fuse.

Although changes in cell shape in general often involve changes in the actin cytoskeleton, it has been established that bending of the neural

Fig. 8.30 Change in cell shape and cell crawling may drive neural folding. Cells at the edge of the neural plate change their shape and appear to crawl along the under surface of the adjacent ectoderm. This may be partly responsible for causing the neural folds to develop.

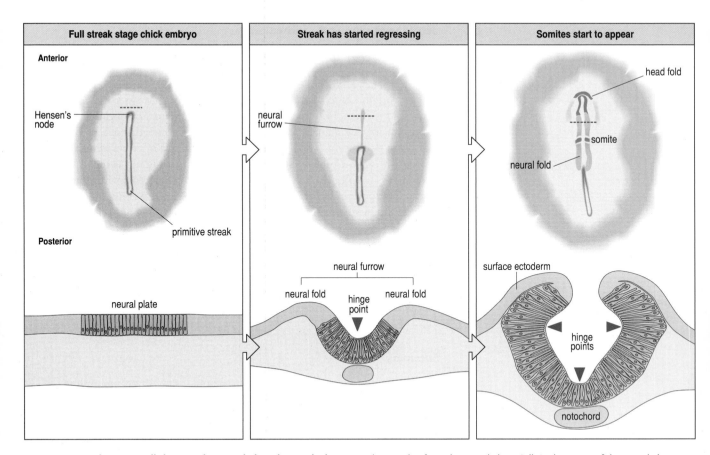

Fig. 8.31 Change in cell shape in the neural plate during chick neurulation. Top: surface views of the chick epiblast. Bottom: cross-sections taken through the epiblast at the sites indicated by the dotted lines in the top diagrams. Cells anterior to Hensen's node become elongated to form the neural plate. Cells in the center of the neural plate become wedge-shaped, defining the 'hinge point' at which the neural plate bends. Additional hinge points with wedge-shaped cells also form on the sides of the furrow. After Schoenwolf, G.C., et al.: 1990.

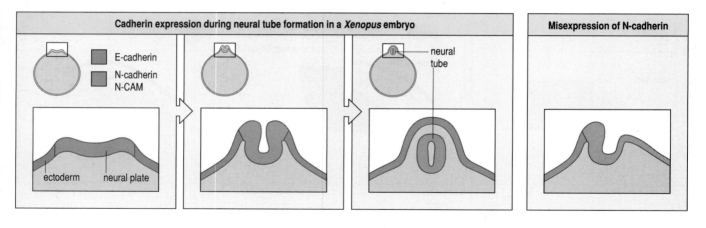

Fig. 8.32 Expression of cell adhesion molecules during neural tube formation in *Xenopus*. The cells of the neural plate express N-cadherin and the adjacent ectoderm expresses E-cadherin. If N-cadherin is misexpressed in the ectoderm on one side, there is a failure of the neural tube to separate at that site (right panel).

plate in the mouse is not dependent on actin microfilaments throughout. Treatment of embryos with cytochalasin B blocks neural fold formation in the head region but not in the spinal cord. By a mechanism that is still unknown, cells at the hinge joints continue to change shape and adopt a wedge-shaped form.

8.14 Changes in the pattern of expression of cell adhesion molecules accompany neural tube formation

The neural tube, which is initially part of the ectoderm, separates from the presumptive epidermis after its formation. This involves changes in cell adhesion. The cells of the neural plate, like the rest of the ectoderm in the chick, initially express the adhesion molecule L-CAM on their surface. However, as the neural folds develop, the neural plate ectoderm begins to express both N-cadherin and N-CAM, whereas the adjacent ectoderm expresses E-cadherin only (Fig. 8.32). These changes in adhesiveness may enable the neural tube to separate from the surrounding ectoderm and allow it to sink beneath the surface, the rest of the ectoderm reforming over it in a continuous layer. The changes in adhesiveness could themselves provide the mechanism for neural tube formation, in a manner similar to that described in Section 8.1 for the sorting out of cells. Support for this view comes from altering the pattern of expression of adhesion molecules in *Xenopus* embryos. Injection of N-cadherin mRNA into the gastrula ectoderm in the region adjacent to one side of the future neural tube, results in the persistence of a continuous layer of cells between the ectoderm and neural tube, and a failure of the neural tissue to separate at that site (see Fig. 8.32).

Summary

In vertebrates, neurulation results in neural tube formation following induction by the mesoderm. Neurulation in mammals, birds, and amphibians results from the folding of the neural plate into a tube, the neural folds fusing in the midline of the embryo. Separation of the neural tube from the adjacent ectoderm requires changes in cell adhesiveness. The formation of the neural folds and their coming together in the midline is apparently driven by changes in cell shape within the tube itself, as well as by forces generated by adjacent tissues.

Cell migration

Cell migration is a major feature of animal morphogenesis, with cells moving over relatively long distances, from one site to another. In this section, we consider two examples, the migrations of the primary mesenchyme cells from the interior of the sea urchin blastula and of the neural crest cells in the chick embryo. Each of these involve interactions between the migrating cells and the substratum, which control the pattern of migration. As a contrast, we also look at the aggregation of individual amebae of the slime mold *Dictyostelium* into a multicellular fruiting body, which provides an example of how chemotaxis and signal propagation can control migration pattern. We leave until Chapter 11 the migration of immature neurons that is so fundamental to the morphogenesis of the nervous system. Other important examples of cell migration that are dealt with in later chapters include muscle cell migration in vertebrate limbs (see Chapter 10), and germ cell migration (see Chapter 12).

8.15 The directed migration of sea urchin primary mesenchyme cells is determined by the contacts of their filopodia with the blastocoel wall

After the primary mesenchyme cells have entered the blastocoel, as described in Section 8.7, they move within the blastocoel to become distributed in a characteristic pattern on the inner surface of the blastocoel wall. The mesenchyme cells become arranged in a ring around the gut in the vegetal region at the ectoderm–endoderm border. Some then migrate to form two extensions toward the animal pole on the ventral (oral) side (Fig. 8.33). The migration path of individual cells varies considerably from embryo to embryo, but their final pattern of distribution is fairly constant. The primary mesenchyme cells later lay down the skeletal rods of the sea urchin endoskeleton by secretion and, as these develop, the distribution of the cells changes, so their pattern is a dynamic one.

The primary mesenchyme cells move over the inner surface of the blastocoel wall by means of fine filopodia, which can be up to 40 μm long and may extend in several directions. Each cell at any one time has, on average, six filopodia, most of which are branched. When filopodia make contact

Fig. 8.33 Migration of primary mesenchyme in early sea urchin development. The primary mesenchyme cells enter the blastocoel at the vegetal pole and migrate over the blastocoel wall by filopodial extension and contraction. Within a few hours they take up a well-defined ring-like pattern in the vegetal region with extensions along the ventral side.

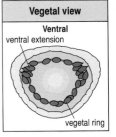

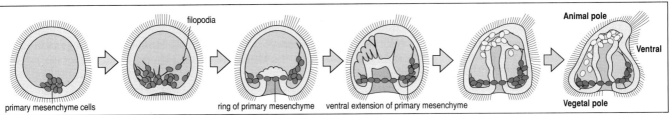

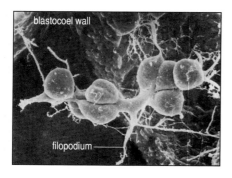

Fig. 8.34 Filopodia of sea urchin mesenchyme. The scanning electron micrograph shows a group of primary mesenchyme cells, some of which have fused together, moving over the blastocoel wall by means of their numerous filopodia, which can extend and contract.
Photograph courtesy of J. Morrill, from Morrill, J.B., et al.: 1985.

with, and adhere to, the blastocoel wall, they contract, drawing the cell body toward the point of contact. Because each cell extends several filopodia (Fig. 8.34), some or all of which may contract on contact with the wall, there seems to be competition between the filopodia, the cell being drawn toward that region of the wall where the filopodia make the most stable contact. The movement of the primary mesenchyme cells therefore resembles a random search for the most stable attachment. As they migrate, their filopodia fuse, forming cable-like extensions.

Analysis of video films of migrating cells suggests that the most stable contacts are made in the regions where the cells finally accumulate, namely the vegetal ring and the two ventro-lateral clusters. Thus, the pattern of contact stability of the inner surface of the blastocoel wall determines the pattern of migration of the cells, but the molecular basis for this adhesion is not known, although glycosaminoglycans and proteoglycans are involved. The surfaces of the cells over which the mesenchyme cells move are covered with a basal lamina, which has been implicated in influencing the contacts between filopodia and the blastocoel wall. For example, there is a strong association between filopodia and a matrix component.

Primary mesenchyme cells introduced by injection at the animal pole move in a directed manner to their normal positions in the vegetal region. This suggests that guidance cues, possibly graded, are globally distributed over the blastocoel wall. Even cells that have already migrated, migrate again to form a similar pattern when introduced into a younger embryo.

8.16 Neural crest migration is controlled by environmental cues and adhesive differences

Neural crest cells of vertebrates have their origin at the edges of the neural folds and first become recognizable during neurulation. When the neural tube closes, they undergo an epithelial to mesenchymal transition (see Section 8.7) and leave the epithelial sheet in the midline, migrating away from it on either side.

The epithelial to mesenchymal transition in vertebrates involves the *slug* gene, which controls the process by which nonmotile epithelial cells become migrating cells. The *slug* gene is related to the *Drosophila snail* gene and is expressed in all migratory neural crest cells. Inhibition of *slug* expression inhibits migration.

Neural crest cells migrate away from the neural tube, giving rise to a wide variety of different cell types that include cartilage in the head, pigment cells in the dermis, the medullary cells of the adrenal gland, glial Schwann cells, and the neurons of both the peripheral and the autonomic nervous systems. Here, we focus on the migration of the crest cells in the trunk region of the chick embryo. We have already discussed the migration of neural crest cells from the hindbrain region into the branchial arches (see Section 4.11).

Various strategies have been employed to follow the migration of neural crest cells. For example, because quail cells have a nuclear marker that

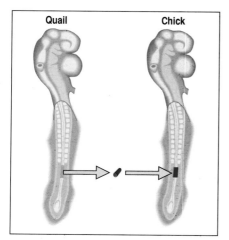

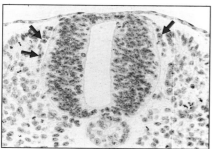

Fig. 8.35 Following cell migration pathways by grafting a piece of quail neural tube to a chick host. A piece of neural tube from a quail embryo is grafted to a similar position in a chick host. The photograph shows migration of the quail neural crest cells (red arrows). Their migration can be followed as quail cells have a nuclear marker that distinguishes them from chick cells.

Photograph courtesy of N. Le Douarin.

distinguishes them from chick cells, grafting a neural tube from a quail embryo into a chick embryo allows the subsequent migration pathways of the quail neural crest cells in the chick embryo to be followed (Fig. 8.35). It is also possible to identify migrating chick neural crest cells by tagging them with labeled monoclonal antibodies, or by labeling them with the dye DiI. The initiation of neural crest migration seems to involve disruption of the basement membrane surrounding the neural tube, allowing the crest cells to escape, and there is a need to regulate cell–cell adhesion.

Multiple subtypes of cadherin are dynamically expressed during these processes. In the chick embryo, for example, the ectoderm initially expresses L-CAM; during neural plate invagination, however, L-CAM expression is replaced by that of N-cadherin. At the same time, cadherin-6B begins to be expressed in the invaginating neural plate, most strongly at the neural crest-generating area. When neural crest cells leave the neural tube, these cadherins become undetectable, and cadherin-7 appears instead. Cadherin-7 expression persists during migration of the crest cells. From these observations, it is proposed that changes in cadherin expression during neural crest development may have a role in the segregation of neural crest cells from the neural tube. If N-cadherin or cadherin-7 are constitutively expressed in cells of the dorsal neural tube by injection of retroviral vectors carrying the cadherin genes, the cells do not turn off N-cadherin expression or they express cadherin-7 prematurely. Under these experimental conditions, the movement of crest cells from the neural tube is dramatically suppressed. Migration also requires loss of adhesion of the cells to the neural tube, and both N-cadherin and E-cadherin (see Section 8.14) are lost from neural crest cells at about the time of migration.

There are two main migratory pathways for neural crest cells in the trunk of the chick embryo (Fig. 8.36). One goes dorso-laterally under the

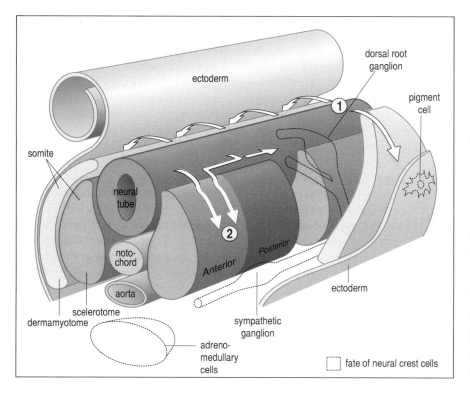

Fig. 8.36 Neural crest migration in the trunk of the chick embryo. One group of cells (1) migrates under the ectoderm to give rise to pigment cells (shown in outline). The other group of cells (2) migrates over the neural tube and then through the anterior half of the somite; the cells do not migrate through the posterior half of the somite. Those that migrate along this pathway give rise to dorsal root ganglia, sympathetic ganglia, and cells of the adrenal cortex, and their future sites are also shown in outline. Those cells opposite the posterior regions of a somite migrate in both directions along the neural tube until they come to the anterior region of a somite. This results in a segmental pattern of migration and is responsible for the segmental arrangement of ganglia.

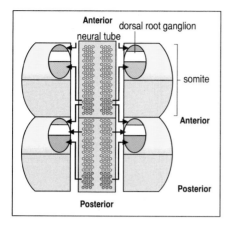

Fig. 8.37 Segmental arrangement of dorsal root ganglia is due to the migration of crest cells through the anterior half of the somite only. Neural crest cells cannot migrate through the posterior (gray) half of a somite. They thus accumulate in the anterior half. The dorsal root ganglion that develops from these cells is made up of crest cells from anterior (orange), adjacent (white), and posterior (red) regions.

ectoderm and over the somites; cells that migrate this way mainly give rise to pigment cells, which populate the skin and feathers. The other pathway is more ventral, primarily giving rise to sympathetic and sensory ganglion cells. Some crest cells move into the somites to form dorsal root ganglia; others migrate through the somites to form sympathetic ganglia and adrenal medulla, but appear to avoid the region around the notochord. Trunk neural crest selectively migrates through the anterior (rostral) half of the somite and not through the posterior (caudal) half. Within each somite, neural crest cells are found only in the anterior half, even when they originate in neural crest adjacent to the posterior half of the somite. This behavior is unlike that of the neural crest cells taking the dorsal pathway, which migrate over the whole dorso-lateral surface of the somite. The anterior migration pathway results in the distinct segmental arrangement of spinal ganglia in vertebrates, with one pair of ganglia corresponding to one pair of somites—one segment—in the embryo (Fig. 8.37). The segmental pattern of migration is due to the adhesive properties of the somites. If the somites are rotated through 180° so that their antero-posterior axis is reversed, the crest cells still migrate through the original anterior halves only. Two members of the Eph family of transmembrane ligands are expressed in the posterior halves of the somites, while crest cells have receptors for these ligands. The interactions between the ligands and receptors could result in the expulsion of the crest cells from the posterior halves of the somites. This would provide a molecular basis for the segmental arrangement of the spinal ganglia.

The neural tube and notochord also both influence neural crest migration. If the early neural tube is inverted through 180°, before neural crest migration starts, so that the dorsal surface is now the ventral surface, one might think that the cells that normally migrate ventrally, now being nearer to their destination, would move ventrally. But this is not the case, and many of the crest cells move upward through the sclerotome in a ventral to dorsal direction, staying confined to the anterior half of each somite. This suggests that the neural tube somehow influences the direction of migration of neural crest cells. The notochord also exerts an influence, inhibiting neural crest cell migration over a distance of about 50 µm, and thus preventing the cells from approaching it.

Many different extracellular matrix molecules have been detected along neural crest migratory pathways, and the neural crest cells may interact with their molecules by means of their cell-surface integrins. Neural crest cells cultured *in vitro* adhere to, and migrate efficiently on, fibronectin, laminin, and various collagens. Blocking the adhesion of the neural crest cells to fibronectin or laminin by blocking the integrin β_1 subunit *in vivo* causes severe deficiencies in the head region but not in the trunk, suggesting that the crest cells in these two regions adhere by different mechanisms, probably involving other integrins. It is striking how neural crest cells in culture will preferentially migrate along a track of fibronectin, although the role of this molecule in guiding the cells in the embryo is still unclear.

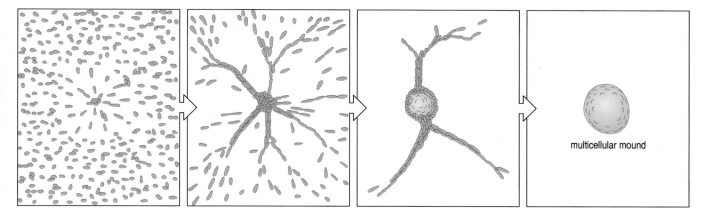

8.17 Slime mold aggregation involves chemotaxis and signal propagation

The myxamebae of the cellular slime mold *Dictyostelium discoideum* feed on bacteria. When a local food source is exhausted, they enter the multi-cellular stage of the slime mold life cycle (see Fig. 6.23). The first phase of this stage is aggregation, which when observed in time-lapse videos presents a dramatic spectacle, the cells streaming toward a center of aggregation like small rivers of cells converging into a lake. As they approach the focus of aggregation, the cells in the streams adhere to each other at their anterior and posterior ends through a membrane glycoprotein that is expressed at this stage. Eventually the cells collect into a compact multi-cellular mound (Fig. 8.38), which develops into a stalked fruiting body (see Fig. 6.23).

The cells move intermittently rather than continuously, in pulses of inward movement. The mechanism of aggregation involves both chemo-taxis of individual cells, and the propagation of the chemotactic signal from one ameba to another. In *Dictyostelium*, cyclic AMP (cAMP) is the chemo-attractant. Amebae respond to an increasing concentration of cAMP, and will thus move up a gradient of cAMP by extending a pseudopod in the direction of the source (Fig. 8.39). Chemotaxis up such a gradient only operates over short distances, much less than 1 mm, because it is difficult to

Fig. 8.38 Aggregation in the cellular slime mold *Dictyostelium*. Aggregating amebae of *D. discoideum* stream toward a focal point, eventually forming a multicellular mound that develops into a fruiting body (not shown). In the cell streams, the normally ameboid cells become bipolar and adhere to each other at either end.

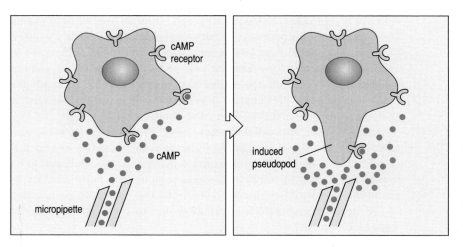

Fig. 8.39 Chemotactic response of slime mold cells. Amebae will move up a gradient of cyclic AMP (cAMP). A localized source of cAMP binds to membrane receptors on the side of the cell facing the source, resulting in the ameba extending a pseudopod toward the source. After Alberts, B., *et al.*: 1989.

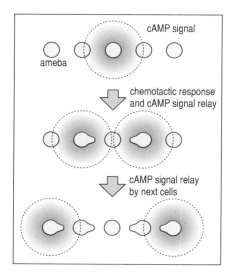

Fig. 8.40 Propagation of a cAMP signal and chemotaxis by slime mold amebae. A cell at the center of the aggregate gives off a pulse of cAMP. This induces pseudopod formation and chemotactic movement in the surrounding cells in the direction of the cAMP source. It also causes the responding cells to produce a pulse of cAMP themselves, and the signal is thus relayed outward. After releasing a pulse of cAMP, a cell becomes refractory to cAMP for a short time. This ensures the signal is propagated outward only.

establish a reliable diffusion gradient of a chemoattractant over larger distances. Yet amebae can aggregate from distances up to 5 mm from the center. This is achieved by the propagation of the chemotactic signal, in a manner rather similar to the conduction of a nerve impulse.

Dictyostelium cells can respond chemotactically to a stable gradient in which there is only a 1% difference in cAMP across the cell. When an ameba receives a pulse of cAMP, the cAMP binds to membrane receptors that stimulate the cell not only to respond chemotactically by moving towards the source, but also to propagate the signal by producing a pulse of cAMP itself. By this relay mechanism, a wave of cAMP is propagated across the field of aggregating amebae (Fig. 8.40). The pulsatory nature of the signal is shown by introducing a micropipette containing cAMP into a field of amebae ready to aggregate: it can act as an aggregation center only if it provides cAMP pulses of the correct frequency. An important feature of the propagation of the signal is that the cells are refractory to cAMP for a short time immediately after they have given out a pulse of cAMP. This ensures that the pulse only propagates outward. The presence of the enzyme phosphodiesterase, which breaks down cAMP, prevents its concentration rising to a level where it saturates the system, which would prevent propagation.

Aggregation of slime mold cells is the best understood example of chemotaxis in a developing system. The chemotactic response is thought to involve local activation of actin polymerization at the leading edge, causing protrusion of the cell in this direction, accompanied by inhibition of such activity over the rest of the cell. We come across further examples of chemotaxis in the development of the nervous system in Chapter 11. However, the signal propagation relay used by the slime mold has not yet been found in any other system.

Summary

Directed cell migration in animal embryos is controlled mainly by interactions with the substratum over which the cells move, although signals from other cells may also play a part. The directed migration of the primary mesenchyme cells in the sea urchin blastula is due to filopodia on these cells exploring their environment, and then drawing the cell to that region of the blastocoel wall where the filopodia make the most stable attachment. Thus, the final position of a mesenchyme cell is determined by the adhesive properties of the basal lamina and ectoderm over which it moves. Similarly, migration pathways of vertebrate neural crest cells are determined by the adhesive properties over which they move. Differences in adhesiveness between the anterior and posterior halves of the somites result in neural crest being prevented from migrating over or through posterior halves. Thus, presumptive dorsal ganglia cells collect adjacent to anterior halves, giving them a segmental arrangement.

Aggregation in the cellular slime mold *Dictyostelium* involves both chemotaxis and signal propagation. The individual amebae extend processes in the direction of an increasing concentration of cAMP, and move toward the source. In addition, the cells respond to a pulse of cAMP by giving out a pulse of cAMP themselves, and this results in propagation of the signal, enabling cells as far away as 5 m to be attracted to the aggregation center.

Summary: directed cell migration of vertebrate neural crest

neural crest cells

migration through anterior
of somites

migration over somites

dorsal ganglion cells

pigment cells

Directed dilation

Hydrostatic pressure can provide the force for morphogenesis in a variety of situations. An increase in hydrostatic pressure inside a spherical sheet of cells causes the sphere to increase in volume. We have already seen how hydrostatic pressure is involved in blastula formation in both the mouse and amphibian. Here, we consider examples of **directed dilation** where the increase in pressure causes an asymmetric change in shape. If the circumferential resistance to pressure in a tube is much greater than the resistance to longitudinal extension, then an increase in the internal pressure causes an increase in length (Fig. 8.41).

After the *Xenopus* notochord has formed, as discussed in Section 8.12, its volume increases threefold, and there is considerable further lengthening as it straightens and becomes stiffer. At this stage, the notochord has become surrounded by a sheath of extracellular material, which restricts circumferential expansion but does allow expansion in the antero-posterior direction. The cells within the notochord develop fluid-filled vacuoles and expand in volume as a result. They thus exert hydrostatic pressure on the notochord sheath, resulting in directed dilation: circumferential expansion of the notochord is prevented by the resistance of the sheath, and this ensures that the increase in volume (dilation) is directed along the notochord's long axis.

The vacuoles in the notochord cells are filled with glycosaminoglycans which, because of their high carbohydrate content, tend to attract water into the vacuoles by osmosis. It is this that produces the hydrostatic pressure that causes the increase in cell volume, and the consequent stiffening and straightening of the notochord.

Changes in the structure of the sheath during the period of notochord elongation fit well with the proposed hydrostatic mechanism. The sheath contains both glycosaminoglycans, which have little tensile strength, and the fibrous protein collagen, which has a high tensile strength; during notochord dilation the density of collagen fibers increases, providing resistance to circumferential expansion. The crucial role of the sheath in dilation and elongation is shown by the fact that if it is digested away, the notochord buckles and folds and the notochord cells, instead of being flat, become rounded.

8.18 Circumferential contraction of hypodermal cells elongates the nematode embryo

During the early development of the nematode there is little change in body shape from the spherical form of the fertilized egg, even during

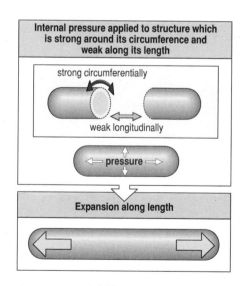

Fig. 8.41 Directed dilation. Hydrostatic pressure inside a constraining sheath or membrane can lead to elongation of the structure. If the circumferential resistance is much greater than the longitudinal resistance, as it is in the notochord sheath, the rod of cells inside the sheath lengthens.

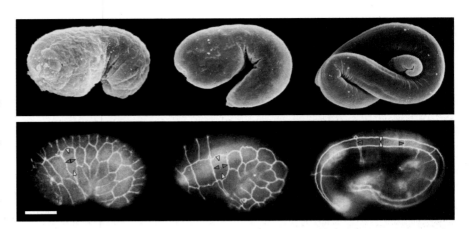

Fig. 8.42 Increase in nematode body length by directed dilation. The change in body shape over 2 hours is illustrated in the top panel. The increase in length is due to circumferential contraction of the hypodermal cells, as shown in the bottom panel. The change in shape of a single cell can be seen in the cell marked with arrows. Scale bar = 10 μm.

Photographs courtesy of J. Priess, from Priess, J.R., et al.: 1986.

gastrulation. After gastrulation, about 5 hours after fertilization, the embryo begins to elongate rapidly along its antero-posterior axis. Elongation takes about 2 hours, during which time the nematode embryo decreases in circumference about threefold and undergoes a fourfold increase in length.

This elongation is brought about by a change in shape of the hypodermal (epidermal) cells that make up the outermost layer of the embryo; their destruction by laser ablation prevents elongation. During embryo elongation, these cells change shape so that, instead of being elongated in the circumferential direction, they become elongated along the antero-posterior axis (Fig. 8.42). Throughout this elongation the hypodermal cells remain attached to each other by desmosome cell junctions. The desmosomes are also linked within the cells by actin-containing fibers that run circumferentially, and these fibers appear to shorten as the cells elongate. The disruption of actin filaments by cytochalasin D treatment blocks elongation, and so it is very likely that their contraction brings about the change in cell shape. Circumferential contraction of the hypodermal cells causes an increase in hydrostatic pressure within the embryo, forcing an extension in an antero-posterior direction. Circumferentially oriented microtubules may also have a mechanical role in constraining the expansion, in the same way as the sheath of the *Xenopus* notochord described above. Increase in nematode body length is thus another example of directed dilation.

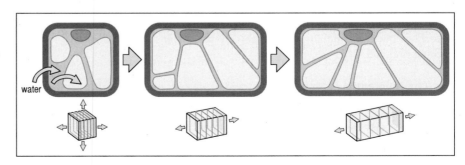

Fig. 8.43 Enlargement of a plant cell. Plant cells expand as water enters the cell vacuoles and thus causes an increase in intracellular hydrostatic pressure. The cell elongates in a direction perpendicular to the orientation of the cellulose fibrils in its cell wall.

water

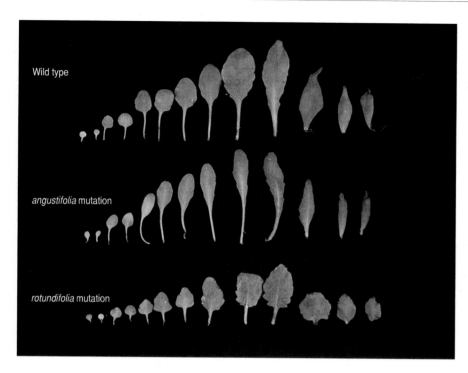

Fig. 8.44 The shape of the leaves of *Arabidopsis* are affected by mutations affecting cell elongation. The *angustifolia* mutation results in less elongation in width, while mutations in *rotundifolia* cause short, fat leaves to develop.

Photograph courtesy of H. Tsukaya, from Tsuge, T. et al.: 1996.

8.19 The direction of cell enlargement can determine the form of a plant leaf

Cell enlargement is a major process in plant morphogenesis, providing up to a 50-fold increase in the volume of a tissue. The driving force for expansion is the hydrostatic pressure—turgor pressure—exerted on the cell wall as the protoplast swells as a result of the entry of water into cell vacuoles by osmosis (Fig. 8.43, opposite). Plant cell expansion involves synthesis and deposition of new cell wall material, and is an example of directed dilation. The direction of cell growth is determined by the orientation of the cellulose fibrils in the cell wall. Enlargement occurs primarily in a direction at right angles to the fibrils, where the wall is weakest. The orientation of cellulose fibrils in the cell wall is thought to be determined by the microtubules of the cell's cytoskeleton, which are responsible for positioning the enzyme assemblies that synthesize cellulose at the cell wall. Plant growth hormones, such as ethylene and gibberellins, alter the orientation in which the fibrils are laid down and so can alter the direction of expansion. Auxin aids expansion by loosening the structure of the cell wall.

The development of a leaf involves a complex pattern of cell division and cell elongation, with cell elongation playing a central part in the expansion of the leaf blade. Two mutations that affect the shape of the blade by affecting the direction of cell elongation have been identified. Leaves of the *Arabidopsis* mutant *angustifolia* are similar in length to the wild type but are much thinner (Fig. 8.44). In contrast, the *rotundifolia* mutations reduce the length of the leaf relative to its width. Neither of these mutations affects the number of cells in the leaf. Examination of the cells in the developing leaf shows that these mutations are affecting the direction of elongation of the enlarging cells.

Summary

Directed dilation results from an increase in hydrostatic pressure and unequal peripheral resistance to this pressure. Extension of the notochord is brought about by directed dilation, in which the notochord increases in volume while its circumferential expansion is constrained by the notochord sheath, forcing it to elongate. Similarly, the nematode embryo elongates after gastrulation due to a circumferential contraction in the outer hypodermal cells that generates pressure on the internal cells, forcing the embryo to extend in an antero-posterior direction. In plants, the direction of cell enlargement determines the shape of leaves. In plant cell enlargement, the direction of elongation is yet another example of directed dilation.

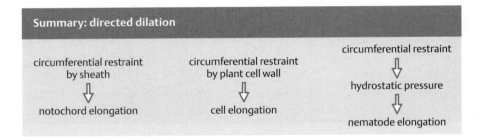

Summary: directed dilation

circumferential restraint by sheath ⟱ notochord elongation

circumferential restraint by plant cell wall ⟱ cell elongation

circumferential restraint ⟱ hydrostatic pressure ⟱ nematode elongation

SUMMARY TO CHAPTER 8

Cells are held together by specific adhesion molecules. Changes in the shape of the embryo and cell migration are due to changes in cell adhesion and to forces generated by the cell. Formation of the blastula results from cell division and cell polarization, and in some cases from the movement of water into the blastocoel. Gastrulation involves major movements of cell sheets so that the future endoderm and mesoderm move inside the embryo to their appropriate positions in relation to the main body plan. Invagination of the mesoderm in both sea urchins and *Drosophila* involves similar changes in cell shape. Convergent extension has a key role in both sea urchin and amphibian gastrulation, and is the result of particular patterns of cell intercalation. Epiboly, the spreading of a multilayered sheet, is due to intercalation. Convergent extension, together with directed dilation, is also involved in notochord formation. Directed dilation causes the elongation of the nematode embryo and the direction of enlargement of plant cells. Changes in cell shape and cell adhesion are responsible for neural tube formation. Directed cell migration of sea urchin mesenchyme and vertebrate neural crest cells is controlled by the adhesiveness of the substratum over which the cells move. Migration and aggregation of slime mold cells is by propagation of a chemotactic signal.

REFERENCES

8.1 Sorting out of dissociated cells demonstrates differences in cell adhesiveness in different tissues

Holder, N., Klein, R.: **Eph receptors and ephrins: effectors of morphogenesis.** *Development* 1999, **126**: 2033–2044.

Steinberg, M.S.: **Does differential adhesion govern self-assembly processes in histogenesis? Equilibrium configurations and the emergence of a hierarchy among populations of embryonic cells.** *J. Exp. Zool.* 1970, **173**: 395–433.

Townes, P., Holtfreter, J.: **Directed movements and selected adhesions of embryonic amphibian cells.** *J. Exp. Zool.* 1955, **128**: 53–120.

Xu, Q., Mellitzer, G., Wilkinson, D.G.: **Roles of Eph receptors and ephrins in segmental patterning.** *Proc. Roy. Soc. B.* 2000, **353**: 993–1002.

Box 8A Cell adhesion molecules

Brown, N.H., Gregory, S.L., Martin-Bermudo, M.D.: **Integrins as mediators of morphogenesis in** *Drosophila.* *Dev. Biol.* **223**: 1–16.

Cunningham, B.A.: **Cell adhesion molecules as morphoregulators.** *Curr. Opin. Cell Biol.* 1995, **7**: 628–633.

Giancotti, F.G., Ruoslahti, E.: **Integrin signaling.** *Science* 1999, **285**: 1028–1032.

Klymkowsky, M.W., Parr B.: **The body language of cells: the intimate connection between cell adhesion and behavior.** *Cell* 1995, **83**: 5–8.

Tepass, U., Truong, K., Godt, D., Ikura, M., Peifer, M.: **Cadherins in embryonic and neural morphogenesis.** *Nat. Rev. Mol. Cell Biol.* 2000, **1**: 91–100.

8.2   Cadherins can provide adhesive specificity

Levine, E., Lee, C.H., Kintner, C., Gumbiner, B.M.: **Selective disruption of E-cadherin function in early** *Xenopus* **embryos by a dominant negative mutant.** *Development* 1994, **120**: 901–909.

Steinberg, M.S., Takeichi, M.: **Experimental specification of cell sorting, tissue spreading, and specific spatial patterning by quantitative differences in cadherin expression.** *Proc. Natl Acad. Sci. USA* 1994, **91**: 206–209.

Takeichi, M., Nakagawa, S., Aono, S., Usui, T., Uemura, T.: **Patterning of cell assemblies regulated by adhesion receptors of the cadherin superfamily.** *Proc. Roy. Soc. B* 2000, **355**: 885–896.

8.3   The asters of the mitotic apparatus determine the plane of cleavage at cell division

Staiger, C., Doonan, J.: **Cell division in plants.** *Curr. Opin. Cell Biol.* 1993, **5**: 226–231.

Strome, S.: **Determination of cleavage planes.** *Cell* 1993, **72**: 3–6.

8.4   Cells become polarized in early mouse and sea urchin blastulas

Fleming, T.P., Johnson, M.H.: **From egg to epithelium.** *Annu. Rev. Cell Biol.* 1988, **4**: 459–485.

Sobel, J.S.: **Membrane–cytoskeletal interactions in the early mouse embryo.** *Semin. Cell Biol.* 1990, **1**: 341–348.

Sutherland, A.E., Speed, T.P., Calarco, P.G.: **Inner cell allocation in the mouse morula: the role of oriented division during fourth cleavage.** *Dev. Biol.* 1990, **137**: 13–25.

Winkel, G.K., Ferguson, J.E., Takeichi, M., Nuccitelli, R.: **Activation of protein kinase C triggers premature compaction in the four-cell stage mouse embryo.** *Dev. Biol.* 1990, **130**: 1–15.

8.5   Ion transport is involved in fluid accumulation in the frog blastocoel

Warner, A.E.: **Physiological approaches to early development.** *Recent Adv. Physiol.* 1984, **10**: 87–123.

8.6   Internal cavities can be created by cell death

Coucouvanis, E., Martin, G.R.: **Signals for death and survival: a two-step mechanism for cavitation in the vertebrate embryo.** *Cell* 1995, **83**: 279–287.

8.7   Gastrulation in the sea urchin involves cell migration and invagination

Davidson, L.A., Koehl, M.A.R., Keller, R., Oster, G.F.: **How do sea urchins invaginate? Using biomechanics to distinguish between mechanisms of primary invagination.** *Development* 1995, **121**: 2005–2018.

Davidson, L.A., Oster, G.F., Keller, R.E., Koehl, M.A.R.: **Measurements of mechanical properties of the blastula wall reveal which hypothesized mechanisms of primary invagination are physically plausible in the sea urchin** *Stronglyocentrotus purpuratus.* *Dev. Biol.* 1998, **204**: 235–250.

Hardin, J., McClay, D.R.: **Target recognition by the archenteron during sea urchin gastrulation.** *Dev. Biol.* 1990, **142**: 86–102.

Ingersoll, E.P., Ettensohn, C.A.: **An N-linked carbohydrate-containing extracellular matrix determinant plays a key role in sea urchin gastrulation.** *Dev. Biol.* 1994, **163**: 359–366.

Miller, J.R., McClay, D.R.: **Changes in the pattern of adherens junction-associated β-catenin accompany morphogenesis in the sea urchin embryo.** *Dev. Biol.* 1997, **192**: 310–322.

Odell, G.M., Oster, G., Alberch, P., Burnside, B.: **The mechanical basis of morphogenesis. I. Epithelial folding and invagination.** *Dev. Biol.* 1981, **85**: 446–462.

8.8   Mesoderm invagination in *Drosophila* **is due to changes in cell shape, controlled by genes that pattern the dorso-ventral axis**

Costa, M., Wilson, E.T., Wieschaus, E.: **A putative cell signal encoded by the** *folded gastrulation* **gene coordinates cell shape changes during** *Drosophila* **gastrulation.** *Cell* 1994, **76**: 1075–1089.

Leptin, M.: *Drosophila* **gastrulation: from pattern formation to morphogenesis.** *Annu. Rev. Cell Biol.* 1995, **11**: 189–212.

Leptin, M.: **Morphogenesis: control of epithelial cell shape changes.** *Curr. Biol.* 1994, **4**: 709–712.

Seher, T.C., Leptin, M.: **Tribbles, a cell-cycle brake that coordinates proliferation and morphogenesis during** *Drosophila* **gastrulation.** *Curr. Biol.* 2000, **10**: 623–629.

Wilson, R., Leptin, M.: **FGF-receptor dependent morphogenesis of the** *Drosophila* **mesoderm.** *Proc. Roy. Soc. B* 2000, **355**: 891–895.

8.9   Dorsal closure in *Drosophila* **and ventral closure in** *C. elegans* **are brought about by the action of filopodia**

Chin-Sang, I.D., Chisholm, A.D.: **Form of the worm: genetics of epidermal morphogenesis in** *C. elegans.* *Trends Genet.* 2000, **16**: 544–551.

Jacinto, A., Wood, W., Balayo, T., Turmaine, M., Martinez-Arias, A., Martin, P.: **Dynamic actin-based epithelial adhesion and cell matching during** *Drosophila* **dorsal closure.** *Curr. Biol.* 2000, **10**: 1420–1426.

8.10   *Xenopus* gastrulation involves several different types of tissue movement

Shih, J., Keller, R.: **Gastrulation in** *Xenopus laevis*: **involution— a current view.** *Dev. Biol.* 1994, **5**: 85–90.

Wacker, S., Grimm, K., Joos, T., Winklbauer, R.: **Development and control of tissue separation at gastrulation in** *Xenopus.* *Dev. Biol.* 2000, **224**: 428–439.

8.11   Convergent extension and epiboly are due to cell intercalation

Keller, R., Shih, J., Sater, A.: **The cellular basis of the convergence and extension of the** *Xenopus* **neural plate.** *Dev. Dyn.* 1992, **193**: 199–217.

Keller, R., Davidson, L., Edlund, A., Elul, T., Ezin, M., Shook, D., Skoglund, P.: **Mechanisms of convergence and extension by cell intercalation.** *Proc Roy. Soc. B* 2000, **355**: 897–922.

8.12 Notochord elongation is caused by cell intercalation

Adams, D.S., Keller, R., Koehl, M.A.: **The mechanics of notochord elongation, straightening, and stiffening in the embryo of *Xenopus laevis*.** *Development* 1990, **100**: 115–130.

Keller, R., Cooper, M.S., D'Anilchik, M., Tibbetts, P., Wilson, P.A.: **Cell intercalation during notochord development in *Xenopus laevis*.** *J. Exp. Zool.* 1989, **251**: 134–154.

8.13 Neural tube formation is driven by cell migration and changes in cell shape

Alvarez, I.S., Schoenwolf, G.C.: **Expansion of surface epithelium provides the major extrinsic force for bending of the neural plate.** *J. Exp. Zool.* 1992, **261**: 340–348.

Davidson, L.A., Keller, R.E.: **Neural tube closure in *Xenopus laevis* involves medial migration, directed protrusive activity, cell intercalation and convergent extension.** *Development* 1999, **126**: 4547–4556.

Schoenwolf, G.C., Smith, J.L.: **Mechanisms of neurulation: traditional viewpoint and recent advances.** *Development* 1990, **109**: 243–270.

Ybot-Gonzalez, P., Copp, A.J.: **Bending of the neural plate during mouse spinal neurulation is independent of actin microfilaments.** *Dev. Dyn.* 1999, **215**: 273–283.

8.14 Changes in the pattern of expression of cell adhesion molecules accompany neural tube formation

Bok, G., Marsh, J.: (eds) *Neural Tube Defects*. Ciba Foundation Symposium 181. John Wiley: Chichester, 1994.

Detrick, R.J., Dickey, D., Kintner, C.R.: **The effect of N-cadherin misexpression on morphogenesis in *Xenopus* embryos.** *Neuron* 1990, **4**: 493–506.

8.15 The directed migration of sea urchin primary mesenchyme cells is determined by the contacts of their filopodia with the blastocoel wall

Ettensohn, C.A.: **Cell movements in the sea urchin embryo.** *Curr. Opin. Genet. Dev.* 1999, **9**: 461–465.

Ettensohn, C.A., McClay D.R.: **The regulation of primary mesenchyme cell migration in the sea urchin embryo: transplantations of cells and latex beads.** *Dev. Biol.* 1986, **117**: 380–391.

Gustafson, T., Wolpert, L.: **Studies on the cellular basis of morphogenesis in the sea urchin embryo. Directed movements of primary mesenchyme cells in normal and vegetalized larvae.** *Exp. Cell Res.* 1999, **253**: 288–295.

Hodor, P.G., Illies, M.R., Broadley, S., Ettensohn, C.A.: **Cell-substrate interactions during sea urchin gastrulation: migrating primary mesenchyme cells interact with and align extracellular matrix fibers that contain ECM3, a molecule with NG2-like and multiple calcium-binding domains.** *Dev. Biol.* 2000, **222**: 181–194.

Malinda, K.A., Ettensohn, C.A.: **Primary mesenchyme cell migration in the sea urchin embryo: distribution of directional cues.** *Dev. Biol.* 1994, **164**: 562–578.

Malinda, K.A., Fisher, G.W., Ettensohn, C.A.: **Four-dimensional microscopic analysis of the filopodial behavior of primary mesenchyme cells during gastrulation in the sea urchin embryo.** *Dev. Biol.* 1995, **172**: 552–566.

8.16 Neural crest migration is controlled by environmental cues and adhesive differences

Bronner-Fraser, M.: **Mechanisms of neural crest migration.** *BioEssays* 1993, **15**: 221–230.

Delannet, M., Martin, F., Bussy, B., Chersh, D.A., Reichardt, L.F., Duband, J.L.: **Specific roles of the $\alpha_V\beta_1$, $\alpha_V\beta_3$ and $\alpha_V\beta_5$ integrins in avian neural crest cell adhesion and migration on vitronectin.** *Development* 1994, **120**: 2687–2702.

Erickson, C.A., Perris, R.: **The role of cell–cell and cell–matrix interactions in the morphogenesis of the neural crest.** *Dev. Biol.* 1993, **159**: 60–74.

Nagawa, S., Takeichi, M.: **Neural crest emigration from the neural tube depends on regulated cadherin expression.** *Development* 1998, **125**: 2963–2971.

Nieto, M.A., Sargent, M.G., Wilkinson, D.G., Cooke, J.: **Control of cell behaviour during vertebrate development by *Slug*, a zinc finger gene.** *Science* 1994, **264**: 835–839.

8.17 Slime mold aggregation involves chemotaxis and signal propagation

Adams, D.S., Keller, R., Koehl, M.A.: **The mechanics of notochord elongation, straightening and stiffening in the embryo of *Xenopus laevis*.** *Development* 1990, **110**: 115–130.

Firtel, R.A., Chung, C.Y.: **The molecular genetics of chemotaxis: sensing and responding to chemoattractant gradients.** *BioEssays* 2000, **22**: 603–615.

Gerisch, G.: **Cyclic AMP and other signals controlling cell development and differentiation in *Dictyostelium*.** *Annu. Rev. Biochem.* 1987, **56**: 853–879.

Sager, B.M.: **Propagation of traveling waves in excitable media.** *Genes Dev.* 1996, **10**: 2237–2250.

Siu, C.H.: **Cell–cell adhesion molecules in *Dictyostelium*.** *BioEssays* 1990, **12**: 357–362.

8.18 Circumferential contraction of hypodermal cells elongates the nematode embryo

Priess, J.R., Hirsh, D.I.: ***Caenorhabditis elegans* morphogenesis: the role of the cytoskeleton in elongation of the embryo.** *Dev. Biol.* 1986, **117**: 156–173.

8.19 The direction of cell enlargement can determine the form of a plant leaf

Jackson, D.: **Designing leaves. Plant morphogenesis.** *Curr. Biol.* 1996, **6**: 917–919.

Tsuge, T., Tsukaya, H., Uchimiya, H.: **Two independent and polarized processes of cell elongation regulate leaf blade expansion in *Arabidopsis thaliana* (L.) Heynh.** *Development* 1996, **122**: 1589–1600.

Cell differentiation

- The control of gene expression
- Models of cell differentiation
- The reversibility of patterns of gene activity

"Considering that we all started off being similar, the diversity of our characteristics is remarkable."

Cellular differentiation describes the process by which embryonic cells become different from one another, acquiring distinct identities and specialized functions. As we have seen in previous chapters, many developmental processes, such as the early specification of the germ layers, involve transient changes in cell form, in patterns of gene activation, and in the molecules synthesized by the cells. However, **cell differentiation** involves the emergence of cell types that have a clear-cut identity in the adult, such as muscle cells, nerve cells, skin cells, and fat cells. In mammals, there are more than 200 clearly recognizable differentiated cell types, such as blood, epidermal, muscle, and nerve cells.

Differentiated cells serve specialized functions and have achieved a terminal and stable state, in contrast with many of the transitory differences in cell state that are characteristic of earlier stages in development. The specified precursors of cartilage and muscle cells have no obvious structural differences from each other and so look the same, but differentiate as cartilage and muscle, respectively, when cultured under appropriate conditions. The differences between the precursor cells at these early stages reflect differences in gene activity and thus the proteins they contain, which control their development.

There are no completely 'undifferentiated' cells, even in the early embryo as, even then, groups of cells differ from each other in their properties and patterns of gene activity. The term 'undifferentiated cells' almost always refers to precursor cells that show no clear structural features to foreshadow their later development. The mesodermal precursors of muscle cells, for example, have no sign of the complex arrangement of contractile filaments that they will develop internally; similarly, at an early stage, the precursors of white blood cells are indistinguishable from those of red blood cells.

As with earlier processes in development, the central feature of cell differentiation is a change in gene expression. This eventually leads to the

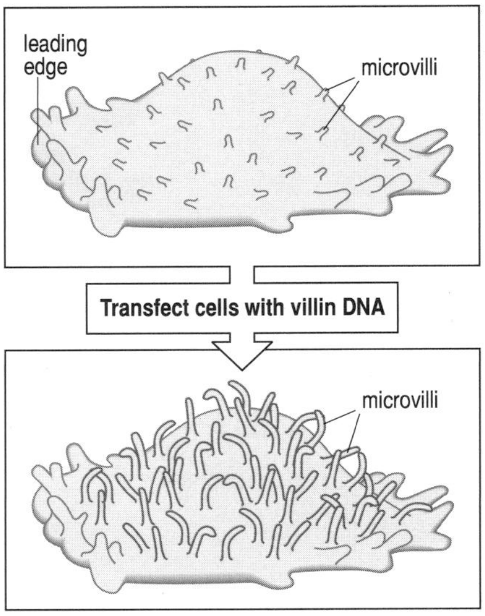

Fig. 9.1 A single protein can change a cell's structure. The protein villin is found in the projections—the microvilli—of the cells lining the gut. When the gene coding for villin is introduced into an epithelial cell line that has few poorly developed microvilli (top panel), the cell becomes covered with microvilli (bottom panel). After Friederich, E., *et al.*: 1990.

production of so-called 'luxury' or cell-specific proteins that characterize a fully differentiated cell: hemoglobin in red blood cells, keratin in skin epidermal cells, and muscle-specific actins and myosins. Naturally, the genes expressed in a differentiated cell include not only those encoding the luxury or cell-specific proteins, but also those for a wide range of 'housekeeping' proteins, such as the glycolytic enzymes involved in energy metabolism.

A differentiated cell is characterized by the proteins it contains. The presence of different proteins can result in considerable structural changes; mature mammalian red blood cells lose their nucleus and become biconcave discs stuffed with hemoglobin, whereas neutrophils, a type of white blood cell, develop a multi-lobed nucleus and a cytoplasm filled with secretory granules. Introduction of a single protein into a cell can dramatically alter its shape. An example of this is provided by the actin-binding protein villin, which is mainly produced in epithelial cells that develop a brush border of microvilli on their apical face; each microvillus is supported by an internal core of actin filaments, and villin is required for the assembly of this core. When a cell line that normally only has sparse and rudimentary microvilli is transfected with DNA coding for villin, the upper face of the cells becomes covered with numerous long microvilli (Fig. 9.1).

In the earliest stages of differentiation, differences between cells are not easily detected and probably consist of subtle alterations caused by a change in activity of a few or perhaps many genes. At this early stage, cells become **determined**, or committed, with respect to their developmental potential (see Section 1.10): mesoderm of somites, for example, can give rise to muscle, cartilage, dermis, and vascular tissue, but not to other cell types. Cells that are determined with respect to their eventual fate will form only the appropriate cell types when grafted to a different site in the embryo; they retain their identity. Once determined for a particular fate, cells pass on that determined state to all their progeny.

Cell differentiation is known to be controlled by a wide variety of external signals, ranging from cell-surface proteins to secreted polypeptide cytokines and molecules of the extracellular matrix. Examples of all of these signals will be encountered in this chapter, but it is important to remember that, while the external signals that stimulate differentiation are often referred to as being instructive, they are in general selective, in the sense that at any stage of development, the number of developmental options open to a cell are very limited (see Section 1.10). There are rarely more than a few options available at any given time; these options are set by the cell's internal state, which in turn reflects its developmental history. External signals cannot, for example, convert an endodermal cell into a muscle or nerve cell, although they can cause cells that are already determined to be either muscle or cartilage to develop into one or the other. Some self-renewing **stem cells** (see Section 1.15) can give rise to a wide range of cell types under the influence of different external signals.

Differentiation is often a gradual process, occurring over successive cell generations, with each generation becoming progressively more differentiated. In many cases, however, there is a conflict between cell division and cell differentiation, with cell-cycle arrest being necessary for cell differentiation. Cell proliferation is most evident before the terminal stage of differentiation, and terminally differentiated cells divide rarely. Some cells, such as skeletal muscle and nerve cells, do not divide at all after they have become fully differentiated. In cells that do divide after differentiation, the differentiated state, like the determined state, is passed on through all subsequent cell divisions to the progeny. As the pattern of gene activity is the key feature in cell differentiation, this raises the question of how a particular pattern of gene activity is first established, and then how it is passed on to daughter cells.

We start this chapter with a consideration of the mechanisms by which a pattern of gene activity can be established, maintained, and inherited at cell division. We then turn to a major question, namely the molecular basis of the specificity of cell differentiation, with muscle cells, blood cells, and neural crest cells providing our main model systems. Finally, we look at the reversibility and stability of the differentiated state.

This chapter concentrates on differentiation in animal cells. As we have seen in Chapter 7, plant cells do not have a permanently determined state, and a single somatic cell can give rise to a whole plant.

The control of gene expression

Every nucleus in the body of a multicellular organism is derived from a single zygotic nucleus in the fertilized egg. But patterns of gene activity in differentiated cells vary enormously from one cell type to another. The egg itself has a pattern of gene activity that is different from that of cells at later stages of embryonic development. This raises the question of what determines the particular pattern of gene activity in a differentiated cell and how it is inherited.

In order to understand the molecular basis of cell differentiation, we first need to determine how a gene can be expressed in a cell-specific manner. Why does a certain gene get switched on in one cell and not in another? We focus here on the regulation of **transcription**, which is the first (and generally the most significant) step in the expression of a gene. Control of protein synthesis can, however, also occur after transcription, for example, at the stages of **RNA splicing** or of **translation**. We have seen an example of translational regulation in early *Drosophila* development, where the nanos protein prevents translation of maternal *hunchback* RNA (see Section 5.3).

Control of transcription is crucial to cell differentiation as it determines which genes are transcribed and therefore which proteins are made by the cell. We have already discussed the importance of the regulation of transcription in early development, for example in relation to patterning in *Drosophila* (see Section 5.14). Here, we focus on the control of transcription in relation to cell differentiation. We start with a brief reminder of the basic mechanisms of transcriptional control in eukaryotic cells.

9.1 Control of transcription involves both general and tissue-specific transcriptional regulators

The transcription of a gene is initiated when RNA polymerase binds tightly to the start site of transcription in the **promoter** region of the gene (see Section 1.8). The polymerase, with its associated proteins, unwinds a short region of the DNA helix and starts synthesizing RNA, using one of the DNA strands as the template. The promoter region, and the other sites in the DNA at which expression of the gene can be regulated, are known as its

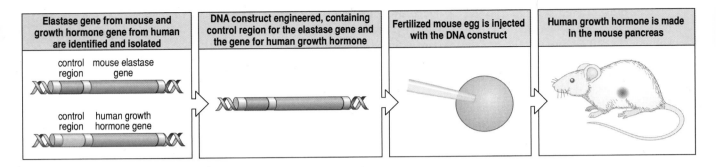

| Elastase gene from mouse and growth hormone gene from human are identified and isolated | DNA construct engineered, containing control region for the elastase gene and the gene for human growth hormone | Fertilized mouse egg is injected with the DNA construct | Human growth hormone is made in the mouse pancreas |

Fig. 9.2 Tissue-specific gene expression is controlled by regulatory regions. The control region of the mouse elastase gene is joined to a sequence of DNA that codes for human growth hormone. This DNA construct is injected into the nucleus of a fertilized mouse egg where it becomes integrated into the genome. When the mouse develops, human growth hormone is made in the pancreas under the control of the mouse elastase promoter. Normally, growth hormone is made only in the pituitary, and elastase only in the pancreas.

control regions, in contrast to the **coding regions**, which encode the amino acid sequence of the protein. In general, transcription of genes in development requires activators to bind to the control regions. There are usually several such activating proteins that interact to form a transcriptional-activating complex.

The importance of control regions in tissue-specific gene expression can be clearly demonstrated by experiments in which the control region of one tissue-specific gene is replaced by the control region of another. For example, in mice, the enzyme elastase is only synthesized in the pancreas, whereas growth hormone is only synthesized in the pituitary gland. The control region of an isolated mouse elastase gene can be joined to the protein-coding region of an isolated human growth hormone gene, and the resulting DNA construct injected into the nucleus of a fertilized mouse egg, where it becomes integrated into the genome (see Box 3C, p. 89). In the transgenic mouse embryos that develop from this egg, human growth hormone can be detected in the pancreas, showing that the human growth hormone gene is now expressed under the control of the mouse elastase promoter (Fig. 9.2). This experiment shows the importance of the promoter control elements in determining where a gene is expressed.

Similar experiments show that **enhancers**, which are additional control sites in the DNA, often located at some distance from the start of transcription, also direct tissue-specific gene expression. The enhancer element from the mouse insulin gene, together with the insulin promoter region can, for example, direct expression of any attached coding region in the islet β cells of the pancreas, the cells in which the insulin gene is normally expressed.

In eukaryotic cells, binding of RNA polymerase onto the correct region of DNA to start transcription requires the cooperation of a set of so-called general transcription factors, which form an initiation complex with the polymerase at the promoter site (Fig. 9.3). One can think of this complex as a transcribing machine. In eukaryotes, most protein-coding genes are transcribed by RNA polymerase II, which forms a multicomponent transcriptional initiation complex with a set of general transcription factors. This complex binds to the promoter at a sequence of DNA known as the TATA box, located near the starting point of transcription. Most of the key genes in development are in an inactive state and require activators to turn them on; these activators are the **gene regulatory proteins** or **transcription factors**.

Gene regulatory proteins bind to specific sites on DNA that together control the specificity of gene activity. Some of these regulatory sites are within the promoter region, adjacent to the TATA box, and are present in similar

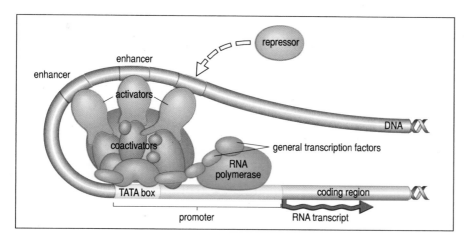

Fig. 9.3 Gene expression is regulated by the coordinated action of gene regulatory proteins that bind to control regions in DNA. The transcription machinery, composed of RNA polymerase and the general transcription factors that bind to the promoter region, is common to many cells. Other regulatory regions, which have to be bound by gene regulatory proteins (activators and repressors) before transcription can occur, are located adjacent to the promoter and usually also further upstream, and sometimes downstream, of the gene. The upstream regulatory regions, such as enhancers, may be many kilobases away from the start point of transcription. After Tijan, R.: 1995.

positions in most protein-coding genes. The enhancer elements, however, vary much more in type and position from gene to gene, and may be thousands of base pairs away from the starting point of transcription. Binding of regulatory proteins at enhancer sites can increase the rate of transcription initiation several hundredfold. All these distant sites are thought to be able to control gene activity because the DNA can form loops, thus bringing enhancers into close proximity with the promoter region. Proteins bound at enhancers can thus make contact with the proteins bound in the promoter region, forming a transcription initiation complex that can initiate transcription at a high rate.

An important class of regulatory proteins are those that link the transcriptional machinery to the proteins that bind to specific sites in the DNA such as enhancer elements. One such linking protein is CBP, which is essential for the co-activation of a variety of genes. CBP has a histone acetyl-transferase activity and may act to decondense **chromatin**—the complex of DNA and histone proteins in the chromosome—to enable transcription to occur. We shall return to the importance of histone acetylation in regulating gene activity in Section 9.3.

The transcriptional regulators that interact with the control regions of eukaryotic genes fall into two main groups: those that are required for the transcription of a wide range of genes, and which are found in a number of cell types; and those that are required for a particular gene or set of genes whose expression is tissue restricted, and which are only found in one or a very few cell types. We shall encounter both types of transcription factor in the following sections, when we look at the regulation of muscle genes in muscle precursor cells and the expression of globin genes in red blood cell precursors.

The expression of any given gene is initiated and maintained by a particular combination of gene regulatory proteins. We have already seen examples of this combinatorial control in the early development of *Drosophila*, where the temporal and regional activation of pair-rule genes is brought about by specific combinations of gap gene proteins binding to the regulatory regions of the pair-rule genes (see Fig. 5.23). The control regions of genes whose activity is restricted to one or a few cell types also appear to be activated by the action of a combination of factors, some of which are present only in those cells.

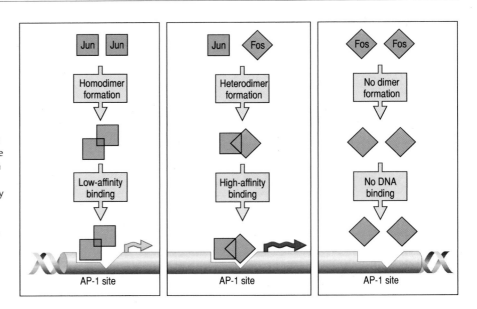

Fig. 9.4 Interaction between the transcription factors Fos and Jun in gene activation. The proteins Fos and Jun form a heterodimer, which binds to sites known as AP-1 sites in DNA, where it acts as a transcriptional regulator. On its own, Jun can form homodimers, which bind to the AP-1 site less strongly than the Fos:Jun dimer. The homodimers can activate gene expression, but less efficiently than the Fos:Jun heterodimer. Fos itself cannot form homodimers or bind to the AP-1 site alone, and so cannot activate transcription at all.

An important feature in the control of transcription is interaction between the gene regulatory proteins themselves and their interactions with other proteins and small molecules. An example of the former is provided by the two ubiquitous transcription factors Fos and Jun, which increase in activity following treatment of cells with growth factors. They bind as a Fos:Jun heterodimer to the AP-1 site, which is present in the control regions of many genes, often in several copies. Fos, but not Jun, can also form a homodimer which will also bind to the AP-1 site, but much less strongly than the Fos:Jun heterodimer (Fig. 9.4). The relative concentrations of Fos and Jun in the cell determine which dimer is preferentially formed and thus whether the AP-1 sites become activated. This simple case illustrates a general feature of transcription factors: they can interact with each other in a variety of ways to increase or decrease their binding affinity for specific control sites in DNA. Interactions between transcription factors can indeed be extremely complicated.

Small differences in the regulatory region of a gene can have significant effects. The transcription factor Pit-1 is needed to activate the genes for three hormones—growth hormone, prolactin, and thyrotropin—each of which is made by a different type of cell in the pituitary gland. But how does Pit-1 turn on the right gene in each cell type without activating the other two? Pit-1, like other transcription factors, has to bind to a regulatory sequence on its target genes and just a small sequence variation between the regulatory elements of the prolactin and growth hormone genes causes Pit-1 to bind very differently to the two. In prolactin-producing cells, Pit-1 is an activator of the prolactin gene, whereas when Pit-1 binds to the growth hormone gene regulatory region in the same cell, it represses it. Just two additional bases in the growth hormone regulatory region seem to underlie this very different response.

9.2 External signals can activate genes

Most of the molecules that act as signals during development are peptides or proteins. These bind to receptors in the cell membrane and do not themselves enter the cell. The signal is relayed to the cell nucleus by a process known as **signal transduction**, of which we already have had examples that include the Wnt/wingless, TGF-β, and hedgehog pathways. This can be a very complex process but the bare essentials of one type of signaling pathway involving a receptor tyrosine kinase are shown in Fig. 9.5. Intracellular signal transduction pathways of this sort involve the sequential activation of protein kinases. Binding of the external signal to a receptor results in the cytoplasmic domain of the receptor becoming phosphorylated. This leads to the activation of the Ras protein at the plasma membrane, which results in the Raf protein binding to it, which in turn results in the phosphorylation and activation of the protein kinase MEK. MEK then phosphorylates another kinase, ERK, which moves into the nucleus. There, it phosphorylates a transcription factor, activating gene expression (Fig. 9.5).

A rather different type of signal transduction pathway is that in which a signal at the cell surface leads to translocation of transcription factors that are stored as inactive cytoplasmic complexes into the nucleus. The activation of the receptor protein Toll in *Drosophila*, leading to dorsal protein entering the nucleus, is an example of this type (see Section 5.5).

Unlike protein hormones and growth factors, steroid hormones and retinoic acid are lipid soluble; they cross the plasma membrane unaided and enter the cell and have a simpler pathway of signal transduction than protein or peptide signals. Once inside the cell, steroids activate gene expression by initially binding to receptor proteins; the complex of steroid and receptor is then able to act as a transcriptional regulator, binding directly to control sites in the DNA to activate (or in some cases repress) transcription. In many cases, the control sites, known generally as steroid response elements, act as enhancers.

As we shall see in Chapter 12, steroid hormones produced in the testis are responsible for the secondary sexual characteristics that make male mammals different from females. In insects, the steroid hormone ecdysone is responsible for metamorphosis (see Chapter 14) and induces differentiation in a wide variety of cells. In these cases, the hormone can turn a whole variety of genes in the cell on and off. An example of the control of tissue-specific expression of a single gene is provided by the steroid hormone estrogen, which causes the chick oviduct to produce the protein ovalbumin, a major component of egg white. The continued presence of estrogen is required for transcription of the ovalbumin gene; when estrogen is withdrawn, ovalbumin mRNA and protein disappear.

The action of steroid hormones in the chick provides an excellent example of the tissue-specificity of the response to these hormones. For example, estrogen activates the ovalbumin gene in the cells of the oviduct, but the ovalbumin gene in liver cells is unaffected. This tissue-specific response in the oviduct cannot easily be accounted for by additional proteins binding to the regulatory regions. Rather, the explanation for the differential response in these two tissues is thought to be some preceding heritable change in the structure of the chromatin. This may, for example, allow transcription factors such as the steroid-hormone receptor complexes to have access to the ovalbumin gene in one cell type but not in the other. Other agents that act through this type of transcription factor complex are thyroid hormone and retinoic acid, which may be a developmental morphogen (see Sections 4.5, 10.5, and Chapter 13).

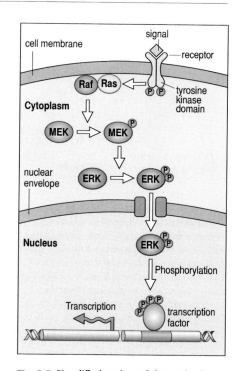

Fig. 9.5 Simplified outline of the Ras/Raf intracellular signaling pathway by which signals received at the cell membrane can alter gene expression. Binding of the signal molecule to the cell-surface receptor sets in train an intracellular cascade of protein phosphorylations. The cytoplasmic portion of the receptor comprises a tyrosine protein kinase, which is stimulated to phosphorylate itself when the signal molecule binds. The phosphorylated receptor then stimulates the activation of the GTP-binding proteins Ras and Raf (details not shown). Their activation leads to the phosphorylation of the protein kinase MEK. This then phosphorylates another protein kinase, ERK, which moves into the nucleus, where it can phosphorylate a transcription factor and so alter gene expression.

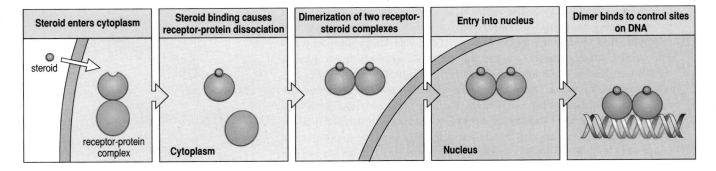

| Steroid enters cytoplasm | Steroid binding causes receptor-protein dissociation | Dimerization of two receptor-steroid complexes | Entry into nucleus | Dimer binds to control sites on DNA |

Fig. 9.6 Steroid hormones control transcription by binding to an intracellular receptor to form a transcriptional regulator. In the case of the glucocorticoid receptor shown here, the receptor is present in the cytoplasm as a complex with another protein. Steroid binding causes the receptor to dissociate from this protein and form a dimer with another steroid-bound receptor. This dimer enters the nucleus, where it binds to control sites on DNA and activates transcription.

Some steroid hormones, such as glucocorticoids, bind to a cytoplasmic receptor protein that is then translocated into the nucleus. For example, binding of a steroid to the glucocorticoid receptor in the cytoplasm results in the dissociation of the receptor from a cytoplasmic protein that keeps it inactive in the absence of steroids (Fig. 9.6). Two steroid–receptor complexes then join to form dimers, which move into the nucleus, bind to DNA, and activate specific genes. Other steroids have receptors that are already bound to their target DNA sequences even when the hormone is absent, but hormone binding is necessary to enable them to activate transcription.

9.3 Maintenance and inheritance of patterns of gene activity may depend on chemical and structural modifications of DNA as well as on regulatory proteins

A central feature of development in general, and cell differentiation in particular, is that some genes are maintained in an active state, while

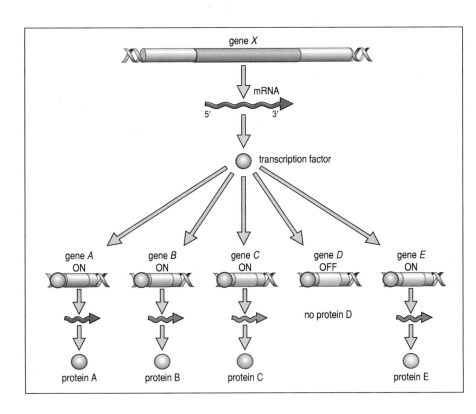

Fig. 9.7 A gene can regulate the activity of other genes when it codes for a transcription factor. Transcription factors can activate or repress the activity of genes by binding to their control regions. In the illustration, the activation of gene *X*, encoding a transcription factor, eventually leads to the production of four new proteins (A, B, C, and E), and the repression of protein D production. After Alberts, B., *et al.*: 1989.

others are repressed and inactive. Moreover, in many cells, including fibroblasts, liver cells, and myoblasts (cells that give rise to muscle cells), a particular pattern of gene activity can be reliably transmitted over many cell
cycles through cell division. It is important to bear in mind that the activity
of one gene can affect the activity of several, often many, other genes if it
codes for a transcription factor, a gene regulatory protein that can switch
other genes on or off (Fig. 9.7, opposite).

Eukaryotic genes, especially those active during development, generally
have complex control regions. These contain binding sites for a variety of
transcription factors, some of which may activate transcription and some
which may repress it. Whether or not an individual transcription factor
affects the activity of a particular gene depends on many factors: whether
the regulatory region of that gene contains a binding site for the transcription factor; what other gene regulatory proteins interact with that gene;
whether or not the transcription factor is phosphorylated; and whether it
binds to proteins or other molecules that can inactivate it. Very subtle
changes in a transcription factor, such as a change in just one amino acid,
can alter its activity by altering any of these properties.

Whether or not a gene is activated thus depends on the precise combination and levels of regulatory factors present. A mechanism for maintaining a differentiated pattern of gene activity could require the continued
presence of these specific gene regulatory proteins. The continued activity
of a gene, in both the differentiated cell itself and in its progeny, would
require the continual presence of the appropriate positive regulatory proteins. Continued inactivity of a gene would require the continual presence
of repressor proteins.

One way of keeping a gene active is for the gene product itself to act as
the positive regulatory protein (Fig. 9.8). All that is then required for the
pattern of gene activity to be maintained is the initial event that first activates the gene; once switched on it remains active. This type of positive

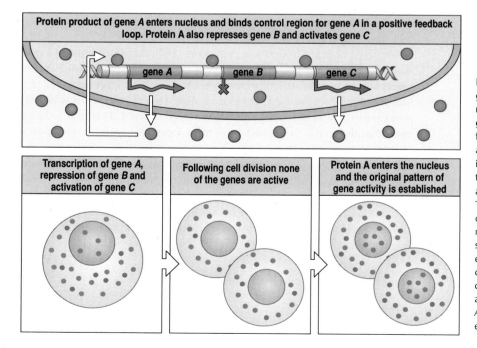

Fig. 9.8 **Continued expression of
gene regulatory proteins could
maintain a pattern of differentiated
gene activity.** Top panel: transcription
factor A is produced by gene A, and acts
as a positive gene regulatory protein for
its own control region. Once activated
therefore, gene A remains switched on
and the cell always contains A.
Transcription factor A also acts on the
control regions of genes B and C to
repress and activate them respectively,
setting up a cell-specific pattern of gene
expression. Bottom panels: after cell
division, the cytoplasm of both
daughter cells contains sufficient
amounts of protein A to reactivate gene
A, and thus maintain the pattern of
expression of genes B and C.

Protein product of gene *A* enters nucleus and binds control region for gene *A* in a positive feedback
loop. Protein A also represses gene *B* and activates gene *C*

gene *A*   gene *B*   gene *C*

Transcription of gene *A*,
repression of gene *B* and
activation of gene *C*

Following cell division none
of the genes are active

Protein A enters the nucleus
and the original pattern of
gene activity is established

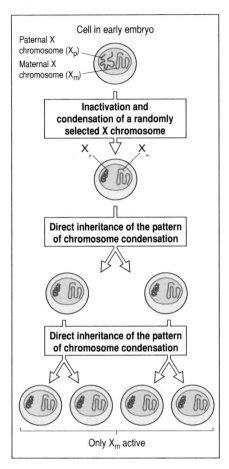

Cell in early embryo

Paternal X chromosome (X$_p$)

Maternal X chromosome (X$_m$)

Inactivation and condensation of a randomly selected X chromosome

X$_p$ X$_m$

Direct inheritance of the pattern of chromosome condensation

Direct inheritance of the pattern of chromosome condensation

Only X$_m$ active

Fig. 9.9 Inheritance of an inactivated X chromosome. In early mammalian embryos one of the two X chromosomes, either the paternal X (X$_p$) or the maternal X (X$_m$), is randomly inactivated. In the figure, X$_p$ is inactivated and this inactivation is maintained through many cell divisions. The inactivated chromosome becomes highly condensed. After Alberts, B., et al.: 1989.

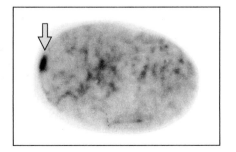

Fig. 9.10 Inactivated X chromosome (the Barr body). The photograph shows the Barr body (arrowed) in the interphase nucleus of a female human buccal cell.

Photograph courtesy of J. Delhanty.

feedback control of gene expression occurs in muscle cell differentiation, where the product of the gene *myoD* acts as an activator of *myoD* expression.

In *Drosophila*, genes that remain active throughout development and maintain the developmental pathway of a region are the selector genes (see Fig. 5.37). These genes encode transcription factors, many of which are homeodomain transcription factors (see Box 4A, p. 117), that act by establishing and maintaining a pattern of gene activity. Such a pattern can be inherited through large numbers of cell divisions if the cytoplasm contains sufficient quantities of the necessary regulatory proteins.

A different mechanism for maintaining and inheriting a pattern of active and inactive genes relies on chemical or structural alterations in the chromosomes. One type of change that could be involved is in the packing state of chromatin, which is the complex of DNA and associated proteins that makes up the chromosomes. Gross changes in chromosome morphology occur at each mitosis. The chromatin becomes packed into a much more compact structure and the chromosomes therefore become visible under the light microscope. In this condensed state, chromosomes are transcriptionally inactive. A change of this kind in localized regions of chromosomes could inactivate genes during development.

Evidence that changes in the packing state of chromatin could play a part in keeping genes inactive over a long period is provided by the phenomenon of X-chromosome inactivation (see Chapter 12). In female mammals, one of the two X chromosomes is inactivated and becomes highly condensed early in embryogenesis. Although it is replicated at each division it remains transcriptionally inactive. This pattern is preserved through many cell divisions in the individual's lifetime (Fig. 9.9). Inactivation is not irreversible, however, as the inactive X chromosome is reactivated during germ cell formation. The chromatin in the inactivated X chromosome is in a different physical state from that in the active one. During interphase, when the other chromosomes become extended, threads of chromatin and are no longer visible under the light microscope, the inactive X chromosome remains condensed into a highly compact form of chromatin—**heterochromatin**—which is not transcribed and is visible in human cells as a Barr body (Fig. 9.10).

Localized changes in chromatin packing could therefore provide a general mechanism for gene inactivation. A reflection of the possible differences in chromatin structure between active and inactive genes is that some active genes have hypersensitive sites, that is, they show a heightened sensitivity to digestion with the enzyme DNase I. This may indicate that the chromatin is packed in a more open structure in genes that can be transcribed, allowing access of the transcription apparatus to the DNA. DNase I digests DNA, but when the chromatin is packed more tightly, the DNA is inaccessible to DNase I, and is therefore protected from digestion.

Chemical modifications in the DNA itself are also associated with a shutdown of transcription. In vertebrates, methylation of cytosine at certain

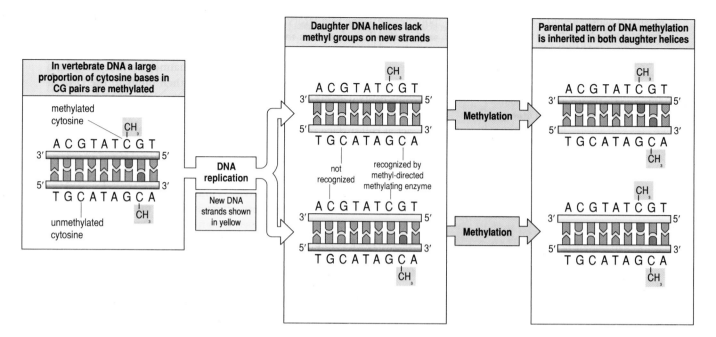

sites in the DNA is correlated with the absence of transcription in those regions. Moreover, the pattern of methylation can be faithfully inherited when the DNA replicates, thus providing a mechanism for passing on a pattern of gene activity to daughter cells (Fig. 9.11). DNA methylase recognizes the presence of a methyl group on a cytosine and methylates the corresponding group on one strand only, after DNA replication. The inactive X chromosome has a pattern of methylation which differs from that of the active X, and so it is likely that methylation also plays an important part in its inactivation.

Histone acetylation is closely linked to gene transcription. The identification of histone acetyltransferases (HATs) and the large multiprotein complexes in which they reside has yielded important insights into how these enzymes regulate transcription. The demonstration that HAT complexes interact with sequence-specific activator proteins illustrates how these complexes target specific genes. In addition to histones, some HATs can acetylate non-histone proteins, suggesting multiple roles for these enzymes. HAT activity is itself regulated by phosphorylation and interaction with other proteins. Histone deacetylases also play an important role by acting as repressors.

Summary

Transcriptional control is a key feature of cell differentiation. Tissue-specific expression of a eukaryotic gene depends on both the promoter and enhancer sequences of its control regions. The combinatorial action of different regulatory proteins on the control regions determines whether a gene is active or not. Small differences in regulatory regions can have significant effects. In some cases, cell-specific expression is due to the correct combination of regulatory proteins only being present in that cell type; in other cases, gene expression may be prevented by the gene being packed away in a state in which it is inaccessible to transcription factors and RNA

Fig. 9.11 Inheritance of the pattern of methylation on DNA. Many cytosine bases occurring in cytosine–guanine (CG) pairs in vertebrate DNA are methylated (a CH_3 group is added). When the DNA is replicated, the new complementary strand lacks methylation but the pattern can be reinstated on the new strand by methylase enzymes that recognize the methylated cytosine of the CG pair on the old DNA strand and methylate the cytosine of the corresponding CG pair on the new strand. After Alberts, B., et al.: 1989.

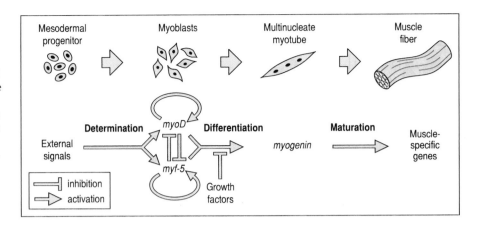

Fig. 9.13 Key features of the differentiation of vertebrate skeletal muscle. External signals initiate muscle differentiation by activating the genes *myoD* and *myf-5*. One of these genes is expressed preferentially (depending on the species) and their activity is mutually inhibitory and self-sustaining. The proteins encoded by these genes activate further genes, such as *myogenin*, which in turn activate expression of muscle-specific genes.

skeletal striated muscle although they have other defects; the most obvious abnormality in these mice is a shortening of the ribs. However, mice lacking both Myf-5 and MyoD not only lack all skeletal muscle, but also the myogenic precursor cells. Thus, these proteins are acting as myogenic determination factors. Indeed, cells in embryos in which mutant *myf-5* and *myoD* genes have been activated but do not make functional Myf-5 or MyoD protein mislocate, and may be integrated into other differentiation pathways such as cartilage and bone. Mice in which the *myogenin* gene has been knocked out lack most skeletal muscle, but precursor myoblasts are present. Myogenin is thus considered to be a myogenic differentiation factor rather than essential for muscle determination.

The MyoD family of transcription factors activate transcription of muscle-specific genes by binding to a nucleotide sequence called the E-box, which is present in the enhancer and/or promoter regions of these genes. MyoD forms a heterodimer with products of the E2.2 gene such as E_{12}, and this binds to the E-box. As the activity of MyoD and its relatives is essential for muscle cell differentiation, we must now look at how the activity of MyoD could be regulated.

9.5 The differentiation of muscle cells involves withdrawal from the cell cycle, but is reversible

Cell proliferation and differentiation of muscle cells are mutually exclusive phenomena. Skeletal myoblasts that are multiplying in culture do not differentiate. Only when proliferation ceases does differentiation begin. In the presence of protein growth factors, myoblasts that are expressing both MyoD and Myf-5 proteins continue to proliferate and do not differentiate into muscle. This means that the presence of MyoD and Myf-5 is not in itself sufficient for muscle differentiation. An additional signal or signals are required. In culture, this stimulus to differentiate can be supplied by the removal of growth factors from the medium. The myoblasts then withdraw from the cell cycle, cell fusion occurs, and differentiation takes place.

There is an intimate relationship between the proteins that control **cell cycle** progression—cell growth and cell division (see Chapter 14)—and the muscle determination and differentiation factors. MyoD and Myf-5 are phosphorylated by cyclin-dependent protein kinases, which become activated at key points in the cycle. Phosphorylation affects the activity of

MyoD and Myf-5 by regulating their degradation, making them more likely to be degraded. Thus, in actively proliferating cells, the levels of the myogenic factors will be kept low. A number of other proteins interact with myogenic factors or with their active partners such as E_{12}, and inhibit their activity. An example is the transcription factor Id, which is present at high concentrations in proliferating cells, and which probably helps to prevent premature activation of muscle-specific genes.

On the other hand, myogenic factors themselves can interfere with the cell cycle. Myogenin activates transcription of members of the p21 family of proteins, which block the cell cycle. This leads the cell to withdraw irreversibly from the cell cycle and to differentiate. MyoD interacts strongly and specifically with cyclin-dependent kinase 4, which inhibits both their activities. The retinoblastoma protein (Rb) is also essential for stable withdrawal from the cycle. In proliferating cells it is phosphorylated, and its dephosphorylation is essential for both withdrawal from the cell cycle and muscle differentiation. It also interacts with Id in inhibiting cell-cycle arrest.

The opposition between cell proliferation and terminal differentiation makes developmental sense. In those tissues where fully differentiated cells do not divide further, it is essential that sufficient cells are produced to make a functional structure, such as a muscle, before differentiation begins. The reliance on external signals to induce differentiation, or to permit a cell already poised for differentiation to start differentiating, is a way of ensuring that differentiation only occurs in the appropriate conditions.

But despite the apparently terminal nature of muscle cell differentiation, muscle cells provide an example of a terminally differentiated cell that can undergo **dedifferentiation**—the loss of differentiated characteristics—re-enter the cell cycle and even give rise to other cell types. The mouse gene *mox1* encodes a homeodomain-containing transcriptional repressor, which is normally expressed in undifferentiated tissue and whose ectopic expression can block muscle cell differentiation. When mouse muscle myotubes in culture were transfected with *mox1*, a significant number of the myotubes cleaved to form both smaller multinucleate myotubes and, more significantly, mononucleated cells that could undergo cell division. When these dividing mononucleated cells were cultured in the appropriate conditions, some expressed markers for other cell types such as cartilage or fat cells. A similar reprogramming of muscle cells can be seen in newt limb regeneration (Section 13.1).

We now turn to hematopoiesis, another well-studied example of differentiation. In this case, we consider how numerous different cell types differentiate from a **pluripotent** stem cell—a stem cell that can give rise to many different types of differentiated cell. We first look at the process of hematopoiesis and at the external signals that influence it, and then consider the control of gene expression in one cell type—the red blood cell.

9.6 All blood cells are derived from pluripotent stem cells

All the blood cells in the adult mammal originate from a population of pluripotent stem cells located in the bone marrow. These stem cells are self-renewing and are the precursors of progenitor cells that become irreversibly committed to one of the hematopoietic lineages at a later stage. Thus, hematopoiesis is in effect a complete developmental system in miniature, in which a single cell—the pluripotent **stem cell**—gives rise to numerous

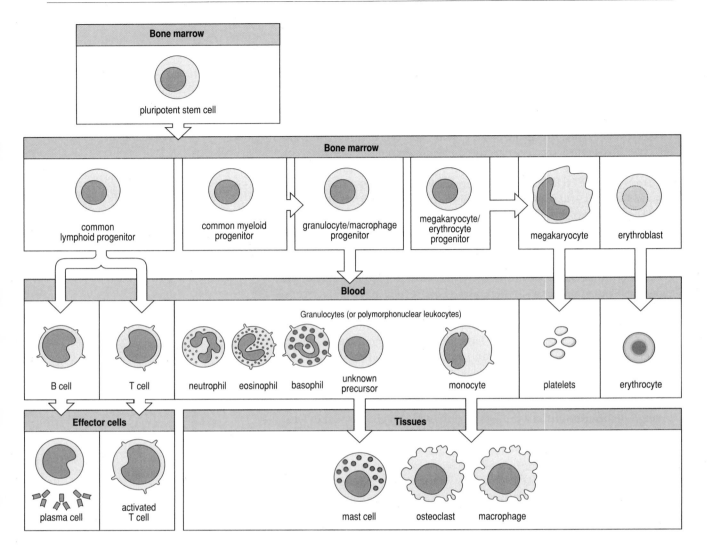

Fig. 9.14 The derivation of blood cells from pluripotent stem cells. The pluripotent stem cells, which are self-renewing, give rise to all of the blood cells, as well as to tissue macrophages, mast cells, and osteoclasts in bone. The stem cells are thought to give rise to uncommitted progenitors, which then split into more committed myeloid, erythroid, and lymphoid precursors. These then give rise to the different blood cell lineages (not all final differentiated cell types are shown). Hematopoietic growth factors influence the proliferation and differentiation of the different lineages. After Janeway, C.A., et al.: 2001.

different cell types. Most of these are short-lived and must be replaced continuously throughout adult life. Hematopoiesis is a particularly well-studied system because it has the advantage of being relatively accessible, as well as being of considerable medical importance. As a measure of complexity of the process, hematopoietic stem cells from fetal liver express almost 200 transcription factors, a similar number of membrane-associated proteins, and almost 150 signaling molecules.

The stem cells for hematopoiesis in mammals are derived from mesoderm in the yolk sac blood islands and a region of the aorta. They colonize a number of definitive blood-forming sites that include the fetal liver, thymus, spleen, and bone marrow. Mammalian blood cells comprise numerous fully differentiated cell types, together with immature cells at various stages of differentiation (Fig. 9.14). The cell types are derived from three main lineages—the erythroid, lymphoid, and myeloid lineages. The first yields the red blood cells, or erythrocytes, and the megakaryocytes, which give rise to blood platelets. The lymphoid lineage produces the lymphocytes, which include the two antigen-specific cell types of the immune system, the B and T lymphocytes. In mammals, B lymphocytes develop in the bone marrow, whereas T lymphocytes develop in the thymus, from precursors that originate from the pluripotent stem cells in the bone marrow and migrate from the bloodstream into the thymus. Both B and T lymphocytes undergo a further terminal differentiation after they encounter antigen. Terminally differentiated B lymphocytes become antibody-secreting plasma cells, while T cells undergo terminal differentiation into at least three functionally distinct effector cell types. The myeloid lineage gives rise to the rest of the white blood cells or leukocytes, which all derive from the bone marrow in adults. These are the eosinophils, neutrophils, and basophils (collectively known as granulocytes or polymorphonuclear leukocytes), mast cells, and monocytes. Monocytes differentiate into macrophages after they have left the bone marrow and entered tissues. Mast cells also reside in tissues.

Within the bone marrow, the different types of blood cells and their precursors are intimately mixed up with connective tissue cells—the bone marrow stromal cells. The pluripotent stem cell itself has not yet been definitively identified within this complex mixture, and its existence is inferred from the ability of bone marrow to reconstitute a complete blood system when transplanted into individuals whose own bone marrow has been destroyed. The key experiment used to show this is the transfusion of a suspension of bone marrow cells into a mouse that has received a potentially lethal dose of X-irradiation. The mouse would normally die as a result of lack of blood cells which, like other proliferating cells, are particularly sensitive to radiation, but transfusion of bone marrow cells reconstitutes the hematopoietic system and allows the mouse to recover.

The hematopoietic stem cell generates progenitor cells that become irreversibly committed to one or other of the various lineages leading to the different blood cell types. There is early commitment to either the myeloid or lymphoid lineage, for example, followed by later commitment to the lineages leading to the differentiated cell types. The hematopoietic system can be considered as a hierarchical system with the pluripotent stem cell at the top (see Fig. 9.14). This generates uncommitted precursors which then become committed to the main lineages, after which they undergo further rounds of proliferation and further commitment, finally undergoing overt differentiation into the different cell types.

Stem cells need to balance the requirement for self-renewal and a commitment to differentiation. Several thousand gene products are present in these cells, many of which they share in common with other cell types. However, an analysis of highly purified stem cells from mouse liver has identified 161 transcription factors, 174 cell-membrane-associated proteins, 28 secreted proteins, and 147 signaling molecules that have not previously been described. The control of stem cell behavior is obviously highly complex.

All this activity occurs in the microenvironment of the bone marrow stroma, and is regulated by external signals provided by the hematopoietic growth factors and other cytokines. These are thought to act mainly as permissive and selective signals, allowing already committed cells of a certain type to proliferate and differentiate. In this way, the numbers of the different types of blood cells can be regulated according to the individual's physiological need. Loss of blood results in an increase in red cell production, whereas infection leads to an increase in lymphocytes and other white blood cells.

9.7 Colony-stimulating factors and intrinsic changes control differentiation of the hematopoietic lineages

Hematopoiesis can be characterized by the activities of a hierarchy of transcription factors, whose overlapping patterns of expression specify the various cell lineages. Some are only expressed in immature cells and are not lineage specific—the proto-oncogene product c-Myb falls into this class. Others, and there are at least 20, have lineage specificity. The transcription factor GATA-2, for example, is present in all the myeloid and erythroid precursors but not in the precursors of the lymphoid system. GATA-1 is more specific, being present in only some of these lineages, and is required for red blood cell differentiation. Its effect can be concentration-dependent: eosinophils form when low levels of GATA-1 are present, whereas red cells and megakaryocytes develop at higher concentrations. However, GATA-1 is also expressed in the testis, which well illustrates the fact that it is the combination of transcription factors, and not any particular one, that regulates the expression of particular genes, and thus cell differentiation. Another key transcription factor for red cells is NF-E2. How then is the activity of all of these factors controlled? Signals delivered by extracellular protein growth factors and differentiation factors have a key role in this.

Studies on blood cell differentiation in culture have identified at least 20 extracellular proteins, known generally as colony-stimulating factors or hematopoietic growth factors, that can affect cell proliferation and cell differentiation at various points in hematopoiesis (Fig. 9.15). Both stimulators and inhibitors have been identified. Not all the factors controlling blood cell production are made by blood cells or stromal cells. The protein erythropoietin, for example, which induces the differentiation of committed red blood cell precursors, is mainly produced by the kidney, in response to physiological signals that indicate red blood cell depletion.

Despite the well-established role of these factors in blood cell proliferation and differentiation, there is evidence that the commitment of a cell to one or other pathway in the myeloid lineage might simply be a chance event. When the daughters of single early progenitor cells (which can give rise to a variety of blood cell types) are cultured separately but under similar conditions, both daughter cells usually give rise to colonies containing the same combination of cell types. But in about 20% of cases, the daughters give rise to dissimilar combinations of cells. The implication is that it may

Hematopoietic growth factors and their target cells	
Type of growth factor	Responding hematopoietic cells
Erythropoietin (EPO)	Erythroid progenitors
Granulocyte colony-stimulating factor (G-CSF)	Granulocytes
Interleukin-4	B cells, T cells
Interleukin-7	Lymphoid stem cells
Interleukin-3	Pluripotent precursor cells, megakaryocytes
Granulocyte-macrophage colony-stimulating factor (GM-CSF)	Pluripotent precursor cells, megakaryocytes
Interleukin-5	B cells, eosinophils
Interleukin-2	T cells
Interleukin-6	T cells, activated B cells, monocytes
Thrombopoietin	Megakaryocytes
Macrophage colony-stimulating factor (M-CSF)	Macrophages, granulocytes

Fig. 9.15 Hematopoietic factors and their target cells.

be an intrinsic property of the cells that a random event at cell division, at various points in the lineage, commits cells to different pathways. The function of growth factors would then be to promote the survival of particular cell lineages.

Among the plethora of growth factors, it is difficult to distinguish those that might be exerting specific effects on differentiation from those that may be required for the survival and proliferation of a lineage or group of lineages. The contrasting functions of three growth factors—granulocyte-macrophage colony-stimulating factor (GM-CSF), macrophage colony-stimulating factor (M-CSF), and granulocyte colony-stimulating factor (G-CSF)—are well established. GM-CSF is generally required for the development of most myeloid cells from the earliest progenitors that can be identified. In combination with G-CSF, however, it tends to stimulate only granulocyte (principally neutrophil) formation from the common granulocyte-macrophage progenitor. This is in contrast to M-CSF which, in combination with GM-CSF, tends to stimulate differentiation of monocytes (macrophages) from the same progenitors (Fig. 9.16).

These growth and differentiation factors have no strict specificity of activity, in the sense of each factor exerting a specific effect on one type of target cell only; rather, they act in different combinations on target cells to produce different results, the outcome also depending on the developmental history of their target.

9.8 Globin gene expression is controlled by distant upstream regulatory sequences

We now focus on one type of differentiated blood cell, the red blood cell or erythrocyte, to consider the transcription factors that are active in these cells during their differentiation, and that control expression of their cell-specific genes. The main feature of red blood cell differentiation is the synthesis of large amounts of the oxygen-carrying protein hemoglobin, which involves the coordinated regulation of two different sets of globin genes.

Vertebrate hemoglobin is a tetramer of two identical α-type and two identical β-type globin chains. The α-globin and β-globin genes belong to different multigene families. Each family consists of a cluster of genes and the two families are located on different chromosomes. In mammals, different members of each family are expressed at various stages of development so that distinct hemoglobins are produced during embryonic, fetal, and adult life. Here, we look at the control of expression of the β-globin genes, as examples of genes that are not only uniquely cell specific, but are also developmentally regulated.

The human β-globin gene cluster contains five genes—ε, γ_G, γ_A, δ, and β (Fig. 9.17). These genes are expressed at different times during development: ε is expressed in the early embryo in the embryonic yolk sac; the two γ genes, which differ by only one amino acid, are expressed in the fetal liver; and the δ and β genes are expressed in erythroid precursors in the adult bone marrow. The protein products of all these genes combine with globins encoded by the α-globin complex to form physiologically different hemoglobins at each of these three stages of development.

The control regions that regulate expression of the β-globin gene cluster are complex and extensive. Each gene has a promoter and control sites

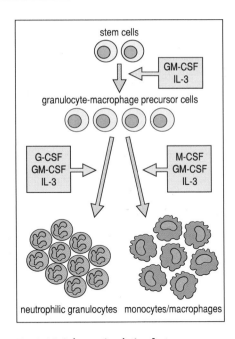

Fig. 9.16 Colony-stimulating factors can direct the differentiation of neutrophils and macrophages. Neutrophils and macrophages are derived from a common granulocyte-macrophage precursor cell. The choice of differentiation pathway can be determined by, for example, the growth factors G-CSF and M-CSF. The growth factors GM-CSF and IL-3 are also required, in combination, to promote the survival and proliferation of cells of the myeloid lineage generally. After Metcalf, D.: 1991.

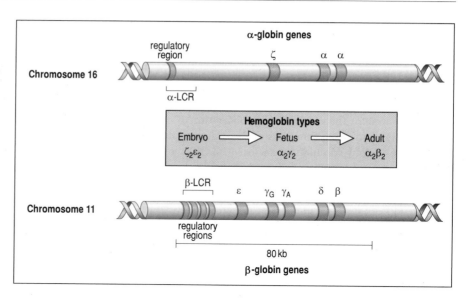

Fig. 9.17 The human globin genes.
Human hemoglobin is made up of two identical α-type globin subunits and two identical β-type globin subunits, which are encoded by α-family and β-family genes on chromosomes 16 and 11, respectively. The composition of human hemoglobin changes during development. In the embryo it is made up of ζ (α-type) and ε (β-type) subunits; in the fetal liver, of α and γ subunits; and in the adult most of the hemoglobin is made up of α and β subunits, but a small proportion is made up of α and δ subunits. The regulatory regions α-LCR and β-LCR are locus control regions involved in switching on the hemoglobin genes at different stages.

immediately upstream of the transcription starting point, and there is also an enhancer downstream (to the 3′ side) of the β-globin gene, which is the last gene in the cluster (Fig. 9.18). But these local control sequences, which contain binding sites for transcription factors specific for erythroid cells, as well as for other more widespread transcriptional activators, are not sufficient to provide properly regulated expression of the β-globin genes.

Fully regulated expression of the β-globin gene cluster depends on a region a considerable distance upstream from the ε gene. This is the locus control region (LCR) (see Fig. 9.17), which stretches over some 10,000 base pairs at between 5000 and 18,000 base pairs from the 5′ end of the ε gene. The LCR confers high levels of expression of any β-family gene linked to it, and is also involved in directing the developmentally correct sequence of expression of the whole β-globin gene cluster in transgenic mice, even though the β-globin gene itself, for example, is around 50,000 base pairs away from the extreme 5′ end of the LCR. A similar control region has been found upstream of the α-globin gene cluster. Indeed, the globin LCRs are

Fig. 9.18 The control regions of the β-globin gene. The β-globin gene is part of a complex of other β-type globin genes and is only expressed in the adult. Binding sites for the relatively erythroid-specific transcription factor GATA-1, as well as for other non-tissue-specific transcription factors, such as NF1 and CP1, are located in the control regions. Upstream of the whole β-globin cluster, in the locus control region (LCR), are additional control regions required for high-level expression and full developmental regulation of the β-globin genes.

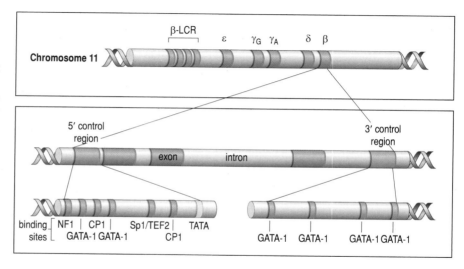

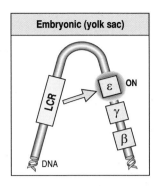 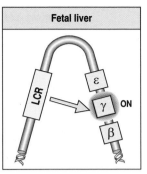 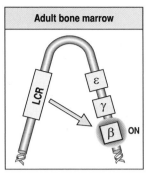

Fig. 9.19 A possible mechanism for the successive activation of β-family globin genes by the LCR during development. The LCR (locus control region) is thought to make contact with the promoter of each gene in succession, at different stages of development, thus controlling their temporal expression. After Crossley, M., *et al.*: 1993.

probably the best-characterized determinants of tissue-specific gene expression discovered so far.

An intriguing feature of globin gene expression is the successive switching on and off of different globin genes during development. An attractive model for the control of globin gene switching envisages an interaction of LCR-bound proteins with proteins bound to the promoters of successive globin genes (Fig. 9.19). The DNA between the LCR region and the globin genes is thought to loop such that proteins binding to the LCR can physically interact with proteins bound to the globin gene promoters. Thus, in erythrocyte precursors in the embryonic yolk sac the LCR would interact with the ε promoter, in fetal liver it would interact with the two γ promoters, and in the adult bone marrow with the β-gene promoter.

9.9 Neural crest cells differentiate into several cell types

We now turn to the neural crest cells, which give rise to a remarkable number of cell types, including neurons and glial cells, cartilage, pigment cells, and chromaffin cells. Neural crest cells arise from the ectoderm (see Section 2.1) and are recognizable as discrete cells when they emerge from the neural tube; this involves an epithelial to mesenchyme transition. Precursor cells within the tube can give rise not only to neural crest but also to neural tube and epidermis. The specification of the neural crest is the result of an inductive interaction between the neural plate and the adjacent presumptive ectoderm, and involves Wnt and BMP signals. The cells then migrate to many sites (Chapter 8). BMPs are required for the initiation of migration.

Differentiation of neural crest cells bears some resemblance to differentiation of the hematopoietic system, in that an apparently uncommitted multipotential population of neural crest stem cells can give rise to a wide variety of cell types. But unlike the blood system, neural crest contributes to several different tissue types. At the time of closure of the neural tube in vertebrate embryos, neural crest cells can be seen on each side. These cells leave the neural epithelium and migrate to give rise to much of the skeleton of the head, the neurons and glia of the peripheral nervous system (which includes the sensory and autonomic nervous systems), endocrine cells, such as the chromaffin cells of the adrenal medulla, and melanocytes—pigmented cells found in the skin and other tissues (Fig. 9.20). The neural crest thus gives rise to both ectodermal and mesodermal types of tissues. This tremendous potential was originally discovered by removing the neural crest from amphibian embryos and noting which structures failed to develop.

Fig. 9.20 Derivatives of the neural crest. Neural crest cells give rise to a wide variety of cell types that include melanocytes, cartilage, glia, and a variety of neurons distinguished by their functional specializations and the neurotransmitters they produce. Cholinergic neurons use acetylcholine as their neurotransmitter, adrenergic neurons principally use noradrenaline (norepinephrine), and peptidergic and serotonergic neurons produce peptide neurotransmitters and serotonin, respectively.

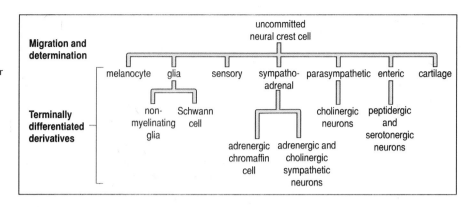

The differentiation of neural crest cells has been studied in detail using quail neural crest transplanted into chick embryos. Quail cells contain a distinctive nucleus that distinguishes them from chick cells. It is therefore possible to track the transplanted quail cells and see which tissues they contribute to and what cell types they differentiate into (see Fig. 8.35). To construct a fate map (see Section 1.10) of the neural crest, small regions of quail neural tube are grafted to equivalent positions in chick embryos of the same age, prior to neural crest migration. The fate map shows a general correspondence between the position of crest cells along the antero-posterior axis and the axial position of the cells and tissues they will give rise to (Fig. 9.21, top two panels). For example, the neural crest cells that give rise to tissues in the facial and pharyngeal regions are derived from crest anterior to somite 5, whereas ganglion cells of the autonomic sympathetic system come from crest posterior to somite 5.

To what extent is the fate of neural crest cells fixed before migration? Many cells are unquestionably stem cells. Single neural crest cells injected with a tracer shortly after they have left the neural tube can be seen to give rise to a number of different cell types, both neuronal and non-neuronal. Also, by changing the position of neural crest before cells start to migrate, it has been shown that the developmental potential of these cells is much greater than their normal fate would suggest (see Fig. 9.21, bottom panel). Quail neural crest from various positions can be grafted to different sites along the antero-posterior axis of chick embryos of the same age. In general, this change in position has little effect on development and the neural crest develops in accordance with its new position. There is, however, some restriction of potential with respect to formation of the skeletal and connective tissues of the head. Neural crest originating posterior to somite 5 cannot give rise to these structures when transplanted into an appropriate position in the chick embryo.

The broad developmental potential of the neural crest is also demonstrated by the behavior of neural crest cells in culture. Almost all of the cell types to which the neural crest can give rise will differentiate in tissue culture. Developmental potential can be studied in cultures derived from single neural crest cells. Most of the clones from cells taken at the time of migration contain more than one cell type, showing that the neural crest cell at the time of culture is multipotent. As neural crest cells migrate, their potential decreases progressively. So both the size of the clones and the diversity of cell types they contain get smaller. Thus, shortly after the beginning of migration, the neural crest is a mixed population of

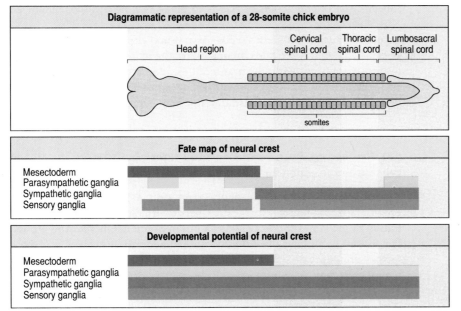

Fig. 9.21 Fate and developmental potential of neural crest cells. The fate map of the neural crest (middle panel) shows that there is a general correspondence between the position of the neural crest cells along the antero-posterior axis and the position of the structure to which these cells give rise (top panel). For example, the crest that gives rise to the mesectoderm of the head comes from the anterior region. By contrast, the developmental potential, with the exception of the presumptive mesectoderm, is very much greater than the presumptive fate (bottom panel).

pluripotent cells, together with cells whose potential is restricted. It is possible to isolate a multipotent stem cell from neural crest that can give rise to neurons, glia, and smooth muscle. Similar multipotent cells have even been isolated from the peripheral nerves of mammalian embryos days after neural crest migration.

What then determines the normal fate of neural crest cells? It is clear that environmental influences they encounter during migration must affect their differentiation. Numerous factors have been identified that affect differentiation and proliferation of neural crest cells in culture, but their role in normal development still remains unclear. So here we focus on just a few examples.

9.10 Steroid hormones and polypeptide growth factors specify chromaffin cells and sympathetic neurons

The chromaffin cells of the adrenal medulla are small cells containing numerous cytoplasmic granules and secrete adrenaline (epinephrine) into the circulation. By contrast, sympathetic neurons primarily secrete noradrenaline (norepinephrine) and are much larger cells, with axons and dendrites. Yet both these cell types arise from the same progenitor cell, which comes from the neural crest. The progenitor cells can be cultured, and under the action of particular inducing substances will differentiate into either sympathetic neurons or chromaffin cells (Fig. 9.22).

The development of chromaffin cells from the progenitor cell requires a high concentration of glucocorticoid hormones, which are synthesized *in vivo* by the cells of the surrounding adrenal cortex. Glucocorticoids inhibit differentiation along the neuronal pathway and also promote the maturation of chromaffin cells. In contrast, differentiation into neurons is induced by two growth factors, fibroblast growth factor (FGF) and nerve growth factor (NGF), which act sequentially. FGF alone can induce neuron differentiation, but the cells will not survive unless NGF is also present.

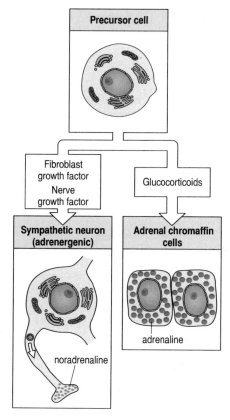

Fig. 9.22 Control of differentiation of neural crest derivatives by external signals. Both adrenergic sympathetic neurons and adrenal chromaffin cells are derived from a common precursor cell. Peptide growth factors and glucocorticoids can specify the pathway of differentiation. After Doupe, A.J., *et al.*: 1985.

Indeed, part of the action of FGF is to make cell survival dependent on NGF.

Differentiated chromaffin cells in culture continue to require the presence of glucocorticoids to maintain the chromaffin phenotype. If glucocorticoids are withdrawn and FGF is added to the culture medium, the cells differentiate into sympathetic neurons. This is just one example of the phenomenon of transdifferentiation, which we shall return to later.

9.11 Neural crest diversification involves signals both for specification of cell fate and selection for cell survival

Experiments on cultured rat neural crest cells have shown that several growth factors can direct their fate. Glial growth factor promotes differentiation of glia and suppresses neuron differentiation, whereas the growth factor BMP-2 promotes the differentiation of neurons. In the embryo, these factors are expressed at sites consistent with their role in determining neural crest fate.

Neural crest cells migrate to give rise to sensory dorsal root ganglia and the autonomic ganglia of the peripheral nervous system. These ganglia contain both neurons and glia, the neurons differentiating first. So what determines whether a neural crest cell gives rise to a neuron or to glia? There is evidence that activation of Notch in neural crest stem cells inhibits neuronal formation and promotes differentiation into glia. There is also evidence that Notch ligands such as Delta are expressed on neuroblasts, the precursors of neurons. Thus, once neuroblasts have formed, under the influence of neurogenic signals such as BMP-2, they may provide a feedback signal that prevents the surrounding cells from developing further as neuroblasts and promotes formation of glia instead.

Two types of spinal ganglia, both segmentally arranged, develop from the neural crest: the dorsal root sensory ganglia, which contain cholinergic neurons, and the sympathetic ganglia of the autonomic nervous system, in which the vast majority of neurons are adrenergic. A surprising find was that early dorsal root sensory ganglia contain cells that can still give rise to neurons of a different type, namely sympathetic neurons. This is shown by grafting fragments of dorsal root ganglia back into the neural crest pathway of younger embryos. Cells from the dorsal root fragment migrate for a second time and some enter sympathetic ganglia, where they develop into adrenergic neurons. Some also migrate to the adrenal medulla and differentiate into adrenergic cells. In marked contrast, developing sympathetic ganglia are never found to contain precursor cells that could give rise to sensory neurons.

An explanation for the absence of prospective sensory cells in sympathetic ganglia is that their survival is dependent on some factor produced by the neural tube. Dorsal root ganglia are in close proximity to the neural tube, but sympathetic ganglia are formed close to the dorsal aorta and far from the neural tube. This would mean that when cells migrate to the site of the sympathetic ganglia, presumptive sensory cells would die. This proposition can be tested by inserting an impermeable barrier between the neural tube and the newly forming dorsal root ganglia. The presumptive sensory cells die and the dorsal root ganglia do not form, whereas the sympathetic ganglia are unaffected. Furthermore, an extract from the embryonic neural tube can rescue the cells *in vivo*, and enable sensory neurons to develop. The active factor is probably brain-derived neurotrophic factor (BDNF). Thus, locally produced factors seem to be involved in selecting particular populations of neural crest cells for survival.

Melanocytes come from neural crest cells that follow a dorsal migration route under the ectoderm. In zebrafish, Wnt signaling induces pigment cell formation from neural crest cells. Mice homozygous for either the *W* (*white spotting*) or *Steel* mutations (genes originally identified by unusual coat color) lack melanocytes in the skin and other tissues. (They are also deficient in hematopoietic stem cells, and their germ cell development is affected.) Although they apparently have the same gross effect, the genes code for two quite different proteins. *W* codes for a cell-surface receptor known as Kit, and is expressed on developing melanoblasts, which are the undifferentiated precursors of skin melanocytes. *Steel*, in contrast, codes for a protein produced by the fibroblasts and keratinocytes that surround the melanoblast, and which is the ligand for Kit. Mutations in either gene block melanocyte differentiation, showing that the signal generated by the interaction of the Steel protein and Kit is essential for the differentiation of melanocytes (Fig. 9.23).

9.12 The epithelia of adult mammalian skin and gut are continually replaced by derivatives of stem cells

Adult vertebrate skin is composed of three layers—the **dermis**, which mainly contains fibroblasts, the protective outer **epidermis**, which mainly contains **keratinocytes**, and the **basal lamina** or basement membrane, composed of extracellular matrix, which separates the epithelial epidermis from the dermis.

Because of the protective function of the epidermis, cells are continuously lost from its outer surface and must be replaced. Cells are replaced from a basal layer of proliferating epidermal cells in contact with the basal lamina. These proliferating cells are derived from stem cells within the basal layer. The stem cells form a special population in the basal layer that multiplies relatively slowly and retains proliferative potential throughout the life of the organism. Stem cells divide asymmetrically, with one daughter remaining a stem cell and the other daughter becoming committed to differentiation. After becoming committed, this cell divides a few times more, and its progeny differentiate further after leaving the basal layer. As they move through the skin they mature, until by the time they reach the outermost layer they are completely filled with keratin and their membranes are strengthened with the protein involucrin. The dead cells eventually flake off the surface (Fig. 9.24).

Cell adhesion molecules play a key role in keratinocyte differentiation. The basal cell layer is attached to the basement membrane by specialized cell junctions that include both **hemi-desmosomes** and focal contacts. The epidermal cells in the upper layers are connected to each other through extensive **desmosomal junctions**. The integrin family of cell adhesion molecules (see Box 8A, p. 255) plays the most important part in keratinocyte development. The expression of integrins is largely confined to the basal layer, where they mediate attachment of cells to the basal lamina at the hemi-desmosome junctions. Stem cells express higher levels of $\alpha_2\beta_1$

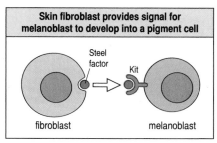

Skin fibroblast provides signal for melanoblast to develop into a pigment cell

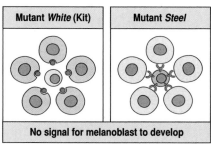

Fig. 9.23 The receptor Kit and its ligand the Steel factor are involved in melanoblast differentiation. When either the gene encoding Kit (the gene *W*) or the *Steel* gene are mutant, melanoblasts fail to differentiate into melanocytes.

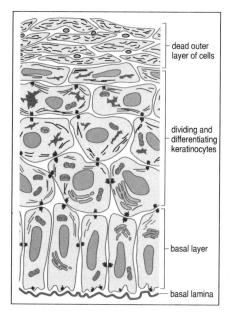

dead outer layer of cells

dividing and differentiating keratinocytes

basal layer

basal lamina

Fig. 9.24 Differentiation of keratinocytes in human epidermis. Descendants of stem cells, which will become keratinocytes, are detached from the basal lamina, divide several times, and leave the basal layer before beginning to differentiate. Differentiation of keratinocytes involves the production of large amounts of the intermediate filament protein keratin. In the intermediate layers, the cells are still large and metabolically active, whereas in the outer epidermal layers, the cells lose their nuclei, become filled with keratin filaments, and their membranes become insoluble owing to deposition of the protein involucrin. The dead cells are eventually shed from the skin surface.

and α3β1 integrins (which are receptors for collagen and laminin) than other cells of the basal layer, and these can be used to determine the cells' locations within the layer. It may be that these high levels of integrins keep the cells in the basal layer. It is likely that components of the basal lamina actively inhibit keratinocyte differentiation; differentiation and migration out of the basal layer only begins when a cell becomes detached from the basal lamina. This is accompanied by a decrease in integrins on the cell surface.

The epithelial cells lining the gut are also continuously replaced from stem cells. In the small intestine, the epithelial cells form a single-layered epithelium. This is highly folded into villi that project into the gut, and crypts that penetrate the underlying connective tissue. Cells are continuously shed from the tips of the villi, while the stem cells that produce their replacements lie near the base of the crypts. The new cells generated by the stem cells move upward toward the tips of the villi. En route, they multiply in the lower half of the crypt (Fig. 9.25).

A single crypt in the small intestine of the mouse contains about 250 cells of four main differentiated cell types. About 150 of the cells are proliferative, dividing about twice a day so that the crypt produces 300 new cells each day; about 12 cells leave the crypt each hour. It is estimated that there is just one stem cell per crypt, but some 30–40 potential stem cells that may acquire stem cell properties following injury or other trauma.

9.13 Many stem cells have a greater potential for differentiation than they normally show

We have already discussed the stem cells that are responsible for hematopoiesis, generation of neural cells, and renewal of the epidermis and the gut lining. In normal circumstances all these stem cells give rise to a limited range of cell types. If placed in a different environment, however, they can undergo a surprising range of differentiation. Ectoderm-derived neural stem cells from mouse embryos, which normally form neurons or glia, developed into blood cells (normally derived from mesodermal cells) when transfused into mice whose bone marrow had been destroyed. Equally surprising, adult mouse bone marrow stromal cells, which contain mesenchymal stem cells that can give rise to muscle cells and blood vessels, have been observed to give rise to neuron-like cells when injected into mouse brain. And highly purified bone marrow hematopoietic stem cells have been shown to give rise to liver cells (hepatocytes) on transplantation.

The primary example of almost completely totipotent stem cells are the **embryonic stem cells** of the inner cell mass, which have already been mentioned in Section 3.11. Mammalian embryonic stem cells (**ES cells**) derived from the inner cell mass can be maintained in culture for long

Fig. 9.25 Epithelial cells lining the mammalian gut are continually replaced. The upper panel shows the villi—epithelial extensions covering the wall of the gut. The lower panel is a detail of a crypt and villus. The base of the crypts contains stem cells. They give rise to cells that proliferate and move upward. These differentiate into epithelium in the projecting villus, and are shed from the tips. The total time of transit is about 4 days. After Alberts, B., *et al.*: 1989.

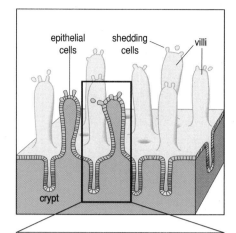

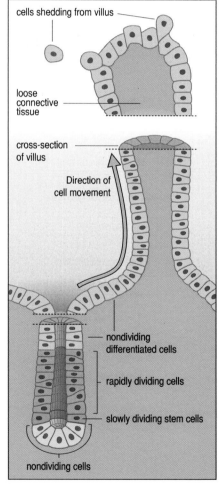

periods, apparently indefinitely. If cultured ES cells are injected into a blastocyst which is then returned to the uterus, they can contribute to all types of cell in the embryo that develops (see Box 3C, p. 89). They can also give rise to many cell types in culture. To maintain their pluripotent state in culture, ES cells must both express the transcription factor Oct-4 and be exposed to the cytokine leukemia inhibitory factor (LIF). If LIF is withdrawn, the ES cells spontaneously aggregate into embryo-like bodies and differentiate into many cell types, including heart muscle, blood islands, neurons, pigmented cells, epithelia, fat cells, and macrophages. By manipulating the culture conditions, particularly in respect of the growth factors present, ES cells can be made to differentiate into a particular cell type. One example is the development of stem cells for the glia, the supporting cells of the nervous system. If aggregates of ES cells are first propagated in a medium of FGF-2, then have EGF added, and are then cultured in FGF-2 and platelet-derived growth factor, they continue to proliferate. But when the growth factors are withdrawn they differentiate into either of two glial cell types—astrocytes or oligodendrocytes. Other combinations of growth factors can lead to the differentiation of ES cells into vascular smooth muscle, fat cells, macrophages or red blood cells. This ability to manipulate ES cells has clear implications for the possible generation of cell types and organs for medical application, particularly when combined with cell cloning.

9.14 Differentiation of cells that make antibodies is irreversible as a result of loss of DNA

Cell differentiation does not usually involve any permanent alteration to the cell's DNA, but is caused chiefly by reversible associations of gene regulatory proteins with DNA and reversible enzymatic modifications of DNA and chromatin proteins that produce changes in the pattern of gene expression. There are, however, rare cases in which the DNA in the differentiated cell becomes irreversibly altered. The only known examples of this in vertebrates are the B and T lymphocytes of the vertebrate immune system. In this case, the changes in gene expression that give these cells their ability to recognize and respond to foreign antigens are accompanied by irreversible DNA rearrangements which involve loss of DNA. We look here at what happens during the differentiation of B lymphocytes—the cells that give rise to antibody-producing plasma cells.

It has been estimated that the human immune system can produce up to 10^{15} different antibody molecules, which vary in their ability to bind different antigens. To encode each antibody separately in the genome is obviously impossible, as it would require a number of genes far greater than any genome could possibly contain. This amazing diversity arises instead from a unique combinatorial mechanism, in which complete antibody genes are assembled during cell differentiation by combining different gene

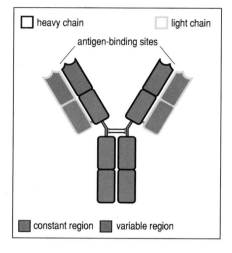

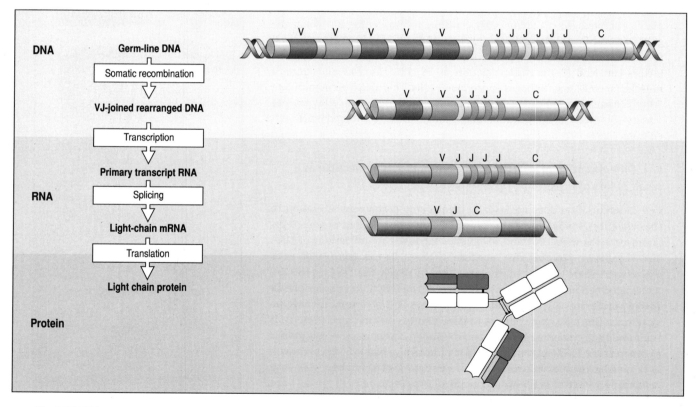

Fig. 9.26 Schematic drawing of an antibody molecule. The molecule is made up of two light chains and two heavy chains. Most of the molecule has a constant structure but the antigen-binding sites are located in the variable region. After Janeway, C.A., *et al.*: 2001.

segments from a relatively small collection (no more than a few hundred) encoded in the genome.

An antibody, or immunoglobulin, molecule is Y-shaped (Fig. 9.26) and is composed of two identical 'light' chains and two identical, larger 'heavy' chains. There are two antigen-binding sites on each antibody, at the ends of each arm of the Y, which are formed by parts of the heavy and light chains. The antigen-binding site determines the antigen specificity of the antibody and varies in structure from one type of antibody molecule to the next. The regions of the light and heavy chains that form the antigen-binding site are known as the variable regions, as they vary in amino acid sequence from one antibody to another. The remainder of the antibody molecule is more constant between different antibodies. It is the variable regions of the genes for light and heavy chains that are assembled by irreversible DNA rearrangements during B-lymphocyte development.

Fig. 9.27 DNA rearrangements generate a functional immunoglobulin light-chain gene. At the immunoglobulin light-chain locus, germline DNA carries a set of V gene segments and a set of J gene segments. During B-lymphocyte development, one V gene segment and one J gene segment are joined at random by somatic recombination to give a DNA sequence coding for a unique light-chain variable region. This is transcribed into RNA, together with the constant (C) region sequence. RNA splicing removes the RNA between the end of the J segment and the beginning of the C region to give a light-chain mRNA, which is then translated into protein. In the antibody molecule (illustrated bottom right), the V, J, and C regions of the light-chain protein are indicated in red, yellow, and blue, respectively. After Janeway, C.A., *et al.*: 2001.

Each differentiated B cell expresses just one type of heavy chain and one type of light chain, and produces an immunoglobulin of a single antigen specificity. Each heavy-chain variable region is encoded by a DNA sequence made up from three different gene segments (called V, D, and J), whereas the light-chain variable region is encoded by just two gene segments (V and J). During the development of a B lymphocyte, several DNA recombination events occur which assemble complete heavy- and light-chain genes from these segments. Because there are multiple copies of the different types of gene segment, many different combinations can be assembled, and thus many different antibody-binding sites generated. The principle of gene assembly can be illustrated for the light-chain gene. During B-lymphocyte differentiation, one V gene segment is moved by a DNA recombination event to lie next to one of the J gene segments. The rearranged variable region, together with the DNA sequence encoding the constant region, is then transcribed into one long RNA, which is spliced to make mRNA encoding the complete light chain (Fig. 9.27, opposite). Millions of B lymphocytes are produced by the immune system each day; by joining up different gene segments in different B cells, considerable antibody diversity can be achieved.

9.15 Programmed cell death is under genetic control

Selective cell death is a normal part of development. It is involved, for example, in the morphogenesis of the vertebrate limb, where interdigital cell death is essential for separating the digits (see Chapter 10). Programmed cell death is particularly important in the development of the nematode (see Chapter 6): of the 1090 somatic cells that come from the egg, 131 die. In the development of the vertebrate nervous system, there is extensive death of neurons. In all these cases, the dying cell undergoes a type of cell suicide known as **apoptosis**, which requires both RNA and protein synthesis and has certain characteristic features. The cytosolic calcium concentration rises and leads to the activation of an endonuclease, which fragments the chromatin. The cell contents remain bounded by a cell membrane throughout, even though the cell breaks up into fragments. The dying cell is finally phagocytosed by scavenger cells. These features distinguish apoptosis from cell death due to pathological damage, where the whole cell tends to swell and eventually burst open (lyse); this type of cell death is known as **necrosis**.

Studies of cell death in the nematode have shown that although different types of cells die, their death is initiated through a common mechanism involving the actions of two genes—*ced-3* and *ced-4*. In nematodes with mutations that inactivate either of these two genes, none of the 131 cells that normally die do so, but appear instead to differentiate in the same way as their sister cells. Such worms appear to survive well and die at the usual age of several weeks.

The nematode's cell death program is under the control of another gene, *ced-9*, which acts, it seems, as a brake on the program (Fig. 9.28). If *ced-9* is inactivated by mutation, many cells that normally do not die will die. And if the *ced-9* gene is made abnormally active by mutation, no cell deaths occur. It appears that in many cells that normally do not die, the ced-9 protein prevents the ced-4 protein from activating the ced-3 protein. The ced-3 protein is a pro-enzyme which, when activated, becomes a caspase, a type of

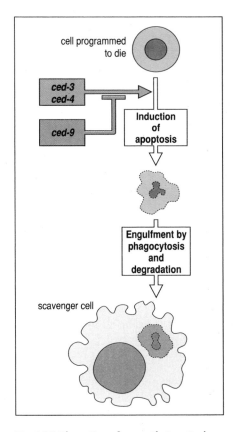

Fig. 9.28 The action of genes that control programmed cell death in the nematode. Programmed cell death requires the action of the genes *ced-3* and *ced-4*. These genes induce apoptosis (programmed cell death). The actions of *ced-3* and *ced-4* are inhibited by the gene *ced-9*. If this gene is active in a cell, the cell will not undergo apoptosis.

protease involved in apoptosis. The mammalian *Bcl-2* genes are similar to *ced-9* and control apoptosis in mammals. The sequence homologies between *Bcl-2* and *ced-9* are so strong that *Bcl-2* introduced into nematodes will function in place of *ced-9*. As in the nematode, the mammalian homolog of *ced-3* encodes a caspase, which initiates a proteolytic cascade that leads to cell death.

Far from programmed cell death being a rare phenomenon, there is the distinct possibility that in animals the cells in all tissues are intrinsically programmed to undergo cell death, and are only prevented from dying by positive control signals from neighboring cells.

Summary

In general, complex combinations of transcription factors control cell differentiation, and their expression and activity are influenced by external signals. Some transcription factors are common to many cell types, but others have a very restricted pattern of expression. Muscle differentiation can be brought about by the expression of genes such as *myoD*, that initiate the differentiation program. The products of these genes are transcription factors, which can bind to the control regions of muscle-specific genes. The protein products of the *myoD* gene family are themselves subject to control by external signals and by interactions with other proteins. The hematopoietic system shows how a single progenitor cell, the stem cell, can both be self-renewing and can also generate a variety of different lineages. The differentiation of the various lineages seems to depend on external signals, although there is evidence for autonomous diversification, and external factors are essential for survival and proliferation of specific cell types. Similar processes also occur in the diversification of the neural crest. Before migration, single crest cells can have a broad developmental potential, and environmental signals can both direct the pathways of neural crest differentiation and promote survival of particular cell types. The epithelial of mammalian gut and skin are replaced by derivatives of stem cells. Differen-

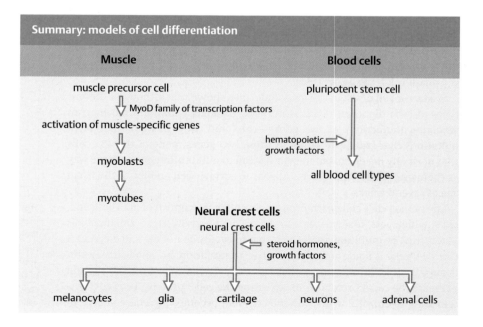

Summary: models of cell differentiation

Muscle

Blood cells

muscle precursor cell

pluripotent stem cell

⇩ MyoD family of transcription factors

activation of muscle-specific genes

hematopoietic ⇨ growth factors

⇩

myoblasts

⇩

myotubes

all blood cell types

Neural crest cells

neural crest cells

⇦ steroid hormones, growth factors

melanocytes   glia   cartilage   neurons   adrenal cells

tiation of cells very rarely involves irreversible change to the DNA. In vertebrates, this occurs in lymphocytes in the assembly of the genes for immunoglobulins and T-cell receptors. Programmed cell death is a common fate for cells, especially during development, and there is evidence that positively acting signals are generally required to prevent programmed cell death and allow cells to continue to survive. Embryonic stem cells and stem cells from other tissues can generate a wide variety of cell types in culture.

The reversibility of patterns of gene activity

In the previous parts of this chapter we have looked at examples of genes being turned on and off during development and have considered some of the mechanisms involved. These chiefly involve the binding of different combinations of gene regulatory proteins to the control regions of genes and structural changes in the chromatin such as acetylation of histone proteins and methylation of the DNA, which are carried out by enzymes. How reversible are these changes, and thus how reversible is cell differentiation? Clearly, there are some cases, such as the developing lymphocyte, where the changes are irreversible. But such physical DNA rearrangements are very rare. In the majority of cases, the gene regulatory mechanisms seem, in principle, to be reversible. One way of finding out whether they can be reversed in practice is to place the nucleus of a differentiated cell in a different cytoplasmic environment, containing a different set of potential gene regulatory proteins.

9.16 Nuclei of differentiated cells can support development of the egg

The most dramatic experiments addressing this question have investigated the ability of nuclei from cells at different stages of development to replace the nucleus of a fertilized egg and to support normal development. If they can do this, it would indicate that no irreversible changes have occurred to the genome during differentiation. It would also show that a particular pattern of nuclear gene activity is determined by whatever transcription factors and other regulatory proteins are being synthesized in the cytoplasm of the cell. Such experiments were initially carried out using the fertilized eggs of amphibians, which are particularly robust to experimental manipulation.

In the unfertilized eggs of *Xenopus*, the nucleus lies directly below the surface at the animal pole. A dose of ultraviolet radiation directed at the animal pole destroys the DNA within the nucleus, and thus effectively removes all nuclear function. These enucleated eggs are then injected with a nucleus taken from a cell at a later stage of development to see whether it can function in place of the inactivated nucleus. The results are striking: in *Xenopus*, nuclei from early embryos and from some types of differentiated cells of larvae and adults, like gut and skin epithelial cells, can replace the egg nucleus and support the development of an embryo. In a small number of cases the embryo has developed into an adult (Fig. 9.29).

Nuclei from adult skin, kidney, heart, and lung cells, as well as from the intestinal cells and myotome of tadpoles, can support the development of a tadpole and even an adult when transplanted into enucleated eggs. However, the success rate with nuclei from cells of adults is very low, with

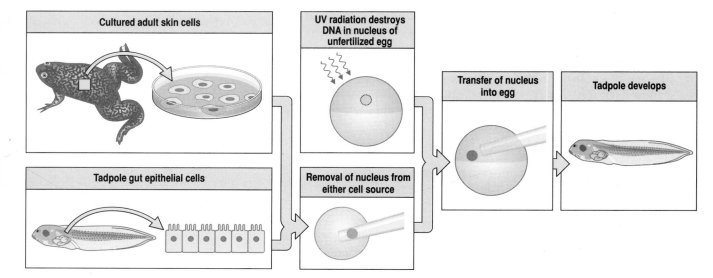

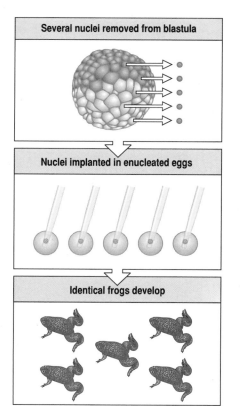

Fig. 9.29 Nuclear transplantation. The nucleus of an unfertilized *Xenopus* egg is rendered nonfunctional by treatment with ultraviolet radiation. A nucleus taken from the epithelial cells lining the tadpole gut or from cultured adult skin cells is transferred into the enucleated egg, and can support the development of an embryo until at least the tadpole stage.

Fig. 9.30 Cloning by nuclear transfer. Nuclei from cells of the same blastula are transferred into enucleated, unfertilized *Xenopus* eggs. The frogs that develop are clones as they all have the same genetic constitution.

only a small percentage of nuclear transplants developing past the cleavage stage. In general, the later the developmental stage the nuclei come from, the less likely they are to be able to support development. Transplantation of nuclei taken from cells of the blastula is much more successful; by transplanting nuclei from the same blastula into several enucleated eggs, a **clone** of genetically identical frogs can be obtained (Fig. 9.30).

These results show that, for a representative range of differentiated cell types, the genes required for development are not irreversibly altered. More importantly, when exposed to the egg's cytoplasmic factors, they behave as the genes in a fertilized egg's nucleus would. So in this sense at least, many of the nuclei of the embryo and the adult are equivalent, and their behavior is determined entirely by factors present in the cell. However, the nuclei of some cell types, such as neurons, are not able to support development in transplantation experiments. This inability is probably related to the fact that these cell types do not undergo mitosis and cell division once fully differentiated, rather than to some structural change in the genome. Even those nuclei that do support development cannot be equivalent in all respects to the egg nucleus, as the adult animals they give rise to are infertile. The reasons for this are not understood, but may be related to special features of germ cell development (see Chapter 12).

What about other organisms? Similar results have been obtained in insects, where nuclei from the blastoderm stage can, when returned to the egg, participate in the formation of a wide range of tissue types. In ascidians, the genetic equivalence of nuclei at different developmental stages can also be demonstrated. In plants, single somatic cells can give rise to fertile adult plants, demonstrating complete reversibility of the differentiated state at the cellular level (see Section 7.3).

In some mammalian species, successful development has been achieved following removal of the egg nucleus with a micropipette and injection of a nucleus from a somatic cell, or by fusion of the somatic cell with the egg cell. In cattle and sheep, nuclei from other embryonic cell cultures can successfully support development when transplanted into an oocyte. A viable lamb, the famous Dolly, has even been produced by transplanting the nucleus from a cell line derived from an adult udder. Mice have been similarly cloned using nuclei from the somatic cumulus cells that surround the recently ovulated egg. Transplantation of nuclei from neurons, however, only supported development to the late blastocyst stage. In earlier experiments in mice, even when nuclei for transplantation came from the two-cell embryo, very few of the treated eggs developed into a blastocyst. This failure to routinely support development does not, however, imply loss of genetic information. It suggests that there are extensive changes in the state of the nuclei, possibly related to the ability to maintain the cell cycle.

One clear exception to the general reversibility of cell differentiation is the irreversible DNA recombination events that occur during the differentiation of B and T lymphocytes in vertebrates (see Section 9.14). Another classic case occurs in the nematode *Parascaris aequorum*, in which only the germ cells retain a full set of chromosomes. In cells destined to become somatic cells, chromosome diminution occurs: the chromosomes fragment, and only a small portion of each original chromosome is found in the somatic cells.

9.17 Patterns of gene activity in differentiated cells can be changed by cell fusion

Nuclear transplantation into eggs, particularly frog eggs, is facilitated by their size and the large amount of cytoplasm. With other cell types, particularly differentiated cells, injection of nuclei into foreign cytoplasm is not possible. It is possible, however, to expose the nucleus of one cell type to the cytoplasm of another type by fusing the two cells together. In this process, which can be brought about by treatment with certain chemicals or viruses, the plasma membranes fuse and nuclei from different cells come to share a common cytoplasm.

A striking example of the reversibility of gene activity after cell fusion is provided by the fusion of chick red blood cells with human cancer cells in culture. Unlike mammalian red blood cells, mature chick red blood cells have a nucleus. Gene activity in the nucleus is, however, completely turned off and it thus produces no mRNA. When a chick red blood cell is fused with a cell from a human cancer cell line, gene expression in the red blood cell nucleus is reactivated, and chick-specific proteins are expressed. This shows that human cells contain cytoplasmic factors capable of initiating transcription in chick nuclei.

Fusion of differentiated cells with striated muscle cells from a different species provides further evidence for reversibility of gene expression in differentiated cells Fig 9.31. Multinucleate striated muscle cells are good partners for cell fusion studies as they are large, and muscle-specific proteins can easily be identified. Differentiated human cells representative of each of the three germ layers can be fused with mouse multinucleate muscle cells, so that the human cell nuclei are exposed to the mouse muscle cytoplasm. This results in muscle-specific gene expression being switched on in the human nuclei. For example, human liver cell nuclei in mouse muscle cytoplasm no longer express liver-specific genes; instead, their muscle-specific genes are activated and human muscle proteins are made.

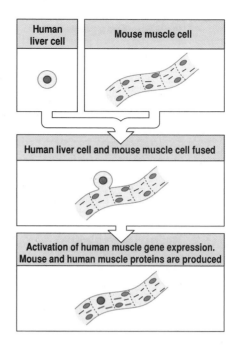

Fig. 9.31 Cell fusion shows the reversibility of gene inactivation during differentiation. A human liver cell is fused with a mouse muscle cell. The exposure of the human nucleus (red) to the mouse muscle cytoplasm results in the activation of muscle-specific genes and repression of liver-specific genes in the human nucleus. Human muscle proteins are made together with mouse muscle proteins. This shows that the inactivation of muscle-specific genes in human liver cells is not irreversible.

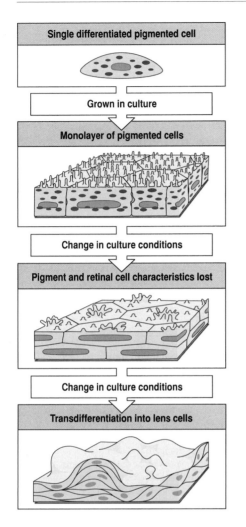

Fig. 9.32 Transdifferentiation of retinal pigment cells. A single pigmented epithelial cell from the embryonic chick retina can be grown in culture to produce a monolayer of pigmented cells. On further culture in the presence of hyaluronidase, serum, and phenylthiourea, they lose their pigment and retinal cell characteristics. If cultured at a high density with ascorbic acid, they differentiate as lens cells and produce the lens-specific protein crystallin. After Okada, T.S.: 1992.

These results clearly show that patterns of gene expression in differentiated cells can be changed, and that gene expression can be controlled by factors synthesized in the cytoplasm. These factors are transcription factors, as we see in subsequent sections, and it leads one to consider that the differentiated state may be maintained by the continuous action of transcription factors on gene activity.

9.18 The differentiated state of a cell can change by transdifferentiation

A fully differentiated cell is generally stable, which is essential if it is to serve a particular function in the mature animal. If it has retained the ability to divide, the cell passes its differentiated state on to all of its descendants. In some long-lived cells such as neurons, which do not divide once they are differentiated, the differentiated state must remain stable for many years. Plant cells maintain their differentiated state while in the plant, but do not maintain it when placed in cell culture.

Under certain conditions, however, differentiated cells are not stable, and this provides yet another demonstration of the potential reversibility of patterns of gene activity. We discussed above the dedifferentiation of muscle cells and, as we see in Chapter 13, cells of one type can change into a different, but related, type in regenerating tissues. The classic example of such a change is regeneration of the lens in the newt eye, which occurs by recruitment of cells from the dorsal region of the iris pigmented epithelium. Another example is the dedifferentiation—the loss of differentiated characteristics—of muscle cells in newt limb regeneration and the ability of these cells to give rise to cartilage. The change of one differentiated cell type into another is known as **transdifferentiation**, and a number of cell types have been found to undergo such a change in tissue culture, particularly when the culture conditions are altered by the addition of chemical agents.

One well-established case of transdifferentiation occurs when pigmented epithelial cells of the embryonic chick retina are exposed to certain culture conditions. The pigment disappears and the cells start to take on the structural characteristics of lens cells, and to produce the lens-specific protein crystallin (Fig. 9.32). Another case of transdifferentiation occurs in the chromaffin cells of the adrenal gland (Fig. 9.33). These small cells are derived from the neural crest and secrete adrenaline (epinephrine) into the blood. In culture, the chromaffin phenotype can be maintained by the addition of glucocorticoids, but when the steroid is removed and nerve growth factor (NGF) is added to the culture medium, the chromaffin cells transdifferentiate into sympathetic neurons. These neurons are larger than chromaffin cells, with dendritic and axonal processes, and they secrete noradrenaline (norepinephrine) instead of adrenaline. This transformation shows that even terminally differentiated cells can change into another type when exposed to the appropriate environmental signals. In both of the cases mentioned above, the transdifferentia-

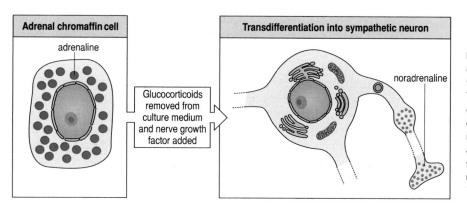

Fig. 9.33 Transdifferentiation of chromaffin cells. Chromaffin cells of the adrenal medulla secrete adrenaline. Their phenotype is maintained in culture in the presence of glucocorticoids. If the steroids are removed and nerve growth factor added, the cells transdifferentiate into sympathetic neurons which secrete noradrenaline. After Doupe, A.J., *et al.*: 1985.

tion is to a developmentally related cell type; pigment cells and lens cells are both derived from the ectoderm and are involved in eye development; sympathetic neurons and chromaffin cells are both derived from the neural crest.

A particularly interesting example of transdifferentiation, in which one cell type transdifferentiates into two different types in succession, is provided by jellyfish striated muscle. When a small patch of striated muscle and its associated extracellular matrix (mesogloea) is cultured, the striated muscle state is maintained. However, if the cultured tissue is treated with enzymes that degrade the extracellular matrix, the cells form an aggregate, and within 1–2 days some have transdifferentiated into smooth muscle cells, which have a different cellular morphology. This is followed by the appearance of a second cell type, nerve cells. This example indicates a role for the extracellular matrix in maintaining the differentiated state of the striated muscle.

All of the above observations suggest that gene activity in some cell types is reversible. Nevertheless, in the majority of cases the differentiated state is remarkably stable.

Summary

Cell differentiation leads to distinguishable cell types whose character and properties are determined by their pattern of gene activity and the proteins they produce. Transplantation of nuclei from differentiated animal cells into fertilized eggs, and cell fusion studies, show that the pattern of gene expression in the nucleus of a differentiated cell can often be reversed, implying that it is determined by factors supplied by the cytoplasm and that no genetic material has been lost. Although the differentiated state of an animal cell *in vivo* is usually extremely stable, some cases of differentiation are reversible. Transdifferentiation of one differentiated cell type into another type has been shown to occur during regeneration and in cells in tissue culture.

SUMMARY TO CHAPTER 9

Cell differentiation leads to distinguishable cell types whose specialized characters and properties are determined by their pattern of gene activity and thus the proteins they produce. The pattern of gene activity in a differentiated cell can be maintained over long periods of time and can be transmitted to the cell's progeny. Maintenance and inheritance of a pattern of gene activity may involve several mechanisms, including the continued action of gene regulatory proteins, changes in the packing state of chromatin, and chemical modifications to DNA.

Differentiation reflects the establishment of a particular pattern of gene expression rather than the loss of genetic material: with the exception of lymphocytes, no irreversible changes have occurred to the DNA in differentiated cells. It is possible to clone both amphibia and mammals by nuclear transplantation from somatic cells into an egg. The differentiated state of an animal cell *in vivo* is usually extremely stable, but in some cases transdifferentiation of one differentiated cell type into another can occur, as during tissue regeneration.

Apart from the use of universal mechanisms of gene regulation, there may be little similarity between the detailed mechanisms of differentiation in different cell types. The understanding of any given differentiation pathway lies in the detail of the external signals, internal signal transduction pathways, gene regulatory proteins, and gene products at work in each particular case.

Induction of cell differentiation by signaling molecules, such as growth factors, requires the execution of a complex program of intracellular events. Different stimuli, or even the same stimulus at different developmental stages, can activate the same internal signaling pathway, yet activate different genes in different cells, because of the cells' own developmental histories. Signal transduction at the cell membrane sets in motion a cascade of events within the cell, which results in the phosphorylation and activation of transcription factors and the consequent switching on and off of gene expression. Stem cells are self-renewing and can generate a wide variety of cell types. Programmed cell death is a common fate of differentiating cells during development, and there is evidence that positively acting signals are generally required to prevent it and allow cells to continue to survive.

REFERENCES

Introduction

Friederich, E., Huet, C., Arpin, M., Louvard, D.: **Villin induces microvilli growth and actin redistribution in transfected fibroblasts**. *Cell* 1989, **59**: 461–475.

Myster, D.L., Duronio, R.J.: **Cell cycle: To differentiate or not to differentiate**. *Curr. Biol.* 2000, **10**: R302–R304.

9.1 Control of transcription involves both general and tissue-specific transcriptional regulators

Cheung, W.L., Briggs, S.D., Allis, C.D.: **Acetylation and chromosomal fucntions**. *Curr. Opin. Cell Biol.* 2000, **12**: 328–333.

Latchman, D.S.: *Eukaryotic Transcription Factors*. 3rd edn. London: Academic Press, 1998.

Mannervik, M., Nibur, Y., Zhang, H., Levine, M.: **Transcriptional coregulators in development**. *Science*, 1999, **284**: 606–609.

Scully, K.M., Jacobsen, E.M., Jepsen, K., Lunyak, V., Viadiu, H., Carriere, C., Rose, D.W., Hooshmand, F., Aggarwal, A.K., Rosenfeld, M.G.: **Allosteric effects of Pit-1 DNA sites on long-term repression in cell type specificiation**. *Science* 2000, **290**: 1127–1131.

Struhl, K.: **Fundamentally different logic of gene regulation in eukaryotes and prokaryotes**. *Cell* 1999, **98**: 1–4.

Tijan, R.: **Molecular machines that control genes**. *Sci. Amer.* 1995, **2**: 336–344.

9.2 External signals can activate genes

Winston, L.A., Hunter, T.: **Intracellular signaling: Putting JAKs on the kinase MAP**. *Curr. Biol.* 1996, **6**: 668–671.

9.3 Maintenance and inheritance of patterns of gene activity may depend on chemical and structural modifications of DNA as well as regulatory proteins

Dillon, N., Grosveld, F.: **Chromatin domains as potential units of eukaryotic gene function**. *Curr. Opin. Genet. Dev.* 1994, **4**: 260–264.

9.4 A family of genes can activate muscle-specific transcription

Arnold, H.H., Winter, B.: **Muscle differentiation: more complexity to the network of myogenic regulators**. *Curr. Opin. Gen. Dev.* 1998, **8**: 539–544.

Naya, F.J., Olson, E.: **MEF2: a transcriptional target for signaling pathways controlling skeletal muscle growth and differentiation**. *Curr. Opin. Cell Biol.* 1999, **11**: 683–688.

Olson, E.N., Arnold, H.H., Rigby, P.W., Wold, B.J.: **Know your neighbours: three phenotypes in null mutant of the myogenic bHLH gene MRF4.** *Cell* 1996, **85**: 1–4.

9.5 The differentiation of muscle cells involves withdrawal from the cell cycle, but is reversible

Novitch, B.G., Mulligan, G.J., Jacks, T., Lassar, A.B.: **Skeletal muscle cells lacking the retinoblastoma protein display defects in muscle gene expression and accumulate in S and G$_2$ phases of the cell cycle.** *J. Cell Biol.* 1996, **135**: 441–456.

Odelberg, S.J., Kolhoff, A., Keating, M.: **Dedifferentiation of mammalian myotubes induced by msx1.** *Cell* 2000, **103**: 1099–1109.

Yun, K., Wold, B.: **Skeletal muscle determination and differentiation: story of a core regulatory network and its context.** *Curr. Opin. Cell Biol.* 1996, **8**: 877–889.

9.6 All blood cells are derived from pluripotent stem cells

Dieterlen-Lievre, F.: **Hematopoiesis: Progenitors and their genetic program.** *Curr. Biol.* 1998, **8**: 727–730.

Morrison, S.J., Shah, N.M., Anderson, D.J.: **Regulatory mechanisms in stem cell biology.** *Cell* 1997, **88**: 287–298.

Phillips, R.L., Ernst, R.E., Brunk, B., Ivanova, N., Mahan, M.A., Deanehan, J.K., Moore, K.A., Overton, G.C., Lemischka, I.R.: **The genetic program of hematopoietic stem cells.** *Science* 2000, **288**: 1635–1640.

9.7 Colony-stimulating factors and intrinsic changes control differentiation of the hematopoietic lineages

D'Andrea, A.D.: **Hematopoietic growth factors and the regulation of differentiative decisions.** *Curr. Opin. Cell Biol.* 1994, **6**: 804–808.

Karin, M., Hunter, T.: **Transcriptional control by protein phosphorylation: signal transmission from the cell surface to the nucleus.** *Curr Biol.* 1995, **5**: 747–757.

Metcalf, D.: **Control of granulocytes and macrophages: molecular, cellular, and clinical aspects.** *Science* 1991, **254**: 529–533.

Ness, S.A., Engel, J.D.: **Vintage reds and whites: combinatorial transcription factor utilization in hematopoietic differentiation.** *Curr. Opin. Genet. Dev.* 1994, **4**: 718–724.

Orkin, S.H.: **Diversification of haematopoietic stem cells to specific lineages.** *Nature Rev. Genet.* 2000, **1**: 57–64.

9.8 Globin gene expression is controlled by distant upstream regulatory sequences

Dillon, N., Grosveld, F.: **Transcriptional regulation of multigene loci: multilevel control.** *Trends Genet.* 1993, **9**: 134–137.

Engel, J.D., Tanimoto, K.: **Looping, linking, and chromatin activity: new insights into β-globin locus regulation.** *Cell* 2000, **100**: 499–502.

Grosveld, F.: **Activation by locus control regions?** *Curr. Opin. Genet. Dev.* 1999, **9**: 152–157.

Orkin, S.: **Development of the hematopoietic system.** *Curr. Opin. Genet. Dev.* 1996, **6**: 597–602.

Tanimoto, K., Liu, Q., Bungert, J., Engel, J.D.: **Effects of altered gene order or orientation of the locus control region on human beta-globin gene expression in mice.** *Nature* 1999, **398**: 344–348.

Wood, W.G.: **The complexities of β-globin gene regulation.** *Trends Genet.* 1996, **12**: 204–206.

9.9 Neural crest cells differentiate into several cell types

Anderson, D.J.: **Genes, lineages and the neural crest.** *Proc. Roy. Soc. Lond. B* 2000, **355**: 953–964.

Garcia-Castro, M., Bronner-Fraser, M.: **Induction and differentiation of the neural crest.** *Curr. Opin. Cell. Biol.* 1999, **11**: 695–698.

Le Douarin, N.M., Dupin, E., Ziller, C.: **Genetic and epigenetic control in neural crest development.** *Curr. Opin. Genet. Dev.* 1994, **4**: 685–695.

9.10 Steroid hormones and polypeptide growth factors specify chromaffin cells and sympathetic neurons

Patterson, P.H.: **Control of cell fate in a vertebrate neurogenic lineage.** *Cell* 1990, **62**: 1035–1038.

Stemple, D.L., Anderson, D.J.: **Lineage diversification of the neural crest: *in vitro* investigations.** *Dev. Biol.* 1993, **159**: 12–23.

9.11 Neural crest diversification involves signals both for specification of cell fate and selection for cell survival

Dorsky, R.I., Moon, R.T., Raible, D.W.: **Environmental signals and cell fate specification in premigratory neural crest.** *BioEssays* 2000, **22**: 708–716.

Fleischman, R.A.: **From white spots to stem cells: the role of the Kit receptor in mammalian development.** *Trends Genet.* 1993, **9**: 285–290.

Le Douarin, N.M., Ziller, C., Couly, G.F.: **Patterning of neural crest derivatives in the avian embryo: *in vivo* and *in vitro* studies.** *Dev. Biol.* 1993, **159**: 24–49.

Morrison, S.J., White, P.M., Zock, C., Anderson, D.J.: **Prospective identification, isolation by flow cytometry, and in vivo self-renewal of multipotent mammalian neural crest stem cells.** *Cell* 1999, **96**: 737–749.

Morrison, S.J., Perez, S.E., Qiao, Z., Verdi, J.M., Hicks, C., Weinmaster, G., Anderson, D.J.: **Transient Notch activation initiates an irreversible switch from neurogenesis to gliogenesis by neural crest stem cells.** *Cell* 2000, **101**: 499–510.

Shah, N.M., Groves, A.K., Anderson, D.J.: **Alternative neural crest cell fates are instructively promoted by TGF-β superfamily members.** *Cell* 1996, **85**: 331–343.

9.12 The epithelia of adult mammalian skin and gut are continually replaced by derivatives of stem cells

Gordon, J.I., Hermiston, M.L.: **Differentiation and self-renewal in the mouse gastrointestinal epithelium.** *Curr. Opin. Cell Biol.* 1994, **6**: 795–803.

Jones, P.H., Harper, S., Watts, F.M.: **Stem cell patterning and fate in human epidermis.** *Cell* 1995, **80**: 1–20.

Poten, C.S., Loeffler, M.: **Stem cells: attributes, cycles, spirals, pitfalls and uncertainties. Lessons for and from the crypt.** *Development* 1990, **110**: 1001–1020.

Sellheyer, K., Bickenbach, J.R., Rothnagel, J.A., Bundman, D., Langley, M.A., Krieg, T., Roche, N.S., Roberts, A.B., Roop, D.R.: **Inhibition of skin development by overexpression of transforming growth factor-β1 in the epidermis of transgenic mice.** *Proc. Natl Acad. Sci. USA* 1993, **90**: 5237–5241.

Slack, J.M.W.: **Stem cells in epithelial tissues.** *Science* 2000, **287**: 1431–1433.

Watt, F.M.: **Terminal differentiation of epidermal keratinocytes.** *Curr. Opin. Cell Biol.* 1989 **1**: 1107–1115.

Watt, F.M., Jones, P.H.: **Expression and function of the**

keratinocyte integrins. *Development (Suppl.)* 1993: 185–192.

9.13 Many stem cells have a greater potential for differentiation than they normally show

Blau, H.M., Brazelton, T.R., Weimann, J.M.: **The evolving concept of a stem cell: entity or function?** *Cell* 2001, **105**: 829–841.

Fuchs, E., Segre, J.A.: **Stem cells: a new lease on life.** *Cell* 2000, **100**: 143–155.

Gage, F.H.: **Mammalian neural stem cells.** *Science* 2000, **287**: 1430–1433.

Morrison, S.J.: **Stem cell potential: Can anything make anything?** *Curr. Biol.* 2001, **11**: R7–R9.

Vogel, G.: **Stem cells: new excitement, persistent questions.** *Science* 2000, **290**: 1672–1674.

Watt, F.M., Hogan, B.L.M.: **Out of Eden: stem cells and their niches.** *Science* 2000, **287**: 1430–1431.

Weissman, I.L.: **Stem cells: units of development, units of regeneration, and units of evolution.** *Cell* 2000, **100**: 157–168.

9.14 Differentiation of cells that make antibodies is irreversible as a result of loss of DNA

Janeway, C.A., Travers, P., Shlomchik, M., Walport, M.: *Immunobiology: The Immune System in Health and Disease.* 5th edn. London/New York: Garland Publishing, 2001.

9.15 Programmed cell death is under genetic control

Ellis, R.E., Yuan, J.Y., Horvitz, H.R.: **Mechanisms and functions of cell death.** *Annu. Rev. Cell. Biol.* 1991, **7**: 663–698.

Jacobson, M.D., Weil, M., Raff, M.C.: **Programmed cell death in animal development.** *Cell* 1997, **88**: 347–354.

Raff, M.: **Cell suicide for beginners.** *Nature* 1998, **396**: 119–122.

Vaux, D.L., Korsmeyer, S.J.: **Cell death in development.** *Cell* 1999, **96**: 245–254.

9.16 Nuclei of differentiated cells can support development of the egg

Gurdon, J.B.: **Nuclear transplantation in eggs and oocytes.** *J. Cell Sci. Suppl.* 1986, **4**: 287–318.

Gurdon, J.B., Colman, A.: **The future of cloning.** *Nature* 1999, **402**: 743–746.

Onishi, A., Iwamoto, M., Akita, T., Mikawa, S., Takeda, K., Awata, T., Hanada, H., Perry, A.C.F.: **Pig cloning by microinjection of fetal fibroblast nuclei.** *Science* **289**: 1188–1190.

Solter, D.: **Mammalian cloning: advances and limitations.** *Science* 2000, **289**: 1188–1190.

Wilmut, I., Schnieke, A.E., McWhir, J., Kind, A.J., Campbell, K.H.S.: **Viable offspring derived from fetal and adult mammalian cells.** *Nature* 1997, **385**: 810–813.

9.17 Patterns of gene activity in differentiated cells can be changed by cell fusion

Blau, H.M.: **How fixed is the differentiated state? Lessons from heterokaryons.** *Trends Genet.* 1989, **5**: 268–272.

Blau, H.M., Baltimore, D.: **Differentiation requires continuous regulation.** *J. Cell Biol.* 1991, **112**: 781–783.

Blau, H.M., Blakely, B.T.: **Plasticity of cell fates: insights from heterokaryons.** *Cell Dev. Biol.* 1999, **10**: 267–272.

9.18 The differentiated state of a cell can change by transdifferentiation

Anderson, D.J.: **Cellular 'neoteny': a possible developmental basis for chromatin cell plasticity.** *Trends Genet.* 1989, **5**: 174–178.

Brockes, J.: **Muscle escapes from a jelly mould.** *Curr. Biol.* 1994, **4**: 1030–1032.

Eguchi, G., Kodama, R.: **Transdifferentiation.** *Curr. Opin. Cell Biol.* 1993, **5**: 1023–1028.

Okada, T.S.: *Transdifferentiation.* Oxford: Clarendon Press, 1992.

Organogenesis

10

- The chick limb
- Insect wings and legs
- The nematode vulva
- Internal organs: heart, blood vessels, lungs, kidneys, and teeth

> "When we were older we made use of what we had learnt when young to make quite new patterns and structures."

So far, we have concentrated almost entirely on the broad aspects of development involved in laying down the basic body plan in various organisms. We now turn to the development of specific organs and structures—**organogenesis**—which is a crucial phase of development in which the embryo at last becomes a fully functioning organism, capable of independent survival.

The development of certain organs has been studied in great detail and they provide excellent models for looking at developmental processes such as pattern formation, the specification of positional information, induction, change in form, and cellular differentiation. In this chapter, we first consider the development of model organ systems in a variety of organisms—the chick limb, insect appendages such as legs and wings in *Drosophila*, and the nematode vulva. We finish by looking at the development of the vertebrate heart and examples of epithelial patterning and morphogenesis in the development of the lungs, kidney, blood vessels, and teeth. We do not deal with other important organs such as the gut, liver, or pancreas because their development is, in general, less well understood, and there is no evidence that different principles are involved.

The cellular mechanisms involved in organogenesis are essentially similar to those encountered in earlier stages of development; they are merely employed in different spatial and temporal patterns. There are, for example, striking similarities between the development of the vertebrate limb and the wings and legs of *Drosophila*, and many of the genes and signaling molecules involved will be familiar from earlier chapters.

The chick limb

The vertebrate embryonic limb is a particularly good system in which to study pattern formation. The basic pattern is initially quite simple, and in the chick embryo at least, the limbs are easily accessible for surgical

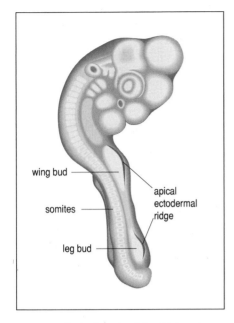

Fig. 10.1 The limb buds of the chick embryo.
Limb buds appear on the flanks of the embryo on the third day of incubation (only limbs on the right side are shown here). They are composed of mesoderm, with an outer covering of ectoderm. Along the tip of each runs a thickened ridge of ectoderm, the apical ectodermal ridge.

manipulation. The limb is a good model for studying cellular interactions within a structure containing a large number of cells, and for elucidating the role of intercellular signaling in development. In chick embryos, the limbs begin to develop on the third day after the egg is laid, when the structures of the main body axis are already well established. The limbs develop from small protrusions—the limb buds—which arise from the body wall of the embryo (Fig. 10.1). By 10 days, the main features of the limbs are well developed (Fig. 10.2). These are the cartilaginous elements (which are later replaced by bone), the muscles and tendons, and the surface structures such as feather buds. The limb has three developmental axes: the **proximo-distal axis** runs from the base of the limb to the tip; the **antero-posterior axis** runs parallel with the body axis (in the human hand it goes from the thumb to the little finger, and in the chick limb from digit 2 to digit 4); the **dorso-ventral axis** runs from the back of the human hand to the palm.

10.1 The vertebrate limb develops from a limb bud

The early limb bud has two major components—a core of loose mesenchymal cells derived from the lateral plate mesoderm, and an outer layer of ectodermal epithelial cells (Fig. 10.3). The skeletal elements and connective tissues of the limb develop from the mesenchymal core, but the limb muscle cells have a separate lineage and migrate into the bud from the somites (see Section 4.2). At the very tip of the limb bud is a thickening in the ectoderm—the **apical ectodermal ridge** or **apical ridge**, which runs along the dorso-ventral boundary (Fig. 10.4). Directly beneath this lies a **progress zone** of rapidly dividing and proliferating undifferentiated mesenchymal cells. It is only when cells leave the progress zone that they begin to differentiate. As the bud grows, the cells start to differentiate and cartilaginous structures begin to appear in the mesenchyme. The proximal part of the limb, that is the part nearest to the body, is the first to differentiate, and differentiation proceeds distally as the limb extends. Among the structures found in the developing limb, the patterning of the cartilage has been the best studied, as this can be stained and seen easily in whole mounts of

Fig. 10.2 The embryonic chick wing.
The photograph shows a stained whole mount of the wing of a chick 10 days after the egg has been laid. By this time, the main cartilaginous elements (e.g. humerus, radius, and ulna) have been laid down. They later become ossified to form bone. The muscles and tendons are also well developed at this stage but cannot be seen in this type of preparation, but the feather buds can be seen. The three developmental axes of the limb are proximo-distal, antero-posterior, and dorso-ventral, as shown in the top panel. Note that the chick wing has only three digits, which have been called 2, 3, and 4. Scale bar = 1 mm.

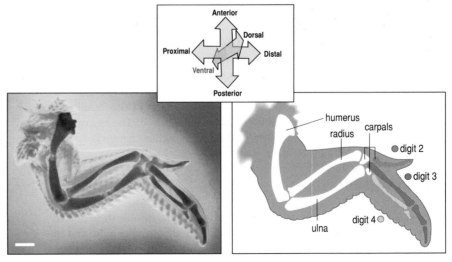

the embryonic limb. The disposition of muscles and tendons is more intricate, and while this can be studied in whole mounts, histological examination of serial sections through the limb may be needed.

The first clear sign of cartilage differentiation is the increased packing of groups of cells, a process known as condensation. The cartilage elements are laid down in a proximo-distal sequence in the chick wing—the humerus, radius, and ulna, the wrist elements (carpals), and then three easily distinguishable digits, 2, 3, and 4 (see Fig. 10.2). Fig. 10.5 compares this sequence in the development of a chick wing with the similar sequence in the development of a mouse forelimb.

The first event in joint formation is the formation of an 'interzone' at the site of the prospective joint, in which the cartilage-producing cells become fibroblast-like and cease producing the cartilage matrix, thus separating one cartilage element from another. At this time, the BMP-related protein GDF-5 is expressed in the presumptive joint region together with Wnt-14. Ectopic misexpression of Wnt-14 can induce the early steps in joint formation, and mice lacking GDF-5 have some of their joints missing. A little later, the musculature begins to function and muscle contraction occurs. At this point high levels of hyaluronan, a component of the proteoglycans that lubricate joint surfaces, are made by the cells of the interzone; the hyaluronan is secreted in vesicles which coalesce to form the joint cavity, which separates the articulating surfaces of the joint from each other. Synthesis of hyaluronan by the cells lining these surfaces depends on the mechanical stimulation that results from muscle contraction, and in the absence of this activity, the joints fuse. Cells isolated from the interzone region and grown in culture respond to mechanical stimulation by increased secretion of hyaluronan when the substratum to which they are attached is stretched.

The chick limb bud at 3 days is about 1 mm wide by 1 mm long, but by 10 days it has grown some tenfold, mostly in length. The basic pattern has been laid down well before then (see Fig. 10.5). But even at 10 days, the limb is still small compared with the newborn limb. The apical ridge disappears as soon as all the basic elements of the limb are in place, and growth occupies most of the subsequent development of the limb, both before and after birth. During the later growth phase, the cartilaginous elements are largely replaced by bone. Nerves only enter the limb after the cartilage has been laid down (see Section 11.12). The problem of limb patterning is to understand how the basic pattern of cartilage, muscle, and tendons is formed in the right places and how they make the right connections with each other.

10.2 Homeobox genes expressed in the lateral plate mesoderm are involved in specifying the position and type of limb

The forelimbs and hindlimbs of vertebrates arise at precise positions along the antero-posterior axis of the body. Transplantation experiments have shown that the lateral plate mesoderm that gives rise to the limb bud mesenchyme becomes determined to form limbs in these exact positions long before limb buds are visible. As the lateral plate mesoderm also becomes regionalized by Hox gene expression along this body axis, it is not difficult to envisage how Hox gene coding could determine prospective forelimb and hindlimb regions, thus specifying the position and type of a limb bud.

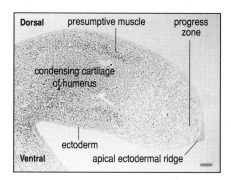

Fig. 10.3 Cross-section through a developing limb bud. The thickened apical ectodermal ridge is at the tip. Beneath the ridge is the progress zone where cells proliferate. Proximal to the progress zone, mesenchyme cells condense and differentiate into cartilage. Presumptive muscle cells migrate into the limb from the adjacent somites. Scale bar = 0.1 mm.

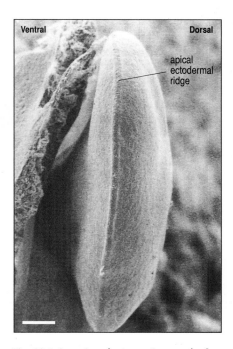

Fig. 10.4 Scanning electron micrograph of a chick limb bud at 4½ days after laying, showing the apical ectodermal ridge. Scale bar = 0.1 mm.

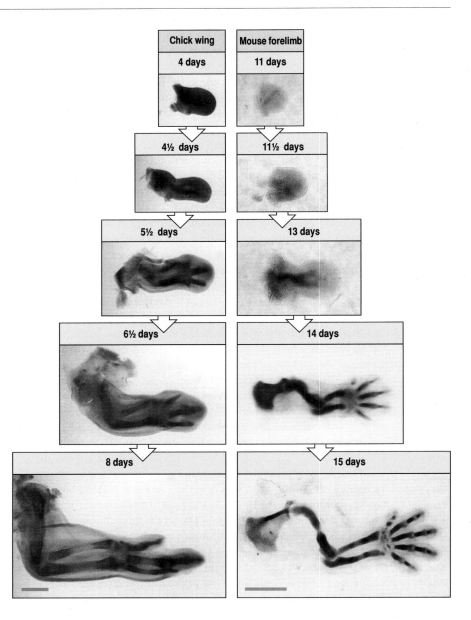

Fig. 10.5 The development of the chick wing and mouse forelimb are similar. The skeletal elements are laid down, as cartilage, in a proximo-distal sequence as the limb bud grows outward. The cartilage of the humerus is laid down first, followed by the radius and ulna, wrist elements, and digits. Scale bar = 1 mm.

This would be analogous to the specification of the different thoracic segments and their appendages in insect development (see Chapter 5).

Proteins of the fibroblast growth factor (FGF) family appear to be key signaling molecules in initiating both forelimb and hindlimb development in vertebrates. These molecules are involved in establishing and maintaining the two main organizing regions of a limb. These are the apical ectodermal ridge, which is essential for limb outgrowth and for correct patterning along the proximo-distal axis of the limb, and the mesodermal polarizing region, which is situated on the posterior side of the limb bud and is crucial for determining pattern along the antero-posterior limb axis, as we shall discuss later. FGF–10 is expressed in limb-forming regions, and knock-out of FGF receptors in the mouse embryo results in embryos lacking limb buds. Wnt genes in the mesoderm play a key role in determining where FGFs are expressed and maintained. The Hox gene coding in the mesoderm could thus determine where limbs develop by specifying where Wnts and thus FGFs are expressed, and thus where an apical ridge and polarizing region develop.

For example, evidence for the combined action of Hox genes in determining the position of the forelimb bud along the antero-posterior axis comes from knock-out mice lacking *Hoxb5* expression, in which the forelimbs develop at a more anterior level. Evidence that Hox genes are involved in specifying the position of the polarizing region comes from mouse embryos that express a transgene of *Hoxb8* in more anterior regions of the embryo than normal. In these mice, an extra polarizing region forms at the anterior margin of the forelimb buds.

The different patterns of Hox gene expression at different positions along the antero-posterior axis of the body could also specify whether the limb will be a forelimb or hindlimb. When an FGF bead is placed in the flank region between fore- and hindlimb buds, it induces an ectopic wing or leg, depending on the exact position along the flank (see Section 10.3). Hox gene expression is found to be respecified in this position, and the new pattern of Hox gene expression corresponds to that normally present in either the wing or the leg region.

The limb-specific proteins that might interpret the Hox gene positional signals are gradually being revealed. Two T-box transcription factors, Tbx4 and Tbx5, which are related to the transcription factor Brachyury (see Section 3.20), are expressed in prospective leg and wing regions respectively, and their expression is controlled by the Hox genes expressed along the antero-posterior axis.

10.3 The apical ectodermal ridge induces the progress zone

The apical ectodermal ridge consists of closely packed columnar epithelial cells, which are linked by extensive gap junctions. Their tight packing gives the ridge a mechanical strength that probably keeps the limb flattened along the dorso-ventral axis; the length of the ridge controls the width of the bud. The progress zone, lying beneath the apical ectodermal ridge, is a region of rapidly proliferating mesenchymal cells, which produces the initial outgrowth of the limb bud. It is also the region where the mesenchymal cells acquire the positional information that determines what structures they will develop into. Perhaps surprisingly, the establishment of the progress zone does not reflect a local increase in cell division in the limb bud region but rather a decrease from a previously high rate of cell proliferation along the rest of the flank.

The apical ectodermal ridge is essential for both the outgrowth of the limb and for its correct proximo-distal patterning. This is due to its role in inducing and maintaining the progress zone. If the ridge is removed from a chick limb bud by microsurgery, there is a significant reduction in growth and the limb is anatomically truncated, with distal parts missing. The proximo-distal level at which the limb is truncated depends on the time at which the ridge is removed (Fig. 10.6). The earlier the ridge is removed, the greater the effect; removal at a later stage only results in loss of the ends of the digits. Proximo-distal patterning thus occurs progressively over time in the progress zone. Following apical ridge removal, cell proliferation in the progress zone is greatly reduced.

Signaling by the apical ectodermal ridge to the underlying mesenchyme

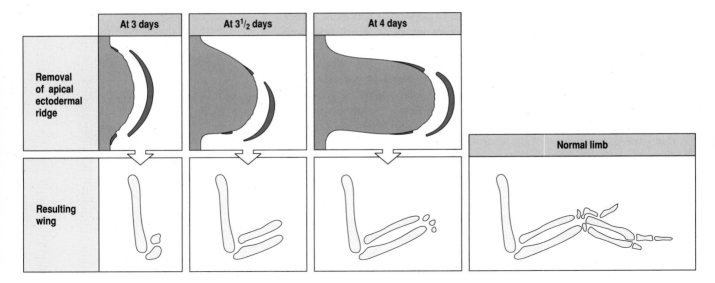

Fig. 10.6 The apical ectodermal ridge is required for proximo-distal development. Limbs develop in a proximo-distal sequence. Removal of the ridge from a developing limb bud leads to truncation of the limb; the later the ridge is removed, the more complete the resulting limb.

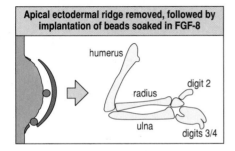

Fig. 10.7 The growth factor FGF-8 can substitute for the apical ectodermal ridge. After the ridge is removed, grafting beads that release FGF-8 into the limb bud results in almost normal limb development.

and induction of a progress zone has been shown by grafting an isolated ridge to the dorsal surface of an early limb bud. The result is an ectopic outgrowth, which may even develop cartilaginous elements and digits. In the chick, a key signal from the ridge is provided by fibroblast growth factors. FGF-8 is expressed throughout the ridge and FGF-4 in the posterior region. FGF-8 can act as a functional substitute for an apical ridge. If the ridge is removed and beads that release the growth factor are grafted into the tip of the limb in its place, more-or-less normal outgrowth of the limb continues. If sufficient FGF-8 is provided to the outgrowing cells, a fairly normal limb develops (Fig. 10.7). Similar results are obtained with FGF-4.

The apical ridge is thus one of the crucial organizing regions of the vertebrate limb. The other is a mesodermal region at the posterior margin of the limb bud which is known as the **polarizing region** or **zone of polarizing activity (ZPA)** (Fig. 10.8). The early limb bud has considerable powers of regulation and pieces of it can generally be removed, rotated, or put in a different position without perturbing the final pattern. But this does not apply to either the apical ridge, as we have just seen, or the polarizing region. In the chick embryo, removal of the anterior half of a limb bud has no effect on development and a normal limb is produced. But removal of the posterior half, which contains the polarizing region, leads to abnormal limb development.

The apical ridge, polarizing region and progress zone are mutually dependent on each other for their maintenance and function. The apical ridge itself is maintained by signals from the progress zone and the polarizing region. One of the crucial signaling molecules is Sonic hedgehog, which is expressed in the polarizing region (see Fig. 10.8). There is a positive feedback loop between the polarizing region and the apical ectodermal ridge: Sonic hedgehog signaling modulates the level of FGF-4 in the posterior ridge, while FGF-4 maintains *Sonic hedgehog* expression. The BMP signaling molecules have a complex role in the limb bud tip. A minimum level of BMP is required to maintain the progress zone, but if the level rises too high the apical ridge regresses. Extra BMP in the form of BMP-soaked beads placed under the ridge causes cell death, whereas if BMP activity is blocked, the ridge both expands anteriorly and its lifetime is extended. A balance is maintained by a complex feedback signaling loop, in which Sonic hedge-hog induces expression of the protein Gremlin, which blocks BMP activity.

The interaction between FGF and Sonic hedgehog can be shown when ectopic limbs are induced in the flank. Localized application of FGF-4 to the flank of a chick embryo, between the wing and leg buds, induces the production of FGF-8 in the ectoderm, and then the ectopic expression of *Sonic hedgehog*. The Sonic hedgehog protein then feeds back to induce expression of the embryo's own gene for FGF-4, resulting in the maintenance of the ridge and the outgrowth of an additional limb bud at this site. Wing buds develop from application of FGF to the anterior flank, whereas leg buds develop from application to the posterior flank (Fig. 10.9).

We now consider how the developing limb is patterned along its three axes: antero-posterior, proximo-distal, and dorso-ventral.

10.4 Patterning of the limb involves positional information

Some aspects of vertebrate limb development fit very well with a model of pattern formation based on positional information. The developing chick limb behaves as if its cells' future development is determined by their position with respect to the main limb axes while they are in the progress zone (Fig. 10.10). The fate of cells along the proximo-distal axis may involve a timing mechanism, their fate being determined by how long they remain in the progress zone. Patterning along the antero-posterior axis is specified by a signal, or signals, emanating from the polarizing region at the posterior margin of the limb bud. The dorso-ventral axis is specified by a signal, or signals, from the ectoderm overlying the progress zone.

Central to the idea of positional information is the distinction between positional specification and interpretation. Cells acquire positional information first and then interpret these positional values according to their developmental history. It is the difference in developmental history (see Section 10.2) that makes wings and legs different, for positional information is signaled in wing and leg buds in the same way. An attractive hypothesis for limb patterning is that a single three-dimensional positional field controls the development of the cells that give rise to all the mesodermal limb elements—cartilage, muscle, and tendons. The specification of all three axes is, as we shall see, linked by molecular signals.

10.5 The polarizing region specifies position along the limb's antero-posterior axis

The polarizing region of a vertebrate limb bud has organizing properties that are almost as striking as those of the Spemann organizer in amphibians. When the polarizing region from an early wing bud is grafted to the anterior margin of another early wing bud, a wing with a mirror-image pattern develops: instead of the normal pattern of digits—2 3 4—the pattern 4 3 2 2 3 4 develops (Fig. 10.11). The pattern of muscles and tendons in the limb shows similar mirror-image changes.

The additional digits come from the host limb bud and not from the graft, showing that the grafted polarizing region has altered the developmental fate of the host cells in the anterior region of the limb bud. The limb bud widens in response to the polarizing graft, which enables the additional digits to be accommodated; widening of a limb bud is always

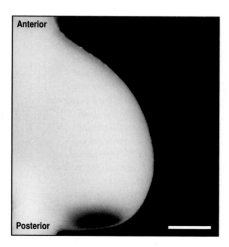

Fig. 10.8 *Sonic hedgehog* **is expressed in the polarizing region of a chick limb bud.** *Sonic hedgehog* is expressed at the posterior margin of the limb bud in the polarizing region and provides a positional signal along the antero-posterior axis. Scale bar = 0.1 mm.

Photograph courtesy of C. Tabin.

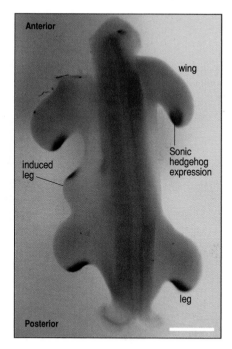

Fig. 10.9 **Result of local application of FGF-4 to the flank of a chick embryo.** FGF-4 is applied to a site close to the hindlimb bud, where it induces outgrowth of an additional hindward limb. The dark regions indicate expression of *Sonic hedgehog*. Scale bar = 1 mm.

Photograph courtesy of J-C. Izpisúa-Belmonte, from Cohn, M.J., et al.: 1995.

Fig. 10.10 Cells acquire their positional value in the progress zone. The progress zone is specified at the distal end of the bud by the apical ectodermal ridge. As the limb bud grows outward, cells in the progress zone proliferate and acquire a positional value. When they leave the zone, cartilage can begin to differentiate and the cartilaginous elements are laid down in a proximo-distal sequence, starting with the humerus. The position of the cells along the antero-posterior axis is specified by a signal from the polarizing region, which is located at the posterior side of the bud.

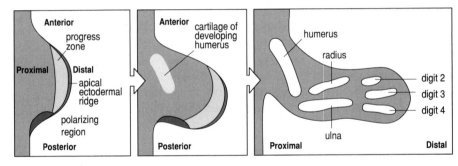

associated with an increase in the extent of the apical ectodermal ridge and the rate of cell division is higher than in the anterior region of a normal bud.

One way that the polarizing region could specify position along the antero-posterior axis is by producing a morphogen that forms a posterior to anterior gradient. The concentration of morphogen could specify the

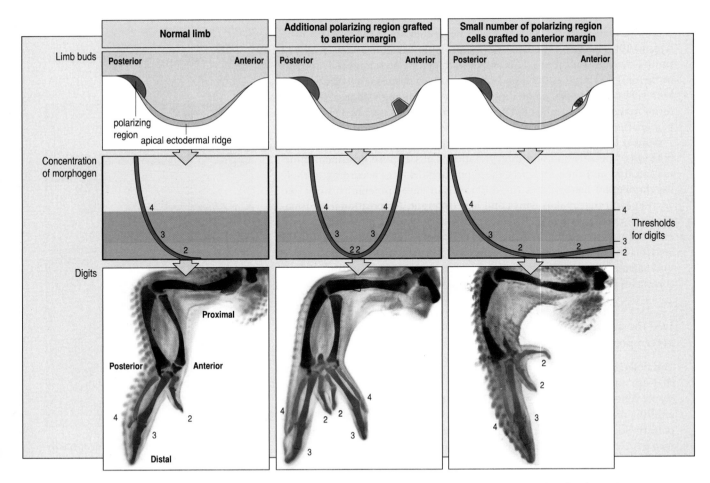

Fig. 10.11 The polarizing region can specify pattern along the antero-posterior axis. If the polarizing region is the source of a graded morphogen, the different digits could be specified at different threshold concentrations of signal (left panels). Grafting an additional polarizing region to the anterior margin of a limb bud (center panels) would result in a mirror-image gradient of signal, and thus the observed mirror-image duplication of digits. The signal from a grafted polarizing region can be attenuated by grafting only a small number of polarizing region cells to the anterior margin of the limb bud (right panels), so that only an extra digit 2 develops.

position of cells along the antero-posterior axis with respect to the polarizing region located at the posterior margin of the limb. Cells could then interpret their positional values by developing specific structures at particular threshold concentrations of morphogen. Digit 4, for example, would develop at a high concentration, digit 3 at a lower one, and digit 2 at an even lower one. According to this model, a graft of an additional polarizing region to the anterior margin would set up a mirror-image gradient of morphogen, which would result in the 4 3 2 2 3 4 pattern of digits that is observed (see Fig. 10.11, center panel).

If the action of the polarizing region in specifying the character of a digit is due to a gradient of signal, then when the signal is weakened, the pattern of digits should be altered in a predictable manner. Grafting small numbers of polarizing region cells to a limb bud results only in the development of an additional digit 2 (see Fig. 10.11, right panel). An analogous result can be obtained by leaving the polarizing region graft in place for a short time and then removing it. When left for 15 hours, an additional digit 2 develops, whereas the graft has to be in place for up to 24 hours for a digit 3 to develop.

But just because a diffusible morphogen model is consistent with many experimental results, this is not sufficient to prove it. Other models suggest that the same results could be obtained by a relay of short-range signals from the polarizing region. For example, digit identity is specified at a surprisingly late stage in limb development, and identity remains labile even when the digit primordia have formed. It now appears that digit identity is specified by the interdigital mesenchyme and requires BMP signaling. There is also evidence that mechanisms other than a diffusible morphogen operate to lay down the initial pattern of cartilage, which is then modified by a signal from the polarizing region, as discussed later.

A persuasive case can, however, be made for the Sonic hedgehog protein as the key component of the natural polarizing signal, whether the gradient of activity is formed by simple diffusion or by other means. The *Sonic hedgehog* gene is expressed in the polarizing region (see Fig. 10.8) and Sonic hedgehog protein is known to be involved in numerous patterning processes elsewhere in vertebrate development, for example in the somites (see Section 4.2) and the neural tube (see Section 11.6). The *Drosophila* homolog of Sonic hedgehog, the hedgehog protein, is a key signaling molecule in the patterning of the segments in the *Drosophila* embryo (see Section 5.16), as well as of *Drosophila* legs and wings. Chick fibroblast cells transfected with a retrovirus containing the *Sonic hedgehog* gene acquire the properties of a polarizing region; they cause the development of a mirror-image limb when grafted to the anterior margin of a limb bud, as will beads soaked in Sonic hedgehog protein. The bead must be left in place for 16 to 24 hours for extra digits to be specified, and the pattern of digits depends on the concentration of Sonic hedgehog protein in the bead. A higher concentration is required for a digit 4 to develop than for a digit 2, for example.

Expression of *Sonic hedgehog* requires the prior expression of the transcription factor dHand. This is produced in the early limb bud and becomes restricted to the posterior margin, thus restricting *Sonic hedgehog* expression to this region also. Overexpression of dHand results in *Sonic hedgehog* expression anteriorly as well as posteriorly, and leads to limb duplications.

What determines the size of the polarizing region? Increasing Sonic

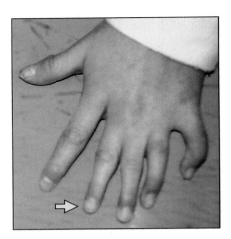

Fig. 10.12 Polydactyly in a human hand. The additional digit (arrowed) resembles the adjacent digit.

Photograph courtesy of R. Winter.

hedgehog levels in the region by local application leads to reduction in *Sonic hedgehog* gene expression and an increase in cell death. This could be the basis of an autoregulatory mechanism to keep signaling more or less constant.

Further evidence for the key role of *Sonic hedgehog* comes from the mouse mutation *extra-toes*, in which there is an additional anterior digit and additional anterior expression of *Sonic hedgehog*. Polydactyly, the development of additional digits, occurs in humans (Fig. 10.12) and this could result from anterior expression of the human version of *Sonic hedgehog* or, as we see below, from mutations in the Hox genes. Mice in which the *Sonic hedgehog* gene has been knocked out develop proximal limb structures but lack distal ones, showing that the gene is not required for the actual initiation of limb development. The loss of distal structures probably reflects a requirement for Sonic hedgehog signaling in order to maintain the apical ectodermal ridge, and thus the progress zone, for a sufficient time for these structures to develop (see Section 10.3).

Several other molecules are also involved in antero-posterior positional signaling in the vertebrate limb bud. Genes for BMP-2 and BMP-4 are expressed in the buds and BMP-2 expression is localized to the polarizing region. Local application of Sonic hedgehog protein induces expression of the BMP growth factors and there is now evidence that these may play a key role in specifying digit identity. If BMP activity is blocked by anti-BMP antibodies following application of Sonic hedgehog to the anterior margin of the bud, additional digits form as expected, but they all have the same identity. Thus Sonic hedgehog may specify digit number but BMP-2 may specify digit identity. Just how the positional values are specified remains unclear.

Retinoic acid is also present at a higher concentration in posterior than anterior regions and, when applied locally to the anterior margin of a chick wing bud, causes formation of extra digits in a mirror-image pattern similar to that obtained by grafting a polarizing region. Retinoic acid induces a new polarizing region by inducing *Sonic hedgehog* expression, so is unlikely to act as a positional signal itself. It is possible, however, that retinoic acid is required for limb bud initiation, as inhibition of retinoic acid synthesis blocks limb bud outgrowth.

10.6 Position along the proximo-distal axis may be specified by a timing mechanism

In contrast to the specification of pattern along the limb's antero-posterior axis, patterning along the proximo-distal axis is less well understood but may be specified by the time cells spend in the progress zone. As the limb bud grows, cells are continually leaving the progress zone. In the forelimb, for example, cells leaving first develop into the humerus and those leaving last form the tips of the digits. If cells could measure the time they spend in the progress zone, for example by counting cell divisions or by increased synthesis or degradation of some molecule, this would give them a positional value along the proximo-distal axis (Fig. 10.13). A timing mechanism of this sort is consistent with the experimental observation that removal of the apical ridge results in a distally truncated limb (see Fig. 10.6). Removal means that cells in the progress zone no longer proliferate, so more distal structures cannot form.

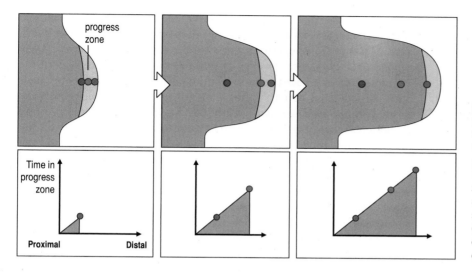

Fig. 10.13 A cell's proximo-distal positional value may depend on the time it spends in the progress zone. Cells are continually leaving the progress zone. If the cells could measure how long they spend in the progress zone, this could specify their position along the proximo-distal axis could be specified. Cells that leave the zone early (red) form proximal structures whereas cells that leave it last (blue) form the tips of the digits.

Evidence for a mechanism based on time comes from killing cells in the progress zone or blocking their proliferation at an early stage by, for example, X-irradiation. The result is that proximal structures are absent but distal ones are present, and can be almost normal. Because many cells in the irradiated progress zone do not divide, the number of cells that leave the progress zone during each unit of time is much smaller than normal. Proximal structures are thus very small, or even absent. As the bud grows out, however, the progress zone becomes repopulated by the surviving cells, so that distal structures such as digits are formed at almost normal size.

10.7 The dorso-ventral axis is controlled by the ectoderm

In the chick wing, there is a well-defined pattern along the dorso-ventral axis; large feathers are only present on the dorsal surface and the muscles and tendons have a complex dorso-ventral organization. Dorsal and ventral compartments are already specified in the ectoderm of the presumptive limb bud regions in early chick embryos, with the apical ridge developing at the boundary between them. Signals from the somites are involved in specifying the dorsal ectoderm. No such compartments are present in the underlying mesoderm.

In the chick, the establishment of the dorso-ventral ectodermal boundary appears to involve signaling from the Notch receptor, modified by the actions of the protein Radical fringe, which is expressed in the presumptive dorsal limb bud ectoderm before formation of the ridge. The chick apical ridge develops at the boundary between cells that express Radical fringe and those that do not. Radical fringe does not appear to be needed for apical ridge formation in mice, however, as the ridge still develops in mice lacking Radical fringe.

The development of pattern along the dorso-ventral axis has been studied by recombining ectoderm taken from left limb buds with mesoderm from right limb buds, so that the dorso-ventral axis of the ectoderm is reversed with respect to the underlying mesoderm. The right and left wing buds are removed from a chick embryo and the ectoderm separated from the mesenchyme. The wing buds are then recombined with the dorso-ventral axis

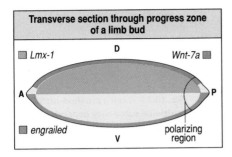

Fig. 10.14 The ectoderm controls dorso-ventral pattern in the developing limb. The gene *Wnt-7a* is expressed in the dorsal ectoderm and *engrailed* (the vertebrate version of the *Drosophila* gene *engrailed*) is expressed in the ventral ectoderm. The gene *Lmx-1* is induced by *Wnt-7a* in the dorsal mesoderm and is involved in specifying dorsal structures.

of the ectoderm opposite to that of the mesoderm. The reconstituted limb buds are then grafted to the flank of a host embryo, with just the dorso-ventral axis of the ectoderm inverted with respect to the enclosed mesenchyme, and allowed to develop. In general, the proximal regions of the limbs that develop have the dorso-ventral polarity of a normal limb: that is their dorso-ventral pattern corresponds to the source of the mesoderm. Distal regions, however, particularly the 'hand' region, have a reversed dorso-ventral axis, with the pattern of muscles and tendons reversed and corresponding to that of the dorso-ventral axis of the ectoderm. The ectoderm can therefore specify dorso-ventral patterning in the limb.

A dorso-ventral pattern is easily seen in the digits of mouse limbs, where ventral surfaces—the palm or paw pads—normally have no fur, whereas dorsal surfaces do. Genes controlling the dorso-ventral axis in vertebrate limbs have been identified from mutations in mice. For example, mutations that inactivate the gene *Wnt-7a*, which encodes a secreted signaling protein of the Wnt family (see Section 3.19), result in limbs in which many of the dorsal tissues adopt ventral fates to give a double ventral limb, the two halves being mirror images. The *Wnt-7a* gene is expressed in the dorsal ectoderm (Fig. 10.14) and this suggests that the ventral pattern may be the ground state and is modified dorsally by the dorsal ectoderm, with Wnt-7a playing a key role in patterning the dorsal mesoderm. Expression of the gene *engrailed* characterizes ventral ectoderm. Mutations that destroy *engrailed* function result in *Wnt-7a* being expressed ventrally as well, giving a double dorsal limb.

One function of Wnt-7a is to induce expression of the LIM homeobox gene *Lmx-1* in the underlying mesenchyme. Mammalian Lmx-1 specifies a dorsal pattern in the mesoderm. Ectopic expression of *Lmx-1* in the ventral mesoderm of the mouse limb results in the cells adopting a dorsal fate, resulting in a mirror-image dorsal limb. Inhibition of *Lmx-1* expression in the chick limb results in the absence of dorsal structures, but the cells do not adopt a ventral fate. Thus other factors are involved in specifying a ventral fate in the chick.

In many *Wnt-7a* mutant mice, posterior digits are lacking, suggesting that *Wnt-7a* is also required for normal antero-posterior patterning. Similar results are observed in chick embryos when dorsal ectoderm is removed. This has led to the suggestion that development along all three limb axes is integrated by interactions between the signals Wnt-7a, FGF-4, and Sonic hedgehog.

10.8 Different interpretations of the same positional signals give different limbs

The positional signals controlling limb patterning are the same in chick wing and leg as well as in different vertebrates, but they are interpreted differently. A polarizing region from a wing bud, for example, will specify additional mirror-image digits if grafted into the anterior margin of a leg bud. However, these digits develop as toes, not wing digits, as the signal is being interpreted by leg cells. Similarly, a polarizing region from a mouse limb bud specifies additional wing structures when grafted into the anterior margin of a chick wing bud. A mouse apical ectodermal ridge grafted in place of a chick apical ectodermal ridge can provide the appropriate signal for early chick limb development. Thus, the signals from

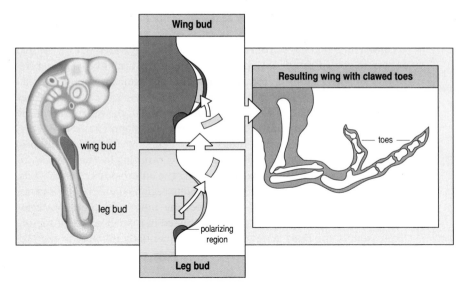

Fig. 10.15 Chick proximal leg bud cells grafted to a distal position in a wing bud acquire distal positional values. Proximal tissue from a leg bud that would normally develop into a thigh, is grafted to the tip of a wing bud. In the progress zone, it acquires more distal positional values and interprets these as leg structures, forming clawed toes at the tip of the wing.

the different polarizing regions are the same, as are the signals from the different apical ridges; the difference in the structures formed by the limbs is a consequence of how the signals are interpreted and thus depends on the genetic constitution and developmental history of the responding cells in the limb bud.

A further demonstration of the interchangeability of positional signals comes from grafting limb bud tissue to different positions along the proximo-distal axis of the bud. If tissue that would normally give rise to the thigh is grafted from the proximal part of an early chick leg bud to the tip of an early wing bud, it develops into toes with claws (Fig. 10.15). The tissue has acquired a more distal positional value after transplantation, but interprets it according to its own developmental program, which is to make leg structures.

The positional signals must thus be interpreted by limb-specific proteins. Two of these are the T-box transcription factors, Tbx4 and Tbx5, which are related to the transcription factor Brachyury (see Section 3.20). They are expressed in the prospective leg and wing regions respectively (see Section 10.2), and then in the leg and wing buds. The transcription factor Pitx1 (related to Otx homeodomain proteins) is expressed in the leg bud only. Misexpression of Tbx5 in the leg results in its developing some wing-like characters. Pitx1 expression in the wing, on the other hand, results in expression of Tbx4 and the development of leg-like distal structures.

10.9 Homeobox genes may also provide positional values for limb patterning

Hox genes specify position along the antero-posterior axis of the vertebrate body (see Section 4.3) and there is now considerable evidence that they may also provide positional values within the limbs. At least 23 different Hox genes are expressed during chick limb development. Attention has been largely focused on the genes of the Hoxa and Hoxd gene clusters, which are related to the *Drosophila Abdominal-B* gene (see Box 4A, p. 117). These sets of genes are expressed in both forelimbs and hindlimbs, whereas the

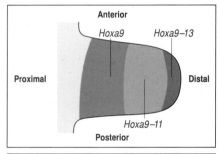

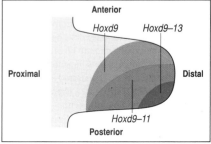

Fig. 10.16 Pattern of Hox gene expression in the chick wing bud. The Hoxa genes (top panel) are expressed in a nested pattern along the proximo-distal axis, *Hoxa13* being expressed most distally, while the Hoxd genes (bottom panel) are expressed in a similarly nested pattern along the antero-posterior axis, *Hoxd13* being expressed most posteriorly.

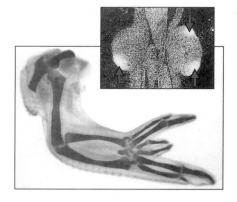

Fig. 10.17 Change in expression of Hoxd genes following a polarizing region graft. Grafting a polarizing region to the anterior margin of a limb bud results in mirror-image expression of the *Hoxd13* genes and a mirror-image duplication of the digits (main panel). The inset shows the expression of *Hoxd13* (arrowed) at the posterior margin in a normal limb bud (inset left). In the operated limb (inset right) *Hoxd13* is expressed at both the anterior and posterior margins.

expression of Hoxb and Hoxc gene clusters is restricted to the forelimb and the hindlimb respectively.

Once a limb bud has started to form, Hox genes expressed within the bud appear to record positional information along the limb axes. The expression of Hox genes during limb development is dynamic, however; the pattern of expression of a particular gene can undergo significant changes as the limb bud grows. Here, we confine our attention to the expression of the Hoxa and Hoxd genes at just one stage of wing development. At this stage, expression of *Hoxd9* through *Hoxd13* is sequentially initiated at the posterior margin of the limb in the progress zone, resulting in a nested pattern of expression of different Hoxd genes along the antero-posterior limb axis. That is, *Hoxd9* expression covers the complete region where the Hoxd genes are expressed, and successive genes occupy more and more posterior regions, so that *Hoxd9–Hoxd13* are all expressed in a small posterior region. Expression of the Hoxa genes is similarly sequentially initiated in the progress zone, and ends up as a nested pattern of expression along the proximo-distal axis (Fig. 10.16).

At the stage of limb development depicted in Fig. 10.16, the proximo-distal sequence of expression of Hoxa genes corresponds to the three main proximo-distal regions of the limb. *Hoxa9* is expressed in the upper limb, where the humerus (or femur in the hindlimb) forms. *Hoxa9–11* are expressed in the lower limb, where the radius and ulna (or tibia and fibula) develop, and *Hoxa9–13* are expressed in the wrist and digits.

If the Hox genes are involved in recording positional information, then experimental manipulations that lead to changes in the pattern of the limb skeletal elements should cause a corresponding change in Hox expression domains. If a polarizing region is grafted to the anterior margin of a wing bud, there is indeed a change to a mirror-image pattern of Hoxd expression (Fig. 10.17). This occurs within 24 hours of grafting, which is about the time required for the polarizing region to exert its effect.

Does alteration of Hox gene expression in the limb lead to homeotic changes, similar to those seen in the vertebrae in Section 4.4? The results of gene knock-out experiments designed to answer this question are complicated, and there is evidently no simple Hox code for the cartilaginous elements of the limb. Some examples of the effect of Hox gene deletion on the mouse forelimb are shown in Fig. 10.18. Knock-outs of individual Hox genes in the mouse do not transform one digit into another. Instead, many bones in the hand are affected in both size and shape, and new elements may even develop. When more than one Hox gene is knocked out at the same time, the effects can be much more severe, and it seems that the Hox genes have an important influence on the growth of the cartilaginous elements. Thus, knock-out of both *Hoxa11* and *Hoxd11* results in the absence of both the radius and ulna. Overexpression of *Hoxa13*, which is normally expressed in the distal region of the limb bud, results in limbs in which the radius and the ulna are reduced in size. This suggests that they are transformed into small wrist-like elements, probably by a change in the control of cell multiplication. Overexpression of *Hoxd13* results in the shortening of the long bones of the leg because of the effects of *Hoxd13* on the rate of cell proliferation in the growing cartilaginous elements. These results show that expression of the Hox genes can control the size of the cartilaginous elements in the limbs at both early and later stages.

Just how the Hox genes determine patterning along the proximo-distal axis remains unclear, but some clues are provided by a class of homeoprotein transcription factors of the PBC family, which combine with Hox proteins to provide transcription factors that modulate recognition of their target genes. One PBC-family protein, Pbx, is present in the nuclei of proximal cells but the cytoplasm of distal ones. Nuclear localization is influenced by the proximal expression of another protein, Meis, as the ectopic expression of Meis distally results in the conversion of distal limb elements to proximal ones. *Meis* expression is, in turn, controlled by Hox genes.

The involvement of Hox genes in human limb development is shown by the phenotype of human Hox gene mutations. A mutation in the human *Hoxd13* gene results in polydactyly and fusion of digits, whereas a mutation in human *Hoxa13* results in anterior and posterior digits that are reduced in size.

10.10 Self-organization may be involved in pattern formation in the limb bud

The development of a pattern of cartilaginous elements along the antero-posterior axis may involve mechanisms other than signaling by the polarizing region. Evidence for this comes, for example, from observing the development of reaggregated limb buds after the mesenchymal cells of early chick limb buds have first been disaggregated and thoroughly mixed to disperse the polarizing region. They are then reaggregated, placed in an ectodermal jacket, and grafted to a site where they can acquire a blood supply, such as the dorsal surface of an older limb. Limb-like structures develop from these grafted buds even though they have developed in the absence of a polarizing region. Several long cartilaginous elements may form in the more proximal regions of these abnormal limbs, although none of the proximal elements can be easily identified with normal structures. More distally, however, reaggregated hindlimb buds develop identifiable toes (Fig. 10.19). The fact that well-formed cartilaginous elements can develop at all in the absence of a discrete polarizing region shows that the bud has a considerable capacity for self-organization. In the digits that develop in hindlimb reaggregates there is no sign of the correlation between Hoxd gene expression and antero-posterior position seen in normal development.

There may therefore be a mechanism in the limb bud that generates a basic pattern—a **prepattern**—of equivalent cartilaginous elements. These elements could then be given their identities and further refined by

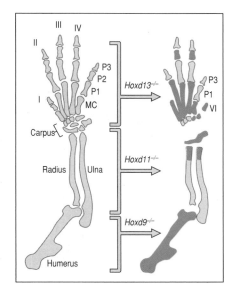

Fig. 10.18 Skeletal defects produced by mutation of Hoxd genes. The genes are inactivated by targeted mutations. The forelimb skeleton of an adult wild-type mouse is shown on the left. The skeletal elements selectively affected by the inactivation of *Hoxd13*, *Hoxd11* or *Hoxd9* singly are shaded in the right-hand diagrams. After Rijli, F.M., Chambon, P.: 1997.

Fig. 10.19 Reaggregated limb bud cells form digits in the absence of a localized polarizing region. Mesodermal cells from a chick leg bud are separated, mixed to disperse all the cells, including the polarizing region, and then reaggregated, placed in an ectodermal jacket, and grafted into a neutral site. Well-formed toes develop distally.

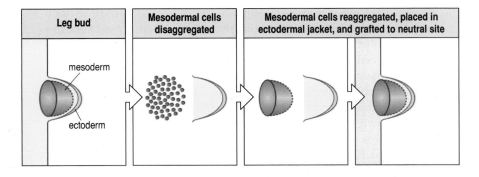

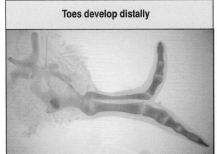

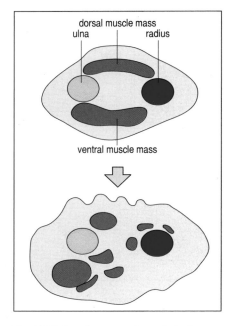

Fig. 10.20 Development of muscle in the chick limb. A cross-section through the chick limb in the region of the radius and ulna shortly after cartilage formation shows presumptive muscle cells present as two blocks—the dorsal muscle mass and the ventral muscle mass. These blocks then undergo a series of divisions to give rise to individual muscles.

response to positional information involving signals such as Sonic hedgehog and the Hox genes. The mechanism for generating the prepattern could be based on a **reaction–diffusion mechanism** (Box 10A). In the wing, for example, a reaction–diffusion or related mechanism could result in a single peak in some morphogen forming in the proximal region of the limb, which would specify a prepattern for the humerus. More distally, alterations in the reaction–diffusion conditions, due to changes in proximo-distal positional information, could give rise to three peaks of the morphogen, giving the cartilaginous elements of the three wing digits. These prepatterns could then be modified by signals specifying antero-posterior and dorso-ventral positional information.

In this reaction–diffusion model, polydactyly in humans could simply result from a chance widening of the limb bud. If a reaction–diffusion mechanism generates a periodic pattern of cartilage elements across the limb as the digits are forming, merely widening the limb bud by some small developmental accident would enable a further digit to develop.

10.11 Limb muscle is patterned by the connective tissue

Limb muscle cells have a different origin from that of the limb mesenchymal cells. If quail somites are grafted into a chick embryo at a site opposite to where the wing bud will develop, the wing that subsequently forms will have muscle cells of quail origin, but all its other cells will be of chick origin. This demonstrates that limb muscle cells have a different lineage to that of limb connective tissues (cartilage and tendons). Cells that give rise to muscle migrate into the limb bud from the somites at a very early stage (see Section 4.2). This migration can be followed by injecting the cDNA for green fluorescent protein into the muscle precursors. After migration, the future muscle cells multiply and initially form a dorsal and a ventral block of presumptive muscle (Fig. 10.20). These blocks undergo a series of divisions to give the final muscle masses. Unlike the cartilage and connective tissue cells, which acquire positional information in the progress zone, the presumptive muscle cells, at least initially, do not acquire positional values, and are all equivalent. However, *Hoxa11* is expressed in the presumptive muscle cells when they enter the limb and is probably induced by the surrounding mesenchyme. When the cells are in the dorsal and ventral masses they express *Hoxa13* in the posterior region. In both cases this pattern of expression is different from that of the mesenchyme in the same regions.

Evidence for the early equivalence of muscle cells comes from experiments where somites are grafted from the future neck region of the early embryo in order to replace the normal wing somites. The muscle cells that develop thus come from neck somites, but a normal pattern of limb muscles still develops. This shows that the muscle pattern is determined by the connective tissue into which the muscle cells migrate, rather than by the muscle cells themselves. A mechanism that could pattern the muscle is based on the prospective muscle-associated connective tissue having surface or adhesive properties that the muscle cells recognize, resulting in the migration of muscle cells to these regions. The pattern of muscle could thus be determined by the pattern of muscle-associated connective tissue, which is presumably specified by mechanisms similar to those that produce the pattern of cartilage. If the pattern of connective tissue changed with time,

Box 10A Reaction-diffusion mechanisms

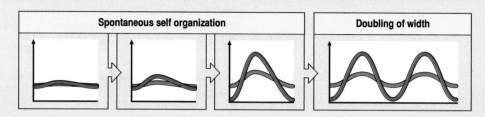

Spontaneous self organization	Doubling of width

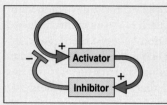

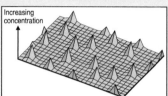

Increasing
concentration

There are some self-organizing chemical systems that spontaneously generate spatial patterns of concentration of some of their molecular components. The initial distribution of the molecules is uniform, but over time the system forms wave-like patterns (see above). The essential feature of such a self-organizing system is the presence of two or more types of diffusible molecules that interact with one another. For that reason, such a system is known as a **reaction–diffusion system**. For example, if the system contains an activator molecule that stimulates both its own synthesis and that of an inhibitor molecule, which in turn inhibits synthesis of the activator, a type of lateral inhibition will occur so that synthesis of activator is confined to one region (see left).

Under appropriate conditions, which are determined by the reaction rates and diffusion constants of the components, a closed system of a certain size can spontaneously develop a spatial pattern of activator with a single concentration peak. If the size of the system is increased, two peaks will eventually develop, and so on. Such a mechanism could thus generate periodic patterns such as the arrangement of digits or the sepals or petals of flowers. When the chemicals can diffuse in two dimensions, the system can give rise to a number of peaks that are somewhat irregularly spaced (see left).

Such a system may underlie some of the patterns of pigmentation that are common throughout the animal kingdom, such as the patterns of spots and stripes seen in the zebra and cheetah (see two bottom panels, left). How these patterns are generated is not yet known, but one possibility is a reaction–diffusion mechanism. Assuming that pigment is synthesized in response to some activator, and that synthesis only occurs at high activator concentrations, some animal color patterns can be mimicked by a reaction–diffusion system in computer simulations.

A characteristic feature of reaction–diffusion patterns is that new intermediate peaks appear as the system grows in size. The angelfish *Pomocanthus semicirculatus* provides a remarkable example of striping that could be generated by a reaction–diffusion mechanism (see below). Juvenile *P. semicirculatus*, less than 2 cm long, have only three dorso-ventral stripes. As the fish grow, the intervals between the stripes increase until the fish is around 4 cm long. New stripes then appear between the original stripes and the stripe intervals revert to those present at the 2 cm stage. As the fish grows larger, the process is repeated. This type of dynamic patterning is what would be expected of a reaction–diffusion mechanism. Computer simulations of reaction–diffusion mechanisms can also generate the patterns seen on a wide variety of mollusc shells. Nevertheless, there is as yet no direct evidence for a reaction–diffusion system patterning any developing organism. Top figure after Meinhardt, H., *et al*.: 1974.

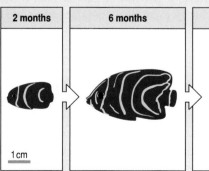

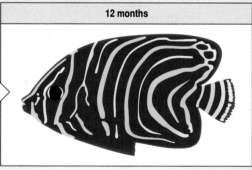

2 months	6 months	12 months

1 cm

the presumptive muscle cells would migrate to the new sites, and this could account for the splitting of the muscle masses.

Presumptive muscle cells proliferate in the limb and express *Pax3*. When *Pax3* is downregulated, cell proliferation stops and the cells differentiate. The overlying ectoderm is required to prevent premature differentiation, and BMP-4 plays a key role in this.

10.12 The initial development of cartilage, muscles, and tendons is autonomous

The pattern of cartilage, tendon, and muscle in the limb may be specified by the same signals, as a polarizing region graft causes the development of a mirror-image pattern of all these elements. Each of the elements develops in its final position and there is little interaction between them. Thus, for example, if just the tip of an early chick wing bud is removed and grafted to the flank of a host embryo, it initially develops into normal distal structures with a wrist and three digits. The long tendon that normally runs along the ventral surface of digit 3 starts to develop in this situation, even though both its proximal end and the muscle to which it attaches are absent. The tendon does not continue to develop, however, because it does not make the necessary connection to a muscle, and so is not put under tension. The mechanism whereby the correct connections between tendons, muscles, and cartilage are established has still to be determined. It is clear, however, that there is little or no specificity involved in making such connections; if the tip of a developing limb is inverted dorso-ventrally, dorsal and ventral tendons may join up with inappropriate muscles and tendons. They simply make connections with those muscles and tendons nearest to their free ends.

10.13 Separation of the digits is the result of programmed cell death

Programmed cell death by apoptosis plays a key role in molding the form of chick and mammalian limbs, especially the digits. The region where the digits form is initially plate-like, as the limb is flattened along the dorso-ventral axis. The cartilaginous elements of the digits develop from the mesenchyme at the correct positions within this plate. Separation of the digits then depends on the death of the cells between these cartilaginous elements (Fig. 10.21). There is evidence for the involvement of BMPs in cell death: if the functioning of BMP receptors is blocked in the developing chick leg, cell death does not occur and the digits are webbed.

Fig. 10.21 Cell death during leg development in the chick.
Programmed cell death in the interdigital region results in separation of the toes. Scale bar = 1 μm.

Photographs courtesy of V. Garcia-Martinez, from Garcia-Martinez, V., et al.: 1993.

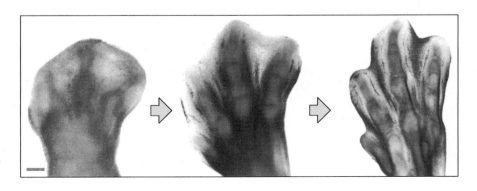

This cell death is a normal programmed part of patterning and cell differentiation (see Chapter 9). The fact that ducks and other waterfowl have webbed feet whereas chickens do not is simply the result of there being less cell death between the digits of waterfowl. If chick limb mesoderm is replaced with duck limb mesoderm, cell death between the digits is reduced and the chick limb develops 'webbed' feet (Fig. 10.22). It is the mesoderm that determines the patterns of cell death both within the mesoderm and the overlying ectoderm. In amphibians, digit separation is not due to cell death, but results from the digits growing more than the interdigital region. That the mesoderm specifies the fate of the overlying epithelium is a general developmental principle.

Programmed cell death also occurs in other regions of the developing limb, such as the anterior margin of the limb bud, between the radius and ulna, and in the wing polarizing region at a later stage. Indeed, it was the investigation of cell death in this region by transplanting it to the anterior margin of the limb bud that led to the discovery that it acts as a polarizing region.

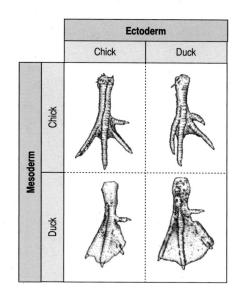

Fig. 10.22 The mesoderm determines the pattern of cell death. The webbed feet of ducks and other water birds form because there is less cell death between the digits than in birds without webbed feet. When the mesoderm and ectoderm of embryonic chick and duck limb buds are exchanged, webbing develops only when duck mesoderm is present.

Summary

Positioning and patterning of the vertebrate limb is largely carried out by intercellular interactions. The position of the limb buds along the antero-posterior axis of the body is probably related to Hox gene expression. There are two key signaling regions within the limb bud. One is the apical ectodermal ridge, which specifies the progress zone in the underlying mesenchyme where cells acquire their positional identity; the second is the polarizing region at the posterior margin of the limb, which specifies pattern along the antero-posterior axis. The signals from the apical ridge are essential for limb outgrowth; one of these signals is probably a fibroblast growth factor. Sonic hedgehog protein is expressed in the polarizing region and may provide a graded positional signal. The dorso-ventral axis is specified by the ectoderm. Hox genes are expressed in a well-defined

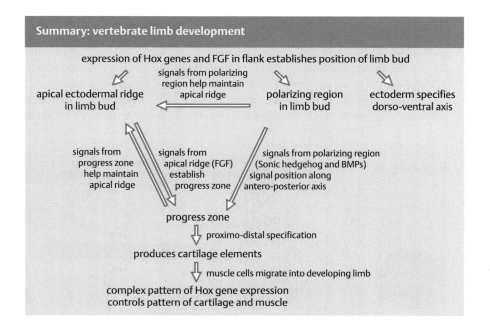

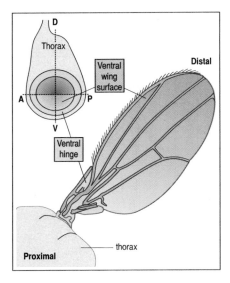

Fig. 10.23 Fate map of the wing imaginal disc of *Drosophila*. The wing disc before metamorphosis is an ovoid epithelial sheet bisected by the boundary between the anterior (A) and the posterior (P) compartments. There is also a compartment boundary between the future ventral (V) and dorsal (D) surfaces of the wing. At metamorphosis, the ventral surface will fold under the dorsal surface (see Fig. 10.24). The disc also contains part of the dorsal region of thoracic segment 2 in the region of wing attachment. After French, V., *et al.*: 1994.

spatio-temporal pattern within the limb bud and together with BMP-2 may provide the molecular basis of positional identity. The limb muscle cells are not generated within the limb but migrate in from the somites and are patterned by the limb connective tissue. Patterning of the cartilaginous elements may involve a self-organizing mechanism. In birds and mammals, separation of the digits is achieved by programmed cell death.

Insect wings and legs

The appendages of the adult *Drosophila*, such as wings, legs and antennae, develop from **imaginal discs**, which are excellent systems for analyzing pattern formation. The discs invaginate from the embryonic ectoderm as simple pouches of epithelium during embryonic development and remain as such until metamorphosis (see Fig. 2.34). In the leg and wing discs, the specification as leg or wing and the basic patterning occur within the embryonic epithelium, at the time when the segments are being patterned and given their identity. The wing and leg discs are initially specified in the embryo as clusters of 20 to 40 cells. During larval development, the discs grow about 1000-fold.

Leg and wing discs are both divided by a compartment boundary that separates them into anterior and posterior developmental regions (see Section 5.15). In the wing disc, a second compartment boundary between the dorsal and ventral regions develops during the second larval instar (Fig. 10.23). When the wing forms at metamorphosis, the future ventral surface folds under the dorsal surface in the distal region to form the double-layered insect wing.

Insect legs are essentially jointed tubes of epidermis and thus have a quite different structure from those of vertebrates. The epidermal cells secrete the hard outer cuticle that forms the exoskeleton. Internally, there are muscles, nerves, and connective tissues. The adult wing is similarly a largely epidermal structure in which two epidermal layers—the dorsal and ventral surfaces—are close together. At metamorphosis, by which time patterning of the imaginal discs is largely complete, each disc undergoes a series of profound anatomical changes to produce a leg or a wing. Essentially, the invaginated epithelial pouch is turned inside out, as its cells differentiate and change shape. In the case of the wing, one surface folds under the other to form the double-layered wing structure (Fig. 10.24).

Although they differ so greatly in appearance, insect wings and legs are developmentally homologous structures, and the strategy of their

Fig. 10.24 Schematic representation of the development of the wing blade from the imaginal disc. Initially, the dorsal and ventral surfaces are in the same plane. At metamorphosis, the sheet folds and extends, so the dorsal and ventral surfaces come into contact with each other.

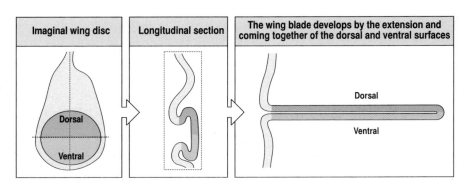

patterning is very similar. Moreover, the mechanisms by which they are patterned, and even the genes involved, show some amazing similarities to the patterning of the vertebrate limb.

Although all imaginal discs superficially look rather similar, they develop according to the segment in which they are located. We shall look later at how the segment-specific character of a particular disc is specified; first we discuss the patterning of the wing and leg discs.

10.14 Positional signals from the antero-posterior and dorso-ventral compartment boundaries pattern the wing imaginal disc

Imaginal discs are essentially sheets of epidermis. The epidermis of body segments, wings, and legs is divided into anterior and posterior compartments—regions of lineage restriction (see Section 5.15). The wing is also divided into dorsal and ventral compartments.

In the wing disc, signaling regions are set up along the compartment boundaries (Fig. 10.25). Cells at the antero-posterior compartment boundary form a signaling region that specifies pattern along the antero-posterior axis of the wing. A cascade of events sets up this signaling center. It begins with the expression of the *engrailed* gene in the posterior compartment of the disc (Fig. 10.25), which reflects the pattern of gene expression in the embryonic parasegment from which the disc derives.

The cells that express *engrailed* also express the segment polarity gene *hedgehog* (see Section 5.16). The maintenance of compartment boundaries in the wing depends partly on communication between compartments. At the compartment boundary, the secreted hedgehog protein acts over about 10 cell diameters and induces adjacent cells in the anterior compartment to express the *decapentaplegic* gene (Fig. 10.25). The hedgehog signal acts through a receptor complex involving the proteins patched and smoothened. The response of the anterior cells to hedgehog involves the transcription factor Cubitus interruptus. This can exist in two forms—a gene activator and a repressor. The activator form is generated in response to hedgehog and both induces transcription of *decapentaplegic* and maintains the compartment boundary, presumably by regulating the expression of an adhesion molecule. The decapentaplegic protein is a member of the TGF-β family of signaling proteins and is related to vertebrate BMP-2 and BMP-4. It is secreted at the compartment boundary (Fig. 10.25) and is probably the positional signal for patterning both the anterior and posterior compartments along the antero-posterior axis.

Fig. 10.25 Establishment of signaling regions in the wing disc at the compartment boundaries. The disc is divided into anterior and posterior compartments and, later in devleopment, into dorsal and ventral compartments. The gene *engrailed* is expressed in the posterior compartment where the cells also express the gene *hedgehog* and secrete the hedgehog protein. Where hedgehog protein interacts with anterior compartment cells, the gene *decapentaplegic* is activated, and decapentaplegic protein is secreted into both compartments, as indicated by the arrows. At the dorso-ventral compartment boundary, the *wingless* gene is activated and the signaling molecule is the protein wingless.

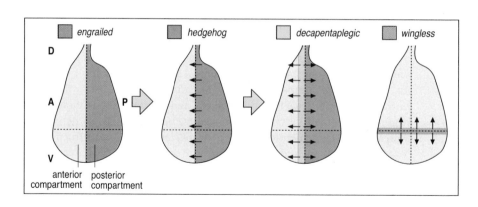

engrailed   hedgehog   decapentaplegic   wingless

anterior compartment   posterior compartment

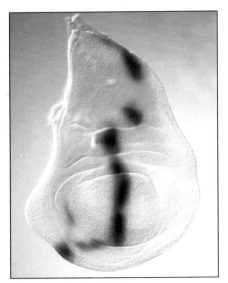

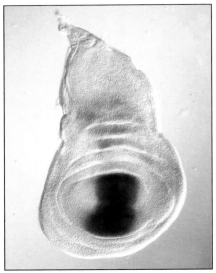

Fig. 10.26 *decapentaplegic* **and** *spalt* **expression in** *Drosophila* **wing disc.** The top panel shows the expression of *decapentaplegic* at the antero-posterior compartment border of the wing blade. The bottom panel shows the expression of *spalt* in the wing blade.

Photographs courtesy of K. Basler, from Nellen, D., et al.: 1996.

There is evidence that the decapentaplegic protein forms a concentration gradient in the wing which provides a long-range signal controlling the localized expression of the genes *spalt* and *omb* in bands overlapping the expression of decapentaplegic (Fig. 10.26). This provides a further level of patterning in the wing. Decapentaplegic protein acts as a morphogen, specifying the expression of *spalt* and *omb* when particular threshold concentrations are reached (Fig. 10.27); low levels induce *omb* whereas higher levels are required to induce *spalt*. The evidence that decapentaplegic is a morphogen comes from several types of experiment. In the first, clones of cells unable to respond to a decapentaplegic signal do not express *spalt* or *omb*. In the second, ectopic expression of *hedgehog* leads to *decapentaplegic* expression and the localized activation of *spalt* and *omb* around the *hedgehog* clone. A third approach is to use temperature-sensitive mutants of *decapentaplegic* so that the signaling activity of the decapentaplegic protein can be reduced by keeping the embryos at raised temperatures. In such mutants, there is a reduction in the region expressing low-threshold genes like *omb* around the source of decapentaplegic, until finally, at very low levels of decapentaplegic, expression of *omb* is lost. Clones expressing low or high levels of decapentaplegic protein can distinguish low-threshold genes, such as *omb*, from high-threshold genes such as *spalt*.

The pattern of veining on the adult wing is one of the ultimate outcomes of this patterning activity. Wing vein L4 forms at the compartment boundary, and it is hedgehog rather than decapentaplegic signaling that defines the position of vein L3 just anterior to it. Vein L2 forms immediately adjacent to cells expressing *spalt*. The presumptive veins may themselves act as boundaries that influence growth in the intervein territories.

The role of the *hedgehog* and *decapentaplegic* genes in wing patterning is further shown by ectopically expressing *hedgehog* in genetically marked cell clones generated at random in wing discs. When such clones form in the posterior compartment they have little effect and development is more or less normal. However, with *hedgehog* clones in the anterior compartment, a mirror-symmetric repeated pattern is produced along the antero-posterior axis in that compartment (Fig. 10.28). If we describe the normal wing pattern from anterior to posterior as 123/45, where the numbers describe the wing veins and / indicates the compartment boundary, then the pattern in the experimental wings can be 123h321123/45, where h indicates the *hedgehog* clone. One can interpret these results in terms of decapentaplegic protein being secreted wherever the hedgehog protein interacts with anterior compartment cells. In a normal wing this interaction is localized at the anterior side of the compartment boundary, but in an experimental wing, the ectopic expression of *hedgehog* sets up new sites of *decapentaplegic* expression in the anterior compartment. This results in new gradients of decapentaplegic protein being formed.

The wing disc is also subdivided into dorsal and ventral compartments, which correspond to the dorsal and ventral surfaces of the adult wing (see Fig. 10.23). These compartments develop after the wing disc is formed— during the second larval instar. The dorsal and ventral compartments were originally identified by cell lineage studies, but they are also defined by expression of the homeotic selector gene *apterous*, which is confined to the dorsal compartment, and which defines the dorsal state. The apterous protein is a structural relative of Lmx-1 from the mouse, which specifies a dorsal pattern in the mesoderm of the mouse limb bud (see Section 10.7).

Fig. 10.27 A model for patterning the antero-posterior axis of the wing disc. The decapentaplgic protein is produced at the compartmental boundary. The schematic cross-section of the wing disc (X–Y) along the antero-posterior axis shows how the presumed gradient in decapentaplegic in both anterior and posterior compartments activates the genes *spalt* and *omb* at their threshold concentrations.

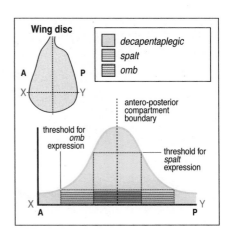

Expression of *apterous* induces the synthesis of the proteins fringe and Serrate. Their actions lead to the Notch receptor protein being activated in a discrete band of cells. The involvement of fringe in setting the dorsal-ventral boundary is similar to that of Radical fringe in the vertebrate limb bud (see Section 10.7). The region of Notch activity thus established marks a compartment boundary, where intense signaling by Notch leads to expression of the *wingless* gene (see Fig. 10.25).

Like the antero-posterior compartment boundary, the dorso-ventral boundary acts as an organizing center, and the secreted protein wingless (see Section 5.16) is the signaling molecule at this boundary. Wingless belongs to the same family as the vertebrate Wnt proteins. It serves a role analogous to decapentaplegic in the antero-posterior patterning system, and activates expression of the genes *achaete*, *Distal-less* and *vestigial* at specific thresholds (Fig. 10.29). These genes and others control the further patterning and growth of the wing.

While both decapentaplegic and wingless act as morphogens whose concentrations provide cells with positional information, the formation of the gradient of morphogen activity is not due to simple diffusion. Studies tracking the movement of the morphogen decapentaplegic labeled with green fluorescent protein show that the decapentaplegic gradient is set up by active transport of the morphogen, rather than by diffusion, and that this involves endocytosis. The gradient is also modified in various ways. Both decapentaplegic and wingless can regulate the expression of their receptors. Decapentaplegic represses the expression of one of its receptors, Thick veins, so that receptor levels become reduced where the morphogen levels are high. More laterally, where the concentration of decapentaplegic is less, the levels of Thick veins are higher. This has two effects. It prevents the spread of decapentaplegic in this region, and the cells reach threshold at lower levels of decapentaplegic. Wing disc cells also have thin actin-based extensions (cytonemes) that project to the antero-posterior signaling boundary. These projections may also be involved in long-range cell–cell communication.

Fig. 10.28 Alteration of wing patterning due to ectopic expression of *hedgehog* and *decapentaplegic*. When *hedgehog* is expressed in a clone of cells in the anterior compartment a new source of decapentaplegic protein is set up. Left panel: in the normal wing, decapentaplegic protein is made at the compartment boundary. The decapentaplegic protein may act as a morphogen, patterning both anterior and posterior compartments, the different veins being specified at threshold concentrations. Right panel: the ectopic expression of *hedgehog* in the anterior compartment results in a new source of decapentaplegic protein and a new wing pattern develops in relation to this new source with veins appearing at threshold concentrations.

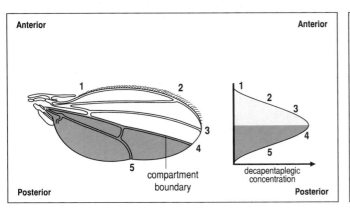

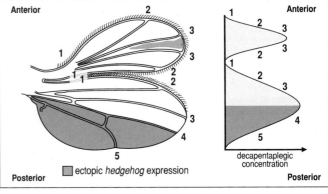

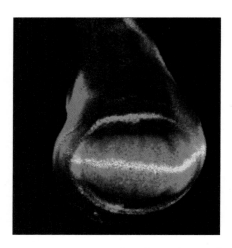

Fig. 10.29 Expression of *wingless* and *vestigial* in the wing disc. *wingless* expression is shown in green, while *vestigial* is red. Both *wingless* and *vestigial* are expressed at the dorso- ventral boundary as indicated by the yellow stripe.

Photograph courtesy of K. Basler, from Zecca, M., et al.: 1996.

As well as the general asymmetry of wing structure, which is due to patterning along the various axes, as discussed above, the individual cells in the adult *Drosophila* wing cuticle become polarized in a proximo-distal direction; this is reflected by the pattern of hairs (trichomes) on the wing cuticle. Each cell produces a single hair, which points distally. The gene *frizzled*, which encodes a transmembrane receptor, has a key role in determining this polarity, as in its absence the hairs do not all point distally but are arranged in characteristic swirling patterns. The mechanism of frizzled action in determining cell polarity is not yet known, but it has been established that the frizzled protein becomes localized to the distal edges of the cells.

10.15 The leg disc is patterned in a similar manner to the wing disc, except for the proximo-distal axis

The easiest way of visualizing a leg imaginal disc is to think of it as a collapsed cone. Looking down on the disc, one can imagine it as a series of concentric rings, each of which will form a proximo-distal segment of leg. A change in the shape of the epithelial cells is responsible for the outward extension of the leg at metamorphosis. The process is rather like turning a sock inside out, with the result that the center of the disc ends up as the distal end, or tip, of the leg (Fig. 10.30). The outermost ring gives rise to the base of the leg, which is attached to the body, and those nearer the center give rise to the more distal structures (Fig. 10.31).

The first steps in the patterning of the leg disc along its antero-posterior axis are the same as in the wing. The *engrailed* gene is expressed in the posterior compartment and induces expression of *hedgehog*. The hedgehog protein induces a signaling region at the antero-posterior compartment border. In the dorsal region of the leg disc, the expression of *decapentaplegic* is induced, as it is in the wing. In the ventral region, however, hedgehog induces expression of *wingless* instead of *decapentaplegic* along the border, and wingless protein acts as the positional signal (Fig. 10.32). The complementary patterns of *wingless* and *decapentaplegic* expression are maintained by mutual inhibition of each other's expression, but there are no dorsal and ventral compartments. The setting up of the proximo-distal axis of the leg disc also involves interactions between wingless and decapentaplegic proteins. The gene *Distal-less*, which marks the distal end of the proximo-distal axis, is expressed where wingless and decapentaplegic expression meet.

Fig. 10.30 *Drosophila* leg disc extension at metamorphosis. The disc epithelium, which is an extension of the epithelium of the body wall, is initially folded internally. At metamorphosis it extends outward, rather like turning a sock inside out. The red arrow in the first panel is the viewpoint that produces the concentric rings shown in Fig. 10.31.

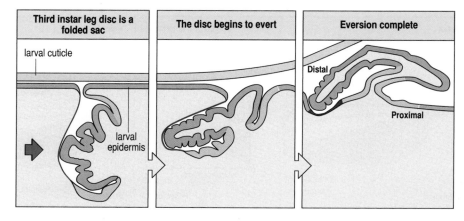

Fig. 10.31 Fate map of the leg imaginal disc of *Drosophila*. The disc is a roughly circular epithelial sheet, which becomes transformed into a tubular leg at metamorphosis. The center of the disc becomes the distal tip of the leg and the circumference gives rise to the base of the leg—this defines the proximo-distal axis. The presumptive regions of the adult leg, such as the tibia, are thus arranged as a series of circles with the future tip at the center. A compartment boundary divides the disc into anterior and posterior regions. After Bryant, P.J.: 1993.

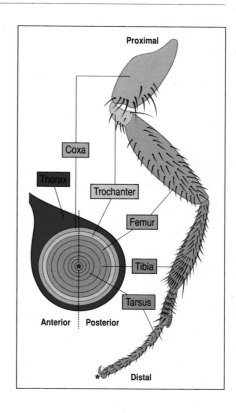

The leg disc is initially divided by the decapentaplegic and wingless signals into a distal domain at the center of the disc, which expresses the gene *Distal-less*, and a proximal domain expressing the gene *homothorax*, both of which encode transcription factors. Their actions lead to the expression of the gene *dachshund*, which also encodes a transcription factor, in a ring between *Distal-less* and *homothorax*, which eventually leads to some overlap of the expression of *dachshund* with *Distal-less* and *homothorax* (Fig. 10.33). Each of these genes is required for the formation of particular leg regions but the expression domains do not correspond precisely with leg segments. Expression of *dachshund*, for example, corresponds to femur, tibia and proximal tarsus. There is no evidence for a proximo-distal compartment boundary; cells expressing *homothorax* and *Distal-less* are, however, prevented from mixing at the interface between their two territories. *Notch* is also expressed in the distal-most cells of each leg segment and Notch signaling is required for joint formation.

10.16 Butterfly wing markings are organized by additional positional fields

The variety of color markings on butterfly wings is remarkable: more than 17,000 species can be distinguished. Many of these patterns are variations on a basic 'ground plan' consisting of bands and concentric eyespots (Fig. 10.34). The wings are covered with overlapping cuticular scales, which are colored by pigment synthesized and deposited by the epidermal cells. How are these patterns specified? Butterfly wings develop from imaginal discs in the caterpillar in a similar way to *Drosophila* wings. Surgical manipulation has shown that the eyespot is specified at a late stage in the development of the wing disc, and that the pattern is dependent on a signal emanating from the center of the spot. A number of the genes that control wing development in *Drosophila*, such as *apterous*, are expressed in the butterfly in a spatial and temporal pattern similar to that in the fly. Thus, as in

Fig. 10.32 Establishment of signaling centers in the antero-posterior compartment of the leg disc, and the specification of the distal tip. The gene *engrailed* is expressed in the posterior compartment and induces expression of *hedgehog*. Where hedgehog protein meets and signals to anterior compartment cells, the gene *decapentaplegic* is expressed in dorsal regions and the gene *wingless* is expressed in ventral regions. Both of these genes encode proteins that are secreted. Expression of the gene *Distal-less*, which specifies the proximo-distal axis, is activated where the wingless and decapentaplegic proteins meet.

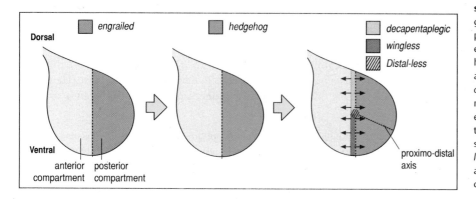

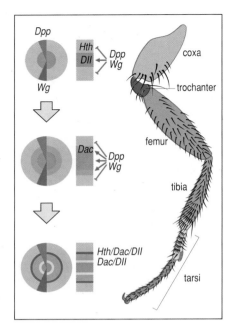

Fig. 10.33 Regional subdivision of the *Drosophila* leg along the proximo-distal axis. The pattern of gene expression in the leg is shown on the right. The stages of gene expression are shown in the two columns on the left, and are viewed as if looking down on the disc. Because of the way in which the leg extends from the disc, the center corresponds to the future tip of the leg, and the more proximal regions to successive rings around it. *decapentaplegic* (dpp) and *wingless* (wg) are initially expressed in a graded manner along the antero-posterior compartment boundary and together induce *Distal-less* (Dll) in the center and repress *homothorax* (hth), which is expressed in the outer region. They then induce *dachshund* (dac) in a ring between Dll and hth. Further signaling leads to these domains overlapping. While hth expression corresponds to proximal regions and Dll to distal regions, there is no simple relation between the other genes and the leg segments. After Milan, M., Cohen, S.M.: 2000.

Drosophila, the shape and structure of the butterfly wing is patterned by a field of positional information. The patterning of the pigmentation, however, involves the establishment of additional fields of positional information.

The expression pattern of *Distal-less* in butterfly wing discs suggests that the mechanism used to delineate the color pattern on the butterfly wing is similar to the mechanism that specifies positional information along the proximo-distal axis of the insect leg. In the butterfly wing, *Distal-less* is expressed in the center of the eyespot, whereas in the *Drosophila* leg disc it is expressed in the central region corresponding to the future tip of the limb. Thus, it is possible that eyespot development and distal leg patterning involve similar mechanisms. The eyespot may be thought of as a proximo-distal pattern superimposed on the two-dimensional wing surface. The center of the eyespot may represent the distal-most positional value, with the surrounding rings representing progressively more proximal positions, as in the leg. The site of the eyespot may be specified with reference to the primary wing pattern (that is, the anterior, posterior, dorsal, and ventral compartments); a secondary coordinate system centered on the eyespot would then be established, with expression of the *Distal-less* gene defining the central focus.

10.17 The segmental identity of imaginal discs is determined by the homeotic selector genes

Patterning of the leg and wing involves similar signals, such as decapentaplegic protein, yet the actual pattern that develops is very different. This implies that the wing and leg discs interpret positional signals in different ways. This interpretation is under the control of the Hox genes and can be illustrated with respect to the leg and antenna. If the Hox gene *Antennapedia*, which is normally expressed in parasegments 4 and 5 (see Section 5.20) and specifies the discs for the second pair of legs, is expressed in the head region, the antennae develop as legs (Fig. 10.35). Using the technique of mitotic recombination (see Box 5B, p. 173) it is possible to generate a clone of *Antennapedia*-expressing cells in a normal antennal disc. These cells develop as leg cells, but exactly which type of leg cell depends on their

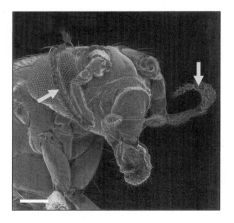

Fig. 10.34 Butterfly wing pattern. Ventral view of a female African butterfly *Bicyclus anynana*, showing wing color pattern and prominent eyespots. Scale bar = 5 mm.

Photograph courtesy of V. French and P. Brakefield.

Fig. 10.35 Scanning electron micrograph of *Drosophila* carrying the *Antennapedia* mutation. Flies with this mutation have the antennae converted into legs (arrowed). Scale bar = 0.1 mm.

Photograph by D. Scharfe, from Science Photo Library.

position along the proximo-distal axis; if, for example, they are at the tip, they form a claw. It is as if the positional values of the cells in the antenna and leg are similar, and the difference between the two structures lies in the interpretation of these values, which is governed by the expression or non-expression of the *Antennapedia* gene. This can be illustrated with respect to the French flag and the Union Jack (Fig. 10.36)—cells develop according to both their position and their developmental history, which determines the genes that they are expressing at any given time. This principle applies also to wing and haltere imaginal discs. Thus, we find a similarity in developmental strategy between insects and vertebrates; both use the same positional information in appendages like legs and wings, and interpret it differently.

The character of a disc and how positional information is interpreted is determined by the Hox genes. Insect legs develop only on the three thoracic segments and not on the abdominal segments and, in *Drosophila*, wings develop only on the second thoracic segment. These adult structures are segment-specific because the particular type of imaginal disc that gives rise to them is only formed by certain parasegments. There are no appendages on abdominal segments in *Drosophila* because the genes required for formation of leg and wing discs are suppressed in the abdomen. The type of disc formed in a particular thoracic segment is typically specified by the action of one of the Hox genes expressed in the segment. Expression of the genes *Antennapedia* and *Ultrabithorax* specifies the second and third pair of legs, respectively. But how does a mutation in *Antennapedia* result in transformation of antennae into legs?

The answer involves the genes *homothorax* and *Distal-less*, which are required in both antenna and leg for the formation of distal and proximal elements, respectively (see Section 10.15). *Distal-less* and *homothorax* also act in combination as selector genes to specify antenna; if *homothorax* is not expressed in the antennal disc, leg structures develop instead. It seems that in the leg imaginal discs, *Antennapedia* prevents *homothorax* and *Distal-less* acting together, and so legs form and not antennae. *Antennapedia* is not normally expressed in the antennal disc, but the dominant *Antennapedia* mutation results in the gene being turned on here as well. As in the leg disc, it then blocks the combined action of *Distal-less* and *homothorax*, resulting in legs and not antennae.

The leg imaginal discs arise from small clusters of ectodermal cells in parasegments 3–6, which contribute to the future thoracic segments of the embryo (Fig. 10.37). In the embryo, each disc initially contains around 25 cells and is formed during growth of the blastoderm. The discs arise at the parasegment boundaries, with anterior and posterior compartments of adjacent parasegments contributing to each disc. In the future second thoracic segment, the leg disc splits off a second disc early in its development, which becomes the wing disc. Similarly, an additional disc, which develops into the haltere, a balancing organ, is formed in the future third thoracic segment.

Mutations in Hox genes in *Drosophila* adults can cause compartment-specific homeotic transformations of halteres into wings. In normal *Drosophila* adults, the wing is on the second thoracic segment and the haltere on the third thoracic segment; they arise from imaginal discs originating at the boundaries of parasegments 4/5 and 5/6, respectively (see Fig. 10.37). In the normal embryo, the *Ultrabithorax* gene, one of the genes of the bithorax

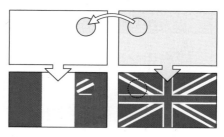

Fig. 10.36 Cells interpret their position according to their developmental history and genetic make-up. For example, if there are two flags which use the same positional information to produce a different pattern, then a graft from one to the other would result in the graft developing according to its intrinsic pattern, but with the pattern appropriate to its new position. This is what happens in imaginal discs.

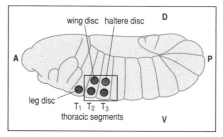

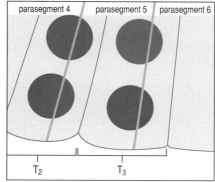

Fig. 10.37 The position in the late *Drosophila* embryo of the imaginal discs that give rise to the adult thoracic appendages. The imaginal discs for the legs, wings, and halteres lie across the parasegment boundaries in the thoracic segments as shown for the wing and haltere discs in the lower panel.

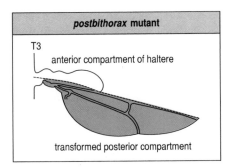

Fig. 10.38 Effects of the *postbithorax* mutation in *Drosophila*. The mutation *postbithorax* acts on the posterior compartment of a haltere, converting it into the posterior half of a wing.

complex (see Box 4A, p. 117), is expressed in parasegments 5 and 6 and is involved in specifying their identity. The *bithorax* mutation (*bx*), which causes the *Ultrabithorax* gene to be misexpressed, can transform the anterior compartment of the third thoracic segment, and thus of the haltere, into the corresponding anterior compartment of the second segment: the anterior half of the haltere thus becomes a wing (see Fig. 5.38). The *postbithorax* mutation (*pbx*), which affects a regulatory region of the *Ultrabithorax* gene, transforms the posterior compartment of the haltere into a wing (Fig. 10.38). If both mutations are present in the same fly, the effect is additive and the result is a fly that has four wings but cannot fly (see Fig. 5.38). Another mutation, *Haltere mimic*, causes a homeotic transformation in the opposite direction: the wing is transformed into a haltere.

As with the antenna and leg, it is possible to generate a mosaic haltere with a small clone of cells containing an *Ultrabithorax* mutation (such as *bithorax*) in the disc; the cells in the clone make wing structures, which correspond exactly to those that would form in a similar position in a wing. It is as if the positional values in haltere and wing discs are identical and all that has been altered in the mutant is how this positional information is interpreted (see Fig. 10.36). In fact, other discs seem to have similar positional fields.

Summary

The legs and wings of *Drosophila* develop from epithelial sheets—imaginal discs—which are set aside in the embryo. Hox genes acting in the parasegments specify which sort of appendage will form and can direct the interpretation of positional information. The leg and wing imaginal discs are divided at an early stage into anterior and posterior compartments. The boundary between the compartments, is a pattern-organizing center and a source of signals that pattern the disc. In the wing disc, expression of *decapentaplegic* is activated by the hedgehog protein at the antero-posterior compartment boundary and the decapentaplegic protein acts as the antero-posterior patterning signal. The wing disc also has dorsal and ventral compartments, with wingless acting as the patterning signal along the dorso-ventral boundary. In the leg disc, the antero-posterior compartment boundary is established in a very similar way, except that hedgehog activates *wingless* instead of *decapentaplegic* in the ventral regions of the leg and the wingless protein acts as the patterning signal in these regions. There is no evidence of dorsal and ventral compartments in the leg disc. The proximo-distal axis of the leg is specified by the interaction between decapentaplegic and wingless proteins, which activates leg-specific genes like *dachshund*. The colorful eyespots on butterfly wings may be patterned by a mechanism similar to that used to organize the proximo-distal axis of the insect leg.

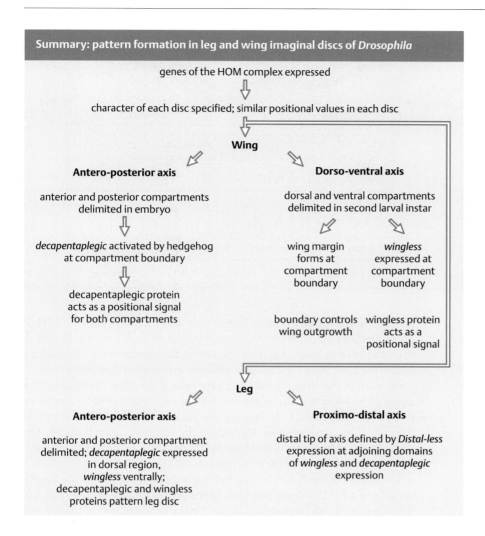

Summary: pattern formation in leg and wing imaginal discs of *Drosophila*

genes of the HOM complex expressed

⇩

character of each disc specified; similar positional values in each disc

⇩

Wing

Antero-posterior axis

anterior and posterior compartments delimited in embryo

⇩

decapentaplegic activated by hedgehog at compartment boundary

⇩

decapentaplegic protein acts as a positional signal for both compartments

Dorso-ventral axis

dorsal and ventral compartments delimited in second larval instar

wing margin forms at compartment boundary

wingless expressed at compartment boundary

boundary controls wing outgrowth

wingless protein acts as a positional signal

Leg

Antero-posterior axis

anterior and posterior compartment delimited; *decapentaplegic* expressed in dorsal region, *wingless* ventrally; decapentaplegic and wingless proteins pattern leg disc

Proximo-distal axis

distal tip of axis defined by *Distal-less* expression at adjoining domains of *wingless* and *decapentaplegic* expression

The nematode vulva

The vulva forms the external genitalia of the hermaphrodite of the nematode worm *Caenorhabditis elegans* and connects with the uterus. Its interest as a model developmental structure arises from the small number of cells—just four, one inducing and three responding—involved in its initial specification, and the fact that more than 40 genes involved in its development have been identified. We deal here with patterning and induction of the vulva at the level of the individual cell. The vulva is an adult structure that develops in the last larval stage. The mature vulva contains 22 cells, with a number of different cell types. It is derived from three of six P cells of ectodermal origin, which persist in a small row aligned antero-posteriorly on the ventral side of the larva. Three of these cells—$P5_p$, $P6_p$, and $P7_p$—give rise to the vulva, each having a well-defined lineage. In discussing the development of the vulva, three distinct cell fates are conventionally distinguished—primary (1°), secondary (2°), and tertiary (3°). The primary and secondary fates refer to different cell types in the vulva, whereas the tertiary fate is non-vulval. $P6_p$ normally gives rise to the primary lineage, whereas $P5_p$ and $P7_p$ give rise to secondary lineages. The other three P cells

Fig. 10.39 Development of the nematode vulva. The vulva develops from three cells, P5$_p$, P6$_p$, and P7$_p$, in the post-embryonic stages of nematode development. Under the influence of a fourth cell, the anchor cell, P6$_p$, undergoes a primary pathway of differentiation that gives rise to eight vulval cells. P6$_p$ is flanked by P5$_p$ and P7$_p$, which each undergo a secondary pathway of differentiation that gives rise to seven cells of different vulval cell type. Three other P cells nearby adopt a tertiary fate and give rise to epidermis.

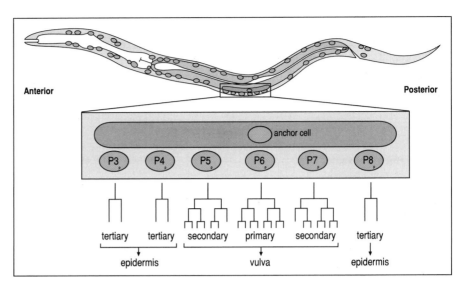

give rise to the tertiary lineage, divide once and then fuse to form parts of the epidermis (Fig. 10.39). Initially, however, all six P cells are equivalent in their ability to develop as vulval cells. One of the questions we address here is how the final three cells become selected as the vulval precursor cells.

Cell fate of the three vulval precursor cells is specified by an inductive signal from a fourth cell, the gonadal anchor cell, which confers a primary fate on the cell nearest to it, P6$_p$, and a secondary fate on the cells lying just beyond it, P5$_p$ and P7$_p$ (see Fig. 10.39). In addition, once induced, the primary cell inhibits its immediate neighbors from expressing a primary fate. P cells that do not receive the anchor cell signal adopt a tertiary fate.

The lineage of the P cells is fixed following induction. If one of the daughter cells of a P cell is destroyed, the other does not change its fate. There is no evidence for cell interactions in the further development of these P cell lineages, and cell fate is thus probably specified by asymmetric cell divisions. The vulva is formed from the 22 cells derived from the three lineages. Actual morphogenesis of the vulva requires the cells to divide, move and fuse in a precise pattern to form seven toroidal rings (Fig. 10.40).

10.18 The anchor cell induces primary and secondary fates in prospective nematode vulva precursor cells

How are just three P cells specified, and how is the primary fate of the central cell made different from the secondary fate of its neighbors? The six P cells are initially equivalent, in that any of them can give rise to vulval tissue. The key determining signal is that provided by the anchor cell, which normally lies above P6$_p$. The vital role of the anchor cell is shown by cell ablation experiments; when the anchor cell is destroyed with a laser microbeam, the vulva does not develop. If all except one P cell is destroyed, its fate depends on how close it is to the anchor cell. If it is very close it adopts a primary fate, if far away a tertiary fate, and at intermediate distances, a secondary fate. This suggests that a graded signal with thresholds specifies the primary and secondary fates.

The anchor cell's signal is the product of the *lin-3* gene, which is a protein similar to the EGF growth factor of vertebrates. Mutations in *lin-3* result in

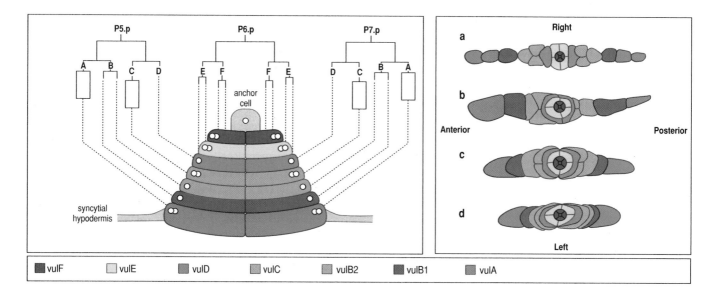

the same abnormal development that follows removal of the anchor cell, and so no vulva is formed. The receptor for the inductive signal is an EGF receptor-like transmembrane tyrosine kinase, encoded by the *let-23* gene (Fig. 10.41). Mutations that inactivate *let-23* also result in no vulva being formed, as the P cells cannot respond to the inductive signal. Direct evidence for a graded signal comes from placing *lin-3* under the control of a heat-shock promoter, so that its activity can be controlled. All but one P cell are then killed and the effect of increasing *lin-3* expression is followed. At low levels of *lin-3* expression, a secondary fate is induced in the remaining P cell, and at high levels a primary fate is the result.

In addition to a graded signal, there is a separate signal from the primary cell that specifies a secondary fate in the adjacent cells. Evidence for this comes from the observation that in genetic mosaics in which the future secondary cells lack the receptor for the *lin-3* signal, development is normal. This shows that the primary cell must be providing a signal for secondary fate, which is not LIN-3 protein. The induction of secondary fate involves a transmembrane receptor, encoded by *lin-12*, which belongs to the Notch family. Thus, two systems pattern early vulval development, perhaps to ensure that it develops correctly.

Fig. 10.40 Formation of the vulva by migration and fusion of specified precursors. The left-hand diagram shows a lateral representation of the seven rings of the vulva at around 39 hours and the cell lineages that give rise to them. The daughter cells of A and C cells fuse. The open circles represent nuclei. At a later stage there will be further fusion of cells in the rings. The right panel shows a ventral view of the developing vulva. The sequence a to d shows morphogenesis of the vulva over a period of 5 hours, from 34 hours. Changes in the shape of the cells and cell fusion give rise to the conical structure seen in the left panel. After Sharma-Kishore, R., *et al.*: 1999.

Fig. 10.41 Cell interactions in vulval development. The anchor cell produces a diffusible signal LIN-3, which induces a primary fate in the precursor cell closest to it by binding to the receptor LET-23. It also induces a secondary fate in the two P cells slightly further away, where the concentration of the signal is lower. The cell adopting a primary fate inhibits adjacent cells from adopting the same fate by a mechanism of lateral inhibition involving LIN-12, and also induces a secondary fate in these cells. A constitutive signal from the epidermis inhibits the development of both primary and secondary fates, but is overruled by the initial inductive signal from the anchor cell.

At least one other signal is involved in establishing the vulva: a signal from the hypodermis—the larval nematode's epidermis—which inhibits adoption of a vulval fate in all the six cells of the equivalence group, and which is overruled by the inductive signal.

Summary

The nematode vulva is derived from just three out of six potentially competent cells, and involves both cell interactions and a well-defined cell lineage. The fates of the three precursor P cells are specified by a graded signal from an adjacent anchor cell; this induces a primary fate in the cell nearest to it, and a secondary fate in the two cells adjacent to the primary cell. These cells then have a well-defined lineage. In addition, the cell adopting a primary fate induces the adjacent cells to develop a secondary fate. The inducing signal (LIN-3) is a protein similar to the vertebrate growth factor EGF, and the second signal involves a Notch-like transmembrane receptor—LIN-12.

Internal organs: heart, blood vessels, lungs, kidneys, and teeth

All the structures discussed so far are external, and this has made them relatively easy to study. We now consider some, mainly vertebrate, internal organs whose development is, as yet, not as well understood. Their development illustrates the principles underlying developmental phenomena we have not yet encountered, such as branching morphogenesis and mesenchyme to epithelium transitions. While numerous genes involved in the development of the organs have been identified, the mechanisms controlling their development are less well understood than those so far considered.

Epithelia are the commonest type of tissue organization found in animals and play a key role in the development of many organs such as the kidney, lung, and blood vessels. Cells in epithelia adhere tightly together to form a sheet, which can be single layered, as in the endothelial lining of capillaries and kidney tubules, or multilayered, as in skin. A common feature of epithelia is that they are separated from underlying tissue by a basal lamina of extracellular matrix. It is common for epithelia involved in organogenesis to be induced in the adjacent mesenchyme—the **mesenchyme to epithelium transition**—and so to form tubes and tubules.

10.19 The development of the vertebrate heart involves specification of a mesodermal tube which is patterned along its long axis

The heart is one of the first structures to form during organogenesis. It is of mesodermal origin and is first established as a single tube consisting of two epithelial layers—the inner **endocardium**, which is a set of endothelial cells, and the outer **myocardium**, which is contractile. During development, this tube becomes divided longitudinally into two chambers, the atrial and ventricular chambers. A two-chambered heart is the basic adult

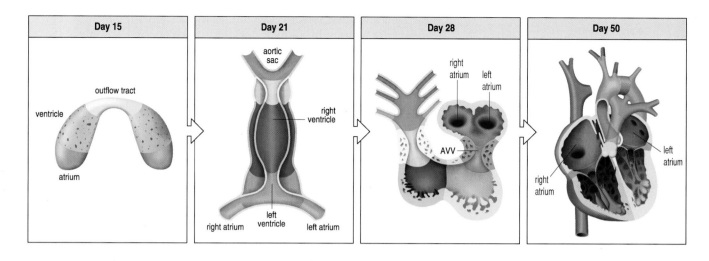

Fig. 10.42 Schematic of human heart development. By day 15 of human embryonic dvelopment, cardiogenic precursors have formed a crescent, as shown in the first panel, in which the main regions of the heart are already specified. The two arms of the crescent fuse along the midline to give a linear heart tube, which is patterned along the antero-posterior axis with the regions and chambers of the mature heart (second panel). After looping (third panel) these regions are disposed approximately in their eventual positions. Later development results in further patterning (fourth panel), and the formation of valves between, for example, the atria and ventricles. AVV, atrioventricular valve region. After Srivastava, D., Olson, E.N.: 2000.

form in fish, but in higher vertebrates, such as birds and mammals, looping and further partitioning give rise to the four-chambered heart.

In the chick and mouse, the lateral plate mesoderm is the major source of heart mesoderm. In the chick, ingression of heart precursor cells through the primitive streak occurs throughout most of the period of streak formation. The heart precursors ingress in a roughly similar antero-posterior order to their eventual position in the heart. Thus, cells close to the streak contribute to anterior structures, and those more lateral to posterior structures such as the atria. Before ingression through the streak, the presumptive heart cells are not yet determined and can give rise to somites if transplanted to the appropriate region. A few hours after ingression, however, the cells are committed and will differentiate into cardiac muscle cells. Shortly afterwards, the antero-posterior patterning of the heart is specified. In *Xenopus*, heart mesoderm is specified by signals from the organizer region (see Section 3.9). In chick, there is some evidence of a role for anterior endoderm in inducing heart tissue.

In all vertebrates, the precursor heart cells come to lie in two patches of lateral plate mesoderm on either side of the midline. They then undergo a complex morphogenetic process and move towards the midline and fuse to form the heart tube (Fig. 10.42). A number of mutations in the zebrafish disrupt this process, leading to two laterally positioned hearts—cardia bifida. One of the genes mutated is *miles apart*, which codes for a receptor that binds lysosphingolipids; sphingosine-1-phosphate is the likely ligand in this case. *Miles apart* is not expressed in the migrating heart cells themselves, but in cells on either side of the midline, and thus may be involved in directing the migration of the presumptive heart cells. Cells appear to be committed to a myocardial or endocardial cell fate early in heart development, as lineage studies of individual cells in the early heart region showed that they gave rise to clones of only one cell type.

There are remarkable similarities—though we should no longer be surprised—between the genes involved in heart development in *Drosophila* and in vertebrates. The homeobox gene *tinman* is required for heart formation in *Drosophila*, and its vertebrate homolog *Nkx-2.5* is expressed in early heart cells and also in neighboring non-heart cells during vertebrate heart development. Mutations in this gene in mouse and human result in heart abnormalities. When vertebrate *tinman*-like genes are expressed in *Drosophila* they can substitute for *Drosophila tinman* and rescue some of the abnormalities caused by lack of normal *tinman* functions. In *Drosophila*, *decapentaplegic* maintains *tinman* expression in the dorsal mesoderm; in vertebrates, BMPs have been shown to induce *Nkx-2.5* and are also able to induce cardiac cell differentiation. Nkx-2.5 or Tinman act together with GATA transcription factors to turn on cardiac-specific gene expression, although GATA transcription factors are not expressed exclusively in cardiac precursors.

There is some evidence that the heart develops as a modular organ in which each anatomical region is controlled by a distinct transcriptional program. The regions are the segments along the heart tube—atria, left ventricle, right ventricle, and ventricular outflow tract in the mammalian heart (see Fig. 10.42). Precursor cells of these regions appear to have separate lineages that develop according to their position along the antero-posterior axis. For example, the regulation of *Nkx-2.5* gene expression is complex and somewhat modular. Seven different activating regions and three repressor regions have been identified in this gene. Activating regions include those that lead to its expression in the entire heart tube, or just in the right ventricle and outflow tract. There is evidence that retinoic acid acts as a signal for antero-posterior positional activity, its source being mesoderm posterior and lateral to the node.

Among the hundreds of known zebrafish mutations, none has eliminated heart tissue completely, although mutations have been found that eliminate ventricle formation, leaving a heart with only one major chamber, the atrium. Thus, it is most likely that many genes are involved in specifying the different heart regions.

The later development of the heart involves asymmetric looping (see Section 3.8) and the formation of the separate chambers—four in the case of the mammalian heart (see Fig. 10.42). Cells of neural crest origin contribute to the outflow tracts; they are essential, for example, to formation of the pulmonary artery and aorta from the single embryonic outflow tract, the truncus arteriosus. Defects in these regions account for 30% of congenital heart defects in humans, some of which are the result of developmental perturbations in neural crest cell specification and migration.

10.20 The vascular system develops from angioblasts and the vessels then undergo angiogenesis

The defining cell type of the vascular system is the endothelial cell which forms the lining of the entire circulatory system, including heart, veins, and arteries. Development of blood vessels starts with **angioblasts** in mesodermal tissues; these are the precursors of the endothelial cells (Fig. 10.43). Angioblasts initially assemble into the main vessels of the vasculature, in the process called **vasculogenesis**. This is then elaborated by the process of **angiogenesis**, which involves vessel extension and branching to form venules, arterioles, and extensive networks of capillaries. Antibodies specific for the endothelial cells of the quail embryo have been used to identify the origin of blood vessels from free angioblasts in quail–chick chimeras. Angioblast differentiation requires the growth factor VEGF (vascular endothelial growth factor) and its receptors. VEGF is also a potent mitogen for endothelial cells.

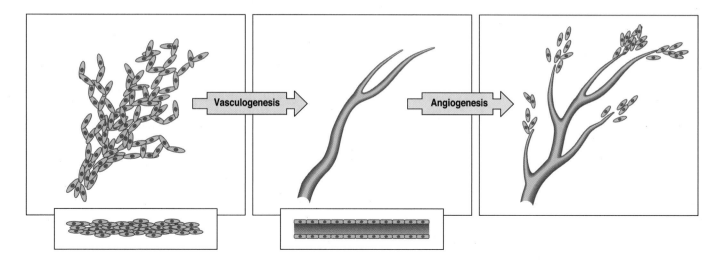

New blood capillaries are formed by sprouting from pre-existing blood vessels (Fig. 10.43). The capillary grows by degradation of the extracellular matrix and proliferation of cells at the tip of the sprout. Cells at the tip extend filopod-like processes which guide and extend the sprout. Many solid tumors produce growth factors that stimulate angiogenesis, and isolated endothelial cells in culture can be stimulated with tumor-conditioned medium to form capillary-like networks.

VEGF stimulates endothelial cells to degrade the surrounding basement membrane matrix and to migrate and proliferate. Mice with only one copy of the *Vegf* gene die before birth because of blood vessel abnormalities. Expression of *Vegf* is induced by lack of oxygen and thus promotes vascularization of an active organ. The Tie family of receptors and their ligands, the angiopoietins, are also required for angiogenesis. If endothelial cells are cultured on beads and then embedded in a fibrin gel, treatment with angiopoietin-1 will induce the formation of capillary sprouts from the cells on the beads.

Arteries and veins are defined by the direction of blood flow as well as by structural and functional differences. The first tubular structures are formed by endothelial cells and the vessels are then covered by pericytes and smooth muscle cells. Arterial and venous endothelial cells are molecularly distinct from the earliest stages of angiogenesis. Ephrin-B2 is expressed in arterial endothelial cells, while ephrin-B4, which is a receptor for ephrin-B2, is present on venous endothelial cells. Proper development of the vascular system involves reciprocal signaling between these membrane-bound molecules.

Fig. 10.43 Vasculogenesis and angiogenesis. Angioblasts assemble into simple tubes during vasculogenesis. These are then elongated during angiogenesis, which involves vessel extension, branching, and growth.

10.21 The tracheae of *Drosophila* and the lungs of vertebrates branch using similar mechanisms

The mammalian lung and the **tracheal system** of *Drosophila* are both involved in the delivery of oxygen from the air to the tissues, and both have elaborate branching patterns. These essentially tubular structures are constructed from an epithelial cell monolayer. The development of both these structures begins with a simple sac or tube from which new branches grow out.

Air enters the tracheal system of the *Drosophila* larva through the spiracles, and oxygen is delivered to the tissues by some 10,000 or so fine branches. This system develops from 20 ectodermal sacs, of about 80 cells each, which invaginate during embryogenesis. Each sac gives rise to six primary tracheal branches. These branches form when a few cells migrate away from the sac in particular directions to form a tubular branch. Remarkably, the whole branching pattern is due to such cell migrations and rearrangements, and no cell proliferation is involved. Secondary branches form by sprouting from the ends of the primary branches and will also branch further.

More than 50 genes required for tracheal branching have been identified in *Drosophila*. *Trachealess* is turned on in the sac, and six domains of *branchless* expression are found in the cells around the sac and determine where the primary branches will form. *Branchless* codes for an FGF-like protein that signals primary branch formation Fig. 10.44. As the primary branch moves towards the source of Branchless, the gene is turned off. Later, *branchless*-expressing cells again stimulate branch formation from the tips of the primary branches. Branchless induces a new set of genes including *sprouty*. The Sprouty protein blocks Branchless signaling to sites further away from the source, and so limits branching to the tip.

Both the trachea and lungs of the mouse develop from the foregut when it is still a simple tube of endoderm. The trachea is unbranched, and is formed by the division of the foregut by a longitudinal septum into two tubes—the trachea and the esophagus. The development of mouse lungs starts with left and right epithelial lung buds, which derive from the endoderm of the foregut. These move into the surrounding mesenchyme, and give rise to millions of subdivisions by successive branching. Early lung development has a reproducible pattern, even in culture. The left primary lung bud sprouts several secondary buds along its lateral face in a precise spatial and temporal sequence. The pattern of branching of the right bud is quite different; for example, the first branch arises from the dorsal surface. Transplantation studies show that the mesenchyme plays a key role in controlling branching and bud formation. Mesenchyme from distal lung buds can induce ectopic buds when grafted next to tracheal endoderm. By contrast, tracheal mesenchyme inhibits outgrowth of lung buds.

Fig. 10.44 Models of branching patterns in the tracheae of *Drosophila* and the mouse lung. Top panels: in *Drosophila*, localized secretion of Branchless protein (blue) attracts the tracheal cells to move towards it, creating a primary branch (left panel). At the top of the branch, the gene *sprouty* is induced, and the sprouty protein (green) inhibits branching further away from the source of Branchless (center panel). Branchless also induces secondary branches (right panel). Bottom panel: in the mouse, FGF-10 (blue) induces outgrowth of a branch of a lung tracheole towards the FGF-10 source. At the tip of the branch, secretion of an inhibitor (green) of *FGF-10* expression (possibly Sonic hedgehog) splits the region of attraction, forming two secondary branches. After Metzger, R.J., Krasnow, M.A.: 1999.

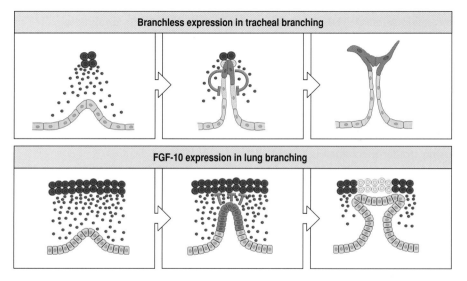

Branchless expression in tracheal branching

FGF-10 expression in lung branching

FGF-10 is the most likely candidate for regulating lung budding as the buds grow towards areas in which FGF-10 is being expressed. Sonic hedgehog is proposed to act as a negative feedback signal that shuts off *FGF-10* gene expression in the mesenchyme as the branch approaches it (Fig. 10.44). Two new buds then sprout towards each of the two lateral domains that continue to secrete FGF-10. Embryos lacking FGF-10 develop neither lung buds nor limbs. However, a bead loaded with FGF-10 does not induce a bud from tracheal endoderm but will stimulate outgrowth when placed near distal lung buds.

10.22 The development of the ureteric bud and mesenchymal tubules of the kidney involves induction

The kidney forms from mesoderm. In mammals and birds, the kidney develops from the metanephros, but two forms of primitive kidney also develop—the pronephros and the mesonephros. The pronephros duct associated with the pronephros differentiates into the Wolffian duct, which gives rise to the ureteric bud; this interacts with the metanephros to form the collecting tubules. The mesonephros also contributes to gonad development (Chapter 12). In amphibians and fish, it is the mesonephros that develops into the functioning kidney.

Tubule formation in the mammalian kidney is an example of organogenesis that involves both induction and a mesenchyme to epithelium transition. The adult kidney develops from a mass of mesenchyme known as the metanephric blastema. This gives rise to the renal tubules and glomeruli—the basic functional units of the kidney—and an epithelial bud, the ureteric bud, which buds off the embryonic ureter and branches to form the collecting system for urine (Fig. 10.45). The ureteric bud induces the mesenchymal cells to condense around it and form epithelial structures that develop into renal tubules. Each tubule elongates to form a glomerulus at one end, where filtration of the blood will occur, while the other end fuses to a collecting duct connecting to the ureter. The intermediate portion of the tubule is convoluted and its epithelial cells become specialized for the resorption of ions from urine. The mammalian kidney is a good developmental system to study as it can develop in organ culture; over a period of about 6 days explanted metanephric mesenchyme will form many glomeruli and tubules, even though a blood supply is absent.

The development of the ureteric bud and the mesenchyme depends on mutual inductive interactions, neither being able to develop in the absence of the other. The mesenchyme causes the ureteric bud to grow and bifurcate, and the bud induces the mesenchyme to differentiate into nephrons. The transcription factor WT1 is necessary for the mesenchyme to be competent to be induced. Mutations in the *WT1* gene are associated with a cancer of the kidney in children known as Wilm's tumor. Glial-derived neurotrophic factor (GDNF) is a likely inducer of ureteric bud growth. It is expressed in the mesenchyme and its receptor, Ret, is present on the bud. Knock-out of the gene for either GDNF or Ret results in the absence of ureteric bud outgrowth. Hepatocyte growth factor is also involved in kidney induction; antibodies to hepatocyte growth factor prevent bud development in culture, and the growth factor can stimulate a cultured kidney cell line to form tubules.

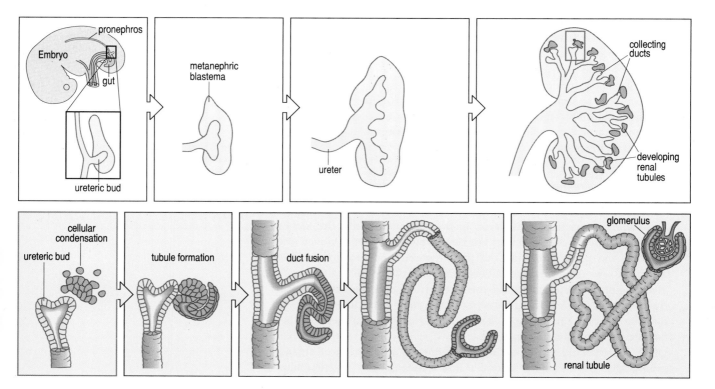

Fig. 10.45 Kidney development from mesenchyme. The kidney develops from a loose mass of mesenchyme, the metanephric blastema, which is induced to form tubules by the ureteric bud. The ureteric bud is itself induced by the mesenchyme to grow and branch to form the collecting ducts of the kidney that connect to the ureter. The mesenchyme cells form cellular condensations that become epithelial tubules, which open into the collecting system formed from the ureteric bud. Each tubule develops a glomerulus, through which waste products are filtered out of the blood.

The requirements for induction of the mesenchyme can be studied by placing mesenchyme on one side of a coarse filter and the ureteric bud on the other. Tubule formation in the mesenchyme is taken as a sign that induction has occurred. Several tissues other than the ureteric bud, such as the neural tube, can also induce tubule formation in the kidney mesenchyme. One of the factors secreted by the ureteric bud that can induce tubule formation is leukemia inhibitory factor (LIF). This cytokine is a good candidate for inducing the mesenchyme to epithelial transition. FGF-2, BMP-7, and Wnt-4 are also involved. A very early response to induction is the expression of *Pax2* in the mesenchyme, which is essential for subsequent tubule development. The expression of *WT1* then increases, the loose mass of mesenchyme cells condenses, and the cells start to express a matrix glycoprotein, syndecan, on their surface. This condensing stage is followed by the formation of distinct cellular aggregates, in which the mesenchymal cells become polarized and acquire an epithelial character. Each aggregate then forms an S-shaped tube, which elongates and differentiates to form the functional unit of a renal tubule and glomerulus. During this transition, the composition of the extracellular matrix secreted by the cells changes: mesenchymal collagen I is replaced by basal lamina proteins, such as collagen IV and laminin, which are typically secreted by epithelial cells. The adhesion molecules (see Box 8A, p. 265) expressed by the cells also change; for example, the N-CAM expressed by the mesenchymal cells is replaced by E-cadherin in the epithelial cells. Integrins are also involved in the epithelial–mesenchyme interactions.

Branching of the bud is again due to interaction with the mesenchyme; FGF-7 is involved in this process and in determining the number of nephrons.

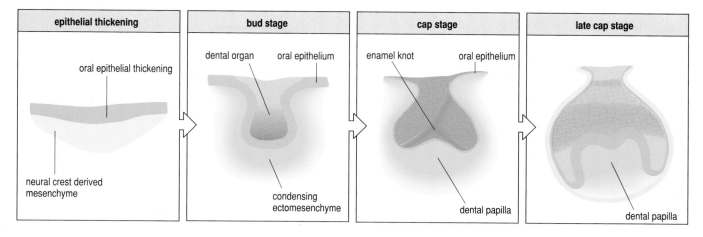

| epithelial thickening | bud stage | cap stage | late cap stage |

Fig. 10.46 Schematic diagram of the stages of tooth development.

10.23 A homeobox gene code specifies tooth identity

Teeth develop from a well-defined series of interactions between the oral epithelium and the mesenchyme of the jaws, which is of neural crest origin. Teeth form from areas of thickened oral epithelium. The epithelium invaginates and the surrounding mesenchyme condenses, forming a tooth primordium or tooth germ (Fig. 10.46). In each primordium, a specialized group of cells—the enamel knot—acts as a signaling center for later development, such as cusp formation. The mesenchyme forms the dental papilla, which gives rise to the pulp and dentine, while the enamel is secreted by the epithelial cells. Tooth primordia that give rise to different types of teeth, such as incisors and molars, are specified at precise positions, and it is this positioning that we will discuss here.

Early tooth development is controlled by reciprocal interactions, by way of secreted signals, between the oral epithelium and the underlying mesenchyme. The sites of tooth formation are determined initially by signals produced by the epithelium. BMP-4 is expressed in the epithelium at all sites of tooth formation, and induces, or maintains, expression of the homeodomain transcription factor Msx-1 in the underlying mesenchyme. BMP-4 is then expressed in the mesenchyme together with Msx-1 at these sites. Sonic hedgehog is also expressed in the epithelium at sites of tooth formation throughout tooth development, whereas Wnt-7b is expressed elsewhere in the oral epithelium except at sites of tooth formation. The reciprocal expression of these latter two signaling molecules establishes the boundaries between the oral epithelium and the dental epithelium that invaginates to form the tooth. As tooth bud development proceeds, the direction of signaling changes so that signals secreted by the mesenchyme direct the development of the epithelium, for example the development of the enamel knot.

In rodents, it has been established that the type of tooth that develops—incisor or molar—is determined by the expression of homeobox genes in the mesenchyme. These homeobox genes, which include *Barx-1, Dlx-1, Dlx-2, Msx-1* and *Msx-2* among others, provide a spatial code in the facial mesoderm (Fig. 10.47), analogous to the Hox gene code along the antero-posterior body axis Thus, *Dlx-2* and *Barx-1* are expressed in mesenchymal cells that will form molars, while *Msx-1* and *Msx-2* are expressed in cells that will form incisors. In mouse mutants lacking both *Dlx-2* and *Dlx-1* (a similar

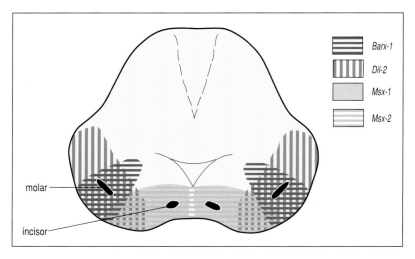

Fig. 10.47 The expression domains of four homeobox genes in the mesoderm of the mandible before initiation of tooth primordia. The positions at which the incisors and molars will develop are indicated in black. After Ferguson, C.A., *et al.*: 2000.

gene that can substitute for *Dlx-2* in development), molars do not develop, but ectopic cartilage structures form at the molar sites.

Summary

Development of internal organs such as the heart, vascular system, lungs, kidneys, and teeth are intimately involved with epithelial patterning and morphogenesis, and a mesenchyme to epithelial transition is common. The heart initially forms as a linear tube made up of an inner endocardium and an outer myocardium. Its later development may be modular and is controlled by genes similar to those that regulate heart formation in *Drosophila*. The vascular system develops from endothelial cells which initially form the large vessels of the vascular system, which then grow and branch in the process of angiogenesis. Branching of the tracheae of *Drosophila* and the vertebrate lung involve similar principles and are controlled by the surrounding mesenchyme. The development of the mammalian kidney involves both mesenchyme to epithelium transition and reciprocal induction. The ureteric bud is induced by the prospective kidney mesenchyme to grow and branch, while the ureteric bud induces the cells in the mesenchyme to condense and form tubules. Teeth form by an invagination of the oral epithelium and the underlying mesenchyme and their positioning and development is controlled by reciprocal signaling between epithelium and mesenchyme. A homeobox gene code in the mesenchyme specifies tooth identity.

SUMMARY TO CHAPTER 10

The development of organs involves similar processes, and in some cases the same genes, as those used earlier in development. The pattern of the vertebrate limb is specified along three axes, at right angles to each other. Signaling molecules provide positional information, and interpretation of this information involves Hox genes. The development of *Drosophila* wings and legs from imaginal discs depends on positional signals generated at compartment boundaries, and some of these signal proteins are similar to those used in vertebrate limb patterning. Lateral inhibition and cell–cell signaling underlies vulva development in the nematode. The development of internal organs such as the vascular system, lungs, and teeth involves epithelial morphogenesis and, in some cases, branching of tubular structures.

GENERAL REFERENCES

Riddle, R.D., Tabin, C.J.: **How limbs develop.** *Sci. Am.* 1999, **280**: 54–59.

Morata, G.: **How** *Drosophila* **appendages develop.** *Nature Rev. Mol. Cell Biol.* 2001, **2**: 89–97.

SECTION REFERENCES

10.1 The vertebrate limb develops from a limb bud

Francis-West, P.H., Parish, J., Lee, K., Archer, C.W.: **BMP/GDF-signalling interactions during synovial joint development.** *Cell Tissue Res.* 1999, **296**: 111–119.

Hartmann, C., Tabin, C.J.: **Wnt-14 plays a pivotal role in inducing synovial joint formation in the developing appendicular skeleton.** *Cell* 2001, **104**: 341–351.

Spitz, F., Duboule, D.: **The art of making a joint.** *Science* 2001, **291**: 1713–1714.

Strom, E.E., Kingsley, D.M.: **CDF5 coordinates bone and joint formation during digit formation.** *Dev. Biol.* 1999, **209**: 11–27.

Wolpert, L.: **Vertebrate limb development and malformations.** *Pediatr. Res.* 1999, **46**: 247–254.

10.2 Homeobox genes expressed in the lateral plate mesoderm are involved in specifying the position and type of limb

Charité, J., De Graaff, W., Shen, S., Deschamps, J.: **Ectopic expression of** *Hoxb-8* **causes duplication of the ZPA in the forelimb and homeotic transformation of axial structures.** *Cell* 1994, **78**: 589–601.

Isaac A., Cohn, M.J., Ashby, P., Ataliotis, P., Spicer, D.B., Cooke, J., Tickle, C.: **FGF and genes encoding transcription factors in early limb specification.** *Mech. Dev.* 2000, **93**: 41–48.

Kawakami, Y., Capdevila, J., Buscher, D., Itoh, T., Rodriguez Esteban, C., Izpisua Belmonte, J.C.: **WNT signals control FGF-dependent limb initiation and AER induction in the chick embryo.** *Cell* 2001, **104**: 891–900.

Tickle, C., Münsterberg, A.: **Vertebrate limb development – the early stages in chick and mouse.** *Curr. Opin. Genet. Dev.* 2001, **11**: 476–481

10.3 The apical ectodermal ridge induces the progress zone

Dudley, A.T., Tabin, C.J.: **Constructive antagonism in limb development.** *Curr. Opin. Genet. Dev.* 2000, **10**: 387–392.

Niswander, L., Tickle, C., Vogel, A., Booth, I., Martin, G.R.: **FGF-4 replaces the apical ectodermal ridge and directs outgrowth and patterning of the limb.** *Cell* 1993, **75**: 579–587.

Pizette, S., Niswander, L. **BMPs negatively regulate structure and function of the limb apical ectodermal ridge.** *Development* 1999, **126**: 883–894.

10.4 Patterning of the limb involves positional information

Cohn, M.J., Tickle, C.: **Limbs: a model for pattern formation within the vertebrate body plan.** *Trends Genet.* 1996, **12**: 253–257.

10.5 The polarizing region specifies position along the limb's antero-posterior axis

Dahn, R.D., Fallon, J.F.: **Interdigital regulation of digit identity and homeotic transformation by modulated BMP signaling.** *Science* 2000, **289**: 438–441.

Drossopoulou, G., Lewis, K.E., Sanz-Ezquerro, J.J., Nikbakht, N., McMahon, A.P., Hofmann, C., Tickle, C.: **A model for anteroposterior patterning of the vertebrate limb based on sequential long- and short-range Shh signalling and Bmp signalling.** *Development* 2000, **127**: 1337–1348.

Lewis, P.M., Dunn, M.P., McMahon, J.A., Logan, M., Martin, J.F., St-Jacques, B., McMahon, A.P.: **Cholestoral modification of sonic hedgehog is required for long-range signaling activity and effective modulation of signaling by Ptc1.** *Cell* 2001, **105**: 599–612.

Riddle, R.D., Johnson, R.L., Laufer, E., Tabin, C.: *Sonic hedgehog* **mediates polarizing activity of the ZPA.** *Cell* 1993, **75**: 1401–1416.

Tickle, C.: **Retinoic acid and chick limb development.** *Development (Suppl.)* 1991, **1**: 113–121.

Yang, Y., Frossopoulou, G., Chuang, P-T., Duprez, D., Marti, E., Bumerot, D., Vargesson, N., Clarke, J., Niswander, L., McMahon, A., Tickle, C.: **Relationship between dose, distance and time in Sonic hedgehog-mediated regulation of anteroposterior polarity in the chick limb.** *Development* 1997, **124**: 4393–4404.

10.6 Position along the proximo-distal axis may be specified by a timing mechanism

Wolpert, L., Tickle, C., Sampford, M.: **The effect of cell killing by X-irradiation on pattern formation in the chick limb.** *J. Embryol. Exp. Morph.* 1979, **50**: 175–193.

10.7 The dorso-ventral axis is controlled by the ectoderm

Altabef, M., Clarke, J.D.W., Tickle, C.: **Dorso-ventral ectodermal compartments and origin of apical ectodermal ridge in developing chick limb.** *Development* 1997, **124**: 4547–4556.

Geduspan, J.S., MacCabe, J.A.: **The ectodermal control of mesodermal patterns of differentiation in the developing chick wing.** *Dev. Biol.* 1987, **124**: 398–408.

Michaud, J.W., Lapointe, F., Le Douarin, N.M.: **The dorsoventral polarity of the presumptive limb is determined by signals produced by the somites and by the lateral somatopleure.** *Development* 1997, **124**: 1453–1463.

Parr, B.A., McMahon, A.P.: **Dorsalizing signal Wnt-7a is required for normal polarity of D-V and A-P axes of mouse limb.** *Nature* 1995, **374**: 350–353.

Riddle, R.D., Ensini, M., Nelson, C., Tsuchida, T., Jessell, T.M., Tabin, C.: **Induction of the LIM homeobox gene** *Lmx1* **by** *Wnt-7a* **establishes dorsoventral pattern in the vertebrate limb.** *Cell* 1995, **83**: 631–640.

Yang, Y., Niswander, L.: **Interaction between the signaling molecule Wnt-7a and Shh during vertebrate limb development: dorsal signals regulate antero-posterior patterning.** *Cell* 1995, **80**: 939–947.

10.8 Different interpretations of the same positional signals give different limbs

Krabbenhoft, K. M., Fallon, J. F.: **The formation of leg or wing specific structures by leg bud cells grafted to the wing bud is influenced by proximity to the apical ridge.** *Dev. Biol.* 1989, **131**: 373–382.

Logan, M., Tabin, C.J.: **Role of Pitx1 upstream of Tbx4 in specification of hindlimb identity.** *Science* 1999, **283**: 1736–1739.

10.9 Homeobox genes may also provide positional values for limb patterning

Davis, A.P., Capecchi, M.R.: **A mutational analysis of the 5′** HoxD genes: dissection of genetic interactions during limb development in the mouse. *Development* 1996, **122**: 1175–1185.

Goff, D.J., Tabin, C.J.: **Analysis of *Hoxd-13* and *Hoxd-11* misexpression in chick limb buds reveals that Hox genes affect both bone condensation and growth.** *Development* 1997, **124**: 627–636.

Nelson, C.E., Morgan, B.A., Burke, A.C., Laufer, E., DiMambro, E., Muytaugh, L.C., Gonzales, E., Tessarollo, L., Parada, L.F., Tabin, C.: **Analysis of Hox gene expression in the chick limb bud.** *Development* 1996, **122**: 1449–1466.

Rijli, F.M., Chambon, P.: **Genetic interactions of Hox genes in limb development: learning from compound mutants.** *Curr. Opin. Genet. Dev.* 1997, **7**: 481–487.

Scott, M.P.: **Hox genes, arms, and the man.** *Nature Genet.* 1997, **15**: 117–118.

Vogt, T.F., Duboule, D.: **Antagonists go out on a limb.** *Cell* 1999, **99**: 563–566.

Yokouchi, Y., Nakazato, S., Yamamoto, M., Goto, Y., Kameda, T., Iba, H., Kuroiwa, A.: **Misexpression of *Hoxa-13* induces cartilage homeotic transformation and changes cell adhesiveness in chick limb buds.** *Genes Dev.* 1995, **9**: 2509–2522.

10.10 Self-organization may be involved in pattern formation in the limb bud

Hardy, A., Richardson, M.K., Francis-West, P.N., Rodriguez, C., Izpisúa-Belmonte, J.C., Duprez, D., Wolpert, L.: **Gene expression, polarising activity and skeletal patterning in reaggregated hind limb mesenchyme.** *Development* 1995, **121**: 4329–4337.

Box 10A Reaction–diffusion mechanisms

Kondo, S., Asai, R.: **A reaction–diffusion wave on the skin of the marine angelfish *Pomocanthus*.** *Nature* 1995, **376**: 765–768.

Meinhardt, H., Gierer, A.: **Pattern formation by local self-activation and lateral inhibition.** *Bioassays.* 2000, **22**: 753–760.

Murray, J.D.: **How the leopard gets its spots.** *Sci. Amer.* 1988, **258**: 80–87.

10.11 Limb muscle is patterned by the connective tissue

Amthor, H., Christ, B., Patel, K.: **A molecular mechanism enabling continuous embryonic muscle growth—a balance between proliferation and differentiation.** *Development* 1999, **126**: 1041–1053.

Hashimoto, K., Yokouchi, Y., Yamamoto, M., Kuroiwa, A.: **Distinct signaling molecules control *Hoxa-11* and *Hoxa-13* expression in the muscle precursor and mesenchyme of the chick limb bud.** *Development* 1999, **126**: 2771–2783.

Knight, B., Laukaitis, C., Akhtar, N., Hotchin, N.A., Edlung, M., Horwitz, A.R.: **Visualizing muscle cell migration *in situ*.** *Curr. Biol.* 2000, **10**: 576–595.

Robson, L.G., Kara, T., Crawley, A., Tickle, C.: **Tissue and cellular patterning of the musculature in chick wings.** *Development* 1994, **120**: 1265–1276.

10.12 The initial development of cartilage, muscles, and tendons is autonomous

Kardon, G.: **Muscle and tendon morphogenesis in the avian hind limb.** *Development* 1998, **125**: 4019–4032.

Ros, M.A., Rivero, F.B., Hinchliffe, J.R., Hurle, J.M.: **Immunohistological and ultrastructural study of the developing tendons of the avian foot.** *Anat. Embryol.* 1995, **192**: 483–496.

10.13 Separation of the digits is the result of programmed cell death

Garcia-Martinez, V., Macias, D., Gañan, Y., Garcia-Lobo, J.M., Francia, M.V., Fernandez-Teran, M.A., Hurle, J.M.: **Internucleosomal DNA fragmentation and programmed cell death (apoptosis) in the interdigital tissue of the embryonic chick leg bud.** *J. Cell. Sci.* 1993, **106**: 201–208.

Zou, H., Niswander, L.: **Requirement for BMP signaling in interdigital apoptosis and scale formation.** *Science* 1996, **272**: 738–741.

10.14 Positional signals from the antero-posterior and dorso-ventral compartment boundaries pattern the wing imaginal disc

Blair, S.S.: **Notch signaling: Fringe really is a glycosyltransferase.** *Curr. Biol.* 2000, **10**: R608–R612.

Diaz-Benjumea, F.J., Cohen, S.M.: **Interaction between dorsal and ventral cells in the imaginal disc directs wing development in *Drosophila*.** *Cell* 1993, **75**: 741–752.

Entchev, E.V., Schwabedissen, A., González-Gaitán, M.: **Gradient formation of the TGF-β homolog Dpp.** *Cell* 2000, **103**: 981–991.

Kim, J., Sebring, A., Esch, J.J., Kraus, M.E., Vorwerk, K., Magee, J., Casroll, B.: **Integration of positional signals and regulation of wing formation and identity by *Drosophila vestigial* gene.** *Nature* 1996, **382**: 133–138.

Lawrence, P.: **Morphogens: how big is the picture?** *Nat. Cell Biol* 2001, **3**: E151–E154.

Lawrence, P.A., Struhl G.: **Morphogens, compartments, and pattern: lessons from *Drosophila*?** *Cell* 1996, **85**: 951–961.

Lecuit, T., Brook, J.W., Ng, M., Calleja, M., Sun, H., Cohen, S.: **Two distinct mechanisms for long-range patterning by Decapentaplegic in the *Drosophila* wing.** *Nature* 1996, **381**: 387–393.

Rauskolb, C., Correia, T., Irvine, K.D.: **Fringe-dependent separation of dorsal and ventral cells in the *Drosophila* wing.** *Nature* 1999, **401**: 476–480.

Smith, J.: **How to tell a cell where it is.** *Nature* 1996, **381**: 367–368.

Strigini, M., Cohen, S.M.: **Formation of morphogen gradients in the *Drosophila* wing.** *Semin. Dev. Biol.* 1999, **10**: 335–344.

Strutt, D.I.: **Asymmetric localization of Frizzled and the establishment of cell polarity in the *Drosophila* wing.** *Mol. Cell* 2001, **7**: 367–375.

Tabata, T.: **Genetics of morphogen gradients.** *Nature Rev. Genet.* 2001, **2**: 620–630.

Teleman, A.A., Cohen, S.M.: **Dpp gradient formation in the *Drosophila* wing imaginal disc.** *Cell* 2000, **103**: 971–980.

Teleman, A.A., Strigini, M., Cohen, S.M.: **Shaping morphogen gradients.** *Cell* 2001, **105**: 559–562.

Vincent, J.P., Lawrence, P.A.: **Developmental genetics. It takes three to distalize.** *Nature* 1994, **372**: 132–133.

10.15 The leg disc is patterned in a similar manner to the wing disc, except for the proximo-distal axis

Brook, W.J., Cohen, S.M.: Antagonistic interactions between wingless and decapentaplegic responsible for dorsal-ventral pattern in the *Drosophila* leg. *Science* 1996, **273**: 1373-1377.

Campbell, G., Tomlinson, A.: Initiation of the proximo-distal axis in insect legs. *Development* 1995, **121**: 619-625.

Milan, M., Cohen, S.M.: Subdividing cell populations in the developing limbs of *Drosophila*: do wing veins and leg segments define units of growth control? *Dev. Biol.* 2000, **217**: 1-9.

Morata, G., Sanchez-Herrero, E.: Pulling the fly's leg. *Nature* 1998, **392**: 657-658.

Wu, J., Cohen, S.M.: Proximal distal axis formation in the *Drosophila* leg: distinct functions of Teashirt and Homothorax in the proximal leg. *Mech. Dev.* 2000, **94**: 47-56.

10.16 Butterfly wing markings are organized by additional positional fields

Brakefield, P.M., French, V.: Butterfly wings: the evolution of development of colour patterns. *BioEssays* 1999, **21**: 391-401.

Brakefield, P.M., Gates, J., Keys, D., Kesbeke, F., Wijngaarden, P.J., Monteiro, A., French, V., Carroll, S.B.: Development, plasticity, and evolution of butterfly eyespot patterns. *Nature* 1996, **384**: 236-242.

Carroll, S.B., Gates, J., Keys, D.N., Paddock, S.W., Panganiban, G.E., Silegue, J.E., Willliams, J.A.: Pattern formation and eyespot determination in butterfly wings. *Science* 1994, **265**: 109-114.

Nijhout, H.F.: *The Development and Evolution of Butterfly Wing Patterns.* Washington: Smithsonian Institution Press, 1991.

North, G., French, V.: Insect wings. Patterns upon patterns. *Curr. Biol.* 1994, **7**: 611-614.

10.17 The segmental identity of imaginal discs is determined by the homeotic selector genes

Carroll, S.B.: Homeotic genes and the evolution of arthropods and chordates. *Nature* 1995, **376**: 479-485.

Morata, G.: How *Drosophila* appendages develop. *Nature Rev. Mol. Cell Biol.* 2001, **2**: 89-97.

Si Dong, P.D., Chu, J., Panganiban, G.: Cooexpression of the homeobox genes *Distal-less* and *homothorax* determines *Drosophila* antennal identity. *Development* 2000, **127**: 209-216.

Vachon, G., Cohen, B., Pfeifle, C., McGuffin, M.E., Botas, J., Cohen, S.M.: Homeotic genes in the Bithorax complex repress limb development in the abdomen of the *Drosophila* embryo through the target gene *Distal-less*. *Cell* 1992, **71**: 437-450.

10.18 The anchor cell induces primary and secondary fates in prospective nematode vulva precursor cells

Grunwald, I., Rubin, G.M.: Making a difference: the role of cell–cell interactions in establishing separate identities for equivalent cells. *Cell* 1992, **68**: 271-281.

Kenyon, C.: A perfect vulva every time: gradients and signaling cascades in *C. elegans*. *Cell* 1995, **82**: 171-174.

Newman, A.P., Sternberg, P.W.: Coordinated morphogenesis of epithelia during development of the *Caenorhabditis*

elegans uterine-vulval connection. *Proc. Natl Acad. Sci. USA* 1996, **93**: 9329-9333.

Sharma-Kishore, R., White, J.G., Southgate, E., Podbilewicz, B.: Formation of the vulva in *Caenorhabditis elegans*: a paradigm for organogenesis. *Development* 1999, **126**: 691-699.

Sundaram, M., Han, M.: Control and integration of cell signaling pathways during *C. elegans* vulval development. *BioEssays* 1996, **18**: 473-480.

10.19 The development of the vertebrate heart involves specification of a mesodermal tube which is patterned along its long axis

Driever, W.: Bringing two hearts together. *Nature* 2000, **406**: 141-142.

Fishman, M.C., Chien, K.R.: Fashioning the vertebrate heart: earliest embryonic decisions. *Development* 1997, **124**: 2099-2117.

Fishman, M.C., Olson, E.N.: Parsing the heart: genetic modules for organ assembly. *Cell* 1997, **91**: 153-156.

Rosenthal, N.: In *Heart Development* (Harvey, R.P., ed.). New York: Academic Press, 1999.

Rosenthal, N., Xavier-Neto, J.: From the bottom of the heart: anteroposterior decisions in cardiac muscle differentiation. *Curr. Opin. Cell Biol.* 2000, **12**: 742-746.

Schwartz, R.J., Olson, E.N.: Building the heart piece by piece: modularity of *cis*-elements regulating Nkx2-5 transcription. *Development* 1999, **126**: 4187-4192.

Srivastava, D., Olson, E.N.: A genetic blueprint for cardiac development. *Nature* 2000, **407**: 221-226.

Stainier, D.Y.R.: Zebrafish genetics and vertebrate heart formation. *Nature Rev. Genet.* 2001, **2**: 39-48.

10.20 The vascular system develops from angioblasts and the vessels then undergo angiogenesis

Risau, W.: Mechanisms of angiogenesis. *Nature* 1997, **386**: 671-674.

Roman, B.L., Weinstein, B.M.: Building the vertebrate vasculature: research is going swimmingly. *BioEssays* 2000, **22**: 882-893.

Wang, H.U., Chen, Z-F., Anderson, D.J.: Molecular distinction and angiogenic interaction between embryonic arteries and veins revealed by ephrin-B2 and its receptor Eph-B4. *Cell* 1998, **93**: 741-753.

10.21 The trachea of *Drosophila* and the lungs of vertebrates branch using similar mechanisms

Hogan, B.L.: Morphogenesis. *Cell* 1999, **96**: 225-233.

Metzger, R.J., Krasnow, M.A.: Genetic control of branching morphogenesis. *Science* 1999, **284**: 1635-1639.

10.22 The development of the ureteric bud and mesenchymal tubules of the kidney involves induction

Barasch, J., Yang, J., Ware, C.B., Taga, T., Yoshida, K., Erdjument-Bromage, H., Tempst, P., Parravicini, E., Malach, S., Aranoff, T., Oliver, J.A.: Mesenchymal to epithelial conversion in rat metanephros is induced by LIF. *Cell* 1999, **99**: 377-386.

Bard, J.B., Davies, J.A., Karavanova, I., Lehtonen, E., Sariola, H., Vainio, S.: Kidney development: the inductive interactions. *Seminars in Cell Dev.* 1996, **7**: 195-202.

Kuure, S., Vuolteenaho, R., Vainio, S.: Kidney morphogenesis: cellular and molecular regulation. *Mech. Dev.* 2000, **94**: 47-56.

Sainio, K., Suvanto, P., Davies, J., Wartiovaara, J., Wartiovaara, K., Saarma, M., Arumae, U., Meng, X., Lindahl, M., Pachnis, V., Sarioia, H.: **Glial-cell-line derived neurotrophic factor is required for bud initiation from uteric epithelium.** *Development* 1997, **124**: 4077–4087.

Schedl, A., Hastie, N.D.: **Cross-talk in kidney development.** *Curr. Opin. Genet. Dev.* 2000, **10**: 543–549.

Stark, K., Vainio, S., Vassileva, G., McMahon, A.P.: **Epithelial transformation of metanephric mesenchyme in the developing kidney regulated by Wnt-4.** *Nature* 1994, **372**: 679–683.

Vainio, S., Muller, U.: **Inductive tissue interactions, cell** signaling, and the control of kidney organogenesis. *Cell* 1997, **90**: 975–978.

10.23 A homeobox gene code specifies tooth identity

Ferguson, C.A., Hardcastle, Z., Sharpe, P.T.: **Development and patterning of the dentition.** *Linn. Soc. Symp.* 2000, **20**: 188–201.

Sarkar, L., Cobourne, M., Naylor, S., Smalley, M., Dale, T., Sharpe, P.T.: **Wnt/Shh interactions regulate ectodermal boundary formation during mammalian tooth development.** *Proc. Natl Acad. Sci. USA* 2000, **97**: 4520–4524.

Development of the nervous system

11

- Specification of cell identity in the nervous system
- The insect compound eye
- Axonal guidance
- Neuronal survival, synapse formation, and refinement

" Special individuals were selected as communicators and they made many connections, which often required them to go on long journeys. "

The nervous system is the most complex of all the organ systems in the animal embryo. In mammals, for example, billions of nerve cells or **neurons** develop a highly organized pattern of connections, creating the neuronal network that makes up the functioning brain and the rest of the nervous system. There are many hundreds of different types of neurons, differing in identity and connectivity, even though they may look quite similar. There are also supporting tissues in the nervous system, known collectively as **glia**; for example, Schwann cells surround the axons of peripheral neurons, and oligodendrocytes and astrocytes ensheath the axons of central neurons. Yet for all its complexity, the nervous system is the product of the same kind of cellular and developmental processes that are involved in the development of other organ systems.

All nervous systems, vertebrate and invertebrate, provide a system of communication through a network of neurons of varying shapes and functions (Fig. 11.1). Each neuron connects with other neurons or target cells at synapses, sites at which the propagation of an action potential induces neurotransmitter release. The **dendrites** and **axon** terminals of individual neurons can be extensively branched; a single neuron can receive as many as 100,000 different inputs. The nervous system can only function properly if the neurons are correctly connected to one another, and thus a central question surrounding neural system development is how the connections between neurons develop with the appropriate specificity. The overall process of nervous system development can be divided up into four major stages: the specification of neural (neuron or glial) cell identity; the migration of neurons and the outgrowth of axons to their targets; the formation of **synapses** with the target cells, which can be other neurons, muscle or gland cells; and the refinement of synaptic connections, through the elimination of axon branches and cell death.

We have already considered some aspects of the development of the vertebrate nervous system, including neural induction (Chapter 4), the

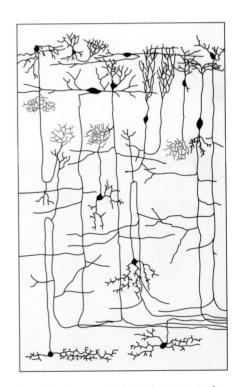

Fig. 11.1 A network of varied neurons in the optic lobe of a bird. Axons and dendrites extend from the cell bodies, which are usually small and rounded, and contain the nucleus.

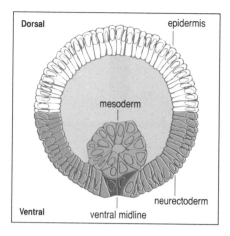

Fig. 11.2 Cross-section of a *Drosophila* early gastrula showing the location of the neurectoderm (pale blue). The mesoderm (red), which originally lies along the ventral midline, has been internalized.

formation of the neural tube (Chapter 8), the segmental arrangement of sensory dorsal root ganglia (Chapter 8), and the differentiation of neurons from neural crest cells (Chapter 9). We will re-examine some of these key steps when we consider the specification of neural cells. The general focus of this chapter is on the mechanisms that control the identity of cells in the nervous system and that pattern their synaptic connections.

Specification of cell identity in the nervous system

We start by considering how neural cells acquire their individual identities. As in other developing systems, nerve cell specification is governed both by external signals and by intrinsic differences generated through asymmetric cell divisions. We first consider neural specification in *Drosophila*, which has revealed many key developmental processes in neurogenesis, and then turn to vertebrates, where similar mechanisms are involved.

11.1 Neurons in *Drosophila* arise from proneural clusters

In *Drosophila*, the presumptive central nervous system is specified at an early stage of embryonic development as two longitudinal stripes of cells, just dorsal to the ventral mesoderm. This region is called the neurogenic zone, or **neurectoderm** (Fig. 11.2), and comprises ectodermal cells with the potential to form neural cells and epidermis. The neurectoderm is sub-divided along the antero-posterior and dorso-ventral axes into a precise orthogonal pattern of **proneural clusters**. Within each cluster, cell–cell interactions direct one cell into a neural precursor or **neuroblast** fate. The rest become epidermal cells. The neuroblasts enlarge, and delaminate into the interior of the embryo to form an invariant pattern of about 30 neurob-lasts per hemisegment. A **hemisegment** is the lateral half of a segment, one on each side of the midline, and is the developmental unit for the nervous system. Each neuroblast that forms within a hemisegment

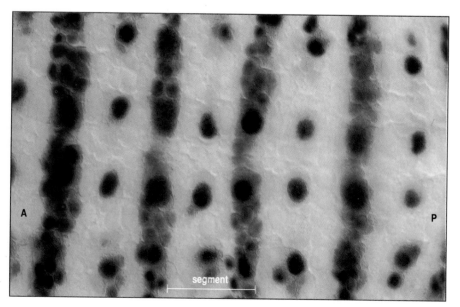

Fig. 11.3 Pattern of expression of the gene *achaete* in the neurectoderm of the *Drosophila* embryo. Within each future segment, there are proneural clusters expressing *achaete* (black). Some of these are formed in the region of *engrailed* expression (purple) at the posterior end of the segment, and others in a more anterior region.

Photograph from Skeath, J.B., Doe, C.Q.: 1996.

expresses a unique set of genes, and divides in a specific lineage to produce neurons and glia.

Neurectodermal cells are given the potential to become neural precursors by the expression of genes known as **proneural genes**. Of particular importance are the genes of the achaete–scute complex. These encode transcription factors, notably the proteins achaete and scute, which have a basic helix-loop-helix DNA-binding motif (the basic region binds DNA and the helix binds to other transcriptional regulatory proteins). These transcription factors form homodimers and heterodimers with each other, and initiate neural specification by turning on transcription of their target genes. Proneural clusters are evident within the neurectoderm when one looks at the pattern of *achaete* gene expression in the early *Drosophila* embryo. The achaete protein first appears after gastrulation in a segmentally repeated pattern. In each future segment, some clusters are formed in the posterior region, where *engrailed* is expressed, whereas others form more anteriorly (Fig. 11.3, opposite). Further proneural clusters can also form elsewhere in the segment under the control of other genes of the achaete–scute complex.

Neuroblast formation occurs in five temporally and spatially distinct waves. Neuroblasts that form in identical positions in different hemisegments at the same time express the same set of genes and follow the same fate; thus each neuroblast acquires a unique fate based on its position within the hemisegment and its time of formation (Fig. 11.4). In *Drosophila*, two groups of genes act along the antero-posterior and dorsoventral axes, respectively, to control both neuroblast formation and fate. The segment polarity genes (see Section 5.16), which include *wingless* and *gooseberry*, divide the neurectoderm transversely into four rows, while

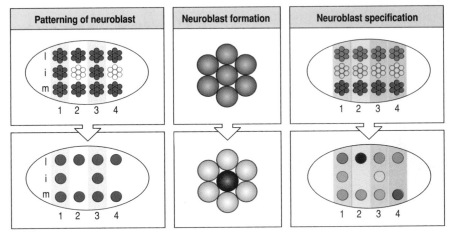

Fig. 11.4 The patterning, formation, and individual specification of neuroblasts in embryonic *Drosophila* neurectoderm. In each hemisegment of the embryo, segment polarity gene expression subdivides the neurectoderm into four transverse stripes (1, 2, 3, 4), while the columnar genes divide it into three longitudinal columns (l, m, i). The patterning of the first 10 proneural clusters formed during the first wave of patterning is shown here. As shown in the left-hand panels, each proneural cluster (top), and thus each neuroblast (bottom) forms within a unique region delineated by the overlapping regions of segment polarity gene and columnar gene expression. The center panels show the formation of a neuroblast (dark red) from a group of initially equipotential cells (light red)—the proneural cluster. The right-hand panels show how the fate of each neuroblast (colored circles in bottom panel) is individually specified while still in the neurectoderm by the different combinations of segment polarity gene (light blue, orange, purple, light green) and columnar gene (green, yellow, red) expression (top panel). These combinations give each neuroblast in each hemisegment a unique identity. Anterior is to the left, and ventral down. Modified from Sneath, J.B., 1999.

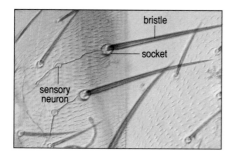

Fig. 11.5 Sensory bristles in *Drosophila*. The bristles respond to external stimuli and stimulate the sensory neurons.

Photograph courtesy of Y. Jan.

genes acting along the dorso-ventral axis divide the hemisegment into three longitudinal columns, thus partitioning the hemisegment into a number of unique regions. Individual proneural clusters are positioned within these unique regions (see Fig. 11.4). The columnar genes include one for an epidermal growth factor (EGF) receptor (DER, the *Drosophila* EGF receptor) and three that code for homeodomain transcription factors—Msh, Ind, and Vnd. Their dorso-ventral sequence of expression is controlled by the dorsal, snail, and decapentaplegic proteins (see Section 5.8).

Shortly after the formation of a proneural cluster in the embryonic neurectoderm, one cell in the cluster starts to express achaete protein at a higher level than the other cells. This will become the neuroblast. Neuroblasts leave the surface layer during germ band extension and eventually give rise to neurons, which will form the ventral nerve cord. The mechanism whereby just one cell is singled out to become a neuroblast is considered in the next section.

The achaete–scute complex also has an important role in the specification of the sensory nervous system of the adult fly. The pattern of sensory bristles (Fig. 11.5) and other sensory organs on the adult cuticle is already present in the imaginal discs in the form of a pattern of **sensory organ precursors** that later give rise to these structures. As in the neurectoderm, proneural clusters are first selected by expression of the achaete–scute complex genes in a precise spatial pattern (Fig. 11.6). This pattern is constructed by the independent action of site-specific regulatory modules distributed along the achaete–scute complex. As with the specification of neuroblasts in the proneural clusters of the neurectoderm, only one of the cells in the cluster expressing achaete–scute adopts a neural fate and becomes a sensory organ precursor.

11.2 Lateral inhibition allocates neural precursors

Not all of the cells in the neurectoderm become neural cells. Initially, all the cells within a proneural cluster are capable of becoming neuroblasts, but one cell, through an apparently random event, then produces a signal that prevents its immediate neighbors from developing further along the neural pathway (Fig. 11.7, top row). This **lateral inhibition** eventually leads to just one cell developing as a neuroblast, with the rest of the cluster becoming epidermal cells. In the grasshopper embryo, if a neuroblast is killed just as it begins to develop, a neighboring cell develops into a neuroblast instead. Once specified, the neuroblasts migrate individually from the superficial cell layer inward and go through a stereotyped series of cell divisions to produce neurons and glia. The sensory organ precursors that give rise to the sensory bristles (sensilla) of the *Drosophila* adult epidermis are also singled out from the proneural clusters in the imaginal discs through lateral inhibition.

The *Notch* and *Delta* genes play a key role in lateral inhibition within the proneural clusters. They both encode transmembrane proteins; the Notch protein is a receptor and Delta protein is its ligand. Activation of Notch by Delta leads to inhibition of the proneural genes in the Notch-bearing cell, which thus loses the ability to give rise to a neuroblast. Small differences in levels of Notch–Delta signaling between the cells in the proneural cluster allow one cell to advance along the pathway to neural specification sooner

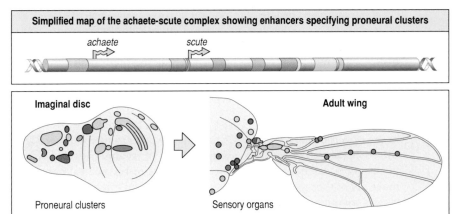

Fig. 11.6 Specification of sensory bristles in the *Drosophila* wing. The location of proneural clusters in *Drosophila* wing imaginal discs is largely under the control of the achaete–scute gene complex, the expression of which is activated at specific sites through different enhancers. Top panel: the position of the various enhancers that determine the site-specific expression of the *achaete* and *scute* genes. These enhancers are located both upstream and downstream of the *achaete* and *scute* genes. Bottom panel: location of the pro-neural clusters in the imaginal disc (left) and of the corresponding sensory bristles in the adult wing (right), with colors corresponding to the enhancer responsible for the site-specific expression of the genes. After Campuzano, S., et al.: 1992.

than the others; it then sends a signal that prevents the less advanced cells from adopting a neural fate.

Initially, all the cells in a proneural cluster can express both Notch and Delta. One cell gains an advantage, for example by starting to express Delta sooner or more strongly than the others. Through interacting with the Notch receptors on its neighbors, it inhibits their further neural development and this also suppresses their production of the Delta protein (Fig. 11.7, bottom row). In this way, the initially equivalent cells of the proneural

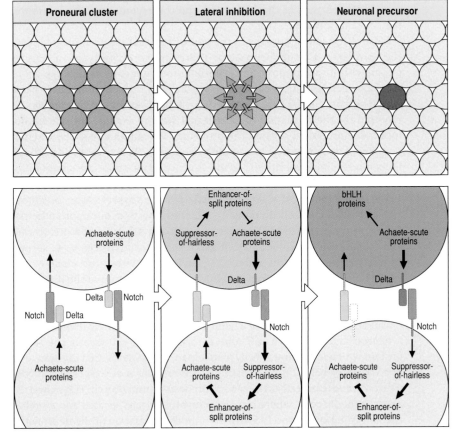

Fig. 11.7 The role of Notch signaling in lateral inhibition in neural cell development. As shown in the top row, the proneural cluster gives rise to a single neuronal precursor cell, the neuroblast or sensory organ precursor, by means of lateral inhibition. The rest of the cells in the cluster become epidermal or support cells. As shown in the bottom panels, Notch signaling between cells of the proneural cluster is initially similar and keeps the cells in the proneural state. An imbalance develops when one cell begins to express higher levels of Delta, the ligand for Notch, and thus activates Notch to a higher level in neighboring cells. This initial imbalance is rapidly amplified by a feedback pathway involving two transcription factors, Suppressor-of-hairless and Enhancer-of-split. Their activities repress the production of achaete–scute proteins and Delta protein in the affected cells. This prevents them from proceeding along the pathway of neuronal development and from delivering inhibitory Delta signals to the prospective neuroblast. After Kandel, E.R. et al.: 2000.

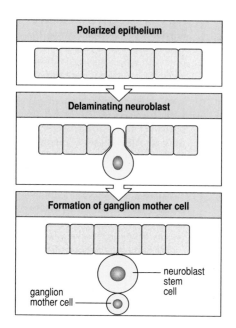

Fig. 11.8 Generation of ganglion mother cell by the asymmetric division of a neuroblast. Once specified, neuroblasts move out of the neurectodermal epithelium and then behave as neural stem cells. Each neuroblast divides asymmetrically to give an apical cell which remains a neuroblast stem cell, and a smaller basal cell, the ganglion mother cell. Adapted from Jan, Y-N., Yan, L.Y.: 2000.

cluster are resolved by lateral inhibition into one neural precursor cell. The other cells go on to develop into epidermal cells and cease expression of *Delta*. If either *Notch* or *Delta* function is inactivated, the neurectoderm makes many more neuroblasts and fewer epidermal cells than normal.

11.3 Asymmetric cell divisions are involved in *Drosophila* neuronal development

In the *Drosophila* nervous system, the majority of neurons are derived from a fixed lineage which involves multiple rounds of asymmetric cell divisions. In the central nervous system, the neuroblasts of the ventral nerve cord move out of the epithelium and then behave as stem cells; each neuroblast divides asymmetrically to give an apical cell which remains a neuroblast stem cell, and a smaller basal cell, the **ganglion mother cell** (Fig. 11.8). This divides again to give two neurons. Two proteins, numb and prospero, are localized in the basal cortex of the neuroblast as it enters mitosis, so after division they are localized mainly in the basal daughter cell. The protein miranda localizes prospero. Numb may act by antagonizing Notch activity, thus making the two daughter cells different in their response to a Notch signal.

A number of other genes are involved in the asymmetric division of central nervous system neuroblasts. For example, in *inscuteable* mutants, the orientation of the mitotic spindle is incorrect, while in *bazooka* mutants, Numb and Prospero are uniformly distributed within the neuroblast. *bazooka* appears to be at the top of a complex cascade of interactions that sets up the polarity of the neuroblast and the asymmetric distribution of the cell determinants.

Most of the sensory organs in the adult *Drosophila* are each formed from the division of a single neuroblast. There are two main types of sensory organ, with invariant positions: external sensory organs, which can act as mechanoreceptors or chemoreceptors, and internal chordotonal organs, which monitor stretching. Expression of the achaete–scute complex is required for the formation of all external sensory organs, but not for the internal chordotonal organs. The proneural clusters that give rise to chordotonal organs instead express the gene *atonal*, which codes for a basic helix-loop-helix transcription factor. Both types of organ are made up of four cells, only one of which is a sensory neuron (Fig. 11.9, first panel).

The development of a sensory bristle from a sensory organ precursor (SOP) involves two cell divisions to generate the four different cells that form the organ. These are a sensory neuron and its associated sheath cell, and two support cells—a hair cell and a socket cell (see Fig. 11.9, second panel). In contrast to the neuroblast, division is within the plane of the epithelium and *inscuteable* is not involved. The orientation of the division is under the control of *frizzled* and *dishevelled*, genes that also control planar polarity in the wing and eye. The first cell division gives two daughter cells (IIa and IIb). IIa divides to give a hair cell and a socket cell; IIb divides to give a neuron and its sheath cell. One determinant of cell fate is the numb protein which is asymmetrically distributed so that at the first division only one daughter cell (IIb) receives numb protein. This cell gives rise to the neuron and sheath cell at the second division. Numb protein is synthesized anew in the IIa cells, where it is again asymmetrically located and specifies the hair cell. Numb protein is essential for the formation of the neuron and

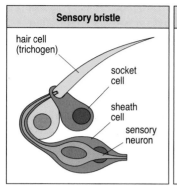

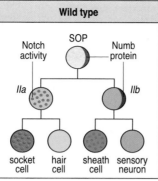

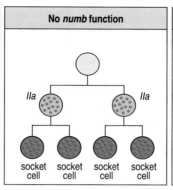

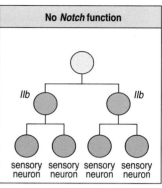

| Sensory bristle | Wild type | No *numb* function | No *Notch* function |

sheath cell; in mutant embryos lacking *numb* expression these cells do not develop and four outer support cells develop instead (see Fig. 11.9, third panel). Numb appears to act by inhibiting Notch signaling; high levels of Notch signaling in the absence of numb lead to all cells developing as socket cells. If Notch activity is absent, all daughter cells develop as neurons (Fig. 11.9, fourth panel).

We have seen many examples of the importance of localization of cytoplasmic determinants in early development. Sensory cell development in *Drosophila* is a very clear example of asymmetric division determining cell specification by distributing localized cytoplasmic determinants such as numb differently to the two daughter cells.

11.4 The vertebrate nervous system is derived from the neural plate

All the cells of the vertebrate central nervous system derive from the neural plate, a region of columnar epithelium induced from the ectoderm on the dorsal surface of the embryo during gastrulation (see Section 4.7), and from sensory placodes in the head region, which contribute to the cranial nerves. Toward the end of gastrulation, the neural plate begins to fold and form the neural tube (Fig. 11.10). Neural crest cells migrate away from the dorsal half of the tube—undergoing an epithelial to mesenchymal transition—and give rise to both the sensory neurons of the peripheral nervous system and to the autonomic nervous system. The cells that remain within the neural tube give rise to the brain and spinal cord, which together comprise the central nervous system. The mechanisms by which cells are specified as neural precursors in the vertebrate neural plate are gradually emerging, and important insights have come from studying the corresponding processes in *Drosophila*.

The neural tube becomes regionalized along the antero-posterior axis, as described in Section 4.8, and this regionalization is particularly clear in the hindbrain region, where the Hox genes are expressed in a well-defined pattern. The expression of these genes is thought to give the cells of the neural tube their positional value along the antero-posterior axis of the hindbrain. The neural tube also becomes patterned along the dorso-ventral

Fig. 11.9 Development of a *Drosophila* sensory bristle, which is made up of four cells, depends on asymmetric cell division. The sensory organ precursor (SOP) contains numb protein on one side only. It undergoes an asymmetric division in which the numb protein is passed on to just one of the daughter cells (IIb). Numb protein inhibits Notch function. This cell gives rise to a sensory neuron and a sheath cell. The cell that does not receive numb protein (IIa) displays Notch activity and gives rise to a hair cell and a socket cell. In mutants lacking numb, both daughters of the sensory organ precursor form socket cells. After Kandel, E.R. *et al.*: 2000.

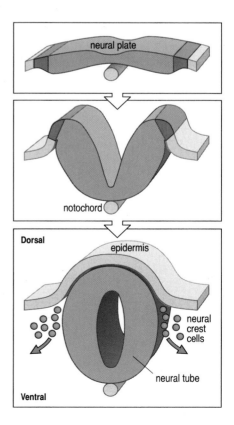

Fig. 11.10 The vertebrate nervous system is derived from the neural plate. The neural tube forms by folding of the neural plate. It sinks beneath the surface and is overlain by the epidermis. The neural tube gives rise to both the brain and the spinal cord. Neural crest cells originating from the line of fusion of the neural folds migrate away from the neural tube to form sensory nerves and the autonomic nervous system (among other structures).

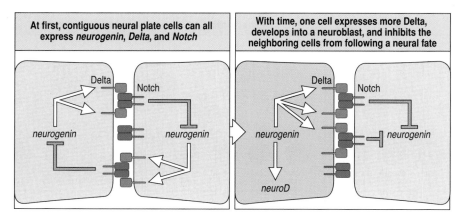

At first, contiguous neural plate cells can all express *neurogenin*, *Delta*, and *Notch*	With time, one cell expresses more Delta, develops into a neuroblast, and inhibits the neighboring cells from following a neural fate

Fig. 11.11 Lateral inhibition specifies single cells as neuronal precursors in the vertebrate nervous system. The *neurogenin* gene is initially expressed in stripes of contiguous cells in the neural plate, and these cells also express the Delta and Notch proteins. The Delta–Notch interaction between adjacent cells mutually inhibits expression of the *neurogenin* gene. When, by chance, one cell expresses more Delta than its neighbors, it inhibits the expression of Delta protein in the neighboring cells and so develops as a neuron, expressing *neurogenin* and *neuroD*.

axis. We next consider the specification of neurons as a class of cells, and then turn to how different types of neurons can be specified in the correct position.

11.5 Specification of vertebrate neuronal precursors involves lateral inhibition

The specification of cells as neuronal precursors in vertebrates involves processes similar to those already described for *Drosophila* (see Sections 11.1 and 11.2). In early *Xenopus* embryos, neurons are not generated uniformly over the neural plate but are initially confined to three longitudinal stripes on each side of the midline. The medial stripe, nearest the midline, which becomes the ventral region of the neural tube, gives rise to motor neurons, whereas the lateral stripes give rise to interneurons and sensory neurons. The *neurogenin* gene, which codes for a basic helix-loop-helix transcription factor, is distantly related to the *achaete* and *scute* genes of *Drosophila*. It is expressed in the cells of the three longitudinal stripes and its activity is required for neural differentiation. Selection of cells to form the stripes involves lateral inhibition, in which the proteins Delta and Notch again play a key role.

Binding of Delta to the Notch protein provides a signal that inhibits neuronal differentiation by inhibiting the synthesis of neurogenin. Initially, all cells in the neural plate are able to make neurogenin, Delta, and Notch, and neural differentiation is prevented by mutual inhibition. When one cell, by chance, starts to express Delta more strongly than its neighbors, the stronger signal delivered to the adjacent cells both prevents neural differentiation in these cells, by shutting down *neurogenin* expression, and represses the synthesis of Delta, stopping the cells from delivering reciprocal inhibition. Synthesis of neurogenin in this one cell then leads to activation of *neuroD*, which encodes a transcription factor required for neuronal differentiation (Fig. 11.11). In support of this model, overexpression of Delta or expression of active mutant Notch, which signals continuously, results in a reduction in the number of neurons. Inhibition of Delta function leads to an increase in the number of neurons, as does overexpression of neurogenin.

11.6 The pattern of differentiation of cells along the dorso-ventral axis of the spinal cord depends on ventral and dorsal signals

There is a distinct dorso-ventral pattern in the developing spinal cord. Future motor neurons are located ventrally, whereas commissural neurons, which send axons across the cord, differentiate primarily in the dorsal region (Fig. 11.12). In addition to the neuronal cell types, there is, in the ventral midline, a group of non-neuronal cells, which forms the **floor plate**. There are also non-neuronal **roof plate** cells in the dorsal midline. Sensory neurons derived from the neural crest cells arise laterally and dorsally and migrate into the dorsal root ganglia (Fig. 11.13). Each cell type is present symmetrically on either side of the midline.

Patterning of the cells in the ventral half of the spinal cord neural tube appears to be controlled by the notochord and floor plate cells. The floor plate can be induced by the notochord, but the relationship between notochord and floor plate does not fit snugly into standard models of induction. One reason is that the presumed inducer—the notochord—produces the same signal as the tissue induced—the floor plate. In this case the signal is Sonic hedgehog. Another reason is that precursor cells at the anterior end of the primitive streak (see Fig. 2.14) contribute to both notochord and floor plate. A substantial proportion of floor plate cells are, however, specified quite late from cells in the neural tube.

The patterning activity of the notochord can be shown by grafting a segment of notochord to a site lateral or dorsal to the neural tube; a second floor plate is induced in neural tube in direct contact with the transplanted notochord, and additional motor neurons are generated. Molecular markers of dorsal cell differentiation are suppressed in the region of the graft. Both the short-range and long-range signals are mediated by Sonic hedgehog protein, whose role as a signal we have already discussed in connection with the development of the vertebrate limb (Chapter 10), and which will be discussed further below.

The dorsal epidermal ectoderm, and later the roof plate, patterns the dorsal half of the neural tube. In addition to the signal provided by Sonic

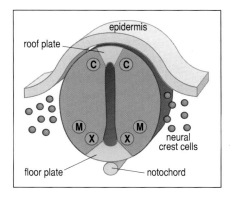

Fig. 11.12 Dorso-ventral organization in embryonic spinal neural tube. In neural tube that develops into spinal cord, a floor plate of non-neuronal cells develops along the ventral midline, and a roof plate of non-neuronal cells along the dorsal midline. Neuronal cells whose phenotype is not known (X) differentiate nearest to the floor plate. Immediately above this region are future motor neurons (M). Commissural neurons (C) differentiate in the dorsal region, near the roof plate.

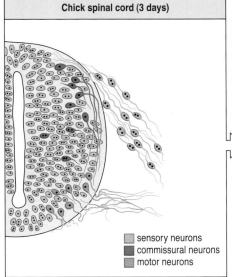
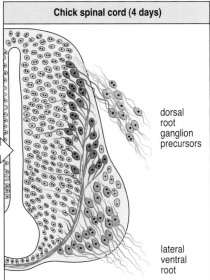

sensory neurons
commissural neurons
motor neurons

dorsal root ganglion precursors

lateral ventral root

Fig. 11.13 Formation of sensory and motor neurons in the chick spinal cord. Three days after an egg has been laid, motor neurons are beginning to form within the chick embryo neural tube. A day later, the motor neurons migrate to form the lateral ventral root of the spinal cord. The sensory neurons that migrate from the dorsal part of the spinal cord are derived from neural crest cells and form the segmental dorsal root ganglia (see Section 8.16). The commissural neurons put out axons towards the ventral midline and cross it.

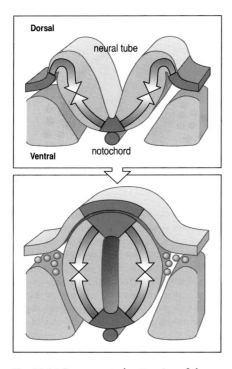

Fig. 11.14 Dorso-ventral patterning of the spinal cord involves both dorsal and ventral signals. As the neural tube forms (top panel), dorsal and ventral signaling occurs. The ventral signal from the floor plate is the Sonic hedgehog protein (red), whereas the dorsal signal (blue) includes BMP-4, which is induced during neural tube formation by the adjacent ectoderm. Ventral and dorsal signals oppose each other's action.

hedgehog from the floor plate and notochord in the ventral midline, there are also dorsal signals. Using genetic knock-out and cell fate mapping, it has been shown that selective removal of the roof plate from the neural tube of the mouse embryo results in the loss of the most dorsal interneurons. BMPs from the ectoderm specify the roof plate, and the signals from the roof plate include BMPs which pattern the dorsal region of the neural tube. BMP-4 and BMP-7 are expressed in the epidermal ectoderm and, in experiments in chick embryos, have been shown to mimic the effect of the dorsalization signal. These BMP signals apparently ensure dorsal differentiation by opposing the action of signals from the ventral region (Fig. 11.14).

The induction of dorsal fate results in the expression of the gene for BMP-4 and of another TGF-β family gene, *Dorsalin-1*, in the dorsal region of the neural tube. The pattern of expression of *Dorsalin-1* is strongly influenced by ventral signals. This system bears a strong similarity to the mechanism of patterning of the dorso-ventral body axis in early *Drosophila* development, where the expression of *decapentaplegic*, another TGF-β family gene, is confined to dorsal regions because its expression is repressed in ventral regions by the nuclear protein, dorsal (see Section 5.8).

Thus, in the spinal cord, Sonic hedgehog protein and BMPs may provide positional signals with opposing actions emanating from the two ends of the dorso-ventral axis. The similarity to the patterning of both the vertebrate somite (see Section 4.2) and the early *Drosophila* embryo again emphasizes how patterning mechanisms have been conserved.

Motor neurons in the spinal cord develop in the ventral region and can be classified on the basis of the position of their cell bodies along the dorso-ventral axis within the cord, and the muscles that their axons innervate. In the chick embryo, motor neurons in the ventral region are subdivided into three main longitudinal columns on each side of the midline—a median column nearest the midline, and a lateral column, which is further divided into two columns, lateral and medial (Fig. 11.15). Motor neurons in the median column send axons to axial and body wall muscles, whereas medial and lateral neurons of the lateral motor column project to ventral and

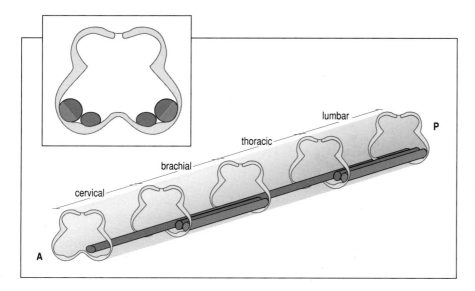

Fig. 11.15 The organization of the motor columns in the developing chick spinal cord. The organization of motor columns along the chick embryo spinal cord at stage 28–29. The median motor column is in blue, the medial lateral column in red and the lateral lateral column in green. The lateral columns are only present in the brachial and lumbar regions, where the limbs will form.

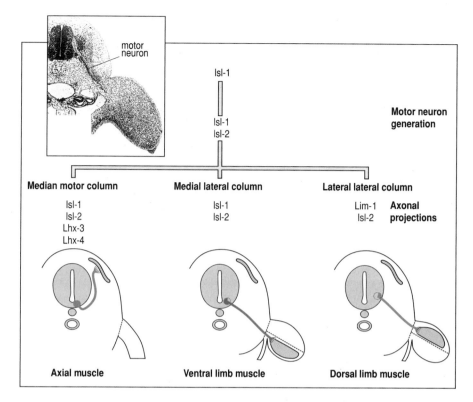

Fig. 11.16 Different motor neurons in the chick spinal cord express different LIM homeodomain proteins. The photograph (inset) shows a transverse section through a chick embryo at the level of the developing wing. Motor neurons can be seen entering the limb. Initially, all developing motor neurons express the LIM proteins Isl-1 and Isl-2. At the time of axon extension, motor neurons that innervate different muscle regions express different combinations of Isl-1, Isl-2, Lim-1, Lhx-3, and Lhx-4.

dorsal limb muscles respectively. The lateral motor column is only present in the brachial and lumbar regions, where the limbs develop.

Motor neurons are not all the same even if they look similar, and each may have a unique identity. Each motor neuron subtype expresses a combination of proteins encoded by the homeobox genes of the LIM family, which may provide the neurons with positional identity and enable their axons to select a particular pathway (Fig. 11.16). The expression of Lim-1 by a lateral set of lateral motor column neurons, for example, ensures that these axons select a dorsal trajectory in the limb. The complementary expression of Lmx-1b in dorsal limb mesenchyme is also involved in controlling axonal outgrowth. If either of these proteins is absent, then the motor axons select dorsal and ventral pathways randomly.

How are neuronal subtypes specified? Secretion of Sonic hedgehog by the notochord and the floor plate controls the specification of five different classes of ventral neurons—motor neurons and four classes of interneurons—in a dose-dependent manner; these different neuronal types can be generated *in vitro* in response to a two- to threefold change in concentration of Sonic hedgehog. Several homeodomain genes are expressed in ventral progenitor cells in response to Sonic hedgehog signaling. These genes can be divided into two classes. Class I are generally repressed by Sonic hedgehog, and include *Pax7, Pax6, Dbx1, Dbx2* and *Irx3*, whereas Class II, such as *Nkx2.2* and *Nkx6.1*, are activated (Fig. 11.17). Class I genes have ventral limits of expression whereas Class II have dorsal limits. Interactions between the regions expressing Class I and Class II proteins establish sharp boundaries between progenitor domains that give rise to the different types of neurons (Fig. 11.17).

Fig. 11.17 Specification of neuronal subtypes in the ventral neural tube. Different neural subtypes (shown as different colors) are specified along the dorso-ventral axis of the ventral neural tube in response to a gradient of Sonic hedgehog (Shh) with its source in the floorplate. Shh protein mediates the repression of the so-called Class I homeodomain protein genes (*Pax7, Pax6, Dbx1, Dbx2* and *Irx3*), whereas Class II homeodomain protein genes (*Nkx2.2* and *Nkx6.1*), are activated. Interactions between Class I and Class II genes pattern this expression further. Five neuronal subtypes are generated from the five domains. MN is motor neurons. Adapted from Jessel, T.M.: 2000.

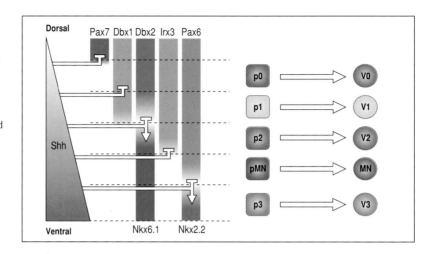

Motor neurons also acquire a regional identity along the antero-posterior axis. Transplantation of a region of the spinal cord from a site adjacent to a limb to a thoracic region leads to expression of both LIM and Hox genes appropriate to the new site. The signals determining their expression come from the adjacent paraxial mesoderm.

11.7 Neurons in the mammalian central nervous system arise from asymmetric cell divisions, then migrate away from the proliferative zone

The neural tube generates a large number of different neuronal and glial cell types. In both brain and spinal cord, all neurons and glia arise from a proliferative layer of epithelial cells lining the lumen of the neural tube called the **ventricular proliferative zone**. Once formed, a neuron never divides, and migrates away from the proliferative zone. Here we look at how neurons of the mammalian cerebral cortex are generated by cell division in the proliferative zone. There is even generation of new neurons in the adult mammalian brain.

The mammalian cerebral cortex is organized into six layers (I–VI), numbered from the cortical surface inward. They each contain neurons with distinctive shapes and connections. For example, large pyramidal cells are concentrated in layer V, and smaller stellate neurons predominate in layer IV. All of these neurons have their origin in the ventricular proliferative zone lining the inside of the neural tube, and migrate outward to their final positions along radial glial cells—greatly elongated cells that extend across the developing neural tube (Fig. 11.18).

The identity of a cortical neuron is specified before it begins to migrate. The layer to which neurons migrate after their birth in the proliferative layer is related to the time when the neuron is born. The neurons of the mammalian central nervous system do not divide once they have become specified, and so a neuron's time of birth is defined by the last mitotic division that its progenitor cell underwent. The newly formed neuron is still immature; later it extends an axon and dendrite processes, and assumes the morphology of a mature neuron. Birth times of individual neurons can be determined by exposing the neural tube to a short pulse of

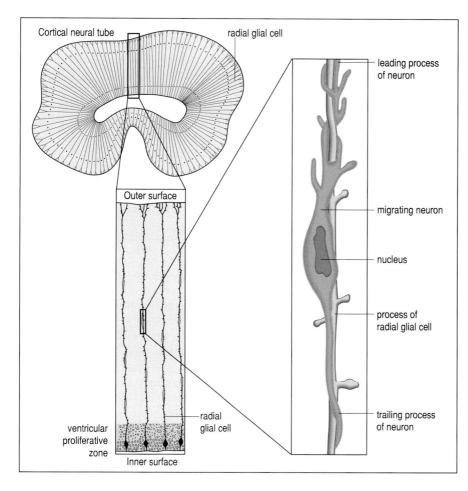

Fig. 11.18 Cortical neurons migrate along radial glial cells. Neurons generated in the ventricular proliferative zone migrate to their final locations in the cortex along radial glial cells that extend right across the wall of the neural tube. After Rakic, P.: 1972.

^{3}H-labelled thymidine (Fig. 11.19). This is incorporated into the neuron's DNA during the S-phase of the cell cycle, when DNA synthesis takes place. A neuron that is formed at the end of a cell cycle during which ^{3}H-thymidine is incorporated can be identified by its heavily labeled DNA, whereas neurons born later have the label diluted out through further rounds of DNA synthesis in the progenitor cells.

Neurons born at early stages of cortical development migrate to layers closest to their site of birth, whereas those born later end up further away, in more superficial layers (see Fig. 11.19). The younger neurons must migrate past the older ones on their way to their correct position. Thus, there is an inside–outside sequence of neuronal differentiation in the neural tube, which gives rise to the cerebral cortex and other layered brain structures, such as the cerebellum. Mutations in the gene *reelin*, which codes for a cerebellar protein, disrupts migration and the neurons seem unable to bypass their predecessors and proper cerebellar cortical layers do not form.

Environmental influences can affect neuronal fate by influencing the character of the progenitor cells. In experiments on the mammalian visual cortex, progenitor cells that produce neurons that would normally migrate to layer VI are transplanted into older embryos. If these cells are transplanted at an early stage in the cell cycle, or undergo further divisions, the

Fig. 11.19 The time at which a neuron is born determines which cortical layer it becomes part of. In the monkey visual cortex, a neuron's time of birth is determined as the time that the precursor cell stops dividing and leaves the mitotic cycle. This figure shows the results of the injection of ³H-labeled thymidine into a developing neural tube at various times during development. The red bars represent the distribution of heavily labeled neurons found after each injection. Neurons born first remain closest to their site of birth—the proliferative ventricular zone. Neurons born later become part of successively higher cortical zones. Cortical zones are numbered from the outer surface inward; layer VI is the layer nearest to the ventricular zone.

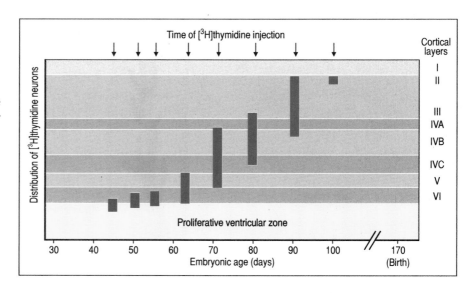

neurons they give rise to migrate to layers II and III instead, which indicates an environmental influence on their fate. However, if the cells are transplanted late in their final division cycle, the neurons derived from them still migrate to layer VI, their normal location. Thus, commitment to a particular neuronal fate occurs late in the division cycle and, once it occurs, the neuron becomes unresponsive to environmental influences on its fate. Both signals from the extracellular environment and intrinsic intracellular restrictions therefore have a role in the specification of neuronal identity in vertebrates.

How is the decision made for a cell to leave the cell cycle and become a neuron? Direct observation of slices of cells from developing brain in culture show that the ventricular epithelial cells that give rise to neurons can divide in either of two different planes. After division of a ventricular epithelial cell parallel to the epithelial surface, which leads to formation of distinct apical and basal cells, the basal cell does not undergo any further division and becomes a neuron; the apical cell behaves as a stem cell and can divide again. In contrast, when the plane of division is at right angles to the epithelium, the two daughter cells have a tendency to remain in the proliferative layer and continue to divide (Fig. 11.20).

Why do daughter cells behave similarly after vertical divisions but differently after horizontal divisions? By analogy with the asymmetric divisions in *Drosophila* sensory cell development, the answer may involve two asymmetrically localized cytoplasmic factors, the Notch-1 protein and the numb protein (the vertebrate equivalents of the *Drosophila* Notch and numb proteins). At cell division, Notch-1 is localized in the basal domain of the mammalian ventricular epithelial cells, so that in horizontal cell divisions the basal cell, which becomes a neuron and migrates away, contains all the Notch-1. In contrast, numb protein is localized at the dividing cell's apical end and cells inheriting it remain in the proliferative layer (see Fig. 11.20). It is not clear, however, that it is these two proteins that determine whether a cell becomes a neuron or remains a stem cell.

There is now evidence for generation of new neurons in the adult mammalian brain. One region where neurons are formed is the well of the lateral ventricle; these neurons contribute to the olfactory system. Another

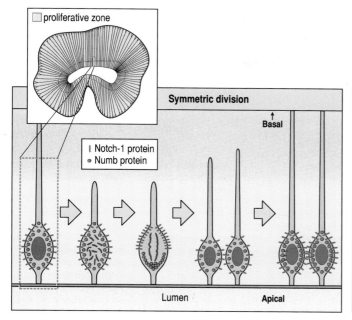

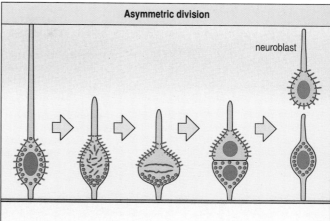

region of neuron formation is the dentate gyrus, where a special type of neuron, the granule cell, is generated. Moreover, neuronal stem cells from adult brain can be made to differentiate into a variety of different cell types (Chapter 9).

Summary

In *Drosophila*, prospective neural tissue is specified early in development as a band of neurectoderm along the dorso-ventral axis, above the mesoderm. Within the neurectoderm, expression of proneural genes, such as those of the achaete–scute complex, provides clusters of neurectoderm cells with the potential to form neural cells. Only one cell of the cluster finally gives rise to a neuroblast because of lateral inhibition, involving the Notch receptor protein and its ligand Delta, which prevents other cells developing as neuroblasts. Development of the spatially patterned sensory bristles on the adult cuticle involves a similar mechanism of lateral inhibition. Sensory organs, each composed of a single sensory neuron and three associated cells, develop from the neural precursor cell by a mechanism involving asymmetric division, which differentially allocates asymmetrically distributed proteins such as numb to daughter cells.

The vertebrate nervous system is derived from the neural plate, which is specified during gastrulation. It contains the cells that form the brain and spinal cord, as well as the neural crest cells that contribute to the peripheral nervous system. Neurons are specified within the neural plate by a mechanism of lateral inhibition, similar to that in *Drosophila*. Patterning of neuronal cell types along the dorso-ventral axis of the spinal cord involves signals from both the ventral and dorsal regions. Secretion of Sonic hedgehog by the notochord and floor plate provides a graded signal that leads to specification of different neuronal types in the ventral region of the spinal cord. The specification of neuronal cell type in the mammalian cortex depends on asymmetric cell divisions and the time when the neurons are

Fig. 11.20 Cortical neurons are formed by asymmetric divisions in the ventricular proliferative zone of the neural tube. Stem cells in the epithelium divide to generate neurons. Divisions at right angles to the epithelial surface give rise to two equal cells that continue to proliferate (left panel). Divisions parallel to the surface give rise to one proliferative cell that remains in the zone and one cell that migrates away to become a neuroblast (right panel). The Notch-1 protein is present on the basal face of dividing ventricular zone cells and is distributed to the basal cell only in horizontal divisions, whereas numb protein is localized in the apical region.

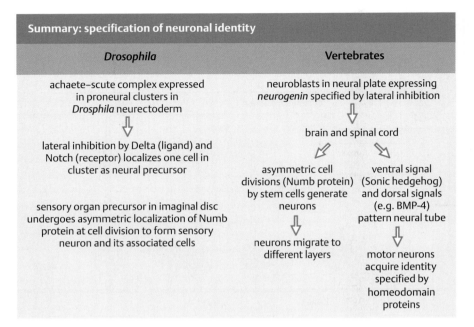

Summary: specification of neuronal identity

Drosophila	Vertebrates
achaete–scute complex expressed in proneural clusters in *Drosphila* neurectoderm	neuroblasts in neural plate expressing *neurogenin* specified by lateral inhibition
⇩	⇩
lateral inhibition by Delta (ligand) and Notch (receptor) localizes one cell in cluster as neural precursor	brain and spinal cord

⬊ ⬊

sensory organ precursor in imaginal disc undergoes asymmetric localization of Numb protein at cell division to form sensory neuron and its associated cells	asymmetric cell divisions (Numb protein) by stem cells generate neurons	ventral signal (Sonic hedgehog) and dorsal signals (e.g. BMP-4) pattern neural tube
	⇩	⇩
	neurons migrate to different layers	motor neurons acquire identity specified by homeodomain proteins

formed, as well as on environmental signals. The earliest formed neurons make the innermost cortical layer, while neurons formed later make successive outer layers.

The insect compound eye

An excellent model system illustrating cell–cell interactions in the development of neural tissues is the eye of *Drosophila*. In this very well studied system, light-sensitive sensory neurons form a highly organized pattern. The nervous system depends on sensory input, such as that from the eye. The *Drosophila* compound eye is composed of about 800 identical photoreceptor organs (**ommatidia**) arranged in a regular hexagonal array (Fig. 11.21). Each ommatidium is made up of eight photoreceptor neurons (R1–R8), together with four overlying cone cells (which secrete the lens) and additional pigment cells. It is these red-pigmented cells that give wild-type *Drosophila* its red eyes. The genetic analysis of ommatidium development has provided one of the best model systems for studying the patterning of a small group of cells, and the basis of their intercellular interactions.

The eye develops from the single layered epithelial sheet of the eye imaginal disc, located at the anterior end of the larva. Specification and patterning of the cells of the ommatidia begins in the middle of the third larval instar. Patterning starts at the posterior of the eye disc and progresses anteriorly, taking about 2 days, during which time the disc grows eight times larger. One of the earliest events in eye differentiation is the formation of a groove, the **morphogenetic furrow**, in the disc. As the furrow sweeps across the epithelium from posterior to anterior, clusters of cells that will give rise to the ommatidia appear behind it, spaced in a hexagonal array. The furrow moves slowly across the disc, at a rate of 2 hours per row of ommatidial clusters. As the morphogenetic furrow moves forward, the

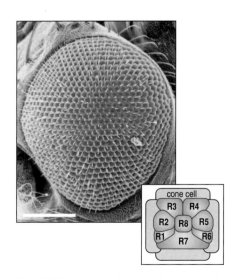

Fig. 11.21 Scanning electron micrograph of the compound eye of *Drosophila*. Each unit is an ommatidium. At the third instar, the ommatidial cluster (inset) is made up of eight photoreceptor neurons (R1–R8) and four cone cells. Scale bar = 50 μm.

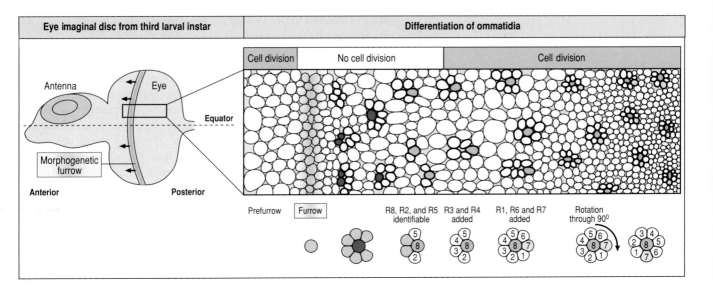

| Eye imaginal disc from third larval instar | Differentiation of ommatidia |

Fig. 11.22 Development of ommatidia in the *Drosophila* compound eye. The compound eye develops from the eye imaginal disc, which is part of a larger disc that also gives rise to an antenna. During the third larval instar, the morphogenetic furrow develops in the eye disc and moves across it in a posterior to anterior direction. Ommatidia develop behind the furrow, the photoreceptor neurons being specified in the order shown, R8 developing first and R7 last. The individual ommatidia are regularly spaced in a hexagonal grid pattern. After Lawrence, P.: 1992.

cells behind it start to differentiate to form regularly spaced ommatidia. The ommatidia are arranged in rows, with each row half an ommatidium out of register with the previous one. This gives the characteristic hexagonal packing arrangement (see Fig. 11.21). The first cells to differentiate are the R8 photoreceptor neurons. These appear as regularly spaced cells in each row, separated from each other by about eight cells. This separation sets the spacing pattern for the ommatidia.

Each R8 cell initiates a cascade of signals that recruits a surrounding cluster of 20 cells to form an ommatidium (Fig. 11.22). R2 and R5 differentiate on either side of it, to form two functionally identical neurons. R3 and R4, which are a slightly different type of photoreceptor, form next. All these cells become arranged in a semi-circle with R8 at the center. R1 and R6 differentiate next and almost complete the circle, which is finally closed by the differentiation of R7 adjacent to R8 (see Fig. 11.22). The clusters rotate 90° so that R7 comes to be closest to the equator of the disc and R3 furthest away (see Fig. 11.22); rotation is in the opposite direction in the dorsal and ventral halves. This polarization of the ommatidium involves the activation of the frizzled protein in R3/R4.

The determination of the equator of the eye results from the specification of the dorsal region of the eye disc by members of the *Iroquois* gene complex, which is followed by specification of the equator itself, involving the actions of Notch, Serrate, Delta and fringe in a similar way to the specification of the dorso-ventral compartment boundary in the wing (Section 10.14). The dorsal and ventral regions of the eye have distinct polarity, with the ommatidia in each half of the eye oriented towards the equator. In the following sections we consider two aspects of eye development: the mechanism that spaces the ommatidia in such a beautifully regular hexagonal array; and the patterning of an individual ommatidium, in particular the patterning of the eight photoreceptor neurons.

11.8 **Signals generated in the *Drosophila* eye disc maintain progress of the morphogenetic furrow and the ommatidia are spaced by lateral inhibition**

The morphogenetic furrow is the result of a wave of signals that initiates the development of ommatidia from the eye disc cells. Its passage across the disc is essential for differentiation of the ommatidia, as mutations that block its progress also block the differentiation of new rows of ommatidia, resulting in a fly with abnormally small eyes. Although there are no anterior and posterior compartments in the eye discs, the cells just behind the furrow can be regarded as resembling posterior cells, similar to the situation in the wing disc posterior compartment (see Section 10.14). They secrete hedgehog protein, which triggers the expression of decapentaplegic, which causes the cells to become competent to form neural tissue. The gene *wingless* also plays a part in eye disc patterning. It is expressed at the lateral edges of the eye disc, and prevents the furrow starting in these regions. We thus see that, even though imaginal discs give rise to very diverse structures, the key signals involved in patterning the leg, wing, and eye are similar, although they have different roles.

The question of how the regular spacing of the ommatidia within the eye is achieved can be answered by considering how the R8 cells are spaced (see Fig. 11.22). This involves a lateral inhibition mechanism (see Section 11.2), in which differentiating cells inhibit the differentiation of those cells immediately adjacent to them. All cells in the eye disc initially have the capacity to differentiate as R8 cells, and as the morphogenetic furrow passes they start to do so. But some inevitably gain a lead and are thus able to inhibit the differentiation of another R8 cell over a range of around three cell diameters. Cells that will give rise to R8 express *atonal*. A candidate for the inhibitor of *atonal*, that spaces the R8 cells, is the scabrous protein, a secreted protein related to the vertebrate fibrinogens (proteins involved in blood clotting). Mutations that inactivate the *scabrous* gene result in reduced spacing between ommatidia, which suggests that the scabrous protein is involved in keeping the ommatidia at the correct distance apart.

11.9 **The patterning of the cells in the ommatidium depends on intercellular interactions**

Within the cluster of cells that will become an ommatidium, there is no constant lineage relationship determining which cells become photoreceptors, for example, and which become pigment cells. Genetic mosaics, in which individual cells in the developing eye are marked, show that any pair of cells in the ommatidium could be sister cells. For example, one daughter cell arising from the division of a precursor cell can become a photoreceptor neuron, whereas the other can become a pigment cell. The patterning of the ommatidium is achieved entirely by inductive interactions between the cells forming the photoreceptor cluster. These interactions occur in a fixed sequence of small steps as each set of cells is recruited into the cluster and becomes specified. R8 differentiates first and induces the specification and differentiation of R2 and R5, which in turn induce R3 and R4, and so on. After photoreceptors have differentiated, the four lens-producing cone cells develop, and finally the surrounding ring of accessory cells. Note that in the insect ommatidium (unlike vertebrate limbs and *Drosophila* appendages) cells are being specified and determined individually, not as groups of cells.

One protein crucial in patterning the ommatidium is the EGF receptor, which in *Drosophila* is called DER (see Section 11.1). A key ligand for DER in ommatidial patterning is the spitz protein. A model for patterning the ommatidium is based both on DER activation and the age of the cells (Fig. 11.23). Spitz is produced by the three earliest-specified and most centrally located cells—R8, R2 and R5. It activates DER on neighboring cells, which recruits R3, R4, R1, R6 and R7 to a photoreceptor fate. The responding cells also secrete the protein argos. This diffuses away and inhibits more distant cells from being activated by spitz, so that no more cells in the prospective ommatidial cluster develop as photoreceptors. The actual character of each photoreceptor may be determined by the age of the cell, with cells passing through a series of 'states', each representing a potential fate. Other signals are, however, involved. The difference between R3 and R4, for example, involves Notch; the cell with the high level of Notch activity is inhibited from becoming R3, and becomes R4.

Specification of R7 requires expression of the genes *sevenless* and *bride-of-sevenless*. When either gene is inactivated, the phenotype is the same: R7 does not develop. Instead, an extra cone cell is formed. *sevenless* codes for a transmembrane receptor tyrosine kinase, and *bride-of-sevenless* is its ligand. Sevenless protein is produced not only by R7 but by other cells in the ommatidium, including lens cells. Thus, expression of sevenless is a necessary, but not sufficient, condition for R7 specification. Using genetic mosaics, it can be shown that for R7 to develop, only R8 need express bride-of-sevenless protein and that this is the signal by which R8 induces R7. Bride-of-sevenless is also an integral membrane protein; it is present on the apical surface of the R8 cell, where it makes contact with R7. A second signal for R7 is provided by the R1/R6 pair, which must activate Notch in the R7 cell.

11.10 Activation of the gene *eyeless* can initiate eye development

The *eyeless* gene is involved in initiating eye development throughout the animal kingdom, even though the resulting eyes are very different in structure. In *Drosophila*, mutations in *eyeless* result in the reduction or complete absence of the fly's compound eye. This gene is expressed in the region of the eye disc anterior to the morphogenetic furrow. Induction of expression of the *eyeless* gene in other imaginal discs results in the development of ectopic eyes, which have been induced in this way in wings, legs, antennae, and halteres. The fine structure of these ectopic eyes is remarkably normal, and distinct ommatidia are present. Therefore, *eyeless* is clearly a key gene for eye development and its expression switches on the requisite developmental pathway. It is estimated that some 2000 genes are eventually activated as a result of *eyeless* activity, and all of them are required for eye morphogenesis. The action of *eyeless* may be similar to that of the Hox genes, as it appears to change the interpretation of positional information in discs in which it is ectopically expressed (see Section 10.17).

The *eyeless* gene is an example of the remarkable conservation of developmental genes throughout the animal kingdom. Its homolog in vertebrates is called *Pax-6*. Like *Drosophila* without *eyeless* function, mice with

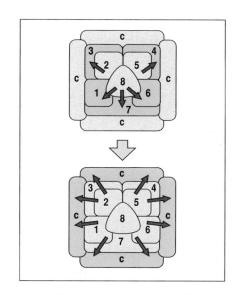

Fig. 11.23 Sequential recruitment of photoreceptors and cone cells during ommatidial development. The three earliest-specified and most centrally located photoreceptor cells—R8, R2 and R5—produce the protein spitz. It activates the *Drosophila* EGF receptor DER on neighboring cells, which recruits R3, R4, R1, R6 and R7 to a photoreceptor fate (top). These also then produce spitz, which interacts with DER on the prospective cone cells (c) to recruit them to the ommatidium. Once cells are determined they secrete the protein argos. This diffuses away and inhibits more distant cells from being activated by spitz, so that no more cells in the prospective ommatidial cluster develop as photoreceptors or cone cells. Adapted from Freeman: 1997.

defective *Pax-6* function have smaller eyes than normal, or no eyes at all. And mutations in the human *Pax-6* gene are responsible for a variety of eye malformations in the disease known as aniridia, because of the partial or complete absence of the iris. Ectopic expression of *Pax-6* in the fly can substitute for *eyeless*, resulting in the development of an ectopic eye. In *Xenopus*, *Pax-6* can induce ectopic eyes when its mRNA is injected into a blastomere at the 16-cell stage.

Summary

The *Drosophila* compound eye contains around 800 individual ommatidia, arranged in a regular hexagonal pattern. It develops from an imaginal disc, which is patterned at a late larval stage, and provides an important model for patterning a developing structure cell by cell.

The regular spacing of the ommatidia is achieved by lateral inhibition. The patterning of the eight photoreceptor neurons (R1–R8) in each ommatidium is due to local cell–cell interactions, in which the photoreceptor cells are specified and differentiate in a strict order. The specification of photoreceptor R7 involves direct contact with R8, which carries the bride-of-sevenless protein on its surface, as well as Notch activation by R1/R6. *eyeless* is a key gene in eye development. In its absence, eyes do not develop or are much reduced and it can induce ectopic eyes in other imaginal discs. It is related to *Pax-6* in vertebrates.

Summary: development of the ommatidia of the *Drosophila* eye

eyeless gene expression required for eye development. Morphogenetic furrow moves across eye disc, associated with hedgehog and decapentaplegic signals

the eight photoreceptors of the future ommatidia begin to develop behind the furrow, R8 developing first and R7 last.
Ommatidia spaced by lateral inhibition

photoreceptors specified by DER activation, other signals, and time;
R7 specification depends on a signal from R8

Axonal guidance

The working of the nervous system depends on discrete neuronal circuits, in which neurons make numerous connections with each other (Fig. 11.24). Here we look at how these connections are set up. The functioning network of neurons is established by migration of immature neurons and by the guided outgrowth of axons toward their target cells. What guides the migrating neurons and controls the growth of axons and the contacts they make with other cells?

We have already discussed one example of extensive cell migration in relation to neural development—the migration of neural crest cells to give rise to the segmental arrangement of dorsal root ganglia (see Section 8.16). In that case, the migration is controlled by signals from the adjacent somites. In this chapter, we deal with axon guidance in the central and

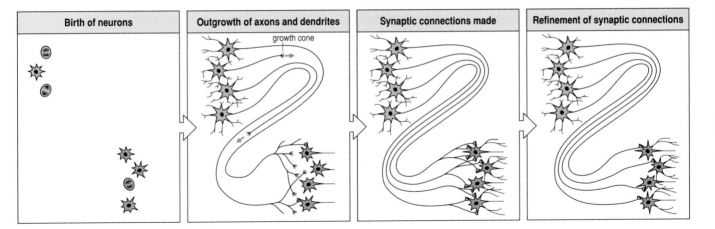

| Birth of neurons | Outgrowth of axons and dendrites | Synaptic connections made | Refinement of synaptic connections |

growth cone

Fig. 11.24 Neurons make precise connections with their targets. Neurons (green) and their target cells usually develop in different locations. Connections between them are established by axonal outgrowth, guided by the movement of the axon tip (growth cone). The initial set of relatively nonspecific synaptic connections are then refined to produce a precise pattern of connectivity. After Alberts, B., *et al.*: 1989.

peripheral nervous system. We first look at mechanisms for guiding axons and evidence for specificity of guidance and connections.

11.11 The growth cone controls the path taken by the growing axon

An early event in the differentiation of a neuron is the extension of its axon by the **growth cone** at the axon tip (Fig. 11.25). The growth cone both moves and senses its environment. As in other cells capable of migration, for example the primary mesenchyme of the sea urchin embryo (see Section 8.15), the growth cone can continually extend and retract filopodia, which help to pull the axon tip forward over the underlying substratum. Between the filopodia, the edge of the growth cone forms thin ruffles—lamellipodia—similar to those on a moving fibroblast (see Box 8B, p. 265). Indeed, in its ultrastructure and mechanism of movement, the growth cone closely resembles the leading edge of a fibroblast crawling over a surface. Unlike a moving fibroblast, however, the extending axon also grows in size, with an accompanying increase in the total surface area of the neuron's plasma membrane. The additional membrane is provided by intracellular vesicles, which fuse with the plasma membrane.

The activity of the growth cone guides axon outgrowth, and is influenced by the contacts the filopodia make with other cells and with the

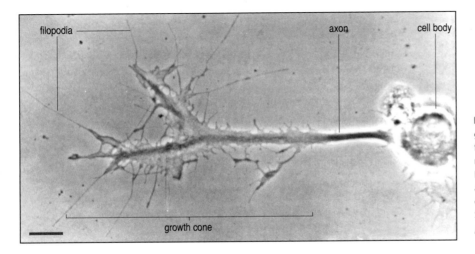

filopodia · · · · · · · · axon · · · · · · · cell body

growth cone

Fig. 11.25 A developing axon and growth cone. The axon growing out from a neuron's cell body ends in a motile structure called the growth cone. Many filopodia are continually extended and retracted from the growth cone to explore the surrounding environment. Scale bar = 10 μm.

Photograph courtesy of P. Gordon-Weeks.

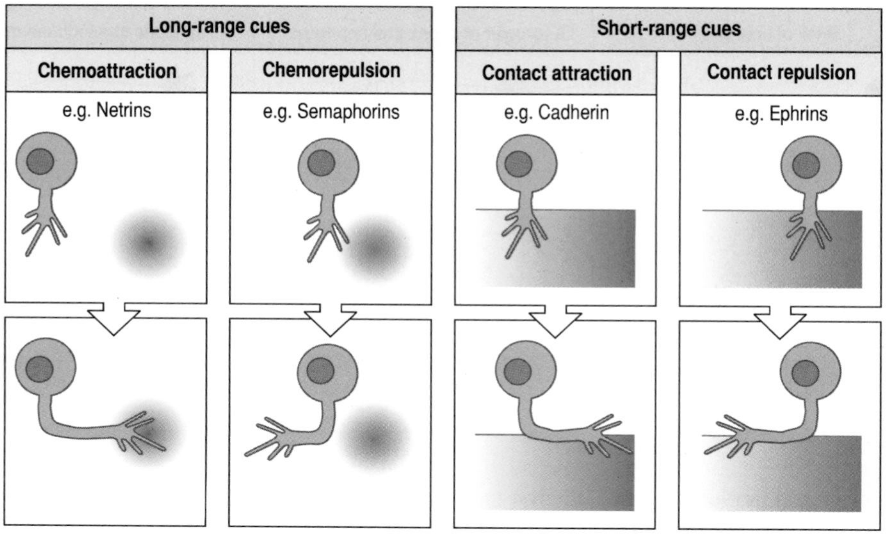

Fig. 11.26 Axon guidance mechanisms. Four types of mechanism can contribute to controlling the direction in which a growth cone moves. Attractant molecules can either be present as a diffusible gradient of chemoattractant, or be bound to the substratum, leading to contact-dependent attraction. Similarly, molecules that repel axon growth can be either bound or diffusible.

extracellular matrix. In general, the growth cone moves in the direction in which its filopodia make the most stable contacts. In addition, diffusible substances can bind to receptors on the growth cone surface, and so influence its direction of migration. Some of these cues promote axon extension, whereas others inhibit it by causing growth cone collapse. The extension and retraction of filopodia involves the assembly and disassembly of the actin cytoskeleton. Members of a family of intracellular signaling proteins, the Ras-related GTPases, are involved in the reorganization of the actin cytoskeleton. Activation of one of these, Rho, causes growth cones to stop extending, whereas Rac and Cdc42, other members of the family, are involved in growth-cone extension. But just how growth cones transduce extracellular signals so as to extend or collapse filopodia is not fully understood.

Axon growth cones are guided by two main types of cue—attractive and repulsive. In addition, cues can act either at long or short range, thus giving four ways in which the growth cone can be guided (Fig. 11.26). Long-range attraction involves diffusible chemoattractants released from the target cells and is, in principle, similar to the chemotaxis of motile cells such as those of slime molds (see Section 8–17). Both chemoattractants and chemorepellents have been identified in the developing nervous system and are discussed later. Short-range guidance is mediated by contact-dependent mechanisms involving molecules bound to other cells or to the extracellular matrix; again, such interactions can be either attractive or repulsive. Cadherins, for example, can act as short-range attractors of neurons whereas ephrins can mediate repulsion on contact.

A number of different extracellular molecules that guide axons have been identified. The **semaphorins** are among the most prominent of the conserved families of axon guidance molecules. There are nine different subfamilies, three of which are secreted whereas the others are associated with the cell membrane. There are two classes of neuronal receptors for semaphorins—the plexins and neuropilins. Axon repulsion is a major theme of semaphorin action. Some, however, can both attract and repel growth cones; which they do will depend on the nature of the neuron and on the semaphorin receptors present in the neuron membrane. Other classes of neuronal guidance molecules, such as **netrins**, are bifunctional, in that they attract some neurons and repel others. The attractant function is mediated by the DCC ('deleted in colorectal cancer') family of receptors, whereas repulsion requires the transmembrane protein Unc-5. Slit proteins repel a variety of axon classes but can also stimulate elongation and branching in sensory axons. Even different regions of the same neuron can respond in opposite ways to the same signal—the apical dendrites of pyramidal neurons in the cortex grow towards a source of semaphorin 3A, whereas their axons are repelled. This difference in response appears to be due to the presence of guanylate cyclase, a component of the signal transduction process, in the dendrites but not in the axon.

11.12 Motor neurons from the spinal cord make muscle-specific connections

An animal's muscular system enables it to carry out an enormous variety of movements, which all depend on motor neurons having made the correct connections with muscles. The development of the chick limb provides a good model for studying how these connections are set up. Motor neurons at different dorso-ventral positions within the spinal cord innervate distinct target muscles within the limb. Those near the midline project to ventral muscle, whereas lateral neurons project to muscles derived from the dorsal muscle mass. Each neuronal subtype expresses a combination of LIM family homeobox genes (see Fig. 11.16), which may provide the neurons with positional identity and enable their axons to select a particular pathway.

The pattern of innervation of the limb muscles of the chick is rather precise (Fig. 11.27). In the chick embryo, the motor neuron axons that enter a developing limb appear to make their connections with particular muscles from an early stage. When the motor axons growing out from the spinal cord first reach the base of the limb, they are all mixed up in a single bundle. At the base of the developing limb, however, the axons separate out and form new nerve tracts containing only those axons that will make connections with particular muscles.

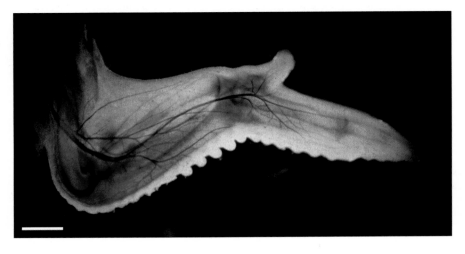

Fig. 11.27 Innervation of the embryonic chick wing. The nerves are stained brown and include both sensory and motor neurons. Scale bar = 1 mm.

Photograph courtesy of J. Lewis.

Evidence that local cues in the limb are responsible for this sorting-out comes from inverting, along the antero-posterior axis, a section of the spinal cord whose motor neurons will innervate the hindlimb. After inversion, motor axons from lumbo-sacral segments 1–3, which normally enter anterior parts of the limb, initially grow into posterior limb regions, but then take novel paths to innervate the correct muscles. So, even when the axon bundles enter in reverse order, the correct relationship between motor neurons and muscles is achieved. Motor axons can thus clearly find their appropriate muscles when their displacement relative to the limb is small. But with a large displacement, this is not the case. A complete dorso-ventral inversion of the limb bud results in the axons failing to find their appropriate targets. In such cases, the axons follow paths normally taken by other neurons. The mesoderm of the limb thus provides local cues for the motor axons and the axons themselves have an identity that allows them to choose the correct pathway.

11.13 Choice of axon pathway depends on environmental cues and neuronal identity

One of the ways in which axons reach a distant target is to make use of 'stepping stones' or 'guide posts' along the pathway. These are localized regions to which the growth cone moves when its filopodia make contact. Arrival at the ultimate target is achieved through a sequence of stages, characterized by arrival at consecutive guide posts. For example, early in the development of the grasshopper, sensory neurons arise in the epithelium at the distal tip of each developing leg. (The grasshopper does not have a separate larval stage.) These neurons extend axons under the epithelium, along a well-defined route within the leg, to make connections with the central nervous system. Because the behavior of the growth cones can be directly observed, it is known that the pathway they follow, although stereotyped, is not absolutely fixed, and they make frequent small corrections to their course.

A distal to proximal gradient of semaphorin defines the general direction of sensory axon outgrowth in the developing grasshopper leg. The numerous fine filopodia of each growth cone continually explore the surrounding substratum. They thus migrate in a proximal direction until the growth cone makes contact with the first of three guidepost cells. These cells, which will themselves become neurons, provide a high-affinity substratum for the migrating growth cone. Growth cones in the vicinity of the guidepost cells often turn sharply toward them as the fine filopodia make contact (Fig. 11.28). The axon then continues along its proximal path until it meets the second guidepost cell. At this site, the growth cone extends branches in both dorsal and ventral directions. Eventually, the dorsal branch is withdrawn, and the axon extends ventrally to make contact with the third guidepost cell, after which it moves toward the central nervous system (Fig. 11.29).

If the first two guidepost cells are removed, there is increased neuronal branching and slower axon movement, but the axon eventually reaches its destination. This indicates the presence of guidance cues in the epithelium over which the growth cones migrate. Local cues of this type provide the basis of axonal outgrowth in a wide variety of systems, including axons that cross the midline of the central nervous system and the specificity of

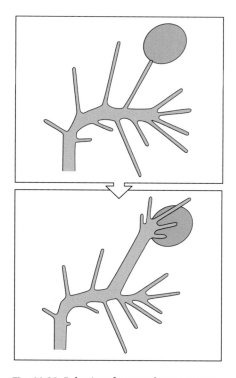

Fig. 11.28 Behavior of a growth cone near a guidepost cell. Contact of a single filopodium with a guidepost cell (red) results in further extension of the growth cone toward it. After O'Connor, T.P., *et al.*: 1990.

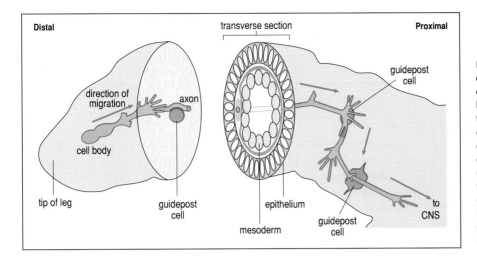

Fig. 11.29 Outgrowth and guidance of a peripheral sensory neuron in the developing grasshopper leg. The cell body of the sensory neuron lies beneath the epithelium at the tip of the developing leg. The axon tip migrates over the inner surface of the epithelium covering the limb. The axon extends until it meets a guidepost cell (red), and then continues to extend to make contact with two more guidepost cells in succession. Eventually, it makes contact with the central nervous system (CNS).

connections between the vertebrate retina and the tectum, which provided much of the early evidence for the presence of local cues.

11.14 Axons crossing the midline are both attracted and repelled

Early in the development of the central nervous system, neurons have to make a decision with respect to the midline—should they cross it or not? And if they do cross it, will the axons cross it again? These choices are very clear in the insect nerve cord and vertebrate spinal cord.

The *Drosophila* midline provides a good example of long-range attraction and repulsion of growth cones. While the development of the nervous system is largely symmetrical about the midline of the body, and proceeds independently on either side, one half of the body needs to know what the other half is doing. So, although some neurons remain entirely within the lateral half of the body in which they were formed, many axons cross the midline of the *Drosophila* embryo, and those that do cross it do not cross back. In vertebrates, axons of commissural neurons also cross the midline of the developing brain and spinal cord and then move away. Similar mechanisms that allow crossing but not recrossing of the midline are involved in both cases.

The protein netrin, which is secreted by glial cells in the midline of the *Drosophila* central nervous system, attracts axons towards it. Some of these axons cross the midline but are then unable to recross it. What allows them to cross but then prevents them from doing so? A large-scale mutant screen in *Drosophila* for genes that control the decision to cross the midline identified two key genes—*robo* (*roundabout*) and *comm* (*commissureless*) (Fig. 11.30). Robo protein functions as a receptor that causes repulsion and whose ligand is the slit protein, which is expressed in the midline. Slit thus prevents axons expressing robo from crossing the midline. However, comm protein, which is also expressed in the midline, appears to function by downregulating robo function, thus allowing the axons to cross the midline, since they are no longer affected by slit. After crossing the midline, the axons again express high levels of robo. Thus they cannot recross because slit prevents them. How robo is upregulated after crossing the midline is not known. A combinatorial code of three types of robo proteins

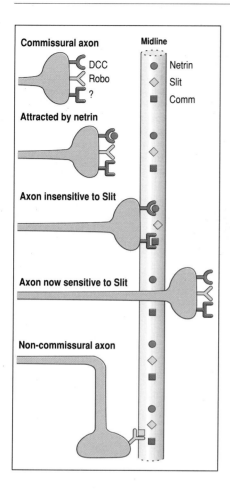

Fig. 11.30 Slit prevents commissural axons from crossing the midline. Commissural axons carry receptors for the proteins netrin (DCC), slit (Robo) and comm, which are expressed in the midline. The commissural axon is attracted to the midline by netrin. Comm downregulates the slit receptor, robo, so that the axon is insensitive to slit and can cross the midline. Having crossed, the axon re-expresses robo, making the axon sensitive to slit again and preventing it from crossing back. Non-commissural axons are repelled by slit and so cannot cross the midline.

acting as receptors for slit is involved in controlling the lateral position of the axons.

Netrin also attracts neurons to the midline in the vertebrate central nervous system. In the spinal cord, commissural neurons, whose axons cross the midline and link the two sides of the cord, develop in the dorsal region and send their axons ventrally along the lateral margin of the cord (see Fig. 11.13). When the axons reach about halfway down the cord, they make an abrupt turn and project to the floor plate in the ventral midline, bypassing the motor neurons (Fig. 11.31, left panel; see also Fig. 11.13, right panel).

Evidence for chemotaxis toward a diffusible chemoattractant produced by the floor plate comes from experiments *in vitro*. When rat dorsal neural tube containing only commissural cells is cultured, there is hardly any outgrowth of the axons, but when it is cultured with a floor plate explant, there is extensive outgrowth of axons toward the floor plate tissue. If the floor plate explant is placed to the side of the dorsal neural tube explant so that it is parallel to its dorso-ventral axis, most of the axons that emerge from the dorsal explant are oriented toward the floor plate (see Fig. 11.31, right panel). The fact that only the axons close to the floor plate tissue grow toward it is good evidence for the presence of a diffusible chemoattractant.

One diffusible factor responsible for this chemotaxis in the spinal cord is netrin-1, whose mRNA is present in floor plate cells. Cells transfected with netrin-1 mRNA mimic the long-range chemoattractant effect of floor plate cells. Knock-out of the mouse *netrin-1* gene, or the gene for one of its receptors, results in abnormal commissural axon pathways (Fig. 11.32). Robo and its ligand slit also appear to be involved in preventing midline re-crossing in vertebrates. Slit is expressed in the floor plate of the vertebrate neural tube but some axons only become responsive to its inhibitory effect after

Fig. 11.31 Chemotaxis can guide commissural axons in the spinal cord. In the chick, spinal cord commissural neurons extend ventrally, and then toward the floor plate cells (left panel). In explants of rat spinal cord cultured together with floor plate tissue, axons that are within 250 μm of the floor plate tissue grow toward it. After Tessier-Lavigne, M., *et al.*: 1991.

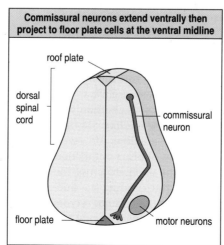

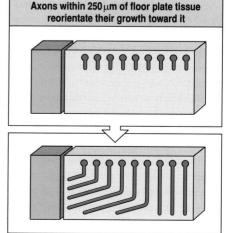

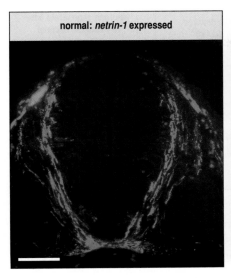
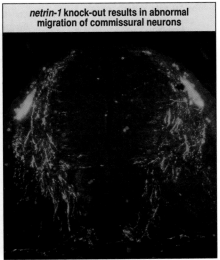

normal: *netrin-1* expressed

netrin-1 knock-out results in abnormal migration of commissural neurons

Fig. 11.32 Effect of *netrin-1* gene knock-out in mice. In mice lacking *netrin-1*, the commissural axons do not migrate toward the floor plate. Scale bar = 0.1 mm.

Photographs courtesy of M. Tessier-Lavigne, from Serafini, T., et al.: 1996.

crossing the midline. *Robo* is expressed on the axons and, as in *Drosophila*, is probably involved in the inhibitory action of slit. Thus, similar mechanisms involving attraction and repulsion, and changes in the level of receptor proteins, are the basis for midline patterning in both the fly and vertebrates.

11.15 Neurons from the retina make ordered connections on the tectum to form a retino-tectal map

The vertebrate visual system has been the subject of intense investigation for many years, and the highly organized projection of neurons from the eye to the brain is one of the best models we have to show how specific neural connections are made. A characteristic feature of the vertebrate brain is the presence of topographic maps. That is, the neurons from one region of the nervous system project in an ordered manner to one region of the brain, so that nearest-neighbor relations are maintained. The most studied topographical projection is that of the optic nerve from the retina into the brain.

The retina develops light-sensitive cells that indirectly activate neurons—the retinal ganglion cells—whose axons are bundled together to form the optic nerve, which connects the retina to a region of the brain called the **optic tectum** in amphibians and birds, and the **lateral geniculate nucleus** and superior colliculus in mammals. In amphibians, the optic nerve from the right eye makes connections with the left optic tectum, while the nerve from the left eye makes connections to the right side of the brain. The optic nerve from each eye is made up of thousands of axons, which connect with the optic tectum in a highly ordered manner: there is a point-to-point correspondence between a position on the retina and one on the tectum. Neurons in the dorsal region of the retina project to the ventral region of the tectum and regions in the anterior (nasal) region of the retina project to the posterior region of the tectum (Fig. 11.33). The development of the projection from the retina to the tectum is initially only reasonably precise, but is fine-tuned later by nerve impulses from the retina.

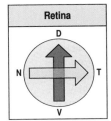
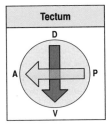

Retina

Tectum

Fig. 11.33 The retina maps onto the tectum. Dorsal (D) retinal neurons connect to ventral (V) tectum and temporal (T) (or posterior) retinal neurons to anterior (A) tectum. N, nasal (anterior) region of the retina.

Fig. 11.34 Retino-tectal connections in amphibians are re-established in the original arrangement after severance of the optic nerve and rotation of the eye. Left panel: neurons in the optic nerve from the left eye mainly connect to right optic tectum and those from the right eye to the left optic tectum. There is a point-to-point correspondence between neurons from different regions in the retina (nasal (N), temporal (T), dorsal (D), ventral (V)) and their connections in the tectum (posterior (p), anterior (a), ventral (v), dorsal (d), respectively). Center and right panels: if one optic nerve of a frog is cut and the eye rotated dorso-ventrally through 180°, the severed ends of the axons degenerate. When the neurons regenerate, they make connections with their original sites of contact in the tectum.

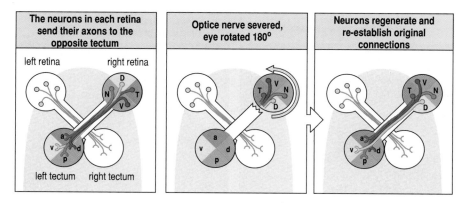

Remarkably, in some lower vertebrates, such as fish and amphibians, the pattern of connections can be re-established with precision when the optic nerve is cut. The ends of the axons distal to the cut die, new growth cones form, and axon outgrowth reforms connections to the tectum. In frogs, even if the eye is inverted through 180° the axons still find their way back to their original sites of contact (Fig. 11.34). However, the animals subsequently behave as if their visual world has been turned upside down: if a visual stimulus, such as a fly is presented above the inverted eye, the frog moves its head downward instead of upward (Fig. 11.35), and can never learn to correct this error.

From such experiments it was suggested that each retinal neuron carries a chemical label that enables it to connect reliably with an appropriately chemically labeled cell in the tectum. This is known as the **chemoaffinity hypothesis** of connectivity. The chemical labels probably do not provide unique lock-and-key interactions. Rather, it is thought that graded spatial distributions of a relatively small number of factors on the tectum provide positional information, which can be detected by the retinal axons. The spatially graded expression of another set of factors on the retinal axons would provide them with their own positional information. The development of the retino-tectal projection could thus, in principle, result from the interaction between these two gradients.

Fig. 11.35 The effect of eye rotation on the visual behavior of a frog. After severance of the optic nerve and rotation of the eye, the retina re-establishes its original connections with the optic tectum (see Fig. 11.34). However, because the eye has been rotated, the image falling on the tectum is upside down compared with normal. When the frog sees a fly above its head, it thinks the fly is below it, and moves its head downward to try to catch it.

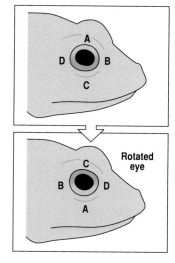

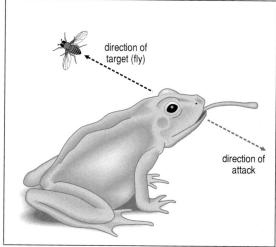

Such gradients have indeed been found in the developing visual system of the chick embryo. An axon-guiding activity based on repulsion has been detected along the antero-posterior axis of the chick tectum. Normally, the temporal (posterior) half of the retina projects to the anterior part of the tectum and the nasal (anterior) retina projects to the posterior tectum. When offered a choice between growing on posterior or anterior tectal cells, temporal axons from an explanted chick retina show a preference for anterior tectal cells (Fig. 11.36). This choice is mediated by repulsion of the axons—as shown by the collapse of their growth cones—by a factor located on the surface of posterior tectal cells.

Ephrins and their receptors (see Section 4.10) may be candidates for the gradients that mediate specificity as they are expressed in reciprocal gradients in retina and tectum. In the chick, the receptor EphA3 is expressed on retinal neurons in a decreasing gradient from temporal to nasal, while ephrin A5 and ephrin A2 are both expressed in the tectum in a gradient that decreases from posterior to anterior. Retinal axons bearing low levels of the receptor move into areas containing high levels of the ligands, whereas retinal axons bearing high levels of receptor stop before they reach areas with high levels of the ligand (Fig. 11.37). The best explanation for this is that binding of ligand to the receptor sends a repulsive signal, and the axon ceases to migrate when this signal reaches a threshold value. The strength of the signal received by the migrating axon will be proportional to the product of the concentrations of receptor and ligand. Thus, the threshold will be reached when there is a high level of both receptor and ligand, but not when there is a low level of receptor, even when ligand level is high. The situation is, however, complicated by the presence of ephrins A2 and A5 in the retina itself. In genetic knock-outs of both ephrins, the mapping of the retina to the superior colliculus is almost, but not completely, abolished.

Summary

Growth cones at the tip of the extending axon guide it to its destination. Filopodial activity at the growth cone is influenced by environmental factors, such as contact with the substratum and with other cells, and can also be guided by chemotaxis. Guidance involves both attraction and repulsion. In the development of motor neurons that innervate vertebrate limb muscles, the growth cones guide the axons so as to make the correct muscle-specific connections, even when their normal site of entry into the limb is disturbed. LIM proteins give motor neurons an identity that determines their pathway. The sensory neurons in the limb of the grasshopper are guided both by the substratum and by special guidepost cells along the migration pathway. Attraction and repulsion control axons crossing the midline in both *Drosophila* and vertebrates. Gradients in diffusible molecules are probably responsible for the directional growth of commissural axons in the spinal cord. Neuronal guidance of axons from the amphibian retina to make the correct connections with the optic tectum involves gradients in cell-surface molecules both on the tectal neurons and on the retinal axons that can promote or repel growth cone approach.

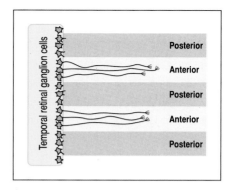

Fig. 11.36 Choice of targets by retinal axons. If pieces of temporal retina are placed next to a 'carpet' of alternating stripes (90 μm wide) of anterior and posterior tectal membranes, the temporal retinal cells only extend axons onto anterior tectal membranes.

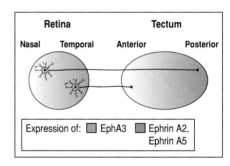

Fig. 11.37 Complementary expression of the ephrins and their receptors in the chick embryo retino-tectal projection. The receptor EphA3 is a receptor tyrosine kinase and ephrins A2 and A5 are its ligands. Temporal retinal neurons, which express high levels of EphA3 on their surface, are repulsed by the posterior tectum, where high levels of ephrins A2 and A5 are also present, but can make connections with the anterior tectum, where ephrins A2 and A5 are scarce or absent. Nasal neurons from the retina make the best contacts in the posterior tectum, as they express low levels of EphA3 and thus are not repulsed by the high levels of ephrins in this area.

Summary: axonal guidance

axonal outgrowth led by growth cone

⇩

attractive and repulsive cues are both short range and long range

motor neurons take
distinct pathways in limb

attraction and repulsion
control midline crossing
in *Drosophila*

retinal axons connect with
tectal neurons to make
retino-tectal map

⇩

connections guided by
gradients; EphA3 on tectum
and ephrins on retinal cells

Neuronal survival, synapse formation and refinement

When axons reach their targets they make specialized connections—synapses—which are essential for signaling between neurons and their target cells. Formation of synapses in the correct pattern is a basic requirement of any developing nervous system. The connections may be made with other nerve cells, with muscles, and also with certain glandular tissues. Here, we focus mainly on the development and stabilization of synapses at the junctions between nerve cells and muscle cells in vertebrates—the **neuromuscular junctions**. There are very many neuronal cell types—hundreds if not thousands—and how they match up to make synapses reliably is a central problem. The molecular mechanisms are only beginning to be known. One class of molecules that might be involved are those of the cadherin family, which could provide a combinatorial code. It is worth noting, however, that pre-synaptic and post-synaptic cells can exchange signals that stimulate and coordinate their mutual differentiation. In the developing cerebellum, for example, granule cells remodel the growth cones of the mossy fibers with which they form synapses; Wnt-7a is a key signal here. Another example is provided by the mouse cerebral cortex, where a variety of Notch-type receptors are expressed and are involved in the extension of both axons and dendrites; activation of Notch by neighboring neurons restricts axon outgrowth and helps to stabilize the system.

Setting up the organization of a complex nervous system in vertebrates involves refining an initially rather imprecise organization by extensive programmed cell death (see Section 9.15). The establishment of a connection between a neuron and its target appears to be essential not only for the functioning of the nervous system but for the very survival of many neurons. Neuronal death is very common in the developing vertebrate nervous system; too many neurons are produced initially, and only those that make appropriate connections survive. Survival depends on the neuron receiving neurotrophic factors, such as nerve growth factor, which are produced by the target tissue and for which neurons compete.

A special feature of nervous system development is that fine-tuning of synaptic connections depends on the interaction of the organism with its environment and the consequent neuronal activity. This is particularly true of the vertebrate visual system, where sensory input from the retina in a period immediately after birth modifies synaptic connections so that the animal can perceive fine detail. Again, this refinement seems to involve competition for neurotrophic factors. We return to this topic after first considering the survival of motor neurons that innervate the developing vertebrate limb.

11.16 Many motor neurons die during limb innervation

Some 20,000 motor neurons are formed in the segment of spinal cord that provides innervation to a chick leg, but about half of them die soon after they are formed (Fig. 11.38). Cell death occurs after the axons have grown out from the cell bodies and entered the limb, at about the time that the axon terminals are reaching their potential targets—the skeletal muscles of the limb. The role of the target muscles in preventing cell death is suggested by two experiments. If the leg bud is removed, the number of motor neurons surviving sharply decreases. When an additional limb bud is grafted at the same level as the leg, providing additional targets for the axons, the number of surviving motor neurons increases.

Survival of a motor neuron may depend on its establishing a functional synapse with a muscle cell. Once a neuromuscular junction is established, the neuron can activate the muscle, and this is followed by the death of a proportion of the other motor neurons that are approaching the muscle cell. Muscle activation by neurons can be blocked by the drug curare, which prevents neuromuscular transmission, and this blocking results in a large increase in the number of motor neurons that survive. A possible explanation is that in the absence of activation, the muscle produces a trophic factor in amounts sufficient to enable many neurons to survive. Once a neuromuscular junction is established and the muscle is activated, production of the trophic factor may be reduced. Neurons that have already established a connection to the muscle cell will survive, while neurons that have not yet reached the cell will die.

Even after neuromuscular connections have been made, some are subsequently eliminated. At early stages of development, single muscle fibers are innervated by axon terminals from several different motor neurons. With time, most of these connections are eliminated, until each muscle fiber is innervated by the axon terminals from just one motor neuron. The continued survival of a neuron that has established a connection must therefore be ensured by additional means. It is now clear that this later elimination process is also dependent on neural activity.

The well-established matching of the number of motor neurons to the number of appropriate targets by these mechanisms suggests that a general mechanism for the development of nervous system connectivity in all parts of the vertebrate nervous system is that excess neurons are generated and only those that make the required connections are selected for survival. This mechanism is well suited to regulating cell numbers by matching the size of the neuronal population to its targets. We now look in more detail at the neurotrophic factors that promote neuronal survival, and then consider the role of neural activity in the elimination of neuromuscular synapses.

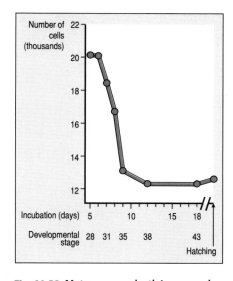

Fig. 11.38 Motor neuron death is a normal part of development in the chick spinal cord. The number of motor neurons innervating a limb decreases by about half before hatching, as a result of programmed cell death during development. Most of the neurons die over a period of 4 days.

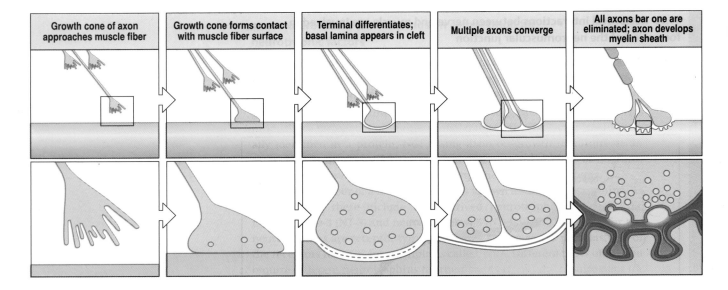

| Growth cone of axon approaches muscle fiber | Growth cone forms contact with muscle fiber surface | Terminal differentiates; basal lamina appears in cleft | Multiple axons converge | All axons bar one are eliminated; axon develops myelin sheath |

Fig. 11.41 Development of the neuromuscular junction. The boxed areas in the top panels are shown at a higher magnification in the bottom panels. From left to right: a growth cone approaches a muscle fiber and forms an unspecialized, but functional, contact on its surface. The terminal then differentiates and basal lamina (red line) appears in the widened synaptic cleft. As the basal lamina appears extrasynaptically, multiple axons converge on the synaptic site. Finally, all axons but one are eliminated. The surviving axon terminal expands and forms a mature neuromuscular junction, with the axon developing a myelin sheath.

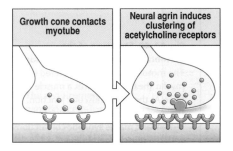

| Growth cone contacts myotube | Neural agrin induces clustering of acetylcholine receptors |

Fig. 11.42 Clustering of acetylcholine receptors during development of the neuromuscular junction. When a growth cone contacts a muscle fiber it releases the protein agrin. This causes acetylcholine receptor clustering in the muscle cell membrane, and localized deposition of basal lamina material.

network-like arrangement. Each branch ends in a swelling, which is in contact with a special endplate region on the muscle fiber. The axon's plasma membrane is separated from the muscle cell's plasma membrane by a narrow cleft (the synaptic cleft) filled with extracellular material (the basal lamina), which is secreted by both the nerve and the muscle cell. The whole structure, comprising the axon terminal plasma membrane, the opposing muscle cell plasma membrane, and the cleft between them, is called a neuromuscular junction or synapse.

Electrical signals cannot pass across the synaptic cleft, and for the neuron to signal to the muscle, the electrical impulse propagated down the axon is converted at the terminal into a chemical signal. This is the release of a neurotransmitter into the synaptic cleft from synaptic vesicles in the axon terminal. Molecules of the neurotransmitter diffuse across the cleft and interact with receptors on the muscle cell membrane, causing the muscle fiber to contract. The neurotransmitter used by motor neurons connecting with skeletal muscles is acetylcholine. Because the signal travels from nerve to muscle, the axon terminal is called the pre-synaptic part of the junction and the muscle cell is the post-synaptic partner (see Fig. 11.40).

Development of a neuromuscular junction is progressive—in the rat it takes about 3 weeks. How does a single axon terminal become selected and develop, and how do acetylcholine receptors become localized in the right place on the post-synaptic muscle cell membrane opposite the terminal? The axon terminals that make contact with the muscle cell are initially unspecialized, but they soon begin to accumulate synaptic vesicles. Initially, several synapses from different axons are made on the same immature muscle cell (or myotube) but, with time, all but one are eliminated. Synaptic transmission begins shortly after a contact is made. The development of a neuromuscular junction is shown in Fig. 11.41. There is some evidence for pre-pattern in the muscle that determines where the synapse will form.

The permanent establishment of a neuromuscular junction relies on an exchange of signals between axon terminal and muscle cell. A key event in establishment is the aggregation of acetylcholine receptors at the post-synaptic membrane. Initially, acetylcholine receptors are present through-

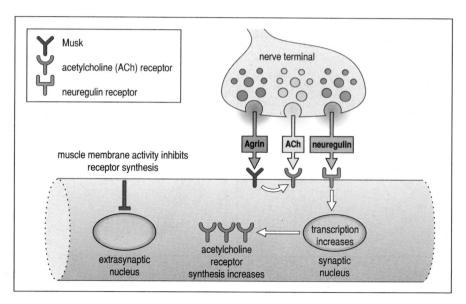

Fig. 11.43 Control of acetylcholine receptor localization and synthesis in a muscle cell membrane by the pre-synaptic axon. Beneath the neuromuscular junction, acetylcholine receptors cluster because of agrin interacting with the receptor protein Musk. Release of neuregulin from the axon stimulates transcription of the genes for the acetylcholine receptor proteins in the nuclei nearest to the junction, and production of acetylcholine receptors. Away from this region, electrical activity in the muscle cell membrane reduces synthesis of the receptor.

out the plasma membrane of the immature muscle fiber, but soon after axon contact they begin to accumulate at the site of contact. A key signal for clustering of the receptors is provided by the protein agrin (Fig. 11.42). Agrin is secreted by motor neurons at the pre-synaptic terminal and induces membrane specialization. It probably acts by activating a tyrosine kinase receptor named Musk on the muscle cell. Knock-out of the genes for either agrin or Musk in mice results in the absence of functional neuromuscular junctions; there is little or no clustering of acetylcholine receptors and consequently no muscle activity.

The innervation of muscle causes not only local clustering of acetyl-choline receptors but also a localized increase in their synthesis. This is due to an increased rate of acetylcholine receptor gene transcription in those muscle cells that lie directly beneath the post-synaptic membrane (Fig. 11.43). Stimulation of acetylcholine receptor gene transcription is thought to be caused by proteins called neuregulins or ARIAs released from the axon terminal. Receptor synthesis elsewhere in the muscle is reduced in response to the enhanced electrical activity in the muscle triggered through the junction.

Neural activity is also crucial in refining the pattern of neuromuscular junctions. Almost all mammalian muscle fibers are innervated by two or more motor axons during the embryonic period. After birth, branches of motor neurons are withdrawn, so that each muscle fiber is finally innerv-ated by a single motor neuron (Fig. 11.44). This change in connectivity is due to competition between synapses. When a motor neuron stimulates a muscle fiber, the activity between other neurons and the muscle fiber is suppressed, and these synapses are eventually eliminated. As we see next, a similar mechanism seems to be involved in refining synaptic connections between neurons.

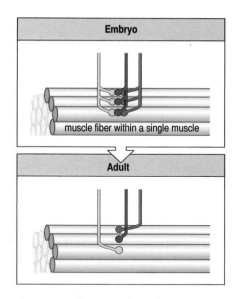

Fig. 11.44 Refinement of muscle innervation by neural activity. Initially, several motor neurons innervate the same muscle fiber. Elimination of synapses means that each fiber is eventually innervated by only one neuron. After Goodman, C.S., Shatz, C.J.: 1993.

11.19 The map from eye to brain is refined by neural activity

We have already considered how axons from the amphibian retina make connections with the tectum so that a retino-tectal map is established. This map is initially rather coarse-grained, in that axons from neighboring cells

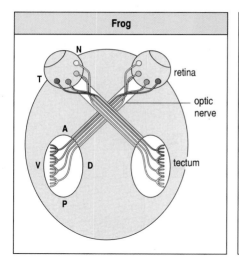

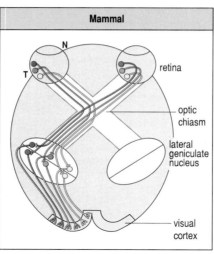

Fig. 11.45 Comparison of amphibian and mammalian visual systems. Left panel: in the frog, neurons from the retina project directly to the optic tectum, neurons from the left retina connecting to the right optic tectum and those from the right retina connecting to the left tectum. Right panel: in mammals with good binocular (stereo) vision, the retinal neurons project to the lateral geniculate nucleus (LGN), from which neurons project to the visual cortex. Neurons from one half of the retina project to the LGN on the same side as the eye, and those from the other half cross over to the opposite LGN at the optic chiasm. For the sake of clarity, the neurons from only one half of the retina are shown. After Goodman, C.S., Shatz, C.J.: 1993.

in the retina make contacts over a large area of the tectum. This area is much larger early on than at later stages of development, when the retino-tectal map is more finely tuned. Fine tuning of the map results, as in muscle, from the withdrawal of axon terminals from most of the initial contacts, and requires neural activity. This requirement is seen particularly clearly in the development of visual connections in mammals.

The mammalian visual system is more complex than that of lower vertebrates (Fig. 11.45). Axons from the retina first connect to the lateral geniculate nucleus, onto which they map in an ordered manner. Input from one half of each eye goes to the opposite side of the brain, whereas input from the other half of each eye goes to the same side of the brain. Neurons from the lateral geniculate nucleus then send axons to the visual cortex. When there is a visual stimulus, the inputs from the retinal axons activate neurons in the lateral geniculate nucleus, which then activate neurons in the corresponding region of the visual cortex. There is thus input from both eyes at the same location in the cortex. The adult visual cortex consists of six cell layers, but we need only focus here on layer 4, which is where many axons from the lateral geniculate make connections.

In the lateral geniculate nucleus, as in the visual cortex, the neurons are arranged in layers. Each layer receives input from retinal axons from either the right or the left eye, but not from both. Thus, from the beginning, inputs from left and right eyes are separated. However, layers in the lateral geniculate nucleus that receive left or right inputs both make connections with layer 4 of the visual cortex. At birth, these inputs overlap and are mixed, but with time the inputs from left and right eyes become separated into blocks of cortical cells about 0.5 mm wide, which are known as **ocular dominance columns** (Fig. 11.46). Adjacent columns respond to the same stimulus in the visual field, one column responding to signals from the left eye, and the next to signals from the right eye. This arrangement is essential for good binocular vision. The columns can be detected and mapped by making electrophysiological recordings. They can also be directly observed by injecting a tracer, such as radioactive proline, into one eye. The tracer is taken up by retinal neurons, transported by the optic nerve to the lateral geniculate, and from there to the visual cortex, where its pattern can be detected by autoradiography. This reveals a striking array of stripes representing the input from one eye (see Fig. 11.46).

Neural activity and visual input are essential for the development and maintenance of the ocular dominance columns. While sensory input is important, spontaneous activity plays a key role. Initial formation of the stripes in non-human primates occurs before visual experience and thus involves spontaneously generated waves of action potentials in the mammalian retina. These waves may act through the release of neurotrophins that can remodel synaptic connections. If neural activity is blocked during development by the injection of tetrodotoxin, ocular dominance columns do not develop, and the inputs from the two eyes into the visual cortex remain mixed. If the input from one eye is blocked, the territory in the visual cortex occupied by the other eye's input expands at the expense of the blocked eye.

The favored explanation for the formation of ocular dominance columns is based on competition between incoming neurons (Fig. 11.47). Because of the initial connections that establish the map, individual cortical neurons can initially receive input from both eyes. Within a particular region, there will therefore be overlap of stimuli originating from the two eyes, and this overlap has to be resolved. Neighboring cells carrying input from the same eye tend to fire simultaneously in response to a visual stimulus; if they both innervate the same target cell, they can thus cooperate to excite it. As in muscle, stimulation of electrical activity in the target cell tends to strengthen the active synapses and suppress those that are not active at the time—cells that fire together, wire together. As there is competition between neurons for targets, this could generate discrete regions of cortical cells that respond only to one eye or the other, and so form the ocular dominance columns. Such a mechanism explains why experimental exposure of animals to continuous strobe lighting after birth, which causes simultaneous firing of neurons in both eyes, prevents the formation of ocular dominance columns.

A possible mechanism for refining connections in response to neuronal activity involves local release of neurotrophins. A certain level of activation, or activation by two axons simultaneously, may induce the release of neurotrophins from the target cells, and only those axons that have been recently active may be able to respond to them. Neural activity is also important in the development of the visual nervous system in vertebrates other than mammals. When neural activity in developing chick or amphibian embryos is blocked with drugs that prevent neuronal firing, no fine-grained retino-tectal map develops.

11.20 Synaptic input can influence the properties of both the post-synaptic cell and the pre-synaptic neuron

The connections made between neurons and muscles in the developing nervous system are remarkably specific. We have already seen this in relation to muscle innervation in the limb and the retino-tectal system. The ability of one synapse to influence the development of its partner is best illustrated with respect to muscle. Mammalian muscle fibers can be divided into fast-twitch and slow-twitch categories which are determined by the

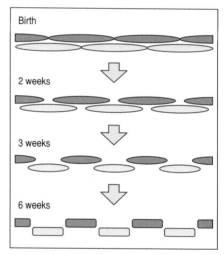

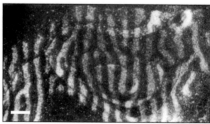

Fig. 11.46 Visualization of ocular dominance columns in the cortex. A radioactive tracer is injected into one eye, from where it is transported to the cortex through the neurons. Tracer injected at birth is broadly distributed in the cortex. Tracer injected at later times becomes confined to alternating columns of cortical cells, representing the ocular dominance columns for that eye, as seen in the photograph. Scale bar = 1 mm. After Kandel, E.R., *et al.*: 1995.

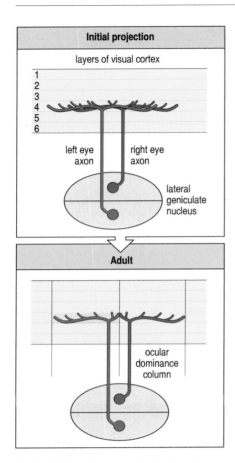

Initial projection

layers of visual cortex

1
2
3
4
5
6

left eye axon   right eye axon

lateral geniculate nucleus

Adult

ocular dominance column

Fig. 11.47 Development of ocular dominance columns. Initially, neurons from the lateral geniculate nucleus, representing projections from both eyes and stimulated by the same visual stimulus, project to the same region of the visual cortex. (Only projection to layer 4 is shown here.) With visual stimulation, the neuronal connections separate out into columns, each representing innervation from only one eye. If stimulation by vision is blocked, ocular dominance columns do not form. After Goodman, C.S., Shatz, C.J.: 1993.

nature of their contractile proteins. Most muscles contain both fast and slow fibers but a few have all slow or all fast fibers. A motor neuron exclusively innervates either slow or fast fibers, even when both are present next to each other in the same muscle.

The neurons do not simply recognize muscle fibers of the appropriate type, but can induce the fiber they innervate to be either slow or fast. Slow and fast muscles can be experimentally cross-innervated before the functional fiber state is fully established so that a nerve that normally innervates a predominantly slow muscle now innervates a muscle that would normally be fast, and vice versa. Under these conditions, the properties of the muscle are partly transformed to correspond to the nerve that innervated it; for example, a normally slow muscle now has some characteristics and molecular characteristics of a fast muscle.

The properties of the presynaptic neuron can also be influenced by the cell with which it makes a synapse. Most neurons of the sympathetic nervous system use noradrenaline (norepinephrine) as a transmitter, but some, such as those that innervate the sweat glands of the footpad, use acetylcholine. This feature is specified by the sweat glands themselves: if contact is not made, the neurons remain adrenergic. This general phenomenon is not unusual, and for many neurons in the brain, specification of the neurotransmitter used is part of the developmental program.

11.21 The ability of mature vertebrate axons to regenerate is restricted to peripheral nerves

In spite of the dynamic nature of the maturing nervous system, and its ability to alter connectivity, the capacity of the vertebrate central nervous system to replace lost connections after neurons have been destroyed is rather small. By contrast, the peripheral nervous system has considerable regenerative capability, even in adults. We have already seen that the optic nerve of lower vertebrates can regenerate. Even so, regeneration involves the regrowth of axons, not the replacement of neurons. If the cell body itself is destroyed, the neuron is not replaced.

Some axons of peripheral neurons in adult vertebrates, such as the motor and sensory axons between the spinal cord and the ends of the limbs, can be hundreds of centimeters long. When such axons are cut they can regenerate. A new growth cone forms at the cut surface and grows down the pathway of the original nerve trunk to make functional connections, leading to an almost complete recovery. In the case of motor neurons, the axon terminal finds the site of the original synapse on the muscle cell by recognizing the basal lamina that fills the synaptic cleft.

The adult central nervous system is not able to regenerate correct axonal and dendritic connections. This is not due to an intrinsic deficit in the neurons but rather is due to the damaged environment, which does not support, or even prevents, regeneration. Although the remaining cell body produces outgrowths, these do not develop further. It seems that the failure of the axons to regenerate is partly the result of an inhibitory effect exerted by central nervous system glial cells, associated with the myelin of nerve fibers. Culturing immature neurons with oligodendrocytes (the source of myelin in the central nervous system) can induce growth cone collapse. In contrast, Schwann cells, which are responsible for myelination in the peripheral nervous system, can promote axon outgrowth, and when Schwann cells are implanted into the central nervous system, axons migrate along tracts of them.

Stem cells have great promise as sources for introducing new neurons and glia into damaged regions of the central nervous system. Embryonic stem cells (Section 9.13), for example, can differentiate into neurons when transplanted into the brain. While such experiments are encouraging, it still needs to be shown that this approach can restore normal function after damage.

Summary

Many neurons in the developing nervous system die. About half of the motor neurons that initially innervate the vertebrate limb undergo cell death; those that survive do so because they make functional connections with muscles. Many neurons depend on neurotrophins, such as nerve growth factor, for their survival, with different classes of neurons requiring different neurotrophins. The formation of a neuromuscular junction involves changes in both pre-synaptic (neuronal) and post-synaptic (muscle cell) membranes once contact is made. The clustering of acetylcholine receptors at the synapse is due to their selective accumulation and localized synthesis in response to signal proteins from the axon terminal. Initially, most mammalian muscle fibers are innervated by two or more motor axons, but neural activity results in competition between synapses so that fibers are eventually innervated by only one motor neuron. Neural activity has a major role in refining the connections between the eye and the brain. In mammals, input from the left and right eyes is required for the development of ocular dominance columns in the visual cortex. These are adjacent columns of cells responding to the same stimulus from left and right eyes, respectively. Formation of the columns, which are essential for binocular vision, is a result of competition for cortical targets by axons carrying visual input from different eyes. Some axons in the peripheral nervous system can regrow when cut, and even make functional synapses, but the central nervous system does not have this ability, probably as a result of local inhibitory factors.

Summary: synapse formation and refinement

axon of motor neuron releases agrin at junction with muscle
⇩
acetylcholine receptors cluster at junction and local synthesis of receptors occurs
⇩
electrical activity of muscle reduces receptor synthesis elsewhere

neurotrophins are required for neuronal survival— about 50% of limb motor neurons die

retino-tectal map refined by neuronal activity, leading to ocular dominance columns

SUMMARY TO CHAPTER 11

The processes involved in the development of the nervous system, with its numerous connections, are similar to those found in other developmental systems. Presumptive neural tissue is specified early in development—during gastrulation in vertebrates, and when the dorso-ventral axis is patterned in *Drosophila*. Within the neurectodermal tissue, the specification of cells that give rise to neural cells involves lateral inhibition. The further development of neurons from neuronal precursors involves asymmetric cell divisions and cell–cell signaling. The patterning of different types of neurons within the vertebrate spinal cord is due to both ventral and dorsal signals. As they develop, neurons extend axons and dendrites. The axons are guided to their destination by growth cones at their tips. Guidance is due to the growth cone's response to attractive and repulsive signals, which may be diffusible or bound to the substratum. Gradients in such molecules can guide the axons to their destination, as in the retino-tectal system of vertebrates. The functioning of the nervous system depends on the establishment of specific synapses between axons and their targets. Specificity appears to be achieved by an initial overproduction of neurons that compete for targets, with many neurons dying during development. Refinement of synaptic connections involves further competition. Neural activity plays a major role in refining connections, such as those between the eye and the brain. The peripheral nervous system has a much greater capacity for regeneration than the central nervous system.

GENERAL REFERENCES

Kandel, E.R., Schwartz, J.H., Jessell, T.H.: *Principles of Neural Science* (4th edn). New York: McGraw-Hill, 2000.

11.1 Neurons in *Drosophila* arise from proneural clusters

Gómez-Skarmeta, J.L., Rodriguez, I., Martinez, C., Culí, J., Ferrés-Marco, D., Beamonte, D., Modolell, J.: **Cis-regulation of *achaete* and *scute*: shared enhancer-like elements drive their coexpression in proneural clusters of imaginal discs.** *Genes Dev.* 1995, **9**: 1809–1882.

Simpson, P.: ***Drosophila* development: a prepattern for sensory organs.** *Curr. Biol.* 1996, **6**: 948–950.

Skeath, J.B.: **At the nexus between pattern formation and cell-type specification: the generation of individual neuroblast fates in the *Drosophila* embryonic central nervous system.** *BioEssays* 1999, **21**: 922–931.

Udolph, G., Lüer, K., Bossing, T., Technau, G.M.: **Commitment of CNS progenitor along the dorso-ventral axes of *Drosophila* neurectoderm.** *Science* 1995, **269**: 1278–1281.

11.2 Lateral inhibition allocates neural precursors

Artavanis-Tsakonas, S., Matsuno, K., Fortini, M.E.: **Notch signaling.** *Science* 1995, **268**: 225–232.

Lawrence, P.A.: *The Making of a Fly*. Oxford: Blackwell Scientific Publications, 1992.

11.3 Asymmetric cell divisions are involved in *Drosophila* sensory organ development

Guo, M., Jan, L.Y., Jan, Y.N.: **Control of daughter cell fates during asymmetric division: interaction of Numb and Notch.** *Neuron* 1996, **17**: 27–41.

Jan, Y-N., Jan, L.Y.: **Polarity in cell division: what frames thy fearful asymmetry?** *Cell* 2000, **100**: 599–602.

Jarman, A.P., Grau, Y., Jan, L.Y., Jan, Y.N.: **atonal is a proneural gene that directs chordotonal organ formation in the *Drosophila* peripheral nervous system.** *Cell* 1993, **73**: 1307–1321.

Knoblich, J.A.: **Asymmetric cell division during animal development.** *Nat. Rev. Mol. Cell. Biol.* 2001, **2**: 11–20.

11.5 Specification of vertebrate neuronal precursors involves lateral inhibition

Chitins, A., Henrique, D., Lewis, J., Ish-Horowitcz, D., Kintner, C.: **Primary neurogenesis in *Xenopus* embryos regulated by a homologue of the *Drosophila* neurogenic gene *Delta*.** *Nature* 1995, **375**: 761–766.

Ma, Q., Kintner, C., Anderson, D.J.: **Identification of *neurogenin*, a vertebrate neuronal determination gene.** *Cell* 1996, **87**: 43–52.

11.6 The pattern of differentiation of cells along the dorso-ventral axis of the spinal cord depends on ventral and dorsal signals

Briscoe, J., Pierani, A., Jessell, T.M., Ericson, J.: **A homeodomain protein code specifies progenitor cell identity and neuronal fate in the ventral neural tube.** *Cell* 2000, **102**: 161–173.

Dodd, J., Jessell, T.M., Placzek, M.: **Coordinate roles for LIM homeobox genes in directing the dorsoventral trajectory of motor axons in the vertebrate limb.** *Cell* 2000, **102**: 161–173.

Ensini M., Tsuchida T.N., Belting, H-G., Jessell, T.M.: **The control of rostrocaudal pattern in the developing spinal cord:**

specification of motor neuron subtype identity is initiated by signals from paraxial mesoderm. *Development* 1998, **125**: 969–982.

Jacob, J., Hacker, A., Guthrie, S.: **Mechanisms and molecules in motor neuron specification and axon pathfinding.** *BioEssays* 2001, **23**: 582–595.

Jessell, T.M.: **Neuronal specification in the spinal cord: inductive signals and transcriptional controls.** *Nat. Rev. Genet.* 2000, **1**: 20–29.

Kania A., Johnson, R.L., Jessell, T.M.: **Coordinate roles for LIM homeobox genes in directing the dorsoventral trajectory of motor axons in the vertebrate limb.** *Cell* 2000, **102**: 161–173.

Roelink, H., Porter, J.A., Chian, C., Tanabe, Y., Chang, D.T., Beachy, P.A., Jessell, T.M.: **Floor plate and motor neuron induction by different concentrations of the amino terminal cleavage product of Sonic hedgehog autoproteolysis.** *Cell* 1995, **81**: 445–455.

Tanabe, Y., Jessell, T.M.: **Diversity and pattern in the developing spinal cord.** *Science* 1996, **274**: 1115–1123.

11.7 Neurons in the mammalian central nervous system arise from asymmetric cell divisions, then migrate away from the proliferative zone

Chenn, A., McConnell, S.K.: **Cleavage orientation and the asymmetric inheritance of Notch-1 immunoreactivity in mammalian neurogenesis.** *Cell* 1995, **82**: 631–641.

D'Arcangelo, G., Curran, T.: *Reeler:* **new tales on an old mutant mouse.** *BioEssays* 1998, **20**: 235–244.

Gage, F.H.: **Mammalian neural stem cells.** *Science* 2000, **287**: 1433–1438.

Kim, H., Schagat, T.: **Neuroblasts: a model for the asymmetric division of cells.** *Trends Genet.* 1996, **13**: 33–39.

McConnell, S.K., Kaznowski, C.E., O'Rourke, N.A., Dailey, M.E., Roberts, J.S.C.: **Neurogenesis, determination and migration during cerebral cortical development.** In *Molecular Basis of Morphogenesis.* Edited by Bernfield, N. New York: Wiley-Liss, 1993: 135–154.

The insect compound eye

Bonini, N.M., Choi, K.W.: **Early decisions in *Drosophila* eye morphogenesis.** *Curr. Opin. Genet. Dev.* 1995, **5**: 507–515.

Strutt, H., Strutt, D.: **Polarity determination in the *Drosophila* eye.** *Curr. Opin. Genet. Dev.* 1999, **9**: 442–446.

11.8 Signals generated in the *Drosophila* eye disc maintain progress of the morphogenetic furrow and the ommatidia are spaced by lateral inhibition

Baonza, A., Casci, T., Freeman, M.: **A primary role for the epidermal growth factor receptor in ommatidial spacing in the *Drosophila* eye.** *Curr. Biol.* 2001, **11**: 396–404.

Greenwood, S., Struhl, G.: **Progression of the morphogenetic furrow in the *Drosophila* eye: the roles of Hedgehog, Decapentaplegic and the Raf pathway.** *Development* 1999, **126**: 5795–5808.

Tomlinson, A., Struhl, G.: **Delta/Notch and Boss/Sevenless signals act combinatorially to specify the *Drosophila* R7 photoreceptor.** *Mol. Cell* 2001, **7**: 487–495.

11.9 The patterning of the cells in the ommatidium depends on intercellular interactions

Domínguez, M., Hafen, E.: **Genetic dissection of cell fate specification in the developing eye of *Drosophila.*** *Cell Dev. Biol.* 1996, **7**: 219–226.

Freeman, M.: **Cell determination strategies in the *Drosophila* eye.** *Development* 1997, **124**: 261–270.

Krämer, H., Cagan, R.L.: **Determination of photoreceptor cell fate in the *Drosophila* retina.** *Curr. Opin. Neurobiol.* 1994, **4**: 14–20.

11.10 Activation of the gene *eyeless* can initiate eye development

Halder, G., Callaerts, P., Gehring, W.J.: **Induction of ectopic eyes by targeted expression of the eyeless gene in *Drosophila.*** *Science* 1995, **267**: 1788–1792.

Gehring, W.J., Kazuho, I.: **Mastering eye morphogenesis and eye evolution.** *Trends Genet.* 1999, **15**: 371–377.

11.11 The growth cone controls the path taken by the growing axon

Tear, G.: **Neuronal guidance: a genetic perspective.** *Trends Genet.* 1999, **15**: 113–118.

Tessier-Lavigne, M., Goodman, C.S.: **The molecular biology of axon guidance.** *Science* 1996, **274**: 1123–1133.

11.12 Motor neurons from the spinal cord make muscle-specific connections

Lance-Jones, C., Landmesser, L.: **Pathway selection by embryonic chick motoneurons in an experimentally altered environment.** *Proc. R. Soc. Lond.* 1981, **214**: 19–52.

Tosney, K.W., Hotary, K.B., Lance-Jones, C.: **Specifying the target identity of motoneurons.** *BioEssays* 1995, **17**: 379–382.

11.13 Choice of axon pathway depends on environmental cues and neuronal identity

Bentley, D., O'Connor, T.P.: **Guidance and steering of peripheral pioneer growth cones in grasshopper embryos.** In *The Nerve Growth Cone.* Edited by Letourneau, P.C., Kater, S.K., Machgno, E.R. New York: Raven, 1992: 265–282.

Polleux, F., Morrow, T., Ghosh, A.: **Semaphorin 3A is a chemoattractant for cortical apical dendrites.** *Nature* 2000, **404**: 567–573.

Van Vactor, D., Lorenz, L.J.: **Neural development: the semantics of axon guidance.** *Curr. Biol.* 1999, **9**: R201–R204.

11.14 Axons crossing the midline are both attracted and repelled

Giger, R.J., Kolodkin, A.L.: **Silencing the siren: guidance cue hierarchies at the CNS midline.** *Cell* 2001, **105**: 1–4

Simpson, J.H., Bland, K.S., Fetter, R.D., Goodman, C.S.: **Short-range and long-range guidance by Slit and its Robo receptors: a combinatorial code of Robo receptors controls lateral position.** *Cell* 2000, **103**: 1019–1032.

11.15 Neurons from the retina make ordered connections on the tectum to form a retino-tectal map

Löschinger, J., Weth, F., Bonhoeffer, F.: **Reading of concentration gradients by axonal growth cones.** *Phil. Trans. R. Soc. Lond. B.* 2000, **355**: 971–982.

Wilkinson, D.G.: **Topographic mapping: organising by repulsion and competition?** *Curr. Biol.* 2000, **10**: R447–R451.

11.16 Many motor neurons die during limb innervation

Oppenheim, R.W.: **Cell death during development of the nervous system.** *Annu. Rev. Neurosci.* 1991, **14**: 453–501.

11.17 Neuronal survival depends on competition for neurotrophic factors

Birling, M.C., Price, J.: **Influence of growth factors on neuronal differentiation.** *Curr. Opin. Cell Biol.* 1995, **7**: 878–884.

Burden, S.J.: **Wnts as retrograde signals for axon and growth cone differentiation.** *Cell* 2000, **100**: 495–497.

Davies, A.M.: **Neurotrophic factors. Switching neurotrophin dependence.** *Curr. Biol.* 1994, **4**: 273–276.

Henderson, C.E.: **Programmed cell death in the developing nervous system.** *Neuron* 1996, **17**: 579–585.

Lindsay, R.M.: **Neuron saving schemes.** *Nature* 1995, **373**: 289–290.

Serafini, T.: **Finding a partner in a crowd: neuronal diversity and synaptogenesis.** *Cell* 1999, **98**: 133–136.

Šestan, N., Artavanis-Tsakonas, S., Rakic, P.: **Contact-dependent inhibition of cortical neurite growth mediated by Notch signaling.** *Science* 1999, **286**: 741–746.

Snider, W.D.: **Functions of the neurotrophins during nervous system development: what the knockouts are teaching us.** *Cell* 1994, **77**: 627–638.

11.18 Reciprocal interactions between nerve and muscle are involved in formation of the neuromuscular junction

Goodman, C.S., Shatz, C.J.: **Developmental mechanisms that generate precise patterns of neuronal connectivity.** *Cell* 1993, **72**: 77–98.

Lin, W., Burgess, R.W., Dominguez, B., Pfaff, S.L., Sanes, J.R., Lee, K-F.: **Distinct roles of nerve and muscle in postsynaptic differentiation of the neuromuscular synapse.** *Nature* 2001, **410**: 1057–1064.

Wallace, B.G.: **Signaling mechanisms mediating synapse formation.** *BioEssays* 1996, **18**: 777–780.

11.19 The map from eye to brain is refined by neural activity

Katz, L.C., Shatz, C.J.: **Synaptic activity and the construction of cortical circuits.** *Science* 1996, **274**: 1133–1138.

11.20 Synaptic input can influence the properties of both the postsynaptic cell and the presynaptic neuron

Buonanno, A., Fields, R.D.: **Gene regulation by patterned electrical activity during neural and skeletal muscle development.** *Curr. Opin. Neurobiol.* 1999, **9**: 110–120.

Francis, N.J., Landis, S.C.: **Cellular and molecular determinants of sympathetic neuron development.** *Annu. Rev. Neurosci.* 1999, **22**: 541–566.

11.21 The ability of mature vertebrate axons to regenerate is restricted to peripheral nerves

Horner, P.J., Gage, F.H.: **Regenerating the damaged central nervous system.** *Nature* 2000, **407**: 963–970.

Terenghi, G.: **Peripheral nerve regeneration and neurotrophic factors.** *J. Anat.* 1999, **194**: 1–14.

Germ cells and sex

- Determination of the sexual phenotype
- The development of germ cells
- Fertilization

"Some of us were singled out at a very early age to found a new colony when we were older."

A great deal of the biology of animals and plants is, not surprisingly, devoted to reproduction and sex. The embryos of all sexually reproducing organisms develop from a single cell, formed by the fusion of a male and a female **gamete** at **fertilization**. The importance of sexual reproduction is that it keeps generating new genotypes by bringing together the genes from two different individuals, male and female, in the fertilized egg. In sexually reproducing organisms, a fundamental distinction is between **germ cells** and somatic cells (see Section 1.2). The former give rise to the gametes—eggs and sperm in animals—whereas somatic, or body, cells make no contribution to the next generation. Thus, a key issue in sexual development is how the germ cells are specified.

In animals, germ cells are usually specified and set aside early in embryonic development. In this chapter we look at sex determination, germ-cell formation, and fertilization, mainly in the mouse, *Drosophila*, and *Caenorhabditis*. It is worth noting that there is great variety in reproduction; for example, some animals can reproduce asexually, such as *Hydra*, which can reproduce by budding, and turtles, whose eggs can develop without being fertilized. Plants, although reproducing sexually, differ from most animals in that their germ cells are not specified early in embryonic development, but during the development of the flowers from the floral meristems.

We begin by considering sex determination. In those animals in which sex is determined genetically, male and female embryos initially look the same, with sexual differences only emerging as a result of the activity of sex-determining genes located on the sex chromosomes. These genes set in motion the processes that result in sexual differences and which involve genes on other chromosomes. Molecular mechanisms of sex determination vary greatly, even between those organisms in which sex is determined genetically. We next consider how germ cells are initially specified in the early embryo and how they differentiate in the gonads. Finally, we look at fertilization and the activation of the egg—the vital step that initiates development.

Determination of the sexual phenotype

In organisms that produce two phenotypically different sexes, sexual development is the result of the modification of a basic developmental program in order that one of the sexes can develop. The early embryo is similar in both males and females, with sexual differences only developing at later stages. In the organisms considered here, somatic sexual phenotype is genetically fixed at fertilization by the chromosomal content of the gametes (the reproductive cells) that fuse to form the fertilized egg. In mammals, for example, sex is determined by the X and Y chromosomes. Males are XY, females are XX.

Even among vertebrates, however, sex is not always determined by which chromosomes are present; in alligators, it is determined by the environmental temperature during incubation of the embryo, and some fish can switch sex as adults in response to environmental conditions. In insects, there is a wide range of different sex-determining mechanisms. Intriguing though these are, we focus here on those organisms in which the genetic and molecular basis of sex determination is best understood— mammals, *Drosophila*, and the nematode—in which sex is determined by chromosomal content, although by quite different mechanisms.

We first consider the determination of the somatic sexual phenotype— the development of the individual as either male or female. We then deal with the determination of the sexual phenotype of the germ cells— whether they become eggs or sperm—and finally consider how the embryo compensates for the difference in chromosomal composition between males and females.

12.1 The primary sex-determining gene in mammals is on the Y chromosome

The genetic sex of a mammal is established at the moment of conception, when the sperm introduces into the egg either an X or a Y chromosome (Fig. 12.1). Eggs contain one X chromosome; if the sperm introduces another X the embryo will be female, if a Y it will be male. The presence of a Y chromosome results in the somatic cells of the embryo's gonads developing into testes rather than into ovaries. The testes secrete Müllerian-inhibiting substance, which suppresses further female development, and the hormone testosterone, which stimulates the development of male reproductive organs. Specification of a gonad as a testis is controlled by a single gene on the Y chromosome, the **sex-determining region of the Y chromosome** (*SRY* in humans and *Sry* in mice), which was formerly known as testis-determining factor.

Evidence that a region on the Y chromosome actively determines maleness first came from two unusual human syndromes: Klinefelter syndrome, in which individuals have two X chromosomes and one Y (XXY), but are still males; and Turner syndrome, in which individuals have just one X chromosome (XO) and are female. Both these types of individual have some abnormalities; those with Klinefelter syndrome are infertile males with

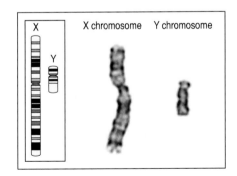

Fig. 12.1 The sex chromosomes in humans. If two X chromosomes (XX) are present, a female develops, whereas the presence of a Y chromosome (XY) leads to development of a male. The inset shows a diagrammatic representation of the banding in the chromosomes, which represent regions of increased chromatin condensation.

Photograph courtesy of Cytogenetic Services.

small testes, whereas females with Turner syndrome do not produce eggs. There are also rare cases of XY individuals who are female, and XX individuals who are phenotypically male. This is due to part of the Y chromosome being lost (in XY females) or to part of the Y chromosome being transferred to the X chromosome (in XX males). This can happen during meiosis in the male germ cells; the X and Y chromosomes pair up, and crossing over can occur between them. Very rarely, this crossing over transfers the *SRY* gene from the Y chromosome onto the X (Fig. 12.2), thus leading to sex reversal.

The sex-determining region alone is sufficient to specify maleness, as shown by an experiment in which the mouse equivalent of the *SRY* gene (*Sry*) was introduced into the eggs of XX mice. These transgenic embryos developed as males, even though they lacked all the other genes on the Y chromosome. The presence of the *Sry* gene, which encodes a transcription factor, resulted in these XX embryos developing testes instead of ovaries. The gene is expressed in the developing gonad just before its differentiation as a testis and activates another transcription factor Sox9. However, these transgenic mice were not completely normal males, as other genes on the Y chromosome are necessary for the development of the sperm. These XX males were therefore infertile.

12.2 Mammalian sexual phenotype is regulated by gonadal hormones

All mammals, whatever their genetic sex, start off as embryos along a sexually neutral developmental pathway. The presence of a Y chromosome causes testes to develop, and the hormones they produce switch the development of all somatic tissues to a characteristic male pathway. In the absence of a Y chromosome, the development of somatic tissues is along the female pathway. Thus, while the sex of the **gonads**—the reproductive organs—is genetically determined, all the other cells in the mammalian body are neutral, irrespective of their chromosomal sex. It does not matter if they are XX or XY, as any future sex-specific development they undergo is controlled by hormones. The primary role of the testis in directing male development was originally demonstrated by removing the prospective gonadal tissue from early rabbit embryos. All the embryos developed as females, irrespective of their chromosomal constitution. To develop as a male, therefore, a testis has to be present. The testis exerts its effect on sexual differentiation of somatic tissues primarily by secreting the hormone testosterone.

The gonads in mammals develop in close association with the **mesonephros**; this is an embryonic kidney that contributes to both the male and female reproductive organs. Associated with the mesonephros on each side of the body are the **Wolffian ducts**, which run down the body to the cloaca, an undifferentiated opening. Another pair of ducts, the **Müllerian ducts**, run parallel to the Wolffian ducts and also open into the cloaca. In early mammalian development, before gonadal differentiation, both sets of ducts are present (Fig. 12.3). In females, in the absence of the testes, the Müllerian ducts develop into the **oviducts** (Fallopian tubes), which transport eggs from the ovaries to the uterus, while the Wolffian ducts degenerate. In males, the expression of *Sry* results in the differentiation of Sertoli cells, which are the somatic cells of the testis and are essential for testis formation and for spermatogenesis, as they retain the germ cells that

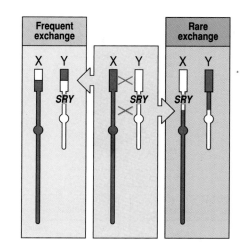

Fig. 12.2 Sex reversal in humans due to chromosomal exchange. At meiosis in male germ cells, the X and Y chromosomes pair up (center panel) and there is crossing over of the distal region (blue cross), which does not affect sexual development (left panel). On rare occasions, crossing over involves a larger segment that includes the *SRY* gene (red cross), so that the X chromosome now carries this male-determining gene (right panel). After Goodfellow, P.N., *et al.*: 1993.

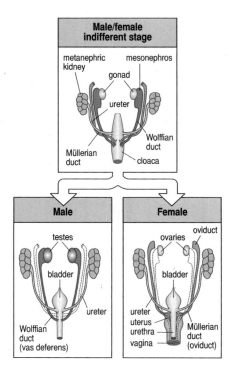

Fig. 12.3 Development of the gonads and related structures in mammals. Top panel: early in development, there is no difference between males and females in the structures that give rise to the gonads and related organs. The future gonads lie adjacent to the mesonephros, which are embryonic kidneys that are not functional in adult mammals. (The true kidney develops from the metanephros, from which the ureter carries urine to the bladder.) Two sets of ducts are present; the Wolffian ducts, which are associated with the mesonephros, and the Müllerian ducts. Both ducts enter the cloaca. Bottom left panel: after testes develop in the male, their secretion of Müllerian-inhibiting substance results in degeneration of the Müllerian duct by programmed cell death, whereas the Wolffian duct becomes the vas deferens, carrying sperm from the testis. Bottom right panel: in females, the Wolffian duct disappears, also by programmed cell death, and the Müllerian duct becomes the oviduct. The uterus forms at the end of the Müllerian ducts. After Higgins, S.J., *et al.*: 1989.

migrate into the gonad. Sertoli cell development involves upregulation of the gene *Sox-9* and the secretion of **Müllerian-inhibiting substance**, which induces regression of the Müllerian duct, largely by apoptosis. The interstitial cell lineage in the testis then differentiates into Leydig cells, which produce testosterone. This causes the Wolffian duct to become the vas deferens, the duct that carries sperm to the penis. There is evidence that the extracellular signaling molecule Wnt-4 represses testosterone production in the undifferentiated gonad, and that one of the functions of the *Sry* gene is to cause the downregulation of *Wnt-4* expression, which occurs when Sertoli cells differentiate. FGF-9 is required for Sertoli cell differentiation and male mice lacking it develop as females.

The main secondary sexual characters that distinguish males and females are the reduced size of mammary glands in males, and the development of a penis and a scrotum in males instead of the clitoris and labia of females (Fig. 12.4). At early stages of embryonic development the genital region of males and females is indistinguishable. Differences only arise after gonad development, as a result of the action of gonadal hormones. For example, in humans, the phallus gives rise to the clitoris in females and the end of the penis in males.

The role of gonadal hormones in sexual development is illustrated by rare cases of abnormal sexual development. Certain XY males develop as phenotypic females in external appearance, even though they have testes and secrete testosterone. They have a mutation that renders them insensitive to testosterone because they lack the testosterone receptor, which is present throughout the body. Conversely, genetic females with a completely normal XX constitution can develop as phenotypic males in external appearance if they are exposed to male hormones during their embryonic development.

Sex-specific behavior is also affected by the hormonal environment because of the effects of hormones on the brain. For example, male rats castrated after birth develop the sexual behavioral characteristics of genetic females.

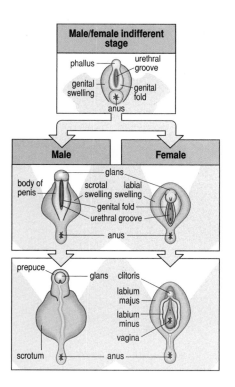

Fig. 12.4 Development of the genitalia in humans. At an early embryonic stage, the genitalia are the same in males and females (top panel). After testis formation in males, the phallus and the genital fold give rise to the penis, whereas in females they give rise to the clitoris and the labia minus. The genital swelling forms the scrotum in males and the labia majus in females.

12.3 The primary sex-determining signal in *Drosophila* is the number of X chromosomes, and is cell autonomous

The external sexual differences between *Drosophila* males and females are mainly in the genital structures, although there are also some differences in bristle patterns and pigmentation, and male flies have a sex comb on the first pair of legs. In flies, sex determination of the somatic cells is cell autonomous, and is therefore specified on a cell-by-cell basis; there is no process resembling the control of somatic cell sexual differentiation by hormones. The pathway of somatic sexual development is the result of a series of gene interactions that are initiated by the primary sex signal, and which act on a binary genetic switch. The end result of this cascade is the expression of just a few effector genes, whose activity controls male or female differentiation of the somatic cells.

Like mammals, fruit flies have two unequally sized sex chromosomes, X and Y, and males are XY and females XX. But these similarities are misleading. In flies, sex is not determined by the presence of a Y chromosome, but by the number of X chromosomes. Thus, XXY flies are female and X flies are male. The chromosomal composition of each somatic cell determines its sexual development. This is beautifully illustrated by the creation of genetic mosaics in which the left side of the animal is XX and the right side X: the two halves develop as female and male, respectively (Fig. 12.5).

In flies, the presence of two X chromosomes results in the production of the protein Sex-lethal, coded for by the gene *Sex-lethal*, which is located on the X chromosome. This leads to female development through the activation of a series of genes that determine the sexual state and then produce the sexual phenotype (Fig. 12.6). The end of the sex-determination pathway is the *transformer* gene, which determines how the mRNA of the *doublesex* gene is spliced; *doublesex* encodes a transcription factor whose activity ultimately produces most aspects of somatic sex. The *doublesex* gene is active in both males and females, but different protein products are produced in the two sexes as a result of sex-specific RNA splicing. Males and females thus express similar but distinct doublesex proteins, which act in somatic cells to induce expression of sex-specific genes, as well as to repress characteristics of the opposite sex. Expression of the *transformer* gene is controlled by *Sex-lethal*. In the presence of Sex-lethal protein, *transformer* RNA is productively spliced and this, together with the transformer-2 protein, leads to the female form of the doublesex protein being made, resulting in female differentiation. Production of the male form of the protein is the default pathway.

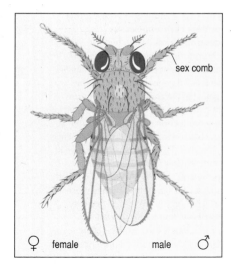

Fig. 12.5 A *Drosophila* female/male genetic mosaic. The left side of the fly is composed of XX cells and develops as a female, whereas the right side is composed of X cells and develops as a male. The male fly has smaller wings, a special structure, the sex comb, on its first pair of legs, and different genitalia at the end of the abdomen (not shown).

Fig. 12.6 Outline of the sex-determination pathway in *Drosophila*. The number of X chromosomes is the primary sex-determining signal, and in females the presence of two X chromosomes activates the gene *Sex-lethal* (*Sxl*). This produces Sex-lethal protein, whereas no Sex-lethal protein is made in males, who have only one X chromosome. The activity of *Sex-lethal* is transduced via the *transformer* gene (*tra*) and causes sex-specific splicing of *doublesex* RNA (*dsx^f*), such that the cells follow a female developmental pathway. In the absence of Sex-lethal protein, the splicing of *doublesex* RNA to give *dsx^m* RNA leads to male development.

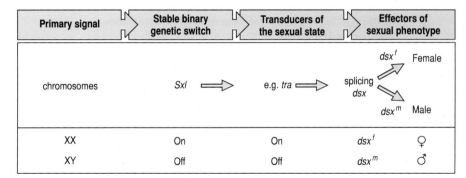

Primary signal	Stable binary genetic switch	Transducers of the sexual state	Effectors of sexual phenotype		
chromosomes	*Sxl* ⟹	e.g. *tra* ⟹	splicing *dsx*	*dsx^f* Female / *dsx^m* Male	
XX	On	On	*dsx^f*	♀	
XY	Off	Off	*dsx^m*	♂	

Fig. 12.7 Production of Sex-lethal protein in *Drosophila* sex determination. When two X chromosomes are present, the early establishment promoter (P$_e$) of the *Sex-lethal* (*Sxl*) gene is activated at the syncytial blastoderm stage in future females, but not in males. This results in the production of Sxl protein. Later, at the blastoderm stage, the maintenance promoter (P$_m$) of *Sxl* becomes active in both females and males, and P$_e$ is turned off. The *Sxl* RNA is only correctly spliced if Sxl protein is already present, which is only in females. A positive feedback loop for Sxl protein production is thus established in females. The continued presence of Sxl protein initiates a cascade of gene activity leading to female development. If no Sxl protein is present, male development ensues. After Cline, T.W.: 1993.

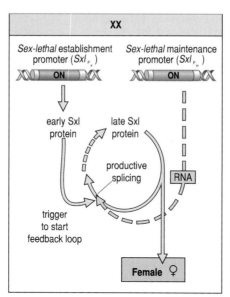
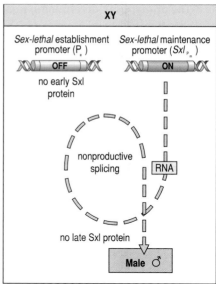

Sex-lethal is only turned on in females with two X chromosomes. In the absence of early *Sex-lethal* expression, male development occurs. Once *Sex-lethal* is activated in females, it remains activated through an autoregulatory mechanism. Early expression of *Sex-lethal* in females occurs through activation of a promoter P$_e$ (e stands for establishment) at about the time of syncytial blastoderm formation. At this stage, Sex-lethal protein is synthesized and accumulates in the blastoderm of female embryos. At the cellular blastoderm stage, another promoter for *Sex-lethal*, P$_m$ (m stands for maintenance), becomes active in both males and females, and P$_e$ is shut off, but the sex is already determined. The *Sex-lethal* RNA transcribed can only be spliced into an mRNA for the Sex-lethal protein if some Sex-lethal protein is already present. Only females contain any Sex-lethal protein, and so only in females is the mRNA productively spliced and more Sex-lethal protein synthesized (Fig. 12.7). This autoregulatory loop at the post-transcriptional level results in Sex-lethal protein being synthesized throughout female development.

How does the number of X chromosomes control these key sex-determining genes? The mechanism in *Drosophila* involves interactions between the products of so-called numerator genes on the X chromosome and of genes on the autosomes, as well as maternally specified factors. In females, the twofold higher level of numerator proteins activates the *Sex-lethal* gene by binding to sites in the P$_e$ promoter, and so overcomes the repression of *Sex-lethal* exerted by genes on the autosomes.

The pathway outlined in Fig. 12.6 is an oversimplification, as *doublesex* does not control all aspects of somatic sexual differentiation in *Drosophila*. There is an additional branch of the pathway downstream of *transformer*, which controls sexually dimorphic aspects of the nervous system and sexual behavior. This branch includes the genes *fruitless*, whose activity has been shown to be necessary for male sexual behavior, and *dissatisfaction*. The *transformer* gene may act directly on *fruitless*.

In other dipteran insects the same general strategy for sex determination is used, but there are marked differences at the molecular level. Only

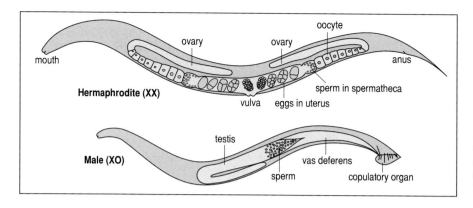

Fig. 12.8 Hermaphrodite and male *C.* elegans. The hermaphrodite has a 'two-armed' gonad and makes both eggs and sperm. The eggs are fertilized internally. The male makes sperm only.

doublesex has been found in dipterans distantly related to *Drosophila*, and *Sex-lethal* has not been found in any other dipteran.

12.4 Somatic sexual development in *Caenorhabditis* is determined by the number of X chromosomes

In the nematode *C. elegans*, the two sexes are hermaphrodite (essentially a modified female) and male (Fig. 12.8), although in other nematodes they are male and female. Hermaphrodites produce a limited amount of sperm early in development, with the remainder of the germ cells developing into oocytes. Sex in *C. elegans* (and other nematodes) is determined by the number of X chromosomes: the hermaphrodite (XX) has two X chromosomes, whereas the presence of just one X chromosome leads to development as a male (XO). The primary sex signal in *C. elegans* acts on the gene *XO lethal* (*xol-1*). With two X chromosomes, *xol-1* expression is low, resulting in the development of a hermaphrodite. *xol-1* is repressed by the SEX-1 protein, which is encoded on the X chromosome and is a key element in 'counting' the number of X chromosomes present.

A cascade of gene activity converts the level of *xol-1* expression into the somatic sexual phenotype (Fig. 12.9). Genes involved include those for nuclear proteins, such as SDC-1, and for a secreted protein, Hermaphrodite-1 (HER-1). At the end of the cascade is the gene *transformer-1 (tra-1)*, which encodes a transcription factor. Expression of transformer-1 protein is both necessary and sufficient to direct all aspects of hermaphrodite (XX) somatic cell development. A gain-of-function mutation in *tra-1* leads to hermaphrodite development in an XO animal, irrespective of the state of any of the regulatory genes that normally control its activity. Mutations that

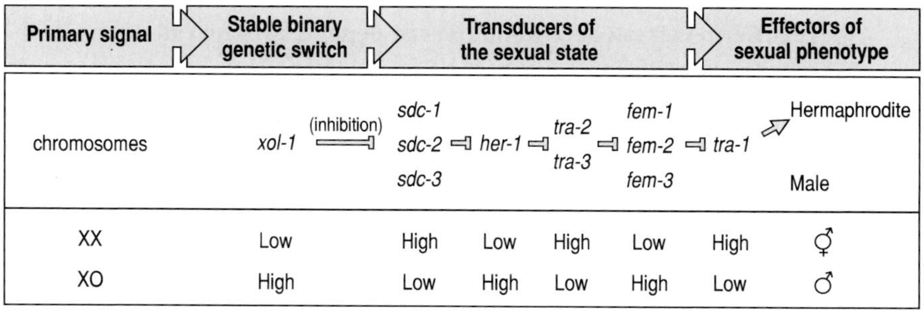

Fig. 12.9 Outline of the somatic sex determination pathway in *C.* elegans. The primary signal for sex determination is set by the number of X chromosomes. When two X chromosomes are present, the expression of the gene *XO lethal* (*xol-1*) is low, leading to hermaphrodite development, whereas *xol-1* is expressed at a high level in males. There is a cascade of gene expression starting from *xol-1* that leads to the gene *transformer-1* (*tra-1*), which codes for a transcription factor. If *tra-1* is active, development as a hermaphrodite occurs, but if it is expressed at low levels, males develop. The product of the *hermaphrodite-1* (*her-1*) gene is a secreted protein, which probably binds to a receptor encoded by *transformer-2* (*tra-2*), inhibiting its function.

Primary signal	Stable binary genetic switch	Transducers of the sexual state			Effectors of sexual phenotype		
chromosomes	xol-1 —(inhibition)⊣	sdc-1 sdc-2 ⊣ her-1 ⊣ sdc-3		tra-2 tra-3	fem-1 ⊣ fem-2 ⊣ tra-1 fem-3		Hermaphrodite ↗ Male
XX	Low	High	Low	High	Low	High	♀
XO	High	Low	High	Low	High	Low	♂

inactivate *tra-1* lead to complete masculinization of XX hermaphrodites. Unlike *Drosophila*, sex determination in *C. elegans* requires cell–cell interactions, as the process involves secreted proteins. The *mab-3* gene, which is related to *Drosophila doublesex*, acts downstream of *tra-1* to promote male-specific development in the peripheral nervous system and the gut. *mab-3* is turned on in males and, early in male development this results in the death of certain neurons associated in the hermaphrodite with egg-laying.

Before considering how the sex of the germ cells is determined and how the imbalance of X-linked genes between the sexes is dealt with in animals, we need to digress briefly to touch upon sex determination in flowering plants.

12.5 Most flowering plants are hermaphrodites, but some produce unisexual flowers

As we have seen, the flowers of angiosperms share a common organization in which the four types of floral organs are arranged in concentric whorls (see Section 7.12). The two inner whorls are the sexual organs—the stamens and carpels. Sepals and petals in the outer whorls are not sexual organs, but may serve to attract pollinators. Stamens produce pollen, which contains the male gametes corresponding to the sperm of animals. The female reproductive structures of flowers are the carpels, which are either free, or are fused to form a compound ovary. Carpels are the site of ovule formation, and each ovule produces an egg cell. Most flowering plants are hermaphrodites, bearing flowers with functional male and female sexual organs. However, not all flowering plants are of this type.

In about 10% of flowering plants, flowers of just one sex are produced. Flowers of different sexes may occur on the same plant, or be confined to different plants. The development of male or female flowers usually involves the selective resorption of either the stamens or pistil after they have been specified and have started to grow.

In maize, male and female flowers develop at particular sites on the shoot. The tassel at the tip of the main stem (see Fig. 7.13) only bears flowers with stamens; the 'ears' at the ends of the lateral branches bear female flowers containing pistils. Sex determination becomes visible when the flower is still small, with the stamen primordia being larger in males and the pistil longer in females. The smaller organs eventually degenerate.

The plant hormone gibberellic acid may be involved in sex determination, as differences in gibberellin concentration are associated with the different sexual organs. In the maize tassel, gibberellin concentration is 100-fold lower than in the developing ears. If the concentration of gibberellic acid is increased in the tassel, pistils can develop.

12.6 Germ-cell sex determination can depend both on cell signals and genetic constitution

The determination of the sex of animal germ cells—that is, whether they will develop into eggs or sperm, makes use of different mechanisms from those used for somatic cell sexual development. In the mouse, the primordial diploid germ cells continue to proliferate for a few days after entering the genital ridge, which is where the gonads form. At this stage, male and female germ cells are indistinguishable. Their future development is determined largely by the sex of the gonad in which they reside, and not by their own chromosomal constitution. Initially, they undergo mitotic division in both males and females. In females, diploid germ cells enter prophase of the first meiotic division in the embryo; they then arrest at this stage until the mouse becomes a sexually mature female. In male embryos, diploid germ cells divide mitotically for some time in the genital ridge, but then also stop dividing, becoming arrested in the G_1 phase of the cell cycle (Fig. 12.10). They start dividing again after birth and enter meiosis some 7 to 8 days after birth.

All mouse germ cells that enter meiosis before birth develop as eggs, whereas those not entering meiosis until after birth develop as sperm. Germ cells, whether XX or XY, that fail to enter the genital ridge and instead end up in adjacent tissues such as the embryonic adrenal gland or mesonephros, enter meiosis and begin developing as oocytes in both male and female embryos. In XX/XY chimeras, XX germ cells that are surrounded by testis cells develop along the spermatogenesis pathway. However, the later development of germ cells that develop in these inappropriate sites is abnormal. There is a little reproductive future for XY germ cells in the ovaries and none for XX germ cells in the testes.

In *Drosophila*, the difference in the behavior of XY and XX germ cells depends initially on the number of X chromosomes, as in somatic cells; the *Sex-lethal* gene again plays an important role, although other elements in the sex-determination pathway may differ from those in somatic cells. As in mammals, both chromosomal constitution and cell interactions are involved in the development of germ-cell sexual phenotype. Transplantation of genetically marked pole cells (see Section 2.5) into a *Drosophila* embryo of the opposite sex shows that male XY germ cells in a female XX embryo become integrated into the ovary and begin to develop as sperm; that is, their behavior is autonomous with respect to their genetic constitution. By contrast, XX germ cells in a testis develop as sperm, showing a role for environmental signals. In neither case, however, are functional sperm produced.

The hermaphrodite of *C. elegans* provides a particularly interesting example of germ-cell differentiation, as both sperm and eggs develop within the same gonad. Unlike the somatic cells, which have a fixed lineage and number (see Section 6.1), the number of germ cells in an adult nematode is indeterminate, with about 1000 germ cells in each 'arm' of the gonad. At hatching of the first-stage larva, there are just two founder germ cells, which proliferate to produce the germ cells. The germ cells are flanked by distal tip cells on each side, and their proliferation is controlled by a signal from these distal tip cells. The distal tip signal is the protein LAG-2, which is homologous to the Delta protein of *Drosophila* (see Section 11.2). The receptor for LAG-2 on the germ cells is probably GLP-1, which is similar both to nematode LIN-12 (see Section 10.18) and *Drosophila* Notch.

In *C. elegans*, entry of germ cells into meiosis from the third larval stage onward is controlled by the distal tip signal. In the presence of this signal, the cells proliferate, but as they move away from it, they enter meiosis and develop as sperm (Fig. 12.11, top). In the hermaphrodite gonad, all the cells that are initially outside the range of the distal tip signal develop as sperm, but cells that later leave the proliferative zone and enter meiosis develop as oocytes (see Fig. 12.11, bottom). The eggs are fertilized by stored sperm as

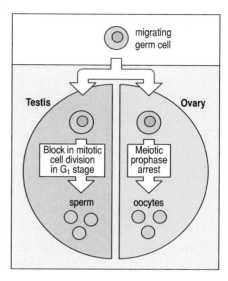

Fig. 12.10 Environmental signals specify germ-cell sex in mammals. Migrating germ cells, whether XX or XY, enter meiotic prophase and start developing as oocytes unless they enter a testis. In the testis, the germ cells receive an inhibitory signal that blocks mitotic division and prevents them entering meiotic prophase.

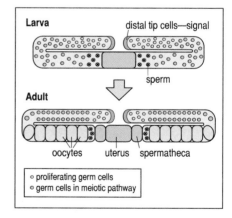

Fig. 12.11 Determination of germ-cell sex in the hermaphrodite nematode gonad. Top: during the larval stage, germ cells in a zone close to the distal tips of the gonad multiply; when they leave this zone in the larval stage they enter meiosis and develop into sperm. Bottom: in the adult, cells that leave the proliferative zone develop into oocytes. The eggs are fertilized as they pass into the uterus. After Clifford, R., *et al.*: 1994.

they pass into the uterus. The male gonad has similar proliferative meiotic regions, but all the germ cells develop as sperm.

Sex determination of the nematode germ cells is somewhat similar to that of the somatic cells, in that the chromosomal complement is the primary sex-determining factor and many of the same genes are involved in the subsequent cascades of gene expression. The terminal regulator genes required for spermatogenesis are called *fem* and *fog*. In hermaphrodites, there must be a mechanism for activating the *fem* genes in some of the XX germ cells, so allowing them to develop as sperm.

12.7 Various strategies are used for dosage compensation of X-linked genes

In each of the animals whose sex determination we have looked at there is an imbalance of X-linked genes between the sexes. One sex has two X chromosomes, whereas the other has one. This imbalance has to be corrected to ensure that the level of expression of genes carried on the X chromosome is the same in both sexes. The mechanism by which the imbalance in X chromosomes is dealt with is known as **dosage compensation**. Failure to correct the imbalance leads to abnormalities and arrested development. Different animals deal with the problem of dosage compensation in different ways (Fig. 12.12).

Mammals, such as mice and humans, achieve dosage compensation by inactivating one of the X chromosomes in females after the blastocyst has implanted in the uterine wall. Once X inactivation is initiated in female embryos, it is maintained in all somatic cells throughout life. In some tissues, the inactive X chromosome can be identified in the nucleus as a Barr body (see Section 9.3) and X inactivation provides an important model for the inheritance of gene expression. Whatever the sex chromosome constitution, for example, XY, XX, XXY, or XXXY, there is only one active X

Fig. 12.12 Mechanisms of dosage compensation. In mammals, *Drosophila*, and *C. elegans* there are two X chromosomes in one sex and only one in the other. Mammals inactivate one of the X chromosomes in females; in *Drosophila* males there is an increase in transcription from the single X chromosome; and in *C. elegans* there is a decrease in transcription from the X chromosomes in hermaphrodites. The result of these different dosage compensation mechanisms is that the level of X chromosome transcripts is approximately the same in males and females.

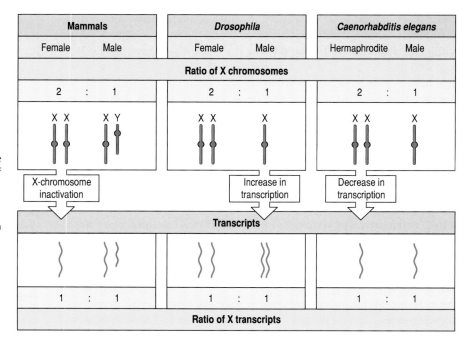

chromosome per somatic cell, with all the other X chromosomes inactivated. During early cleavage stages, both X chromosomes are active and the first X inactivation occurs in the extra-embryonic tissues, when the paternal X is exclusively inactivated. Later, at gastrulation, random X inactivation occurs in the cells of the embryo. As we see below, the inactive X becomes reactivated during germ-cell development. And in the early embryos of mice cloned from a differentiated cell nucleus, both X chromosomes are initially active and then normal X inactivation occurs.

X inactivation is dependent on a small region of the X chromosome, the inactivating center. It contains a proposed switch gene, *Xist*, which produces a non-coding RNA. Inactivation seems to be caused by the chromosome being coated by *Xist* RNA, which is expressed by the inactive X but not by an active X chromosome. If the *Xist* gene is introduced into another chromosome, that chromosome becomes silenced. This inactivation is correlated with methylation of the DNA (see Section 9.3).

Dosage compensation in *Drosophila* works in the opposite way to that in the mouse; instead of repression of the 'extra' X activity in females, transcription of the X chromosome in males is increased nearly twofold. A set of male-specific genes controls dosage compensation, and these are repressed in females by Sex-lethal protein, thereby preventing excessive transcription of the X chromosome. The increased activity in males is regulated by the primary sex-determining signal, which results in the dosage compensation mechanism operating when the *Sex-lethal* gene is 'off'. In females, where *Sex-lethal* is 'on', a cascade of gene activity turns off the dosage compensation mechanism.

In *C. elegans*, dosage compensation is achieved by reducing the level of X chromosome expression in XX animals to that of the single X chromosome in XO males. This involves a cascade of gene interactions set off by the primary sex-determining signal. The key event in initiating nematode dosage compensation is thought to be expression of the protein SDC-2, which occurs only in hermaphrodites. SDC-2 is a very large protein which forms a complex specifically with the X chromosomes and reduces transcription, possibly by causing condensation of the chromatin. SDC-2 also represses *her-1* on the sex-determination pathway.

Summary

The development of early embryos of both sexes is very similar. A primary sex-determining signal sets off development toward one or the other sex, and in mammals, *Drosophila*, and *C. elegans*, this signal is determined by the chromosomal complement of the fertilized egg. In mammals, the *Sry* gene on the Y chromosome is responsible for the embryonic gonad developing into a testis and producing hormones that determine male sexual characteristics. The sexual phenotype of the somatic cells is determined by the gonadal hormones. In *C. elegans* and *Drosophila*, the primary sex-determining signal is the number of X chromosomes. In *Drosophila*, the gene *Sex-lethal* is turned on in females but not in males, in response to this signal. In both cases, this results in further gene activity in which sex-specific RNA splicing is involved. In *C. elegans*, the gene *XO lethal* is turned off in hermaphrodites and on in males, eventually leading to sex-specific expression of the gene *transformer-1*, which determines the sexual phenotype. Somatic sexual differentiation in *Drosophila* is cell-autonomous and is controlled by the number of X chromosomes; in *C. elegans*, cell–cell interactions are also involved. Most flowering plants are hermaphrodites, having flowers with both male and female organs.

In mammals, signals from the gonads determine whether the germ cells develop into oocytes or sperm. Male germ cells in *Drosophila* develop along the sperm pathway even in an ovary, but female germ cells develop along the sperm pathway when placed in a testis. Most *C. elegans* adults are hermaphrodites, and produce both sperm and eggs from the same gonad.

Various strategies of dosage compensation are used to correct the imbalance of X chromosomes between males and females. In female mammals, one of the X chromosomes is inactivated; in *Drosophila* males the activity of the single X chromosome is upregulated; and in *C. elegans* the activity of the X chromosomes in XX hermaphrodites is downregulated to match that from the single X chromosome in males.

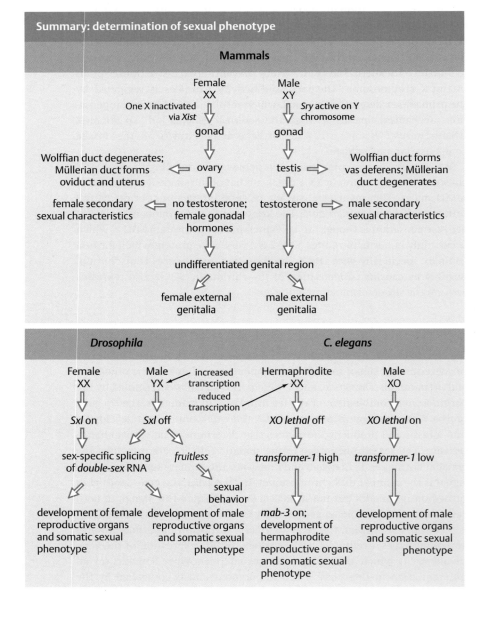

The development of germ cells

In all but the simplest animals, the cells of the germ line are the only cells that can give rise to a new organism. So, unlike somatic cells, which eventually all die, germ cells can in a sense outlive the bodies that produce them. They are therefore very special cells. The outcome of germ-cell development is either a male gamete (the sperm in animals), or a female gamete (the egg). The egg is a particularly remarkable cell, as it is the cell from which all the cells in an organism are derived. In species whose embryos receive no nutrition from the mother after fertilization, the egg must also provide everything necessary for development, as the sperm contributes virtually nothing to the organism other than its chromosomes.

Animal germ cells typically differ from somatic cells during early development by dividing less often. Later, they are the only cells to undergo meiosis. In many animals, primordial germ cells are formed in locations that seem to protect them from the inductive signals specifying the fate of somatic cells. In *C. elegans*, there is evidence for mechanisms that generally repress transcription in germ cells compared to somatic cells.

Germ cells are specified very early in some animals, although not in mammals, by cytoplasmic determinants present in the egg. We therefore start our discussion of germ-cell development by looking at the specification of primordial germ cells by special cytoplasm—the **germ plasm**. The localization of this special cytoplasm in the insect egg during oogenesis has been particularly well studied, and many of the genes involved are known.

In plants, there is no early specification of germ cells, as a single somatic plant cell can give rise to a sexually mature plant. The germ cells are specified late in development, during the development of the flower from the floral meristem (see Section 7.12).

In animals, once the primordial germ cells are specified, they migrate into the gonads, somatic structures that usually develop some distance away from the site of germ cell origin. Once within the gonads, the germ cells differentiate as either male or female gametes.

12.8 Germ-cell fate can be specified by a distinct germ plasm in the egg

In flies, nematodes, and frogs, molecules localized in specialized cytoplasm in the egg are involved in specifying the germ cells. The clearest example of this is in *Drosophila*, where primordial germ cells become distinct at the posterior pole of the egg about 90 minutes after fertilization, several hours before cellularization of the rest of the embryo (see Section 2.5). The cytoplasm at the posterior pole is distinguished by large organelles, the polar granules, which contain both proteins and RNAs. That there is something special about this posterior cytoplasm—the so-called **pole plasm**—is demonstrated by two key experiments. First, if the posterior end of the egg is irradiated with ultraviolet light, which destroys the pole plasm activity, no germ cells develop. Second, if pole plasm of an egg is transferred to the anterior pole of another embryo, the nuclei that become surrounded by the pole plasm are specified as germ cells (Fig. 12.13). If they are then transplanted into the future genital region, they develop as functional germ cells.

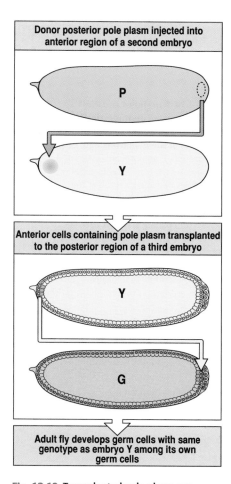

Fig. 12.13 Transplanted pole plasm can induce germ-cell formation in *Drosophila*. Pole plasm from a fertilized egg of genotype P (pink) is transferred to the anterior end of an early cleavage stage embryo of genotype Y (yellow). After cellularization, cells containing pole plasm induced at the anterior end of embryo Y are transferred to the posterior end (a site from which germ cells can migrate into the gonad) of another embryo, of genotype G (green). The adult fly that develops from embryo G contains germ cells of genotype Y as well as those of G.

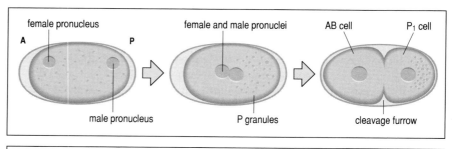

Fig. 12.14 P granules and PIE-1 protein become asymmetrically distributed to germline cells during cleavage of the nematode egg. Before fertilization, P granules are distributed throughout the egg. After fertilization, P granules become localized at the posterior end of the egg. At the first cleavage, they are only included in the P_1 cell (top panel), and thus become confined to the P cell lineage. The PIE-1 protein is only present in P cells. All germ cells are derived from P_4, which is formed at the fourth cleavage.

In the nematode, a germ-cell lineage is set up at the end of the fourth cleavage division, with all the germ cells being derived from the P_4 blastomere (see Section 6.2). The P_4 cell is derived from three stem-cell like divisions of the P_1 cell. At each of these divisions, one daughter produces somatic cells whereas the other divides again to produce a somatic cell progenitor and a P cell. The egg contains P granules in its cytoplasm which become asymmetrically distributed after fertilization, and are subsequently confined to the P-cell lineage (Fig. 12.14). The association of germ-cell formation with the P granules suggested that they might have a role in germ-cell specification, and at least one P-granule component, the product of the *pgl* gene, has been shown to be necessary for germ-cell development. PGL may specify germ cells by regulating some aspect of mRNA metabolism. The gene *pie-1* is involved in maintaining the stem-cell property of the P blastomeres. It encodes a nuclear protein that is expressed maternally and is not a component of P granules; the PIE-1 protein is only present in the germline blastomeres, and represses new transcription of zygotic genes in these blastomeres until it disappears at around the 100-cell stage. This general repression may protect the germ cells from the actions of transcription factors that promote development into somatic cells.

There is also evidence for germ plasm in *Xenopus* eggs. After fertilization, distinct yolk-free patches of cytoplasm aggregate at the yolky vegetal pole.

When the blastomeres at the vegetal pole cleave, this cytoplasm is distributed asymmetrically so that it is retained only in the most vegetal daughter cells, from which the germ cells are derived. Ultraviolet irradiation of the vegetal cytoplasm abolishes the formation of germ cells, and transplantation of fresh vegetal cytoplasm into an irradiated egg restores germ-cell formation. At gastrulation, the germ plasm is located in cells in the floor of the blastocoel cavity, among the cells that give rise to the endoderm. Cells containing the germ plasm are, however, not yet determined as germ cells and can contribute to all three germ layers if transplanted to other sites. At the end of gastrulation, the primordial germ cells are determined, and migrate out of the presumptive endoderm and into the genital ridge.

There is no evidence for germ plasm being involved in germ-cell formation in the mouse or other mammals. Germ-cell specification in the mouse involves cell–cell interactions, as cultured embryonic stem cells can, when injected into the inner cell mass, give rise to both germ cells and somatic cells (see Box 3C, p. 89). The earliest that germ cells have been identified in the mouse is midway through gastrulation, when, as a result of their high alkaline phosphatase activity, they can be identified in the extra-embryonic region posterior to the primitive streak (see Fig. 3.13). These cells have their origin in the proximal epiblast (the region of the epiblast nearest to the site of implantation) and BMP-4 is necessary for their specification. As in *Xenopus*, germ cells in mammals migrate to the genital ridge, which is where the gonads form.

12.9 Pole plasm becomes localized at the posterior end of the *Drosophila* egg

In Chapter 5, we saw how the main axes of the *Drosophila* egg are specified by the follicle cells in the ovary, and how the mRNAs for proteins such as bicoid and nanos become localized in the egg (see Section 5.7). The pole plasm also becomes localized at the posterior end of the egg under the influence of the follicle cells.

Several maternal genes are involved in pole plasm formation in *Drosophila*. Mutations in any of at least eight genes result in the affected homozygous individual being 'grandchildless'. Its heterozygous offspring lack a proper pole plasm, and although they may develop normally in other ways, they lack germ cells and are therefore sterile. One of these eight genes is *oskar*, which plays a central role in the organization and assembly of the pole plasm (Fig. 12.15); of the genes involved in pole plasm formation, *oskar* is the only one to have its mRNA localized at the posterior pole. The signal for localization is contained in the 3' untranslated region of the mRNA. Staufen protein is required for the localization of *oskar* mRNA and may act

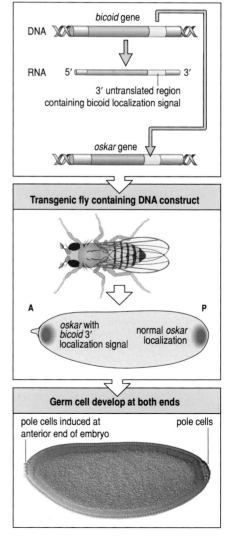

Fig. 12.15 The gene *oskar* is involved in specifying the germ plasm in *Drosophila*. In normal eggs, *oskar* mRNA is localized at the posterior end of the embryo, whereas *bicoid* mRNA is at the anterior end. The localization signals for both *bicoid* and *oskar* mRNA are in their 3' untranslated regions. By manipulating the *Drosophila* DNA, the localization signal of *oskar* can be replaced by that of *bicoid* (top panel). A transgenic fly is made containing the modified DNA. In its egg, *oskar* becomes localized at the anterior end (middle panel). The egg therefore has *oskar* mRNA at both ends, and germ cells develop at both ends of the embryo, as shown in the photograph (bottom panel). Thus, *oskar* alone is sufficient to initiate the specification of germ cells.

Photograph courtesy of R. Lehmann.

by linking the mRNA to the microtubule system that is polarized along the antero-posterior axis (see Section 5.7). When the region of the *oskar* gene that codes for the 3′ localization signal is replaced by the *bicoid* 3′ localization signal, flies made trangenic with this DNA construct have *oskar* mRNA localized at both the anterior and posterior ends of the egg (Fig. 12.15).

12.10 Germ cells migrate from their site of origin to the gonad

In many animals, germ cells develop at some distance from the gonads, and only later migrate to them, where they differentiate into eggs or sperm. The reason for this separation of site of origin from final destination is not known, but it may be a mechanism for excluding germ cells from the general developmental upheaval involved in laying down the body plan, or a mechanism for selecting the healthiest of the germ cells, namely those that survive migration. Their pathway of migration is controlled by the environment, and in *Xenopus*, for example, germ cells transplanted to the wrong place in the blastula do not end up in the gonad. In zebrafish, however, the primordial germ cells are specified in random positions relative to the embryonic axis.

The vertebrate gonad develops from the mesoderm lining the abdominal cavity, which is known as the **genital ridge**. The primordial germ cells migrate there from distant sites. In *Xenopus*, the primordial germ cells originate in the endoderm (which forms the gut) and migrate to the future gonad along a cell sheet that joins the gut to the genital ridge. Only a small number of cells start this journey, dividing about three times before arrival, so that about 30 germ cells colonize the gonad. The number of primordial germ cells that arrive at the genital ridge of the mouse, by a very similar pathway (Fig. 12.16), is about 2500. In chick embryos, the pattern of migration is different: the germ cells originate at the head end of the embryo, and most arrive at their destination via the blood vessels, leaving the bloodstream at the hindgut and then migrating along the epithelial sheet. In *Drosophila*, prospective germ cells are carried inside the embryo to the midgut during gastrulation, and they then migrate through the gut to the site at which the gonads develop. The gene *nanos*, one of the posterior group genes (see Section 5.3), is required for their correct migration.

In the mouse gastrula, germ cells first become detectable, by staining for alkaline phosphatase, posterior to the primitive streak. They become incorporated into the hindgut and then migrate into the adjacent connective tissue. Two genes involved in controlling proliferation of migrating germ cells are *White spotting* (*W*) and *Steel*, which we have already met in relation to melanocyte differentiation (see Section 9.11). Mutations that inactivate either of these genes cause a decrease in germ-cell numbers. *White spotting* codes for the cell-surface receptor Kit, which is expressed in the migrating germ cells. Its ligand, the Steel protein, is expressed in the cells along which the germ cells migrate. The requirement for these two genes indicates that the migrating cells are continuously receiving signals from the surrounding tissue. Once they arrive at the gonad, the germ cells begin to differentiate into sperm or eggs, as described in the next section.

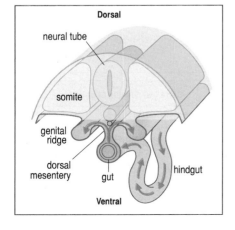

Fig. 12.16 Pathway of primordial germ-cell migration in the mouse embryo. In the final stage of migration, the cells move from the gut tube into the genital ridge, via the dorsal mesentery. After Wylie, C.C., *et al.*: 1993.

12.11 Germ-cell differentiation involves a reduction in chromosome number

Germ cells have to reduce their chromosome number by half during gamete formation, so that at fertilization the diploid chromosome number is reinstated. Primordial germ cells are diploid, and reduction from the diploid to the haploid state occurs during **meiosis** (Fig. 12.17). Meiosis comprises two cell divisions, in which the chromosomes are replicated before the first division, but not before the second, so that their number is reduced by half.

Development of eggs and sperm follow different courses, even though they both involve meiosis. The development of the egg is known as **oogenesis**. The main stages of mammalian oogenesis are shown in Fig. 12.18 (left panel). In mammals, germ cells undergo a small number of proliferative mitotic cell divisions as they migrate to the gonad. In the case of developing eggs, the diploid oogonia continue to divide mitotically for a short time in the ovary. After entry into meiosis, the primary oocytes become arrested in the prophase of the first meiotic division. They never proliferate again; thus, the number of oocytes at this embryonic stage is the total number of eggs the female mammal ever has. In humans, most of these oocytes degenerate before puberty, leaving about 40,000 out of an original 6 million. In mammals and many other vertebrates, oocyte development may be held in suspension after birth for months (mice) or years (humans). The female becomes sexually mature at puberty, when the oocytes undergo further development, growing in size up to 1000-fold. In mammals, meiosis resumes at the time of ovulation, proceeds as far as the metaphase of the second meiotic division, where it becomes arrested again, and is completed after fertilization.

The strategy for **spermatogenesis**—the production of sperm—is different. Diploid germ cells that give rise to sperm do not enter meiosis in the embryo, but become arrested at an early stage of the mitotic cell cycle in the embryonic testis. They resume mitotic proliferation after birth. Later, in the sexually mature animal, spermatogonial stem cells give rise to several generations of differentiating spermatocytes, which undergo meiosis, each forming four haploid spermatids that mature into sperm (see Fig. 12.18, right panel). Thus, unlike the fixed number of oocytes in female mammals, sperm continue to be produced throughout life.

In *Drosophila*, there is a continuous production of both eggs and sperm from a population of stem cells. Oogenesis begins with the division of a stem cell (see Fig. 5.10), and so there is no intrinsic limitation to the number of eggs that a female fly can produce. *Drosophila* oogenesis and

Fig. 12.17 Meiosis produces haploid cells. Meiosis reduces the number of chromosomes from the diploid to the haploid number. Only one pair of homologous chromosomes is shown here for simplicity. Before the first meiotic division the DNA replicates, so that each chromosome entering meiosis is composed of two identical chromatids. The paired homologous chromosomes (known as a bivalent) undergo crossing over and recombination, and align on the meiotic spindle at the metaphase of the first meiotic division. The homologous chromosomes separate, and each is segregated into a different daughter cell at the first cell division. There is no DNA replication before the second meiotic division. The daughter chromatids of each chromosome separate and segregate at the second cell division. The chromosome number of the resulting daughter cells is thus halved.

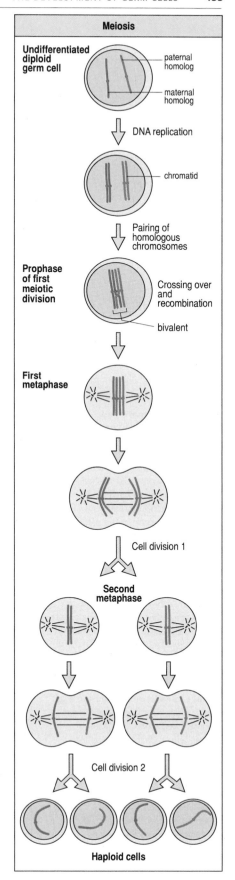

Meiosis

Undifferentiated diploid germ cell — paternal homolog — maternal homolog

DNA replication

chromatid

Pairing of homologous chromosomes

Prophase of first meiotic division — Crossing over and recombination — bivalent

First metaphase

Cell division 1

Second metaphase

Cell division 2

Haploid cells

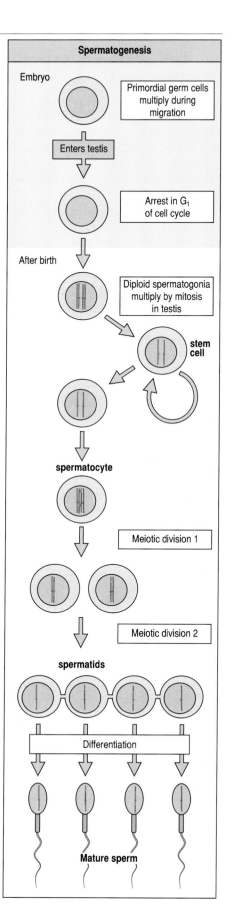

Fig. 12.18 Oogenesis and spermatogenesis in mammals. Left panel: after the germ cells that form oocytes enter the embryonic ovary, they divide mitotically a few times and then enter the prophase of the first meiotic division. No further cell multiplication occurs. Further development occurs in the sexually mature adult female. This includes a 100-fold increase in mass, the formation of external cell coats, and the development of a layer of cortical granules located under the oocyte plasma membrane. In each cycle, a group of follicles starts to grow, oocyte growth and maturation follows; a few eggs are ovulated but most degenerate. Eggs continue to mature in the ovary under hormonal influences, but become blocked in the second metaphase of meiosis, which is only completed after fertilization. Polar bodies are formed at meiosis (see Box 2A, p. 31). Right panel: germ cells that develop into sperm enter the embryonic testis and become arrested at the G₁ stage of the cell cycle. After birth, they begin to divide mitotically again, forming a population of stem cells (spermatogonia). These give off cells that then undergo meiosis and differentiate into sperm. Sperm can therefore be produced indefinitely.

the localization of cytoplasmic determinants in the egg have already been discussed in Section 5.7.

12.12 Oocyte development can involve gene amplification and contributions from other cells

Eggs vary enormously in size among different animals, but they are always larger than the somatic cells. A typical mammalian egg is about 0.1 mm in diameter, a frog's egg about 1 mm, and a hen's egg about 5 cm (see Fig. 3.1). To achieve these large sizes—and even the mammalian egg has a mass 100-fold greater than that of a typical mammalian somatic cell—a variety of mechanisms have evolved, some organisms using several of them together. One strategy is to increase the overall number of gene copies in the developing oocyte, as this proportionately increases the amount of mRNA that can be transcribed, and thus the amount of protein that can be synthesized. This strategy is generally adopted by vertebrate oocytes, which are arrested in the prophase of the first meiotic division and thus have double the normal diploid number of genes, as do all mitotic cells at this stage. Transcription and oocyte growth continue while the oocyte is in the arrested stage. Another strategy, adopted by insects and amphibians, is to produce many extra copies of those genes whose products are needed in large quantities in the egg. Thus, the ribosomal RNA genes of amphibians are amplified during oocyte development from hundreds to millions. In insects, the genes that encode the proteins of the egg membrane, the chorion, become amplified in the surrounding follicle cells.

A different strategy is for the oocyte to rely on the synthetic activities of other cells. In insects, the nurse cells adjacent to the oocyte, which are its sister cells, make and deliver many mRNAs and proteins to the oocyte (see Fig. 5.10). Yolk proteins in birds and amphibians are made by liver cells, and are carried by the blood to the ovary, where the proteins enter the oocyte by endocytosis and become packaged in yolk platelets. From an early stage, the oocyte is polarized, and the yolk platelets accumulate at the vegetal pole. In *C. elegans*, yolk proteins are made in interstitial cells, and in *Drosophila* in the fat body, and then transported to the oocytes.

12.13 Genes controlling embryonic growth are imprinted

Like all other cells in an animal, mature germ cells have completed a program of differentiation. Yet, at the end of that process they must have a genome that, after fertilization, is capable of controlling the development of the embryo; of all the cells in the body, they are the only ones with genomes that are passed on to future generations. Their genomes therefore have to revert to a state from which all the cells of the organism can be derived, and so there must be no permanent alterations in the genetic constitution. However, recent studies on mammals have shown that certain genes in eggs and sperm are programmed to be switched off during development. Evidence for this comes from the different contributions of the maternal and paternal genomes to the development of the embryo.

Mouse eggs can be manipulated by nuclear transplantation to have either two paternal genomes or two maternal genomes, and can be reimplanted into a mouse for further development. The embryos that result are known as **androgenetic** and **gynogenetic** embryos, respectively. Although both

Fig. 12.19 Paternal and maternal genomes are both required for normal mouse development. A normal biparental embryo has contributions from both the paternal and maternal nuclei in the zygote after fertilization (left panel). Using nuclear transplantation, an egg can be constructed with two paternal or two maternal nuclei from an inbred strain. Embryos that develop from an egg with two maternal genomes—gynogenetic embryos (center panel)—have underdeveloped extra-embryonic structures. This results in development being blocked, although the embryo itself is relatively normal and well developed. Embryos that develop from eggs with two paternal genomes—androgenetic embryos (right panel)—have normal extra-embryonic structures, but the embryo itself only develops to a stage where a few somites have formed.

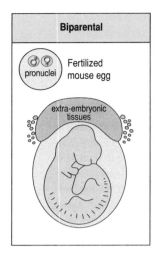
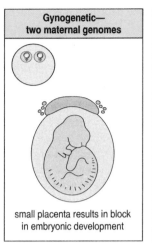
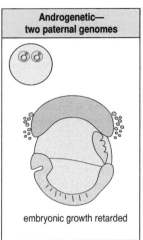

Biparental	Gynogenetic—two maternal genomes	Androgenetic—two paternal genomes

pronuclei / Fertilized mouse egg

extra-embryonic tissues

small placenta results in block in embryonic development

embryonic growth retarded

kinds of embryo have a diploid number of chromosomes, their development is abnormal. The embryos with two paternal genomes have well-developed extra-embryonic tissues, but the embryo itself is abnormal, and does not proceed beyond a stage at which several somites are present. By contrast, the embryos with diploid maternal genomes have relatively well-developed embryos, but the extra-embryonic tissues—placenta and yolk sac—are poorly developed (Fig. 12.19). These results clearly show that both maternal and paternal genomes are necessary for normal mammalian development: the two parental genomes function differently in development, and are required for the normal development of both the embryo and the placenta. This is the reason that mammals cannot be produced parthenogenetically, by activation of an unfertilized egg.

Such observations show that the paternal and maternal genomes are modified, or **imprinted**, during germ-cell differentiation. The paternal and maternal genomes contain the same set of genes, but the imprinting process turns off certain genes in either the sperm or egg, so that they are not expressed during development. For example, some of the genes necessary for yolk sac and placenta development are inactivated in the maternal genome, whereas some of those required for development of the embryo are turned off in the paternal genome. Imprinting implies that the affected genes carry a 'memory' of being in a sperm or an egg.

Imprinting in mammals is a reversible process, whereby modification of the same genes in either the sperm or egg can lead to differences in gene expression in the diploid cells of the embryo. The reversibility requirement is important, because any of the chromosomes may eventually end up in male or female germ cells during development. Inherited imprinting is probably erased during early germ-cell development, and imprinting is later established afresh during germ-cell differentiation.

Fig. 12.20 Imprinting of genes controlling embryonic growth. In mouse embryos, the paternal gene for insulin-like growth factor 2 (*Igf-2*) is on, but the gene on the maternal chromosome is off. In contrast, the *Igf-2r* gene is on in the maternal genome and off in the paternal genome. The product of this gene tends to inhibit growth, as it is involved in degrading IGF-2. *H19* may regulate *Igf-2*.

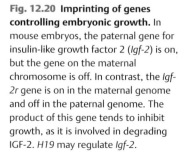
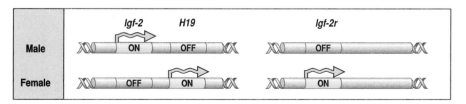

	Igf-2	*H19*	*Igf-2r*	
Male	ON	OFF	OFF	
Female	OFF	ON	ON	

The imprinted genes affect not only early development but also the later growth of the embryo. Further evidence that imprinted genes are involved in the growth of the embryo comes from studies on chimeras made between normal embryos and androgenetic or gynogenetic embryos. When inner cell mass cells from gynogenetic embryos are injected into normal embryos, growth is retarded by as much as 50%. But when androgenetic inner cell mass cells are introduced into normal embryos, the chimera's growth is increased by up to 50%. The imprinted genes on the male genome thus significantly increase the growth of the embryo.

At least 40 imprinted genes have been identified in mice, some of which are involved in growth control. The insulin-like growth factor IGF-2 is required for embryonic growth; its gene, *Igf-2*, is imprinted in the maternal genome—that is, it is turned off, so that only the paternal gene is active. Direct evidence for the *Igf-2* gene being imprinted comes from the observation that when sperm carrying a mutated, and thus defective, *Igf-2* gene fertilize a normal egg, small offspring result. This is because only very low levels of IGF-2 are produced from the imprinted maternal gene and this is not enough to make up for the loss of IGF-2 expression from the paternal genome. In contrast, if the non-functional mutant gene is carried by the egg's genome, development is normal, with the required IGF-2 activity being provided from the normal paternal gene.

Imprinting of *Igf-2* is probably due to imprinting of the *H19* gene, which regulates expression of *Igf-2*. *H19* is imprinted in the opposite direction: it is expressed in the maternal chromosome but not in the paternal one. The gene for the so-called receptor for IGF-2 protein—*Igf-2r*—is also imprinted in this direction; it is turned off in the paternal genome and turned on in the maternal genome (Fig. 12.20). In fact, *Igf-2r* does not encode the receptor for IGF-2 (the true receptor for IGF-2 is probably the IGF-1 receptor), but a protein that is required to degrade IGF-2 protein; expression of this protein tends to reduce growth by controlling the amount of IGF-2 available.

A possible evolutionary explanation for the reciprocal imprinting of genes that control growth is that the reproductive strategies of the father and mother are different—that is, paternal imprinting promotes growth whereas maternal imprinting reduces it. The father wants to have maximal growth for his own offspring, and this can be achieved by having a large placenta, as a result of producing growth hormone, whose production is stimulated by IGF-2. The mother, who may mate with different males, benefits more by spreading her resources over all her offspring, and so wishes to prevent too much growth in any one embryo. Thus, a gene that promotes embryonic growth is turned off in the mother and a gene that tends to reduce it is turned on. Paternal genes expressed in the offspring could be selected to extract more resources from mothers, because an offspring's paternal genes are less likely to be present in the mother's other children.

Imprinting occurs during germ-cell differentiation, and so a mechanism is required both for maintaining the imprinted condition throughout development and for wiping it out during the next cycle of germ-cell development. A possible mechanism for maintaining imprinting is DNA methylation (see Fig. 9.11). Evidence that DNA methylation is required for imprinting comes from transgenic mice in which methylation is impaired. In these mice, the *Igf-2* and *Igf-2r* genes are no longer imprinted. In the cloning of mammals using donor nuclei from differentiated cells, these nuclei will not have the appropriate genes imprinted and this could account for the large number of failures and abnormalities.

Summary

In many animals, germ cells are specified by localized cytoplasmic determinants in the egg. Localization of cytoplasmic determinants is controlled by cells surrounding the oocyte. Germ cells in mammals are an exception, and are specified by intercellular interactions at a later embryonic stage. Once the germ cells are determined, they migrate from their site of origin to the gonads, where further development and differentiation takes place. In the gonads, the diploid germ cells undergo meiosis, eventually producing haploid eggs and sperm. The number of oocytes in a female mammal is fixed before birth, whereas sperm production in male mammals is continuous throughout adult life. Eggs are always larger than somatic cells, and some are very large indeed. In order to achieve their increased size, specialized cells surrounding the developing oocyte may provide some of the constituents, such as yolk; in addition, some genes producing materials required in large amounts may be amplified in the oocyte.

Both maternal and paternal genomes are necessary for normal mammalian development. Embryos with diploid maternal or paternal genomes develop abnormally. The purely maternal ones are deficient in extra-embryonic structures, whereas the embryo proper fails to develop in purely paternal embryos. Certain genes in eggs and sperm are imprinted, so that the activity of the same gene is different depending on whether it is of maternal or paternal origin. Several imprinted genes are involved in growth control of the embryo.

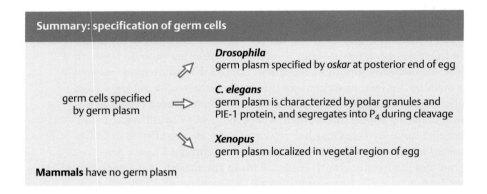

Summary: specification of germ cells

germ cells specified by germ plasm

Drosophila
germ plasm specified by *oskar* at posterior end of egg

C. elegans
germ plasm is characterized by polar granules and PIE-1 protein, and segregates into P_4 during cleavage

Xenopus
germ plasm localized in vegetal region of egg

Mammals have no germ plasm

Fertilization

Fertilization—the fusion of egg and sperm—is the trigger that initiates development. The membranes of the egg and sperm fuse, the sperm nucleus enters the egg cytoplasm, becoming the sperm **pronucleus**. In mammals and many other animals, fertilization triggers the completion of meiosis in the egg and the set of maternal chromosomes retained in the egg becomes the egg pronucleus. The sperm and egg pronuclei form the zygotic nucleus. At fertilization, the egg is activated to start dividing and embark

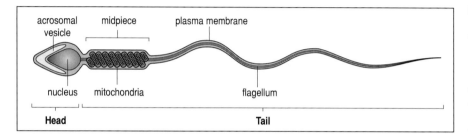

Fig. 12.21 A human sperm. The acrosomal vesicle at the anterior end of the sperm contains enzymes that are used to digest the protective coats around the egg. The plasma membrane on the head of the sperm contains various specialized proteins that bind to the egg coats and facilitate entry. The sperm moves by its single flagellum, which is powered by mitochondria. The overall length from head to tail is about 60 μm.

on its developmental program. Fertilization can be either external, as in frogs, or internal, as in *Drosophila*, mammals, and birds. Of all the sperm released by a male animal, only one fertilizes each egg. In many animals, including mammals, sperm penetration activates a blocking mechanism in the egg that prevents any further sperm entering the egg—the so-called block to **polyspermy**. This is necessary because if more than one sperm nucleus enters the egg there will be additional sets of chromosomes and centrosomes, resulting in abnormal development. There are a variety of mechanisms to ensure that only one sperm nucleus contributes to the zygote. In some animals, such as birds, many sperm penetrate the egg but all but one are destroyed in the cytoplasm.

Both eggs and sperm are structurally specialized for fertilization. The specializations of the egg are directed to preventing fertilization by more than one sperm, whereas those of the sperm are directed to facilitating penetration of the egg. Eggs are usually surrounded by several protective layers, and the eggs of many organisms have a layer of cortical granules just beneath the plasma membrane. The mammalian egg is bounded by a plasma membrane with cortical granules beneath it, and around it the zona pellucida, which serves to make the egg impenetrable to more than one sperm. All sperm are motile cells, typically designed for activating the egg, and at the same time delivering their nucleus into the egg cytoplasm. They essentially consist of a nucleus, mitochondria to provide an energy source, and a flagellum for movement. The anterior end is highly specialized, aiding penetration (Fig. 12.21). The sperm of the nematode *C. elegans* and some other invertebrates is unusual as it more closely resembles a normal cell and moves by ameboid motion. In hermaphrodites there are about 300 sperm, and almost all will fertilize an egg.

12.14 Fertilization involves cell-surface interactions between egg and sperm

After sperm have been deposited in the mammalian female reproductive tract, they undergo a process known as **capacitation** which facilitates fertilization by removing certain inhibitory factors. There are very few mature eggs—usually one or two in humans and about 10 in mice—waiting to be fertilized, and less than a 100 of the millions of sperms deposited actually reach these eggs. The sperm has to penetrate several physical barriers to enter the egg (Fig. 12.22). In mammalian eggs, the first barrier is a layer of cumulus cells, embedded in a sticky mass of hyaluronic acid. Hyaluronidase activity on the surface of the sperm head helps it to penetrate this layer. The sperm next encounters the **zona pellucida**, a layer of glycoproteins surrounding the egg. This also acts as a physical barrier, but sperm are

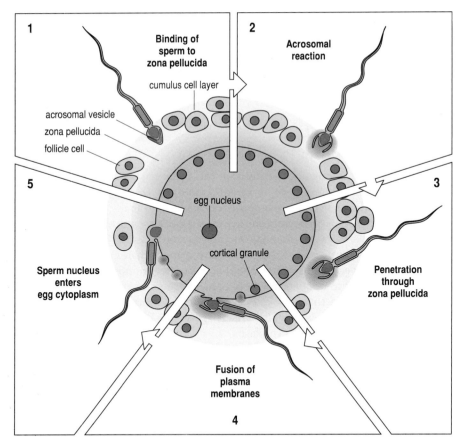

Fig. 12.22 Fertilization of a mammalian egg. After penetrating the follicle-derived cumulus cell layer, the sperm binds to the zona pellucida (1). This triggers the acrosomal reaction (2), in which enzymes are released from the acrosomal vesicle and break down the zona pellucida. This enables the sperm to penetrate the zona pellucida (3), and bind to the egg plasma membrane. The plasma membrane of the sperm head fuses with the egg plasma membrane (4). This activates the egg, causing a release of cortical granules and the sperm nucleus then enters the egg (5). After Alberts, B., *et al.*: 1989.

helped to penetrate it by the **acrosomal reaction**—the release of the contents of the acrosomal vesicle located in the sperm head. The zona pellucida contains a receptor for species-specific binding of the sperm, in the form of the glycoprotein ZP3. On the sperm head is β1,4-galactosyltransferase—an adhesion molecule that may bind to ZP3. When the sperm binds to ZP3, the contents of the acrosome are released by exocytosis. The enzymes released include β-*N*-acetylglucosaminidase, which breaks down the oligosaccharide side chains on the zona pellucida glycoproteins, and a protease called acrosin. These enzymes allow the sperm to approach the egg plasma membrane.

The acrosomal reaction also exposes proteins on the sperm surface that can bind to the egg membrane and are involved in the fusion of sperm and egg membranes. A potential component is the protein fertilin, which may bind to an integrin-like receptor on the egg plasma membrane. A key egg receptor for the sperm is the protein CD9. Interaction of a sperm with CD9 could initiate sperm and egg fusion. In the sperm of many invertebrates, such as the sea urchin, the acrosomal reaction results in the extension of a rod-like acrosomal process. This forms by the polymerization of actin and facilitates contact with the egg membrane.

It is possible to fertilize human eggs in culture, and then transfer a very early embryo to the mother. This technique of *in vitro* fertilization has been of great help to couples who have, for a variety of reasons, difficulty in conceiving. It is even possible to fertilize a human egg by injecting a single intact sperm directly into the egg.

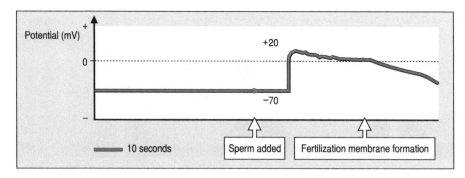

Fig. 12.23 Depolarization of the sea urchin egg plasma membrane at fertilization. The resting membrane potential of the unfertilized sea urchin egg is –70 mV. At fertilization, it changes rapidly to + 20 mV, and then slowly returns to the original value. This depolarization may provide a fast block to polyspermy.

12.15 Changes in the egg membrane at fertilization block polyspermy

Although many sperm attach to the coats surrounding the egg, it is important that only one sperm fuses with the egg plasma membrane and delivers its nucleus into the egg cytoplasm. There are thus mechanisms to prevent more than one sperm entering. The main block to polyspermy in many animals is brought into play as soon as the first sperm fuses with the plasma membrane. This stimulates the release of the cortical granules which, in mammals, contain enzymes that block further sperm from binding to the zona pellucida. In sea urchins, which we shall consider in more detail, the enzymes contribute to the formation of an impenetrable fertilization membrane around the fertilized egg.

In the sea urchin egg, a rapid block to polyspermy is triggered by a transient depolarization of the egg plasma membrane, caused by sperm–egg fusion. The electrical membrane potential across the plasma membrane goes from –70 mV to + 20 mV within a few seconds of sperm entry (Fig. 12.23). The membrane potential slowly returns to its original level while the fertilization membrane, which is made from the cortical granules and vitelline membrane, is formed. If depolarization is prevented, polyspermy occurs, but how depolarization blocks polyspermy is not yet clear. (In mouse fertilization there is no change in membrane potential.) The slower cortical reaction in the sea urchin is better understood. Sperm entry into the sea urchin egg results in the initiation of a wave of calcium release and leads to the cortical granules, which lie just beneath the plasma membrane in the mature egg, releasing their contents to the outside of the plasma membrane by exocytosis. This results in the vitelline membrane lifting off the plasma membrane. The cortical granule contents contribute to the vitelline membrane to form the fertilization membrane, and also provide a hyaline layer between it and the egg's plasma membrane (Fig. 12.24). Both

Fig. 12.24 The cortical reaction at fertilization in the sea urchin. The egg is surrounded by a vitelline membrane, which lies outside the plasma membrane. Membrane-bound cortical granules lie just beneath the egg plasma membrane. At fertilization, the cortical granules fuse with the plasma membrane, and some of the contents are extruded by exocytosis. These join with the vitelline membrane to form a tough fertilization membrane, which then lifts off the egg surface and prevents further sperm entry. Other cortical granule constituents give rise to a hyaline layer, which surrounds the egg under the fertilization membrane.

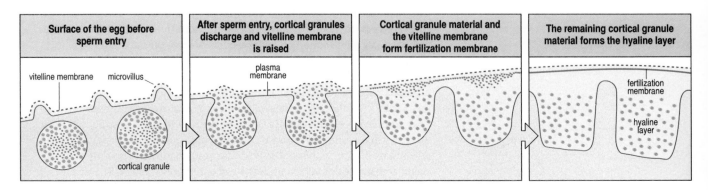

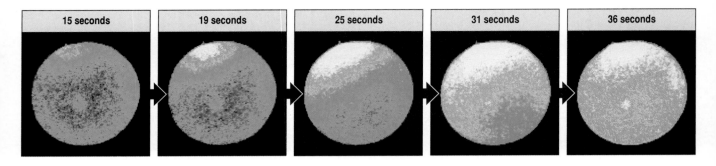

Fig. 12.25 Calcium wave at fertilization. A series of images showing an intracellular calcium wave at fertilization in a sea urchin egg. The fertilizing sperm has fused just to the left of the top of the egg, and triggered the wave. Calcium ion concentration is monitored with a calcium-sensitive fluorescent dye, using confocal fluorescence microscopy. Calcium concentration is shown in false color: red is the highest concentration, then yellow, green, and blue. Times shown are seconds after sperm entry.

Photographs courtesy of M. Whitaker.

these structures prevent sperm from binding to the egg plasma membrane.

12.16 A calcium wave initiated at fertilization results in egg activation

The activation of the egg at fertilization initiates a series of events that result in the start of development. For example, in the sea urchin egg, there is a several-fold increase in protein synthesis, and there are often changes in egg structures, such as the cortical rotation that occurs in amphibian eggs (see Sections 3.2 and 3.3). But the main events are that the egg, which has been blocked at a stage in meiosis, now completes meiosis, whereupon the egg and sperm nuclei fuse to form the diploid zygotic genome, and the fertilized egg enters mitosis. In mice and humans, the pronuclear membranes disappear before the pronuclei come together.

Fertilization and egg activation in mammals and sea urchins are associated with an explosive release of free Ca^{2+} ions within the egg, producing a wave of calcium that travels across it (Fig. 12.25). The calcium release is triggered by sperm entry, and is both necessary and sufficient to initiate the onset of development. The wave starts at the point of sperm entry and crosses the egg at a speed of 5–10 micrometers per second. In all mammals, oscillations in calcium concentration occur for several hours after fertilization. The mechanism for the Ca^{2+} release at fertilization is not known, but it is possible that the sperm introduces a specific protein factor that initiates Ca^{2+} release after granule fusion.

The sharp increase in free Ca^{2+} is crucial for egg activation. Eggs in a variety of animals are activated if the Ca^{2+} concentration in the egg cytosol is artificially increased, for example, by direct injection of Ca^{2+}. Conversely, preventing calcium increase by injecting agents that bind it, such as the calcium chelator EGTA, blocks activation. It has been known for many years that *Xenopus* eggs can be activated simply by prodding them with a glass needle; this is due to a local influx in calcium at the site of insertion that triggers a calcium wave. Calcium initiates development of the fertilized egg by its action on proteins that control the cell cycle.

The unfertilized *Xenopus* egg is maintained in the metaphase of the second meiotic division by the presence of high levels of a protein complex, maturation-promoting factor (MPF), one of whose components is the protein cyclin. For the egg to complete meiosis and to start dividing mitotically, the level of MPF activity must be reduced (Fig. 12.26). The calcium wave results in the activation of the enzyme, calmodulin-dependent protein kinase II. The activity of this kinase results in the degradation of the cyclin component of MPF; and so the egg completes meiosis. The pronuclei

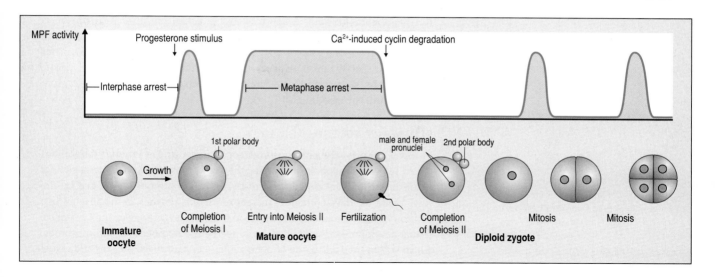

then fuse, and the zygote moves on to the next stage of its cell cycle, which is entry into mitosis.

Summary

The fusion of sperm and egg at fertilization stimulates the egg to start dividing and developing. Both sperm and egg have specialized structures relating to fertilization. The initial binding of sperm to the mammalian egg is mediated by cell-surface molecules, and leads to the release of the contents of the sperm acrosome, which facilitates the penetration of the sperm through the layers surrounding the egg and allows it to reach the egg plasma membrane. A block to polyspermy allows only one sperm to fuse with the egg and deliver its nucleus into the egg cytoplasm. In sea urchins, the first block is rapid and partial, and the second results from the release of egg cortical granule contents to the exterior to form an impenetrable

Fig. 12.26 Profile of maturation-promoting factor (MPF) activity in early *Xenopus* development. The immature *Xenopus* oocyte cell cycle is arrested. On receipt of a hormonal progesterone stimulus it enters, and completes, the first meiotic division, with the formation of the first polar body. It enters the second meiotic division, but becomes arrested again in metaphase. The egg is laid at this point. At fertilization, the calcium wave leads to completion of meiosis, and the second polar body is formed. The zygote starts to cleave rapidly by mitotic divisions. Maturation-promoting factor (MPF) rises sharply just before each division of the meiotic and mitotic cell cycles, remains high during mitosis, and then decreases abruptly and remains low between successive mitoses.

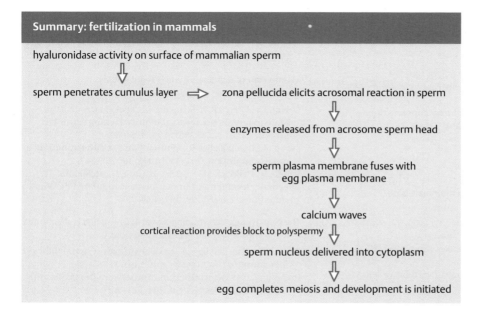

Summary: fertilization in mammals

hyaluronidase activity on surface of mammalian sperm

⇩

sperm penetrates cumulus layer ⟹ zona pellucida elicits acrosomal reaction in sperm

⇩

enzymes released from acrosome sperm head

⇩

sperm plasma membrane fuses with egg plasma membrane

⇩

calcium waves

cortical reaction provides block to polyspermy ⇩

sperm nucleus delivered into cytoplasm

⇩

egg completes meiosis and development is initiated

fertilization membrane. A key role in egg activation after fertilization is played by the release of free calcium ions into the cytosol, which spread in a wave from the site of sperm fusion. In mammals, and most other vertebrates, fertilization triggers the completion of the second meiotic division; the sperm and egg haploid pronuclei give rise to the zygote nucleus, and the egg divides.

SUMMARY TO CHAPTER 12

In many animals, the chromosomal constitution of the embryo determines which sex will develop. In mammals, the Y chromosome is male-determining; it specifies the development of a testis, and the hormones produced by the testis cause the development of male sexual characteristics. In the absence of a Y chromosome, the embryo develops as a female. In *Drosophila* and *C. elegans*, sexual development is initially determined by the number of X chromosomes, which sets in train a cascade of gene activity. In mammals, the somatic sexual phenotype is determined by cell–cell interactions; in the fly, somatic cell sexual differentiation is cell autonomous. In many animals, the future germ cells are specified by localized cytoplasmic determinants in the egg. Germ cells in mammals are unusual, as they are specified entirely by cell–cell interactions, and the same is true of flowering plants, in which the germ cells are only specified at a late stage, as the flowers develop. Most flowering plants are hermaphrodites. In animals, the development of germ cells into sperm or egg depends both on chromosomal constitution and interactions with the cells of the gonad. At fertilization, fusion of sperm and egg initiates development, and there are mechanisms to ensure that only one sperm enters an egg. Both maternal and paternal genomes are required for normal mammalian development, as some genes are imprinted; for such genes, whether they are expressed or not during development depends on whether they are derived from the sperm or the egg. Animals use a variety of strategies of dosage compensation to correct the imbalance in the number of X chromosomes in males and females.

GENERAL REFERENCES

Crews, D.: **Animal sexuality**. *Sci. Amer.* 1994, **270**: 109–114.

Marsh, J., Goodie, J. (eds): *Germline Development*. Ciba Symposium 182. Chichester: John Wiley, 1994.

Zarkower, D.: **Establishing sexual dimorphism: conservation amidst diversity?** *Nat. Rev. Genet.* 2001, **2**: 175–185.

SECTION REFERENCES

12.1 **The primary sex-determining gene in mammals is on the Y chromosome**

Capel, B. **The battle of the sexes**. *Mech. Dev.* 2000, **92**: 89–103.

Goodfellow, P.N., Lovell-Badge, R.: *SRY* **and sex determination in mammals**. *Annu. Rev. Genet.* 1993, **27**: 71–92.

Koopman, P.: **The genetics and biology of vertebrate sex determination**. *Cell* 2001, **105**: 843–847.

Schafer, A.J., Goodfellow, P.N.: **Sex determination in humans**. *BioEssays* 1996, **18**: 955–963.

12.2 **Mammalian sexual phenotype is regulated by gonadal hormones**

Colvin, J.S., Green, R.P., Schmahl, J., Capel, B., Ornitz, D.M.: **Male-to-female sex reversal in mice lacking fibroblast growth factor 9**. *Cell* 2001, **104**: 875–889.

Swain, A., Lovell-Badge, R.: **Mammalian sex determination: a molecular drama**. *Genes Dev.* 1999, **13**: 755–767.

Vainio, S., Heikkila, M., Kispert, A., Chin, N., McMahon, AP.: **Female development in mammals is regulated by Wnt-4 signalling**. *Nature* 1999, **397**: 405–409.

12.3 **The primary sex-determining signal in *Drosophila* is the number of X chromosomes, and is cell autonomous**

Cline, T.W.: **The *Drosophila* sex determination signal: how do flies count to two?** *Trends Genet.* 1993, **9**: 385–390.

Hodgkin, J.: **Sex determination compared in *Drosophila* and *Caenorhabditis***. *Nature* 1990, **344**: 721–728.

12.4 Somatic sexual development in *Caenorhabditis* is determined by the number of X chromosomes

Cline, T.W., Meyer, B.J.: **Vive la difference: males vs. females in flies vs. worms.** *Ann. Rev. Genet.* 1996, **30**: 637–702.

Raymond, C.S., Shamu, C.E., Shen, M.M., Seifert, K.J., Hirsch, B., Hodgkin, J., Zarkower, D.: **Evidence for evolutionary conservation of sex-determining genes.** *Nature* 1998, **391**: 691–695.

12.5 Most flowering plants are hermaphrodite, but some produce unisexual flowers

Irisa, E.N.: **Regulation of sex determination in maize.** *BioEssays* 1996, **18**: 363–369.

12.6 Germ-cell sex determination can depend both on cell signals and genetic constitution

McLaren, A.: **Signaling for germ cells.** *Genes Dev.* 1999, **13**: 373–376.

Seydoux, G., Strome, S.: **Launching the germline in *Caenorhabditis elegans*: regulation of gene expression in early germ cells.** *Development* 1999, **126**: 3275–3283.

12.7 Various strategies are used for dosage compensation of X-linked genes

Avner, P., Heard, E.: **X-chromosomes inactivation: counting, choice and initiation.** *Nat. Rev. Genet.* 2001, **2**: 59–67.

Davis, T.L., Meyer, B.J.: **SDC-3 co-ordinates the assembly of a dosage compensation complex on the nematode X chromosome.** *Development* 1997, **124**: 1019–1031.

Meller, V.H.: **Dosage compensation: making 1X equal 2X.** *Trends Cell Biol.* 2000, **10**: 54–59.

Meyer, B.J.: **Sex in the worm: counting and compensating X-chromosome dose.** *Trends Genet.* 2000, **16**: 247–253.

12.8 Germ-cell fate can be specified by a distinct germ plasm in the egg

Matova, N., Cooley, L.: **Comparative aspects of animal oogenesis.** *Dev. Biol.* 2001, **231**: 291–320.

McLaren, A.: **Establishment of the germ cell lineage in mammals.** *J. Cell. Physiol.* 2000, **182**: 141–143.

Mello, C.C., Schubert, C., Draper, B., Zhang, W., Lobel, R., Priess, J.R.: **The PIE-1 protein and germline specification in *C. elegans* embryos.** *Nature* 1996, **382**: 710–712.

Williamson, A., Lehmann, R.: **Germ cell development in *Drosophila*.** *Annu. Rev. Cell Dev. Biol.* 1996, **12**: 365–391.

12.9 Pole plasm becomes localized at the posterior end of the *Drosophila* egg

Micklem, D.R., Adams, J., Grunert, S., St. Johnston, D.: **Distinct roles of two conserved Staufen domains in oskar mRNA localisation and translation.** *EMBO J.* 2000, **19**: 1366–1377.

12.10 Germ cells migrate from their site of origin to the gonad

Dixon, K.E.: **Evolutionary aspects of primordial germ cell formation.** In *Germline Development.* Ciba Symposium 182. Chichester: John Wiley, 1994: 92–120.

Wylie, C. **Germ cells.** *Cell* 1999, **96**: 165–174.

12.11 Germ-cell differentiation involves a reduction in chromosome number

De Rooij, D.G., Grootegoed, J.A.: **Spermatogonial stem cells.** *Curr. Opin. Cell Biol.* 1998, **10**: 694–701.

Metz, C.B., Monroy, A. (eds.): *Biology of the Sperm. Biology of Fertilization*, vol. 2. Orlando, Florida: Academic Press, 1985.

12.12 Oocyte development can involve gene amplification and contributions from other cells

Browder, L.W.: *Oogenesis.* New York: Plenum Press, 1985.

de Rooij, D.G., Grootegoed, J.A.: **Spermatogonial stem cells.** *Curr. Opin. Cell Biol.* 1998, **10**: 694–701.

Spradling, A.: **Developmental genetics of oogenesis.** In Drosophila *Development*. Edited by Bate, M., Martinez-Arias, A. New York: Cold Spring Harbor Laboratory Press, 1993: 1–69.

12.13 Genes controlling embryonic growth are imprinted

Reik, W., Walter, J.: **Genomic imprinting: parental influence on the genome.** *Nat. Rev. Genet.* 2001, **2**: 21–32.

12.14 Fertilization involves cell-surface interactions between egg and sperm

Singson, A.: **Every sperm is sacred: fertilization in *Caenorhabditis elegans*.** *Dev. Biol.* 2001, **230**: 101–109.

Wasserman, P.M.: **Mammalian fertilization: molecular aspects of gamete adhesion, exocytosis, and fusion.** *Cell* 1999, **96**: 175–183.

12.15 Changes in the egg membrane at fertilization block polyspermy

Ohelndieck, K., Lennarz, W.J.: **Role of the sea urchin egg receptor for sperm in gamete interactions.** *Trends Biochem. Sci.* 1995, **20**: 29–33.

Tian, J., Gong, H., Thomsen, G.H., Lennarz, W.J.: ***Xenopus laevia* sperm–egg adhesion is regulated by modifications in the sperm receptor and the egg vitelline envelope.** *Dev. Biol.* 1997, **187**: 143–153.

12.16 A calcium wave initiated at fertilization results in egg activation

Fissore, R.A., *et al.*: **Sperm induced calcium oscillations. Isolation of the Ca^{2+}-releasing component(s) of mammalian sperm extracts: the search continues.** *Mol. Hum. Reprod.* 1999, **5**: 189–192.

Swann, K., Parrington, J.: **Mechanism of Ca^{2+} release at fertilization in mammals.** *J. Exp. Zool.* 1999, **285**: 267–275.

Regeneration

13

- Limb regeneration
- Regeneration in *Hydra*

"Even when some of us were removed by force, others took our place."

Many of the cells in the adult body, such as muscle and nerve cells, are the same cells as were originally generated during embryonic development. But some tissues, such as blood and epithelia, are continually being replaced by stem cells (see, for example, Section 9.6). We have also seen many examples of the capacity of the embryo to self-regulate when parts of it are removed or rearranged (see, for example, Sections 3.5 and 6.1). Here we look at the related phenomenon of **regeneration** in adult organisms. Regeneration is the ability of the fully developed organism to replace tissues, organs and appendages by growth or remodeling of somatic tissue. Plants have remarkable powers of regeneration: a single somatic plant cell can give rise to a complete new plant. Some animals also show great ability to regenerate: small fragments of animals such as starfish, planarians (flatworms), and *Hydra* can give rise to a whole animal (Fig. 13.1). The ability of animals like *Hydra* and planarians to regenerate may be related to their ability to reproduce asexually. A remarkable case of regeneration is claimed for ascidians, whose blood cells alone have been reported to give rise to a

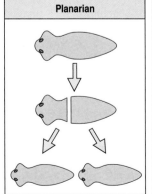
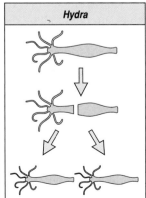
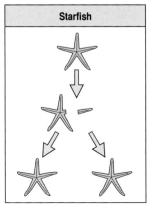

Fig. 13.1 Regeneration in some invertebrate animals. A planarian, *Hydra*, and a starfish all show remarkable powers of regeneration. When parts are removed or a small fragment isolated, a whole animal can be regenerated.

Fig. 13.2 The capacity for regeneration in urodele amphibians. The emperor newt can regenerate its dorsal crest (1), limbs (2), retina and lens (3 and 4), jaw (5), and tail (not shown).

fully functional organism. But just why some animals can regenerate while others are unable to do so is not clear.

Among vertebrates, newts and other urodele amphibians show a remarkable capacity for regeneration (Fig. 13.2). The newt lens, for example, regenerates from the pigmented epithelium of the iris (Fig. 13.3). Some insects and other arthropods can also regenerate lost appendages, such as legs. The regenerative powers of mammals are much more restricted. The mammalian liver can regenerate if a part of it is removed, the antlers of male deer regenerate each year, and fractured bones can mend by a regenerative process. But mammals cannot regenerate lost limbs, although they do have a limited capacity to replace the ends of digits, and nematodes and rotifers cannot regenerate at all.

The issue of regeneration raises several major questions. Why are some animals able to regenerate and others not? What is the origin of the cells that give rise to the regenerated structures? What mechanisms pattern the regenerated tissue and how are these related to the patterning processes that occur in embryonic development? We will focus on two systems in which regeneration has been intensively studied: regeneration of the whole animal in *Hydra*, and limb regeneration in insects and amphibians. Plant regeneration was discussed briefly in Chapter 7 and that of the nervous system in Chapter 11.

A distinction can be drawn at the outset between two types of regeneration. In one—**morphallaxis**—there is little new growth, and regeneration occurs mainly by the repatterning of existing tissues and the re-establishment of boundaries. Regeneration in *Hydra* is a good example of morphallaxis. By contrast, regeneration in the newt limb, for example, depends on the growth of new, correctly patterned structures, and this is known as **epimorphosis**. Both types of regeneration can be illustrated with reference to the French flag pattern (Fig. 13.4). In morphallaxis, new boundary regions are first established and new positional values are specified in relation to them; in epimorphosis, new positional values are linked to growth from the cut surface. In both cases, the origin of the progenitor cells for the regenerated tissue is a key issue.

Limb regeneration

Urodele amphibians such as newts and axolotls show a remarkable capacity for regenerating body structures such as tails, limbs, jaws, and the lens of the eye (see Fig. 13.2). Regeneration of all these structures involves new growth and is therefore an epimorphic type of regeneration. Damaged insect legs and other appendages can also regenerate. This has been studied

Fig. 13.3 Lens regeneration. Removal of the lens from the eye of a newt results in regeneration of a new lens from the dorsal pigmented epithelium of the iris.

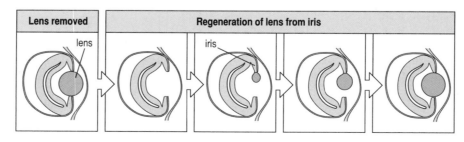

mainly in the larval cockroach, which has relatively large legs that are easy to manipulate. Regeneration of a structure such as an adult vertebrate limb, which contains a variety of fully differentiated cell types in a highly organized arrangement, raises a central question relating to the origin of the cells that give rise to the regenerated structure: are there special reserve cells or do cells dedifferentiate and change their character? As we shall see, the fully differentiated cells of the mature vertebrate limb return to the cell cycle, dedifferentiate, and then redifferentiate into different cell types. Regeneration of insect appendages will also be considered.

13.1 Vertebrate limb regeneration involves cell dedifferentiation and growth

Following amputation of a newt limb, there is a rapid migration of epidermal cells over the wound surface, which is essential for subsequent outgrowth, and a **blastema** begins to form, which will give rise to a regenerated limb (Fig. 13.5). The blastema is formed from cells beneath the wound epidermis which lose their differentiated character and start to divide. As the limb regenerates over a period of weeks, these cells differentiate into cartilage, muscle, and connective tissue. The blastemal cells are derived locally from the mesenchymal tissues of the stump, close to the site of amputation. They particularly come from the dermis but also from the cartilage and muscle. This raises the question as to whether cells differentiating into cartilage and muscle in the blastema are remaining true to type or whether, for example, multinucleate skeletal muscle cells in the stump undergo dedifferentiation and then give rise to other cell types, such as cartilage, during regeneration. In other words, is transdifferentiation (see Section 9.18) occurring during limb regeneration, as it occurs in the transdifferentiation of pigmented iris epithelial cells into lens cells in the regenerating eye lens of a newt (see Fig. 13.3)? The answer, at least in the newt, is yes. If cultured newt limb muscle myotubes, which are multinucleate and have stopped dividing, are labeled in culture with a retrovirus expressing alkaline phosphatase, and are introduced into regenerating limbs, strongly labeled mononucleate cells can be observed in blastemas after 1 week. The majority of the myotubes give rise to mononucleate cells. These mononucleate cells proliferate, and there is some evidence that they later give rise to cartilage as well as new muscle.

These results reinforce the view that the early blastema is an environment that causes cells to dedifferentiate. The cells in this special environment then act as progenitor cells for the regenerating limb. This is further illustrated by the striking observation that if iris epithelium cells are grafted to different sites on the newt's body, they maintain their differentiated state. However, if they are grafted into a limb blastema they differentiate into lens cells.

The ability of muscle cells to re-enter the cell cycle is a special feature of newt limb regeneration, since mature muscle cells normally never divide (Chapter 9). A general feature of vertebrate muscle differentiation is the withdrawal of the muscle precursor cells from the cell cycle after myoblast fusion, which involves the dephosphorylation of the protein product of the *Rb* gene (see Section 9.5). Cultured mouse muscle cells lacking the Rb protein can re-enter the cell cycle. The regenerating newt cells contain Rb protein but it is inactivated by phosphorylation, so the cells can re-enter the

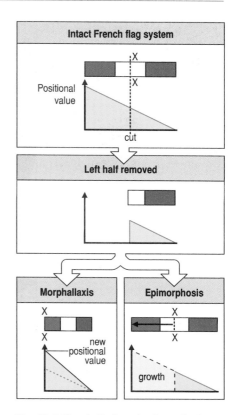

Fig. 13.4 Morphallaxis and epimorphosis. A pattern such as the French flag may be specified by a gradient in positional value (see Fig. 1.22). If the system is cut in half it can regenerate in one of two ways. In regeneration by morphallaxis, a new boundary is established at the cut and the positional values are changed throughout. In regeneration by epimorphosis, new positional values are linked to growth from the cut surface.

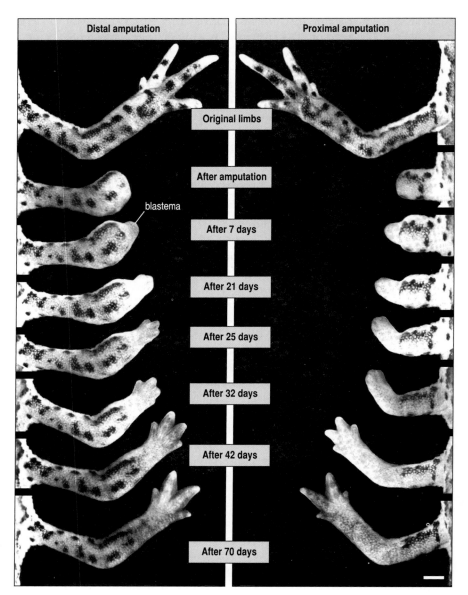

Distal amputation | **Proximal amputation**

Original limbs

After amputation

blastema

After 7 days

After 21 days

After 25 days

After 32 days

After 42 days

After 70 days

Fig. 13.5 Regeneration of the forelimb in the red-spotted newt *Notophthalmus viridescens.* The left panel shows the regeneration of a forelimb after amputation at a distal (mid-radius/ulna) site. The right panel shows regeneration after amputation at a proximal (mid-humerus) site. At the top, the limbs are shown before amputation. Successive photographs were taken at the times shown after amputation. Note that the blastema gives rise to structures distal to the cut. Scale bar = 1 mm.

cell cycle and divide. The ability of newt muscle cells to enter the cell cycle is signaled by local activation of thrombin, a proteolytic protein that is more familiar as part of the blood clotting cascade. The regeneration of the newt lens from the dorsal margin of the iris correlates with thrombin activity in that region.

Growth of the blastema is dependent on its nerve supply (Fig. 13.6) and the overlying wound epidermis, which may play a role similar to the apical ectodermal ridge in limb development (Section 10.3). In limbs in which the nerves have been cut before amputation, a blastema forms but fails to grow. The nerves have no influence on the character or pattern of the regenerated structure; it is the amount of neural innervation, not the type of nerve that matters. Nerve cells, therefore, seem to be providing some essential growth factor. Members of the neuregulin family of growth factors are likely candidates and their local application can result in regeneration of denervated limbs.

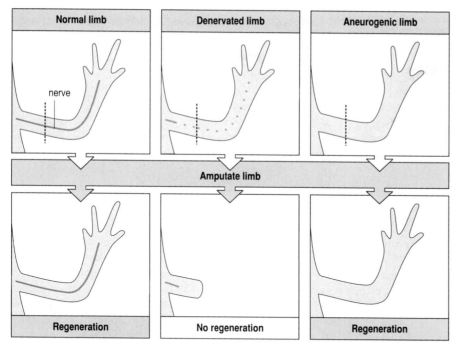

Fig. 13.6 Innervation and limb regeneration. Normal limbs require a nerve supply to regenerate (left panels). Limbs denervated prior to amputation will not regenerate (center panels). However, limbs that have never been innervated, because the nerve was removed during development, can regenerate normally in the absence of nerves (right panels).

An interesting phenomenon, as yet unexplained, is that if embryonic limbs are denervated very early in their development and so have never been exposed to the influence of nerves, they can regenerate in the complete absence of any nerve supply (see Fig. 13.6, right panel). If such an aneurogenic limb is innervated, it rapidly becomes dependent on nerves for regeneration. This evidence suggests that the dependence on the nerve is imposed on the limb after the ingrowth of the nerve.

13.2 The limb blastema gives rise to structures with positional values distal to the site of amputation

Regeneration always proceeds in a direction distal to the cut surface, allowing replacement of the lost part of the limb. If the hand is amputated at the wrist, only the carpals and digits are regenerated, whereas if amputation is through the middle of the humerus, everything distal to the cut (including the distal humerus) is regenerated. Positional value along the axis is therefore of great importance. The blastema has considerable morphogenetic autonomy. If it is transplanted to a neutral location that permits growth, such as the dorsal crest of a newt larva or even the anterior chamber of the eye, it gives rise to a regenerate appropriate to the position from which it was taken.

The growth of the blastema and the nature of the structures it gives rise to are dependent on the site of the amputation and not on the nature of the more proximal tissues. The limb is not, however, simply 'trying' to replace missing parts. This was shown in a classic experiment in which the distal end of a newt limb that had been amputated at the wrist was inserted into the belly of the same animal so as to establish a blood supply to it. The limb was then cut mid-humerus. Both surfaces regenerated distally, even though the part attached to the belly already had a radius and ulna (Fig. 13.7).

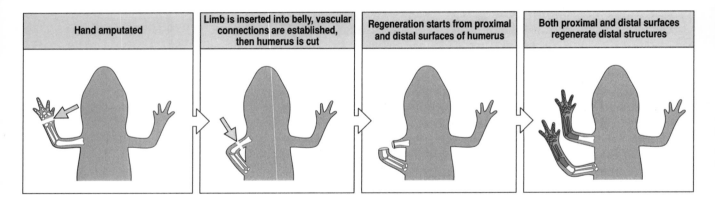

| Hand amputated | Limb is inserted into belly, vascular connections are established, then humerus is cut | Regeneration starts from proximal and distal surfaces of humerus | Both proximal and distal surfaces regenerate distal structures |

Fig. 13.7 Limb regeneration is always in the distal direction. The distal end of a limb is amputated and the limb inserted into the belly. Once vascular connections are established, a cut is made through the humerus. Both cut surfaces regenerate the same distal structures even though, in the case of one of the regenerating limbs, distal structures are already present.

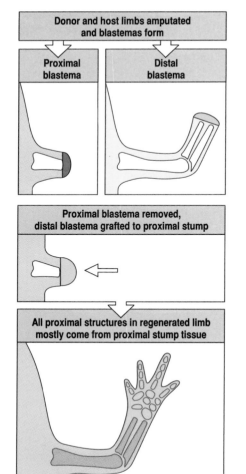

Donor and host limbs amputated and blastemas form

Proximal blastema

Distal blastema

Proximal blastema removed, distal blastema grafted to proximal stump

All proximal structures in regenerated limb mostly come from proximal stump tissue

The blastema is much larger—it can be ten times larger in terms of cell numbers—than the embryonic limb bud. This size makes it most unlikely that there are signals that diffuse right across the blastema. Instead, regeneration can best be understood in terms of an adult limb having a set of positional values along its proximo-distal axis, which are set up during embryonic development (see Section 10.6). The regenerating limb in some way reads the positional value at the site of the amputation and then regenerates all positional values distal to it. Epimorphic regeneration involves the retention of embryonic processes, like the ability to specify new positional values.

The ability of cells to recognize a discontinuity in positional values is illustrated by grafting a distal blastema to a proximal stump. In this experiment, the forelimb's stump and blastema have different positional values, corresponding to shoulder and wrist, respectively. The result is a normal limb in which structures between the shoulder and wrist have been generated by **intercalary growth**, predominantly from the proximal stump, while the cells from the wrist blastema mostly give rise to the hand (Fig. 13.8).

During limb development in the embryo, the Hoxa genes are expressed with temporal and spatial co-linearity along the proximo-distal axis, as we saw in Chapter 10. In the regenerating axolotl limb, however, two Hoxa genes from the 3′ and 5′ ends of the complex are expressed in the stump cells at the same time—24 to 48 hours after amputation. This suggests that the most distal region of the blastema is specified first and that regeneration involves intercalation of positional values between this distal region and the stump. The timing of expression of the Hoxd complex is similar to that in embryonic development, *Hoxd-8* being switched on earlier than *Hoxd-11*, although the early expression of *Hoxd-11* would fit better with the intercalation model as it is in the most distal region of the limb. *Sonic hedgehog* is expressed in a small region at the posterior margin of the blastema but its function is not known. However, limb duplication induced by treatment with retinoic acid is also preceded by anterior expression of *Sonic hedgehog*.

While the nature of the positional values is not known, there is evidence

Fig. 13.8 Proximo-distal intercalation in limb regeneration. A distal blastema grafted to a proximal stump results in intercalation of all the structures proximal to the distal blastema. Almost all of the intercalated region comes from the proximal stump.

that cell-surface properties are involved. When mesenchyme from two blastemas from different proximo-distal sites are confronted in culture, the more proximal mesenchyme engulfs the distal (Fig. 13.9, left panels), whereas mesenchyme from two blastemas at similar sites maintains a stable boundary. This behavior is suggestive of a graded difference in cell adhesiveness along the axis, with adhesiveness being highest distally. The cells in the distal explant remain more tightly bound to each other than do the cells from the proximal blastema, which thus spread to a greater extent. This difference in adhesiveness is also suggested by the behavior of a distal blastema when grafted to the dorsal surface of a proximal blastema, such that their mesenchymal cells are in contact. Under these conditions, the distal blastema moves during limb regeneration to end up at the site from which it originated (see Fig. 13.9, right panels). This suggests that its cells adhere more strongly to the regenerated wrist region than they do to the proximal region to which it was transplanted. Transplantation of a shoulder-level blastema to a shoulder stump does not mobilize the stump tissue, but leads to a normal distal outgrowth from the shoulder blastema. These experiments suggest that proximo-distal positional values in urodele limb regeneration are encoded as a graded property, probably in part at the cell surface, and that cell behavior relevant to axial specification—growth, movement, and adhesion—is a function of the expression of this property, relative to neighboring cells.

Maintaining the continuity of positional values by intercalation is a fundamental property of regenerating epimorphic systems and we consider it further in relation to the cockroach leg, below. Even normal regeneration by outgrowth from a blastema could be considered the result of intercalation between the cells at the level of amputation and those with the most distal positional values, as specified by the wound epidermis. It is not clear to what extent blastemal cells inherit a particular positional value, for example, from their differentiated precursors, and to what extent they are subject to signals that induce the appropriate expression of positional value. The precise relationship between Hox gene expression and positional identity is not understood, either for limb embryonic development or regeneration.

Although mammals cannot regenerate whole limbs, many, including young children, can regenerate the ends of their digits. In mice and children, the level from which digits are able to regenerate is limited to the base of the claw or nail, respectively. This probably reflects the presence of stem cells in the nail bud rather than cell dedifferentiation. In mammalian limb development, the homeobox gene *Msx-1* is expressed in the progress zone of the embryonic limb bud, and in mice it continues to be expressed in the tips of the digits even after birth. Since the region in which digit regeneration in mice can occur corresponds with that of *Msx-1* expression, this gene may be required for the generation of new positional values.

13.3 Retinoic acid can change proximo-distal positional values in regenerating limbs

We have already seen that retinoic acid is present in developing vertebrate limbs and how experimental treatment with retinoic acid can alter positional values in the developing chick limb (see Section 10.5). It also has striking effects on regenerating amphibian limbs.

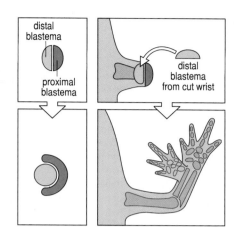

Fig. 13.9 Cell-surface properties vary along the proximo-distal axis. Left panels: when mesenchyme from distal and proximal blastemas is placed in contact in culture, the proximal mesenchyme engulfs the distal mesenchyme, which has greater adhesion between its cells. Right panels: if a distal blastema (in this case from a cut wrist) is grafted to the dorsal surface of a more proximal blastema, the regenerating wrist blastema will move distally to a position on the host limb that corresponds to its original level and regenerates a hand.

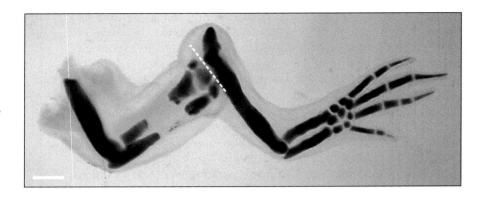

Fig. 13.10 Retinoic acid can proximalize positional values. A forelimb amputated at the level of the hand, as indicated by the dotted line, and then treated with retinoic acid, regenerates structures corresponding to a cut at the proximal end of the humerus. Scale bar = 1 mm.

Exposing a regenerating limb to retinoic acid results in the blastema becoming proximalized; that is, the limb regenerates as if it had originally been amputated at a more proximal site. For example, if a limb is amputated through the radius and ulna, treatment with retinoic acid will result not only in the regeneration of the elements distal to the cut, but also in the production of an extra complete radius and ulna. The effect of retinoic acid is dose dependent, and with a high dose it is possible to regenerate a whole extra limb, including part of the shoulder girdle, on a limb from which only the hand has been amputated (Fig. 13.10). Retinoic acid can therefore alter the proximo-distal positional value of the blastema, making it more proximal. Retinoic acid can also, under some experimental conditions, shift positional values along the anterior–posterior axis in a posterior direction.

In untreated regenerating limbs, endogenous retinoic acid is present in a distinct pattern, although there is no direct evidence that it is involved in regeneration. There is an antero-posterior gradient of retinoic acid in the blastema, and there is also a higher concentration in distal blastemas than in proximal blastemas, suggesting a proximo-distal gradient as well. The wound epidermis is a strong source of retinoic acid.

Retinoic acid is known to act through a variety of receptors. There are a number of different types of receptors in the limb but only one of them (δ_2) is involved in changes in positional values. By constructing a chimeric

Fig. 13.11 Retinoic acid proximalizes the positional value of individual cells. Some of the cells of a newt distal blastema are transfected by a chimeric receptor, through which retinoic acid receptor function can be activated by thyroxine. This blastema is grafted to a proximal stump and treated with thyroxine. During intercalary growth, the transfected cells, which have been labeled, move proximally because their positional values have been proximalized by the activation of the retinoic acid receptor. The photographs illustrate proximalization of the transfected cells. Scale bar = 0.5 mm.

Photographs from Pecorino, L.T., et al.: 1996.

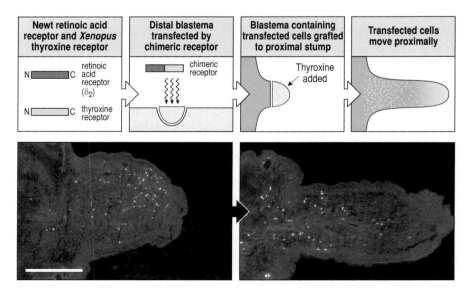

normal limb

Fig. 13.12 Retinoic acid can induce additional limbs in regenerating tail of a frog, *Rana temporaria*. Treatment of the regenerating tail with retinoic acid at the time when hindlimbs are developing, results in the appearance of additional hindlimbs in place of a regenerated tail. Scale bar = 5 mm.

Photograph courtesy of M. Maden.

receptor from a δ_2 retinoic acid receptor and a thyroxine receptor, the retinoic acid receptor can be activated selectively by thyroxine, and the effects of its activation studied experimentally (Fig. 13.11). In these experiments, cells in the distal blastema are transfected with the chimeric receptor, which is then grafted to a proximal stump and treated with thyroxine. The transfected cells behave as if they have been treated with retinoic acid. The result is a movement of the transfected blastemal cells to more proximal regions in the intercalating regenerate. This shows that activation of the retinoic acid pathway can proximalize the positional values of the cells, which then respond by translocation to more proximal sites.

Another remarkable effect of retinoic acid is its ability to bring about a homeotic transformation of tails into limbs in tadpoles of the frog *Rana temporaria*. If the tail of a tadpole is removed it will regenerate. Treatment of regenerating tails with retinoic acid at the same time as hindlimbs are developing results in the appearance of additional hindlimbs in place of a regenerated tail (Fig. 13.12). We have, as yet, no satisfactory explanation for this result, but it has been speculated that the retinoic acid alters the antero-posterior positional value of the regenerating tail blastema to that of the site along the antero-posterior axis where hindlimbs would normally develop.

13.4 Insect limbs intercalate positional values by both proximo-distal and circumferential growth

The legs of some insects such as the cockroach can regenerate. The structure of insect legs is different from those of vertebrates as the key structural feature is the ectodermally derived external cuticle. Nevertheless, intercalation of missing positional values, as already described for the proximo-distal axis of the amphibian limb (see Section 13.2), seems to be a general property of regenerating epimorphic systems. When cells with disparate positional values are placed next to one another, intercalary growth occurs in order to regenerate the missing positional values. Intercalation is particularly clearly illustrated by limb regeneration in the cockroach. Intercalation also occurs in *Drosophila* leg and wing imaginal discs.

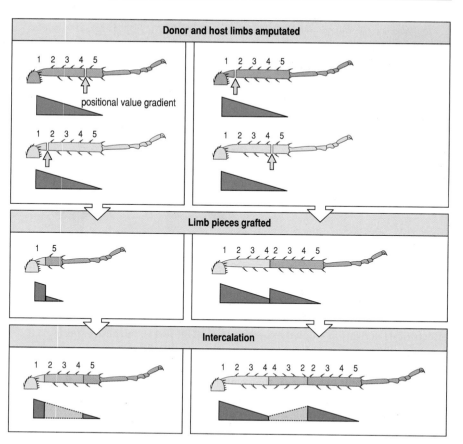

Fig. 13.13 Intercalation of positional values by growth in the regenerating cockroach leg. Left panels: when a distally amputated tibia (5) is grafted to a proximally amputated host (1), intercalation of the positional values 2–4 occurs, irrespective of the proximo-distal orientation of the grafts, and a normal tibia is regenerated. Right panels: when a proximally amputated tibia (1) is grafted to a distally amputated host (4), however, the regenerated tibia is longer than normal and the regenerated portion is in the reverse orientation to normal, as judged by the orientation of surface bristles. The reversed orientation of regeneration is due to the reversal in positional value gradient. The proposed gradient in positional value is shown under each figure. After French, V., *et al.*: 1976.

A cockroach leg is made up of a number of distinct segments, arranged along the proximo-distal axis in the order coxa, femur, tibia, tarsus. Each segment seems to contain a similar set of proximo-distal and circumferential positional values, and will intercalate the missing positional values. When a distally amputated tibia is grafted onto a host tibia that has been cut at a more proximal site, localized growth occurs at the junction between graft and host, and the missing central regions of the tibia are intercalated (Fig. 13.13, left panels). In contrast to amphibian regeneration, there is a predominant contribution from the distal piece. As in the amphibian, however, regeneration is a local phenomenon and the cells are indifferent to the overall pattern of the tibia. Thus, when a proximally cut tibia is grafted onto a more distal site, making an abnormally long tibia, regenerative intercalation again restores the missing positional values, making the tibia even longer (Fig. 13.13, right panels). The regenerated portion is in the reverse orientation to the rest of the limb, as indicated by the direction in which the bristles point, suggesting that the gradient in positional values also specifies cell polarity, as in insect body segments (see Chapter 5). These results also show that when cells with non-adjacent positional values are placed next to each other, the missing values are intercalated by growth to provide a set of continuous positional values.

A similar set of positional values is present in each segment of the limb. Thus, a mid-tibia amputation, when grafted to the mid-femur of a host, will heal without intercalation. But grafting a distally amputated femur onto a proximally amputated host tibia results in intercalation, largely femur in

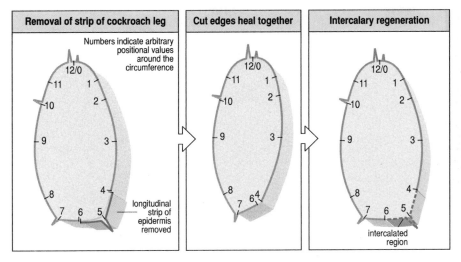

Fig. 13.14 Circumferential intercalation in the cockroach leg. The leg is seen in transverse section. When a piece of cockroach ventral epidermis is removed (left panel), the cut edges heal together (center panel). When the insect molts and the cuticle regrows, circumferential positional values are intercalated (right panel). The positional values are arranged around the circumference of the leg, rather like the hours on a clock face. After French, V., *et al.*: 1976.

type. There must be other factors making each segment different, rather like the segments of the insect larva.

Intercalary regeneration also occurs in a circumferential direction. When a longitudinal strip of epidermis is removed from the leg of a cockroach, normally nonadjacent cells come into contact with one another, and intercalation in a circumferential direction occurs after molting (Fig. 13.14). Cell division occurs preferentially at sites of mismatch around the circumference. One can treat positional values in the circumferential direction as a clock face, with values going continuously 12, 1, 2, 3 . . . 6 . . . 9 . . . 11. As in the proximo-distal axis, there is intercalation of the missing positional values.

Summary

Urodele amphibians can regenerate amputated limbs and tails. Stump tissue at the site of amputation first dedifferentiates to form a blastema, which then grows and gives rise to a regenerated structure. The dedifferentiated cells of the blastema are the progenitors of the regenerate. Regeneration is usually dependent on the presence of nerves, but limbs that have never been innervated are capable of regeneration. Regeneration

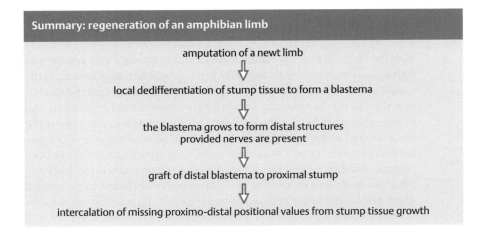

Summary: regeneration of an amphibian limb

amputation of a newt limb

⇩

local dedifferentiation of stump tissue to form a blastema

⇩

the blastema grows to form distal structures provided nerves are present

⇩

graft of distal blastema to proximal stump

⇩

intercalation of missing proximo-distal positional values from stump tissue growth

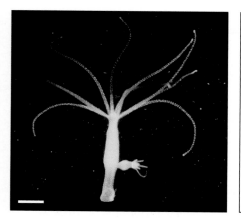
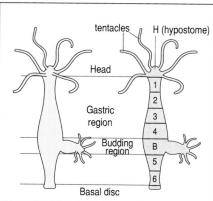
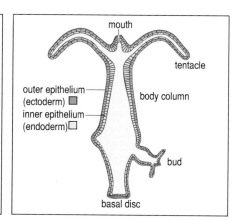

Fig. 13.15 *Hydra*. This freshwater coelenterate (photograph, left panel) has a head with tentacles and mouth at one end and a sticky foot at the other. It can reproduce by budding. For the purposes of grafting experiments, the body column is divided into a series of regions, as indicated in the center panel. The body wall is made up of two epithelial layers, corresponding to the ectoderm and endoderm found in other organisms featured in this book (right panel). Scale bar = 1 mm.

Photograph courtesy of W. Müller, from Müller, W.A.: 1989.

always gives rise to structures with positional values more distal than those at the site of amputation. When a blastema is grafted to a stump with different positional values, proximo-distal intercalation of the missing positional values occurs. Retinoic acid proximalizes the positional values of the cells of the blastema. Insect limbs can also regenerate, intercalation of positional values occurring in both proximo-distal and circumferential directions.

Regeneration in *Hydra*

Hydra is a freshwater coelenterate consisting of a hollow tubular body about 0.5 cm long, with a head region at one end (the distal end) and a basal region at the other (proximal) end, with which it can stick to surfaces (Fig. 13.15). The head consists of a small conical hypostome where the mouth opens, surrounded by a set of tentacles, which are used for catching the small animals on which *Hydra* feeds. Unlike most of the animals discussed so far in this book, which have three germ layers, *Hydra* has only two. The body wall is composed of an outer epithelium, which corresponds to the ectoderm, and an inner epithelium, which corresponds to the endoderm. These two layers are separated by a basement membrane. There are about 20 different cell types in *Hydra*, which include nerve cells, secretory cells, and nematocysts that are used to capture prey.

13.5 *Hydra* grows continuously but regeneration does not require growth

Well-fed *Hydra* are in a dynamic state of continuous growth and pattern formation. The cells of both epithelial layers proliferate steadily and, as the tissues grow, cells are displaced along the body column toward the head or foot (Fig. 13.16). In order for an adult *Hydra* to maintain a constant size, excess cells must be continually lost. Cell loss occurs at the tips of the tentacles and at the basal disc of the foot bud. Most of the excess cell production is taken up by the asexual budding of new *Hydra* from the body column. Budding occurs about two-thirds of the way down the body column; the body wall evaginates by a morphogenetic change in cell shape, generated locally by the cells in the bud region, to form a new column that develops a head at the end and then detaches as a small new *Hydra*.

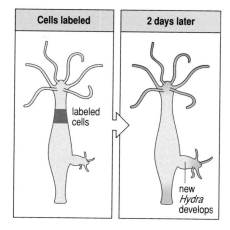

Fig. 13.16 Growth in *Hydra*. The cells in the body column of a *Hydra* are continually dividing and being displaced. If a group of cells in region 1 are labeled (left panel), 5 days later the labeled cells have been displaced to the tentacles and basal disc, where they are lost, and to the budding zone where they form a new *Hydra* (right panel).

Continuous growth in *Hydra* means that cells are continually changing their relative positions and are forming new structures as they move up or down the body column. Moreover, new *Hydra* are generated asexually by budding from the body wall. There must, therefore, be mechanisms for repatterning cells during this dynamic process. It is these mechanisms that give *Hydra* its remarkable capacity for regeneration.

If the body column of a *Hydra* is cut transversely, the lower piece will regenerate a head but the upper piece will regenerate a foot. Thus, what structure the cells regenerate at a cut surface depends on their relative position within the regenerating piece. The cut surface nearest the original head end forms a head—this shows that *Hydra* has a well-defined overall polarity. This polarity is maintained even in small pieces of the body. This can be seen when a short piece is cut out of the body column; the distal end regenerates a head while the proximal end becomes the basal disc.

Regeneration in *Hydra* does not require growth and is thus said to be morphallactic. When a short fragment of the column regenerates, there is no initial increase in size and the regenerated animal will be a small *Hydra*. Only after feeding will the animal return to a normal size. The lack of a growth requirement for regeneration is shown in heavily irradiated *Hydra*; no cell divisions occur in these animals but they can still regenerate more or less normally.

13.6 The head region of *Hydra* acts both as an organizing region and as an inhibitor of inappropriate head formation

At the beginning of this century, it was shown that grafting a small fragment of the hypostome region of a *Hydra* into the gastric region of another *Hydra* induced a new head, complete with tentacles, and a body axis (Fig. 13.17). Similarly, transplantation of a fragment of the basal region induced a new body column with a basal disc at its end. *Hydra* therefore have two organizing regions, one at each end. The hypostome and the basal disc act as organizing regions (like the Spemann organizer in amphibians and the polarizing regions in vertebrate limb buds). These organizing regions at the ends give *Hydra* its overall polarity.

Grafting experiments also show that, as part of its organizing function, the hypostome produces an inhibitor of head formation whose effectiveness drops with distance from the head (Fig. 13.18). This inhibition normally prevents inappropriate head formation in the intact animal. When a body piece from just below the head (region 1 in Fig. 13.15) is grafted into the gastric region, it rarely induces a new head and is usually simply absorbed into the body. But if the head of the host is removed at the time of grafting, the graft can then induce a new axis and head. This suggests that the removal of the head results in the loss of some factor that is inhibiting head formation. This inhibitory effect falls off with distance from the head:

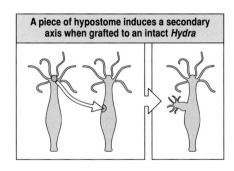

Fig. 13.17 The hypostome can induce a new head and body in *Hydra*. When an excised fragment of hypostome is grafted into the gastric region of another, intact *Hydra*, it can induce the formation of a complete new secondary axis with a head and tentacles.

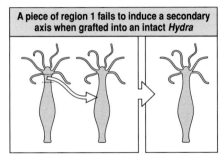

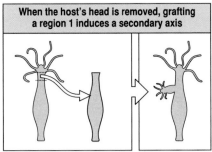

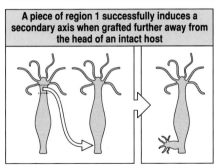

Fig. 13.18 The head region of *Hydra* produces an inhibitory signal that falls off with distance. Region 1 does not induce a head when grafted into the gastric region of an intact *Hydra* (top panel), indicating the presence of an inhibitory signal in the host. If the host's head is removed and a piece of region 1 is then grafted into the host, a secondary axis is induced (middle panel), indicating that the head region is the source of the inhibitory signal. Region 1 can induce a new axis in the foot region of an intact *Hydra* because the inhibitory signal grows weaker further away from the head (bottom panel).

when a region 1 is grafted to near the foot it can induce a head, even when the original head is still in place (see Fig. 13.18, bottom panel). These experiments suggest that the formation of extra heads is normally prevented in *Hydra* by a lateral inhibition mechanism (see Sections 1.14 and 11.2) acting through a gradient of inhibitory signal with its highest concentration at the head end. An opposing gradient inhibiting foot regeneration appears to be produced by the basal disc. These gradients are dynamic and when, for example, the head is removed, the concentration of the inhibitor falls.

13.7 Head regeneration in *Hydra* can be accounted for in terms of two gradients

One can account for the results of most of the regeneration experiments in *Hydra* in terms of an interaction between a gradient of head inhibitor and a gradient of positional value that determines the character of the different regions along the body column. The gradient in positional value appears to determine both head-inducing ability and resistance to inhibition. A gradient in resistance to inhibition can be recognized by differences in the level of inhibitor required to suppress head formation by different regions of the body. This gradient in resistance decreases with distance from the head and is thought to represent a gradient of positional value with a high point at the head end. Thus, there is insufficient inhibitor near the foot to prevent a region 1 transplant forming a head when transplanted to this site, but sufficient inhibitor to prevent a region 5 from doing so.

The gradient in head-inducing ability also runs from a high point at the head end to a low point at the basal end. Evidence that this is determined by a gradient in positional value is provided by the difference in time required for different regions to acquire head-inducing properties after amputation. Region 1 from an intact *Hydra* will not induce an axis when transplanted into the gastric region of another *Hydra*. If, however, the head of the donor *Hydra* is amputated, the region 1 can induce a new axis if taken about 6 hours after head removal (Fig. 13.19). The further down the axis the amputation is made, the longer it takes for the remaining cells to acquire head-like inducing properties. A region 5 can take up to 30 hours.

A simple model for these gradients assumes that the head inhibitor is a secreted factor, made by the head, that diffuses down the body column and is degraded at the basal end. The gradient in positional value is assumed to be an intrinsic property of cells. In this model, both gradients are linear—their values decrease at a constant rate with distance from the head. We further assume that, provided the level of inhibitor is greater than the threshold set by the positional value, head regeneration is inhibited. Removal of the head results in the concentration of inhibitor falling, as the inhibitor is degraded and cannot be replaced. The decrease in inhibitor concentration is greatest at the cut end, and when the inhibitor falls below the threshold concentration set by the local positional value, the positional value increases to that of the head end (Fig. 13.20). Thus, the first key step in this morphallactic regeneration, when the head region is removed, is the specification of a new head region at the cut surface.

When the positional value has increased to that of a normal head region, the cells start to make inhibitor and so prevent head formation in other body regions. The level of inhibitor will always first fall below the threshold

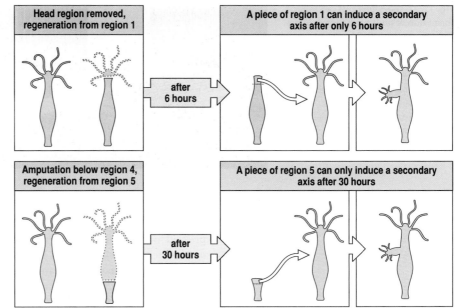

Fig. 13.19 The time needed to acquire head-like inducing properties following amputation increases with distance from the head. Top panels: a region 1 can induce a secondary axis if grafted into an intact *Hydra* 6 hours after head removal from the host. Bottom panels: if amputation is made lower down, it can take up to 30 hours for the region's cells to acquire head-like inducing properties.

for head inhibition where the positional value is highest, thus maintaining polarity. Once a new head has been specified and the inhibitory gradient re-established, the gradient in positional value also returns to normal, but this can take more than 24 hours. Morphallactic regeneration results in a smaller *Hydra* which, after feeding, will eventually grow back to a normal size

Addition of diacylglycerol—an intracellular second messenger in many signal transduction systems—to the medium in which *Hydra* is growing, increases positional value throughout the body column and can cause ectopic head formation, producing a multiheaded *Hydra* (Fig. 13.21). Diacylglycerol is produced in the phosphatidylinositol signaling pathway; this pathway is affected by lithium, which when added to the medium can cause ectopic foot ends to form, apparently by lowering positional values throughout the body column. The molecular nature of the inhibitor remains to be discovered, but there is evidence for peptide signals.

13.8 Genes controlling regeneration in *Hydra* are similar to those expressed in animal embryos

Hox genes, which control body pattern in many animals, are also involved in patterning *Hydra*. Some Hox genes are expressed along the body axis of coelenterates in a regional pattern and may be involved in specifying head–

Fig. 13.20 A simplified model for head regeneration in *Hydra*. This model assumes that there are two gradients running from head to foot. One is a diffusible molecule (I) that inhibits head formation and is produced by the head. The other is a gradient of positional value (P) that is an intrinsic property of the cells. When the head is removed, the concentration of inhibitor falls at the cut surface (graph 2) until a threshold is reached at which the positional value increases to that of the head region (graph 3). This then re-establishes the inhibitor gradient (graph 4). The overall gradient in positional value takes a much longer time to return to normal (graph 5).

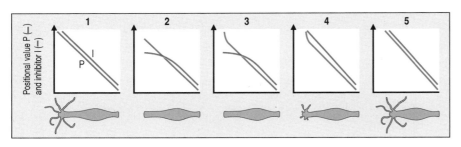

Fig. 13.21 Diacylglycerol can induce formation of multiple heads in *Hydra*. Diacylglycerol is an intracellular second messenger in many signal transduction systems. When added to the medium in which a *Hydra* is growing, it increases the positional value of cells and can cause ectopic head formation, producing a multiheaded *Hydra*. Scale bar = 1 mm.

Photograph courtesy of W. Müller, from Müller, W.A.: 1989.

foot positional values. A homeobox gene of the paired type related to the *Drosophila* gene *aristaless*, which is expressed in the head of the fly, is expressed in *Hydra* endoderm at the time of head determination. *Cnox-1* and *Cnox-2*, which are related to an anterior member of the Hox complex of *Drosophila*, are expressed early and late, respectively. The *Hydra* homolog, *budhead*, of vertebrate *HNF-3b*, which is expressed in the organizer region of vertebrates (see Section 3.20), is present in developing heads. A *Hydra* homolog of *goosecoid* is expressed just above where tentacles will appear. When injected into an early *Xenopus* embryo it can induce a partial secondary axis. This suggests that the roles of those genes in tissue that can act as an organizer have been retained over many millions of years of evolution. In addition, the *Hydra* homolog of the T-box gene *Brachyury* is expressed in the hypostome and in buds is expressed at an early stage in the region of the future hypostomes. Its function is not known, but its presence has evolutionary implications as *Hydra* has no mesoderm, the usual tissue in which *Brachyury* is expressed and suggests that the head corresponds to the proximal end, the blastopore, of other animals.

Wnt and β-catenin are both present in *Hydra*, and the Wnt signaling pathway is of particular importance. The *Hydra* homolog of β-catenin, Hyβ-Cat, shows a high degree of conservation in its role in the Wnt pathway, as injection of Hyβ-Cat mRNA into ventral *Xenopus* blastomeres at the eight-cell stage was able to induce complete secondary body axes. These

Fig. 13.22 Wnt expression in *Hydra* and in a regenerating tip at 1 hour and 48 hours.

Photographs courtesy of T.W. Holstein

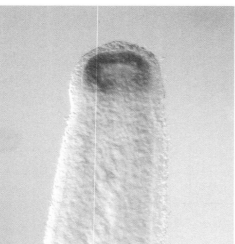
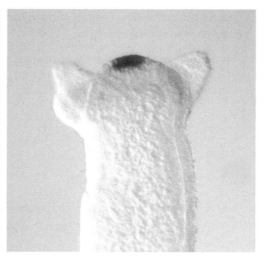

contained the anterior-most structures, such as eyes and cement gland, which were indistinguishable from those induced by *Xenopus* β-catenin. *In situ* hybridization experiments revealed that expression of HyWnt, the *Hydra* Wnt homolog, is restricted to the apical tip of the body axis in adult polyps; this apical tissue is the *Hydra* head organizer (Fig. 13.22, opposite). During head regeneration, both *Hydra* catenin and Wnt genes were expressed at the regenerating tip within an hour after head removal. During budding, a significant upregulation of Hyβ-Cat expression occurred in a ring-like domain in the prospective budding zone before tissue evagination started. This high expression level was maintained until bud detachment. These results demonstrate that Wnt signaling is involved in axis formation in *Hydra* and support the idea that it played a key part in the evolution of axial differentiation in early multicellular animals.

Summary

Hydra grows continually, losing cells from its ends and through the formation of buds. Two organizing regions, one at the head end and one at the basal end, pattern the body and maintain polarity. If transplanted to another site, a head region can induce a new body axis and head. If the body of an intact *Hydra* is severed in two, it regenerates by morphallaxis, which does not require new growth. Regeneration initially involves the respecification of cells at the cut end as 'head', leading to the establishment of an organizing region. The head region produces an inhibitor that prevents other regions forming a head. The concentration of inhibitor decreases with distance from the head. There is also a gradient in positional value that determines the threshold at which head inhibition occurs. When the head is removed, the inhibitor level in the rest of the body falls, and a new head region develops where the positional value is highest, thus maintaining polarity.

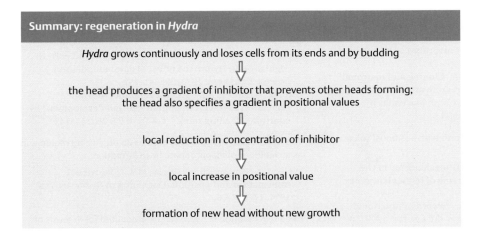

SUMMARY TO CHAPTER 13

Regeneration is the ability of an adult organism to replace a lost part of its body. The capacity to regenerate varies greatly among different groups of organisms and even between different species within a group. Mammals have very limited powers of regeneration, whereas newts can regenerate limbs, jaws, and the lens of the eye. Regeneration of limbs in animals such as amphibians may in part reflect the retention or reactivation of embryonic mechanisms. The coelenterate *Hydra* regenerates by morphallaxis: when its head is removed, the level of inhibitor normally produced by the head falls, resulting in a new head being specified at the cut end, without any growth having occurred. Regeneration of limbs in amphibians and insects is by epimorphosis: a blastema forms by dedifferentiation of stump cells and the blastemal cells divide and differentiate to regenerate distal structures. Intercalation of positional values can occur when normally non-adjacent values are placed next to each other. In these examples of regeneration the cellular mechanisms are related to those operating in normal development.

REFERENCES

Introduction

Goss, R.J.: *Principles of Regeneration*. London and New York: Academic Press, 1969.

Rinkevich, B., Shlemburg, Z., Fishelson, L.: **Whole-body protochordate regeneration from totipotent blood cells.** *Proc. Natl Acad. Sci. USA* 1995, **92**: 7695–7699.

13.1 Vertebrate limb regeneration involves cell dedifferentiation and growth

Brockes, J.P.: **Amphibian limb regeneration: rebuilding a complex structure.** *Science* 1997, **276**: 81–87.

Gardiner, D.M., Carlson, M.R., Roy, S.: **Towards a functional analysis of limb regeneration.** *Semin. Cell Dev. Biol.* 1999, **10**: 385–393.

Kumar, A., Velloso, C.P., Imokawa, Y., Brockes, J.P.: **Plasticity of retrovirus-labelled myotubes in the newt limb regeneration blastema.** *Dev. Biol.* 2000, **218**: 125–136.

Tanaka, E.M., Gann, A.A.F., Gates, P.B., Brockes, J.P.: **Newt myotubules re-enter the cell cycle by phosphorylation of the retinoblastoma protein.** *J. Cell Biol.* 1997, **136**: 155–165.

Tanaka, E.M., Drechel, D.N., Brockes, J.P.: **Thrombin regulates S-phase re-entry by cultured newt myotubes.** *Curr. Biol.* 1999, **9**: 792–799.

Wang, L., Marchioni, M.A., Tassava, R.A.: **Cloning and neuronal expression of a type III newt neuregulin and rescue of denervated nerve-dependent newt limb blastemas by rhGGF2.** *J. Neurobiol.* 2000, **43**: 150–158.

13.2 The limb blastema gives rise to structures with positional values distal to the site of amputation

Gardiner, D.M., Bryant, S.V.: **Molecular mechanisms in the control of limb regeneration: the role of homeobox genes.** *Int. J. Dev. Biol.* 1996, **40**: 797–805.

Gardiner, D.M., Carlosn, M.R.J., Roy, S.: **Towards a functional analysis of limb regeneration.** *Semin. Cell Dev. Biol.* 1999, **10**: 385–393.

Reginelli, A.D., Wang, Y.Q., Sassoon, D., Muneoka, K.: **Digit tip regeneration correlates with Msx-1 (Hox7) expression in fetal and newborn mice.** *Development* 1995, **121**: 1065–1076.

Stocum, D.L.: **A conceptual framework for analyzing axial patterning in regenerating urodele limbs.** *Int. J. Dev. Biol.* 1996, **40**: 773–783.

Torok, M.A., Gardiner, D,M., Shubin, N.H., Bryant, S.V.: **Expression of HoxD genes in developing and regenerating axolotl limbs.** *Dev. Biol.* 1998, **200**: 225–233.

13.3 Retinoic acid can change proximo-distal positional values in regenerating limbs

Bryant, S.V., Gardiner, D.M.: **Retinoic acid, local cell–cell interactions and pattern formation in vertebrate limbs.** *Dev. Biol.* 1992, **152**: 125.

Maden, M.: **The homeotic transformation of tails into limbs in Rana temporaria by retinoids.** *Dev. Biol.* 1993, **159**: 379–391.

Pecorino, L.T., Entwistle, A., Brockes, J.P.: **Activation of a single retinoic acid receptor isoform mediates proximo-distal respecification.** *Curr. Biol.* 1996, **6**: 563–569.

Scadding, S.R., Maden, M.: **Retinoic acid gradients during limb regeneration.** *Dev. Biol.* 1994, **162**: 608–617.

13.4 Insect limbs intercalate positional values by both proximo-distal and circumferential growth

French, V.: **Pattern regulation and regeneration.** *Phil. Trans. Roy. Soc. Lond., Biol. Sci.* 1981, **295**: 601–617.

13.5 *Hydra* grows continuously but regeneration does not require growth

Hicklin, J., Wolpert, L.: **Positional information and pattern regulation in Hydra: the effect of gamma-radiation.** *J. Embryol. Exp. Morph.* 1973, **30**: 741–752.

Otto, J.J., Campbell, R.D.: **Tissue economics of Hydra: regulation of cell cycle, animal size and development by controlled feeding rates.** *J. Cell Sci.* 1977, **28**: 117–132.

13.6 The head region of *Hydra* acts both as an organizing region and as an inhibitor of inappropriate head formation

Wolpert, L., Hornbruch, A., Clarke, M.R.B.: **Positional information and positional signaling in Hydra.** *Am. Zool.* 1974, **14**: 647–663.

13.7 Head regeneration in *Hydra* can be accounted for in terms of two gradients

Bosch, T.C.G., Fujisawa, T.: **Polyps, peptides and patterning.** *BioEssays* 2001, **23**: 420–427.

Hassel, M., Albert, K., Hofheinz, S.: **Pattern formation in Hydra vulgaris is controlled by lithium-sensitive process.** *Dev. Biol.* 1993, **156**: 362–371.

MacWilliams, W.: **Hydra transplantation phenomena and the mechanism of Hydra head regeneration. II. Properties of the head activation.** *Dev. Biol.* 1983, **96**: 239–257.

Müller, W.A.: **Pattern formation in the immortal** *Hydra*. *Trends Genet.* 1996, **12**: 91–96.

13.8 Genes controlling regeneration in *Hydra* are similar to those expressed in animal embryos

Broun, M., Sokol, S., Bode, H.R.: **Cngsc, a homologue of goosecoid, participates in the patterning of the head, and is expressed in the organizer region of Hydra.** *Development* 1999, **126**: 5245–5254.

Galliot, B.: **Conserved and divergent genes in apex and axis development in cnidarians.** *Curr. Opin. Genet. Dev.* 2000, **10**: 629–637.

Hobmayer, B., Rentzsch, F., Kuhn, K., Happel, C.M., Cramer von Laue, C., Snyder, P., Rothbächer, U., Holstein, T.W.: **WNT signalling molecules act in axis formation in the diploblastic metazoan *Hydra*.** *Nature* 2000, **407**: 186–189.

Shenk, M.A., Gee, L., Steele, R.E., Bode, H.R.: **Expression of** *Cnox-2,* **a Hom/Hox gene, is suppressed during head formation in** *Hydra*. *Dev. Biol.* 1993, **160**: 108–118.

Technau, U., Bode, H.R.: **HyBra1, a *Brachyury* homologue, acts during head formation in *Hydra*.** *Development* 1999, **126**: 999–1010.

Growth and post-embryonic development

14

- ■ Growth
- ■ Molting and metamorphosis
- ■ Aging and senescence

> " Growing up was not just about getting bigger, but also about some dramatic changes that made us almost unrecognizable. "

Development does not stop once the embryonic phase is complete. Most, but by no means all, of the growth in animals and plants occurs in the post-embryonic period, when the basic form and pattern of the organism has already been established. The basic patterning is done on a small scale, over dimensions usually less than a millimeter. In many animals, the embryonic phase is immediately succeeded by a free-living larval or immature adult stage. In others, such as mammals, considerable growth occurs during a late embryonic or fetal period, while the embryo is still dependent on maternal resources. Growth then continues after birth. Growth is a central aspect of all developing systems, determining the final size and shape of the organism and its parts. Animals with a larval stage not only grow in size, but may also undergo molting, in which the outer skeleton is shed, and **metamorphosis**, in which the larva is transformed into the adult form. Metamorphosis often involves a radical change in form and the development of new organs.

We first consider the roles of intrinsic growth programs and of extracellular factors such as growth hormones in controlling both embryonic and post-embryonic growth. This is followed by a discussion of metamorphosis in insects and amphibians. Finally, we look at what might be considered an abnormal aspect of post-embryonic development—aging.

Growth

Growth is defined as an increase in the mass or overall size of a tissue or organism; this increase may result from cell proliferation, cell enlargement without division, or by accretion of extracellular material, such as bone matrix or even water (see Fig. 14.1). In early embryonic development, there is little growth during cleavage and blastula formation, and cells get smaller at each cleavage division. However, growth begins in different animals at different stages. In the chick, for example, it occurs in the region anterior to the node, during primitive streak regression, whereas in *Xenopus* it starts after gastrulation.

In animals, the basic body pattern is laid down when the embryo is still small; in humans, all organs are less than 1 cm in their maximum dimension when their pattern is determined. The growth program—that is, how much an organism or an individual organ grows and how it responds to external factors like hormones—may also be specified at an early stage in development. Overall growth of the organism mainly occurs in the period after the basic pattern of the embryo has been established; there are, however, many examples where earlier organogenesis involves localized growth, as in the vertebrate limb bud (see Chapter 10), and the developing nervous system (see Chapter 11). Different rates of growth in different parts of the body, or at different times, during early development profoundly affect the shape of organs and the organism.

Unlike the situation in animals, where the embryo is essentially a miniature version of the free-living larva or adult, plant embryos bear little resemblance to the mature plant. Most of the adult plant structures are generated after germination by the shoot or root meristems, which have a capacity for continual growth (see Chapter 7). In woody plants, the cambial layer in the trunk, branches and roots also retains proliferative capacity. It can give rise to the main tissues of the plant axis, enabling trees to increase in girth year after year.

14.1 Tissues can grow by cell proliferation, cell enlargement, or accretion

Although growth often occurs through an increase in the number of cells, this is only one of three main means of growth (Fig. 14.1). There is also a significant amount of cell death in many growing tissues, and the overall growth rate is determined by the rates of cell death and cell proliferation.

A second strategy is growth by cell enlargement—that is, by individual cells increasing their mass and getting bigger. Once differentiated, skeletal and heart muscle cells and neurons never divide again, although they do increase in size. Neurons grow by the extension and growth of axons and dendrites, whereas muscle growth involves an increase in mass, as well as the fusion of satellite cells to pre-existing muscle fibers to provide new nuclei. Cell enlargement is also a major feature of plant growth, as we shall see. And differences in size between closely related species of *Drosophila* are partly the result of the larger species having larger cells. Some growth is through a combination of cell proliferation and cell enlargement. For example, lens cells are produced by cell division from a proliferative zone for an extended period, while their differentiation involves considerable cell enlargement.

The third growth strategy, accretionary growth, involves an increase in the volume of the extracellular space, which is achieved by secretion of large quantities of extracellular matrix by cells. This occurs in both cartilage and bone, where most of the tissue mass is extracellular.

Certain vertebrate tissues, including the hematopoietic system tissues (see Section 9.6) and epithelia (see Section 9.12), are continually renewed throughout an animal's lifetime by cell division and differentiation from a stem cell population. The end products of this type of proliferative system,

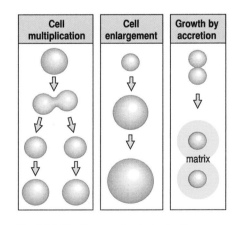

Fig. 14.1 The three main strategies for growth in vertebrates. The most common mechanism is cell proliferation—growth followed by division. A second strategy is cell enlargement, in which cells increase their size without dividing. A third strategy is to increase size by accretionary growth, such as matrix secretion.

such as mature red blood cells and keratinocytes, are themselves incapable of division and die.

14.2 Cell proliferation can be controlled by an intrinsic program

When a eukaryotic cell duplicates itself it goes through a fixed sequence of events called the **cell cycle**. The cell grows in size, the DNA is replicated, and these replicated chromosomes then undergo mitosis, and become segregated into two daughter nuclei. Only then can the cell divide to form two daughter cells, which can go through the whole sequence again. During cleavage of the fertilized egg, as in *Xenopus* and the mouse, there is no cell growth, and cell size decreases with each division, but in other proliferating cells, the cytoplasmic mass must double in preparation for cell division.

The standard eukaryotic mitotic cell cycle is divided into well-marked phases. At the M phase, mitosis and cell cleavage give rise to two new cells. The rest of the cell cycle, between one M phase and the next, is called interphase. Replication of DNA occurs during a defined period in interphase, the S phase. Preceding S phase is a period known as G_1 (the G stands for gap), and after it another interval known as G_2, after which the cells enter mitosis (Fig. 14.2). G_1, S phase, and G_2 collectively make up interphase, the part of the cell cycle during which cells synthesize proteins and grow, as well as replicating their DNA. Particular phases of the cell cycle are absent in some cells: during cleavage of the fertilized egg, G_1 and G_2 are virtually absent; in meiosis (see Section 12.11) there is no DNA replication at the second division; and in insect salivary glands there is no M phase, as the DNA replicates repeatedly without mitosis or cell division, leading to the formation of giant polytene chromosomes.

Growth factors and other signaling proteins play a key role in controlling cell growth and proliferation. Studies of cell cultures show that growth factors are essential for cells to multiply, with the particular growth factor or factors required depending on the cell type. When somatic cells are not proliferating they are usually in a state known as G_0, into which they withdraw after mitosis (see Fig. 14.2). Growth factors enable the cell to proceed out of G_0 and progress through the cell cycle. Numerous growth factors that can control cell proliferation have been discovered, but in general their precise roles in normal development are not yet known. Some exceptions include erythropoietin, which promotes proliferation of red blood cell precursors (see Section 9.7). The role of hormones is considered later.

A striking discovery has been the recognition that cells must receive signals, such as growth factors, not only for them to divide, but also simply to survive. In the absence of all growth factors, cells commit suicide by **apoptosis**, as a result of activation of an internal cell death program (see Section 9.15). There is a significant amount of cell death in all growing tissues, so that overall growth rate depends on the rates of both cell death and cell proliferation.

The timing of events in the vertebrate cell cycle is controlled by a set of 'central' timing mechanisms. One set of proteins, known as the **cyclins**, controls the passage through key transition points in the vertebrate cell cycle. Cyclin concentrations oscillate during the cell cycle, and these oscillations correlate with transitions from one phase of the cycle to the next (see Section 12.16). Cyclins act by forming complexes with, and helping to activate, cyclin-dependent kinases. These phosphorylate proteins that

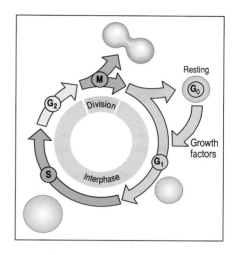

Fig. 14.2 The eukaryotic cell cycle. After mitosis (M), the daughter cells can either enter a resting phase (G_0), in which they effectively withdraw from the cell cycle, or proceed through G_1 to the phase of DNA synthesis (S). This is followed by G_2, and then by mitosis. Cell growth occurs throughout G_1, S, and G_2. The decision to enter G_0 or to proceed through G_1 may be controlled by both intracellular status and extracellular signals such as growth factors. Cells such as neurons and skeletal muscle cells, which do not divide after differentiation, are permanently in G_0.

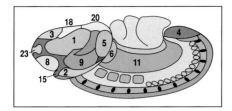

Fig. 14.3 Domains of mitosis in the Drosophila blastoderm. Mitotic domains that divide at the same time are indicated by the various colors. The numbers indicate the order in which various regions undergo mitosis at the 14th cycle. Not all domains are visible in this view. After Edgar, B.A., *et al.*: 1994.

trigger the events of each phase, such as DNA replication in S phase or mitosis in M phase.

In general, the mechanisms controlling the pattern of cell division in the embryo are poorly understood. One example of intrinisic control comes from *Drosophila*, where the early cell cycles are under the control of genes that pattern the embryo; these exert their effect by their influence on the cyclins. The first cell cycles in the *Drosophila* embryo are represented by rapid and synchronous nuclear divisions, without any accompanying cell divisions. This creates a syncytial blastoderm (see Fig. 2.30). There are virtually no G phases, just alternations of DNA synthesis (S phase) and mitosis. But at cycle 14, there is a major transition to a different type of cell cycle, a transition similar to the mid-blastula transition in frogs (see Section 3.17). At cycle 14 and subsequently, the cell cycles have a well defined G_2 phase, and the blastoderm becomes cellularized. After the 17th or 18th cycle, cells in the epidermis and mesoderm stop dividing, and differentiate. This cessation of proliferation is caused by the exhaustion of maternal cyclin E originally laid down in the egg, which is required for progression through the cell cycle.

At the 14th cell cycle, distinct spatial domains with different cell-cycle times can be seen in the *Drosophila* blastoderm (Fig. 14.3). This patterning of cell cycles is produced by a change in the synthesis and distribution of a protein phosphatase called string, which exerts control on the cell cycle by dephosphorylating and activating a cyclin-dependent kinase. In the fertilized egg, the string protein is of maternal origin and is uniformly distributed. It therefore produces a synchronized pattern of nuclear division throughout the embryo. After cycle 13, the maternal string protein disappears and zygotic string protein becomes the controlling factor.

Zygotic *string* gene transcription occurs in a complex spatial and temporal pattern. Only cells in which the *string* gene is expressed enter mitosis. This results in variation in the rate of cell division in different parts of the blastoderm, which ensures that the correct number of cells are generated in different tissues. The pattern of zygotic *string* gene expression is controlled by transcription factors encoded by the early patterning genes, such as the gap and pair-rule genes and those patterning the dorso-ventral axis. One exception to the rule that expression of *string* leads to cell proliferation is the presumptive mesoderm, which is the first domain in which *string* is expressed but the tenth to divide. The failure to divide is due to the expression of the gene *tribble* in this region, as tribble protein degrades the string protein. The delay is necessary to allow ventral furrow formation and thus mesoderm invagination, as cell division inhibits ventral furrow formation (see Section 8.8). Thus, the cell cycles in *Drosophila* development provide a good example of how genes can control the pattern of cell divisions.

In contrast to this intrinsic program of early cell division in the *Drosophila* embryo, the cell proliferation and growth that occurs in the imaginal discs that give rise to adult structures is modulated by cell–cell interactions, that is, by extracellular signals, and is not closely linked to pattern formation.

In early mammalian embryos, cell proliferation times vary with developmental time and place. The first two cleavage cycles in the mouse last about 24 hours, and subsequent cycles take about 10 hours each. After implantation, the cells of the epiblast proliferate rapidly; cells in front of the primitive streak have a cycle time of just 3 hours, but the signals involved in this proliferation have not yet been identified.

14.3 Organ size can be controlled by external signals and intrinsic growth programs

Both intrinsic growth programs and systemic circulating factors can determine organ size, and their relative importance in the development of different organs varies a great deal. Some organs control their own size. For example, if several fetal thymus glands are transplanted into a developing mouse embryo, each grows to its normal size, showing that the thymus follows an intrinsic growth program. If the same experiment is done with spleens, however, each spleen grows much less than normal so that the final total mass of the spleens is equivalent to one normal spleen. For the spleen, therefore, systemic factors produced by the spleen itself are a major controlling factor in its growth. For muscle, the protein myostatin may be a systemic negative regulator of growth, as mutation in the *myostatin* gene in mice leads to a significant increase in muscle mass; the number of muscle fibers and their size are both increased. In butterflies, there is evidence for interactions between imaginal discs; the removal of hindwing discs results in larger forewings and forelegs.

Even when they are capable of division, the cells of many adult tissues do not divide, or divide infrequently. Cells can, however, be induced to divide by injury or other stimuli, such as a reduction in the mass of the tissue. Liver cells divide relatively infrequently but if, for example, two thirds of a rats's liver is removed, the cells in the remaining third proliferate and restore the liver to its normal size within a few weeks. This restorative capacity indicates the presence of factors in the circulation that control cell proliferation. Removal of a kidney leads to an increase in size of the remaining kidney, but in this case the growth is mainly the result of cell enlargement rather than cell proliferation. And in Section 13.4, we saw how intercalary growth is stimulated to replace missing positions of amputated limbs when cells with disparate positional values are put next to each other.

Patterning of the embryo occurs while the organs are still very small. For example, human limbs have their basic pattern established when they are less than 1 cm long. Yet, over the years, the limb grows to be at least one hundred times longer. How is this growth controlled? It appears that each of the cartilaginous elements in the limb has its own individual growth program. In the chick wing, the initial size of the cartilaginous elements in the long bones—the humerus and the ulna—are similar to the elements in the wrist (Fig. 14.4). Yet, with growth, the humerus and ulna increase many times in length compared with the wrist bones. These growth programs are specified when the elements are initially patterned and involve both cell multiplication and matrix secretion. Each skeletal element follows its growth program even when grafted to neutral sites, provided that a good blood supply is established.

A classic illustration of an intrinsic growth program comes from grafting limb buds between large and small species of salamanders of the genus *Ambystoma*. A limb bud from the larger species grafted to the smaller species initially grows slowly, but eventually ends up its normal size, which is much larger than any of the limbs of the host (Fig. 14.5). Whatever the

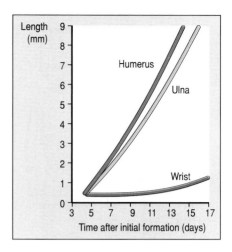

Fig. 14.4 Comparative growth of the cartilaginous elements in the embryonic chick wing. When first laid down, the cartilaginous elements of the humerus, ulna, and wrist are the same size, but the humerus and ulna then grow much more than the wrist element.

Fig. 14.5 The size of limbs is genetically programmed in salamanders. An embryonic limb bud from a large species of salamander, *Ambystoma tigrinum*, grafted to the embryo of a smaller species, *Ambystoma punctatum*, grows much larger than the host limbs—to the size it would have grown in *Ambystoma tigrinum*.

Photograph from Harrison, R.G.: 1969.

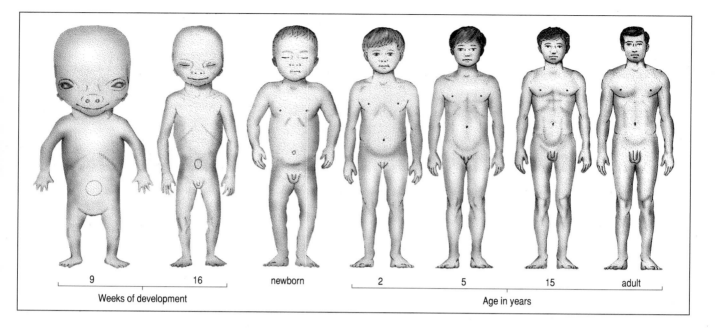

9   16   newborn   2   5   15   adult

Weeks of development   Age in years

Fig. 14.6 Different parts of the human body grow at different rates. At 9 weeks of development, the head is relatively large but, with time, other parts of the body grow much more than the head. After Gray, H.: 1995.

circulatory factors, such as hormones, that influence growth, the intrinsic response of different tissues is therefore crucial. Such intrinsically patterned differential growth can affect the overall form of an organism considerably, as can be seen in Fig. 14.6, even though in many cases circulating hormones are also required for growth.

14.4 Organ size may be determined by absolute dimension rather than cell number or size

Evidence that animals can monitor the linear dimensions of their organs comes from organisms with either less or more than the diploid number of chromosomes—that is haploid and polyploid organisms. For a given cell type, cell size is usually proportional to ploidy, so haploid cells are half the volume of diploid cells, and tetraploid cells twice the volume. Animals such as salamanders with unusual ploidy grow to a normal size, and tetraploid salamanders have only half the number of cells. Tetraploid mouse embryos compensate for larger cells by having fewer cells, but usually die before birth. In *Drosophila*, the final size of mosaics of haploid and diploid cells is normal, the haploid region containing smaller but more numerous cells. Polyploid plants are larger than normal, but in mosaics of diploid and polyploid cells it is possible to find the sort of size compensation seen in animals.

The growth of the *Drosophila* wing provides an excellent model for this mechanism of growth control. The wing disc (see Section 10.14) contains about 40 cells, and this grows in the larva to about 50,000 cells before metamorphosis. A striking feature is that mosaics of slow-growing *Minute* cells (see Box 5B, p. 173) and normal cells result in competition between these cell types, so that the wing is made almost entirely of normal cells. The clonal descendants of a single cell can contribute from a tenth to as much as a half of the wing. That cell proliferation does not equal growth is clearly demonstrated by blocking cell division in either an anterior or posterior compartment, in which case the cells become larger but wing size is normal.

We have already seen how decapentaplegic and wingless proteins play a key role in patterning the limb, and if the production of either protein is defective, growth is stunted. However, these proteins should not necessarily be thought of as growth factors. Rather, as one model proposes, local growth may depend on reading the steepness of the morphogen concentration gradients. As long as the gradient is sufficiently steep, the cells will grow and divide. This would account for growth not depending on cell size or cell number.

14.5 Growth of mammals is dependent on growth hormones

Human growth during the embryonic, fetal, and post-natal periods provides a good model for mammalian growth. The human embryo increases in length from 150 mm at implantation to about 50 cm over the nine months of gestation. During the first 8 weeks after conception, the embryonic body does not increase greatly in size, but the basic human form is laid down in miniature. The greatest rate of growth occurs at about 4 months, when the embryo grows as much as 10 cm per month. Growth after birth follows a well-defined pattern (Fig. 14.7, left panel). During the first year after birth, growth occurs at a rate of about 2 cm per month. The growth rate then declines steadily until the start of the characteristic adolescent growth spurt at about 11 years in girls and 13 years in boys (see Fig. 14.7, right panel). In pygmies, this adolescent growth spurt does not occur, hence their characteristic short stature.

Growth of the different parts of the body is not uniform, and different organs grow at different rates. At 9 weeks of development, the head of a human embryo is more than a third of the length of the whole embryo, whereas at birth it is only about a quarter. After birth, the rest of the body grows much more than the head, which is only about an eighth of the body length in the adult (see Fig. 14.6).

The maternal environment plays an important role in controlling fetal growth. This is well illustrated by crossing a large shire horse with a much smaller Shetland pony. When the mother is a shire mare, the newborn foal is similar in size to a normal shire foal, but when the mother is a Shetland, the newborn is much smaller. However, with growth after birth, the

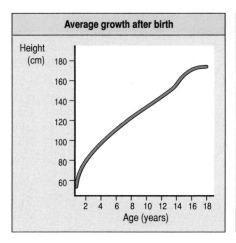

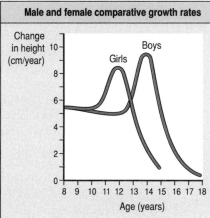

Fig. 14.7 Normal human growth. Left panel: an average growth curve for a human male after birth. Right panel: comparative growth rates of boys and girls. There is a growth spurt at puberty in both sexes, which occurs earlier in girls.

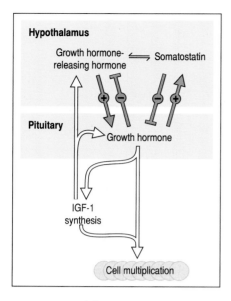

Fig. 14.8 Growth hormone production is under the control of the hypothalamic hormones. Growth hormone is made in the pituitary gland and is secreted. Growth hormone-releasing hormone from the hypothalamus promotes growth hormone synthesis, while somatostatin inhibits it. Growth hormone controls its own release by negative feedback signals to the hypothalamus. Growth hormone causes the synthesis of the insulin-like growth factor IGF-1 and this promotes the production of growth hormone.

offspring of both these crosses become similar in size, and achieve a final size intermediate between shires and Shetlands.

The amount of nourishment an embryo receives can have profound effects in later life. In early intrauterine life, undernutrition tends to produce small but normally proportioned animals. In contrast, undernutrition during the post-natal growth period leads to selective organ damage. For example, rats that are undernourished immediately after weaning have normal skeletal growth, but the liver and kidneys do not grow normally and are permanently small. Epidemiological studies in humans have shown that small size at birth is associated with increased death rates from cardiovascular diseases and non-insulin-dependent diabetes. The mechanisms underlying these long-term effects are not understood, but the studies emphasize the importance of adequate growth for normal development. In one study, rats were given a low-protein diet during the preimplantation period of development of the embryo and this resulted in reduced cell numbers in both inner cell mass and trophectoderm lineages.

Embryonic growth is dependent on growth factors. We have seen that fibroblast growth factors (FGFs) control cell proliferation in the progress zone of the developing chick limb, and are required for proliferation and outgrowth of the bud (see Section 10.2). Evidence for the role of other growth factors in embryonic growth comes from the technique of gene knock-out (see Box 4B, p. 122). Insulin-like growth factors 1 and 2 (IGF-1 and IGF-2), which are single-chain protein growth factors that closely resemble insulin and each other in their amino-acid sequence, have a key role not only in post-natal mammalian growth, but also in growth during embryonic development. Newborn mice lacking a functional *Igf-2* gene develop relatively normally, but weigh only 60% of the normal newborn body weight. Mice in which the *Igf-1* gene has been inactivated are also growth retarded, and IGF-1 also has an important role in post-embryonic growth. Both factors and their receptors are present as early as the eight-cell stage of mouse development. IGF-2 seems to be essential for early embryonic growth, after which the action of IGF-1 becomes dominant. Both proteins act via the IGF-1 receptor; the IGF-2 receptor by contrast reduces the concentration of IGF-2 by enabling it to be degraded. The genes for IGF-2 and its receptor are two of the genes that are imprinted in mammals; they are inactivated in maternal and paternal germ cells, respectively (see Section 12.13).

Growth hormone, which is synthesized in the pituitary gland, is essential for the post-embryonic growth of humans and other mammals. Within the first year of birth, secretion of growth hormone by the pituitary gland commences. A child with insufficient growth hormone grows less than normal, but if growth hormone is given regularly, normal growth is restored. In this case, there is a catch-up phenomenon, with a rapid initial response that tends to restore the growth curve to its original trajectory.

Production of growth hormone is under the control of two hormones produced in the hypothalamus: growth hormone-releasing hormone, which promotes growth hormone synthesis and secretion, and somatostatin, which inhibits its production and release. Growth hormone produces many of its effects by inducing the synthesis of IGF-1 (Fig. 14.8) and, to a lesser extent, IGF-2. Post-natal growth, as well as embryonic growth, is largely due to the actions of these insulin-like growth factors, and complex hormonal regulatory circuits control their production.

As can be seen from measurements of human growth after birth (see Fig. 14.7, left panel), the growth rate decreases until puberty, when there is a sudden growth spurt. This sharp increase is due to secretion of gonadotropins; these cause the increased production of the steroid sex hormones, which in turn cause increased production of growth hormone. This growth spurt occurs a couple of years earlier in girls than in boys. We next look at the growth of some individual organs.

14.6 Growth of the long bones occurs in the growth plates

An important aspect of post-embryonic vertebrate growth is the growth of the long bones of the limbs (humerus, femur, radius, and ulna). The long bones are initially laid down as cartilaginous elements (see Section 10.1) and then become ossified. The early growth of these elements involves both cell proliferation and matrix secretion in a well-defined pattern. In both fetal and post-natal growth, the cartilage is replaced by bone in a process known as **endochondral ossification**, in which ossification starts in the centers (diaphyses) of the long bones and spreads outward (Fig. 14.9). There are also secondary ossification centers at each end of the bones (the epiphyses). The adult long bones thus have a bony shaft with cartilage confined to the articulating surfaces at each end, and to two internal regions near each end, the **growth plates**, in which growth occurs. In the growth plates, the

Fig. 14.9 Growth plates and endochondral ossification in the long bone of a vertebrate. The long bones of vertebrate limbs increase in length by growth from cartilaginous growth plates. The growth plates are cartilaginous regions that lie between the epiphysis of the future joint and the central region of the bone, the diaphysis. In the figure, bone has already replaced cartilage in the diaphysis, and more bone is being added at the growth plates. Within the growth plates, cartilage cells multiply in the proliferative zone, then mature and undergo hypertrophy (cell enlargement). They are then replaced by bone, which is laid down by special cells called osteoblasts. Secondary sites of ossification are located within the epiphyses. After Wallis, G.A.: 1993.

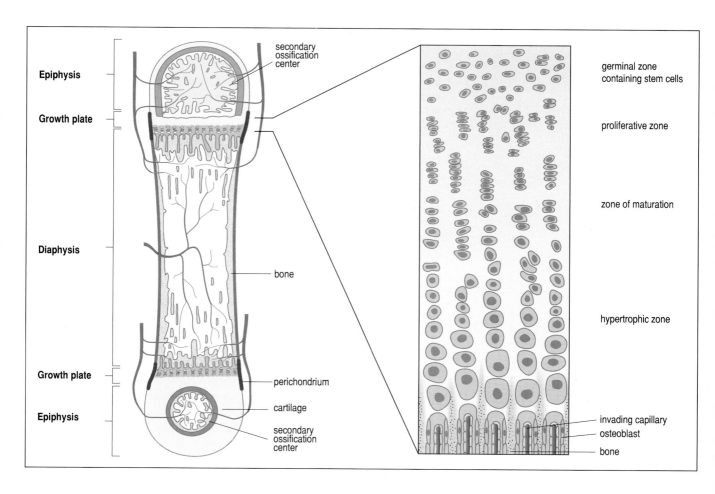

cartilage cells are usually arranged in columns, and various zones can be identified. Just next to the bony epiphysis is a narrow germinal zone, which contains stem cells. Next is a proliferative zone of cell division, followed by a zone of maturation, and a hypertrophic zone, in which the cartilage cells increase in size. Finally, there is a zone in which the cartilage cells die and are replaced by bone laid down by osteoblasts. There is a strong similarity to the development of skin, where basal stem cells give rise to dividing cells, which differentiate into keratinocytes and finally die (see Section 9.12).

The cartilage cells in the growth plates are controlled so that they stop dividing and enlarge to form the scaffolding for bone; the mechanism of this has been studied in mice. It involves a signaling molecule called Indian hedgehog, which belongs to the large and widespread family of proteins that includes mammalian Sonic hedgehog and *Drosophila* hedgehog proteins. Indian hedgehog is expressed in the pre-hypertrophic cells of the growth plate. Its target is a protein that is related to parathyroid hormone and is present in the region around the articulating surfaces. Indian hedgehog promotes the proliferation of cartilage cells and prevents their hypertrophy. Lack of Indian hedgehog protein in mice results in decreased chondrocyte proliferation and increased hypertrophy, and thus short, stubby limbs. The receptor for Indian hedgehog is expressed in the perichondrium and not in the cartilage itself, and the perichondrium is closely involved in regulating chondrocyte proliferation and differentiation. *Wingless*-related genes like *Wnt-5a* and *Wnt-4* are involved in this regulation.

Growth hormone affects bone growth by acting on the growth plates. The cells in the germinal zone have receptors for growth hormone, and growth hormone is probably directly responsible for stimulating these stem cells to proliferate. Further growth, however, is probably mediated by IGF-1, whose production in the growth plate is stimulated by growth hormone. Thyroid hormones are also necessary for optimal bone growth; they act both by increasing the secretion of growth hormone and IGF-1, and by stimulating hypertrophy of the cartilage cells. FGF is also important for bone growth; the genetic defect that gives rise to achondroplasia (short-limbed dwarfism) is a dominant mutation in the FGF receptor-3, whose normal function is to limit, rather than promote, bone formation.

The rate of increase in the length of a long bone is equal to the rate of new cell production per column multiplied by the mean height of an enlarged cell. The rate of new cell production depends both on the time cells take to complete a cycle in the proliferative zone, and the size of this zone. Different bones grow at different rates, and this can reflect the size of the proliferative zone, the rate of proliferation, and the degree of cell enlargement in the growth plate.

In view of the complexity of the growth plate, it is remarkable that human bones in limbs on opposite sides of the body can grow for some 15 years independently of each other, and yet eventually match to an accuracy of about 0.2%. This is achieved by having many columns of cells in each plate, so that growth variations are averaged out. Growth of a bone ceases when the growth plate ossifies, and this occurs at different times for different bones. Ossification of growth plates occurs in a strict order in different bones and can therefore be used to provide a measure of physiological age.

14.7 Growth of vertebrate striated muscle is dependent on tension

The number of striated (skeletal) muscle fibers in vertebrates is determined during embryonic development. Once differentiated, striated muscle cells lose the ability to divide. Post-embryonic growth of muscle tissue results from an increase in individual fiber size, both in length and girth. The number of myofibrils within the enlarged muscle fiber can increase more than 10-fold. Additional nuclei for the much-enlarged cell are provided by the fusion of satellite cells with the fiber. Satellite cells, which are undifferentiated cells lying adjacent to the differentiated muscle, also act as a reserve population of stem cells that can replace damaged muscle.

The increase in the length of a muscle fiber is associated with an increase in the number of sarcomeres, the functional contractile unit. For example, in the soleus muscle of the mouse leg, as the muscle increases in length, the number of sarcomeres increases from 700 to 2300 at 3 weeks after birth. This increase in number seems to depend on the growth of the long bones putting tension on the muscle through its tendons. If the soleus muscle is immobilized by placing the leg in a plaster cast at birth, sarcomere number increases slowly over the next 8 weeks, but then increases rapidly when the cast is removed (Fig. 14.10). One can thus see how bone and muscle growth are mechanically coordinated.

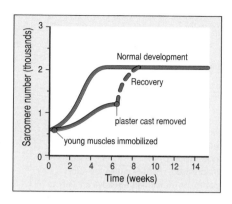

Fig. 14.10 The growth in length of the muscles attached to the long bones of the mouse leg depends on the tension provided by long bone growth. The length of a muscle is related to the number of sarcomeres—the basic contractile unit. If the limb is immobilized by a plaster cast in a position in which there is no tension on the muscle, there is little increase in muscle length until the cast is removed. There is then a rapid increase in length.

14.8 Cancer can result from mutations in genes that control cell multiplication and differentiation

Cancer can be regarded as a major perturbation of normal cellular behavior that results from certain mutations in somatic cells. Creating and maintaining tissue organization requires strict controls on cell division, differentiation, and growth. In cancer, cells escape from these normal controls, and proceed along a path of uncontrolled growth and migration that can kill the organism. There is usually a progression from a benign localized growth to malignancy in which the cells **metastasize**—migrate to many parts of the body where they continue to grow.

The cells most likely to give rise to cancer are those that are undergoing continual division. Because they replicate their DNA frequently, they are more likely than other cells to accumulate mutations that arise from errors in DNA replication. In almost all cancers, the cancer cells are found to have a mutation in one or more genes. Not all mutations give rise to a cancer, however, and particular genes in which mutation can contribute to cancer formation have been identified in humans and other mammals. These genes are known as **proto-oncogenes**; when such a gene undergoes mutation it becomes an **oncogene**. In some cases, the presence of a single oncogene is capable of making a cell cancerous. At least 70 proto-oncogenes have been identified in mammals.

Proto-oncogenes are genes that are involved in cell proliferation, differentiation, and migration. We have seen how extracellular signals provided by growth factors and other signal molecules play a key role in the regulation of cell differentiation and cell proliferation. Many proto-oncogenes encode growth factors and other signaling molecules, or their receptors, or proteins involved in the intracellular response to such signals. An altered version of any of these proteins produced by its mutant oncogene can cause permanent activation of the cell cycle, causing the cell and

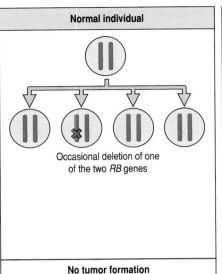
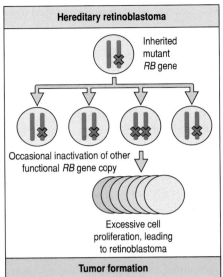

Normal individual	Hereditary retinoblastoma
Occasional deletion of one of the two *RB* genes	Inherited mutant *RB* gene
	Occasional inactivation of other functional *RB* gene copy
	Excessive cell proliferation, leading to retinoblastoma
No tumor formation	**Tumor formation**

Fig. 14.11 The *retinoblastoma* (*RB*) gene is a tumor suppressor gene. If only one copy of the *RB* gene is lost or inactivated, no tumor develops (left panel). In individuals already carrying an inherited mutant *RB* gene, if the other copy of the gene is lost or inactivated in a cell, that cell will generate a retinal tumor (right panel). Such individuals are thus at a much greater risk of developing retinoblastoma, and usually do so at a young age.

its progeny to divide indefinitely. Such mutations have a dominant effect on the cell, as only one copy of the gene needs to be mutated to produce the change that can lead to malignancy.

There is another group of genes in which mutations can also lead to cancer. These are the **tumor suppressor genes**, in which inactivation or deletion of both copies of the gene is required for a cell to become cancerous. The classic example of a tumor caused by the loss of such a gene is the childhood tumor retinoblastoma, which is a tumor of retinal cells. Although retinoblastoma is normally very rare, there are families in which an inherited predisposition to it is determined by a single gene, and they have been the means of identifying the gene responsible for this cancer. The inherited defect in some of these families turns out to be a deletion of a particular region on one of the two copies of chromosome 13. This on its own does not cause the cells to be cancerous. However, if any retinal cell also acquires a deletion of the same region on the other copy of chromosome 13, a retinal tumor develops. The gene in this region that is responsible for susceptibility to retinoblastoma is known as the *retinoblastoma* (*RB*) gene. Both copies of the *RB* gene must be lost or inactivated for a cell to become cancerous (Fig. 14.11), and hence *RB* is regarded as a tumor suppressor gene. The *RB* gene encodes a protein that is involved in the regulation of the cell cycle. We have seen in Section 9.5 that the RB protein plays a role in the normal development of muscle, where it is involved in the withdrawal of muscle precursor cells from the cell cycle.

Another tumor suppressor gene is familiar to us as a gene first identified as a segment polarity gene in *Drosophila*. The gene *patched* codes for a transmembrane protein that is part of the hedgehog signaling pathway (see Section 5.16). In humans, mutations that inactivate or delete both copies of the human homolog of *patched* are associated with a variety of epithelial cell tumors. Loss of just one copy of the homolog leads to some skeletal abnormalities.

A major feature of cancer is the failure of the cells to differentiate properly. The majority of cancers—over 85%—occur in epithelia. This is not surprising when it is recalled that many epithelia (such as the epidermis and the lining of the gut) are constantly being renewed by division and differentiation of stem cells (see Section 9.12). In normal epithelia, cells generated by stem cells continue to divide for a little time until they undergo differentiation, when they stop dividing. By contrast, cancerous epithelial cells continue to divide, although not necessarily more rapidly, and usually fail to differentiate. Another feature of cancer cells, unlike developing cells, is that when they divide, they are genetically unstable; the gain or loss of chromosomes is common in solid tumors.

This failure of cancer cells to differentiate is also clearly seen in certain leukemias—cancers of white blood cells. All blood cells are continually renewed from a pluripotent stem cell in the bone marrow, by a process in which steps in differentiation are interspersed with phases of cell proliferation. The pathway eventually culminates in terminal cell differentiation and a complete cessation of cell division (see Section 9.6). Several types of leukemia are caused by cells continuing to proliferate instead of differentiating. These cells become stuck at a particular immature stage in their normal development, a stage that can be identified by the molecules expressed on their cell surfaces. Cancers that result from a failure of cell differentiation could possibly be cured by the addition of a factor that could promote differentiation.

A number of developmental genes that we have considered elsewhere are also involved in cancer. For example, the first known mammalian member of the Wnt family was an oncogene, and abnormal expression of this signaling protein can block cell differentiation. Wnt genes are known to be involved in colorectal cancers. Mutations in the hedgehog family can also lead to tumor formation. The cell adhesion molecule E-cadherin is down-regulated in cancers that spread. Mutations in the Notch pathway can lead to a block in differentiation, and therefore also result in a cancer. Members of the TGF-β family are involved in tumor suppression.

Rarely, cancers can develop without any alteration in the cell's genetic material. The clearest examples are **teratocarcinomas**, solid tumors that spontaneously arise from germ cells. Teratocarcinomas are unusual tumors, in that they can contain a bizarre mixture of differentiated cell types. Spontaneous teratocarcinomas usually occur in the ovary or testis, and are derived from the germ cells. In the mouse ovary, accidental activation of an unfertilized egg results in its development *in situ* to the stage of epiblast formation; this epiblast then gives rise to a tumor. Similarly, if the epiblast of an early mouse embryo is transplanted to any site in the body of an adult mouse in which it receives a good blood supply, it gives rise to a teratocarcinoma.

There is good reason to believe that teratocarcinomas are not caused by genetic alterations. A mouse inner cell mass placed in culture gives rise to embryonic stem cells (ES cells), which can grow indefinitely in culture (see Box 3C, p. 89). These cells maintain their embryonic character, and when put back into the inner cell mass of another embryo, contribute normally to many different tissues, including the germ line, to produce a chimeric mouse. However, when the same ES cells are placed under the skin of an adult mouse, they develop into a teratocarcinoma (Fig. 14.12). Transgenic mice containing tissues derived from ES cells do not have an increased probability of forming tumors, yet those same cells consistently form tumors when grafted into adult mice. This indicates that

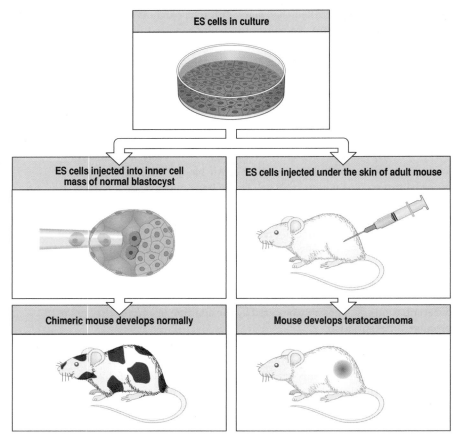

Fig. 14.12 Embryonic stem (ES) cells can develop normally or into a tumor, depending on the environmental signals they receive. Cultured ES cells originally obtained from an inner cell mass of a mouse contribute to a healthy chimeric mouse when injected into an early embryo (bottom left panels), but if injected under the skin of an adult mouse, the same cells will develop into a teratocarcinoma (bottom right panels).

the teratocarcinoma must be the result of the inner mass cells receiving the wrong developmental signals, and not of a genetic change.

14.9 Hormones control many features of plant growth

Unlike the protein growth hormones of animals, plant hormones are typically small molecular weight organic molecules. Auxin (indole-3-acetic acid) is one of the main regulators of plant growth, and is implicated in a large number of developmental processes, including growth toward a light source, tissue polarity, vascular tissue differentiation, and apical dominance, which is the phenomenon of the suppression of the growth of lateral buds immediately below the apical bud. Apical dominance is caused by a diffusible inhibitor of bud outgrowth produced by the shoot apex, as shown by putting an excised apex in contact with an agar block, which can then, on its own, inhibit lateral growth. The inhibitor is auxin. It is produced by the apical bud, transported down the stem, and suppresses the outgrowth of buds that fall within its sphere of influence. If the apical bud is removed, apical dominance is also removed, and lateral buds start to grow (Fig. 14.13). Application of auxin to the cut tip replaces the suppressive effect of the apical bud.

Another family of plant hormones, the gibberellins, regulate stem elongation and have some similar effects to auxin. Cytokinins, which can stimulate cell proliferation in culture, are derivatives of adenine. The chemical nature of plant hormones is very different from those of animals;

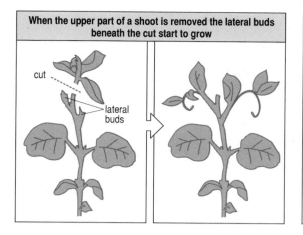

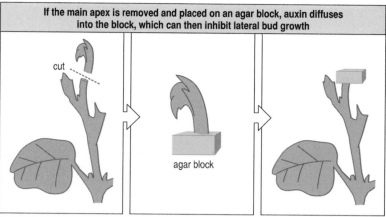

Fig. 14.13 Apical dominance in plants. Left panels: the growth of lateral buds is inhibited by the apical meristem above them. If the upper part of the stem is removed, lateral buds start growing. Right panels: an experiment to show that apical dominance is due to inhibition of lateral bud outgrowth by a substance secreted by the apical region. This substance is the plant hormone auxin.

nevertheless, they are thought to act through specific hormone-binding receptors and to stimulate intracellular signal transduction.

Plant growth is affected by a variety of environmental factors, such as temperature, humidity, and light. A seedling grown in the dark takes on a characteristic etiolated form in which chloroplasts do not develop, there is extensive elongation of internodes, and the leaves do not expand. The effect of light on plant growth (photomorphogenesis) is mediated through a family of intracellular receptor proteins called phytochromes, which respond to red light and regulate many aspects of plant development and growth.

14.10 Cell enlargement is central to plant growth

Plant growth is routinely achieved by cell division in meristems and organ primordia, followed by irreversible cell enlargement, which achieves most of the increase in size. Controlled cell enlargement can result in up to a 50-fold increase in the volume of a tissue. The driving force for expansion is the hydrostatic pressure exerted on the cell wall when the protoplast swells as a result of water entry into the cell vacuoles by osmosis (see Fig. 8.43). Cell expansion involves synthesis and deposition of new wall material, and is accompanied by an increase in both protein and RNA synthesis.

The direction of cell expansion is determined by the orientation of the cellulose fibrils in the cell wall, and is an example of directed dilation (discussed in Chapter 8 in relation to notochord expansion, see Fig. 8.41). Enlargement primarily occurs in a direction at right angles to the fibrils, where the wall is weakest. The orientation of cellulose fibrils in the cell wall is thought to be determined by the microtubules of the cell's cytoskeleton, which are responsible for positioning the enzyme assemblies that synthesize cellulose at the cell wall.

Mutations that affect cell expansion can significantly reduce elongation of the root. Plant growth hormones, such as ethylene and gibberellic acid, alter the orientation in which the fibrils are laid down, and so can alter the direction of expansion. Auxin aids expansion, perhaps by activating expansins, cell wall proteins that act by loosening the structure of the cell wall.

Summary

Growth in both animals and plants mainly occurs after the basic body plan has been laid down and the organs are still very small. Organ size can be controlled both by external signals and by intrinsic growth programs. Organ size can be determined by absolute dimensions rather than cell number or size. In animals, growth can occur by cell multiplication, cell enlargement, and secretion of large amounts of extracellular matrix. A plant's growth in size is both by cell enlargement and cell division. In mammals, insulin-like growth factors (IGFs) are required for normal embryonic growth, and they also mediate the effects of growth hormone after birth. Human post-natal growth is largely controlled by growth hormone, which is made in the pituitary gland. Growth in the long bones occurs in response to growth hormone stimulation of the cartilaginous growth plates at either end of the bone. Cancer is the result of the loss of growth control and differentiation. Plant growth is dependent on auxins, gibberellins, and other growth hormones.

Molting and metamorphosis

Many animals do not develop directly from an embryo into an 'adult' form, but into a larva from which the adult eventually develops by metamorphosis. The changes that occur at metamorphosis can be rapid and dramatic, the classic examples being the metamorphosis of a caterpillar into an adult butterfly, a maggot into a fly, and a tadpole into a frog. Another striking example of metamorphosis is the transformation of the pluteus stage larva of the sea urchin into the adult. In some cases it is hard to see any resemblance between the animal before and after metamorphosis. The adult fly does not resemble the maggot at all, because adult structures develop from the imaginal discs and so are completely absent from the larval stages (see Chapters 5 and 10). In frogs, the most obvious external changes at metamorphosis are the regression of the tadpole's tail and the development of limbs, although many other structural changes occur. In some insects, the entire body plan is transformed, with most larval tissues undergoing cell death as the adult tissues develop from the imaginal discs and histoblasts. In all arthropods, increase in size in larval and pre-adult stages requires shedding of the external cuticle, which is rigid, a process known as molting.

A number of features distinguish early embryogenesis from molting, metamorphosis, and other aspects of post-embryonic development. Whereas the signal molecules in early development act over a short range and are typically protein growth factors, many signals in post-embryonic development are produced by specialized endocrine cells, and include both protein and non-protein hormones. The synthesis of these hormones is orchestrated by the central nervous system in response to environmental cues, and there are complex interactions between the endocrine glands and their secretions.

14.11 Arthropods have to molt in order to grow

Arthropods have a rigid outer skeleton, the cuticle, which is secreted by the epidermis. This makes it impossible for the animals to increase in size gradually. Instead, increase in body size takes place in steps, associated with the loss of the old outer skeleton and the deposition of a new larger one. This process is known as **ecdysis** or **molting**. The stages between molts are known as instars. *Drosophila* larvae have three instars and molts. The increase in overall size between molts can be striking, as illustrated in Fig. 14.14 for the tobacco hornworm.

At the start of a molt, the epidermis separates from the cuticle in a process known as apolysis, and a fluid (molting fluid) is secreted into the space between the two (Fig. 14.15). The epidermis then increases in area by cell multiplication or cell enlargement, and becomes folded. It begins to secrete a new cuticle, and the old cuticle is partly digested away, eventually splits, and is shed.

Molting is under hormonal control. Stretch receptors that monitor body size are activated as the animal grows, and this results in the brain secreting prothoracicotropic hormone. This activates the pro-thoracic gland to release the steroid hormone ecdysone, which is the hormone that causes molting. The action of ecdysone is counteracted by juvenile hormone, produced by the corpus allatum. A similar hormonal circuit controls metamorphosis, and is described more fully in the next section.

14.12 Metamorphosis is under environmental and hormonal control

When an insect larva has reached a particular stage it does not grow and molt any further, but undergoes a more radical metamorphosis into the adult form. Metamorphosis occurs in many animal groups other than arthropods, including amphibians. In both insects and amphibians, environmental cues, such as nutrition, temperature, and light, as well as the animal's internal developmental program, control metamorphosis through their effects on neurosecretory cells in the brain. There are two groups of hormone-producing cells, one of which promotes metamorphosis, and the other inhibits it. Metamorphosis occurs when the inhibition, which is predominant in the larval stage, is overcome in response to environmental cues. The signals produced by the two sets of endocrine cells control the development of all the cells involved in metamorphosis. Larval tissues such as gut, salivary glands and certain muscles undergo programmed cell death. The imaginal discs now develop into rudimentary adult appendages like wings, legs, and antennae. The nervous system is also remodeled.

In insects, temperature and light cues stimulate neurosecretory cells in the larva's central nervous system to release signals that act on a neurosecretory release site behind the brain, which then secretes

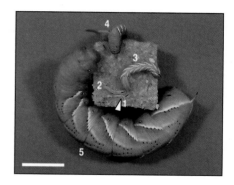

Fig. 14.14 Growth and molting of the tobacco hornworm caterpillar (*Manduca sexta*). The caterpillar goes through a series of molts. The tiny hatchling (1)—indicated by the arrow—molts to become a caterpillar (2), and then undergoes three further molts (3, 4, and 5). The increase in size between molts is about twofold. The caterpillars are sitting on a lump of caterpillar food. Scale bar = 1 cm.
Photograph courtesy of S.E. Reynolds.

Fig. 14.15 Molting and growth of the epidermis in arthropods. The cuticle is secreted by the epidermis. At the start of molting, the cuticle separates from the epidermis—apolysis—and a fluid is secreted between them. The epidermis grows, becomes folded, and begins to secrete a new cuticle. Enzymes weaken the old cuticle, which is shed.

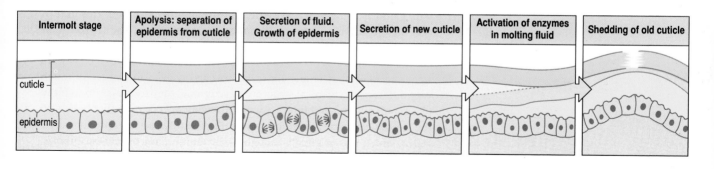

| Intermolt stage | Apolysis: separation of epidermis from cuticle | Secretion of fluid. Growth of epidermis | Secretion of new cuticle | Activation of enzymes in molting fluid | Shedding of old cuticle |

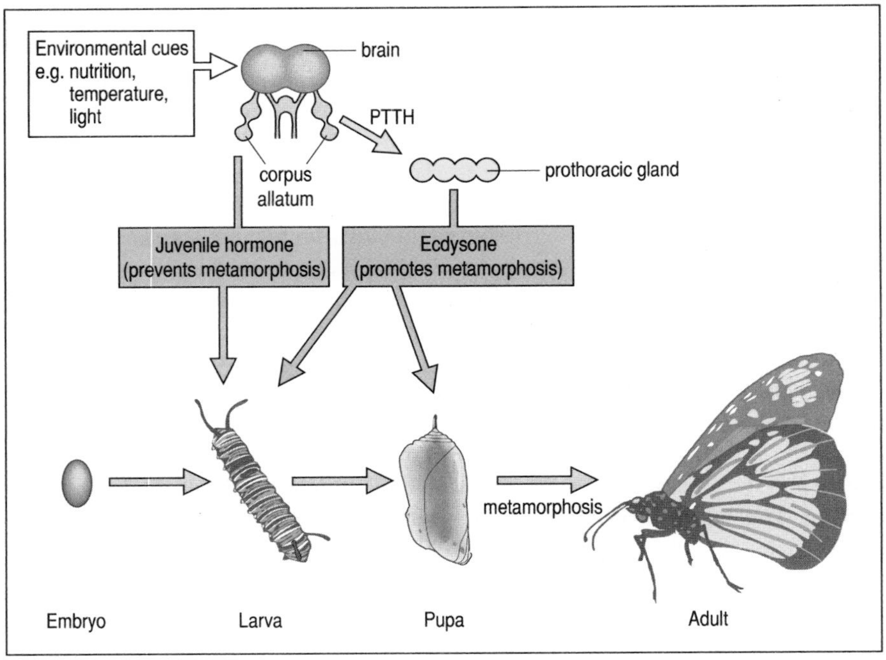

Fig. 14.16 Insect metamorphosis. The corpus allatum of a butterfly larva secretes juvenile hormone, which inhibits metamorphosis. In response to environmental changes, such as an increase in light and temperature, the corpus allatum of the final instar larva begins to secrete prothoracicotropic hormone (PTTH). This acts on the prothoracic gland to stimulate the secretion of ecdysone, the hormone that overcomes the inhibition by juvenile hormone and causes metamorphosis. After Tata, J.R.: 1998.

prothoracicotropic hormone. This acts on the prothoracic gland to stimu-late the production of ecdysone. It is ecdysone that promotes meta-morphosis, as demonstrated by its ability to induce premature meta-morphosis in fly larvae. However, the action of ecdysone can be counter-acted by another hormone—juvenile hormone—produced by the corpus allatum, an endocrine gland located just behind the brain. As its name implies, juvenile hormone maintains the larval state. In butterflies, a pulse of ecdysone in the final instar larva triggers the beginning of pupa forma-tion, and another pulse some days later initiates the later stages of meta-morphosis (Fig. 14.16).

Ecdysone crosses the plasma membrane, where it interacts with intracel-lular ecdysone receptors that belong to the steroid hormone receptor superfamily. These receptors are gene regulatory proteins, which are acti-vated by binding their hormone ligand (see Section 9.2). The hormone–receptor complex binds to the regulatory regions of a number of different genes, inducing a new pattern of gene activity characteristic of metamorphosis.

In amphibians, in response to nutritional status and environmental cues, such as temperature and light, the neurosecretory cells of the hypo-thalamus release corticotropin-releasing hormone, which acts on the pitu-itary gland, causing it to release thyroid-stimulating hormone. This in turn acts on the thyroid gland to stimulate the secretion of the thyroid hor-mones that bring about metamorphosis (Fig. 14.17). The thyroid hormones are the iodo-amino acids thyroxine (T_4), and tri-iodothyronine (T_3). They are signaling molecules of ancient origin, occurring even in plants. Although very different in chemical structure to ecdysone, they too pass through the plasma membrane and interact with intracellular receptors that belong to the steroid hormone receptor superfamily. The pituitary also produces pro-lactin, which was originally thought to be an inhibitor of metamorphosis;

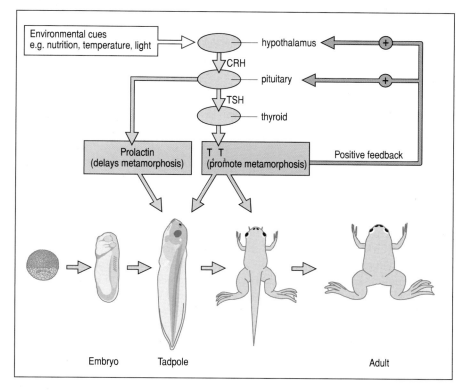

Fig. 14.17 Amphibian metamorphosis. Changes in the environment, such as an increase in nutritional levels, cause the secretion of corticotropin-releasing hormone (CRH) from the hypothalamus, which acts on the pituitary to release thyroid-stimulating hormone (TSH). This in turn acts on the thyroid glands to stimulate secretion of the thyroid hormones thyroxine (T_4) and tri-iodothyronine (T_3), which cause metamorphosis. The thyroid hormones also act on the hypothalamus and pituitary to maintain synthesis of CRH and TSH. After Tata, J.R.: 1998.

however, overexpression of prolactin does not prolong tadpole life but reduces tail resorption.

A striking feature of the hormones that stimulate metamorphosis is that as well as affecting a wide variety of tissues, they affect different tissues in different ways, their effects varying from the subtle to the gross. In the tadpole limb, for example, thyroid hormones promote development and growth, whereas they cause cell death and degeneration in the tail. Fast muscles are the first to go, and the notochord later collapses. Metamorphosis also leads to changes in the responsiveness of cells to other signals; for example, in *Xenopus*, estrogen can only induce the synthesis of vitellogenin, a protein required for the yolk of the egg, after metamorphosis. Each tissue has its own response to the hormones causing metamorphosis, and some of these effects can be reproduced in culture. When excised *Xenopus* tadpole tails are exposed to thyroid hormones in culture, for example, they cause cell death and complete tissue regression.

There are alterations in the expression of many genes—several hundred at least—during *Drosophila* metamorphosis. In *Drosophila*, because of the special characteristics of polytene chromosomes, it is possible to see some changes in gene activity that occur during metamorphosis. Cells in some tissues of the larva grow and repeatedly pass through S phase without undergoing mitosis and cell division. The cells become very large and can have several thousand times the normal complement of DNA. In salivary gland cells, many copies of each chromosome are packed side by side to form giant polytene chromosomes. When a gene is active, the chromosome at that site expands into a large localized 'puff', which is easily visible (Fig. 14.18). The puff represents the unfolding of chromatin and the associated

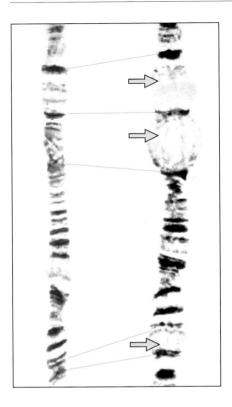

Fig. 14.18 Gene activity seen as puffs on the polytene chromosomes of *Drosophila*. A region of a chromosome is shown from a young third instar (left), and from an older larva (right), after ecdysone has induced puffs at three loci (arrowed).

Photograph courtesy of M. Ashburner.

transcriptional activity. When the gene is no longer active the puff disappears. During the last days of larval life, a large number of puffs are formed in a precise sequence, a pattern that is under the direct influence of ecdysone.

Summary

Arthropod larvae grow by undergoing a series of molts in which the rigid cuticle is shed. Metamorphosis during the post-embryonic period can result in a dramatic change in the form of an organism. In insects, it is hard to see any resemblance between the animal before and after metamorphosis, whereas in amphibians, the change is somewhat less dramatic. Environmental and hormonal factors control metamorphosis. In both insects and amphibians there are two sets of hormonal signals, one promoting, the other delaying, metamorphosis. Thyroid hormones cause metamorphosis in amphibians, and ecdysone does the same in insects. In *Drosophila*, gene activity during metamorphosis can be monitored by localized puffing on the giant polytene chromosomes of the salivary gland cells.

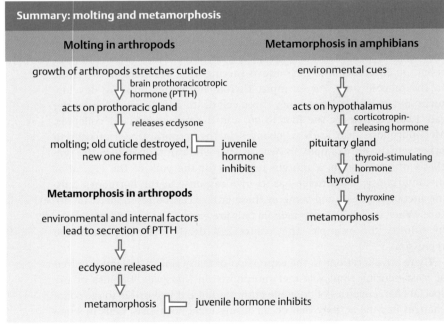

Aging and senescence

Organisms are not immortal, even if they escape disease or accidents. With aging—the passage of time—**senescence** occurs. Senescence is the increase of impairment of physiological functions with age, resulting in a decreased ability to deal with a variety of stresses, and an increased susceptibility to disease. This phenomenon raises many unanswered questions as to its underlying mechanisms, but we can at least consider some general questions, such as whether senescence is part of an organism's post-embryonic developmental program or whether it is simply the result of wear and tear.

Although individuals may vary in the time at which particular aspects of aging appear, the overall effect is summed up as an increased probability of dying in most animals, including humans, with increased age. This life pattern, which is illustrated in relation to *Drosophila* (Fig. 14.19), is typical of many animals, but there is little evidence that aging contributes to mortality in the wild; more than 90% of wild mice die during their first year. There are, however, exceptions, such as the Pacific salmon, in which death does not come after a process of gradual aging, but is linked to a certain stage in the life cycle, in this case to spawning.

One view of senescence is that it is the outcome of an accumulation of damage that eventually outstrips the ability of the body to repair itself, and so leads to the loss of essential functions. For example, some old elephants die of starvation because their teeth have worn out. Nevertheless, there is clear evidence that senescence is under genetic control, as different animals age at vastly different rates, as shown by their different life spans (Fig. 14.20). An elephant, for example, is born after 21 months' embryonic development, and at that point shows few, if any, signs of aging, whereas a 21-month-old mouse is already well into middle age and beginning to show signs of senescence. This genetic control of aging can be understood in

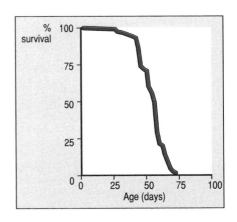

Fig. 14.19 Aging in *Drosophila*. The probability of dying increases rapidly at older ages.

Longevity and time to attain reproductive maturity at puberty for various mammals			
	Maximum life span (months)	**Length of gestation (months)**	**Age at puberty (months)**
Man	1440	9	144
Finback whale	960	12	—
Indian elephant	840	21	156
Horse	744	11	12
Chimpanzee	534	8	120
Brown bear	442	7	72
Dog	408	2	7
Cattle	360	9	6
Rhesus monkey	348	5.5	36
Cat	336	2	15
Pig	324	4	4
Squirrel monkey	252	5	36
Sheep	240	5	7
Gray squirrel	180	1.5	12
European rabbit	156	1	12
Guinea-pig	90	2	2
House rat	56	0.7	2
Golden hamster	48	0.5	2
Mouse	42	0.7	1.5

Fig. 14.20 Table showing life span, length of gestation, and age at puberty for various mammals.

terms of the 'disposable soma theory', which puts it into the context of evolution. The disposable soma theory proposes that natural selection tunes the life history of the organism so that sufficient resources are invested in maintaining the repair mechanisms that prevent aging at least until the organism has reproduced and cared for its young. Thus mice, which start reproducing when just a few months old, need to maintain their repair mechanisms for much less time than do elephants, which only start to reproduce when around 13 years old. In most species of animals in the wild, few individuals live long enough to show obvious signs of senescence and senescence need only be delayed until reproduction is complete.

14.13 Genes can alter the timing of senescence

The maximum recorded life spans of animals show dramatic differences in length (see Fig. 14.20). Humans can live as long as 120 years, some owls 68 years, cats 28 years, *Xenopus* 15 years, mice 3½ years, and the nematode about 25 days. Mutations in genes that affect life span have been identified in both *C. elegans* and humans, and may give clues to the mechanisms involved; both the ability to resist damage to DNA and the effects of oxygen radicals are important.

A *C. elegans* worm that hatches as a first instar larva in an uncrowded environment with ample food grows to adulthood and can survive for 25 days. However, in crowded conditions and when food is short, the animal enters a larval state known as the **dauer** larval state, where it neither eats nor grows nor reproduces until food becomes available again. This state can last for 60 days. An insulin/IGF-1 system in *C. elegans* plays an important role in controlling life span. Mutations that lower the level of DAF-2 expression, which encodes a receptor in this pathway, can cause the animal to remain youthful and live twice as long as normal. The pathway also regulates entry into the dauer state. It is not known exactly how entering the dauer state or mutations in the IGF-1 pathway affect longevity, but there may be a lowering of metabolic rate. Surprisingly, restoring the DAF-2 pathway to neurons alone was sufficient to specify wild-type life span, which suggests that the nervous system is a key regulator of longevity.

A similar system regulates aging in *Drosophila* and mutations in the insulin/IGF-1 pathway almost double the life span. The flies, rather like the dauer larvae, enter into a state of reproductive diapause. These effects on life span may be due to resistance to oxidative stress.

There is evidence that oxidative damage accelerates aging and that reactive oxygen radicals are key players. Reduction in food intake increases life span in other animals; rats on a minimal diet live about 40% longer than rats allowed to eat as much as they like, which is thought to be partly due to a reduced exposure to free radicals. These are formed during the oxidative breakdown of food. They are highly reactive, and can damage both DNA and proteins. A long-lived rodent species generates less reactive oxygen than the laboratory mouse, and a mutation in the mouse gene $p66^{SHC}$ that renders the animal resistant to oxygen radicals increases life span by 30%.

Humans that are homozygous for the recessive gene defect known as Werner's syndrome show striking effects of premature aging. There is growth retardation at puberty, and by their early twenties, those affected by this syndrome have gray hair and suffer from a variety of illnesses, such as heart disease, that are typical of old age. Most die before the age of 50. Fibroblasts taken from Werner's syndrome patients undergo fewer cell divisions in culture before becoming senescent and dying than do fibroblasts from unaffected people of the same age (see Section 14.14). The gene affected in Werner's syndrome has been isolated, and is thought to encode a protein involved in unwinding DNA. Such unwinding is required for DNA replication, DNA repair, and gene expression. The inability to carry out DNA repair properly in Werner's syndrome patients could subject the genetic material to a much higher level of damage than normal. The link between Werner's syndrome and DNA thus fits with the possibility that aging is linked to the accumulation of damage in DNA.

14.14 Cell senescence

One might think that when cells are isolated from an animal, placed in culture, and provided with adequate medium and growth factors, they would continue to proliferate almost indefinitely. But this is not the case. For example, mammalian fibroblasts—connective tissue cells—will only go through a limited number of cell doublings in culture; the cells then stop dividing, however long they are cultured (Fig. 14.21). For normal fibroblasts, the number of cell doublings depends both on the species and the age of the animal from which they are taken. Human fibroblasts taken from a fetus go through about 60 doublings, those from an 80-year-old about 30, and those from an adult mouse about 12–15. When the cells stop dividing, they appear to be healthy, but are stuck at some point in the cell cycle, often G_0. Cells taken from patients with Werner's syndrome, who show an acceleration of many features of normal aging, make significantly fewer divisions in culture than normal cells. However, it is far from clear how this behavior of cells in culture is significant for the aging of the organism and to what extent it reflects the way the cells are cultured.

A feature shared by senescent cells in culture and *in vivo* is shortening of the telomeres. These are the repetitive DNA sequences at the ends of the chromosomes that preserve chromosome integrity and ensure that chromosomes replicate themselves completely without loss of informational

Fig. 14.21 Vertebrate fibroblasts can only go through a limited number of divisions in culture. Fibroblasts placed in culture are subcultured until they stop growing (top panels). The number of cell doublings in culture before they stop dividing is related to their maximum age, as indicated by the figures in brackets on the graph (bottom panel).

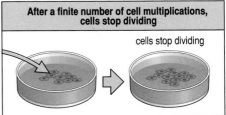

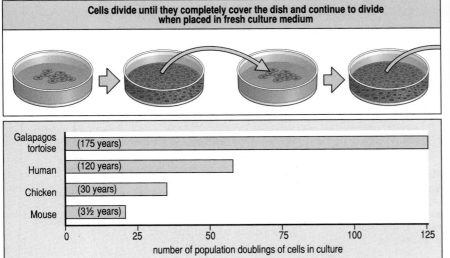

DNA at the ends. The length of the telomeres is reduced in older human cells. Telomere length is found to decrease at each DNA replication, suggesting that they are not completely replicated at each cell division and this may be related to senescence. If the enzyme telomerase, which maintains telomere length and is normally absent from cells in culture, is expressed in those cells, senescence in culture does not occur. However, telomere shortening appears not to be a major cause of aging in somatic cells, as certain rodent cells such as Schwann cells can, under appropriate conditions, proliferate indefinitely and their telomeres do not control replication. Cloning can restore the capacity of mouse cells to divide and enable them even to divide in culture more than normal.

Summary

Aging is largely caused by damage to cells, particularly by reactive oxygen, but is also under genetic control. Many normal cells in culture can only undergo a limited number of cell divisions, which correlate with their age at isolation and the normal life span of the animal from which they came. Genes that can increase life span have been identified in *C. elegans* and *Drosophila* and may act by increasing the animals' resistance to oxidative stress.

SUMMARY IN CHAPTER 14

The form of many animals is laid down in miniature during embryonic development, and they then grow in size, keeping the basic body form, although different regions grow at different rates. Growth may involve cell multiplication, cell enlargement, and the laying down of extracellular material. In vertebrates, structures have an intrinsic growth program, which is under hormonal control. Arthropod larvae grow by molting and both they and other animals, such as frogs, undergo metamorphosis, which is under hormonal control. Cancer can be viewed as an aberration of growth, as it usually results from mutations that lead to excessive cell proliferation and failure of cells to differentiate. The symptoms of aging appear mainly to be caused by damage to cells which acumulates over time, and aging is also under genetic control. In plants, all adult structures are derived by growth from specialized regions called meristems.

REFERENCES

14.1 Tissues can grow by cell proliferation, cell enlargement, or accretion

Goss, R.J.: *The Physiology of Growth*. New York: Academic Press, 1978.

14.2 Cell proliferation can be controlled by an intrinsic program

Edgar, B.: **Diversification of cell cycle controls in developing embryos**. *Curr. Opin. Cell Biol.* 1995, **7**: 815–824.
Edgar, B.A., Lehner, C.F.: **Developmental control of cell cycle regulators: a fly's perspective**. *Science* 1996, **274**: 1646–1652.
Follette, P.J., O'Farrell, P.H.: **Connecting cell behavior to patterning: lessons from the cell cycle**. *Cell* 1997, **88**: 309–314.
Raff, M.C.: **Size control: the regulation of cell numbers in animal development**. *Cell* 1996, **86**: 173–175.

14.3 Organ size can be controlled by external signals and intrinsic growth programs

Conlon, I., Raff, M.: **Size control in animal development**. *Cell* 1999, **96**: 235–244.
Lee, S-J., NcPherron, A.C.: **Myostatin and the control of skeletal muscle mass**. *Curr. Opin. Genet. Dev.* 1999, **9**: 604–607.
Wolpert, L.: **The cellular basis of skeletal growth during development**. *Brit. Med. Bull.* 1981, **37**: 215–219.

14.4 Organ size may be determined by absolute dimension rather than cell number or size

Day, S., Lawrence, P.A.: **Measuring dimensions: the regulation of size and shape**. *Development* 2000, **127**: 2977–2987.

Su, T.T., O'Farrell, P.H.: **Size control: cell proliferation does not equal growth**. *Curr. Biol.* 1998, **8**: R687–R689.

14.5 Growth of mammals is dependent on growth hormones

Baker, J., Liu, J.P., Robertson, E.J., Efstratiades, A.: **Role of insulin-like growth factors in embryonic and postnatal growth**. *Cell* 1993, **75**: 73–82.

Barker, D.J.: **The Wellcome Foundation Lecture, 1994: The fetal origins of adult disease**. *Proc. Roy. Soc. Lond.* 1995, **262**: 37–43.

Brook, C.G.D.: *A Guide to the Practice of Paediatric Endocrinology*. Cambridge: Cambridge University Press, 1993.

Heyner, S., Garside, W.T.: **Biological actions of IGFs in mammalian development**. *BioEssays* 1994, **16**: 55–57.

Kwong, W.Y., Wild, A.E., Roberts, P., Willis, A.C., Fleming, T.P.: **Maternal undernutrition in the postimplantation period of rat development causes blastocyst abnormalities and programming of postnatal hypertension**. *Development* 2000, **127**: 4195–4202.

14.6 Growth of the long bones occurs in the growth plates

Deng, C., Wynshaw-Boris, A., Zhou, F., Kuo, A., Leder, P.: **Fibroblast growth factor receptor 3 is a negative regulator of bone growth**. *Cell* 1996, **84**: 911–921.

Karp, S.J., Schipani, E., St Jacques, B., Hunzelman, J., Kronenberg, H., McMahon, A.P.: **Indian hedgehog coordinates endochondral bone growth and morphogenesis via parathyroid hormone related protein dependent and independent pathways**. *Development* 2000, **127**: 543–548.

Kember, N.F.: **Cell kinetics and the control of bone growth**. *Acta Paediatr. Suppl.* 1993, **391**: 61–65.

Ohlsson, C., Isgaard, J., Tomell, J., Nilsson, A., Isaksson, O.G., Lindahl, A.: **Endocrine regulation of longitudinal bone growth**. *Acta Paediatr. Suppl.* 1993, **391**: 33–40.

Roush, W.: **Putting the brakes on bone growth**. *Science* 1996, **273**: 579.

Wallis, G.A.: **Bone growth: coordinating chondrocyte differentiation**. *Curr. Biol.* 1996, **6**: 1577–1580.

14.7 Growth of vertebrate striated muscle is dependent on tension

Schultz, E.: **Satellite cell proliferative compartments in growing skeletal muscles**. *Dev. Biol.* 1996, **175**: 84–94.

Williams, P.E., Goldspink, G.: **Changes in sarcomere length and physiological properties in immobilized muscle**. *J. Anat.* 1978, **127**: 450–468.

14.8 Cancer can result from mutations in genes that control cell multiplication and differentiation

Hunter, T.: **Oncoprotein networks**. *Cell* 1997, **88**: 333–346.

Rabbitts, T.H.: **Chromosomal translocations in human cancer**. *Nature* 1994, **372**: 143–149.

Sawyers, C.L., Denny, C.T., Witte, O.N.: **Leukemia and the disruption of normal hematopoiesis**. *Cell* 1991, **64**: 337–350.

Shilo, B.Z.: **Tumor suppressors. Dispatches from patched**. *Nature* 1996, **382**: 115–116.

Taipale, J., Beachy, P.A.: **The hedgehog and Wnt signalling pathways in cancer**. *Nature* 2001, **411**: 349–353.

14.9 Hormones control many features of plant growth

Raven, P.H., Evert, R.F., Eichhorn, S.E.: *The Biology of Plants* (5th edn). New York: Worth, 1992.

14.10 Cell enlargement is central to plant growth

Cosgrove, D.J.: **Plant cell enlargement and the action of expansins**. *BioEssays* 1996, **18**: 533–540.

Hauser, M.T., Morikami, A., Benfey, P.N.: **Conditional root expansion mutants of *Arabidopsis***. *Development* 1995, **121**: 1237–1252.

14.11 Arthropods have to molt in order to grow

Reynolds, S.E., Samuels, R.I.: **Physionomy and biochemistry of insect molting fluid**. *Adv. Insect Physical* 1996, **26**: 157–232.

14.12 Metamorphosis is under environmental and hormonal control

Elinson, R.P., Remo, B., Brown, D.D.: **Novel structural element identified during tail resorption in *Xenopus laevis* metamorphosis: lessons from tailed frogs**. *Dev. Biol.* 1999, **215**: 243–252.

Huang, H., Brown, D.D.: **Prolactin is not a juvenile hormone in *Xenopus laevis* metamorphosis**. *Proc. Natl Acad. Sci. USA.* 2000, **97**: 195–199.

Tata, J.R.: *Hormonal Signaling and Postembryonic Development*. Heidelberg: Springer, 1998.

Thummel, C.S.: **Flies on steroids—*Drosophila* metamorphosis and the mechanisms of steroid hormone action**. *Trends Genet.* 1996, **12**: 306–310.

White, K.P., Rifkin, S.A., Hurban, P., Hogness, D.S.: **Microarray analysis of *Drosophila* development during metamorphosis**. *Science* 1999, **286**: 2179–2184.

Aging and senescence

Kirkwood, T.B.L., Austad, S.N.: **Why do we age?** *Nature* 2000, **408**: 233–245.

Orr, W.C., Sohal, R.S.: **Extension of life-span by overexpression of superoxide dismutase and catalase in *Drosophila melanogaster***. *Science* 1994, **263**: 1128–1130.

14.13 Genes can alter the timing of senescence

Kenyon, C.: **A conserved regulatory system for aging**. *Cell* 2001, **105**: 165–168.

Weindruck, R.: **Caloric restriction and aging**. *Sci. Amer.* 1996, **274**: 46–52.

Yu, C.E., Oshima, J., Fu, Y.H., Wijsman, E.M., Hisama, F., Alisch, R., Matthews, S., Nakura. J., Miki, T., Ouais, S., Martin, G.M., Mulligan, J., Schellenberg, G,D.: **Positional cloning of the Werner's syndrome gene**. *Science* 1996, **272**: 258–262.

14.14 Cell senescence

Shay, J.W., Wright, W.E.: **When do telomeres matter?** *Science* 2001, **291**: 839–840.

Sherr, C.J., DePinho, R.A.: **Cellular senescence: mitotic clock or culture shock**. *Cell* 2000, **102**: 407–410.

Evolution and development

15

- Modification of development in evolution
- Changes in the timing of developmental processes during evolution
- Evolution of development

Evolutionary biology and developmental biology have rather different histories and approaches. Whereas studies of development are based largely on model systems and experiments, evolutionary biology uses comparative methodology to describe and research explanations of past and present patterns of diversity and change. Evolution is a one-off historical process that we can only get at indirectly. The evolution of multicellular forms of life is the result of changes in development, and this in turn is entirely due to changes in genes that control cell behavior in the embryo. It is also true, as has been suggested, that nothing in biology makes sense unless viewed in the light of evolution. Certainly, it would be very difficult to make sense of many aspects of development without an evolutionary perspective. For example, in our discussion of vertebrate development we saw how, despite different modes of very early development, all vertebrate embryos develop through a rather similar phylotypic stage, after which their development diverges again (see Fig. 3.1). This shared phylotypic stage, which is the embryonic stage after neurulation and the formation of the somites, is a stage through which some distant ancestor of the vertebrates passed. It has persisted ever since, to become a fundamental characteristic of the development of all vertebrates, whereas the stages before and after the phylotypic stage have evolved differently in different organisms. Genetically based changes in development that generated new adult forms better adapted to their environment and more successful modes of reproduction were selected for during evolution. Genetic variability resulting from mutations, sexual reproduction, and genetic recombination is present in all populations of the organisms we have dealt with in this book, and provides new phenotypes upon which selection can act. Changes in the control regions of genes are fundamental.

Throughout this book we have emphasized the conservation of some developmental mechanisms at the cellular and molecular level among distantly related organisms. The widespread use of the Hox gene complexes and of the same few families of protein signaling molecules provide excellent examples of this. It is this basic similarity in molecular mechanisms that has made developmental biology so exciting in recent years; it has meant that discoveries in one animal, in particular in *Drosophila*, have had important implications for understanding development in other animals. It seems that when a useful developmental mechanism evolved, it was used again and again in very different organisms, and at different times and places in the same organism. Wingless signaling, for example, is already present in simple multicellular animals such as the coelenterates, which arose early in animal evolution. But if the genes themselves are so similar,

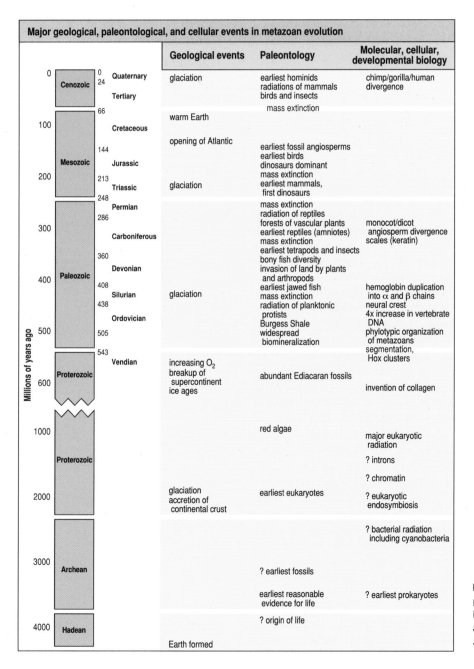

Fig. 15.1 Major geological, paleontological, and cellular events in the evolution of multicellular animals. Modified from Gerhart, J. and Kirschner, M.: 1997.

how do they control cell behavior so that the wide variety of animal forms develops? The answer lies both in the spatial pattern and the timing with which these developmental genes are expressed, and in the target genes that they influence, which control cell behavior. These differences are largely due to differences that arise both in the control regions of the developmental genes themselves, which determine how they are regulated in space and time, and in the control regions of potential target genes, which determine the targets that the developmental genes will influence. In a way, evolution has been lazy; having 'discovered' a good way of developing pattern and form, it just tinkered with it to make new animals.

We have already considered the early development of a wide variety of organisms and found some similarities, as well as a number of differences. Here, we mainly confine our attention to two phyla—the chordates, which include the vertebrates, and the arthropods, which include the insects and crustaceans. We focus on those differences that distinguish the members of a large group of related animals, such as the vertebrates or the insects, from each other. A general view of the evolution of animals over time and their relation to geological periods is shown in Fig. 15.1 (opposite). It is generally accepted that all multicellular animals have evolved from a common ancestor, which itself evolved from a single-celled organism.

We will look at the relationship between **ontogeny**—the development of the individual organism—and **phylogeny**—the evolutionary history of the species or group: why, for example, do all vertebrate embryos pass through an apparently fish-like phylotypic stage that has structures resembling gill slits? We then discuss the many variations that occur on the theme of a basic segmented body plan: what determines the different numbers and positions of paired appendages, such as legs and wings, in different groups of segmented organisms? We consider the timing of developmental events, and how simple alterations in timing, and variations in growth, can have major effects on the shape and form of an organism. In all cases, we ultimately want to understand the changes in the developmental processes and the genes controlling them that have resulted in the extraordinary variety of multicellular animals. This is an exciting area of study in which many problems remain to be solved. Finally, we take a brief look at the evolution of development itself.

Modification of development in evolution

Comparisons of embryos of related species have suggested an important generalization about development: the more general characteristics of a group of animals (that is those shared by all members of the group) appear earlier in their embryos than the more specialized ones, and arose earlier in evolution. In the vertebrates, a good example of a general characteristic would be the notochord, which is common to all vertebrates and is also found in other chordate embryos. The phylum Chordata comprises three subphyla—the vertebrates, the cephalochordates, such as amphioxus (Fig. 15.2), and the urochordates, such as the ascidians (see Chapter 6). They all have, at some embryonic stage, a notochord, flanked by muscle, and a dorsal neural tube. Paired appendages, such as limbs, which develop later, are special characters that develop only in the vertebrates, and differ in form among different vertebrates. All vertebrate embryos pass through a

Fig. 15.2 The cephalochordate amphioxus compared to a hypothetical primitive vertebrate similar to a present-day lamprey. The overall construction of the organisms is very similar, with a dorsal nerve cord, a more ventral axial skeleton known as the notochord, and a ventral digestive tract. Both animals have gills in the pharyngeal region, structures that are designed to capture food or remove oxygen from the water. The notochord extends right to the anterior end of amphioxus, but the vertebrate has a prominent head at the anterior end, extending beyond the notochord. In more advanced vertebrates, the notochord is present only in the embryo, being replaced by the vertebrae. Modified from Finnerty, J.R.: 2000.

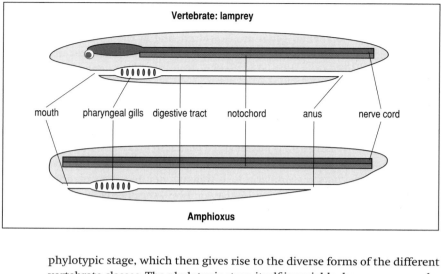

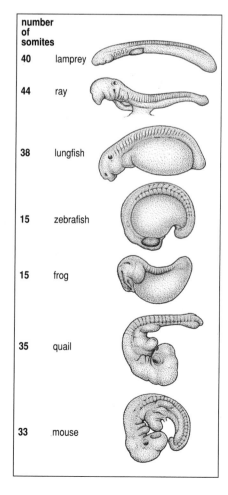

Fig. 15.3 Vertebrate embryos at the phylotypic (tailbud) stage show wide variation in somite number. Somite number is given in parentheses. Not to scale. Modified from Richardson, M.K., *et al.*: 1997.

phylotypic stage, which then gives rise to the diverse forms of the different vertebrate classes. The phylotypic stage itself is variable, however, as can be seen in Fig. 15.3, which shows that at the phylotypic or tailbud stage, the number of somites can be very different, and that some organs such as limb buds are at different stages of development. The development of the different vertebrate classes before the phylotypic stage is also highly divergent, because of their very different modes of reproduction; some developmental features that precede the phylotypic stage are evolutionarily highly advanced, such as the formation of a trophoblast and inner cell mass by mammals. This is an example of a special character that developed late in vertebrate evolution, and is related to the nutrition of the embryo through a placenta, rather than a yolky egg.

In this part of the chapter, we consider the modifications that have occurred to a variety of embryonic structures during evolution, including the basic body plan and the limbs. We start by looking at the branchial arches in vertebrates.

15.1 Embryonic structures have acquired new functions during evolution

If two groups of animals that differ greatly in their adult structure and habits (such as fishes and mammals) pass through a very similar embryonic stage, this could indicate that they are descended from a common ancestor and, in evolutionary terms, are closely related. Thus, an embryo's development reflects the evolutionary history of its ancestors. Structures found at a particular embryonic stage have become modified during evolution into different forms in the different groups. In vertebrates, one good example of this is the evolution of limbs from the embryonic fin-like structures of a fish ancestor, which is discussed in the next section. Division of the body into segments, which then diverge from each other in structure and function, is a common feature in the evolution of both vertebrates and arthropods. An example are the branchial arches and clefts that are present in all vertebrate embryos (see Fig. 2.9), including humans. These segmented structures are not the relics of the gill arches and gill slits of an adult fish-like ancestor, but of structures that would have been present in the embryo of the fish-like ancestor as developmental precursors to gill slits and arches. During evolution, the branchial arches have given rise both to the gills of the primitive jawless fishes and, in a later modification, to jaws (Fig. 15.4). When the ancestor of land vertebrates left the sea, gills were no longer required, but the embryonic structures that gave rise to them persisted, perhaps because their loss would have affected the vasculature. With time they became modified, and in mammals, including humans, they now give rise to various structures in the face and neck (Fig. 15.5), many of which are derived from neural crest cells which migrate into the branchial arches early in development (see Fig. 4.27). The cleft between the first and second branchial arches provides the opening for the eustachian tube, and endo-dermal cells in the clefts give rise to a variety of glands, such as the thyroid and thymus.

Evolution rarely generates a completely novel structure out of the blue. New anatomical features usually arise from modification of an existing structure. One can therefore think of much of evolution as a tinkering with existing structures, which gradually fashions something different. A nice example of a modification of an existing structure is provided by the evolu-tion of the mammalian middle ear. This is made up of three bones that transmit sound from the eardrum (the tympanic membrane) to the inner ear. In the reptilian ancestors of mammals, the joint between the skull and the lower jaw was between the quadrate bone of the skull and the articular bone of the lower jaw, which were also involved in transmitting sound (Fig. 15.6). During mammalian evolution, the lower jaw became just one bone, the dentary, with the articular no longer attached to the lower jaw. By changes in their development, the articular and the quadrate bones in mammals were modified into two bones, the malleus and incus, whose function was now to transmit sound from the tympanic membrane to the inner ear. The quadrate is evolutionarily and developmentally **homologous** with the dorsal bone of the first arch, and the stapes with that of the second. **Homology** here refers to a morphological or structural similarity due to a common ancestry.

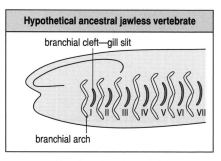

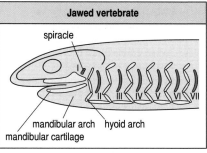

Fig. 15.4 Modification of the branchial arches during the evolution of jaws in vertebrates. The ancestral jawless fish had a series of at least seven gill slits—branchial clefts—supported by cartilaginous or bony arches. Jaws developed from a modification of the first arch to give the mandibular arch, with the mandibular cartilage of the lower jaw and the hyoid arch behind it.

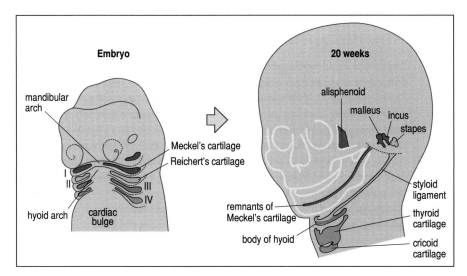

Fig. 15.5 Fate of branchial arch cartilage in humans. In the embryo, cartilage develops in the branchial arches, which gives rise to elements of the three auditory ossicles, the hyoid, and the pharyngeal skeleton. The fate of the various elements is shown by the color coding. After Larsen, W.J.: 1993.

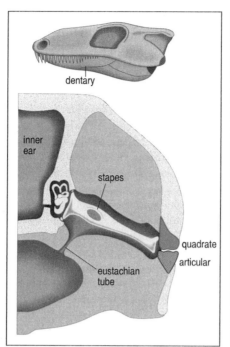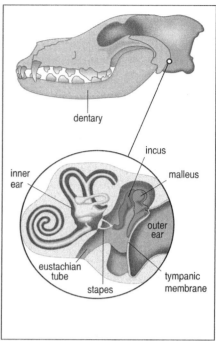

Fig. 15.6 Evolution of the bones of the mammalian middle ear. The articular and quadrate bones of ancestral reptiles (left panel) were part of the lower jaw articulation. Sound was transmitted to the inner ear via these bones and their connection to the stapes. When the lower jaw of mammals became a single bone (the dentary), the articular bone became the malleus, and the quadrate bone the incus of the middle ear, acquiring a new function in transmitting sound to the inner ear from the tympanic membrane (right panel). The eustachian tube forms between branchial arches I and II. After Romer, A.S.: 1949.

Another example of modification of a pre-existing structure is provided by the evolution of the vertebrate kidney. In birds and mammals, three kidney-like structures—metanephros, mesonephros, and pronephros— appear during development. The pronephros and mesonephros are transitory and the functional kidney develops from the metanephros. However, as discussed in Section 12.2, the mesonephros plays a key role in the development of the gonads, giving rise to the somatic cells of the testis and ovary. In lower vertebrates, such as fish and amphibians, the pronephros acts as the functional kidney in the immature juvenile stages, but the mesonephros is the functional kidney in the adult. Thus, in birds and mammals the embryonic kidneys of their ancestors have persisted as embryonic structures, but have been modified to provide structures essential to the development of the gonad.

15.2 Limbs evolved from fins

The limbs of tetrapod vertebrates are special characters that develop after the phylotypic stage. Amphibians, reptiles, birds, and mammals have limbs, whereas fish have fins. The limbs of the first land vertebrates evolved from the pelvic and pectoral fins of their fish-like ancestors. The basic limb pattern is highly conserved in both the forelimbs and hindlimbs of all tetrapods, although there are differences both between forelimbs and hindlimbs, and between different vertebrates. Limbs evolved from fins, but how fins evolved is far less clear. Nevertheless, the development of these appendages made use of signaling molecules like Sonic hedgehog and fibroblast growth factor and of transcription factors such as the Hox proteins, which were already being used to pattern the body.

The fossil record suggests that the transition from fins to limbs occurred in the Devonian period, between 400 and 360 million years ago. The

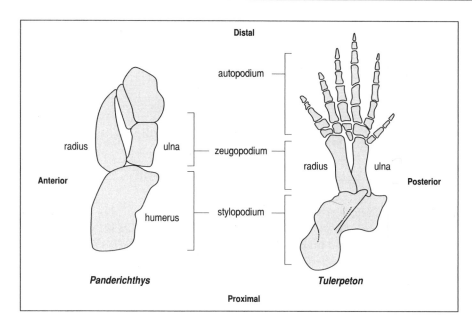

Fig. 15.7 The fin to limb transition. In the lobe-like fin of the Devonian fish *Panderichthys*, there were proximal elements corresponding to the humerus (the stylopodium), radius, and ulna (the zeugopodium), but no distal elements. The Devonian tetrapod *Tulerpeton* has similar proximal elements, but has also developed digits.

transition probably occurred when the fish ancestors of the tetrapod vertebrates living in shallow waters moved onto the land. The fins of Devonian lobe-finned fishes, such as *Panderichthys*, are probably ancestral to tetrapod limbs, an early example of which is the limb of the Devonian tetrapod *Tulerpeton* (Fig. 15.7). The proximal skeletal elements corresponding to the humerus, radius, and ulna of the tetrapod limb are present in the ancestral fish, but there are no structures corresponding to digits. How did digits evolve? Some insights have been obtained by examining the development of fins in a modern fish, the zebrafish *Danio*.

The fin buds of the zebrafish embryo are initially similar to tetrapod limb buds, but important differences soon arise during development. The proximal part of the fin bud gives rise to skeletal elements, which are homologous to the proximal skeletal elements of the tetrapod limb. There are four main proximal skeletal elements in a zebrafish fin, which arise from the subdivision of a cartilaginous sheet (Fig. 15.8). The essential difference between fin and limb development is in the distal skeletal elements. In the zebrafish fin bud, an ectodermal fin fold develops at the distal end of the bud and fine bony fin rays are formed within it. These rays have no relation to anything in the vertebrate limb.

As in the tetrapod limb bud (see Chapter 10), the key signaling gene *Sonic hedgehog* is expressed at the posterior margin of the zebrafish fins and the expression pattern of Hoxd and Hoxa genes is similar to that in tetrapods. However, in the later stages of bud development, the Hox genes are only expressed in the posterior part of the fin bud and there is no distal

Fig. 15.8 The development of the pectoral fin of the zebrafish *Danio*. Left panel: the pectoral girdle and fin fold. Middle panel: four proximal cartilaginous elements and distal fin rays. Right panel: four proximal bony elements supporting the distal fin rays in the adult fish.

Photographs courtesy of D. Duboule, from Sordino, P., et al.: 1995.

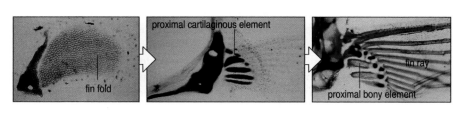

Fig. 15.9 Regions of Hox gene expression in the chick hindlimb and the zebrafish pectoral fin. Left panel: in the zebrafish fin bud, an apical ectodermal fold extends out from the underlying mesoderm. *Hoxd12* remains expressed in the mesoderm, which gives rise to the proximal cartilaginous elements. Right panel: in the chick leg, the mesoderm grows extensively and *Hoxd11* is expressed in the early bud and, additionally, more distally at later stages. After Coates, M.I.: 1995.

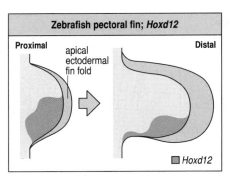

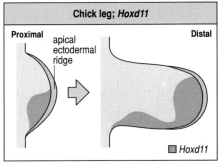

expression where the fin rays develop. On the other hand, in the tetrapod limb bud, an additional domain of Hox gene expression occurs in the distal region that gives rise to the digits (Fig. 15.9). If zebrafish fin development reflects that of the primitive ancestor, then tetrapod digits are novel structures, whose appearance is correlated with a new domain of Hox gene expression. However, they may have evolved from the distal recruitment of the same developmental mechanisms and processes that generate the radius and ulna. There are, as discussed in Chapter 10, mechanisms in the limb for generating periodic cartilaginous structures such as digits. It is likely that such a mechanism was involved in the evolution of digits by an extension of the region in which the embryonic cartilaginous elements form, together with the establishment of a new pattern of Hox gene expression in the more distal region.

Differences in fore- and hindlimb development are related to differences in Hox genes, which are involved in establishing positional differences in the lateral plate mesoderm from which the limb develops. For example, loss of function of *Hoxb5* causes the forelimb to form in a more anterior position. In limb initiation, *Hoxb9* and *Hoxc9* are expressed only in the hindlimb, whereas *Hoxd9* is expressed in both limbs. This staggering of Hox gene expression in the lateral plate mesoderm is thought to be a key feature in limb evolution, possibly related to regionalization of the gut. Another key role of the Hox genes is to specify limb identity. Limb loss is quite common, and a good example of the developmental changes involved is the python. Snakes have hundreds of similar vertebrae in their backbones, as can be seen in the skeleton of a python embryo. Snakes have no forelimbs, but pythons have a pair of hindlimb rudiments at the junction between the rib-bearing thoracic vertebrae, and vertebrae with shorter, forked ribs. In four-legged vertebrates such as mice, expression of *Hoxb5* and *Hoxc8* is confined to a short trunk region. In the python embryo, however, these genes are expressed along the whole body as far as the pelvic rudiment (Fig. 15.10). The expansion of these Hox-expression domains is thought to underlie the expansion of rib-bearing vertebrae and the loss of forelimbs in snake evolution. Hindlimb buds begin to develop, but the signals associated with apical ridge and polarizing region signaling (see Chapter 10) are not activated.

The great range of anatomical specializations in the limbs of mammals (Fig. 15.11) is due to changes both in limb patterning and in the differential growth of parts of the limbs during embryonic development, but the basic underlying pattern of skeletal elements is maintained. If one compares the forelimb of a bat and a horse, one can see that although both retain the

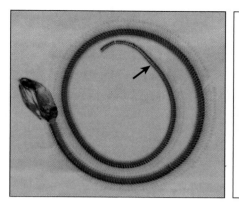

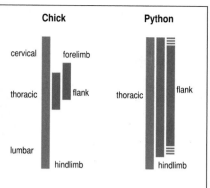

Chick | Python

cervical — forelimb

thoracic — flank

lumbar

hindlimb

thoracic — flank

hindlimb

Fig. 15.10 Comparison of Hox gene expression in python and chick embryos. The photograph shows a skeleton of a python embryo at 24 days incubation stained with Alcian blue and Alizarin red. The arrow marks the position of hindlimb rudiments, which have been removed in this preparation. Note the similarity of the vertebrae anterior to the arrow. The right panel shows a schematic comparison of domains of expression of *Hoxb5* (green), *Hoxc8* (blue) and *Hoxc6* (red) in chick and python embryos. The expansion of the *Hoxc8* and *Hoxc6* domains in the python correlates with the expansion of thoracic identity in the axial skeleton and flank identity in the lateral plate mesoderm. Adapted from Cohn, M.J., Tickle, C.: 1999.

basic pattern of limb bones, it has been modified to provide a specialized function in each. In the bat, the limb is adapted for flying: the digits are greatly lengthened to support a membranous wing. In the horse, the limb is adapted for running: in the forelimb, lateral digits are reduced, the central metacarpal (a hand bone in humans) is lengthened, and the radius and ulna are fused for greater strength. The role of differential growth rates and the loss of skeletal elements in the evolution of the horse's limb, and cases of limb reduction, are considered later. In some cases, differences in limb size and the length of the skeletal elements is present from a very early embryonic stage, as in the development of the kiwi and the bat (Fig. 15.12).

A feature of limb evolution is that while reduction in digit number is common—there are only three in the chick wing and reduction is common in lizards—species with more than five digits are very rare. It seems that there is a developmental constraint on evolving more than five different kinds of digits. This may be due to the Hox genes providing only five discrete genetic programs for giving a digit an identity. In limbs with polydactyly, at least two of the digits are the same (see Section 10.5), and there are still only five different kinds of digits. This may be the reason why in an animal with an additional distinctive digit-like element, such as the giant panda's 'thumb', this digit is in fact a modified wrist bone.

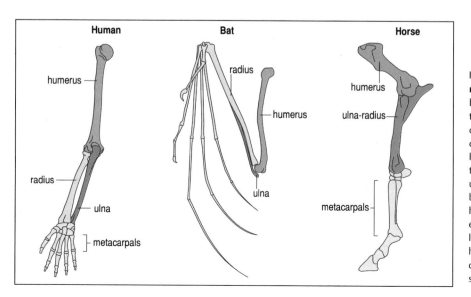

Human | Bat | Horse

humerus

radius

ulna

metacarpals

radius

humerus

ulna

humerus

ulna-radius

metacarpals

Fig. 15.11 Diversification of mammalian limbs. The basic pattern of bones in the forelimb is conserved throughout the mammals, but there are changes in the proportions of the different bones, as well as fusion and loss of bones. This is seen particularly in the horse limb, in which the radius and ulna have become fused into a single bone, and the central metacarpal (a hand bone in humans) is greatly elongated. In addition, there has been loss and reduction of the digits in the horse. In the bat wing, by contrast, the digits have become greatly elongated to support the membranous wing.

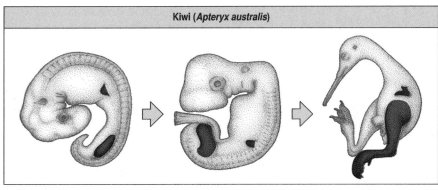

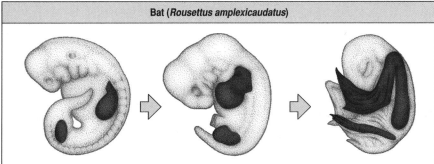

Fig. 15.12 Developmental expression of adult limb size. Top row: the flightless kiwi (*Apteryx australis*) has relatively short forelimbs and large hindlimbs at all stages of development, even in the early embryo. Bottom row: in the bat *Rousettus amplexicaudatus*, which has large forelimbs and smaller hindlimbs when adult, the forelimb bud is large at all stages, relative to the hindlimb. Adapted from Richardson, M.K.: 1999.

15.3 Vertebrate and insect wings make use of evolutionarily conserved developmental mechanisms

Vertebrate and insect wings are not homologous but have some superficial similarities; they have similar functions yet are very different in structure. The insect wing is a double-layered epithelial structure, whereas the vertebrate limb develops mainly from a mesenchymal core surrounded by ectoderm. However, despite these great anatomical differences, there are striking similarities in the genes and signaling molecules involved in setting up the axes and so patterning insect legs, insect wings, and vertebrate limbs (Fig. 15.13, and see Chapter 10). Patterning along the antero-posterior axis of all these appendages uses signals encoded by *hedgehog*-related genes, and by members of the TGF-β family, such as decapentaplegic (in insects) and BMP-2 (in vertebrates). It is remarkable that the dorsal surface of the insect wing is characterized by expression of the gene *apterous*, whereas the related gene *Lmx-1* is expressed in the dorsal mesenchyme of the vertebrate limb. A *fringe*-like gene is involved in the specification of the boundary between dorsal and ventral regions in both insect and bird wings.

All these relationships suggest that, during evolution, a mechanism for patterning and setting up the axes of appendages appeared in some common ancestor of insects and vertebrates. Subsequently, the genes and signals involved acquired different downstream targets so that they could interact with different sets of genes, yet the same set of signals retain their organizing function in these very different appendages. The individual genes involved in specifying the limb axes are probably more ancient than either insect or vertebrate limbs.

Another example illustrating the conservation of the genetic machinery for insect appendages is provided by the gene *Distal-less*, which is also

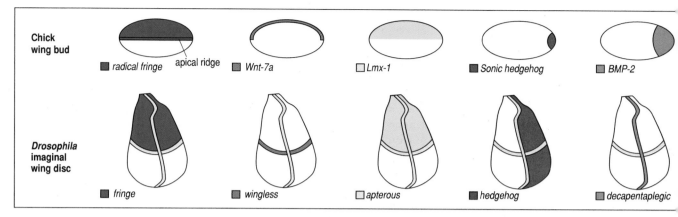

Fig. 15.13 Comparison of developmental signals in the chick wing bud and *Drosophila* wing imaginal disc. First column: the chick wing bud (top) is shown with the distal end facing. The double line bisecting it represents the apical ridge. The chick apical ectodermal ridge forms at the boundary between the dorsal cells, which express *radical fringe*, and the ventral cells, which do not. The dorso-ventral boundary in the insect wing disc also forms at the boundary between *fringe* and non-*fringe* cells (bottom). In the insect wing imaginal disc, the future dorsal and ventral regions are in the same plane. The vertical double lines represent the antero-posterior compartment border, and the horizontal double lines represent the dorso-ventral compartment border. Second column: the dorsal region of the chick wing is specified by *Wnt-7a* in the ectoderm, whereas *wingless* is expressed at the dorso-ventral insect wing margin. Third column: the gene *Lmx-1* is expressed in the dorsal region of chick wing bud mesoderm, whereas the *Drosophila* gene *apterous*, to which it is structurally related, specifies the dorsal region of the insect wing. Fourth and fifth columns: *Sonic hedgehog* in the chick wing and *hedgehog* in *Drosophila* are expressed in posterior regions and both induce expression of genes of the TGF-β family—*BMP-2* and *decapentaplegic*, respectively.

expressed along the proximo-distal axis of a wide variety of developing appendages in other animals, including annelid parapodia and the tube feet of sea urchins. It is also expressed during vertebrate limb outgrowth.

All this emphasizes how evolution has used a wonderfully modifiable system for setting up axes and specifying positional values.

15.4 Hox gene complexes have evolved through gene duplication

Hox genes play a key role in development both in vertebrates and insects. By comparing the organization and structure of the Hox genes in insects and vertebrates, we can determine how one set of important developmental genes has changed during evolution.

A major general mechanism of evolutionary change has been gene duplication and divergence. Tandem duplication of a gene, which can occur by a variety of mechanisms during DNA replication, provides the embryo with an additional copy of the gene. This copy can diverge in its nucleotide sequence and acquire a new function and regulatory region, so changing its pattern of expression and downstream targets without depriving the organism of the function of the original gene. The process of gene duplication has been fundamental in the evolution of new proteins and new patterns of gene expression; it is clear, for example, that the different hemoglobins in humans (see Fig. 9.17) have arisen as a result of gene duplication.

One of the clearest examples of the importance of gene duplication in developmental evolution is provided by the Hox gene complexes. As we have seen, the Hox genes are members of the homeobox gene family, which is characterized by a short 180 base pair motif, the homeobox, which encodes a helix-turn-helix domain that is involved in transcriptional regulation (see Box 4A, p. 117). Two features characterize all known Hox genes:

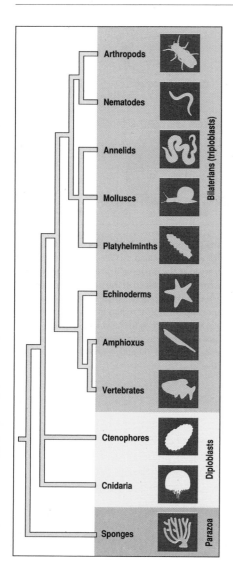

Fig. 15.14 A metazoan family tree. The tree shown here is based on rDNA and Hox gene sequences. Adapted from Ferrier, D.E.K., Holland, P.W.H.: 2001.

the individual genes are organized into one or more gene clusters or complexes, and the order of expression of individual genes along the antero-posterior axis is usually the same as their sequential order in the gene complex.

The animal kingdom can be divided into three main groups—the Bilateria, the Cnidaria and Ctenophora (the coelenterates), and the Parazoa (the sponges). The largest group is the Bilateria, animals that show bilateral symmetry across the main body axis. This group includes vertebrates and other chordates, arthropods, annelids, molluscs, and nematodes (Fig. 15.14). These animals are triploblasts, as they have the three major germ layers—endoderm, mesoderm, and ectoderm. The diploblasts—the Cnidaria and Ctenophora—do not have bilateral symmetry and have only two germ layers, lacking mesoderm. The Parazoa are considered to be a quite separate group from the rest of the multicellular animals. The pathway of Hox gene evolution, which is central to the laying down of the body plan, is relatively clear for the Bilateria, but its origin is less clear. Hox genes have been identified in the Cnidaria, such as *Hydra*, where the gene *Cnox2* is homologous to the Hox2 genes. One model suggests that the duplication and divergence of a 'ProtoHox' cluster of just five genes gave rise to all the later Hox gene complexes, and to the related 'ParaHox' genes, which are found outside the classical Hox clusters. Invertebrates have one Hox complex that specifies pattern along the antero-posterior axis, whereas vertebrates have four, suggesting two further rounds of duplication of the whole complex. In several fish there has been one further round of duplication: the zebrafish has seven Hox clusters, one having been lost after duplication. The Japanese puffer fish *Fugu rubripes* has four Hox clusters, but many genes have been lost, and one cluster bears little relation to any in other vertebrates. It appears that the spatial colinearity of the Hox genes along the chromosome and their expression along the main body axis was confined to the neural tube in the ancestral chordates that gave rise to the vertebrates, and it was later extended to other tissues, thus allowing increased complexity of body plans.

Comparing the Hox genes of a variety of species, it is possible to reconstruct the way in which they are likely to have evolved from a simple set of seven genes in a common ancestor of arthropods and vertebrates (Fig. 15.15). Amphioxus, which is a vertebrate-like chordate, has many features of a primitive vertebrate: it possesses a dorsal hollow nerve cord, a notochord, and segmental muscles that derive from somites. It has only one Hox gene cluster, and one can think of this cluster as most closely resembling the common ancestor of the four vertebrate Hox gene complexes—Hoxa, Hoxb, Hoxc, and Hoxd (see Fig. 15.15). It is possible that both the vertebrate and *Drosophila* Hox complexes evolved from a simpler ancestral complex by gene duplication. In *Drosophila*, the duplications could have given *abdominal-A* (*abd-A*), *Ubx*, and *Antennapedia* (*Antp*). In vertebrates, the Hox genes are arranged in four separate clusters, each of which is on a different chromosome, and are not linked to each other. These separate clusters probably arose from duplications of whole chromosomal regions, in which new Hox genes had already been generated by tandem duplication. For example, sequence comparisons suggest that the multiple mouse Hox genes related to the *Drosophila Abdominal-B* (*Abd-B*) gene do not have direct homologs within the *Drosophila* bithorax complex (see Fig. 15.15); they probably arose by tandem duplication from an ancestral gene after the split of the insect

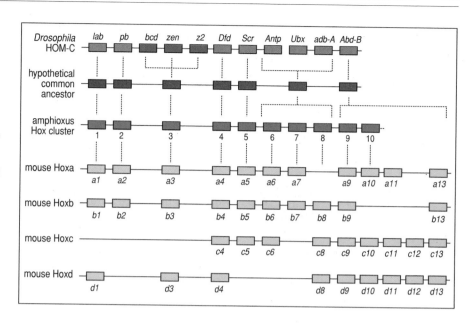

Fig. 15.15 Gene duplication and Hox gene evolution. A suggested evolutionary relationship between the Hox genes of a hypothetical common ancestor and *Drosophila* (an arthropod), amphioxus (a cephalochordate), and the mouse (a vertebrate). Duplications of genes of the ancestral set (red) could have given rise to the additional genes in *Drosophila* and amphioxus. Two duplications of the whole cluster in a chordate ancestor of the vertebrates could have given rise to the four separate Hox gene complexes in vertebrates. There has also been a loss of some of the duplicated genes in vertebrates.

and vertebrate lineages, but before duplication of the whole cluster in vertebrates. We now consider the role of these genes in evolution of the axial body plan.

15.5 Changes in specification and interpretation of positional identity have generated the elaboration of vertebrate and arthropod body plans

Hox genes are, as we have seen, key genes in the control of development and are expressed regionally along the antero-posterior axis of the embryo. The apparent conserved nature of the Hox genes, and of certain other genes homologous with them, in animal development has led to the concept of the **zootype**. This defines both the overall form and the pattern of expression of these key genes along the antero-posterior axis of the embryo, which is present in all animals.

Multicellular organisms are thought to have originated about 1500 million years ago, with the earliest generally accepted fossil of a multicellular animal dating from about 600 million years ago. There are currently about 35 extant animal phyla, each with its own distinctive basic body plan, and all of these had evolved by the end of the Cambrian period, around 500 million years ago. No new basic body plans with clear antero-posterior and dorso-ventral axes have evolved since then, although body plans within the different phyla have been modified and elaborated to give organisms as different as fish and mammals within the chordates. In the following discussion on the elaboration of body plans within phyla, we focus on just two phyla—the chordates, which include the vertebrates, and the arthropods, which include the insects and the crustaceans—because we have a good understanding of the development of some of the members of both these phyla.

The role of the Hox genes is to specify positional identity in the embryo, rather than the development of any specific structure. These positional values are interpreted differently in different embryos to influence how the cells in a region develop into, for example, segments and appendages. The Hox genes exert this influence by their action on the genes controlling the development of these structures. Changes in the downstream targets of the Hox genes can thus be a major source of change in evolution. In addition, changes in the pattern of Hox gene expression along the body can have important consequences. An example is a relatively minor modification of the body plan that has taken place within vertebrates. One easily distinguishable feature of pattern along the antero-posterior axis in vertebrates is the number and type of vertebrae in the main anatomical regions—cervical (neck), thoracic, lumbar, sacral, and caudal (see Fig. 4.13). The number of vertebrae in a particular region varies considerably among the different vertebrate classes—mammals, with rare exceptions, have seven cervical vertebrae, whereas birds can have between 13 and 15. How does this difference arise? A comparison between the mouse and the chick shows that the domains of Hox gene expression have shifted in parallel with the change in number of vertebrae (see Fig. 4.13). For example, the anterior boundary of *Hoxc6* expression in the mesoderm in mice and chicks is always at the boundary of the cervical and thoracic regions. Moreover, the *Hoxc6* expression boundary is also at the cervical–thoracic boundary in geese, which have three more cervical vertebrae than chickens, and in frogs, which only have three or four cervical vertebrae in all. The changes in the spatial expression of *Hoxc6* correlate with the number of cervical vertebrae. Other Hox genes are also involved in the patterning of the antero-posterior axis, and their boundaries also shift with a change in anatomy.

Changes in the targets of genes such as Hox genes probably play a key role in the differences between different animals. There may be many, perhaps very many, downstream targets of the Hox genes (Chapter 5). For example, *Ubx* makes an appendage, the *Drosophila* wing, different from the haltere, and tests have shown that six of the twelve genes known to be involved in patterning the wing are under the control of *Ubx*. But only some of these genes are controlled by *Ubx* in the butterfly wing. Working out the control mechanisms is further complicated by evidence that the amount of Hox protein present in the cell is important. Knock-outs of *Hoxa3* or *Hoxd3* in the mouse give quite different phenotypes, and it seems reasonable to assume that this reflects the differences in the proteins encoded by these genes. However, when the protein-coding region of *Hoxa3* was exchanged for that of *Hoxd3*, the mice lacking Hoxa3 protein developed normally; so it may simply be that a sufficient amount of Hox protein, such as Hoxd3, rather than a particular type, needs to be present, in this case at least.

An example of how subtle variations in body form can be controlled by Hox genes comes from insects. There are around 150,000 species of flies and all use roughly the same set of developmental genes. Diversity in form can reflect different patterns of gene activity, but it is a difficult question in this case as to how the same sets of proteins can generate such subtle morphological diversity. In the genus *Drosophila*, different species have different patterns of non-sensory bristles—trichomes—on the femur of the second leg. *D. melanogaster* has a small naked patch, whereas *D. simulans* has a large one and *D. virilis* no naked patch at all (Fig. 15.16). *Ubx* represses trichome development, and the extent or presence of the naked patch is determined by subtle differences in the regulation of *Ubx* expression in this region. These differences are due to small differences in the regulatory region of *Ubx* in the different species, and this variation provides the evolutionary basis for the variation in pattern.

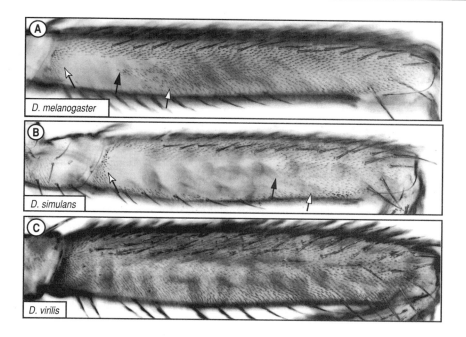

Fig. 15.16 Trichome patterns on the posterior second femur vary among Drosophila species. Top panel, *Drosophila melanogaster*. Middle panel, *D. simulans*. Bottom panel, *D. virilis*. The arrow and arrowhead in the top two panels indicate the extent of the patch of naked cuticle. From Stern: 1998.

Changes in patterns of Hox gene expression can also help explain the evolution of arthropod body plans. Insects and crustaceans are distinct groups of arthropods that have evolved from a common arthropod ancestor, which probably had a body composed of more or less uniform segments. A comparison of Hox gene expression in an insect, the grasshopper, with that in a crustacean, the brine shrimp *Artemia*, shows which body regions in these two present-day arthropods are homologous with each other, and how the different body plans might have evolved from the ancestral body plan. Such a comparison shows that both the pattern of Hox gene expression and the body regions to which particular Hox genes relate have changed during the evolution of these two groups (Fig. 15.17).

The grasshopper has a pattern of Hox gene expression similar to that in *Drosophila*. As with *Drosophila*, the Hox genes *Antp*, *Ubx*, and *abd-A* specify distinct segment types in the thorax and abdomen, but are expressed in

Fig. 15.17 A comparison of body plans and Hox gene expression in two arthropods. A comparison of Hox gene expression in an insect, the grasshopper, with that in a crustacean, the brine shrimp *Artemia*, shows that both the pattern of Hox gene expression and the body regions to which particular Hox genes relate have changed during the evolution of these two groups of arthropods from their common ancestor. In *Artemia*, the three Hox genes *Antennapedia*, *Ultrabithorax*, and *abdominal-A* are all expressed throughout the thorax, where most of the segments are similar. However, the expression of these genes in the thorax and abdomen of the grasshopper is different. They have overlapping and distinct patterns of expression that reflect the regional differences in the insect thorax. The gene *Abdominal-B* is expressed in the genital regions of both animals, indicating that these two regions are homologous. After Akam, M.: 1995.

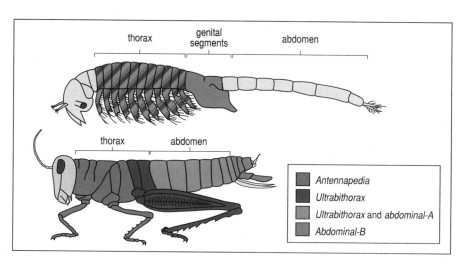

■	*Antennapedia*
■	*Ultrabithorax*
■	*Ultrabithorax* and *abdominal-A*
■	*Abdominal-B*

overlapping domains, and segment types are defined by combinatorial expression (see Section 5.19). In the brine shrimp, however, these genes are all expressed together in a thoracic region that is composed of uniform segments. This suggests that the thorax of *Artemia* might be homologous to the whole insect thorax and much of the insect abdomen. Thus, in the evolution of the grasshopper and the brine shrimp from their common ancestor, the changes in body plan have resulted in part from spatial changes in the expression of particular Hox genes, although changes in their downstream targets have also been involved.

These comparisons clearly show that Hox genes do not specify particular structures but simply provide a regional identity. How that identity is interpreted to produce a particular morphology is the role of genes acting downstream from the Hox genes, and there may be many of these. The head region of different arthropods, for example, can be very different, and spiders do not seem to have a head at all. Their body is divided into a front and a back region, the front end having segments that bear the spider's fangs, pedipalps, and four pairs of walking legs; it acts as a head and thorax combined. In spite of these differences, the pattern of expression of the Hox genes in this region, which include *orthodenticle* and *labial*, is very similar in spiders and insects, showing how much morphology can change due to changes in downstream targets. Further examples of the role of Hox genes are provided by arthropod appendages, which we will look at next.

15.6 The position and number of paired appendages in insects is dependent on Hox gene expression

Insect fossils display a variety of patterns in the position and number of their paired appendages—principally the legs and wings. Some insect fossils have legs on every segment, whereas others only have legs in a distinct thoracic region. The number of abdominal segments bearing legs varies, as does the size and shape of the legs. Wings arose later than legs in insect evolution. Wing-like appendages are present on all the thoracic and abdominal segments of some insect fossils, but are restricted to the thorax in others. To understand how these different patterns of appendages arose during evolution we need to look at how the different patterns of appendages develop in two orders of modern insects, the Lepidoptera (butterflies and moths) and the Diptera (flies, including *Drosophila*).

The basic pattern of Hox gene expression along the antero-posterior axis is the same in all present-day insect species that have been studied. Yet the larvae of Lepidoptera have legs on the abdomen as well as on the thorax, and the adults have two pairs of wings, whereas the Diptera, which evolved later, have no legs on the abdomen in the larva or adult, and only one pair of wings, with the second pair of wings having been modified into halteres. How are these differences related to differences in Hox gene activity in the two groups of insects?

In *Drosophila*, products of the bithorax gene complex suppress appendage formation in the abdomen by repressing the expression of the *Distal-less* gene. This suggests that the potential for appendage development is present in every segment, even in flies, and is actively repressed in the abdomen. It thus seems likely that the ancestral arthropod from which insects evolved had appendages on all its segments. During the embryonic development of Lepidoptera, the bithorax complex genes *Ubx* and *Abd-B* are turned off in the ventral parts of the abdominal segments; this results both in *Distal-less* being expressed and in legs developing on the abdomen in the larva. The presence or absence of legs on the abdomen is thus determined by whether or not a particular Hox gene is expressed there. This shows that changes in the pattern of Hox gene expression have played a key role in evolution. The Hox genes can also determine the nature of an appendage: we have seen how mutations can convert legs into antenna-like structures and an antenna into a leg (see Section 10.17).

It has been proposed that insect wings originated as outgrowths from the first proximal leg segment. However, Hox genes do not seem to have been involved, as *Antp*, which is expressed in the second thoracic wing-bearing segment of *Drosophila*, is not required for wing development. It is likely that wings were originally present on all thoracic and abdominal segments and that Hox genes have repressed and modified their development in all segments except the scond thoracic. For example, the differences between forewings and hindwings of insects with two pairs of wings, such as butterflies, are probably regulated by the *Ubx* gene.

It is clear that in the course of evolution, rather than new genes appearing, new regulatory interactions between the bithorax complex proteins and genes involved in leg and wing development have evolved.

15.7 The basic body plan of arthropods and vertebrates is similar, but the dorso-ventral axis is inverted

From paleontological, molecular, and cellular evidence, all multicellular animals are presumed to descend from a common ancestor, and the broad similarities in the pattern of Hox gene expression in vertebrates and arthropods, whose evolution diverged hundreds of millions of years ago, are taken as good supporting evidence for this concept. However, a comparison between the body plans of arthropods and chordates reveals an intriguing difference. In spite of many similarities in their basic body plan—both have an anterior head, a nerve cord running anterior to posterior, a gut, and appendages—the dorso-ventral axis of vertebrates is inverted when compared to that of arthropods. The most obvious manifestation of this is that the nerve cord runs ventrally in arthropods, but dorsally in vertebrates (Fig. 15.18, left panels).

One explanation for this, first proposed in the 19th century, is that during the evolution of the vertebrates from their common ancestor with the arthropods, the dorso-ventral axis was turned upside-down, so that the ventral nerve cord of the ancestor became dorsal. This startling idea has recently found some support from molecular evidence showing that the same genes are expressed along the dorso-ventral axis in both insects and vertebrates, but in inverse directions. This inversion may have been dictated by the position of the mouth. The mouth defines the ventral side, and a change in the position of the mouth away from the side of the nerve cord would have resulted in the reversal of the dorso-ventral axis in relation to the mouth. The position of the mouth is specified during gastrulation, and it is not difficult to imagine how its position could have moved.

We have seen in Chapters 3 and 5 that the patterning of the dorso-ventral axis in vertebrates and insects involves intercellular signaling. In *Xenopus*, the protein chordin is one of the signals that specifies the dorsal region, whereas the growth factor BMP-4 specifies a ventral fate. In *Drosophila*, the

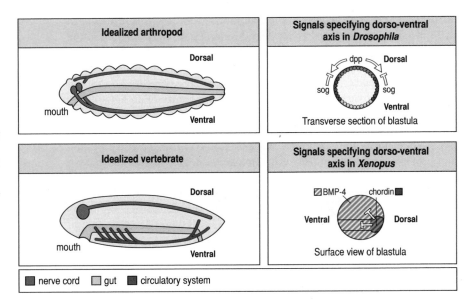

Fig. 15.18 The vertebrate and *Drosophila* dorso-ventral axes are related but inverted. In arthropods, the nerve cord is ventral whereas in vertebrates it is dorsal—dorsal and ventral being defined by the position of the mouth. In *Drosophila* and *Xenopus*, the signals specifying the dorso-ventral axis are similar, but are expressed in inverted positions. The protein chordin, a dorsal specifier in vertebrates, is related to sog, which is a ventral specifier in *Drosophila*, and the vertebrate ventral specifier BMP-4 is related to *Drosophila* decapentaplegic (dpp), which specifies dorsal. After Ferguson, E.L.: 1996.

pattern of gene expression is reversed: the protein decapentaplegic, which is closely related to BMP-4, is the dorsal signal, and the protein short gastrulation (sog), which is related to chordin, is the ventral signal (see Fig. 15.18, right panels). These signaling molecules are experimentally interchangeable between insects and frogs. Chordin can promote ventral development in *Drosophila*, and decapentaplegic protein promotes ventral development in *Xenopus*. The molecules and mechanisms that set up the dorso-ventral axes in the two groups of animals are thus strikingly similar, strongly suggesting that the divergence in the body plans of the arthropods and vertebrates involved an inversion of this axis, by movement of the mouth during the evolution of the vertebrates. Thus our zootype, and our image of the common ancestor of chordates and arthropods, is an animal in which the establishment of the two major body axes, anteroposterior and dorso-ventral, was similar to that in present-day animals.

The establishment of segmentation provides other examples of similar mechanisms in body plan development in chordates and arthropods. The gene *engrailed*, for example, which is expressed in the posterior compartment of each segment in *Drosophila*, is also expressed in the posterior half of each of the first eight somites of amphioxus. And as we saw in Chapter 5, initially similar parasegments can be diversified during development to give segments with different structures and functions.

Summary

The development of an embryo provides insights into the evolutionary origin of the animal. Groups of animals that pass through a similar embryonic stage are descended from a common ancestor and have evolved modifications of gene regulatory circuits. During evolution, the development of structures can be altered so that they acquire new functions, as has happened in the evolution of the mammalian middle ear from a reptilian jaw bone, which itself evolved from the branchial arch. The limbs of tetrapod vertebrates evolved from fins, with the digits as a novel feature. The development of vertebrate and insect limbs involves the same set of pattern-establishing genes, reflecting the evolution of limb development from an ancestral mechanism for specifying body appendages. The basic body plan of all animals is defined by patterns of Hox gene expression that provide positional identity, the interpretation of which has changed in evolution. The Hox genes themselves have undergone considerable evolution by gene duplication and divergence and are a good example of Darwin's description of evolution as "descent with modification". Comparison of patterns of dorso-ventral gene expression suggests that during the evolution of the vertebrates the dorso-ventral axis of an invertebrate ancestor was inverted.

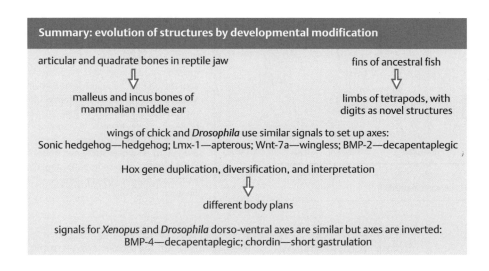

Summary: evolution of structures by developmental modification

articular and quadrate bones in reptile jaw   fins of ancestral fish
⇩   ⇩
malleus and incus bones of mammalian middle ear   limbs of tetrapods, with digits as novel structures

wings of chick and *Drosophila* use similar signals to set up axes:
Sonic hedgehog—hedgehog; Lmx-1—apterous; Wnt-7a—wingless; BMP-2—decapentaplegic

Hox gene duplication, diversification, and interpretation
⇩
different body plans

signals for *Xenopus* and *Drosophila* dorso-ventral axes are similar but axes are inverted:
BMP-4—decapentaplegic; chordin—short gastrulation

Changes in the timing of developmental processes during evolution

In the previous part of the chapter we have focused on changes in spatial patterning that have occurred during evolution. But changes in the timing of developmental processes can also have major effects. In this part of the chapter, we look at some examples of how changes in the timing of growth and sexual maturation can affect animal form and behavior.

15.8 Changes in growth can alter the shapes of organisms

Many of the changes that occur during evolution reflect changes in the relative dimensions of parts of the body. We have seen how growth can alter the proportions of the human baby after birth, as the head grows much less than the rest of the body (see Fig. 14.6). The variety of face shapes in the different breeds of dog, which are all members of the same species, descended from the grey wolf, also provides a good example of the effects of differential growth after birth. All dogs are born with rounded faces; some keep this shape, but in others the nasal regions and jaws elongate during growth. The elongated face of the baboon is also the result of growth of this region after birth.

Because individual structures such as bones can grow at different rates, the overall shape of an organism can be changed substantially during evolution by heritable changes in the duration of growth that leads to an

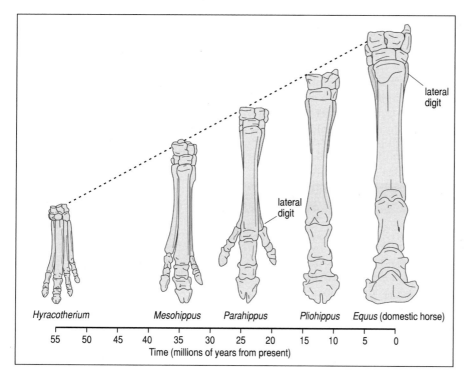

Fig. 15.19 Evolution of the forelimb in horses. *Hyracotherium*, the first true equid, was about the size of a large dog. Its forefeet had four digits, of which one (the third digit in anatomical terms) was slightly longer, as a result of a faster growth rate. All digits were in contact with the ground. As equids increased in size, the lateral digits lost contact with the ground, as the result of the relatively greater increase in length of metacarpal 3 (the hand bone of the third digit). At a later stage, the lateral digits became even shorter, because of a separate genetic change. After Gregory, W.K.: 1957.

increase in the overall size of the organism. In the horse, for example, the central digit of the ancestral horse grew faster than the digits on either side, so that it ended up longer than the lateral digits (Fig. 15.19). As horses continued to increase in overall size during evolution, this discrepancy in growth rates resulted in the relatively smaller lateral digits no longer touching the ground because of the much greater length of the central digit. At a later stage in evolution, the now-redundant lateral digits became reduced even further in size because of a separate genetic change.

The mathematical analysis of the relative growth of parts of an organism during development is known as **allometry**. It has been found that the

Fig. 15.20 Differential growth of body regions in the ant. As the ant grows, the width of the head (*y*) increases much faster than that of the abdomen (*x*) and so the head becomes proportionally larger. The relationship fits the equation $y = bx^a$, where *a* is the slope of the curve when log *y* is plotted against log *x*, and *b* is a constant.

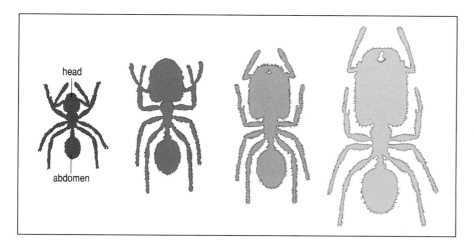

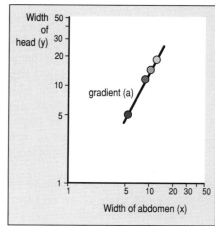

Fig. 15.21 Heterochrony in salamanders. In terrestrial species of the salamander *Bolitoglossa* (top panel), the foot is larger, has longer digits, and is less markedly webbed than in those that live in trees (bottom panel). This difference can be accounted for by foot growth ceasing at an earlier stage in the arboreal species. Scale bar = 1 mm.

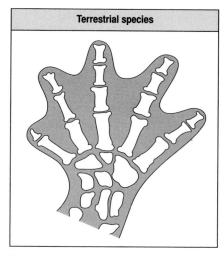

Terrestrial species

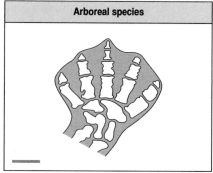

Arboreal species

difference in growth rates of two structures in an organism can be determined from their final dimensions. There is often a mathematical relationship between two structures of lengths x and y, such that $y = bx^a$, where a and b are constants. So when one plots log y against log x, a straight line of slope a is obtained. The slope of the line indicates how much faster y grows than x. An example of this relationship is provided by the relative growth of the abdomen and head in an ant (Fig. 15.20). As the ant grows, the head becomes relatively much larger.

15.9 The timing of developmental events has changed during evolution

Differences among species in the time at which developmental processes occur relative to one another, and relative to their timing in an ancestor, can have dramatic effects on both the structure and behavior of an organism. Differences in the feet of members of a genus of tropical salamanders illustrate the effect of a change in developmental timing on both the morphology and ecology of different species. Many species of the salamander genus *Bolitoglossa* are arboreal (tree living), rather than typically terrestrial, and their feet are modified for climbing on smooth surfaces. The feet of the arboreal species are smaller and more webbed than those of terrestrial species, and their digits are shorter (Fig. 15.21). These differences seem to be mainly the result of the development and growth of the foot ceasing at an earlier stage in the arboreal species than in the terrestrial species. The term used to describe such differences in timing is **heterochrony**.

We saw above that the loss of limbs in the python was related to changes in Hox gene expression along the body axis. Limb reduction and loss in both snakes and whales also involves changes in timing of developmental processes. In general, limb reduction occurs gradually during evolution, with limb structures being lost in a distal to proximal sequence. In the evolution of the whales, fossil evidence shows that the progressive loss of the hindlimbs occurred over many millions of years. Some modern whales still retain internal remnants of a femur and tibia. Flightless birds, such as the kiwi, also show some limb reduction: kiwi wings have just one digit. These evolutionary changes can be partly understood in terms of changes in timing during limb development, such as a reduced rate of growth of the bud and early cessation of growth. The complete loss of the limbs in snakes and legless lizards can be partly understood in terms of a change in the timing of cell death and the disappearance of the apical ridge in the bud. Although the limb bud begins to develop in some legless species, cell death in the apical ridge starts at an early stage, followed by death of the cells in the bud itself, which generally results in the absence of limbs. But in the python, for example, the proximal elements of the limb still develop.

Some of the clearest examples of heterochrony come from alterations in the timing of onset of sexual maturity in organisms with larval stages. The acquisition of sexual maturity by an animal while still in the larval stage is a process that goes under the name **neoteny**. The development of the

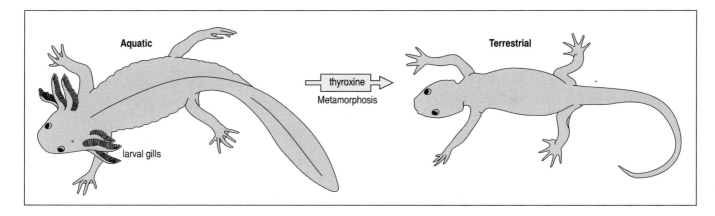

Fig. 15.22 Neoteny in salamanders. The sexually mature Mexican axolotl retains larval features, such as gills, and remains aquatic. This neotenic form metamorphoses into a typical terrestrial adult salamander if treated with thyroxine, which it does not produce, and which is the hormone that causes metamorphosis in other amphibians.

animal, although not its growth, is retarded in relation to the maturation of the reproductive organs. This occurs in the Mexican axolotl, a type of salamander; the larva grows in size and matures sexually, but does not undergo metamorphosis. The sexually mature form remains aquatic and looks like an overgrown larva. However, the axolotl can be induced to undergo metamorphosis by treatment with the hormone thyroxine (Fig. 15.22).

Larval stages may have evolved as a result of heterochrony. For example, if we assume that frog ancestors developed directly into adults, a change in the timing of events in the post-neurula stages, together with structural modifications, could have led to the interposition of a feeding tadpole stage. One of the changes required would have been a delay in the development of limbs.

Some modern frogs have evolved with direct development to the adult by a loss of the larval stage and the acceleration of the development of adult features. Frogs of the genus *Eleutherodactylus*, unlike the more typical amphibians, *Rana* and *Xenopus*, develop directly into an adult frog and there is no aquatic tadpole stage, the eggs being laid on land. Typical tadpole features, such as gills and cement glands, do not develop, and prominent limb buds appear shortly after the formation of the neural tube (Fig. 15.23). In the embryo, the tail is modified into a respiratory organ. Such direct development requires a large supply of yolk to the egg in order to support development through to an adult without a tadpole feeding stage. This increase in yolk may itself be an example of heterochrony, involving a longer or more rapid period of yolk synthesis in the development of the egg.

Most sea urchins have a larval stage that takes a month or more to become adult. But there are some species in which direct development has

Fig. 15.23 Development of the frog *Eleutherodactylus*. Typical frogs, such as *Xenopus* or *Rana*, lay their eggs in water and develop through an aquatic tadpole stage, which undergoes metamorphosis into the adult. Frogs of the genus *Eleutherodactylus* lay their eggs on land and the frog hatches from the egg as a miniature adult, without going through an aquatic free-living larval stage. The embryonic tail is modified as a respiratory organ.

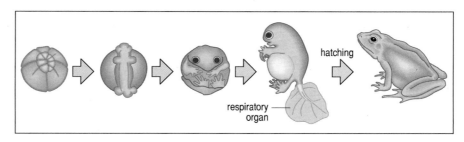

evolved so that they no longer go through a functional larval stage. Such species have large eggs and, as a result of their very rapid development, become juvenile urchins within 4 days. This has required changes in early development so that the directly developing embryo gives rise to a larval stage that lacks a gut and cannot feed, and which metamorphoses rapidly into the adult form.

15.10 Evolution of life histories has implications for development

Animals and plants have very diverse life histories: small birds breed in the spring following their birth, and continue to do so each year until their death; Pacific salmon breed in a suicidal burst at 3 years of age; and oak trees require 30 years of growth before producing acorns, which they eventually produce by the thousand. In order to understand the evolution of such life histories, evolutionary ecologists consider them in terms of probabilities of survival, rates of reproduction, and optimization of reproductive effort. These factors have important implications for the evolution of developmental strategies, particularly in relation to the rate of development. For example, a characteristic feature of many animal life histories is the presence of a feeding larval stage that is distinct from, and usually simpler in form, than the sexually mature adult, and which feeds in a different way. The evolutionary significance of this developmental strategy is not only that it provides the organism with both a means of dispersal and a means of obtaining nutrition before it becomes an adult, but that it also allows the organism to exploit different environments for feeding. Next, we consider two other issues in the evolution of life histories that impinge directly on development—selection for speed of development and selection for egg size.

Life histories help us to understand the evolution of long germ band insects, such as *Drosophila*, which are of more recent origin than short germ band insects. Compared with short germ band insects, such as grasshoppers, which have no larval stage and develop directly into small, immature adult-like forms, *Drosophila* develops very rapidly into a feeding larva. *Drosophila* develops into a feeding larva within 24 hours, whereas a grasshopper takes 5–6 days. It is easy to imagine conditions in which there would have been a selective advantage to insects whose larvae begin to feed as quickly as possible, and it is also likely that embryos are more vulnerable than adults. Thus it is likely that the evolution of the complex developmental mechanisms of long band insects—the system for setting up the whole antero-posterior axis in the egg (see Section 5.18)—resulted from selection pressure for rapid development, perhaps related to the ability to feed on fast-disappearing fruit.

Egg size can also be best understood within the context of life histories. If we assume that the parent has limited energy resources to put into reproduction, the question is how should these resources best be invested in making gametes, particularly eggs; is it more advantageous to make lots of small eggs or a few large ones? In general, it seems that the larger the egg, and thus the larger the offspring at birth, the better the chances of offspring survival, suggesting that in most circumstances an embryo needs to give rise to a hatchling as large as possible and as quickly as possible. Why then, do some species lay many small eggs capable of rapid development? One possible answer is that the parental investment in large eggs may reduce the parents' chance of surviving. Another probable answer is that these species have evolved a strategy that is especially successful in variable conditions where populations can suddenly crash. Again, development of a larva enables early feeding or dispersal to new sites.

Summary

Changes in the timing of developmental processes that have occurred during evolution can alter the form of the body, for if different regions grow faster than others, their size is proportionately increased if the animal gets larger. The decrease in size of the lateral digits of horses is partly due to this type of developmental change. Speed of development and egg size also have important evolutionary implications. Changing the time at which an animal becomes sexually mature can result in adults with larval characteristics. Some animals, such as frogs and sea urchins, which usually have a larval form, have evolved species that develop directly into the adult without a larval stage.

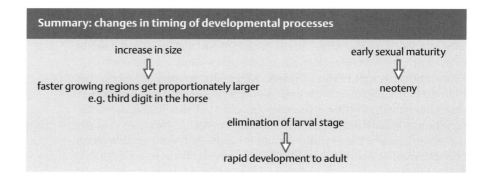

Evolution of development

How did development itself evolve? What is the origin of the egg and how did processes like pattern formation and gastrulation evolve? What follows are some possible answers, admittedly highly speculative.

15.11 How multicellular organisms evolved from single-celled ancestors is still highly speculative

If we consider the four basic processes of development—spatial organization, differentiation, change in form, and growth—it seems not unreasonable to suggest that all were present in the eukaryotic cell that gave rise to multicellular organisms. The cell cycle, for example, can be taken as a model for several key processes in development, as it involves a temporal program of gene activity and highly organized movements, as of the chromosomes at mitosis. Single-celled organisms can undergo a primitive form of differentiation, as in spore formation and the generation of the different mating types in yeast. Thus, given the existence of gene regulatory circuits, transmembrane signal transduction, and cell motility, very little new was required for multicellularity and embryonic development. Indeed, the evolution of the cell is the most remarkable triumph of evolution.

What was the origin of multicellularity and the embryo? One possibility,

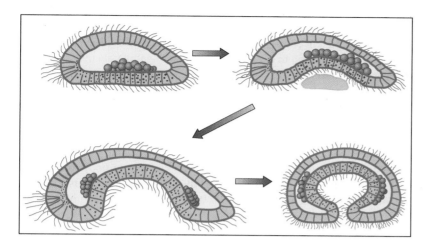

Fig. 15.24 A possible scenario for the development of the gastrula. A colonial protozoan in the form of a hollow multicellular sphere could have settled on the sea bottom and developed a gut-like invagination to aid feeding. Based on Jaegerstern, G.: 1956.

and it is highly speculative, is that a mutation resulted in the progeny of a single-celled organism not separating after cell division, leading to a loose colony of identical cells that occasionally fragmented to give new 'individuals'. One advantage of a colony might originally have been that when food was in short supply, the cells could feed off each other, and so the colony survived. This could have been the origin of both multicellularity and the requirement for cell death in multicellular organisms. The egg may subsequently have evolved as the cell fed by other cells; in sponges, the egg phagocytoses neighboring cells.

Once multicellularity evolved, it opened up all sorts of new possibilities, such as cell specialization for different functions. Some cells could specialize in providing motility, for example, and others in feeding, as we see in the endoderm of *Hydra*. The origin of pattern formation, that is, cell differentiation in a spatially organized arrangement, is not known, but may have depended on gradients set up by external influences. How gastrulation evolved is also unknown, but it is not implausible to consider a scenario in which a hollow sphere of cells, the common ancestor of all multicellular animals, changed its form to assist feeding (Fig. 15.24). This ancestor may, for example, have sat on the ocean floor, ingesting food particles by phagocytosis. A small invagination developing in the body wall could have promoted feeding by forming a primitive gut. Ciliary movement could have swept food particles more efficiently into this region, where they would be phagocytosed. Once the invagination formed, it is not too difficult to imagine how it could eventually extend right across the sphere, fuse with the other side, and form a continuous gut, which would be the endoderm. At a later stage in evolution, cells migrating inside between the gut and the outer epithelium would give rise to the mesoderm.

Summary

The embryo arose during the evolution of multicellular organisms from single-celled organisms, which have most of the cellular properties required by embryonic development. Multicellularity, having arisen, might have persisted originally because a colony of cells could provide a source of food for some of the members in times of scarcity. The egg might have evolved as the cell that is fed by other cells.

SUMMARY TO CHAPTER 15

Many developmental processes have been conserved during evolution. While many questions relating to evolution and development remain unanswered, it is clear that development reflects the evolutionary history of ancestral embryos. All vertebrate embryos pass through a phylotypic developmental stage, although there can be considerable divergence both earlier and later in development. The signals involved in patterning of vertebrate and arthropod appendages, as well as of the dorso-ventral axis, show remarkable similarity and conservation. The pattern of Hox gene expression along the body axis of vertebrates and arthropods is conserved, and changes in body plan reflect changes in both Hox gene expression and their downstream targets. Changes in gene regulatory regions were crucial. Alterations in the timing of developmental events have played an important role in evolution. Such changes can alter the overall form of an organism as a result of differences in growth rates of different structures, and can also result in sexual maturation at larval stages. The origins of multicellularity and the embryo from a single-celled ancestor are still highly speculative.

GENERAL REFERENCES

Carroll, S.: **Chance and necessity: the evolution of morphological complexity and diversity.** *Nature* 2001, **409**: 1102–1109.

Finnerty, J.R.: **Head start.** *Nature* 2000, **408**: 778–780.

Gerhart, J., Kirschner, M.: *Cells, Embryos and Evolution.* Malden, MA: Blackwell Science, 1997.

Graham, A.: **The evolution of the vertebrates—genes and development.** *Curr. Opin. Genet. Dev.* 2000, **10**: 624–628.

Pennisi, E., Roush, W.: **Developing a new view of evolution.** *Science* 1997, **277**: 34–37.

Raff, R.A.: *The Shape of Life.* University of Chicago Press, 1996.

Raff, R.A.: **Evo-devo: the evolution of a new discipline.** *Nature Rev. Genet.* 2000, **1**: 74–79.

Richardson, M.K.: **Vertebrate evolution: the developmental origins of adult variation.** *BioEssays* 1999, **21**: 604–613.

Richardson, M.K., Allen, S.P., Wright, G.M., Raynaud, A., Hanken, J.: **Somite number and vertebrate evolution.** *Development* 1998, **125**: 151–160.

SECTION REFERENCES

15.1 Embryonic structures have acquired new functions during evolution

Romer, A.S.: *The Vertebrate Body.* Philadelphia: W.B. Saunders, 1949.

15.2 Limbs evolved from fins

Capdevila, J., Izpisua Belmonte, J.C.: **Perspectives on the evolutionary origin of tetrapod limbs.** *J. Exp. Zool. (Mol. Dev. Evol.)* 2000, **288**: 287–303.

Coates, M.I., Cohn, M.J.: **Fins, limbs, and tails: outgrowths and axial patterning in vertebrate evolution.** *BioEssays* 1998, **20**: 371–381.

Cohn, M.J., Tickle, C.: **Developmental basis of limblessness and axial patterning in snakes.** *Nature* 1999, **399**: 474–479.

Ruvinsky, I., Gibson-Brown, J.J.: **Genetic and developmental bases of serial homology in vertebrate limb evolution.** *Development* 2000, **127**: 5211–5244.

Sordino, P., van der Hoeven, F., Duboule, D.: **Hox gene expression in teleost fins and the origin of vertebrate digits.** *Nature* 1995, **375**: 678–681.

15.3 Vertebrate and insect wings make use of evolutionarily conserved developmental mechanisms

Panganiban, G., Irvine, S.M., Lowe, C., Roehl, H., Corley, L.S., Sherbon, B., Grenier, J.K., Fallon, J.F., Kimble, J., Walker, M., Wray, G.A., Swalla, B.J., Martindale, M.Q., Carroll, S.B.: **The origin and evolution of animal appendages.** *Proc. Natl Acad. Sci. USA* 1997, **94**: 5162–5166.

15.4 Hox gene complexes have evolved through gene duplication

Brooke, N.M., Gacia-Fernandez, J., Holland, P.W.H.: **The ParaHox gene cluster is an evolutionary sister of the Hox gene cluster.** *Nature* 1998, **392**: 920–922.

Ferrier, D.E.K., Holland, P.W.H: **Ancient origin of the Hox gene cluster.** *Nature Rev. Genet.* 2001, **2**: 33–34.

Valentine, J.W., Erwin, D.H., Jablonski, D.: **Developmental evolution of metazoan bodyplans: the fossil evidence.** *Dev. Biol.* 1996, **173**: 373–381.

15.5 Changes in specification and interpretation of positional identity have generated the elaboration of vertebrate and arthropod body plans

Akam, M.: **Hox genes and the evolution of diverse body plans.** *Phil. Trans. Roy. Soc. Lond. B* 1995, **349**: 313–319.

Averof, M., Akam, M.: **Hox genes and the diversification of insect and crustacean body plans.** *Nature* 1995, **376**: 420–423.

Averof, M.: **Origin of the spider's head.** *Nature* 1998, **395:** 436–437.

Duboule, D.: **A Hox by any other name.** *Nature* 2000, **403:** 607–610.

Slack, J.M., Holland, P.W., Graham, C.F.: **The zootype and phylotypic stage.** *Nature* 1993, **361:** 490–492.

Stern, D.L.: **A role of** *Ultrabithorax* **in morphological differences between** *Drosophila* **species.** *Nature* 1998, **396:** 463–466.

15.6 The position and number of paired appendages in insects is dependent on Hox gene expression

Carroll, S.B., Weatherbee, S.D., Langeland, J.A.: **Homeotic genes and the regulation and evolution of insect wing number.** *Nature* 1995, **375:** 58–61.

Weatherbee, S.D., Carroll, S.: **Selector genes and limb identity in arthropods and vertebrates.** *Cell* 1999, **97:** 283–286.

Weatherbee, S.D., Nijhout, H.F, Grunnert, L.W., Halder, G., Galant, R., Selegue, J., Carroll, S.: **Ultrabithorax function in butterfly wings and the evolution of insect wing patterns.** *Curr. Biol.* 1999, **9:** 109–115.

15.7 The body plan of arthropods and vertebrates is similar, but the dorso-ventral axis is inverted

Arendt, D., Nübler-Jung, K.: **Dorsal or ventral: similarities in fate maps and gastrulation patterns in annelids, arthropods and chordates.** *Mech. Dev.* 1997, **61:** 7–21.

Davis, G.K., Patel, N.H.: **The origin and evolution of segmentation.** *Trends Biochem. Sci.* 1999, **24:** M68–M72.

Holland, L.Z., Kene, M., Williams, N.A., Holland, N.D.: **Sequence and embryonic expression of the amphioxus** *engrailed* **gene (AmphiEn): the metameric pattern of transcription resembles that of its segment-polarity homolog in** *Drosophila*. *Development* 1997, **124:** 1723–1732.

Holley, S.A., Jackson, P.D., Sasai, Y., Lu, B., De Robertis, E., Hoffman, F.M., Ferguson, E.L.: **A conserved system for dorso-ventral patterning in insects and vertebrates involving** *sog* **and** *chordin*. *Nature* 1995, **376:** 249–253.

Müller, M., von Weizsäcker, E., Campos-Ortega, J.A.: **Expression domains of a zebrafish homologue of the** *Drosophila* **pair-rule gene** *hairy* **correspond to primordia of alternating somites.** *Development* 1996, **122:** 2071–2078.

15.8 Changes in growth can alter the shapes of organisms

Huxley, J.S.: *Problems of Relative Growth*. London: Methuen & Co. Ltd., 1932.

15.9 The timing of developmental events has changed during evolution

Alberch, P., Alberch, J.: **Heterochronic mechanisms of morphological diversification and evolutionary change in the neotropical salamander** *Bolitoglossa occidentales* **(Amphibia: Plethodontidae).** *J. Morphol.* 1981, **167:** 249–264.

Lande, R.: **Evolutionary mechanisms of limb loss in tetrapods.** *Evolution* 1978, **32:** 73–92.

Raynaud, A.: **Developmental mechanism involved in the embryonic reduction of limbs in reptiles.** *Int. J. Dev. Biol.* 1990, **34:** 233–243.

Wray, G.A., Raff, R.A.: **The evolution of developmental strategy in marine invertebrates.** *Trends Evol. Ecol.* 1991, **6:** 45–56.

15.10 Evolution of life histories has implications for development

Partridge, L., Harvey, P.: **The ecological context of life history evolution.** *Science* 1988, **241:** 1449–1455.

15.11 How multicellular organisms evolved from single-celled ancestors is still highly speculative

Jaegerstern, G.: **The early phylogeny of the metazoa. The bilaterogastrea theory.** *Zool. Bidrag. (Uppsala)* 1956, **30:** 321–354.

Wolpert L.: **Gastrulation and the evolution of development.** *Development* (Suppl.) 1992, 7–13.

Wolpert, L.: **From egg to adult to larva.** *Evol. Dev.* 1999, **1:** 3–4.

Glossary

The **acrosomal reaction** is the release of enzymes and other proteins from the acrosomal vesicle of the sperm head that occurs once a sperm has bound to the outer surface of the egg. It helps the sperm to penetrate the outer layers of the egg.

Actin filaments or microfilaments are one of the three principal protein filaments of the cytoskeleton. They are involved in cell movement and changes in cell shape. Actin filaments are also part of the contractile apparatus of muscle cells.

In plant leaves, the **adaxial–abaxial** axis runs from the upper surface to the lower surface (dorsal to ventral).

Adhesion molecules bind cells to each other and to the extracellular matrix. The main classes of adhesion molecules important in development are the cadherins, the immunoglobulin superfamily, and the integrins.

The **allantois** is a set of extra-embryonic membranes that develops in some vertebrate embryos. In bird and reptile embryos it acts as a respiratory surface, while in mammals its blood vessels carry blood to and from the placenta.

An **allele** is a particular version of a gene. In diploid organisms two alleles of each gene are present, which may or may not be the same.

Allometry is the mathematical study of the relative growth of parts of an organism.

The **amniotic sac** is an extra-embryonic membrane in birds, reptiles, and mammals, which forms a fluid-filled sac that encloses and protects the embryo. It is derived from extra-embryonic ectoderm and mesoderm.

An **androgenetic** embryo is an embryo in which the two sets of homologous chromosomes are both paternal in origin.

Angioblasts are the mesodermal precursor cells that will give rise to blood vessels.

Angiogenesis is the process by which small blood vessels sprout from the larger vessels.

The **animal region** of an egg is the end of the egg where the nucleus resides, usually away from the yolk. The most terminal part of this region is the **animal pole**, which is directly opposite the vegetal pole at the other end of the egg. In *Xenopus* the pigmented animal half is called the **animal cap**.

The **animal–vegetal axis** runs from the animal to the vegetal pole in an egg or early embryo.

The **anterior visceral ectoderm** is an extra-embryonic tissue in the early mouse embryo that is involved in inducing anterior regions of the embryo.

The **antero-posterior axis** defines which is the 'head' end and which is the 'tail' end of an animal. The head is anterior and the tail posterior. In the vertebrate limb, this axis runs from the thumb to little finger.

Anticlinal cell divisions are divisions in planes at right angles to the outer surface of a tissue.

The **apical ectodermal ridge** or **apical ridge** is a thickening of the ectoderm at the distal end of the developing chick and mammalian limb bud. Signals from the ridge specify the progress zone in the underlying mesoderm.

An **apical meristem** is the region of dividing cells at the tip of a growing shoot or root.

Apoptosis or programmed cell death is a type of cell death that occurs widely during development. In programmed cell death, a cell is induced to commit 'suicide', which involves fragmentation of the DNA and shrinkage of the cell. These apoptotic cells are removed by the body's scavenger cells and, unlike necrosis, their death does not cause damage to surrounding cells.

The **archenteron** is the cavity formed inside the embryo when the endoderm and mesoderm invaginate during gastrulation. It forms the gut.

The **area opaca** is the outer dark area of the chick blastoderm.

The **area pellucida** is the central clear area of the chick blastoderm.

Asymmetric cell divisions or **asymmetric divisions** are cell divisions in which the daughter cells are different from each other because some cytoplasmic determinant(s) have been distributed unequally between them.

The **axes** (singular **axis**) of an organism define its polarity in various directions. In animals the two main axes are the antero-posterior axis, which runs from the head to the tail, and the dorso-ventral axis, which runs at right angles to this, from back to underside, with the mouth defining the ventral or underside. The main axis of a plant runs from shoot tip to root tip and is called the apical-basal axis.

Axons are long cell processes of neurons that conduct nerve impulses away from the cell body. The end of an axon forms contacts (synapses) with other neurons, muscle cells, or glandular cells.

The **basal lamina** is a sheet of extracellular matrix that separates an epithelial layer from the underlying tissues. For example, the epidermis of the skin is separated from the dermis by a basal lamina.

In amphibian limb regeneration, a **blastema** is formed from the dedifferentiation and proliferation of cells beneath the wound epidermis, and gives rise to the regenerated limb.

The **blastocoel** is the fluid-filled cavity that develops in the interior of a blastula.

The **blastocyst** stage of a mammalian embryo corresponds in form to the blastula stage of other animal embryos, and is the stage at which the embryo implants in the uterine wall.

A **blastoderm** is a post-cleavage embryo composed of a solid layer of cells rather than a spherical blastula, as in early chick and **Drosophila** embryos. The chick blastoderm is also known as the **blastodisc**.

Blastomeres are the cells derived from cleavage of the early embryo.

The **blastopore** is the slit-like or circular invagination on the surface of amphibian and sea urchin embryos where the mesoderm and endoderm move inside the embryo at gastrulation.

The **blastula** stage in animal development is the outcome of cleavage. The blastula is a hollow ball of cells, composed of an epithelial layer of small cells enclosing a fluid-filled cavity—the blastocoel.

The **body plan** describes the overall organization of an organism, for example the position of the head and tail, and the plane of bilateral symmetry, where it exists. The body plan of most animals is organized around two main axes, the antero-posterior axis and the dorso-ventral axis.

The functional maturation of sperm after they have been deposited in the female reproductive tract is known as **capacitation**.

Catenins are intracellular proteins that interact with the cytoplasmic tails of cadherin molecules and form a link between the cadherins and the cell's cytoskeleton. Free cytoplasmic catenins are also invovled in specifying the dorsal side of **Xenopus** embryos.

Cell adhesiveness is the property of cells that enables them to attach to each other and to a substratum.

The effects of a gene are **cell-autonomous** if they only affect the cell the gene is expressed in.

The **cell cycle** is the sequence of events by which a cell duplicates itself and divides in two.

During **cell differentiation**, cells become functionally and structurally different from one another and become distinct cell types, such as muscle or blood cells.

Cell-lineage restriction occurs when all the descendants of a particular group of cells remain within a 'boundary' and never mix with an adjacent group of cells. Compartment boundaries in insect development are boundaries of lineage restriction.

Cell motility refers to the ability of cells to change shape and move.

The **chemoaffinity hypothesis** proposes that each retinal neuron carries a chemical label that enables it to connect reliably with an appropiately labeled cell in the optic tectum.

A **chimeric** organism or tissue (a chimera) is made up of cells from two or more different sources, and thus of different genetic constitutions.

The **chorion** is the outermost of the extra-embryonic membranes in birds, reptiles, and mammals. It is involved in respiratory gas exchange. In birds and reptiles it lies just beneath the shell. In mammals it is part of the placenta and is also involved in nutrition and waste removal. The chorion of insect eggs has a different structure.

Chromatin is the material of which chromosomes are made. It is composed of DNA and protein.

Cleavage occurs after fertilization and is a series of rapid cell divisions without growth that divides the embryo up into a number of small cells.

A **clone** is a collection of genetically identical cells derived from a single cell by repeated cell division, or the genetically identical offspring of a single individual produced by asexual reproduction.

The **coding region** of a gene is that part of the DNA that encodes a polypeptide or functional RNA.

The correspondence between the order of Hox genes on a chromosome and their temporal and spatial order of expression in the embryo is known as **co-linearity**.

Induction of cell differentiation in some tissues depends on a **community effect**, in that there have to be a sufficient number of responding cells present for differentiation to occur.

Compaction of the mouse embryo occurs during early cleavage. The blastomeres flatten against each other and microvilli become confined to the outer surface of the ball of cells.

Compartments are discrete areas of an embryo that contain all the descendants of a small group of founder cells and which show cell-lineage restriction. Cells in compartments respect the compartment boundary and do not cross over into an adjacent compartment. Compartments tend to act as discrete developmental units.

Competence is the ability of a tissue to respond to an inducing signal. Embryonic tissues only remain **competent** for a limited period of time.

The **control region** of a gene is the region to which regulatory proteins bind and so determine whether or not the gene is transcribed.

Convergent extension is the process by which a sheet of cells changes shape by extending in one direction and narrowing—converging—in a direction at right angles to the extension.

Cortical rotation occurs immediately after an amphibian egg is fertilized. The egg cortex rotates with respect to the underlying cytoplasm, toward the point of sperm entry.

A **cotyledon** is the part of the plant embryo that acts as a food storage organ.

Cyclins are proteins that periodically rise and fall in concentration during the cell cycle and are involved in controlling progression through the cycle.

Cytoplasmic localization is the nonuniform distribution of some factor or determinant in a cell's cytoplasm, so that when the cell divides, the determinant is unequally distributed to the daughter cells.

The long-lived **dauer** larvae of *C. elegans* are a response to starvation conditions and neither eat nor grow until food is again available.

Dedifferentiation is loss of the structural characteristics of a differentiated cell, which may result in the cell then differentiating into a new cell type.

Dendrites are extensions from the body of a nerve cell that receive stimuli from other nerve cells.

The **dermatome** is the region of the somite that will give rise to the dermis.

The **dermis** of the skin is the connective tissue beneath the epidermis, from which it is separated by a basal lamina.

The **dermomyotome** is the region of the somite that will give rise to both muscle and dermis.

Desmosomes or **desmosomal junctions** are specialized cell junctions between epithelial cells. Adhesion at these junctions is mediated by cadherins.

Determinants are cytoplasmic factors (e.g. proteins and RNAs) in the egg and in embryonic cells that can be asymmetrically distributed at cell division and so influence how the daughter cells develop.

Determination implies a stable change in the internal state of a cell such that its fate is now fixed, or **determined**. A determined cell will follow that fate when grafted into other regions of the embryo.

Diploid cells contain two sets of homologous chromosomes, one from each parent, and thus two copies of each gene.

Directed dilation is the extension of a tube-like structure at each end of a cell due to hydrostatic pressure, the direction of extension reflecting greater circumferential resistance to expansion.

A **dominant** allele is one that determines the phenotype even when present in only a single copy.

A **dominant-negative mutation** inactivates a particular cellular function by the production of a defective RNA or protein molecule that blocks the normal function of the gene product.

Dorsalized embryos develop much increased dorsal regions at the expense of ventral regions.

The **dorso-ventral** axis defines the relation of the upper surface or back (dorsal) to the under surface (ventral) of an organism or structure. The mouth is always on the ventral side.

Dosage compensation is the mechanism that ensures that although the number of X chromosomes in males and females is different, the level of expression of X-chromosome genes is the same in both sexes. Mammals, insects, and nematodes all have different dosage compensation mechanisms.

Ecdysis is a type of molting in arthropods in which the external cuticle is shed to allow for growth.

The **ectoderm** is the germ layer that gives rise to the epidermis and the nervous system.

Embryogenesis is the process of development of the embryo from the fertilized egg.

Embryology is the study of the development of an embryo.

The **embryonic-abembryonic axis** in the mammalian blastocyst runs from the site of attachment of the inner cell mass—the **embryonic pole**—to the opposite pole.

The **embryonic ectoderm** is the name given to the mouse epiblast once it has developed into an epithelial sheet.

Embryonic stem cells (ES cells) are derived from the inner cell mass of a mammalian embryo, usually mouse, and can be indefinitely maintained in culture. When injected into another blastocyst, they combine with the inner cell mass and can potentially contribute to all the tissues of the embryo.

The **endocardium** is the inner endothelial layer of the developing heart.

Endochondral ossification is the replacement of cartilage with bone in the growth plate of vertebrate embryonic skeletal elements, such as those that give rise to the long bones of the limbs.

The **endoderm** is the germ layer that gives rise to the gut and associated organs, such as the lungs and liver in vertebrates.

The **endo-mesoderm** in *Xenopus* is a layer of cells at the boundary between endoderm and mesoderm.

The **endosperm** in higher plant seeds is a nutritive tissue that serves as a source of food for the embryo.

Enhancers are DNA sequences to which regulatory proteins bind to control the time and place of transcription of a gene. Enhancers can be many thousands of base pairs away from the gene's coding region.

Ephrins and their receptors are cell-surface molecules involved in delimiting compartments in rhombomeres and in axonal guidance.

The **epiblast** of mouse and chick embryos is a group of cells within the blastocyst or blastoderm, respectively, that gives rise to the embryo proper. In the mouse, it develops from cells of the inner cell mass.

Epiboly is the process during gastrulation in which the ectoderm extends to cover the whole of the embryo.

The **epidermis** in vertebrates, insects, and plants is the outer layer of cells that forms the interface between the organism and its environment. Its structure is quite different in the different organisms.

Epimorphosis is a type of regeneration in which the regenerated structures are formed by new growth.

An **epithelium to mesenchymal transition** involves cells leaving an epithelial layer and becoming a loose mass of mesenchyme cells which can migrate individually.

Extra-embryonic ectoderm in mammals contributes to the formation of the placenta.

Extra-embryonic membranes are membranes external to the embryo proper that are involved in protection and nutrition of the embryo.

The **fate** of cells describes what they will normally develop into. By marking cells in the embryo, a **fate map** of embryonic regions can be constructed. However, having a particular normal fate does not imply that a cell could not develop differently if placed in a different environment.

Fertilization is the fusion of sperm and egg to form the zygote.

The **floor plate** of the developing neural tube is composed of non-neural cells and is involved in patterning the ventral part of the neural tube.

A **floral meristem** is a region of dividing cells at the tip of a shoot that gives rise to a flower.

The individual parts of a flower develop from **floral organ primordia** generated by the floral meristem.

Follicle cells surround the oocyte and nurse cells during egg development in *Drosophila*.

The **gametes** are the cells that carry the genes to the next generation—in animals they are the eggs and sperm.

A **ganglion mother cell** is formed by division of a neuroblast in *Drosophila* and gives rise to neurons.

Gap genes are zygotic genes coding for transcription factors expressed in early *Drosophila* development that subdivide the embryo into regions along the antero-posterior axis.

The **gastrula** is the stage in animal development when the endoderm and mesoderm of the blastula move inside the embryo.

Gastrulation is the process in animal embryos in which the endoderm and mesoderm move from the outer surface of the embryo to the inside, where they give rise to internal organs.

Gene knock-out refers to the complete inactivation of a particular gene in an organism by means of genetic manipulation.

Gene regulatory proteins are proteins that bind to control regions in DNA and help to switch genes on and off.

The **genital ridge** in vertebrates is the region of mesoderm lining the abdominal cavity from which the gonads develop.

The **genotype** is a description of the exact genetic constitution of a cell or organism in terms of the alleles it possesses for any given gene.

The **germ band** is the name given to the ventral blastoderm of the early *Drosophila* embryo, from which most of the embryo will eventually develop.

Germ cells are those cells that give rise to eggs and sperm.

The **germ layers** refer to the regions of the early animal embryo that will give rise to distinct types of tissue. Most animals have three germ layers—ectoderm, mesoderm, and endoderm.

Germ plasm is the special cytoplasm in some animal eggs, such as those of *Drosophila*, that is involved in the specification of germ cells.

Glia are supporting cells of the nervous system, such as Schwann cells.

The **gonads** are the reproductive organs of animals.

Growth is an increase in size, which occurs by cell multiplication, increase in cell size and deposition of extracellular material.

Axons of developing neurons extend by means of a **growth cone** at their tip. The growth cone both crawls forward on the substrate and senses its environment by means of filopodia.

Growth of vertebrate long bones occurs at the cartilaginous **growth plates**. The cartilage grows and is eventually replaced by bone by the process of endochondral ossification.

A **gynogenetic** embryo is one in which the two sets of homologous chromosomes are both maternal in origin.

Haploid cells are derived from diploid cells by meiosis and contain only one set of chromosomes (half the diploid number of chromosomes), and thus contain only one copy of each gene. In most animals the only haploid cells are the gametes—the sperm or egg.

A **haploid germ cell** is a germ cell that has undergone meiosis and so only has one set of chromosomes.

Hematopoiesis is the process by which all the blood cells are derived from a pluripotent stem cell. This occurs mainly in the bone marrow.

Hemi-desmosomes are specialized junctions between epithelial cells and the underlying basal lamina. They involve integrins.

Hensen's node is a condensation of cells at the anterior end of the primitive streak in chick and mouse embryos. Cells from the node give rise to the notochord. It corresponds to the Spemann organizer in amphibians.

Heterochromatin is the state of chromatin in regions of the chromosome that are so condensed that transcription is not possible.

Heterochrony is an evolutionary change in the timing of developmental events. A mutation that changes the timing of a developmental event is called a **heterochronic** mutation.

Heterotaxis is the state in which internal organs of normal and inverted symmetry are present in the same animal.

A diploid individual is **heterozygous** for a given gene when it carries two different alleles of that gene.

The **homeobox** is a region of DNA in homeotic genes that encodes a DNA-binding domain called the **homeodomain**. Genes containing this motif are known generally as **homeobox genes**. The homeodomain is present in a large number of transcription factors that are important in development, such as the products of the Hox genes and the Pax genes.

Homeosis is the phenomenon in which one structure is transformed into another, homologous, structure. An example of a **homeotic transformation** is the development of legs in place of antennae in *Drosophila* as a result of mutation in a **homeotic gene**.

A **homeotic mutation** is a mutation that gives rise to a homeotic transformation—the transformation of one structure into another.

Homeotic selector genes in **Drosophila** are genes that specify the identity and developmental pathway of a group of cells. They encode homeodomain transcription factors and act by controlling the expression of other genes. Their expression is required throughout development. The *Drosophila* gene *engrailed* is an example of a homeotic selector gene.

Homologous genes share significant similarity in their nucleotide sequence and are derived from a common ancestral gene.

Homologous recombination is the recombination of two DNA molecules at a specific site of sequence similarity.

Homology refers to morphological or structural similarity due to common ancestry.

A diploid individual is **homozygous** for a given gene when it carries two identical alleles of that gene.

Hox genes are a particular family of homeobox-containing genes that are present in all animals (as far as is known) and are involved in patterning the antero-posterior axis. They are clustered on the chromosomes in one or more gene complexes.

The **hypoblast** in the early chick embryo is a sheet of cells that covers the yolk and gives rise to extra-embryonic structures such as the stalk of the yolk sac.

Imaginal discs are small sacs of epithelium present in the larva of *Drosophila* and other insects, which at metamorphosis give rise to adult structures such as wings, legs, antennae, eyes, and genitalia.

A gene is said to be **imprinted** when it is expressed differently (either active or inactive) in the embryo depending on whether it is derived from the mother or father.

Induction is the process whereby one group of cells signals to another group of cells in the embryo and so affects how they will develop.

An **inflorescence** in plants is a flowerhead—a flowering shoot. Shoots that can bear flowers develop as a result of the conversion of a vegetative apical meristem into an **inflorescence meristem**.

Ingression is the movement of individual cells from the outside of the embryo into the blastocoel.

Initials are cells in the meristems of plants that are able to divide continuously, giving rise both to dividing cells that stay within the meristem and to cells that leave the meristem and go on to differentiate.

The **inner cell mass** of the early mammalian embryo is derived from the inner cells of the morula, which form a discrete mass of cells in the blastocyst. Some of the cells of the inner cell mass give rise to the embryo proper.

In situ **hybridization** is a technique used to detect where in the embryo particular genes are being expressed. The mRNA that is being transcribed is detected by its hybridization to a labeled single-stranded complementary DNA probe.

In animals in which the larva goes through successive phases of growth and molting before developing into an adult, the phase between each molt is known as an **instar**.

Integrins are a class of cell adhesion molecules by which cells attach to the extracellular matrix.

Intercalary growth can occur in animals capable of epimorphic regeneration when two pieces of tissue with different positional values are placed next to each other. The intercalary growth replaces the intermediate positional values.

An **internode** is that portion of a plant stem between two nodes (sites at which a leaf or leaves form).

Invagination is the local inward movement of a sheet of embryonic epithelial cells to form a bulge-like structure, as in early gastrulation of the sea urchin embryo.

Involution is a type of cell movement that occurs at the beginning of amphibian gastrulation, when a sheet of cells enters the interior of the embryo by rolling in under itself.

Keratinocytes are differentiated epidermal skin cells that produce keratin, eventually die and are shed from the skin surface.

Knock-out, see gene knock-out.

The **lateral geniculate nucleus** is the region of the brain in mammals where the axons from the retina terminate.

The **lateral plate mesoderm** in vertebrate embryos lies lateral and ventral to the somites and gives rise to the tissues of the heart, kidney, gonads, and blood.

Lateral inhibition is the mechanism by which cells inhibit neighboring cells from developing in a similar way to themselves.

A cell's **lineage** is the sequence of cell divisions that give rise to that cell.

Lineage restriction, see cell-lineage restriction.

In **long-germ development** the blastoderm gives rise to the whole of the future embryo, as in *Drosophila*.

The **marginal zone** of an amphibian embryo is the belt-like region of presumptive mesoderm at the equator of the late blastula.

Maternal-effect mutations are mutations in the genes of the mother which affect the development of the egg and later the embryo.

Maternal factors are proteins and RNAs that are deposited in the egg by the mother during oogenesis. The production of these maternal proteins and RNAs is under the control of so-called **maternal genes**.

Medio-lateral intercalation of cells occurs during convergent extension in amphibian gastrulation. The sheet of cells narrows and elongates by cells pushing in sideways between their neighbors.

Meiosis is a special type of cell division that occurs during formation of sperm and eggs, and in which the number of chromosomes is halved from diploid to haploid.

In **mericlinal chimeras** a genetically marked cell gives rise to a sector of an organ or of a whole plant.

Meristems are groups of undifferentiated, dividing cells that persist at the growing tips of plants. They give rise to all the adult structures—shoots, leaves, flowers, and roots.

Mesenchyme describes loose connective tissue, usually of mesodermal origin, whose cells are capable of migration; some epithelia of ectodermal origin, such as the neural crest, undergo an epithelial to mesenchymal transition.

A **mesenchyme to epithelium transition** occurs when loose mesenchyme cells aggregate and then form an epithelium like a tube, as in kidney development.

The **mesoderm** is the germ layer that gives rise to the skeleto-muscular system, connective tissues, the blood, and internal organs such as the kidney and heart.

The **mesonephros** in mammals is an embryonic kidney that contributes to the male and female reproductive organs.

Messenger RNA (mRNA) is the RNA molecule that specifies the sequence of amino acids in a protein. It is produced by transcription from DNA.

Metamorphosis is the process by which a larva is transformed into an adult. It often involves a radical change in form, and the development of new organs, such as wings in butterflies and limbs in frogs.

Metastasis is the movement of cancer cells from their site of origin to invade underlying tissues and to spread to other parts of the body. Such cells are said to **metastasize**.

Microfilaments, see actin filaments.

Micromeres are small cells that result from unequal cleavage during early animal development.

The **mid-blastula transition** in amphibian embryos is when the embryo's own genes begin to be transcribed, cleavages become asynchronous, and the cells of the blastula become motile.

Mitosis is the division of the nucleus during cell division resulting in both daughter cells having the same diploid complement of chromosomes as the parent cell.

Molting is the shedding of an external cuticle when arthropods grow, and its replacement with a new one.

A **morphogen** is any substance active in pattern formation whose spatial concentration varies and to which cells respond differently at different threshold concentrations.

Morphallaxis is a type of regeneration that involves repatterning of existing tissues without growth.

Morphogenesis refers to the processes involved in bringing about changes in form in the developing embryo.

The **morphogenetic furrow** in *Drosophila* eye development moves across the eye disc and initiates the development of the ommatidia.

A **morula** is the very early stage in a mammalian embryo when cleavage has resulted in a solid ball of cells.

The **Müllerian duct** runs adjacent to the Wolffian duct in the mammalian embryo and becomes the oviduct in females.

Müllerian-inhibiting substance is secreted by the developing testis and induces regression of the Müllerian ducts in males.

The **myocardium** is the outer contractile layer of the developing heart.

Myoplasm is special cytoplasm in ascidian eggs involved in the specification of muscle cells.

The **myotome** is that part of the somite that gives rise to muscle.

Necrosis is a type of cell death due to pathological damage in which cells break up, releasing their contents.

Negative feedback is a type of regulation in which the end-product of a pathway or process inhibits an earlier stage and thus keeps the level of end-product constant.

Neoteny is the phenomenon in which an animal acquires sexual maturity while still in larval form.

Netrins are extracellular proteins involved in axon guidance.

The **neural crest cells** in vertebrates are derived from the edge of the neural plate. They migrate to different regions of the body and give rise to a wide variety of tissues, including the autonomic nervous system, the sensory nervous system, pigment cells, and some cartilage of the head.

Neural folds, neural plate, neural tube, see neurulation.

The **neurectoderm** is embryonic ectoderm with the potential to form neural cells and epidermis.

A **neuroblast** is an embryonic cell that will give rise to neural tissue (neurons and glia).

The **neuromuscular junction** is the specialized area of contact between a motor neuron and a muscle fiber, where the neuron can stimulate muscle activity.

Neurons or nerve cells are the electrically excitable cells of the nervous system, which convey information in the form of electrical signals.

Neurotrophins are proteins that are necessary for neuronal survival, such as nerve growth factor.

A **neurula** is the stage of vertebrate development at the end of gastrulation when the neural tube is forming.

Neurulation in vertebrates is the process in which the ectoderm of the future brain and spinal cord—the **neural plate**—develops folds (**neural folds**) and forms the **neural tube**.

The **Nieuwkoop center** is a signaling center on the dorsal side of the early *Xenopus* embryo. It forms in the vegetal region as a result of cortical rotation.

A **node** in a plant is that part of the stem at which leaves and lateral buds form. In avian and mammalian embryos the term usually refers to Hensen's node or its equivalent.

The effects of a gene are **non cell-autonomous** or **non-autonomous** if they affect cells other than the cell the gene is expressed in.

The **notochord** in vertebrate embryos is a rod-like cellular structure that runs from head to tail and lies centrally beneath the future central nervous system. It is derived from mesoderm.

Nurse cells surround the developing oocyte in *Drosophila* and synthesize proteins and RNAs that are to be deposited in it.

Alternate **ocular dominance columns** in the visual cortex respond to the same visual stimulus from either the left or right eye.

Insect compound eyes are composed of hundreds of individual phtoreceptor organs, the **ommatidia** (singular **ommatidium**).

Many of the genes involved in cell regulation can be mutated into **oncogenes**, which cause cells to become cancerous.

Ontogeny refers to the development of an individual organism.

An **oocyte** is an immature egg.

Oogenesis is the process of egg formation in the female.

The **optic tectum** is the region of the brain in amphibians and birds where the axons from the retina terminate.

An **organizer** or **organizing region** is a signaling center that directs the development of the whole embryo or of part of the embryo, such as a limb. In amphibians, the organizer usually refers to the Spemann organizer.

Organogenesis is the development of specific organs such as limbs, eyes, and heart.

The **oviduct** in female birds and mammals transports the eggs from the ovaries to the uterus.

The **pair-rule genes** in *Drosophila* are involved in delimiting parasegments. They are expressed in transverse stripes in the blastoderm, each pair-rule gene being expressed in alternate parasegments.

Genes within a species that have arisen by duplication and divergence are called **paralogs**. Examples are the Hox genes in vertebrates, which comprise several **paralogous subgroups** made up of **paralogous genes**.

Parasegments in the developing *Drosophila* embryo are independent developmental units that give rise to the segments of the larva and adult.

Pattern formation is the process by which cells in a developing embryo acquire identities that lead to a well ordered spatial pattern of cell activities.

Pax genes encode transcriptional regulatory proteins that contain both a homeodomain and another protein motif, the paired motif.

P elements are transposable DNA elements found in *Drosophila*. They are short sequences of DNA that can become inserted in different positions within a chromosome and can also move to other chromosomes. This property is exploited in the technique of **P-element-mediated transformation** for making transgenic flies.

P granules are granules that become localized to the posterior end of the fertilized egg of *C. elegans*.

Periclinal cell divisions are divisions in a plane parallel to the surface of the tissue.

In **periclinal chimeras** in plants, one of the three meristem layers has a genetic marker which distinguishes it from the other two.

The **phenotype** is the observable or measurable characters and features of a cell or an organism.

Phyllotaxy is the way the leaves are arranged along a shoot.

Phylogeny is the evolutionary history of a species or group.

Vertebrate embryos pass through a developmental stage known as the **phylotypic stage** at which the embryos of the different vertebrate groups closely resemble each other. This is the stage at which the embryo possesses a distinct head, a neural tube, and somites.

A **plasmid** is a small circular DNA molecule that replicates independently of the bacterial genome.

A **pluripotent** stem cell is one that can give rise to many different types of differentiated cell.

The **pluteus** is the larval stage of the sea urchin.

Polar bodies are formed during meiosis in the developing egg. They are small cells containing a haploid nucleus and take no part in embryonic development.

In the developing chick and mouse limb buds, the **polarizing region** at the posterior margin of the bud produces a signal specifying position along the antero-posterior axis.

Pole cells give rise to the germ cells in *Drosophila* and are formed at the posterior end of the blastoderm.

Pole plasm is the cytoplasm at the posterior end of the *Drosophila* egg which is involved in specifying germ cells.

Polyspermy is the entry of more than one sperm into the egg.

Positional information in the form, for example, of a gradient of an extracellular signaling molecule, can provide the basis for pattern formation. Cells acquire a **positional value** that is related to their position with respect to the boundaries of the given field of positional information. The cells then interpret this positional value according to their genetic constitution and developmental history, and develop accordingly.

Posterior dominance or **posterior prevalence** is the process whereby the more posteriorly expressed Hox genes can inhibit the action of more anteriorly expressed Hox genes when they are expressed in the same region.

The **posterior marginal zone** of the chick embryo is a dense region of cells at the edge of the blastoderm that will give rise to the primitive streak.

Post-translational modification of a protein involves changes in the protein after it has been synthesized. The protein can, for example, be enzymatically cleaved, glycosylated or acetylated.

A basic pattern generated automatically in a structure is known as a **prepattern**. It may subsequently be modified during development.

The **pre-somitic mesoderm** is the unsegmented mesoderm between the node (in chick embryos) and the already formed somites. It will form somites from its anterior end as the node regresses.

Primary embryonic induction is the induction of the whole body axis, as demonstrated by transplantation of the Spemann organizer in amphibians.

The **primitive ectoderm** or epiblast is the part of the inner cell mass in the mammalian blastocyst that gives rise to the embryo proper.

The **primitive endoderm** in mammalian embryos is that part of the inner cell mass that contributes to extra-embryonic membranes.

The **primitive streak** of the chick embryo is a strip of cells that extends inward from the posterior marginal zone and is the forerunner of the antero-posterior axis. During gastrulation, cells move through the streak into the interior of the blastoderm. The primitive streak in the mouse embryo has a similar function to that in the chick.

Minute undifferentiated outgrowths that will give rise to a structure such as a tooth, leaf, flower or floral organ are known as **primordia** (singular **primordium**).

Programmed cell death, see apoptosis.

In chick and mouse limb buds the cells in the **progress zone** at the tip of the bud proliferate and acquire positional values.

The **promeristem** is the central region of the meristem that contains cells capable of continued division—the initials.

The **promoter** is a region of DNA close to the coding sequence to which RNA polymerase binds to begin transcription of a gene.

Genes that promote a neural fate in neurectoderm cells are called **proneural** genes.

Proneural clusters are small clusters of cells within the neurectoderm in which one cell will eventually become a neuroblast.

A **pronucleus** is the haploid nucleus of sperm or egg after fertilization but before nuclear fusion and the first mitotic division.

A **proto-oncogene** is a gene that is involved in regulation of cell proliferation which can, when mutated into an oncogene, or expressed under abnormal control, cause cancer.

The **proximo-distal axis** of a limb or other appendage runs from the point of attachment to the body (proximal) to the tip of the limb (distal).

Radial intercalation occurs in a multilayered ectoderm of an amphibian gastrula when cells intercalate in a direction perpendicular to the surface, so thinning and extending the cell sheet.

Reaction-diffusion mechanisms produce self-organizing patterns of chemical concentrations which could underlie periodic patterns.

A **recessive** mutation is a mutation in a gene that only changes the phenotype when both copies of the gene carry the mutation.

Redundancy refers to an apparent absence of an effect when a gene that is normally active during development is inactivated. It is assumed that other pathways exist which can substitute for the missing gene action.

Regeneration is the ability of a fully developed organism to replace lost parts.

Regulation is the ability of the embryo to develop normally even when parts are removed or rearranged. Embryos that can regulate are called **regulative**.

The **rhombomeres** are a sequence of compartments of cell-lineage restriction in the hindbrain of chick and mice embryos.

RNA splicing is the process in eukaryotic cells in which introns are cut out of newly transcribed RNA which is rejoined to make a continuous coding messenger RNA or a functional structural RNA.

The **roof plate** of the developing neural tube is composed of non-neural cells and is involved in patterning the dorsal part of the tube.

The **sclerotome** is that part of a somite that will give rise to the cartilage of the vertebrae.

Segmentation is the division of the body of an organism into a series of morphologically similar units or **segments**.

Segment polarity genes in *Drosophila* are involved in patterning the parasegments and segments.

Selector genes in *Drosophila* determine the activity of a group of cells, and their continued expression is required to maintain that activity.

Semaphorins are extracellular proteins involved in axon guidance.

A **semi-dominant** mutation is a mutation that affects the phenotype when just one allele carries the mutation but where the effect on the phenotype is much greater when both alleles carry the mutation.

Senescence is the impairment of function associated with aging.

A **sensory organ precursor** is an ectodermal cell that will give rise to a sensory bristle in the adult **Drosophila** epidermis.

The **sex-dermining region of the Y chromosome (SRY)** determines maleness by specifying the gonad as a testis.

Short-germ development characterizes those insects in which most of the segments are formed sequentially by growth. The blastoderm itself only gives rise to the anterior segments of the embryo.

Signal transduction is the process by which a cell converts an extracellular signal, usually at the cell membrane, into a response, which is often a change in gene expression.

A **signaling center** is a localized region of the embryo that exerts a special influence on surrounding cells and thus determines how they develop.

In the rare condition **situs inversus** in humans there is complete mirror-image reversal of the position of the internal organs.

Somatic cells are any cells other than germ cells. In most animals, the somatic cells are diploid.

Somites in vertebrate embryos are segmented blocks of mesoderm lying on either side of the notochord. They give rise to body and limb muscles, the vertebral column, and the dermis.

A **specification map** shows how the tissues of an embryo will develop when placed in a simple culture medium.

A group of cells is called **specified** if when isolated and cultured in a neutral medium they develop according to their normal fate.

The **Spemann organizer** is a signaling center on the dorsal side of the amphibian embryo. Signals from this center can organize new antero-posterior and dorso-ventral axes.

Spermatogenesis is the production of sperm.

A **stem cell** is a type of undifferentiated cell that is both self renewing and also gives rise to differentiated cell types. They are found in some adult tissues. *See also* **embryonic stem cells**.

A **synapse** is the specialized point of contact where a neuron communicates with another neuron or a muscle cell.

A **syncytium** is a cell with many nuclei in a common cytoplasm. Cell walls do not develop during nuclear division within the very early **Drosophila** embryo. This gives rise to the **syncytial blastoderm** in which the nuclei are arranged around the periphery of the embryo.

In leeches and other annelids, **teloblasts** give rise to the segmental structures.

The **teloplasm** is the cytoplasm in the eggs of leeches and other annelids which is involved in the specification of the blastomere that will give rise to the teloblasts.

Teratocarcinomas are solid tumors that arise from germ cells and which can contain a mixture of differentiated cell types.

A **threshold concentration** is that concentration of a chemical signal or morphogen that can elicit a particular response from a cell. A specific response to a chemical signal that only occurs above or below a particular threshold concentration of the signal is known as a **threshold effect**.

Totipotency is the capacity of a cell to develop into any of the cell types found in that particular organism. Such a cell is called **totipotent**.

The **tracheal system** of insects is a system of fine tubules that deliver air (and thus oxygen) to the tissues.

When a gene is active its DNA sequence is copied, or **transcribed**, into a complementary RNA sequence, a process known as **transcription**.

A **transcription factor** is a regulatory protein required to initiate or regulate transcription of a gene into RNA. Transcription factors act within the nucleus of a cell by binding to specific regulatory regions in the DNA.

Transdifferentiation is the process by which a differentiated cell can differentiate into a different cell type, such as pigment cell to lens.

Transfection is the technique by which mammalian and other animal cells are induced to take up foreign DNA molecules. The introduced DNA sometimes becomes inserted permanently into the host cell's DNA.

Transgenic organisms have had their genetic constitution deliberately changed by genetic engineering. A new gene—a **transgene**—may have been introduced or a particular gene may have been inactivated.

The process by which messenger RNA directs the order of amino acids in a protein during protein synthesis at ribosomes is known as **translation**. The messenger RNA is said to be **translated** into protein.

A **transposon** is a DNA sequence that can become inserted into a different site on the chromosome, either by the insertion of a copy of the original sequence or by excision and reinsertion of the original sequence.

A **trichome** is a hair-bearing cell in plant epidermis.

The **trophectoderm** is the outer layer of cells of the early mammalian embryo. It gives rise to extra-embryonic structures such as the placenta.

Tumor suppressor genes are genes that can cause a cell to become cancerous when both copies of the gene have been inactivated.

Vasculogenesis is the formation of blood vessels.

When an embryo is treated so that the amount of endoderm is increased at the expense of ectoderm it is **vegetalized**.

The **vegetal region** of an egg is the most yolky region, and is the region from which the endoderm will develop. The most terminal part of this region is called the **vegetal pole**, and is directly opposite the animal pole.

Embryos that are **ventralized** are deficient in dorsal regions and have much increased ventral regions.

The **ventricular proliferative zone** is a layer of dividing cells lining the lumen of the neural tube, from which neurons and glia are formed.

The **visceral endoderm** is derived from the primitive endoderm that develops on the surface of the egg cylinder in the mammalian blastocyst.

The **vitelline membrane** is an extracellular layer surrounding the eggs of the sea urchin and other animals. In the sea urchin it gives rise to the fertilization membrane.

Wolffian ducts are ducts associated with the mesonephros in mammalian embryos, They become the vas deferens in males.

The **yolk sac** is an extra-embryonic membrane in birds and mammals. In the chick embryo it surrounds the yolk.

The **zona pellucida** is a layer of glycoprotein surrounding the mammalian egg which serves to prevent polyspermy.

The **zone of polarizing activity** is another name for the polarizing region at the posterior margin of the limb bud.

The **zootype** refers to a pattern of expression of Hox genes and certain other genes along the antero-posterior axis of the embryo that is characteristic of all animal embryos.

The **zygote** is the fertilized egg. It is diploid and contains the chromosomes of both the male and female parents.

Zygotic genes are those present in the fertilized egg and which are expressed in the embryo itself.

Index